# Computational Methods for Agricultural Research:
## Advances and Applications


Hercules Antonio do Prado
*Brazilian Agricultural Research Corporation & Catholic University of Brasilia, Brazil*

Alfredo Jose Barreto Luiz
*Brazilian Agricultural Research Corporation, Brazil*

Homero Chaib Filho
*Brazilian Agricultural Research Corporation, Brazil*



Information Science REFERENCE

**INFORMATION SCIENCE REFERENCE**
Hershey · New York

| | |
|---|---|
| Director of Editorial Content: | Kristin Klinger |
| Director of Book Publications: | Julia Mosemann |
| Acquisitions Editor: | Lindsay Johnston |
| Development Editor: | Joel Gamon |
| Typesetter: | Deanna Jo Zombro |
| Production Editor: | Jamie Snavely |
| Cover Design: | Lisa Tosheff |

Published in the United States of America by
Information Science Reference (an imprint of IGI Global)
701 E. Chocolate Avenue
Hershey PA 17033
Tel: 717-533-8845
Fax: 717-533-8661
E-mail: cust@igi-global.com
Web site: http://www.igi-global.com



Library of Congress Cataloging-in-Publication Data

Computational methods for agricultural research : advances and applications / Hercules Antonio do Prado, Alfredo Jose Barreto Luiz, and Homero Chaib Filho, editors.
p. cm.
Includes bibliographical references and index.
ISBN 978-1-61692-871-1 (hardcover) -- ISBN 978-1-61692-873-5 (ebook) 1. Agriculture--Research--Data processing. 2. Agricultural informatics. I. Prado, Hercules Antonio do. II. Luiz, Alfredo Jose Barreto, 1963- III. Filho, Homero Chaib.
S540.D38C66 2010
630.72--dc22

2010035367

British Cataloguing in Publication Data
A Cataloguing in Publication record for this book is available from the British Library.

*For our friend and remarkable scientist, Fernando Luís Garagorry, who has introduced innovative data analysis techniques in connection with agricultural statistics.*

*For Orfeo Apolo Droguett Affin (in memoriam) who, against all odds, became a philosopher and taught us that "there is nothing impossible in Science" as well as nothing was impossible in his life.*

*Hércules, Alfredo, and Homero*

# Editorial Advisory Board

# Table of Contents

# Detailed Table of Contents

them correspond to mathematical concepts, while the third one gives the terrestrial distance between national centers of gravity of the products.



The present chapter discusses an analysis of land cover change based on the enhanced transition matrix, and provides a detailed interpretation of the results of the case study chosen: a protected area in the region of Madrid (central Spain) between 1990 and 2000. The method adopted introduces concepts and indicators to study the statistical information contained in the transition matrix in greater depth than with the traditional approach. The chapter describes all the formal, theoretical and practical aspects of the method, along with the preparation of the data and the analysis of the results. The method proved to be useful for synthesizing the land changes observed and identifying systematic transitions.



This chapter discusses how to take advantage of computational models to analyze and extract useful information from time series of climate data and remote sensing images. This kind of data has been used for researching on climate changes, as well as to help on improving yield forecasting of agricultural crops and increasing the sustainable usage of the soil. The authors present three techniques based on the Fractal Theory, data streams and time series mining: the FDASE algorithm, to identify correlated attributes; a method that combines intrinsic dimension measurements with statistical analysis, to monitor evolving climate and remote sensing data; and the CLIPSMiner algorithm applied to multiple time series of continuous climate data, to identify relevant and extreme patterns. The experiments with real data show that data mining is a valuable tool to help agricultural entrepreneurs and government on monitoring sugar cane areas, helping to make the production more useful to the country and to the environment.

This work assessed the capacity of the self-organizing map, an unsupervised artificial neural network, to aid the process of territorial design through visualization and clustering methods applied to a multivariate geospatial temporal dataset. The method was applied in the case study of Sergipe's institutional regional partition (Territories of Identity). Results have shown that the proposed method can improve the exploratory spatial-temporal analysis capacity of policy makers that are interested in territorial typology. A new partition for rural planning was elaborated and confirmed the coherence of the Territories of Identity.



Zoning based on climate and soil characteristics does not represent a full ecological-economic zoning procedure, as the Brazilian law establishes, but it is a positive step forward toward a sustainable use of natural resources. The Embrapa Temperate Climate Research Center has developed, with the collaboration of several entities from Brazil and Uruguay, a zoning procedure based on climate and soil for planting eucalyptus in southern region of Rio Grande do Sul State covering a total area greater than 3,5 million hectares. Three eucalyptus species were considered: Eucalyptus grandis W. Hill ex Maiden, Eucalyptus dunnii Maiden and Eucalyptus globulus Labill. Data were processed and integrated through statistical procedures and by using a GIS. The potential surface area for forestry surpasses 900,000 hectares, but a homogeneous exploitation of that territory is limited by legal restrictions.



The planning of agricultural cultivation and harvesting is a complex task. However, this area of study is still relatively young. This chapter focuses on the tactical and operational planning for sugar cane cultivation and harvesting which determines the best moment to harvest the fields, maximizing the total profit given by the sugar content within the cane. It considers resources such as cutting and transport crews, processing capacities in sugar cane mills, the use of maturation products and the application of vinasse on harvested fields. The mixed integer programming model extends the classical Packing formulation, adding a network flow structure to represent the harvest scheduling. All experiments were performed with real-world instances provided by a Brazilian sugar cane producer. The suggested solutions are discussed in terms of quality and use in practice.

In recent years Europe has seen important developments in the application of mathematical programming in the field of agricultural economics. This is due in part to the need for analytical tools to evaluate the impact of the Common Agricultural Policy (CAP) changes put in place since 1992 in the agricultural sector (the McSharry Reform), and also due to the appearance of positive mathematical programming (PMP) as devised by Howitt (1995). This technique, which can be applied when information is limited, gets round one of the greatest drawbacks of linear programming applications, i.e., the difficulty in obtaining realistic solutions when applying this technique in agricultural economics. PMP avoids this problem by way of the calibration of the model in a base year. In Spain, the PROMAPA model (Programación Matemática para el Análisis de Políticas Agrarias), a PMP model for analysing agricultural policy changes, was developed at the Universidad Politécnica de Madrid and the Consejo Superior de Investigaciones Cientificas. On the basis of the PROMAPA model results, in this chapter the authors will present the analysis of the impact of the new CAP reform (Council Regulation (EC) 1782/2003) adopted in Spain in 2006 on Spanish agriculture. They will analyse as well the impact of some variants of the measures of this reform that could be or are going to be adopted in Spain at some future date. Before presenting this analysis the authors will introduce first, a brief review of the applications and recent developments of PMP in the context of agricultural policy analysis; then, the mathematical basis of calibration in PMP and the method of calibration used in their application; and finally a description of PROMAPA, highlighting the formulation of the principal features of the CAP reform.



The present chapter provides a brief explanation on some aspects involved in the development of models and mathematical-modelling simulations, to show their benefits to the decision-making process in the environmental impact assessment of agriculture. Aspects concerning the agroecosystems were also presented toward the sustainability of Brazilian agricultural production systems. Some applications which have been developed in Brazil were pointed out, as well as a specific case study conducted at the Guarani aquifer recharge area located in Ribeirão Preto, São Paulo state, in order to show the influence of input data on the results provided by CMLS94 simulator.

land use, and social context. The perspectives of different stakeholders are represented as agents that make decisions on whether or not to buy flood insurance. The authors demonstrate that agent-based models can be important for policy issues in general, and for sustainable development policy issues in particular, by aiding stakeholder communication and learning, thereby increasing the chances of reaching robust decisions. The agent-based approach enables the highlighting and communication of distributional effects of policy changes at the micro-level, as illustrated by several graphical representations of outputs from the model.


*Pedro Luiz Pizzigatti Corrêa, Agricultural Automation Laboratory, Polytechnic School of the University of Sao Paulo, EPUSP, Brazil*
*Mariana Aparecida Carvalhaes, Brazilian Agricultural Research Corporation, EMBRAPA Middle-North, Brazil*
*Antonio Mauro Saraiva, Agricultural Automation Laboratory, Polytechnic School of the University of Sao Paulo, EPUSP, Brazil*
*Fabrício Augusto Rodrigues, Agricultural Automation Laboratory, Polytechnic School of the University of Sao Paulo, EPUSP, Brazil*
*Elisângela Silva da Cunha Rodrigues, Agricultural Automation Laboratory, Polytechnic School of the University of Sao Paulo, EPUSP, Brazil*
*Ricardo Luis de Azevedo da Rocha, Laboratory of Languages and Adaptive Techniques, Polytechnic School of the University of Sao Paulo, EPUSP, Brazil*

Computational modeling techniques for species geographic distribution are critical to support the task of identifying areas with high risk of loss of Biodiversity. These tools can assist in the conservation of Biodiversity, in planning the use of non-inhabited regions, in estimating the risk of invasive species, in the proposed reintroduction programs for species and even in planning the protecting endangered species. Furthermore, such techniques can help to understand the effects of climate change and other changes in the geographical distribution of species. This chapter presents concepts related to the species distribution modeling and algorithms based on Neural Networks and Maximum Entropy as alternatives for modeling of species distribution. The algorithms were integrated into the open source tool called openModeller used by biologists and other researchers in this area. A case study of modeling the distribution of babaçu (Orbignya phalerata) in the Piauí State - Brazil is presented, evaluating the potential distribution of this species used to produce bioenergy. Fifty models were generated and merged the ten models with best accuracy for each algorithm. The results show that the models obtained by both are consistent. The models obtained with Maximum Entropy seem to reflect best the reality, considering the occurrence pattern of babaçu as a secondary species.

Agriculture is vulnerable to the interannual climate variability and to its unpredictability, in such a way that most agricultural decisions taken within the time horizon of several months are made in a conservative manner, supposing a near-pessimist scenario. The improvement of climate prediction may help the strategic view, mitigating unwanted impacts and taking advantage of favorable conditions. This chapter presents the development of an Artificial Neural Network (ANN) model for seasonal precipitation forecast based on climate indices, focusing on the practical aspects of selecting the best predictors, defining ANN architecture, data handling and ANN training and validation. The study case is the Pardo/ Mogi-Guaçu rivers watershed in Brazil, which is characterized by intense sugarcane plantation for both ethanol and sugar industries. The results demonstrate how the methodology for seasonal precipitation forecast based on ANN can be particularly helpful, with the use of available time series of climate indices.



The present chapter reports on a practical case of decision making in agricultural TWW reuse. The studied site concerns the Cebala TWW irrigated perimeter in Tunisia, located in North Africa. Crucially though in this perimeter is that TWW is under-exploited. Thus, the proposed research dwells on the future of this perimeter which could oscillate between hope and deadlock. In case of hope, shall we improve the quality of the resource and possibly move to the tertiary treatment? And in case of deadlock, shall we give up the TWW reuse or substitute it by a conventional resource? To make an evaluation of this experience, the authors first gave a general overview of the descriptive study in a dashboard. Contingent valuation method was used to assess qualitative data and gauge the intangible effects. Then, the authors presented a systemic multicriteria method deployed to make decision-making process easier. Such a method or technique, which is grounded on Lp metrics, is called Compromise Programming.

# Preface

This book emphasizes scientific computing and applications in agriculture. It is said in one of its chapters that Mathematics is the stairs given to man by God to reach the infinity. Currently, we are living in the information age and we can not climb the steps of this stairs without information technology and quantitative methods. The Mathematics part of the statement has generalized to scientific computing, even considering its more basic philosophical foundations.

The chapters cover a wide range of applications and techniques from different countries, groups and scientific institutions. Although some of them are still in an early stage and better suited for academic or research environment, others are almost ready to be used in real conditions of rural areas. The richness of this book is precisely its diversity, which is capable of stimulating convergence among different areas of knowledge and may result in unexpected advances, both in the area of computational methods and in agricultural research.

**Chapter 1** has an introductory nature and aims at giving us an account about a successful agricultural research institution that owes an important part of its triumph to a strong investment in scientific computing. This chapter is a testimony from people who played important roles in the creation of the organization and in the adoption of scientific computing as a support for agricultural research.

Next, **Chapter 2** addresses particularly the study of the spatial motion of agricultural activities. The methodology based on statistical descriptive methods associated with concepts from elementary physics and from geographical processing allows the evaluation of the spatial concentration and the dynamics of agricultural products as well as visualizing the center of mass of crops productions. This methodology can be used in order to give an overall summary of the spatial changes along the years. The procedure also allows studying the agricultural development and is particularly relevant to countries where substantial changes in the rural space use are ongoing, as it occurs in Latin America.

Working in a similar domain, **Chapter 3** presents, in a didactic way, a method to assess changes in land cover using the approach of the transition matrix. The method allows the evaluation of gains and losses, swap, and net change in different types of land cover and also allows the identification of systematic transitions in a region between two points in time. The method is a powerful tool for evaluating trends in agriculture within the land use policies that seek to environmental sustainability.

**Chapter 4** presents an approach to analyze, monitor, and discover knowledge from remote sensing images associated to climate data. The techniques are based on the Fractal Theory, data streams, and time series mining: the FDASE algorithm; a method that combines intrinsic dimension measurements with statistical analysis; and the CLIPSMiner algorithm applied to multiple continuous time series of climate data. The monitoring tool presented permits that, instead of spending hours analyzing graphs and charts, the specialists may count on methods that spot the regions and the periods where they should pay more attention during the decision making process.

Also handling digital images, **Chapter 5** provides the theoretical and practical basis to estimate crop areas using statistical objective sampling and remote sensing data. Formulas are presented and can be applied to other land covers that occupy large portions of surface, such as forest, water bodies and urban areas. The estimate of crop areas is of great potential, since it allows predictions provided with quantification of the error. The method can be performed at different scales, from the smallest, as municipalities, to the larger, such as countries.

**Chapter 6** describes how to use the Geographic Information Systems - GIS to combine data of weather and soil in order to define areas of potential suitable for the cultivation of sugarcane. The system generates a map of the distribution of the risk of frost and, as a result, the crop management can be designed in a way that sugarcane achieves physiological maturity before the period of increased climate risk, decreasing the chances of low productivity. This application of GIS shows potential to become a powerful tool for management of rural environment and of public policy.

**Chapter 7** assesses the capacity of the self-organizing map, an unsupervised artificial neural network, for aiding the process of territorial planning through methods of visualization and clustering applied to a multivariate set of geospatial and temporal data. Population growth and hence the demand for quality public services and projects for regional infrastructure, require from public managers greater flexibility in decision-making so that they are observing the requirements of fairness, efficiency and effectiveness. Therefore, territorial typology provided by this method offers an important feedback for the development of a collective solution to issues relating to rural development.

To define the best area for commercial cultivation of eucalyptus in a certain region, **Chapter 8** uses satellite imagery, mapping and geo-referenced data on climate, the soil and watersheds, beyond the topography, with the aim of establishing forest policies considering socioeconomic issues. Specifically, this study allows evaluating the aptitude of a specific region, according to the environmental requirements for each species of eucalyptus, the soil capacity of water storage and the risk of frost.

**Chapter 9** addresses the complex task of planning cultivation and harvesting of sugar cane with focus on tactical and operational aspects which determine the best way to manage fields, maximizing profits. The proposed mixed integer programming model extends the classical Packing formulation, adding a network flow structure to represent the harvest scheduling. Experiments performed with real-world instances ensure a realistic presentation of the processes. The presented algorithms are promising to solve, in reasonable time, even larger problems as they appear in practice.

In agricultural economics, one of the greatest drawbacks of mathematical programming models to evaluate agricultural processes is to calibrate the model in a base year. That is because it is difficult, if not impossible, to introduce in the models all the variables affecting farmer's decisions to obtain reliable results. **Chapter 10** presents a way to get the calibration of these models even using limited information. By using the dual form of the original model, this methodology allows to reproduce the situation existing in a baseline situation of the unit modeled (e.g. farm or region). This method, called the Positive Mathematical Programming, is currently being used in a great number of analyses of new agricultural policies in Europe.

**Chapter 11** provides a brief explanation of some aspects involved with the development of models and mathematical-modeling simulations, to show their benefits to the decision-making process in the environmental impact assessment of agriculture. It shows that there are many types of mathematical methods to deal with decision making in this field of applications. The main models and simulators mentioned on the literature, as well as their uses, are outlined. The chapter also presents emerging trends in implementation of mathematical-modeling on agricultural research.

A linear approach is applied in **Chapter 12** to optimize crop rotation scheduling under some ecological restrictions. The work includes detailed description of three optimization problems that are solved in order to put together the ecological criteria with technical constraints specific of crops. To achieve its objectives, the authors use the technique of column generation to allow the construction of rotation plans in an iterative scheme. The results indicate that the methodology is appropriate and an incorporate new features as they become necessary or can discard those with less importance to a region or farming system specific.

**Chapter 13** discusses the important process of leaching of pesticides and their risk of groundwater contaminating. It presents the basic concepts of simulation models and shows the mathematical development to prepare the model to assess the risk of water contamination by pesticides. The chapter shows the efficiency of the Pesticide Leaching Models (PLM) in use in the European Community to define policies controlling the use of pesticides, and their possible use by the authorities responsible for managing the risk of groundwater contamination.

**Chapter 14** presents a simulation model structured into five modules, in which the use of multi-agent approach has the main role. It is a remarkable example of the combination of several techniques for the simulation of agricultural areas subject to flooding that offers the possibility to evaluate different strategies to be adopted according to the severity of the occurrence of floods.

Another use of models is demonstrated in **Chapter 15** that presents a fairly didactic approach of modeling technique to evaluate the spatial distribution of species. An application is made on data from Babaçu (Orbignya spp.), a palm tree native from Amazon region. Modeling techniques using neural networks and elements of descriptive statistics, together with the concept of maximum entropy, are used to evaluate the degree of loss of biodiversity. The use of the concept of ecological niche is very important in this context because it may allow assessment of the effects experienced by some species due to changes underway in the region.

The knowledge of precipitation conditions is so valuable information for agricultural management purposes that justify the continuous development and improvement of methods for forecasting. **Chapter 16** presents the development of an Artificial Neural Network (ANN) model for seasonal precipitation forecast based on climate indices, focusing on the practical aspects of selecting the best predictors, defining ANN architecture, data handling and ANN training and validation. Climate predictions can lead decisions towards a strategic view, aiming at minimizing unwanted impacts and taking advantage of favorable conditions. This approach can be successfully used to provide site-specific precipitation forecasts, valuable for the agriculture-related decisions.

Although the reuse of water, especially rainwater, has historically had a wide use in various regions of the world, there is still resistance to the reuse of wastewater, especially those used for agricultural irrigation. This important topic is approached in **Chapter 17** that aims at demonstrating the economic feasibility of reuse of water resources in agriculture, which can be one more element in environmental protection and aid to regions with water scarcity. The use of descriptive methods, based on the Lp metric and multicriteria approach is called compromise programming This methodology proves to be quite useful to define the reuse of wastewater or not, describing reuse as an option to be promoted and evaluated.

Facing another challenge, the focus of **Chapter 18** is in technological solutions to guarantee a reliable bovine traceability, what is very important for both producers and consumers of beef cattle. It is presented a decision support system dedicated to animal geolocation and to the analysis of sanitary problems. The system architecture is designed in three layers: acquisition, data management and spatial decision support. It has potential use to governmental institutions and farmers in case of a recall, saving time and costs apart from collaborating to achieve the food security.

Finally, aiming at contributing to the construction of a life cycle inventory database for agricultural production systems, **Chapter 19** proposes an integrated view for a data-construction methodology for agri-environmental assessment.

We would feel rewarded if this volume would help in improving the bridge between the academic people and agriculture practitioners community in benefit of the improvement of the food production in the world.

# Acknowledgment

Overcoming the challenge of organizing a book like the one we bring here would not be possible without the help of many professionals. From the decision of editing it until having the book published, we received the worthwhile support from people for whom we owe the success of this enterprise.

First, we are grateful for our Editorial Board that helped in defining the topics comprised in the book, in identifying prospective chapter authors, and in pointing out reviewers able to find among the many good proposals received those that compose the final manuscript.

Next, we would like to express our gratitude to Edilson Ferneda, Eduardo Delgado Assad, Eliane Gonçalves Gomes, Eliseu Roberto de Andrade Alves, Fernando Luis Garagorry, Geraldo da Silva e Souza, Paulo Martins Engel, Sandro da Silva Camargo, and Otávio Valentim Balsadi that, along with the editors, evaluated the proposals and helped in selecting the final book content.

Finally, we are grateful to the IGI Global team, particularly to Joel Gamon, who provided the fundamental support to develop, step by step, this project.

*Hercules Antonio do Prado*
*Brazilian Agricultural Research Corporation & Catholic University of Brasilia, Brazil*

*Alfredo Jose Barreto Luiz*
*Brazilian Agricultural Research Corporation, Brazil*

*Homero Chaib Filho*
*Brazilian Agricultural Research Corporation, Brazil*

# Chapter 1
# Scientific Computing in the Context of a Successful Agricultural Research Enterprise

**Geraldo da Silva e Souza**
*Brazilian Agricultural Research Corporation - Secretariat for Strategic Planning, Brasília, Brazil*

**Eliseu Roberto de Andrade Alves**
*Brazilian Agricultural Research Corporation - Secretariat for Strategic Planning, Brasília, Brazil*

## ABSTRACT

*Embrapa (Brazilian Agricultural Research Corporation), a governmental agricultural research institution from Brazil, is a case of successful organizational innovation that has as main characteristics: a public corporation model of organization; scale of operation at national level; spatial decentralization; specialized research units; enhanced training and remuneration of human resources and a vision of an agriculture based on science and technology. Moreover, from the beginning the organization has always been result oriented. Among the structural and political issues that led this enterprise to reach a well succeeded position, the authors argue that the strong application of scientific computing is the underlying reason that enabled high quality results achieved in research, development, and innovation. All of these reasons are presented in the next sections.*

## THE SUPPORT OF THE FEDERAL GOVERNMENT

This support has been critical to the survival of Embrapa. In the early years, it took the form of the federal government having understood the importance of technology for the development of agriculture. Once the results proved Embrapa could be profitable as an option for the government, the battle for budget support remained, but it takes place in an environment where the corporation is one of the priorities of government, both in the sphere of executive and the National Congress.

In the first twelve years of its existence, Embrapa was a promise: of bold and modern design, but still a promise. During those twelve years, huge investments were made in the training of human resources and infrastructure - about six billion dollars in 2008 value. The federal government paid for this investment based on the promise that Embrapa could be for the modernization Brazilian agriculture. Without the support from

DOI: 10.4018/978-1-61692-871-1.ch001

the federal government, Embrapa would not have been possible.

But Embrapa´s management has always been aware of the risk that the lack of achievements represented. For this, it led the research centers in a portfolio of research with short-term goals and to the conclusion of research already in progress. Moreover, it also gave special attention to the dissemination of existing results. And the media had a key role in creating the image Embrapa has. It is clear that the media not only operates on top of achievements, but also upon a consistent promise, provided it is not for long. In the early years of life of the corporation, the media bore the promise consistently, even in light of few existing results. This support was crucial to create a favorable image in society and in government. So, it has to be registered how important the development of competence to relate with the press, was for the success of the Embrapa. This relationship helped the government to justify the investment, over a period of lean achievements.

In a period of many macroeconomic imbalances and non-orthodox policies to deal with then, it is surprising that Embrapa´s budget support did not falter, which only shows the decisive support the federal government gave to it. At the stage where Embrapa was only a promise, the expenditures of Embrapa evolved linearly until 1982. This was essential to consolidate its image. From 1982 onwards, the corporation was no longer a promise, and its success will explain the government's continued investments.

## SCALE AND DECENTRALIZATION

Many wished that Embrapa would be small and only coordinated a research program run by the existing institutes and universities. This option was rejected because it was soon realized that, in a country of continental dimensions, the success of Embrapa depend on its size and an accumulated critical mass of researchers, diverse for talent, and branched throughout the national territory. It was understood that Embrapa needed to have the scale as large as Brazil and that it needed to have its own research network, so it could be direct responsible for the results, allowing it to be well known and evaluated on its own merits. This model would also allow for it to seek cooperation with universities, research institutes, private sector and overseas in a position of equals. Being large, diverse and decentralized, Embrapa would have conditions to represent the federal government in an area as important as agriculture and receive priority, both in the allocation of resources and with regard to institutional development.

It was very important for Embrapa to have a presence throughout the national territory. This presence helped to attract sympathy of the state governments and the National Congress. Embrapa has a marked presence in the Federal District, where are located its Headquarters and: Embrapa Technological Information, Embrapa Cerrados, Embrapa Vegetables, Embrapa Genetic Resources & Biotechnology, Embrapa Agroenergy, Embrapa Technology Transfer. The units in the Federal District are an important window of Embrapa. Being in the proximity of power, they have had important role in helping establish and solidify the image of the corporation near the central power and the international market.

## A CONCENTRATE ORGANIZATION MODEL FOR THE RESEARCH UNITS

Embrapa research units are distributed throughout the national territory and are specialized in products, natural resources or themes. For example, farmers know that the unit responsible for maize research is the Embrapa Maize & Sorghum, located in Sete Lagoas, Minas Gerais. Maize producers know where to go with demands for information and results, which will give them ownership in the center, providing help with the political leadership and the economic area of government. Similarly, researchers have the exact notion of their responsibilities, and no ambiguities regarding goals and actions needed. Even more, there are strong ties of solidarity and spirit of corps, as if all employees are committed to having a winning team. Thus, Embrapa´s model has aimed to facilitate and encourage interaction - researcher - farmer and researcher - society. It had an important role in preventing the dispersal of efforts, which is one of his pillars, and is an objective way of identifying priorities for research.

## HUMAN RESOURCES

The human resources policy is one of the main reasons for Embrapa success. The policy aims to develop the human capital of the corporation and it is from this capital that Embrapa derives its success. To stimulate creativity and by creating an environment that encourages coexistence is another specific role of this policy. The policy is based on the following specific points:

1. The establishment of a career that stimulates the desire to study and progress. It has three levels (called positions), where the level I is for someone who holds only a bachelor degree, level II for those with master's and, finally, level III is for those who hold Ph.D. degree. Each level has several sublevels;
2. A level of salary that allows the researcher to have a dignified living with his family based on the salary Embrapa pays him;
3. A retirement plan, with voluntary membership, paid by Embrapa and the employee, to support old age. The plan aims to supplement the retirement of the public social security;
4. A health plan paid by Embrapa and the employees, with two purposes: support the researchers and his families for expenses in health care, and preserve the health of researchers, which is the most important capital of the corporation.
5. A series of activities to stimulate the researcher to accumulate knowledge and experience, allowing him or her to be productive and to love the Embrapa. There is a complex system of promotion on merit, based on individual, group and the research unity. The aphorism, ingrained in Embrapa, says that every year it has to deliver technologies to justify to society the investments made and produce better-trained researchers. Thus, there are two products that Embrapa has to deliver - researchers always more competent and technologies.
6. A training program at graduate and post-graduate levels, that meets both the interests of the corporation and researchers, and which seeks to train them at the same levels of the best centers of advanced countries;
7. The corporation recognizes that the technology generated incorporates the effort of all its employees. Thus, the training program is available for everyone, but graduate training focuses on researchers, but it is not exclusive of them;
8. Each research unit has a critical mass of researchers. It is organized around a specific target audience, a clear main problem to solve and the team's responsibility towards society. Each unit is in itself an instrument of concentrated effort in research and provides a work environment that encourages human development, creativity and sense of usefulness to society;
9. Embrapa seeks to stimulate the researcher to be an entrepreneur in his field, to seek resources, to interact with the outside world and ensure the dissemination of technology. Commands another aphorism: research results in the drawer of the researcher, means that; it did not result in anything or the researcher is not good enough, or both;
10. Embrapa´s communication program aims to provide accountability for work, actions to disseminate research results, giving the

corporation visibility and transparency and valuing its employers. This communication program is organized around many ceremonies, some connected with the anniversary of the corporation and the research units, carried out in other seminars, symposiums in Brazil and abroad, and they all provide opportunities for learning and enhancing the employers. Thus, the communication program is also considered part of the human resource development policy, but with independent living;

11. Although the corporation is always looking for opportunities to improve its human capital, one has to plan for the future and the principle of orderly replacement has to prevail. In Embrapa´s case the goal is to maintain an average age of 45 years old for the PhDs, imagining the following guideline: on the average a researcher should be finish his Ph.D. work around 30 years old, which would leave him with a horizon of around 30 years of productive work. Half of this is 15. So, 15 years should be added to 30, comprising 45 years. Thus, on average, a young doctor has 15 years of work alongside senior researchers. A complementary strategy is for the creation of conditions that would allow for competent and outstanding retiring researchers to continue to do some kind of work with Embrapa. There is much to be done in this regard.
12. In the beginning, an enormous effort was made to integrate the researchers to the spirit of Embrapa, in training courses, meetings and direct communication from the direction of the corporation. As the corporation was young, a world of opportunity was opened to those who joined the effort. The leadership did not have time to stratify on a separate level. For this and by having the support of society, there were no major difficulties to develop the spirit Embrapa. Today, the fame of Embrapa helps to develop the spirit Embrapa, but we should not overlook this point.

In research, it is natural over time that seniority develops, that is how leadership solidifies, founded on knowledge and recognition. If these leaders do not have the ability to integrate with new researchers as part of their work, they will be disappointed to learn that they have no help to develop their careers. This disappointment can be transformed into rancor, leading quickly to an anti Embrapa attitude. This kind of problem cannot be solved by employing more democratic procedures. What should be done is to find mechanisms for promotion for those who can work in teams and spread their knowledge. Procedures have no place in making a competent researcher share more than what he has already reported in text or oral communication. However, a competent researcher has much more to offer.

## SCIENTIFIC COMPUTING

Scientific computing is the field of study concerned with constructing mathematical models and quantitative analysis techniques and using computers to analyze and solve scientific problems.

Much of agricultural research is based on decision under uncertainty. Modeling in such a context depends heavily on quantitative methods and ultimately on scientific computing. Sciences like Statistics, Econometrics, Operational Research and Biometrics among others are of importance to scientific computing.

With the increasing availability of computer resources new techniques of analysis have been developed to analyze data in general from which agricultural research has much to benefit. Research data accumulated for years can now be assessed via data mining and experimentation optimized with the help of simulation.

Scientific computing allows complex modeling of world markets of commodities and the evaluation of trends and demand and supply elasticities of ultimate importance for the implementation of public policies.

In Brazil of today we experiment a revolution in scientific computing. This goes from teaching in elementary schools to research in general. Agriculture is no exception. Indeed Agriculture has been a leading sector in the usage of scientific computing. As early as 1976, close to its very foundation, Embrapa – the leading state company responsible for applied agricultural research in Brazil, had already realized the importance of scientific computing in agricultural research, and created a quantitative methods department – DMQ, embracing four divisions: statistics, operations research and software analysis. The main objective of this administrative structure was to provide continuous support to experimental research, mimicking the operations of other successful institutions like Rothamstead in Great Britain. The new DMQ was filled with the best possible in human resources. The department was so successful that its development spilled over other research groups in Brazil. Indeed DMQ's research and support activities seeded the graduate programs of Statistics, Operations Research and Computer Science in the University of Brasília. In the 80's all research groups in the country engaged in agricultural research were associated with Embrapa in general and with DMQ in particular.

Today, DMQ, regrettably, no longer exists, but its inheritances are still present. From DMQ were originated Embrapa Agriculture Informatics, located inside the University of Campinas in São Paulo, a leading educational institution. This new entity certainly has less statistics and operational research but is still competent in the development of support software to agricultural applications.

Although critical to the analysis and development of agricultural research, scientific computing and its roll of disciplines, remains a bit obscure to a big chunk of the population of agronomists, foresters, agricultural engineers, zootechnicians, veterinarians and administrators responsible for research in agriculture. This is a symptom felt at the market and at the universities as well. This is the truth for Brazil and elsewhere. Quantitative methods require proper education. In essence they are pure art, servicing science and society, but they require sensitivity and intelligence for proper understanding and recognition of their necessity. When Embrapa was founded these features were abundant in its administration. It is of crucial importance to maintain agricultural research quality at a high level.

## OTHER IMPORTANT FACTORS

Without any further discussion we are going to nominate other factors that were important for Embrapa success. They are close integration to international research, ample exposure to the media and, at the first ten years, some research results that impressed Brazilian society such as the development of modern agriculture in the savannah region, nitrogen fixation in soybean to the point of 100% fixation of the plant needs and technologies that reduce the application of agro toxic and protect soils against erosion.

# Chapter 2
# Concentration and Dynamics of the Brazilian Agriculture

**Fernando Luís Garagorry**
*Embrapa - Secretariat for Strategic Management, Brazil*

**Homero Chaib Filho**
*Embrapa Cerrados, Brazil*

## ABSTRACT

*This chapter examines the measurement of motion in agriculture. Not only in Brazil, but in several Latin-American countries, important changes have been observed in the geographical distribution of the agricultural activity. They may take different forms, such as occupation of new areas or reallocation of particular products to areas that were already under some form of agriculture. Besides, in any year, for a given level of geographical subdivision, the distribution of a product shows spatial concentration; but the places which concentrate a substantial portion of the total production may change from year to year, and different subsets of the total territory occupied by a product may move at different speeds. Three distance measures will be introduced in order to assess the motion of individual products; two of them correspond to mathematical concepts, while the third one gives the terrestrial distance between national centers of gravity of the products.*

## INTRODUCTION

In many countries, particularly in South America, it has long been known that agriculture moves over the territory. Of course, in some places, the same products have been grown for many decades. Nevertheless, two facts can be easily deduced from available data: 1) new areas have been opened to agriculture; 2) some areas have shown a decrease in the presence of some products, which have moved to other places. Therefore, independently of the forces that may have caused de changes (market prices, urban developments, roads opening, population policies, etc.), the geographical distribution of many products has changed over the years. As a result, the subject of motion in agriculture can be a matter of systematic study.

In this chapter, some basic computational techniques, which are being used to study the motion taking place in the Brazilian agriculture, will be

DOI: 10.4018/978-1-61692-871-1.ch002

presented. Initially, it seems adequate to mention that the word "dynamics", as used in this chapter, is taken directly from physics, originally described in terms such as "the branch of mechanics concerned with the motion of bodies under the action of forces" (Oxford University Press, 1996). The only difference is that some basic ideas from classical dynamics are being applied to geographical "bodies" related to agricultural products. In this sense, the initial idea is to measure the motion, which can be deduced from agricultural statistics, by means of some concept of distance.

In order to study the motion of a product over the Brazilian territory, it is convenient, at some points, to consider some subsets of the whole area that it occupies. That is, although the motion of 100% of the product is routinely evaluated, some parts, such as 25, 50 or 75% have also been considered. In other words, the "bodies" whose motion is to be studied, in this chapter, are defined by sets of geographical subdivisions which hold the "first" 25, 50 or 75% of a product, besides the whole set occupied by 100% of the product. In this regard, the subject of geographical concentration entered from the beginning in the study of agricultural motion. In general, for any product, it can be said that although 100% of it can be spread over a large part of the Brazilian territory, only relatively small sets of geographical subdivisions hold the "first" 25, 50 or 75%. And, along the years, the distances travelled by these geographical sets are usually very different.

The main objective of this chapter is to present some measures of geographical concentration and distance, in order to assess the territorial changes that have taken place in the Brazilian agriculture. For the most part, the distances refer to some mathematical concept; but, also, a terrestrial distance (in kilometers), which measures the changes in location of the centers of gravity of a product, will be presented. The results obtained with these techniques will be illustrated with examples of a few products. On the overall, this type of work corresponds to a stage of diagnosis of territorial changes; no effort will be made in order to analyze the causes of these changes.

## BACKGROUND

The fact that the agricultural area of a country changes over time has long been a subject of study in Latin America. In many cases, the studies have explored the concept of agricultural frontier (Comisión Económica para América Latina, 1983). This approach has been particularly relevant in the case of the Brazilian agriculture (Mueller, 1983; Sawyer, 1983; Homem de Melo, 1983; Mueller & Martha Júnior, 2008). In this type of studies, there is a special interest in the description of changes in policies that resulted in the expansion of the agricultural area (internal migrations, credit programs, roads opening, irrigation projects, etc.). Nevertheless, Sawyer (1983) mentions that, for some products, the expansion to new areas may take place at the same time as a retraction from traditional areas. Therefore, even if characterized on largely qualitative terms, the idea that agriculture may move from some places to others, at least with reference to a single product, was already present in these studies. In general, the emphasis was placed on the historical, sociological and political aspects which caused the changes in the agricultural frontier. In any case, there was no systematic effort to provide some form of measurement of the extent of the observed changes.

In a broad sense, the study of changes in the spatial distribution of the agricultural activity could be placed under the general framework known as land use and cover change (LUCC), which is the subject of a substantial amount of work, mainly in Europe and the United States. So far, several scientific meetings and three complete issues of international journals have been devoted to the matter (Veldkamp & Lambin, 2001; Verburg & Veldkamp, 2005; Verburg et al., 2006). An impressive array of computational methods has been used to study observed changes or to produce scenarios of future developments in land use, in rural as well as urban areas. In particular, a number of metrics have been explored in order to assess the changes observed within a particular area (for instance, city growth). Nevertheless, it seems that the central subject addressed in this chapter, namely the study of agricultural motion, has remained outside the main issues covered by the LUCC studies.

A study conducted at a Government agency (Diniz, 1995) shows changes in the percent distributions of a few agricultural products, in the states and the five regions of Brazil, up to 1994 for grains and 1990 for cattle, but no joint measurement of the changes is provided. A similar starting point will be taken below, with the addition of some distances which summarize the magnitude of the changes.

A preliminary form of a technique to be shown in this chapter, that is, the changes in the centers of gravity, was used to study the motion of eight agricultural products with respect to the Amazon biome of Brazil (Simon & Garagorry, 2005). In that paper, only the changes in latitude were examined, while a center of gravity is defined by its latitude and longitude. The paper showed that the percentage contributions of cattle and soybean, as compared to their totals in Brazil, were increasing in the Amazon biome, while other products had a more erratic behavior.

More recently, a research project was conducted at the Brazilian Agricultural Research Corporation (Embrapa), in order to study the evolution of 35 agricultural products. For almost all products, the study covered the period 1975-2003. The list of products included 25 crops (both annual and perennial), 6 animals and 2 products from animals (cow milk and hen eggs). Besides the usual computations related to changes in volume (e.g., annual change rate), the project concentrated on the measurement of spatial changes in the geographical distribution of the products. A report on the main results is available in Portuguese (Garagorry & Chaib Filho, 2008). This chapter is largely based on that work, with the extension of the period studied up to 2005.

A follow-up project is being executed, with the addition of 14 products related to agro-energy, plus bushfires and deforestation. Although for the 49 products the data come from agricultural statistics, for the last two items they come from satellite images which, after processing, produce area data. The main idea is that the same techniques can be applied to the study of motion over the national territory, independently of the subject matter, as soon as area data are available.

## STATISTICAL DATA

Annual data were derived from those published by the Brazilian Institute of Geography and Statistics (IBGE), in the Municipal Agricultural Production (PAM) and the Municipal Livestock Survey (PPM) series, for 1975-2005. They are available both from the Agrotec data base at the Brazilian Agricultural Research Corporation (Embrapa) and, for the years since 1990, from the IBGE website (Instituto Brasileiro de Geografia e Estatística, 2009). These data are obtained, at the level of municipalities, by a panel method, from estimates of public and private agents involved in production, trade, research, inspection, technical assistance, financial support and other activities. Although they are not the result of rigorous statistical surveys, the resulting series are the best available sources of information. In the case of crops, there exist data for production, planted area, harvested area and commercial value; for animals, only livestock data are available. In this chapter, only data for production, in the case of crops, and livestock, for animals, will be used; for simplicity, they will be generically referred to as "volume" data.

In this chapter, the period of 1975 to 2005 will be considered. Since two states were created after 1975 (Mato Grosso do Sul and Tocantins), a reallocation of municipalities was performed in order to obtain a situation as if the new states existed prior to 1975. For instance, the state of Tocantins was separated from Goiás, which was and remains in the Center-West region of the country, while the new state was added to the North region of the official subdivision of Brazil. Naturally, although this procedure causes some discrepancies with old publications of IBGE, where the new states were not mentioned, it allows for meaningful comparisons over a period starting in 1975; for instance, in this chapter, the North region includes the state of Tocantins, as if it had existed, at least, since 1975.

The official Territorial Division of Brazil contains regions, states, meso-regions and micro-regions, besides the municipalities. Although the data are kept at the level of municipality, they are unsuitable for the study of the dynamics of agriculture over a period of more than a few years, in view of the frequent creation of new municipalities. Therefore, as a standard procedure, the municipal data were aggregated into their corresponding micro-regions, which constitute the geographical level immediately above that of municipality. The number of micro-regions is fixed at 558 for the whole country.

Some of the techniques presented in this chapter could be applied to any level of subdivision; nevertheless, this chapter will concentrate on the levels of region and micro-region. The section related to centers of gravity will consider all the 557 micro-regions in the mainland, with the exclusion of the remote island of Fernando de Noronha which, in any case, has little agricultural significance.

The results will be shown only for the years of 1975, 1985, 1995 and 2005. Although calculations and analyses are routinely updated on an annual basis, the four selected years will be sufficient to illustrate the main types of results. Rice production will be taken as the basic reference to show the methodology, which has so far been applied to the study of more than 50 products. At some points, a few results for other products (crops or animals) will be shown, in order to further illustrate the methodology.

According to the official Territorial Division of Brazil, the country is divided into five regions: North (N), North-East (NE), South-East (SE), South (S) and Center-West (CW). As a starting point, Table 1 shows the totals of rice production (in metric tons), for Brazil and the five regions, in the selected years.

The usual analysis of these data starts with the calculation of geometric annual rates of change, for the whole period or for some sub-period. As a simple example, from the direct comparison of the values for 1975 and 2005, without adjusting an exponential statistical model, the following annual rates of change are obtained for Brazil and the five regions: 1.78% (BR), 4.41% (N), -0.28 (NE), -4.46% (SE), 3.07% (S) and 1.73% (CW). That type of analysis will not be pursued in this chapter. The only point to be mentioned is that, if the same annual rates of change were obtained for the five regions, there would not be any reason to explore a geographical reallocation of the product, at least at the regional level. In that case, a change in volume (positive or negative) could have taken place, but there would be no motion (or dynamics, as used in this chapter). The fact

*Table 1. Rice production (t) in Brazil and the regions*

| Year | Brazil | N | NE | SE | S | CW |
|---|---|---|---|---|---|---|
| 1975 | 7,781,538 | 406,239 | 1,292,059 | 1,424,996 | 2,946,965 | 1,711,279 |
| 1985 | 9,024,555 | 858,660 | 1,138,697 | 1,556,854 | 3,949,764 | 1,520,580 |
| 1995 | 11,226,064 | 1,125,430 | 1,732,323 | 990,563 | 5,954,577 | 1,423,171 |
| 2005 | 13,192,863 | 1,481,872 | 1,189,173 | 363,030 | 7,295,967 | 2,862,821 |

that, in the case of rice, the rates were very different indicates that there was some type of interregional motion of the product.

## REGIONAL LEVEL

### Concentration

In order to study concentration and dynamics at the regional level, the usual approach starts with the determination of the percent contribution of each region to the national total, for each year. The resulting distributions, derived from the values in Table 1, are shown in Table 2, together with their concentration index (Gini).

The increasing values of the Gini's index summarize a clear path in the evolution of the concentration. In fact, in this example, it could be easily detected by visual inspection of the distributions; for instance, the joint contribution of the South and Center-West regions increases along the period, from a value close to 60%, in 1975, to 77%, in 2005. In other cases, a visual assessment is not as clear-cut, and a measure of concentration is helpful to detect some structural changes.

### Dynamics

Independently of the increase in concentration, there are other changes in the distributions; for instance, the contribution of the Center-West region falls from 1975 to 1995, and then increases nine percentage points from 1995 to 2005. Figure 1 presents a graph which shows the changes in the percent contribution of each region.

*Table 2. Rice production distributions (%) in each year and Gini's concentration index*

| Year | N | NE | SE | S | CW | Gini |
|---|---|---|---|---|---|---|
| 1975 | 5.22 | 16.60 | 18.31 | 37.87 | 21.99 | 0.35 |
| 1985 | 9.51 | 12.62 | 17.25 | 43.77 | 16.85 | 0.37 |
| 1995 | 10.03 | 15.43 | 8.82 | 53.04 | 12.68 | 0.47 |
| 2005 | 11.23 | 9.01 | 2.75 | 55.30 | 21.70 | 0.59 |

*Figure 1. Changes in the regional contributions to the national totals of rice production*

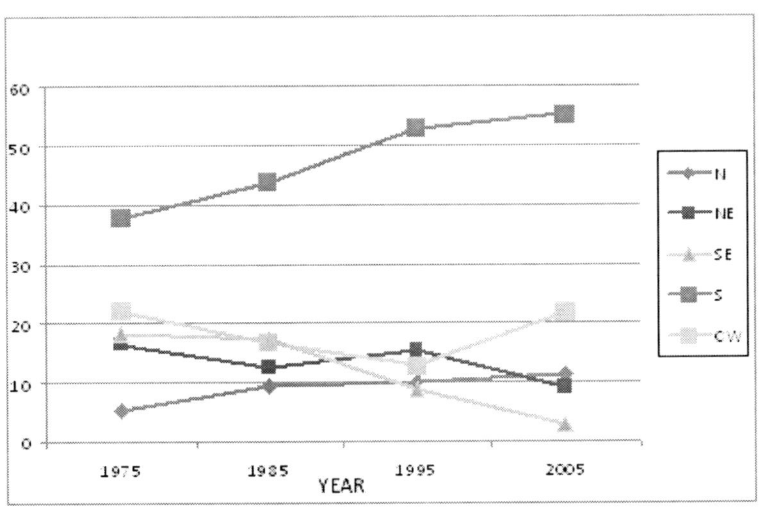

Table 3 shows one type of distance that has been used to evaluate the changes in the distributions corresponding to each pair of the selected years (denoted as "initial year" and "final year"). It is a special form of the well-known $L^1$ distance used in mathematics, which has been standardized in order to take values between 0 and 1. For simplicity, the same name $L^1$ will be used. Namely, if $f_{sk}$ and $f_{tk}$ are the percent contributions of region $k$ ($k = 1, \ldots, 5$) in the years $s$ and $t$, the $L^1$ distance between the distributions in these two years was calculated with the following formula:

$$d(s,t) = \left(\sum_{k=1}^{5} \left| f_{sk} - f_{tk} \right| \right) \Big/ 200$$

In each case, the $L^1$ distance shown in Table 3 provides a lower bound for a comparison with the maximum possible distance that could have been theoretically reached in a specific period. For instance, in the period from 1975 to 2005, considering as fixed the distribution for 1975, the maximum distance would be calculated as follows:

- the minimum contribution in 1975, of 5.22%, happened in the North region;
- the distribution which would have a maximum distance from that in 1975 is the one with 100% in the North region and 0% in the other regions;
- the $L^1$ distance between these two distributions is given by $d = {}^{2}(100 - 5.22)/200 = 0.95$;
- since 0.23/0.95 = 0.24 = 24%, this means that the distribution for 2005 moved away from that in 1975 by a distance which is 24% of the theoretically maximum possible.

*Table 3. Rice production: $L^1$ distance between the distributions corresponding to each pair of selected years*

| **Initial year** | **Final year** 1985 | 1995 | 2005 |
|---|---|---|---|
| 1975 | 0.10 | 0.20 | 0.23 |
| 1985 | — | 0.13 | 0.18 |
| 1995 | — | — | 0.12 |

As seen in this example, the usual situation is that the original $L^1$ distance provides a good estimate that can be interpreted as a percentage of the theoretically maximum distance. In most cases, for a given initial distribution of agricultural production, a theoretical counterpart which would be at a maximum distance is totally unrealistic. For instance, in the case of rice, it is difficult to imagine a situation where the total production would be concentrated exclusively in the North region. In any case, a distance of 0.23, as that shown in Table 3, from 1975 to 2005, can be interpreted as an indication that there was a substantial change in spatial distribution, at the regional level.

## MICRO-REGIONAL LEVEL

In order to study concentration and dynamics at the micro-regional level, the following starting steps have been followed:

1. in each year, for each product, the micro-regions with statistical record of the product were ordered by "volume" (that is, by production in the case of crops and by livestock for animals) and the percent contribution to the annual total, of each micro-region, was determined;
2. the percent contributions were accumulated and each micro-region was allocated to one of the quarters; namely, if ACPCT denotes the accumulated percentage up to a micro-region, the following rules were applied:
   - if ACPCT ≤ 25 then the micro-region was in the first quarter (Q1);
   - else, if ACPCT ≤ 50 then it was in the second quarter (Q2);

- ◦ else, if ACPCT ≤ 75 then it was in the third quarter (Q3);
- ◦ else, it was in the fourth quarter (Q4).

In this way (even in the case of an exact equality of ACPCT with one of the cutting values, or quartiles), it was ensured that Q4 was formed by the minimum number of micro-regions which were sufficient to reach 25% of the total volume, the union of Q4 and Q3 was formed by the minimum number of micro-regions sufficient to attain 50%, and so forth.

Additionally, for convenience, the following groups (sets of micro-regions) were defined:

- group 25 (or G25), which coincides with Q4;
- group 50 (or G50), which is the union of Q3 and Q4;
- group 75 (or G75), which is the union of Q2, Q3 and Q4;
- group 100 (or G100) which is the set of all the micro-regions with record.

## Concentration

Table 4 shows the resulting distributions of micro-regions, into the four quarters, for rice production in the selected years. As it can be seen, in any year, very few micro-regions are in Q4, and most of then are in Q1. By construction, unless there is an exact equality of ACPCT at the 25 or 75% quartiles, since the micro-regions are discrete units, usually the joint production of those in Q4 is more than 25%, while that of those in Q1 is slightly less than 25% of total production. The high values of the Gini's index merely confirm what is clearly visible in the frequency distributions; nevertheless, the trend in that index is useful to indicate a well defined increase in concentration along the years.

As it can be seen in Table 4, 11 micro-regions were sufficient to account for 25% of total production in 1975, while only 4 were required in 2005.

*Table 4. Rice production: distribution of micro-regions into quarters and Gini's concentration index*

| Year | Q1 | Q2 | Q3 | Q4 | Total | Gini |
|---|---|---|---|---|---|---|
| 1975 | 395 | 59 | 24 | 11 | 489 | 0.809 |
| 1985 | 400 | 63 | 21 | 7 | 491 | 0.829 |
| 1995 | 423 | 44 | 11 | 5 | 483 | 0.888 |
| 2005 | 411 | 21 | 9 | 4 | 445 | 0.924 |

In a similar way, while 94 micro-regions formed G75 in 1975, only 34 were sufficient to reach 75% of total production in 2005.

The fact that few micro-regions form G25, G50 or G75, as compared to the number of those in Q1, is not a peculiar attribute of rice production. Quite to the contrary, it is the general situation with regard to the spatial distribution of any agricultural product, in terms of micro-regions. Table 5 provides a small example with several products, showing their distributions into quarters in 2005. The "weight" indicates the percentage of micro-regions for each product, in relation to the total number of micro-regions.

Table 6 shows the percent distributions corresponding to the frequencies in Table 5, together with their weighted mean, where the weights are those in Table 5.

*Table 5. Example of contingency table for several products, with their weight, in 2005*

| Product | Quarter | | | | | |
|---|---|---|---|---|---|---|
| | Q1 | Q2 | Q3 | Q4 | Total | Weight |
| Cattle | 412 | 85 | 36 | 22 | 555 | 24.77 |
| Grapes | 196 | 6 | 1 | 1 | 204 | 9.10 |
| Melon | 99 | 3 | 1 | 1 | 104 | 4.64 |
| Rice | 411 | 21 | 9 | 4 | 445 | 19.86 |
| Soybean | 218 | 26 | 11 | 4 | 259 | 11.56 |
| Pigs | 399 | 106 | 38 | 11 | 554 | 24.72 |
| Wheat | 94 | 13 | 8 | 5 | 120 | 5.35 |
| **Total** | 1,829 | 260 | 104 | 48 | 2,241 | 100.00 |

*Table 6. Distribution (%) of micro-regions in the quarters, for seven products, and weighted mean distribution, in 2005*

| Product | Quarter | | | |
|---|---|---|---|---|
| | Q1 | Q2 | Q3 | Q4 |
| Cattle | 74.23 | 15.32 | 6.49 | 3.96 |
| Grapes | 96.08 | 2.94 | 0.49 | 0.49 |
| Melon | 95.19 | 2.88 | 0.96 | 0.96 |
| Rice | 92.36 | 4.72 | 2.02 | 0.90 |
| Soybean | 84.17 | 10.04 | 4.25 | 1.54 |
| Pigs | 72.02 | 19.13 | 6.86 | 1.99 |
| Wheat | 78.33 | 10.83 | 6.67 | 4.17 |
| **Weighted mean** | 81.62 | 11.60 | 4.64 | 2.14 |

A few products have an old and wide presence in the national territory (that is, they have statistical records in almost all the 558 micro-regions); in their case, and for a few other products, it may happen that, in some years, slightly over 25% of the micro-regions with record are required to reach 75% of their national volume. Therefore, with regard to concentration, a kind of empirical (statistical) result can be stated as follows: in general, with only a few exceptions, in any year, 75% of the total volume of a product can be reached with less than 25% of the micro-regions with record of the product. The last row of Table 6 illustrates this conclusion. Results of a similar type can be stated for groups 25 and 50. Some cases of extreme concentration are not rare; for instance, in the case of grapes, one micro-region (Caxias do Sul, in the state of Rio Grande do Sul) has always been responsible for well over 35% of total production; and similar situations have been registered with melon (in Mossoró, state of Rio Grande do Norte) and papaw (in Porto Seguro, state of Bahia), in more recent years.

The important point to be mentioned here is that the geographical distributions into quarters, similar to those in Tables 4 and 5, indicate different types of agriculture. For instance, in the upper quarter (Q4), modern forms of commercial agriculture predominate, while in the lower quarter (Q1) different forms of subsistence or introductory agriculture are to be found. As an example of "introductory agriculture", the case of soybean may be mentioned; that is, some micro-regions which belonged to Q1 in 1975 (or, in fact, which did not have any record of soybean production at that time, but appeared in Q1 a few years later), were firmly in Q4 by 2005. This type of analysis has clear implications for agricultural research, extension services, industrial processing, logistics and so on.

## Dynamics

Spatial changes at the micro-regional level have been studied for the four groups defined above. Of particular interest have been groups 25 and 75. In the case of rice production, Table 7 shows the changes of micro-regions that happened in G75 between 1975 and the other selected years. That is, in this example, G75 is the "body" whose motion is to be studied. As mentioned before, G75 was formed by 94 micro-regions in 1975. Of this total, only 57 remained in 1985 (column "a"); 37 left (column "b") and were replaced by 34 new ones (column "c"). Of course, the interpretation is the same in the other years. The subset of micro-regions in column "a" has been called the persistent part of the comparison; it corresponds to the intersection of the set of micro-regions which formed G75 in 1975 with the set of the micro-regions in that group in the other selected year. It is not implied that the 37 micro-regions which, in 1985, had left the initial set do not produce any rice; they were simply replaced by 34 new ones because these produced larger quantities. In this way, in any year, the sum of the numbers in columns "a" and "b" is always 94, corresponding to the initial situation. For the year 2005, the sum of the numbers in columns "a" and "c" is 34, as mentioned before, in the comments following Table 4. That is, along the years, some

*Table 7. Rice production: dynamics of G75 between 1975 and the other selected years*

| | Subset | | | Total | Persist- | Distance | |
|---|---|---|---|---|---|---|---|
| Year | b | a | c | (a+b+c) | ence | Cantor | $L^1$ |
| 1985 | 37 | 57 | 34 | 128 | 0.45 | 0.55 | 0.42 |
| 1995 | 60 | 34 | 26 | 120 | 0.28 | 0.72 | 0.53 |
| 2005 | 75 | 19 | 15 | 109 | 0.17 | 0.83 | 0.65 |

micro-regions in the initial G75 move to column "b", and are replaced by new ones whose numbers appear in column "c". This process defines the motion of G75.

In order to have some measure of the motion that has taken place, two simple mathematical concepts of distance have been used. The first one is obtained from Jaccard's matching coefficient (Anderberg, 1973). This coefficient is calculated as follows: $p = a / (a + b + c)$; in this chapter, it is being called "persistence" only to emphasize that it is a matching (intersection) between two sets at different points in time. Its complement to unity, that is, $d = 1 - p = (b + c) / (a + b + c)$, defines a metric which does not seem to have a standard name; sometimes, it has been called Cantor's distance, because it appears, in measure theory, as a distance between two sets, and Cantor was the creator of set theory. This distance is based exclusively on counting the geographical units involved (in this example, they are micro-regions); it does not use any additional information, such as the percent contribution of each unit. Nevertheless, besides being simple to calculate, it has a natural interpretation. For instance, a Cantor's distance of 0.83, between 1975 and 2005, as shown in Table 7, can be interpreted in the sense that there was a geographical change of the order of 83% of the maximum possible. This maximum, which is 1, would be reached in a case where all the micro-regions in the initial G75 were replaced by new ones. Therefore, in this example with rice, there was a substantial change, from 1975 to 2005, in the geographical distribution of G75.

The second distance, which is a variant of the $L^1$ distance already introduced, takes into account the contribution of each micro-region. Having chosen a group (such as G75, in the example of Table 7), in each selected year, the volumes contributed by all the micro-regions in the group are added, and the percent share of each one, in that total, is determined. Therefore, in each year, a percent distribution is obtained. In order to calculate the distance between two years, the union of the sets of micro-regions in the respective groups is formed, and each distribution is extended, if required, over that union. The extension of a distribution is made by assigning zeros to micro-regions which, originally, are only in the other distribution. In this way, the $L^1$ distance can be calculated as before, with the adjustment that the summation in the formula runs over the union of the two groups of micro-regions. Another way of interpreting the use of the $L^1$ distance, at the micro-regional level, is that there is a vector space of 558 dimensions, where each micro-region is associated with a fixed dimension (that is, with a fixed component of the vectors). Once a group, such as G75, has been chosen, the percent contributions to the total in that group are placed in the appropriate components of a vector, in each year, and the remaining components are filled with zeros. In this interpretation, the summation in the formula for $L^1$ runs over the 558 components. In practice, the first interpretation has been used in the computer programs.

The interpretation of a value of 0.65 for $L^1$, as in the last row of Table 7, is the same as before: in the $L^1$ metric, 75% of rice production has moved, from 1975 to 2005, by a distance which is greater than or equal to 65% of the theoretically maximum possible. In any case, this lower bound of 65% captures a substantial dislocation of G75 over the entire period. Although, in each case, the exact percentage of the $L^1$ distance, with respect to the maximum possible, can be easily calculated and, in this way, a corrected distance can be obtained, the original form has been used in order to facilitate some types of comparison. In any way, in most cases, it is difficult to imagine that the theoretically maximum distance could ever be attained. Therefore, the $L^1$ distance, as shown in Table 7, provides an acceptable measure of the magnitude of the motion that has taken place.

In some cases, the contribution of the persistent part decreases rapidly along the years. The case of rice, as shown in Table 8, illustrates a situation where that did not happen. For instance, the 19 micro-regions, which remained in the persistent part in 2005 (Table 7), contributed with 29.83% of total production, in 1975, and with 52.18%, in 2005.

Table 9 gives the Cantor's distances for G75 of rice production, for any pair of selected years. Of course, the numbers in the row corresponding to the initial year of 1975 coincide with those in Table 7. The numbers in the diagonal show that, in each decade, there were substantial changes in micro-regions, between the respective initial and final year. Increasing values in the rows of the initial years of 1975 and 1985 indicate progressive departures from the respective starting geographical distributions. This may not always be the case, and some forms of "return" towards a starting distribution can be detected in this way, particularly if smaller time intervals are used.

Table 10 shows a brief example of a different form of analysis. In fact, calculations as those producing the results shown in Tables 7 to 9 have been conducted for the four groups. In Table 10, only the results for the comparison between 1975 and 2005 are shown. Of course, the numbers in the row of G75 coincide with those in Table 7 for the year of 2005.

In order to fix the ideas, only the column of Cantor's distance will be considered. Usually, the distance corresponding to G100 is smaller than those for the other groups, and this is not different in the case of rice. The most interesting part comes from the comparison of the other three groups. For instance, in this example, the distance decreases from G25 to G50, and increases from G50 to G75, reaching almost the value that it has in G25. There is no general rule relating to this type of behavior and, for each product, these changes indicate a peculiar form of spatial evolution, at the level of micro-regions. Nevertheless, for many products, including rice, the value for G25 is larger than those for the other groups and, whenever this has happened, it has called the attention to the need of a more detailed inspection. In fact, at a first approach, it may be thought that the micro-regions in G25, which are clearly associated with some form of advanced, commercial agriculture, should also display a rather static

*Table 8. Rice production: percent contribution of the subsets of G75, with 1975 as initial year; the persistent part appears in the initial year (ia) and in the final one (fa)*

| Final year | Percentage of contribution | | | |
|---|---|---|---|---|
| | b | ia | fa | c |
| 1985 | 17.63 | 57.39 | 58.66 | 16.35 |
| 1995 | 34.46 | 40.56 | 58.88 | 16.20 |
| 2005 | 45.19 | 29.83 | 52.18 | 22.83 |

*Table 9. Rice production: Cantor's distances between two years, for G75*

| Initial | Final year | | |
|---|---|---|---|
| year | 1985 | 1995 | 2005 |
| 1975 | 0.55 | 0.72 | 0.83 |
| 1985 | — | 0.50 | 0.71 |
| 1995 | — | — | 0.55 |

*Table 10. Rice production: persistence and distances for the different groups, between 1975 and 2005*

| Group | b | a | c | a+b+c | Persist. | Cantor | L[1] |
|---|---|---|---|---|---|---|---|
| G25 | 9 | 2 | 2 | 13 | 0.15 | 0.85 | 0.68 |
| G50 | 25 | 10 | 3 | 38 | 0.26 | 0.74 | 0.61 |
| G75 | 75 | 19 | 15 | 109 | 0.17 | 0.83 | 0.65 |
| G100 | 66 | 423 | 22 | 511 | 0.82 | 0.17 | 0.57 |

behavior. But, as found in the case of rice, soybean, cotton and many other products, several micro-regions, which were in the upper quarter in 1975, were overtaken by others along the years. There are several causes that, for each product, can be associated with these changes (e.g., population policy favoring internal migrations, agricultural policy to stimulate a large increase in production, attractive international prices, release of new varieties adapted to the tropical climate); as a result, new areas have been opened to commercial agriculture, and the production in some of the "new" micro-regions, in more recent years, can be much greater than that in the "old" ones.

Figure 2 shows a type of statistical map that has been used in order to illustrate the changes that have taken place, for a given group, between two years, at the micro-regional level. Instead of showing two maps, side-by-side, one for each year, which makes rather difficult a visual comparison, the type of map used in Figure 2 is the result of the superimposition of the map for the final year over that for the initial year. The painted areas in Figure 2 correspond to the numbers in row 2005, columns "b", "c" and "a", of Table 7, according to the following labeling: 1) label "1975" refers to the 75 micro-regions which were present in 1975 but not in 2005 (Table 7, column "b"); 2) label "2005" refers to the 15 micro-regions which were present in 2005 but not in 1975 (Table 7, column "c"); 3) label "1975/2005" refers to the persistent part, that is, the 19 micro-regions which were present both in 1975 and 2005 (Table 7, column "a"). This type of map shows precisely the persistent part as the intersection of the painted areas corresponding to each year, as if two different maps had been prepared and then superimposed; that is, 75 + 19 = 94 micro-regions in 1975, and 19 + 15 = 34 micro-regions in 2005, with 19 in the intersection.

## CENTERS OF GRAVITY

The concept of center of gravity is useful to provide another visual and quantitative description of agricultural dynamics. Strictly speaking, in the present situation, it would be a center of mass, since no gravitational field is involved; nevertheless, the expression has been widely used in statistics, usually in the sense of some form of weighted mean. In fact, a weighted mean as that presented in the last row of Table 6 is called a center of gravity by the French school of data analysis (Volle, 1997).

In order to use the concept of center of gravity in the description and measurement of movements in agriculture, the following steps were followed:

- the latitude and longitude of a point, called a centroid, was determined in each micro-region, using a GIS package;
- for each product and each one of the selected years, the volume (that is, crop production or animal livestock) in a micro-region was allocated to the centroid, as the mass to be used in the calculation;
- each center of gravity (given by latitude and longitude) was determined with a program used in geodesic computing;

*Figure 2. Dynamics of rice production: comparison of G75 in 1975 and 2005*

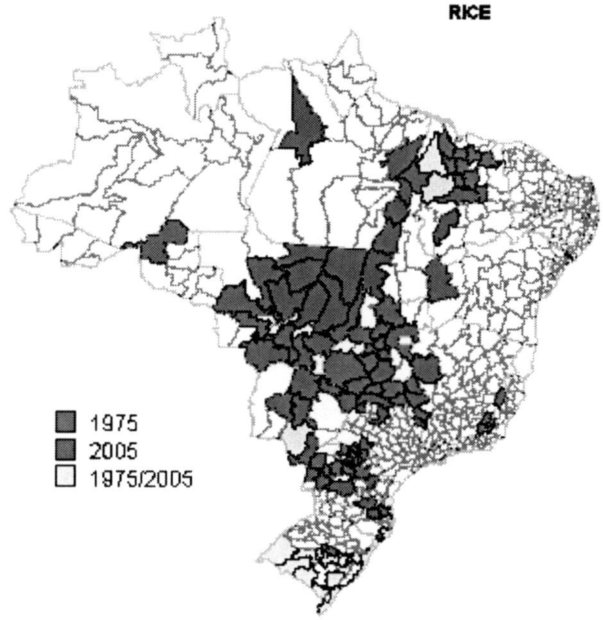


- the resulting points were plotted by a GIS package;
- the terrestrial distance between each pair of centers of gravity was calculated by another program used in geodesic computing;
- finally, a simple statistics, called curvature, was calculated; this is useful to assess the presence of spatial trends in the locations of the centers of gravity, and to detect outliers that might have been caused by some special events.

The central point here is that the center of gravity provides a simple weighted mean of the geographical distribution of a product. In what follows, only national centers of gravity will be considered (that is, they take into account the spatial distribution of the volume in the whole country, excluding the remote island of Fernando de Noronha); but they may be determined for any geographical area made up of micro-regions, such as one of the five regions or a state. Figure 3 shows the national centers of gravity for rice production, in the selected years.

Table 11 gives the terrestrial distance (in km) between any two centers of gravity shown in Figure 3. Increasing values in the lines corresponding to the "initial years" of 1975 and 1985 indicate a progressive departure from the geographical distributions which existed in these years. This fact does not always happen, and some products show a more erratic behavior. Nevertheless, this fact alone is not sufficient to characterize a specific geographic trend in the motion of the centers of gravity. In order to assist in the automatic detection of some outliers or other forms of erratic behavior, a simple measure of curvature was introduced.

*Table 11. Terrestrial distance (km) between two centers of gravity, in the selected years*

| Initial | Final year | | |
|---|---|---|---|
| year | 1985 | 1995 | 2005 |
| 1975 | 129 | 201 | 285 |
| 1985 | — | 79 | 193 |
| 1995 | — | — | 126 |

*Figure 3. Rice production: national centers of gravity*

The whole period from 1975 to 2005 will be considered, so as to illustrate the calculation of the index of curvature. According to the triangle inequality, the sum of the distances in the diagonal of Table 11 has to be greater than or equal to the distance from 1975 to 2005. If equality holds, then the centers of gravity are on a geodesic line (it can be interpreted as a "straight line" on the earth's surface); some allowance has to be made for rounding errors, since the computer program calculates the distances in meters, and they have been rounded to kilometers. In the present example, the index of curvature is calculated as follows:

$$curv(1975, 2005) = 1 - 285 / (129 + 79 + 126) = 0.15.$$

The value of the index of curvature is always between 0 and 1. If it is zero, the centers of gravity are on a geodesic line. In general terms, a value of 0.15, as in the example, is not to be interpreted as showing an erratic behavior; quite to the contrary, it indicates some form of geographic trend, which agrees with what can be seen on Figure 3.

Table 12 presents other examples of distances between the centers of gravity corresponding to the years shown in the column headings, as well as the resulting indexes of curvature for the whole period from 1975 to 2005. According to these results (allowing for rounding to two decimal places), in the cases of cattle, melon and soybean the centers of gravity have followed practically perfect geodesic paths.

Figures 4 to 8 show the motion of the centers of gravity for these products. Additional comments may help to clarify the analysis.

In the case of cattle, a clear trend shows that the center of gravity is moving towards the north-west, which agrees with the increasing activity of cattle growing in the states of Mato Grosso and Pará, in areas that correspond to the "Legal Amazon" and, even more strictly, to the Amazon biome (Simon & Garagorry, 2005).

In the case of grapes, there are two great regional poles: a) one in the south, in the micro-region of Caxias do Sul (state of Rio Grande do Sul), which can be easily related to Italian immigration at the end of the nineteenth and beginning of the twentieth century; b) one in the north-east, in the

*Table 12. Other examples of terrestrial distances (km) and index of curvature for the period 1975-2005*

| Product | 1975/1985 | 1985/1995 | 1995/2005 | 1975/2005 | Curvature |
|---|---|---|---|---|---|
| Cattle | 111 | 139 | 192 | 434 | 0.02 |
| Grapes | 60 | 363 | 191 | 495 | 0.19 |
| Melon | 873 | 746 | 21 | 1,600 | 0.02 |
| Soybean | 311 | 246 | 381 | 938 | 0.00 |
| Wheat | 272 | 58 | 25 | 249 | 0.30 |

*Figure 4. Cattle livestock: national centers of gravity*

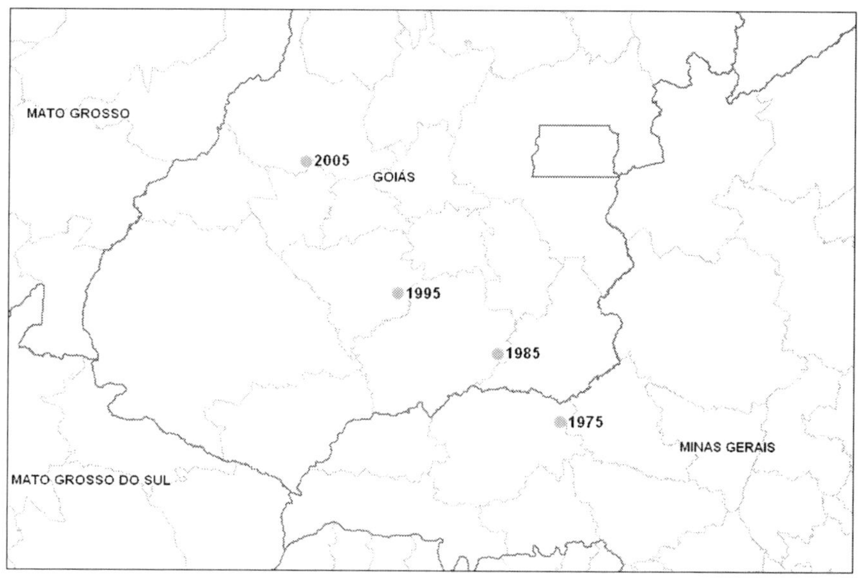


neighboring micro-regions of Petrolina (state of Pernambuco) and Juazeiro (state of Bahia), both over the São Francisco river, where large irrigation projects have been implemented after 1980. The map on Figure 5, where the center of gravity for 1985 was to the south of that for 1975, prompted to a more detailed analysis; that is, a curvature of 0,19, which could be accepted, as it is, showing a definite trend, nevertheless may call the attention to the need of a more careful assessment. In fact, the review of statistical records, together with documents of that time, showed that, in 1985, there was an exceptionally high harvest of grapes in the state of Rio Grande do Sul (Mello et al., 2008). Taking that into account, it was found that, if the point for the year of 1985 were disregarded, as being an outlier, the remaining three centers of gravity were on a geodesic path, moving from the South towards the North-East region. In fact, the terrestrial distance between the centers for 1975 and 1995 was of 304 km; since $d(1975, 1995) + d(1995, 2005) = 304 + 191 = 495 = d(1975, 2005)$, where $d(s, t)$ denotes the terrestrial distance between the centers of gravity corresponding to the years $s$ and $t$, then the three points are, for practical purposes, over a geodesic line.

The case of melon illustrates one of the most striking examples of dynamics in the Brazilian agriculture. Separate results obtained from the spatial distributions over geographical subdivisions (at the regional, state and micro-regional levels) had pointed to a substantial displacement

*Figure 5. Grapes production: national centers of gravity*

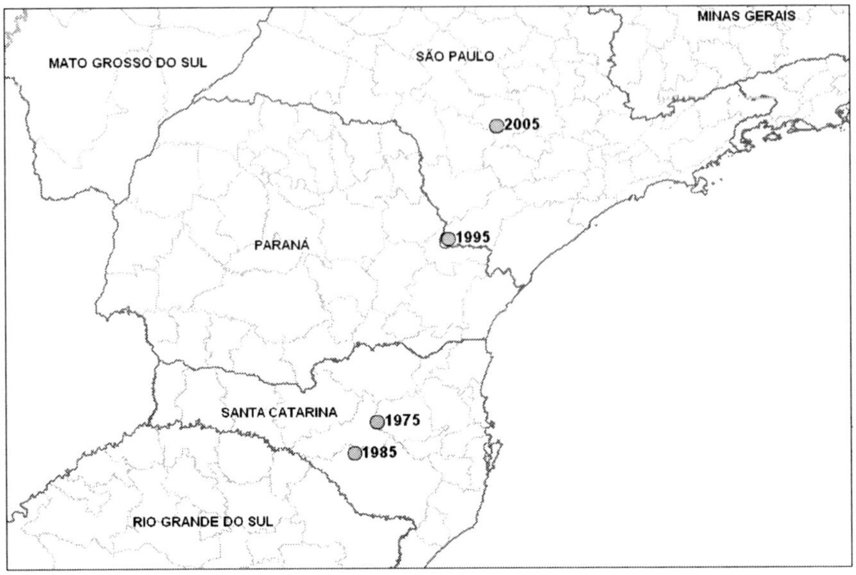


of this product. The distance of 1,600 kilometers between the centers of gravity for 1975 and 2005 summarize this motion. The high values of the terrestrial distances shown in Table 12, from 1975 to 1985, and from 1985 to 1995, together with a small value from 1995 to 2005, show that melon production achieved most of its displacement within the period of only two decades.

The center of gravity of soybean has moved on a geodesic path towards the north. This has been associated with the introduction of tropical varieties and good export prices over the years. Even when yearly centers of gravity are plotted,

*Figure 6. Melon production: national centers of gravity*

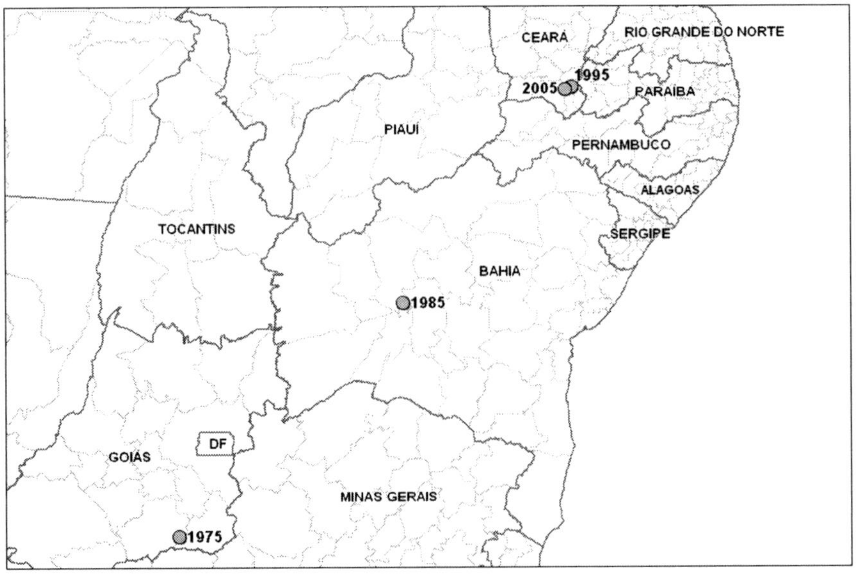

*Figure 7. Soybean production: national centers of gravity*

there is a consistent alignment, with very small departures from a geodesic line.

In the case of wheat, a curvature of 0.30 merely confirms the lack of an overall spatial trend in the motion of its center of gravity. Nevertheless, Figure 8 shows a process of "return to the origin", and some plausible arguments have been presented in order to explain this particular trajectory. Initially, which can be represented by the situation in 1975, as a winter crop, wheat was grown almost exclusively in the South region. But, starting at that year, some successive development programs have been implemented, in order to stimulate agricultural expansion in the Center-West region,

*Figure 8. Wheat production: national centers of gravity*

and wheat has been one of the crops included in this policy. Nevertheless, the main possibility to grow wheat in the Center-West region rests heavily on irrigation, and this has been considered attractive by only a small part of the farmers. In fact, there are strong indications that many farmers, in the Center-West, have been overburdened by debts associated to the attempt of introducing new crops to that region. Therefore, the expansion of wheat to the Center-West has declined. At the same time, new varieties, well-adapted to the southern winter, have been introduced, and the South region has been increasing its contribution to the total wheat production. This brief comment was included here only to suggest the difficulties involved when trying to explain the possible causes of a particular spatial evolution; this point will be addressed below.

## FUTURE RESEARCH DIRECTIONS

### Density and Yield

This discussion applies, particularly, to the study at the micro-regional level. Originally, the micro-regions appear on a nominal scale, where they are identified by an official code and a name. In order to talk about quarters (or about the first micro-region, or the best ten, and so forth), it was necessary to create an ordinal scale. In the previous discussion, the volume was used to order the micro-regions. But two other orderings have also been used. In all cases, the volume was accumulated, since an additive variable was required for that purpose, and volume was the most natural to consider. The additive variable might have been commercial value or land used (such as harvested area), but these were not available for all products. The fact is that two steps are involved: 1) to order the geographical units; 2) to accumulate the volume over that ordering. As can be expected, although the additive variable was always the volume, different stories result, depending on the ordering that has been chosen.

The two other orderings that have been explored result from the consideration of density and yield. In the case of density, the volume in each micro-region was divided by the respective total area. In fact, in many cases, when volume ordering was used, some micro-regions, with a long tradition of producing a particular item, and with good agricultural conditions for that purpose, fell in the lower quarters; their volume was small, as compared to that in other places, just because their area was small. Conversely, some micro-regions, with huge total areas (particularly in the North and Center-West regions), could appear in the upper quarters; in fact, although their volume contribution was important, they could have used only a minor portion of their area. This distorted the analysis of some products, particularly in the case of horticulture, winter cereals and milk. Therefore, it was considered necessary to relate, at least on a first approach to that problem, the volume with some proxy for the area involved. As a result, the use of density, with respect to the total area of the micro-regions, has proved to be very insightful, and is being employed in connection with other types of studies. With regard to concentration and dynamics, the numbers and maps which result from the density ordering are different from those obtained when starting from the volume ordering. In particular, the new maps tend to show some natural clusters, made up of neighboring small micro-regions, such as in the traditional wheat producing part of Rio Grande do Sul.

On the subject of density, it is clear that a division by the total area of the micro-regions provides only a very crude approximation to land use. In principle, it would be better to relate to available agricultural land (and, even much better, to land which is most adequate for specific products). But, in fact, data are not available, or reliable, on the general subject of land suitability. Therefore, the present research effort goes in the direction of establishing "exclusion zones". This implies subtracting from the total area of a micro-region those areas that cannot, or should not, be used for agriculture; it includes urban areas, Indian reservations, national parks, mountain areas, and many others. To this end, satellite technology, legal statutes, different sorts of maps, and other sources of information are being used. Although several results are already available, much remains to be done in this direction.

With respect to yield, it is clear that it constitutes a traditional and central subject of agricultural research. But the relationship between production (volume) and yield may be very complex, when a study reaches the geographical level of rather small subdivisions. At the micro-regional level, some frequent situations have been found, for many products. For instance: 1) very low statistical correlation between production and yield; 2) micro-regions with important volume but very low yield; 3) micro-regions with high yield but very small contribution to total volume. Therefore, the ordering of micro-regions by yield, and the subsequent accumulation of volume, has led to results that are very different from those obtained when ordering by volume or by density. In particular, the spatial changes measured while working with yield are frequently much greater than those measured when starting from the volume ordering. In general, yield values change very rapidly, from year to year and from place to place. In summary, the study of spatial changes involving yield is more complex and subtle than that restricted to volume data.

At present, yield can be determined for crops and for cow milk (since the number of milked cows is available). In this area, some exercises have been conducted in two research directions: 1) identification of micro-regions where a small yield increase may lead to a large increase in volume; 2) identification of micro-regions with very small yield but which have been using a large amount of area. In both cases, the underlying concern was to identify micro-regions where large responses could be obtained, with the products that were being used or with some substitutes, in order to minimize the opening of new areas to agriculture. Although, initially, very simple statistical techniques have been used, there is a wide room to apply more sophisticate data mining tools.

In the case of large animals, particularly cattle, which are grown almost exclusively on pasture lands, the data on grasslands are available only from agricultural censuses. Nevertheless, some measure of yield or of land use (such as cattle heads per hectare), even using some form of interpolation from census data, would be useful. In this sense, some exercises have been carried out. Presently, the main concern is to identify degraded land, that is being occupied with livestock, with very low density, but that, eventually, could be put to a better use. Again, the leading approach is to identify areas that are presently under some form of agriculture, but where an overall increase in production could be obtained, **without opening new lands**. This has become especially relevant in view of the expansion of areas used for sugar cane (and other crops associated with "bioenergy").

In any case, it should be clear that yield monitoring deserves the upmost interest of agricultural research institutions. This is particularly important for the following reasons, among others:

- **Yield dynamics.** As mentioned above, yield statistical estimates show significant changes from year to year and from place to place; although, in some cases, they may be related to natural causes, the fact is that a quick assessment of these changes can provide useful information to agricultural research.
- **Yield gap.** Usually, in developing countries, there exists a huge gap between the theoretical yield that could possibly be reached, according to research station results, and that deduced from agricultural statistics (such as the micro-regional data);

of course, it is expected that yield research estimates would generally be higher than those resulting from wide-area statistics; but the assessment and interpretation of these gaps should provide useful information to identify new research opportunities, even if the gaps cannot be directly attributed to a lack of adequate agricultural technology.

## Forecasting

The Brazilian Ministry of Agriculture has recently published its fourth forecasting report for some important products (Ministério da Agricultura ..., 2008). Besides, the Ministry of Agriculture and Embrapa have been working, for several years, in cooperation with the Organisation for Economic Co-operation and Development (OECD) and the Food and Agriculture Organization of the United Nations (FAO), as part of a worldwide effort to produce agricultural outlooks. That activity tries to provide support to high-level decision making, mainly in connection with agricultural and trade policy. Even if forecasting errors may be quite large at a horizon of 10 years, which is typically the case, the fact is that this type of outlook is required for policy formulation. In the case of Brazil, for the most part, these forecasts refer to totals for the country, covering production, consumption and foreign trade, and are based on data for, at least, 30 years. Forecasts for a few products have been presented at the state level, and they refer to states that are leading producers.

The situation becomes more complex when finer geographical subdivisions are considered, such as micro-regions. In fact, as shown in this chapter, some micro-regions, which were leading producers some 30 years ago, have been overtaken by others, in different states. Some of the main soybean producing micro-regions in recent years (e.g., Alto Teles Pires, in the state of Mato Grosso), did not have any record of the product 30 years ago. Therefore, forecasts involving micro-regions will have to take into account land-use changes over a shorter period, and a base series of 30 years may be meaningless. In this sense, a research project is being conducted at Embrapa, with the objective of producing forecasts for, initially, 5 years, on the basis of data from 1990 to 2006. For each of the 49 products being studied, the geographical entities, where the forecasts are being made, are clusters of micro-regions which, all together, have contributed with at least 75% of the total national volume over the period of 2002-2006.

As can be realized, there are some serious technical difficulties which, for most products, will probably prevent the formulation of forecasts beyond the five-year horizon. On the one hand, the base series cannot be too long, in view of the dynamics of agriculture; on the other hand, a base series of 17 years, as adopted, is too short for the time-series forecasting techniques which are currently available, and the initial results show large forecasting errors. In any case, this type of exercise has proved useful, in order to bring to the research environment a concern with land use, as captured by the agricultural statistics. In particular, soil and climate information has to be presented with reference to micro-regions, the "exclusion zones" have to be taken into account, and then discuss if there is available land where the usually optimistic forecasts can make sense. In other words, the standard forecasting techniques do not take into account any restriction; when constraints, such as available area, are brought into the discussion, the forecasts have to be reassessed. Clearly, many of these issues could be handled by some computer programs, but much remains to be done. This includes both theoretical developments (e.g., constrained time-series forecasting) as well practical applications of diffusion processes (e.g., the possibility of increasing production in neighboring micro-regions, that had not been included in the initial clusters). In any case, all these comments on forecasting refer to a single product. The situation becomes much more complex when several products are considered. Of course, some products can be grown sequentially, along the year; but others compete for the same land.

## Multi-Product Interaction

A simple example may be helpful to suggest the type of problems that are faced when several products are considered simultaneously. Over the ten-year period from 1998 to 2007 (the last year with available annual data, at the time of this writing), in the state of São Paulo, sugar cane showed a large increase in planted area and production, while cattle livestock and milk production declined. Even if annual data for pasture land are not available, all estimates, and other sources of information, strongly indicate that large portions of grassland are being occupied by sugar cane. Therefore, although aggregate data for these three products, at the national level, indicate an increase in total volumes, along that period, this does not mean that the same has happened in all the states. Within the states, some products are competing for the same land.

Also, some forms of synergy have been described, where the development of a product helps that of another one; this has been the case, for instance, of soybean grown under irrigation in central parts of Brazil, that is followed by irrigated corn in the same harvesting year (Alves et al, 1999). In that example, soybean appears as the "driver", which has justified most of the investment; but, since climate conditions and other factors are not limiting (temperature, sunlight hours, manpower, etc.), then the irrigation equipment and some of the machinery can be used with more than one crop in the same year.

More recently, a wide concern has erupted over the questions related to competition between food and biofuels production. At a minimum, different products may compete for the same land (the case of recent sugar cane expansion is in point); but also, there may appear renewed pressures in the sense of opening new lands to agriculture. Therefore, there exists a wide research area in the direction of improving the use of land that has already been under some form of agriculture (e.g., through products substitution, yield increase, and, in general, some type of "intensification"), instead of adding new lands, which can cause all sorts of environmental impacts.

Two exercises have been conducted with 20 products, at the micro-regional level, in the "cerrado" (savanna) area of central Brazil (Garagorry & Chaib Filho, 2008b, 2008c), over the 1975-2003 period. They were intended to obtain joint measures of agricultural change, with simple ranking techniques, which allowed for the introduction of an extension of the Cantor's distance presented before, either for clusters of products or for clusters of micro-regions. In the first case — clusters of products —, the change referred to motion (that is, dynamics, as used in this chapter); it was shown that, over the sub-periods of 1975-1985, 1985-1995 and 1995-2003, different groups of products moved more than others, over the set of micro-regions. In the second case — clusters of micro-regions —, the change referred to the agricultural profiles (in the case, a vector of 20 ranks associated to each micro-region), and it was closer to most published work on land-use change; it was shown that, over the same sub-periods already mentioned, the profiles of different groups of micro-regions changed more than others.

As future research directions, two lines are being considered. One refers to the application of multi-person game theory, where products would be the players. Initially, the idea is not to show that, in a given year, one coalition (that is, a group of products) won, probably in the sense that it used more than 50% of available land. The idea is to explore if solution concepts from game theory, such as the Shapley value or the Banzhaf-Coleman index of power (Owen, 1982), can capture some important changes that took place between the initial and final year of a period, such as a decade.

The other line to be explored is related to the application of computational techniques, such as multi-agent simulation (Parker et al., 2003), where products could be the agents, in order to describe the changes in land use that have happened over a period of several years, probably at the state level.

In any case, the comments presented so far, and the main focus of this chapter, have concentrated on the description of something that has happened, allowing for possible extensions in the form of forecasts. But, there is a clear interest on the formulation of explanations for the spatial changes that have been observed and measured.

## Explanation

Once the basic findings related to the dynamics of Brazilian agriculture have been presented, as in this chapter, one immediate question has been frequently raised: why did it happen? To be sure, the initial intention has been to extract information which was hidden in the statistical data. As a possible elaboration of agricultural statistics, the main tools, including distances, statistical maps and centers of gravity, would fall within a broad view of descriptive statistics. Besides, everybody knows that agriculture changes over time; and, in developing countries, particularly in Latin America, it is clear that it also changes over the space. Nevertheless, as soon as people realize that the agricultural motion can be described and measured, a wide research area appears: the identification of causes that could explain the motion that has been described.

A detailed discussion of this subject would fall outside the objectives of this chapter. In the case of the Brazilian agriculture, several factors have been mentioned, as part of a research project conducted at Embrapa, depending on each product, such as: internal migrations, opening of new roads, release of new varieties, outbreak of diseases, agricultural policies, international prices, agricultural credit, large irrigation projects. Although, for a given product, some of these factors seem most relevant, it is generally very difficult to produce a scientific explanation (e.g., with statistical models).

This difficulty is, to some extent, related to the scarcity of, or limitations to obtain, relevant data for the covariables. A preliminary exercise, at the state level, with longitudinal (panel) statistics methods, suggested that agricultural credit could be one of the main factors to explain the expansion of some products to new areas. Other computational techniques, such as multi-agent simulation, may be useful to identify relevant factors. In any case, it seems clear that a driving force is that agriculture will move to any place where farmers can obtain some economic reward, unless public policies can be enforced so as to limit the expansion.

## Forces

Having the annual national centers of gravity for a product, preliminary exercises have been conducted in order to describe, in physical terms, the forces that cause their motion. This has been made through a discrete version of the second law of motion, with a time-step of one year. Besides, annual centers of gravity at a lower geographical level (such as region or state) can be determined, and the same method can be applied so as to find the forces related to their motion. Now, the national center can be obtained from the regional or state centers (after assigning to them the whole volume in their respective areas), and the force at the national level should be a type of vector resultant of the forces which act at a lower level. This type of analysis may be helpful in order to identify, at a lower geographical level, some factors that can be behind the physical forces which are associated to the motion observed at the national level. In particular, for instance, different covariables may be relevant to explain the motion observed in different regions. In any case, some work remains to be done with respect to estimation and composition of forces which are spread over a vast territory.

## CONCLUSION

In this chapter, only an introductory view of a systematic study of agricultural motion has been presented. The main point was that, if agriculture moves, then this motion can be measured by some distance concept. The work on the general subject of agricultural motion started with a few products, which was later extended to 35 and, then, to 49 products. Besides, the same techniques are being used to study territorial changes in grasslands (with data from agricultural censuses), as well as bushfires (biweekly data) and deforestation (annual data). This activity allowed for a combination of elementary techniques of different origins. That orderly assemblage of simple statistical, geographical and geodesic techniques, in order to assess spatial changes in agriculture, is being called agrodynamics, by Embrapa's researchers involved in the studies. Certainly, other distances can be employed; for instance, distances of the $L^1$-type have been employed in combination with ranks.

With regard to the subject of concentration, which could be treated by itself, in isolation from the dynamics of agriculture, it entered from the beginning as part of the study of agricultural motion, because it was clear that, for any product, some subsets of its total volume (in terms of geographical subdivisions involved) moved faster than others. Therefore, although the study of agricultural dynamics can be conducted without any reference to concentration, it was considered useful to provide some measure of that, in order to stress an additional fact: even though, in general, the value of the index of geographical concentration of a product does not change much over time, the geographical subdivisions which concentrate an important part of the total volume may change substantially. Besides Gini's concentration index, that of Theil is also being used frequently; only the first one was presented here because, with all the products studied so far, both change with similar patterns.

It was shown that different groups of micro-regions, defined from the quarters obtained after ordering the micro-regions by volume and, then, accumulating the volume, move at different speeds, as follows from the distance measures used with the groups. The centers of gravity have also been calculated for each quarter (not shown in this chapter), and it was found that the centers of different quarters also moved at very different speeds. In many cases, the centers corresponding to the two upper quarters (Q3 and Q4) move faster, as measured by the terrestrial distance, than that of the first quarter. This agrees with what was found for the groups. That is, for many products, the sets of micro-regions which have the largest contributions to volume tend to change more frequently than those made up with micro-regions which, individually, have smaller contributions.

Within the general field of agriculture, the techniques presented in this chapter can be applied to other areas, such as agricultural employment or machinery, or to country-level data on production or foreign trade (e.g., the set of ten countries which, in each year, are the main producers of certain crop may change along the years). More generally, similar techniques can be applied to other fields, in order to describe and measure spatial changes.

## REFERENCES


Alves, E., Souza, G. S., & Garagorry, F. L. (1999). Evolution of the corn yield. *Brazilian Review of Agricultural Economics and Rural Sociology*, *37*(1), 183–202.

Anderberg, M. R. (1973). *Cluster analysis for applications*. New York: Academic Press.

Comisión Económica para América Latina (Ed.). (1983). *Expansión de la frontera agropecuaria y medio ambiente en América Latina*. Madrid, Spain: Centro Internacional de Formación en Ciencias Ambientales.

Diniz, C. C. (1995). *A dinâmica regional recente da economia brasileira e suas perspectivas*. Brasília, DF: Instituto de Pesquisa Econômica Aplicada. (Texto para discussão Nº 375.)

Garagorry, F. L., & Chaib Filho, H. (2008a). *Elementos de agrodinâmica*. Retrieved June 29, 2009, from http://www.embrapa.br/publicacoes/tecnico/RELFIN1.pdf

Garagorry, F. L., & Chaib Filho, H. (2008b). Avaliação conjunta da agrodinâmica de 20 produtos no Cerrado. In *Anais do 2º Simpósio Internacional de Savanas Tropicais: Desafios e estratégias para o equilíbrio entre sociedade, agronegócio e recursos naturais*. Brasília, DF: Embrapa Cerrados. 1 CD.

Garagorry, F. L., & Chaib Filho, H. (2008c). Mudanças no perfil agrícola das microrregiões do Cerrado. In *Anais do 2º Simpósio Internacional de Savanas Tropicais: Desafios e estratégias para o equilíbrio entre sociedade, agronegócio e recursos naturais*. Brasília, DF: Embrapa Cerrados. 1 CD.

Homem de Melo, F. (1983). La producción de biomasa y el crecimiento de la superficie de cultivo en el Brasil. In Comisión Económica para América Latina (Ed.), *Expansión de la frontera agropecuaria y medio ambiente en América Latina* (pp. 105-138). Madrid, Spain: Centro Internacional de Formación en Ciencias Ambientales.

Instituto Brasileiro de Geografia e Estatística. (2009). *Sistema IBGE de Recuperação Automática*. Retrieved June 29, 2009, from http://www.sidra.ibge.gov.br

Mello, L. M. R., Garagorry, F. L., & Chaib Filho, H. (2008). Spatial dynamics of grapes production in Brazil between 1975 and 2005. In *31st World Congress of Vine and Wine* (file PIII02.pdf, pp. 1-7).

Ministério da Agricultura. Pecuária e Abastecimento. (2009). *Projeções do agronegócio: Brasil 2008/2009 a 2018/19*. Brasília, DF: Ministério da Agricultura, Pecuária e Abastecimento.

Mueller, C. C. (1983). El Estado y la expansión de la frontera agrícola en la Amazonia. In Comisión Económica para América Latina (Ed.), *Expansión de la frontera agropecuaria y medio ambiente en América Latina* (pp. 37-78). Madrid, Spain: Centro Internacional de Formación en Ciencias Ambientales.

Mueller, C. C., & Martha Júnior, G. B. (2008) A agropecuária e o desenvolvimento socioeconômico recente do Cerrado. In F. G. Faleiro & A. L. Farias Neto (Eds.), *Savanas: Desafios e estratégias para o equilíbrio entre sociedade, agronegócio e recursos naturais* (pp. 105-169). Planaltina, DF: Embrapa Cerrados.

Owen, G. (1982). *Game Theory*. New York: Academic Press.

Oxford University Press. (1996). *Reader's Digest Oxford complete wordfinder*. Pleasantville, NY: The Reader's Digest Association, Inc.

Parker, D. C., Manson, S. M., Janssen, M. A., Hoffmann, M. J., & Deadman, P. (2003). Multi-agents systems for the simulation of land-use and land-cover change: A review. *Annals of the Association of American Geographers. Association of American Geographers*, *93*(2), 314–337. doi:10.1111/1467-8306.9302004

Rome, Italy: Ministero delle Politiche Agricole Alimentari e Forestali. 1 CD.

Sawyer, D. R. (1983). Ocupación y desocupación de la frontera agrícola en el Brasil: Un ensayo de interpretación estructural y espacial. In Comisión Económica para América Latina (Ed.), *Expansión de la frontera agropecuaria y medio ambiente en América Latina* (pp. 79-104). Madrid, Spain: Centro Internacional de Formación en Ciencias Ambientales.

Simon, M. F., & Garagorry, F. L. (2005). The expansion of agriculture in the Brazilian Amazon. *Environmental Conservation*, *32*(3), 203–212. doi:10.1017/S0376892905002201

Veldkamp, A., & Lambin, E. F. (2001). Predicting land-use change. *Agriculture Ecosystems & Environment*, *85*(1-3), 1–6. doi:10.1016/S0167-8809(01)00199-2

Verburg, P. H., Rounsevell, M. D. A., & Veldkamp, A. (2006). Scenario-based studies of future land use in Europe. *Agriculture, Ecosystems & Environment, 114*(***), 1-6.

Verburg, P. H., & Veldkamp, A. (2005). Introduction to the special issue on spatial modeling to explore land use dynamics. *International Journal of Geographical Information Science*, *19*(2), 99–102. doi:10.1080/13658810410001713362

Volle, M. (1997). *Analyse des données*. Paris: Economica.

## KEY TERMS AND DEFINITIONS

**Agrodynamics:** The study of the agricultural activity's motion over a territory.

**Center of Gravity:** As used in this chapter, it is a point over the earth surface, given by its latitude and longitude, which is obtained as a weighted mean of a set of statistical measurements distributed over a territory. In terms of physics, it is really a center of mass, since no gravitational field is involved; nevertheless, the term is widely used in statistics, particularly with respect to different forms of weighted means.

**Distance:** A function which assigns a non-negative number to any pair of points in a set, and which usually satisfies the axioms required for a metric, in mathematics. In more general terms, depending on the type of application, it can include dissimilarities or other statistical concepts that may not satisfy all the axioms of a metric (e.g., the triangle inequality), but which are useful to provide a measure of the separation between two points.

**Dynamics:** As used in this chapter, the term is derived directly from physics, and it means the same as agrodynamics. Although in agriculture, economics and many other fields, it usually refers to any type of change in a variable, in this chapter it is used exclusively with reference to changes in location or geographical area, along the years. Therefore, it includes the time and space dimensions.

**Geodesic Computing:** The use of computer programs that take into account the shape of the earth. It includes the determination (in terms of latitude and longitude) of particular points over the earth surface and the calculation of the terrestrial distance between pairs of such points. Depending on the program used, other quantities, besides the distance, such as the azimuths, may be obtained; these are useful to better analyze the direction of a motion between two points and to characterize the force that may have caused the motion.

**Land Use Change:** Any change in the use of a geographical area. When restricted to agricultural land use, it may include changes in the level of production in traditional areas (e.g., through yield increase), without a clear dislocation of the activity.

**Spatial Concentration:** In this chapter, it refers to the concentration of a statistical distribution defined over some territorial subdivisions of a geographical space.

# Chapter 3
# Land Cover Change:
## Statistical Indexes Using the Enhanced Transition Matrix

**Carlos Pérez-Hugalde**
*Universidad Politécnica de Madrid, Spain*

**Patricia Delgado-Pérez**
*Universidad Politécnica de Madrid, Spain*

**Raúl Romero-Calcerrada**
*Universidad Rey Juan Carlos, Spain*

## ABSTRACT

*The present chapter discusses an analysis of land cover change based on the enhanced transition matrix, and provides a detailed interpretation of the results of the case study chosen: a protected area in the region of Madrid (central Spain) between 1990 and 2000. The method adopted introduces concepts and indicators to study the statistical information contained in the transition matrix in greater depth than with the traditional approach. The chapter describes all the formal, theoretical and practical aspects of the method, along with the preparation of the data and the analysis of the results. The method proved to be useful for synthesizing the land changes observed and identifying systematic transitions.*

## INTRODUCTION

While changes in land cover may be the result of natural processes, such as in the case of forest fires, today and in recent history, most such changes can be attributed to human activity (Briassoulis, 2000). They are, in fact, one of the main indications of the economic and social development-related pressures exerted on the land. These changes have important environmental consequences, including loss of biodiversity, soil degradation and climate alterations. Land-cover changes may follow different patterns depending on the socio-economic processes (urban development, intensified farming, deforestation) taking place in an area. For this reason, the analysis of the nature and intensity of these changes has received significant attention in economic, geographic and ecological studies. Moreover, it has become one of the most relevant tools for recognizing the elements that generate friction between human societies and the ecosystems that support them (Murray et al., 2005)

DOI: 10.4018/978-1-61692-871-1.ch003

and for defining ways to ensure more sustainable long-term development.

In a matter of only a few decades, survey- and sampling-based land cover and land use analysis (Peña et al., 2005) has given way to geographic information system (SIG) imaging and techniques to study and visualize land cover change (Mast et al., 1997). The many methods at hand for detecting change can be grouped under two main categories: (1) methods that use images or bands from different dates and a variety of techniques to detect change (Lu et al., 2004; Coppin et al., 2004); and (2) methods based on the comparison of two previously classified images to analyze change a posteriori (Van Oort, 2007). The data used here, taken from the Corine Land Cover Project, are of the second type.

The purpose of the present study is to report on the analysis of land cover change based on an enhanced transition matrix, i.e., a transition matrix including the improvements and procedures proposed by Pontius, Shusas and McEachern (2004) and Pontius and Malizia (2004). The methodology deployed makes no pretence about originality. The aim, rather, is to provide a detailed and didactic explanation of the method to further the understanding of its fundamentals and utility, as well as a conscientious interpretation of the results of the case study conducted. The area studied is protected land area in the southwestern region of Madrid, in Central Spain, while the period, 1990-2000, was chosen on the grounds of data availability.

The problems associated with land cover change are more sensitive in areas subject to some type of environmental protection. More than that, in many cases the mere fact that change takes place in these areas is an indication that the protective measures are ineffective. The study of such lands is, then, of particular interest and justifies the present choice. European Union Directives 79/409/EEC and 92/43/EEC establish a framework for nature conservation policies. The former, commonly known as the *Bird Directive*, aims to conserve, maintain or restore a sufficient diversity and area of habitats to protect wild birds naturally occurring in Europe. To this end it promotes the delimitation of Special Protection Areas for birdlife (SPA). The second, the *Habitat Directive*, which deals with the conservation of the natural habitats of wild fauna and flora, constitutes the EU's core legislation on biodiversity. The main purpose of this directive, which includes and expands on its predecessor, is to contribute to guaranteeing biodiversity by conserving the natural habitats of wild fauna and flora in Europe. To this end, it promotes the designation of Special Areas of Conservation (SAC) and the creation of the Natura 2000 Network, which encompasses both SPAs and SACs.

While as a rule protected areas are associated with lands whose natural characteristics virtually preclude farming or forestry, in Spain this is not exactly the case. The conservation value of agrarian ecosystems is regarded to be high in light of both the diversity of the systems in question and the type of farming involved, which is characteristically extensive (Suárez et al., 1997). Nonetheless, while agriculture is one of the productive activities most closely related to nature conservation, concurrent intensification and abandonment over the past few decades has determined the co-existence of areas of high natural value, managed more or less traditionally, with others of scant environmental value, such as intensive farming (De Juana et al., 1993). These changes have lowered the biological value of large expanses of farmland, with a concomitant decline in the population of certain species of birds protected world-wide and the endangerment of many habitats and species (Suárez et al., 1997).

The chapter is structured as follows. Next section provides a physical and socio-economic description of the area studied. The statistics given in this chapter were drawn from Corine Land Cover Project cartographic data. This source and the data formatting procedures used in the present study are explained in the Data section. Then we address the Method, Results and Conclusions sections.

## AREA STUDIED

SPA No. 56, denominated *Encinares del río Alberche y el río Cofio*, was catalogued as a Special Protection Area for birdlife by the regional government of Madrid in 1990 due to the presence there of at least five protected species of birds of prey. This SPA, located in the southwest part of the region, covers 83,156 ha and encompasses 19 municipalities. While some of these fall only partially within the limits of the protected area, this survey covers the land pertaining to all 19, for a total of 97,226.10 ha.

Thanks to its diversified geomorphology and climate, the area can support a wide variety of more or less humanized and closely inter-related ecosystems. The most prominent include: pine groves that constitute a nesting grounds for several species of highly endangered bird species such as the Iberian imperial eagle, the black vulture and the black stork; holm oak groves that house a wide range of diverse species and are important feeding areas for several birds of prey, especially the groves in meadows, regarded to be one of the most environmentally valuable agri-systems on the Iberian peninsula; thickets with brush and underbrush, most of which, substituting for potential plant life, are the result of the degradation of natural woodlands or constitute colonization stages of abandoned fields or barren land; small oak and chestnut groves, which while not particularly large are important for their singularity and sustainable use; river copses, associated with the courses of brooks, rivers and films of water, forming a network of corridors that inter-connect ecosystems; pasture lands in mountainous municipalities, regarded to be the most important areas for small game in an area where such hunting is a relevant activity; and cultivated fields, with a patchwork of permanent crops, predominantly vineyards, and extensive fields of annual herbaceous crops, alongside more intensive farming (irrigated vegetable farming) in Villa del Prado, 1 of the 19 municipalities.

Socio-economically speaking, this SPA conforms to a rural population pattern in which a small number of workers manage vast expanses of land. The population density (42.6 inhabitants/km2 in 2001) is much lower than in the non-metropolitan municipalities of the Community of Madrid (108 inhabitants/km2 in 2001). The population pyramid is characterized by a small base, significant narrowing in certain economically active age groups (between 35 and 50) and a large population of people over the age of 65.

The three most prevalent primary sector activities are livestock raising in mountainous municipalities, vineyards in the southwest and truck farming in Villa de Prado. According to the socio-economic survey conducted in the SPA by Martín and Martínez (1998), however, the sector is characterized by substantial structural imbalances, including: the ageing business community; part-time farming; near absence of professional associations; the visible lack of medium-sized farms and a bipolar predominance of micro- and macro-farming, along with scattered land ownership. Manufacturing is marginal in the area's economy. Its population growth is associated with growth in the service sector, which is only relevant in the municipalities with larger populations.

## DATA

Periodically updated sources of information, collected and compiled with uniform methodologies to ensure comparability, are requisite to any study of change in a given place over time. These requirements are met by the CLC90 and CLC00 cartographic databases, developed by the European Environment Agency's Corine Land Cover (CLC) Project, from which the statistical information for this study was drawn. Processing these data is no simple matter. The procedure is explained below for its possible utility for readers interested in such techniques.

The Corine Land Cover Project is part of the CORINE (Coordination of Information of the Environment) Program, launched on 27 June 1985 by virtue of Council Decision 85/338/EEC. The primary aim of the project is to collect and continually update numerical and geographic data to create a consistent and comparable European database on land cover and use on a scale of 1:100,000 (Instituto Geográfico Nacional, 2003).

The land cover database referred to above as CLC90, while dated in 1990, was created between the late nineteen eighties and early nineteen nineties. The preliminary work to update the CLC database (I&CLC2000 Project) was undertaken in 1999.

The revised CLC90 database and the new country level CLC00 database, dated in 2000, are two of the products generated by the I&CLC2000 Project. Both are available in vector coverage file format for export to Arc-info (.e00) as polygons (area defined to contain a single category of land cover) with a minimum mappable area of 25 ha. They are based on a single land cover nomenclature applied across Europe and a clearly defined production methodology.

The CLC00 nomenclature contains 85 categories, while the CLC90 has only 64. In order to compare the information in the two databases, the nomenclature common to both, i.e., the least detailed or CLC90 classification, was used. Consequently, the first step was to convert the CLC00 code to CLC90 nomenclature, pursuant to the instructions in the document published under reference Instituto Geográfico Nacional (2003).

With both country level databases in CLC90 nomenclature, Arc-view software was used to generate shapefiles of the area studied. The result was the land cover for that area in 1990 and 2000. These data were supplemented with attribute tables specifying the respective category code and total area for each polygon identified.

The change in cover in this 10-year period was found by intersecting the 1990 and 2000 shapefiles. As above, this was supplemented with an attribute table which in this case identified the new polygon created after intersection, along with the area and codes under which it was classified in each of the two years. In other words, a polygon with the same code in 1990 as in 2000 defined an area whose cover remained unchanged in the period studied. A polygon with different codes in 1990 and 2000 defined an area whose cover changed in those 10 years.

Of the 64 classes in the nomenclature used, the area studied contained only 33, which were in turn regrouped into the eight categories shown in Table 1. These categories are: artificial surfaces (ART): continuous urban fabric, road and rail networks, residential suburbs, mineral extraction sites, construction and excavation sites, and sport and leisure facilities; water bodies (AGU): natural or artificial inland waters; mixed crops (MIX): complex cultivation patterns, land principally occupied by agriculture with significant areas of natural vegetation and annual crops or pastures with forest cover; herbaceous crops (HER): permanently irrigated and non-irrigated land, fallow land and vegetable crops (open field or under plastic); woody crops (LEÑ): irrigated and non-irrigated; pastures (PAS): pastures and natural grasslands; forests (BOS); and scrub vegetation and open spaces (VAR): sclerophyllous vegetation, transitional woodland-scrub, sands, bare rocks and burnt areas.

## METHOD

The method used to analyze change is based on what is referred to here as the enhanced transition matrix. In this procedure, the statistical information provided by the classic transition matrix is analyzed in greater depth, further to the concepts, calculations and indicators developed by Pontius, Shusas and McEachern (2004) and Pontius and Malizia (2004). These concepts and tools are described in detail in this section. The aim sought is to provide a rigorous and didactic explanation

*Table 1. Categories analyzed and CLC90 nomenclature correlates*

| CLC90 NOMENCLATURE | | | | | | Category analyzed |
|---|---|---|---|---|---|---|
| **Level 1** | **Level 2** | **Level 3** | **Level 4** | **Level 5** | **Code** | |
| **Artificial surfaces** | Urban fabric | Continuous urban fabric | | | 11100 | **Artificial surfaces (ART)** |
| | | Discontinuous urban fabric | Scattered urban development | | 11210 | |
| | | | Detached and/or landscaped housing | | 11220 | |
| | Mine, dump and construction sites | Mineral extraction sites | | | 13100 | |
| | | Construction sites | | | 13300 | |
| | Artificial, non- agricultural vegetated areas | Sport and leisure facilities | | | 14200 | |
| **Agricultural areas** | Arable land | Non-irrigated arable land | | | 21100 | **Herbaceous crops (HER)** |
| | | Permanently irrigated land | Herbaceous crops | | 21210 | |
| | | | Other irrigated areas | | 21220 | |
| | Permanent crops | Vineyards | | | 22100 | **Woody crops (LEÑ)** |
| | | Fruit trees and berry plantations | Non-irrigated | | 22210 | |
| | | | Irrigated | Other fruit trees | 22223 | |
| | | Olive groves | | | 22300 | |
| | Pastures | Pastures | | | 23100 | **Pastures (PAS)** |
| | | Natural grasslands | Other grasslands | | 32120 | |
| | Heterogeneous agricultural areas | Complex cultivation patterns | Annual crops with meadows and/or pastures | | 24210 | **Mixed crops (MIX)** |
| | | | Of permanent crops | | 24220 | |
| | | | Annual crops associated with permanent crops | | 24230 | |
| | | Land principally occupied by agriculture, with significant areas of natural vegetation | | | 24300 | |
| | | Agro-forestry areas | | | 24400 | |
| **Forest and semi-natural areas** | Forests | Broad-leaved forest | Perennials and gall oaks | Perennial sclerophylls | 31111 | **Forests (BOS)** |
| | | | Deciduous and turkey oaks | | 31120 | |
| | | | Other plantation broadleaves | | 31130 | |
| | | Coniferous forest | Evergreens | | 31210 | |
| | | Mixed forest | | | 31300 | |
| | Scrub and/or herbaceous vegetation associations. | Sclerophyllous vegetation | Large expanses of dense or fairly dense thicket | | 32310 | **Scrub vegetation and open spaces (VAR)** |
| | | | Sub-shrub or sparsely populated scrubland | | 32320 | |
| | | Transitional woodland-scrub | | | 32400 | |
| | Open spaces with little or no vegetation | Beaches, dunes, sands | | | 33100 | |
| | | Bare rocks | Barren soil | | 33220 | |
| | | Burnt areas | | | 33400 | |

*continues on following page*

*Table 1. continued*

| CLC90 NOMENCLATURE | | | | | | Category analyzed |
|---|---|---|---|---|---|---|
| **Level 1** | **Level 2** | **Level 3** | **Level 4** | **Level 5** | **Code** | |
| **Water bodies** | Inland waters | Water bodies | Lakes and lagoons | | 51210 | **Water bodies (AGU)** |
| | | | Reservoirs | | 51220 | |

of the fundamentals and utility of the method. First item defines the different types of change and reviews the inter-relationships for each of the categories analyzed. A similar explanation is given in the next item for the entire landscape studied. Last item defines the indicators designed to distinguish systematic from random transitions.

## The Transition Matrix

The information in the databases is displayed in a cross-tabulation or transition matrix to analyze the changes in land cover between time 1 and time 2. Such matrices table the categories in the initial year against the same categories in the end year, in the same order, and distribute the area of the landscape studied across each pair of categories for the two years. Assume that the number of categories to be analyzed is $k$ and that an index is identified for each whose value varies, naturally, from 1 to $k$. This is illustrated in Table 2, whose body consists in $k$ rows for time 1, and $k$ columns for time 2. The generic symbol for table elements is $p_{ij,}$ whose value is expressed as a proportion of the area of the landscape studied. An element located off the central diagonal, such as $p_{ij}$ where $i$ and $j$ represent different categories, shows the area in category $i$ in time 1 but in category $j$ in time 2. Consequently, it denotes the proportion of the landscape that at the beginning of the period was classified in category $i$ and at the end in category $j$, known as the *transition* from category $i$ to $j$. A value on the central diagonal, $p_{jj}$ for instance, indicates the area in category $j$ that did not change over the period studied, denominated the *persistence* of category $j$. Pontius, Shusas et al. (2004) maintain that many of the methods taught or used to analyze land use land cover change "fail to account for persistence in a manner capable of detecting important signals of land change" (p. 252). Nonetheless, persistence dominates most landscapes, including those where authors claim that the change is important and/or large" (p. 253). Hence, the precise definition and accurate evaluation of persistence is essential to land change studies. Be it said in passing that if the categories are further sub-divided, the persistence rate declines because transitions may be recorded between new classes originally pertaining to the same category.

The body of the table therefore shows the transitions between categories observed during the period and the persistence of each category, all expressed in terms of a proportion of the total area of the landscape studied. Row $j$, in turn, shows the distribution by category in time 2 of the total portion in category $j$ in time 1, i.e., the area that remained unchanged, or $p_{jj}$, plus the sum

*Table 2. Transition matrix to compare land cover on two different dates. Values in proportion of area studied*

| | | Time 2 category | | | | | Time 1 total | Loss |
|---|---|---|---|---|---|---|---|---|
| | | **1** | **…** | ***j*** | **…** | ***k*** | | |
| Time 1 category | 1 | $p_{11}$ | | $p_{1j}$ | | $p_{1k}$ | $p_{1+}$ | $p_1$-$p_{11}$ |
| | … | | | | | | | |
| | $j$ | $p_{j1}$ | | $p_{jj}$ | | $p_{jk}$ | $p_{j+}$ | $p_{j+}$-$p_{jj}$ |
| | … | | | | | | | |
| | $k$ | $p_{k1}$ | | $p_{kj}$ | | $p_{kk}$ | $p_{k+}$ | $p_{k+}$-$p_{kk}$ |
| Time 2 total | | $p_{+1}$ | | $p_{+j}$ | | $p_{+k}$ | 1 | |
| Gain | | $p_{+1}$-$p_{11}$ | | $p_{+j}$-$p_{jj}$ | | $p_{+k}$-$p_{kk}$ | | |

of the areas initially in category *j* but found in other categories (generically, *i*) at the end of the period. That total area, symbolized as $p_{j+}$, is given in row *j* under the column *Time 1 total.* Similarly, the information in column *j* shows the distribution by category in time 1 of the total proportion in category *j* in time 2, i.e., the area that remained unchanged, or $p_{jj}$, as above, plus the sum of the areas in category *j* in time 2 but initially in other categories (generically, *i*). That total area, symbolized as $p_{+j}$, is given in column *j* in the row labeled *Time 2 total.*

The above is a description of the basic structure of a transition matrix. On the grounds of these elements, other aggregates of interest for land change analysis can be defined and calculated. The *gross gain* of category *j* ($Gain_j$) is defined to be the area that changed during the period to category *j*, i.e., the proportion of the area of the landscape that was not in category *j* in time 1 but was in that category in time 2. It can be calculated as the total portion in category *j* in time 2, less persistence for that category: $Gain_j = p_{+j} - p_{jj}$, where, as noted earlier, $p_{+j}$ is the sum of $p_{ij}$, with *i* varying from 1 to *k*, or the sum under column *j*. *Gross loss* of category *j* ($Loss_j$) is defined to be the proportion that was in category *j* in time 1 but not at the end of the period. It can be calculated as the category *j* total in time 1, less persistence for that category: $Loss_j = p_{j+} - p_{jj}$, where $p_{j+}$ is the sum of the elements in row *j*. Further to Pontius and Malizia (2004), the transition matrix used here also has an additional column showing gross gains and an additional row showing gross loss, as in Table 2.

Category *j* may be involved in land cover changes in two compatible ways: it can cover new areas or withdraw from areas previously covered. The sum of these two magnitudes gives an overall indication of category dynamism and the pressures to which it is subjected, which translate into losses or gains. The aggregate is known as *total* category *j change*: $Total_j = Gain_j + Loss_j$.

Total change, by definition, can be broken down into gross gain and gross loss. This aggregate might also be usefully broken down under a second criterion, however. During the period, the area attributed to each category may have expanded or contracted. That change, the absolute value of the change in the amount of area covered by category *j*, is known as the *net change* of category *j* ($Net_j$). The net change can be calculated as the absolute value of the difference between the gross gain and gross loss, or as the absolute value of the difference between the area in *j* in time 1 and time 2: $Net_j = |Gain_j - Loss_j| = |p_{+j} - p_{j+}|$.

Net change can also be calculated as the difference between the maximum and minimum gross gain and gross loss: $Net_j = \max(Gain_j, Loss_j) - \min(Gain_j, Loss_j)$.

The part of the total change in *j* that is not net change, i.e., the change in the location of category *j*, is denominated category *j swap* ($Swap_j$). In the following explanation of this concept, the gross grain in category *j* is assumed to be greater than its gross loss. The area lost can be regarded to be offset by an equal amount of area gained: that offset is interpreted to be a change in the spatial location of the category, which withdraws from certain polygons and occupies space in others. The area swapped in this case is twice the gross loss and includes both the area lost and the part of the gross gain that offsets the gross loss. As a rule, swaps are calculated as $Swap_j = 2\min(Gain_j, Loss_j)$

In summary, total change is broken down, according to one criterion, into gross loss and gross gain, and according to another, into change in amount and change in location:

$$Gain_j + Loss_j = Total_j = Net_j + Swap_j \quad (1)$$

For an example of a transition matrix based on real data, see Table 3 in the Results section below. Using the data in that table, the total change in each category and its breakdown as in Equation 1 were calculated and are given in Table 4, likewise in Results. Since net changes are given in absolute values, the symbols + and - alongside the figures

*Table 3. Transition matrix, 1990-2000, for eight categories defined in SPA 56*

| | | Category in 2000 | | | | | | | | Total 1990 | Loss |
|---|---|---|---|---|---|---|---|---|---|---|---|
| | | ART | AGU | MIX | HER | LEÑ | PAS | BOS | VAR | | |
| Category in 1990 | ART | 2.673 | 0.000 | 0.002 | 0.001 | 0.000 | 0.002 | 0.001 | 0.020 | 2.699 | 0.026 |
| | AGU | 0.000 | 1.012 | 0.000 | 0.000 | 0.000 | 0.000 | 0.000 | 0.000 | 1.012 | 0.000 |
| | MIX | 0.119 | 0.000 | 14.328 | 0.017 | 0.000 | 0.003 | 0.003 | 0.184 | 14.654 | 0.326 |
| | HER | 0.190 | 0.000 | 0.488 | 5.597 | 0.001 | 0.066 | 0.001 | 0.503 | 6.846 | 1.249 |
| | LEÑ | 0.035 | 0.000 | 0.001 | 0.001 | 6.549 | 0.023 | 0.004 | 0.320 | 6.933 | 0.383 |
| | PAS | 0.188 | 0.000 | 0.003 | 0.033 | 0.001 | 14.053 | 0.004 | 0.446 | 14.727 | 0.674 |
| | BOS | 0.036 | 0.000 | 0.007 | 0.001 | 0.003 | 0.002 | 16.330 | 0.173 | 16.553 | 0.223 |
| | VAR | 0.214 | 0.000 | 0.037 | 0.031 | 0.005 | 0.007 | 0.320 | 35.963 | 36.576 | 0.613 |
| Total 2000 | | 3.455 | 1.012 | 14.867 | 5.680 | 6.560 | 14.155 | 16.662 | 37.610 | 100.000 | 3.494 |
| Gain | | 0.781 | 0.000 | 0.539 | 0.083 | 0.011 | 0.102 | 0.332 | 1.647 | 3.494 | |

*Note.* Values in percentage of area studied. Categories: ART: artificial surface AGU: water bodies. MIX: mixed crops. HER: herbaceous crops. LEN: woody crops. PAS: pastures. BOS: forests. VAR: scrub vegetation and open spaces.

in the *Net* column in Table 4 respectively denote net gain or loss.

## Synthesis for the Land Area Analyzed: Category Aggregation

All the changes are defined at the category level. The synthesis of all types of change in the landscape studied as a whole, i.e., with no category distinctions, is calculated below. The overall gross gain, expressed as subscript M, is obtained as the sum of the gross gain in all the categories:

$$Gain_M = \sum_{j=1}^{k} Gain_j$$

The gross loss is calculated analogously:

$$Loss_M = \sum_{j=1}^{k} Loss_j \, .$$

Since the total change in a category includes both its gross gains and losses and one category can only gain at the expense of another, to calcu-

*Table 4. Gains, losses, total change, swap and net change extrapolated from Table 3*

| | Gain | Loss | Total Change | Swap | Net Change | |
|---|---|---|---|---|---|---|
| ART | 0.78 | 0.03 | 0.81 | 0.05 | 0.76 | (+) |
| AGU | 0.00 | 0.00 | 0.00 | 0.00 | 0.00 | |
| MIX | 0.54 | 0.33 | 0.86 | 0.65 | 0.21 | (+) |
| HER | 0.08 | 1.25 | 1.33 | 0.17 | 1.17 | (-) |
| LEÑ | 0.01 | 0.38 | 0.39 | 0.02 | 0.37 | (-) |
| PAS | 0.10 | 0.67 | 0.78 | 0.20 | 0.57 | (-) |
| BOS | 0.33 | 0.22 | 0.55 | 0.44 | 0.11 | (+) |
| VAR | 1.65 | 0.61 | 2.26 | 1.23 | 1.03 | (+) |
| TOTAL | 3.49 | 3.49 | 3.49 | 1.38 | 2.11 | |

*Note.* Values in percentage of the area studied. Categories: ART: artificial surface. AGU: water bodies. MIX: mixed crops. HER: herbaceous crops. LEN: woody crops. PAS: pastures. BOS: forests. VAR: scrub vegetation and open spaces.

late the total change in the landscape studied the sum of total changes by categories must be divided by two to avoid counting each change twice, first as a loss and then as a gain: $Total_M = \frac{1}{2}\sum_{j=1}^{k} Total_j$ .

Naturally, the total change may also be obtained for the gain side as $Total_M = \sum_j Gain_j$, or from the loss side as $Total_M = \sum_j Loss_j$. The reasoning used to calculate the total change can be applied to the net change and swap, whereby: $Net_M = \frac{1}{2}\sum_{j=1}^{k} Net_j$ and $Swap_M = \frac{1}{2}\sum_{j=1}^{k} Swap_j$ .

$Net_{M,}$ the quantitative change during the period in all the categories as a whole, is equal to the total net gain of all the categories with net gains, or the total net loss of all the categories with net losses. In light of the foregoing relationships, the following equalities hold for total change in the landscape as a whole:

$$Gain_M = Loss_M = Total_M = Net_M + Swap_M \qquad (2)$$

As noted, the values of total change, gross gain and gross loss are equal when calculated for the landscape as a whole, but not for a specific category. Nonetheless, as in the case of categories, total change is broken down into net change and swaps (compare Equations 2 and 1). The synthesis of changes for the entire area studied is given in the *Total* row in Table 4.

## Transition Analysis

The elements off the central diagonal in Table 2, $p_{ij}$, provide information on transitions as a proportion of the total area. When evaluating these transitions, the widely varying relative weights of these categories must be borne in mind, inasmuch as the transitions of greatest quantitative importance are likely to involve the categories covering the largest areas in the landscape studied. Nonetheless, due to social or economic processes, when one type of cover, such as artificial surfaces, grows, it may not be at the expense of all the other types of cover at random, but preferentially at the expense of one or a few specific types. Similarly, when one type of cover, such as forest land, loses ground, it may be replaced preferentially by one or a few specific types. Hence the need to develop a method able to detect and identify any such systematic transitions. Although not addressed in the present chapter, the ultimate aim, once such transitions are identified, i.e., once the land change patterns are known, would be to associate these patterns with the underlying causes.

Transitions may be evaluated from two complementary angles: as gains or losses. Beginning with the former, gains in category *j*, for instance, consist in transitions to that category from all others, i.e., for all the off-diagonal elements in column *j* ($p_{ij}$ for $i=1,\ldots,k;\ i\neq j$). The decision as to when a transition can be regarded to be systematic or non-random can be based on a comparison of the value observed to the expected value, assuming that category *j* replaces all other categories randomly. Taking the gross gain of *j* to be a fixed amount, the expected values for category *j* transitions are calculated from the following expression:

$$g_{ij} = Gain_j \frac{p_{i+}}{(1 - p_{j+})} \qquad (3)$$

for any category *i* other than *j*. The numerator in the fraction in Equation 3 is the proportion of the landscape studied in category *i* in time 1, and the denominator is the proportion that is not category *j* in time 1. The quotient is, then, the ratio of the area in category *i* to the area not in category *j* in time 1. For a fuller understanding of this equation, imagine that the areas lost by other categories to generate the gain observed in category *j* is divided into a large number of area units, and that in time 1 these units are distributed randomly across the non-$j$ area in the landscape studied. According to probability theory, the proportion lost by each category should be very similar to the ratio of the area in that category to the total area not in category $j$ in time 1. That is precisely what is calculated by Equation 3. Consequently, $g_{ij}$ is the expected area gained by $j$ at the expense of $i$ on the assumption that $j$ replaces all other categories randomly in time 1 to reach the gain observed, or to put it more succinctly: the expected transition from $i$ to $j$, assuming gains are random.

Gains are analyzed in Table 5 in the Results section below. The first value in each cell, shown in bold typeface, is the observed value of the transition, $p_{ij}$. The italicized second value is the expected value, $g_{ij}$. The third value is the *deviation* of the observed from the expected value, $p_{ij}$-$g_{ij}$. The deviation for $p_{ij}$ may be regarded to measure the degree to which this transition fails to comply with the assumption that the gain in category $j$ is randomly distributed. Where the deviation is nil, the degree of non-compliance is also nil. If the deviation is greater than zero, the area gained by category $j$ at the expense of category $i$ is greater than the expected value. A substantial difference would be an indication of bias in the transition from $i$ to $j$: when $j$ gains it tends to be at the expense of $i$. If the deviation is less than zero, the area gained by category $j$ at the expense of $i$ is less than the expected value. This means that when $j$ gains, it tends not to replace category $i$. The fourth value in each cell is the above deviation expressed as a proportion of the expected value:

$$tg_{ij} = \frac{p_{ij} - g_{ij}}{g_{ij}} \tag{4}$$

This ratio, which draws from the chi-square test, helps to reach a quantitative evaluation of the above deviation. It expresses the deviation in transition $p_{ij}$ as the number of times of the expected value.

Since the area that remains unchanged in a given category should not affect this analysis, the expected values for the elements on the diagonal are made to concur with the observed values. The totals for the expected and observed values in time 2 concur (see Table 5) because the persistence and gain values are fixed. On the contrary, the expected value in the *Time 1 total* column includes the sum of the expected values of the transitions for the row category in question, which need not concur with the sum of the respective observed transitions. The former is the total that the categories should have had in the initial year, assuming randomly distributed gains. Similarly, the expected value in the *Loss* column shows the expected loss for the respective row category on the assumption that the rest of the categories replace all the others randomly.

The description of transitions as losses is symmetrical to the above discussion of gains, but inverting the functions of gains and losses and rows and columns. While somewhat reiterative, the explanation is provided in full for greater precision. Losses in category $i$, for instance, consist in transitions from that category to all others, i.e., for all the off-diagonal elements in row $i$ ($p_{ij}$ for $j$=1,…,$k$; $j$≠$i$). The decision as to when a transition can be regarded to be systematic can be based on a comparison of the value observed to the expected value under the assumption that category $i$ is replaced by all other categories randomly. The values expected for transitions from category $i$, based on the assumption that gross loss in category $i$ is a fixed amount, are calculated from the following expression:

$$l_{ij} = Loss_i \frac{p_{+j}}{(1 - p_{+i})} \tag{5}$$

for any category $j$ other than $i$. The numerator in the fraction in Equation (5) is the proportion of the landscape studied in category $j$ in time 2, and the denominator is the proportion that is not

*Table 5. Analysis of transitions as gains, 1990-2000*

| GAIN | | Category in 2000 | | | | | | | | Total 1990 | Loss |
|---|---|---|---|---|---|---|---|---|---|---|---|
| | | ART | AGU | MIX | HER | LEÑ | PAS | BOS | VAR | | |
| Category in 1990 | ART | **2.673** | **0.000** | **0.002** | **0.001** | **0.000** | **0.002** | **0.001** | **0.020** | **2.699** | **0.026** |
| | | *2.673* | *0.000* | *0.017* | *0.002* | *0.000* | *0.003* | *0.011* | *0.070* | *2.777* | *0.104* |
| | | 0.000 | 0.000 | -0.015 | -0.002 | 0.000 | -0.001 | -0.010 | -0.050 | -0.078 | -0.078 |
| | | 0.000 | 0.000 | -0.879 | -0.767 | 0.000 | -0.369 | -0.929 | -0.712 | -0.028 | -0.749 |
| | AGU | **0.000** | **1.012** | **0.000** | **0.000** | **0.000** | **0.000** | **0.000** | **0.000** | **1.012** | **0.000** |
| | | *0.008* | *1.012* | *0.006* | *0.001* | *0.000* | *0.001* | *0.004* | *0.026* | *1.059* | *0.047* |
| | | -0.008 | 0.000 | -0.006 | -0.001 | 0.000 | -0.001 | -0.004 | -0.026 | -0.047 | -0.047 |
| | | -1.000 | 0.000 | -1.000 | -1.000 | -1.000 | -1.000 | -1.000 | -1.000 | -0.044 | -1.000 |
| | MIX | **0.119** | **0.000** | **14.328** | **0.017** | **0.000** | **0.003** | **0.003** | **0.184** | **14.654** | **0.326** |
| | | *0.118* | *0.000* | *14.328* | *0.013* | *0.002* | *0.018* | *0.058* | *0.381* | *14.917* | *0.589* |
| | | 0.002 | 0.000 | 0.000 | 0.004 | -0.001 | -0.015 | -0.056 | -0.196 | -0.263 | -0.263 |
| | | 0.014 | 0.000 | 0.000 | 0.285 | -0.831 | -0.830 | -0.956 | -0.516 | -0.018 | -0.446 |
| | HER | **0.190** | **0.000** | **0.488** | **5.597** | **0.001** | **0.066** | **0.001** | **0.503** | **6.846** | **1.249** |
| | | *0.055* | *0.000* | *0.043* | *5.597* | *0.001* | *0.008* | *0.027* | *0.178* | *5.909* | *0.312* |
| | | 0.135 | 0.000 | 0.445 | 0.000 | 0.000 | 0.058 | -0.026 | 0.325 | 0.937 | 0.937 |
| | | 2.452 | 0.000 | 10.306 | 0.000 | 0.502 | 7.058 | -0.963 | 1.827 | 0.159 | 3.001 |
| | LEÑ | **0.035** | **0.000** | **0.001** | **0.001** | **6.549** | **0.023** | **0.004** | **0.320** | **6.933** | **0.383** |
| | | *0.056* | *0.000* | *0.044* | *0.006* | *6.549* | *0.008* | *0.028* | *0.180* | *6.871* | *0.321* |
| | | -0.021 | 0.000 | -0.043 | -0.005 | 0.000 | 0.014 | -0.024 | 0.140 | 0.062 | 0.062 |
| | | -0.380 | 0.000 | -0.982 | -0.791 | 0.000 | 1.744 | -0.869 | 0.780 | 0.009 | 0.193 |
| | PAS | **0.188** | **0.000** | **0.003** | **0.033** | **0.001** | **14.053** | **0.004** | **0.446** | **14.727** | **0.674** |
| | | *0.118* | *0.000* | *0.093* | *0.013* | *0.002* | *14.053* | *0.059* | *0.382* | *14.720* | *0.667* |
| | | 0.069 | 0.000 | -0.090 | 0.019 | -0.001 | 0.000 | -0.055 | 0.064 | 0.007 | 0.007 |
| | | 0.585 | 0.000 | -0.965 | 1.475 | -0.478 | 0.000 | -0.933 | 0.166 | 0.001 | 0.011 |
| | BOS | **0.036** | **0.000** | **0.007** | **0.001** | **0.003** | **0.002** | **16.330** | **0.173** | **16.553** | **0.223** |
| | | *0.133* | *0.000* | *0.104* | *0.015* | *0.002* | *0.020* | *16.330* | *0.430* | *17.034* | *0.704* |
| | | -0.097 | 0.000 | -0.097 | -0.014 | 0.001 | -0.018 | 0.000 | -0.257 | -0.481 | -0.481 |
| | | -0.728 | 0.000 | -0.929 | -0.925 | 0.548 | -0.919 | 0.000 | -0.597 | -0.028 | -0.684 |
| | VAR | **0.214** | **0.000** | **0.037** | **0.031** | **0.005** | **0.007** | **0.320** | **35.963** | **36.576** | **0.613** |
| | | *0.294* | *0.000* | *0.231* | *0.033* | *0.004* | *0.044* | *0.145* | *35.963* | *36.714* | *0.750* |
| | | -0.080 | 0.000 | -0.194 | -0.002 | 0.001 | -0.037 | 0.174 | 0.000 | -0.137 | -0.137 |
| | | -0.271 | 0.000 | -0.841 | -0.056 | 0.184 | -0.849 | 1.200 | 0.000 | -0.004 | -0.183 |
| Total 2000 | | **3.455** | **1.012** | **14.867** | **5.680** | **6.560** | **14.155** | **16.662** | **37.610** | **100.000** | **3.494** |
| | | *3.455* | *1.012* | *14.867* | *5.680* | *6.560* | *14.155* | *16.662* | *37.610* | *100.000* | *3.494* |
| | | 0.000 | 0.000 | 0.000 | 0.000 | 0.000 | 0.000 | 0.000 | 0.000 | 0.000 | 0.000 |
| | | 0.000 | 0.000 | 0.000 | 0.000 | 0.000 | 0.000 | 0.000 | 0.000 | 0.000 | 0.000 |

*continues on following page*

*Table 5. continued*

| GAIN | Category in 2000 | | | | | | | | Total 1990 | Loss |
|---|---|---|---|---|---|---|---|---|---|---|
| | ART | AGU | MIX | HER | LEÑ | PAS | BOS | VAR | | |
| Gain | **0.781** | **0.000** | **0.539** | **0.083** | **0.011** | **0.102** | **0.332** | **1.647** | **3.494** | |
| | *0.781* | *0.000* | *0.539* | *0.083* | *0.011* | *0.102* | *0.332* | *1.647* | *3.494* | |
| | 0.000 | 0.000 | 0.000 | 0.000 | 0.000 | 0.000 | 0.000 | 0.000 | 0.000 | |
| | 0.000 | 0.000 | 0.000 | 0.000 | 0.000 | 0.000 | 0.000 | 0.000 | 0.000 | |

*Note.* Four values per cell: 1: observed; 2: expected; 3: deviation; 4: deviation as a multiple of the expected value. First three values in percentage of area studied. Categories: ART: artificial cover. AGU: water bodies. MIX: mixed crops. HER: herbaceous crops. LEN: woody crops. PAS: pastures. BOS: forests. VAR: scrub vegetation and open spaces.

in category $i$ in time 2. The quotient is, then, the ratio of the area in category $j$ to the area not in category $i$ in time 2. For a fuller understanding of this equation, imagine that the areas gained by the other categories to generate the loss observed in category $i$ is divided into a large number of area units, and that in time 2 these units are distributed randomly across the non-$i$ area in the landscape studied. The proportion gained by each category should be very similar to the ratio of the area in that category to the area not in category $i$ in time 2. That is precisely what is calculated by Equation (5). Consequently, $l_{ij}$ represents the expected loss of $i$ in favor of $j$ under the assumption that $i$ is randomly replaced by the remaining categories in time 2 to generate the observed loss, i.e., the expected transition from $i$ to $j$ under the random loss hypothesis.

Losses are analyzed in Table 6 in the Results section below. The first value in each cell, shown in bold typeface, is the observed value of the transition, $p_{ij}$. The italicized second value is the expected value, $l_{ij}$. The third value is the deviation of the observed from the expected value, $p_{ij}$-$l_{ij}$. The deviation for $p_{ij}$ measures the degree to which this transition fails to conform to the assumption that $i$ losses are randomly distributed. If the deviation is greater than zero, the area lost by category $i$ to category $j$ is greater than the expected value. A substantial difference would be an indication of bias in the transition from $i$ to $j$: when the area in $i$ declines, it tends to be replaced by $j$. If the deviation is less than zero, the area lost by category $i$ to $j$ is less than the expected value. This means that when the area in $i$ declines, it tends not to be replaced by category $j$. The fourth value in each cell is the above deviation expressed in terms relative to the expected value:

$$tl_{ij} = \frac{p_{ij} - l_{ij}}{l_{ij}} \quad (6)$$

*Note.* Four values per cell: 1: observed; 2: expected; 3: deviation; 4: deviation as a multiple of the expected value. First three values in percentage of area studied. Categories: ART: artificial cover. AGU: water bodies. MIX: mixed crops. HER: herbaceous crops. LEN: woody crops. PAS: pastures. BOS: forests. VAR: scrub vegetation and open spaces.

It can be interpreted in exactly the same way as the $tg_{ij}$ ratio in Equation (4).

Here as well, since the area that remains unchanged in any given category should not affect this analysis, the expected values for the elements on the diagonal are made to concur with the observed values. The totals for the expected and observed values in time 1 concur (see Table 6) because the persistence and loss values are fixed. On the contrary, the expected value in the *Time 2 total* column includes the sum of the expected values of the transitions for the column category in question, which need not concur with the sum of the respective observed transitions. The former

*Table 6. Analysis of transitions as losses, 1990-2000*

| LOSS | | Category in 2000 | | | | | | | | Total 1990 | Loss |
|---|---|---|---|---|---|---|---|---|---|---|---|
| | | ART | AGU | MIX | HER | LEÑ | PAS | BOS | VAR | | |
| Category in 1990 | ART | **2.673** | **0.000** | **0.002** | **0.001** | **0.000** | **0.002** | **0.001** | **0.020** | **2.699** | **0.026** |
| | | *2.673* | *0.000* | *0.004* | *0.002* | *0.002* | *0.004* | *0.005* | *0.010* | *2.699* | *0.026* |
| | | 0.000 | 0.000 | -0.002 | -0.001 | -0.001 | -0.002 | -0.004 | 0.010 | 0.000 | 0.000 |
| | | 0.000 | -1.000 | -0.487 | -0.634 | -0.764 | -0.467 | -0.831 | 0.992 | 0.000 | 0.000 |
| | AGU | **0.000** | **1.012** | **0.000** | **0.000** | **0.000** | **0.000** | **0.000** | **0.000** | **1.012** | **0.000** |
| | | *0.000* | *1.012* | *0.000* | *0.000* | *0.000* | *0.000* | *0.000* | *0.000* | *1.012* | *0.000* |
| | | 0.000 | 0.000 | 0.000 | 0.000 | 0.000 | 0.000 | 0.000 | 0.000 | 0.000 | 0.000 |
| | | 0.000 | 0.000 | 0.000 | 0.000 | 0.000 | 0.000 | 0.000 | 0.000 | 0.000 | 0.000 |
| | MIX | **0.119** | **0.000** | **14.328** | **0.017** | **0.000** | **0.003** | **0.003** | **0.184** | **14.654** | **0.326** |
| | | *0.013* | *0.004* | *14.328* | *0.022* | *0.025* | *0.054* | *0.064* | *0.144* | *14.654* | *0.326* |
| | | 0.106 | -0.004 | 0.000 | -0.005 | -0.025 | -0.051 | -0.061 | 0.040 | 0.000 | 0.000 |
| | | 8.012 | -1.000 | 0.000 | -0.230 | -0.989 | -0.945 | -0.960 | 0.279 | 0.000 | 0.000 |
| | HER | **0.190** | **0.000** | **0.488** | **5.597** | **0.001** | **0.066** | **0.001** | **0.503** | **6.846** | **1.249** |
| | | *0.046* | *0.013* | *0.197* | *5.597* | *0.087* | *0.187* | *0.221* | *0.498* | *6.846* | *1.249* |
| | | 0.144 | -0.013 | 0.292 | 0.000 | -0.086 | -0.122 | -0.220 | 0.005 | 0.000 | 0.000 |
| | | 3.149 | -1.000 | 1.481 | 0.000 | -0.987 | -0.648 | -0.995 | 0.009 | 0.000 | 0.000 |
| | LEÑ | **0.035** | **0.000** | **0.001** | **0.001** | **6.549** | **0.023** | **0.004** | **0.320** | **6.933** | **0.383** |
| | | *0.014* | *0.004* | *0.061* | *0.023* | *6.549* | *0.058* | *0.068* | *0.154* | *6.933* | *0.383* |
| | | 0.020 | -0.004 | -0.060 | -0.022 | 0.000 | -0.035 | -0.065 | 0.166 | 0.000 | 0.000 |
| | | 1.435 | -1.000 | -0.987 | -0.945 | 0.000 | -0.609 | -0.947 | 1.077 | 0.000 | 0.000 |
| | PAS | **0.188** | **0.000** | **0.003** | **0.033** | **0.001** | **14.053** | **0.004** | **0.446** | **14.727** | **0.674** |
| | | *0.027* | *0.008* | *0.117* | *0.045* | *0.052* | *14.053* | *0.131* | *0.295* | *14.727* | *0.674* |
| | | 0.160 | -0.008 | -0.114 | -0.012 | -0.051 | 0.000 | -0.127 | 0.151 | 0.000 | 0.000 |
| | | 5.912 | -1.000 | -0.973 | -0.272 | -0.983 | 0.000 | -0.970 | 0.511 | 0.000 | 0.000 |
| | BOS | **0.036** | **0.000** | **0.007** | **0.001** | **0.003** | **0.002** | **16.330** | **0.173** | **16.553** | **0.223** |
| | | *0.009* | *0.003* | *0.040* | *0.015* | *0.018* | *0.038* | *16.330* | *0.100* | *16.553* | *0.223* |
| | | 0.027 | -0.003 | -0.032 | -0.014 | -0.015 | -0.036 | 0.000 | 0.073 | 0.000 | 0.000 |
| | | 2.914 | -1.000 | -0.813 | -0.927 | -0.833 | -0.957 | 0.000 | 0.726 | 0.000 | 0.000 |
| | VAR | **0.214** | **0.000** | **0.037** | **0.031** | **0.005** | **0.007** | **0.320** | **35.963** | **36.576** | **0.613** |
| | | *0.034* | *0.010* | *0.146* | *0.056* | *0.064* | *0.139* | *0.164* | *35.963* | *36.576* | *0.613* |
| | | 0.180 | -0.010 | -0.109 | -0.025 | -0.060 | -0.132 | 0.156 | 0.000 | 0.000 | 0.000 |
| | | 5.310 | -1.000 | -0.749 | -0.449 | -0.923 | -0.952 | 0.954 | 0.000 | 0.000 | 0.000 |
| Total 2000 | | **3.455** | **1.012** | **14.867** | **5.680** | **6.560** | **14.155** | **16.662** | **37.610** | **100.000** | **3.494** |
| | | *2.817* | *1.054* | *14.893* | *5.759* | *6.797* | *14.534* | *16.982* | *37.166* | *100.000* | *3.494* |
| | | 0.638 | -0.042 | -0.026 | -0.079 | -0.237 | -0.379 | -0.320 | 0.445 | 0.000 | 0.000 |
| | | 0.227 | -0.040 | -0.002 | -0.014 | -0.035 | -0.026 | -0.019 | 0.012 | 0.000 | 0.000 |

*continues on following page*

*Table 6. continued*

| LOSS | Category in 2000 | | | | | | | | Total 1990 | Loss |
|---|---|---|---|---|---|---|---|---|---|---|
| | ART | AGU | MIX | HER | LEÑ | PAS | BOS | VAR | | |
| Gain | **0.781** | **0.000** | **0.539** | **0.083** | **0.011** | **0.102** | **0.332** | **1.647** | **3.494** | |
| | *0.143* | *0.042* | *0.564* | *0.162* | *0.247* | *0.480* | *0.652* | *1.202* | *3.494* | |
| | 0.638 | -0.042 | -0.026 | -0.079 | -0.237 | -0.379 | -0.320 | 0.445 | 0.000 | |
| | 4.447 | -1.000 | -0.046 | -0.489 | -0.957 | -0.788 | -0.491 | 0.370 | 0.000 | |

is the total that the categories should have had in time 2, assuming random losses. Similarly, the expected value in the *Gain* row shows the expected gain for the respective column category on the assumption that the rest of the categories are replaced by all the others randomly.

## RESULTS

While in the foregoing theoretical discussion the areas were expressed as a proportion of the total area studied, the results are given here in percentage.

The distribution of the SPA area by category in 1990 is shown in the *Total 1990* column, and for 2000 in the *Total 2000* row. In both years, the category with the highest percentage was scrub vegetation and open spaces, which accounted for approximately 37% of the area studied, followed by forests, pastures and mixed crops, with percentages ranging from 14.2 to 16.7%. Woody and herbaceous crops covered from 5.7 to 6.9% of the area. The categories with the lowest relative weight were artificial surfaces, with around 3%, and water bodies, with 1.1%. The order of the categories by size remained stable over the 10-year period, except for pastures and mixed crops: in 1990 the former covered slightly more area than the latter, while the order was inverted in 2000.

Persistence prevailed in 96.5% of the area. In other words, 96.5% of the SPA area underwent no change in category in the 10 years studied. The distribution of this total across the categories as shown in the diagonal in Table 3 indicates that the level of persistence is closely associated with the weight of each category in the area studied.

The area changing category in the period, i.e., the total change in the area studied, came to 3.49% of the land area (Table 4), or 3,393.2 ha. Most of this change, 60.5% in all, was net change and involved 2.11% of the total SPA area. This is the amount of area acquired by the categories expanding in the period at the expense of the categories with net losses. Swaps amounted to 39.5% of the total change, affecting 1.38% of the area studied. This is the area lost by categories in some places but offset by their gains in others.

The discussion of the results in Table 4 revolves around net change. The categories with net losses in the period were herbaceous (1.17%) and woody (0.37%) crops and pastures (0.57%). The swap levels in these categories was low or very low, with net change consistently over 73% of the total change and coming to nearly 95% in the woody crops. These categories were clearly receding, with gross losses barely mitigated by gross gains. Despite their scant weight in the area, the case of herbaceous crops is important because their net loss, at 17% of the area covered by these crops in 1990, was higher than for any other category.

The non-artificial categories with net gains were scrub vegetation and open spaces (1.03%), forests (0.11%) and mixed crops (0.21%). These categories also showed the highest swap rates, ranging from 0.44 to 1.23% of the SPA area. In other words, these are very dynamic categories whose gross gains offset their visible gross losses. The changes involving forests are of particular interest. This category showed a net gain of 0.11%

in the period. According to Table 4, however, that net change was the result of a gross loss due to deforestation that came to twice the net gain. This 0.22% loss was offset by a gross gain (reforestation) of 0.33%. While initially encouraging, such figures must be interpreted with caution, however. The losses may well have primarily affected traditional forest, whereas the gains would most likely have been the result of recent repopulation, a type of cover whose environmental quality is presumably lower. In other words, if the analysis is limited to net change, very significant information on the relevance and nature of the changes taking place in each category may be lost.

Artificial surfaces grew substantially in the period, with a net gain of 0.76% of the total area and a low swap level: practically no gross losses were recorded. The relevance of the expansion of artificial surfaces is illustrated by the high net change figure, which amounted to 28% of the area in the category in 1990.

The most prominent transitions in Table 3, ranked from highest to lowest, were: the 0.5% (489 ha) of the total landscape yielded by herbaceous crops to scrub vegetation and the 0.49% (475 ha) to mixed crops; the 0.45% (434 ha) gain in scrub vegetation at the expense of pastures and the 0.32% (312 ha) that replaced woody crops, and the 0.32% (311 ha) yielded by this category to forests; the 0.21% (208 ha) gain in artificial surfaces from scrub vegetation, the 0.19% (185 ha) from herbaceous crops and the 0.19% (182 ha) from pastures. These eight transitions accounted for nearly 70% of the total change observed in the area studied.

The analysis of the transitions from the standpoint of gains is based on the data in Table 5, whose interpretation rests on the principles set out in Method section. The deductions drawn from a review of the table are listed below.

Gains in artificial surfaces came non-randomly at the expense of herbaceous crops (0.19% compared to the expected 0.05%, or 2.45 times more than expected) and pastures (0.19% compared to 0.12% or 0.58 times more than expected). While the greatest gain in artificial surfaces was at the expense of scrub vegetation, this gain was 0.27 times lower than expected under the random gain hypothesis (0.21% compared to the expected 0.29%).

Herbaceous crops were also systematically replaced by mixed crops, yielding over 10 times the terrain that would be expected in a random gain process (0.49% compared to 0.04%). In this case, this transition was also the largest in absolute terms of all the transitions involving gains in this category.

Scrub vegetation area growth also involved non-random replacement of herbaceous (0.5% compared to 0.18%, 1.8 times more than the expected value) and woody (0.32% compared to 0.18%, 0.8 times more than expected) crops.

Forest areas, in turn, replaced scrub vegetation systematically (0.32% compared to 0.15%, or 1.2 times more than expected).

The growth of any of the non-herbaceous crop categories except forests entailed the non-random replacement of herbaceous crops, as deduced from the elements in the *HER* row in Table 5, synthesized in the *Loss* column. While the loss of area devoted to herbaceous crops came to 1.25%, the total that the other categories would be expected to gain from herbaceous crops under the random gain hypothesis was only 0.31%. Therefore, herbaceous crops lost three times more than expected. This finding synthesizes all the above results for herbaceous crops.

The analysis of transitions viewed as losses is based on the data in Table 6. The deductions drawn from a review of the table are set out below.

When the area of non-artificial surfaces declined, these surfaces were non-randomly replaced by artificial surfaces. This can be seen in each element of the ART column, as summarized in the gains row. Artificial areas were observed to grow by 0.78%. The total that the other categories would be expected to yield to artificial surfaces under the random loss hypothesis was 0.14%.

Therefore, artificial surfaces grew 4.5 times more than expected.

When the area attributed to non-scrub vegetation categories declined it was non-randomly covered by scrub vegetation (1.6% compared to 1.2%, or 37% more than expected).

Losses in the area covered by scrub vegetation were systematically replaced by artificial surfaces (0.21% compared to 0.03%, 5.3 times more than expected) and forests (0.32% compared to 0.16%, 95% more than expected).

## CONCLUSION

The method provides for a highly structured synthesis of the changes recorded for each category and the implications for the landscape. The findings show the importance of considering swaps when analyzing change. In the present case study, for instance, swaps accounted for a substantial portion of the total change in the SPA. Moreover, their omission in certain categories (forests, mixed crops) would be tantamount to omitting most of the change recorded.

For transitions, the method records the inter-category changes observed as the fraction or the percentage of the total landscape area, as in classic transition matrices. But it also delivers information that can be used to determine which of these transitions can be regarded to be systematic and therefore not the result of gains or losses occurring at random across the area of the landscape studied. These two types of information are obviously complementary. The method does not, however, specify how they should be integrated or the importance to be attached to each. These issues are left to the analyst's discretion. For instance, the method may identify systematic transitions whose absolute value renders them quantitatively irrelevant, for transitions are evaluated in terms of the expected value. But where is the cut-off? When should a transition be regarded to be irrelevant? In a similar vein, what deviation between the observed and expected values should be defined as indicative of systematic change? Statistical indicators obviously fail to solve the problem entirely, whereas a full understanding of the natural and socio-economic environments is essential to drawing genuinely useful knowledge from land cover change information.

Lastly, a word is in order on the number of categories studied, for the larger the number, the greater is the complexity in interpreting the findings. In this regard, the eight categories analyzed in the present survey would appear to constitute the upper limit.

## REFERENCES

Briassoulis, H. (2000). Analysis of Land Use Change: Theoretical and Modeling Approaches. In S. Loveridge (Ed.), *The Web Book of Regional Science* (www.rri.wvu.edu/regscweb.htm). Morgantown, WV: Regional Research Institute, West Virginia University.

Coppin, P., Jonckheere, I., Nackaerts, K., & Muys, B. (2004). Digital change detection methods in ecosystem monitoring: a review. *International Journal of Remote Sensing*, *25*(9), 1565–1596. doi:10.1080/0143116031000101675

De Juana, E. (Ed.). (1990). *Áreas importantes para las aves en España*. Madrid: Monografías de la Sociedad Española de Ornitología, n.º 3.

Instituto Geográfico Nacional. (2003). *Generación de la Base de Datos de cambios Corine Land Cover y estadísticas por Km2. Proyecto Image & Corine 2000*. Madrid: Ministerio de Fomento, Área de Teledetección del Instituto Geográfico Nacional.

Lu, D., Mausel, P., Brondizios, E., & Moran, E. (2004). Change detection techniques. *International Journal of Remote Sensing*, *25*(12), 2365–2407. doi:10.1080/0143116031000139863

Martín Lou, M. A., & Martínez Vega, J. (1998). *Situación socioeconómica de la ZEPA Encinares del río Alberche y río Cofio (Comunidad de Madrid)*. Madrid: Instituto de Economia y Geografía (CSIC).

Mast, J. N., Veblent, T. T., & Hodgson, M. E. (1997). Tree invasion within a pine/grassland ecotone: An approach with historic aerial photography and GIS modelling. *Forest Ecology and Management*, *93*(3), 181–194. doi:10.1016/S0378-1127(96)03954-0

Murray, I., Blàzquez, M., & Rullan, O. (2005). Los cambios en la cobertura de la Tierra. Una revisión bibliográfica desde la geografía. *Biblio 3W, Revista Bibliográfica de Geografía y Ciencias Sociales, Universidad de Barcelona, 10*, 571. Retrieved Mars 15 2005, from http://www.ub.es/geocrit/b3w-571.htm

Peña, J., Poveda, R. M., Bonet, A., Bellot, J., & Escarré, A. (2005). Cartografía de las coberturas y usos del suelo de la Marina Baixa (Alicante) para 1956, 1978 y 2000. *Investigaciones Geográficas*, *37*, 93–107.

Pontius, R. G., Jr., & Malizia, N. R. (2004). Effect of category aggregation on map comparison. In Egenhofer, M.J., Freksa, C., & Miller, H.J. (Ed.), *GIScience 200*, (LNCS 3234, pp. 251-268). Berlin Heildelberg: Springer-Verlag.

Pontius, R. G. Jr, Shusas, E., & McEachern, M. (2004). Detecting important land changes while accounting for persistence. *Agriculture Ecosystems & Environment*, *100*, 251–268. doi:10.1016/j.agee.2003.09.008

Suárez, F., Oñate, J., Malo, J., & Peco, B. (1997). Las políticas agroambientales y de conservación de la naturaleza en España. *Economía Agraria*, *179*, 267–296.

Van Oort, P. A. J. (2006). Interpreting the change detection error matrix. *Remote Sensing of Environment*, *108*(1), 1–8. doi:10.1016/j.rse.2006.10.012

## ADDITIONAL READING

Alo, C. A., & Pontius, R. G. (2008). Identifying systematic land-cover transitions using remote sensing and GIS: the fate of forests inside and outside protected areas of Southwestern Ghana. *Environment and Planning. B, Planning & Design*, *35*(2), 280–295. doi:10.1068/b32091

Anav, A., Ruti, P. M., Artale, V., & Valentini, R.Modelling the effects of land-cover changes on surface climate in the Mediterranean region. *Climate Research*, *41*(2), 91–104.

Calvo-Iglesias, M. S., Fra-Paleo, U., Crecente-Maseda, R., & Diaz-Varela, R. A. (2006). Directions of change in land cover and landscape patterns from 1957 to 2000 in agricultural landscapes in NW Spain. *Environmental Management*, *38*(6), 921–933. doi:10.1007/s00267-005-0276-1

Cots-Folch, R., Martínez-Casasnovas, J. A., & Ramos, M. C. (2009). Agricultural trajectories in a Mediterranean mountain region (Priorat, NE Spain) as a consequence of vineyard conversion plans. *Land Degradation & Development*, *20*(1), 1–13. doi:10.1002/ldr.856

Detsis, V., Ntasiopoulou, G., Chalkias, C., & Efthimiou, G. Recent Insular Mediterranean Landscape Evolution: A Case Study on Syros, Greece. *Landscape Research, 35*(3), 361-381.

Garcia-Romero, A., Munoz, J., Andres, N., & Palacios, D.Relationship between climate change and vegetation distribution in the Mediterranean mountains: Manzanares Head valley, Sierra De Guadarrama (Central Spain). *Climatic Change*, *100*(3-4), 645–666. doi:10.1007/s10584-009-9727-7

Garcia Ruiz, J. M., La Santa, T., Ruiz Flano, P., Ortigosa, L., White, S., & Gonzalez, C. (1996). Land-use changes and sustainable development in mountain areas: a case study in the Spanish Pyrenees. *Landscape Ecology*, *11*(5), 267–277. doi:10.1007/BF02059854

Gellrich, M., Baur, P., Koch, B., & Zimmermann, N. E. (2007). Agricultural land abandonment and natural forest re-growth in the Swiss mountains: A spatially explicit economic analysis. *Agriculture Ecosystems & Environment*, *118*(1-4), 93–108. doi:10.1016/j.agee.2006.05.001

Gellrich, M., & Zimmermann, N. E. (2007). Investigating the regional-scale pattern of agricultural land abandonment in the Swiss mountains: A spatial statistical modelling approach. *Landscape and Urban Planning*, *79*(1), 65–76. doi:10.1016/j.landurbplan.2006.03.004

Hersperger, A. M., & Burgi, M. How Do Policies Shape Landscapes? Landscape Change and its Political Driving Forces in the Limmat Valley, Switzerland 1930-2000. *Landscape Research,35*(3), 259-279. doi:10.1080/01426391003743561

Kayhko, N., & Skanes, H. (2006). Change trajectories and key biotopes--Assessing landscape dynamics and sustainability. *Landscape and Urban Planning*, *75*(3-4), 300–321. doi:10.1016/j.landurbplan.2005.02.011

Kerr, J. T., & Cihlar, J. (2003). Land use and cover with intensity of agriculture for Canada from satellite and census data. *Global Ecology and Biogeography*, *12*(2), 161–172. doi:10.1046/j.1466-822X.2003.00017.x

Kuzera, K., & Pontius, R. G. (2008). Importance of matrix construction for multiple-resolution categorical map comparison. *Giscience & Remote Sensing*, *45*(3), 249–274. doi:10.2747/1548-1603.45.3.249

MacDonald, D., Crabtree, J. R., Wiesinger, G., Dax, T., Stamou, N., & Fleury, P. (2000). Agricultural abandonment in mountain areas of Europe: Environmental consequences and policy response. *Journal of Environmental Management*, *59*(1), 47–69. doi:10.1006/jema.1999.0335

Manandhar, R., Odeh, I. O. A., & Pontius, R. G. (2010). Analysis of twenty years of categorical land transitions in the Lower Hunter of New South Wales, Australia. *Agriculture Ecosystems & Environment*, *135*(4), 336–346. doi:10.1016/j.agee.2009.10.016

Martinez-Falero, E., & Gonzalez-Alonso, S. (1995). *Quantitative techniques in landscape planning*. London: Boca Raton, Fla.

Millington, J. D. A., Perry, G. L. W., & Romero-Calcerrada, R. (2007). Regression techniques for examining land use/cover change: A case study of a mediterranean landscape. *Ecosystems (New York, N.Y.)*, *10*(4), 562–578. doi:10.1007/s10021-007-9020-4

Munsi, M., Malaviya, S., Oinam, G., & Joshi, P. K.A landscape approach for quantifying land-use and land-cover change (1976-2006) in middle Himalaya. *Regional Environmental Change*, *10*(2), 145–155. doi:10.1007/s10113-009-0101-0

Paquette, S., & Domon, G. (2004). Changing ruralities, changing landscapes: exploring social recomposition using a multi-scale approach. *Journal of Rural Studies*, *19*(4), 425–444. doi:10.1016/S0743-0167(03)00006-8

Pontius, J., Gil, R., & Schneider, L. C. (2001). Land-cover change model validation by an ROC method for the Ipswich watershed, Massachusetts, USA. *Agriculture Ecosystems & Environment*, *85*(1-3), 239–248. doi:10.1016/S0167-8809(01)00187-6

Pontius, R. G. (2002). Statistical methods to partition effects of quantity and location during comparison of categorical maps at multiple resolutions. *Photogrammetric Engineering and Remote Sensing*, *68*(10), 1041–1049.

Pontius, R. G., Boersma, W., Castella, J. C., Clarke, K., de Nijs, T., & Dietzel, C. (2008). Comparing the input, output, and validation maps for several models of land change. *The Annals of Regional Science*, *42*(1), 11–37. doi:10.1007/s00168-007-0138-2

Pontius, R. G., & Connors, J. (2009). Range of Categorical Associations for Comparison of Maps with Mixed Pixels. *Photogrammetric Engineering and Remote Sensing*, *75*(8), 963–969.

Pontius, R. G., Huffaker, D., & Denman, K. (2004). Useful techniques of validation for spatially explicit land-change models. *Ecological Modelling*, *179*(4), 445–461. doi:10.1016/j.ecolmodel.2004.05.010

Pontius, R. G., & Petrova, S. H. (2010). Assessing a predictive model of land change using uncertain data. *Environmental Modelling & Software*, *25*(3), 299–309. doi:10.1016/j.envsoft.2009.09.005

Pontius, R. G., Thontteh, O., & Chen, H. (2008). Components of information for multiple resolution comparison between maps that share a real variable. *Environmental and Ecological Statistics*, *15*(2), 111–142. doi:10.1007/s10651-007-0043-y

Pontius, R. G., Walker, R., Yao-Kumah, R., Arima, E., Aldrich, S., & Caldas, M. (2007). Accuracy assessment for a simulation model of Amazonian deforestation. *Annals of the Association of American Geographers. Association of American Geographers*, *97*, 677–695. doi:10.1111/j.1467-8306.2007.00577.x

Rabbinge, R., & van Diepen, C. A. (2000). Changes in agriculture and land use in Europe. *European Journal of Agronomy*, *13*(2-3), 85–99. doi:10.1016/S1161-0301(00)00067-8

Rocchini, D., Perry, G. L. W., Salerno, M., Maccherini, S., & Chiarucci, A. (2006). Landscape change and the dynamics of open formations in a natural reserve. *Landscape and Urban Planning*, *77*(1-2), 167–177. doi:10.1016/j.landurbplan.2005.02.008

Romero-Calcerrada, R., & Perry, G. L. W. (2004). The role of land abandonment in landscape dynamics in the SPA 'Encinares del rio Alberche y Cofio, Central Spain, 1984-1999. *Landscape and Urban Planning*, *66*(4), 217–232. doi:10.1016/S0169-2046(03)00112-9

Sluiter, R. (2005). *Mediterranean land cover change: Modelling and monitoring natural vegetation using GIS and remote sensing* (pp. 17–144). Nederlandse Geografische Studies.

Southworth, J., Munroe, D., & Nagendra, H. (2004). Land cover change and landscape fragmentation--comparing the utility of continuous and discrete analyses for a western Honduras region. *Agriculture Ecosystems & Environment*, *101*(2-3), 185–205. doi:10.1016/j.agee.2003.09.011

Veldkamp, A., & Lambin, E. F. (2001). Predicting land-use change. *Agriculture Ecosystems & Environment*, *85*, 1–6. doi:10.1016/S0167-8809(01)00199-2

Verburg, P. H., Kok, K., Pontius, R. G., & Veldkamp, A. (2006). Modeling land-use and land-cover change. *Land-Use and Land-Cover Change: Local Processes and Global Impacts*, 117-135.

Verburg, P. H., Veldkamp, A., & Fresco, L. O. (1999). Simulation of changes in the spatial pattern of land use in China. *Applied Geography (Sevenoaks, England)*, *19*(3), 211–233. doi:10.1016/S0143-6228(99)00003-X

Wang, Y., & Zhang, X. (2001). A dynamic modeling approach to simulating socioeconomic effects on landscape changes. *Ecological Modelling*, *140*(1-2), 141–162. doi:10.1016/S0304-3800(01)00262-9

Yilmaz, R.Monitoring land use/land cover changes using CORINE land cover data: a case study of Silivri coastal zone in Metropolitan Istanbul. *Environmental Monitoring and Assessment*, *165*(1-4), 603–615. doi:10.1007/s10661-009-0972-z

Zampieri, M., & Lionello, P.Simple statistical approach for computing land cover types and potential natural vegetation. *Climate Research*, *41*(3), 205–220. doi:10.3354/cr00846

## KEY TERMS AND DEFINITIONS

**Gross Gain of A:** Is the total land that has become category A from time 1 to time 2.

**Gross Loss of Category A:** Is the area that has ceased to be of category A from time 1 to time 2.

**Net Change of A:** Is the change in the amount of land cover area occupied by A from time 1 to time 2.

**Persistence of Category A is the Stable Area:** Land cover A doesn't change of category from time 1 to time 2.

**Swap of Category A:** Corresponds to the areas gained and lossed by A that can be considedered as compensating each other. That is, it is the part of total change implied in change of location of land cover A.

**Total Change of Category A:** Is the sum of the gross loss and gross gain of the category. It is the area that has had a change that affects the category A, either because this category has experienced a gain or because it has had a loss.

**Transition from Category A to Category B:** Is the area that was clasified in category A in time 1 but in category B in time 2.

# Chapter 4

# Mining Climate and Remote Sensing Time Series to Improve Monitoring of Sugar Cane Fields

**Luciana A. S. Romani**
*University of Sao Paulo at Sao Carlos, Brazil & Embrapa Agriculture Informatics at Campinas, Brazil*

**Elaine P. M. de Sousa**
*University of Sao Paulo at Sao Carlos, Brazil*

**Marcela X. Ribeiro**
*Federal University of Sao Carlos, Brazil*

**Ana M. H. de Ávila**
*University of Campinas, Brazil*

**Jurandir Zullo Jr.**
*University of Campinas, Brazil*

**Caetano Traina Jr.**
*University of Sao Paulo at Sao Carlos, Brazil*

**Agma J. M. Traina**
*University of Sao Paulo at Sao Carlos, Brazil*

## ABSTRACT

*This chapter discusses how to take advantage of computational models to analyze and extract useful information from time series of climate data and remote sensing images. This kind of data has been used for researching on climate changes, as well as to help on improving yield forecasting of agricultural crops and increasing the sustainable usage of the soil. The authors present three techniques based on the Fractal Theory, data streams and time series mining: the FDASE algorithm, to identify correlated attributes; a method that combines intrinsic dimension measurements with statistical analysis, to monitor evolving climate and remote sensing data; and the CLIPSMiner algorithm applied to multiple time series of continuous climate data, to identify relevant and extreme patterns. The experiments with real data show that data mining is a valuable tool to help agricultural entrepreneurs and government on monitoring sugar cane areas, helping to make the production more useful to the country and to the environment.*

## INTRODUCTION

Brazil is an agricultural country. It is a leading corn and soybeans producer and the main sugar cane and coffee producer. According to official data (IBGE, 2007), the agribusiness segment contributed with 23.3% of the national Gross Domestic Product (GDP), 42% of exports and 37% of jobs in 2007. Significant advances have been made in determining suitable areas for agricultural

DOI: 10.4018/978-1-61692-871-1.ch004

crops development through agricultural zoning program, developed by the Brazilian Ministry of Agriculture (Rossetti, 2001).

The Brazilian agricultural zoning program aims at reducing agriculture losses caused by two climatic-associated risks: dry spells during the reproductive stage and excessive rainfall during the harvesting periods. This official program defines planting calendars for the main crops in the country, which have been calculated to achieve risk rates lower than 20% regarding climate problems, based on climate data and agrometeorological methods. However, after planting the crop, it is extremely important to monitor crop yields.

Reliable estimates of agricultural production are powerful tools to guide producers on issues related to planting and also to assist agribusiness in operating and marketing sectors. They may generate reliable data to support the government in the decision making process, aimed at reducing negative impacts on the economy or to take advantage of favorable situations in the weather and in the agricultural market.

In addition, the crop forecasts performed by agrometeorological agents in the country appear as an effective mechanism for protection of domestic production. This occurs because they generate a frame capable of preventing or reducing the reaction caused by speculative estimates from external agents, often owned by competing countries in the international market.

Developing countries, like Brazil, usually do not have a well-distributed network of meteorological stations. Therefore, monitoring the weather conditions is one of the main difficulties that decision makers must face. In particular, we observed that data from surface networks are not enough to appropriately solve small-scale variability. Such variability includes strong gradients of rainfall intensity, the small size of cells and a short precipitation cycle time, which are characteristics of convective rainfall, responsible for much of precipitation that occurs in tropical regions. Moreover, experts are becoming increasingly concerned about the negative impacts of the meteorological conditions and of the natural disasters.

Using remote sensing data is an alternative to more conventional methods, because the sensors have an excellent spatial and temporal coverage. These sensors also make it possible to obtain continuous information from the country land, with spatial resolution of few kilometers and temporal data in order of minutes. However, measurements obtained from remote sensors are indirect and, therefore, it is necessary to develop models to relate the features available in the satellite spectral channels to parameters associated with the required information.

In this scenario, several satellites are being used to assist in land monitoring and climate forecasting. The NOAA (National Oceanic and Atmospheric Administration) satellites, originally designed to act as meteorological satellites, have been widely used for vegetation monitoring on both regional and global levels and, more recently, to monitor agricultural crops. The AVHRR (Advanced Very High Resolution Radiometer) on board of the NOAA satellites includes two channels in the red and near-infrared spectra, which are adopted in studies of the vegetation. The frequency of AVHRR imaging varies from 2 to 4 images from the same place per day, increasing the probability of obtaining good quality images throughout the development cycle of commercial crops.

According to Fontana & Berlato (1998), the inclusion of spectral variable into agricultural models of monitoring aims at estimating parameters that cannot be fully represented in agrometeorological models. When spectral information is acquired with higher frequency throughout the production cycle, it becomes possible to assess the intrinsic crop parameter evolution more accurately, such as for the biomass and leaf area index (LAI), which may have closer correlation with the crop yield. In this sense, the orbital sensors having high temporal resolution are an important source of spectral information.

Sugar cane has become increasingly strategic in the Brazilian economy due to the replacement of fossil fuels with renewable energy sources, such as the ethanol. Another important economic issue is the sugar production due to the increasing imports stimulated by Asiatic countries, like China, which are steadily increasing the demand for these products. Thus, we can expect a strong pressure on exports driven by China in the next years.

In Brazil, the sugar cane is the main agricultural crop used to produce ethanol. The country has a privileged position to support the growing international demand for sugar and anhydrous ethanol for fuel. With two main producing regions and alternate crops, Brazil is able to maintain its worldwide market presence throughout the year. In fact, this agricultural commodity has a strategic importance to the national economy. Therefore, there is an evident need for accurate crop prediction techniques that would help the production planning and the marketing strategy for domestic and foreign markets.

Besides being important to the current agribusiness, sugar cane also has influence in the climate changes. According to the fourth Assessment Report (AR4) of Intergovernmental Panel on Climate Change (IPCC), temperature and precipitation should increase in the planet due to natural and anthropogenic effects (IPCC, 2007). Therefore, researches have been pursued to forecast the changes in the temperature patterns to define methods to reduce the emission of greenhouse gases and to adapt agricultural crops to the new conditions of higher temperatures. An alternative to reduce greenhouse gases emission is to replace fossil fuels with renewable sources.

Under such scenario, the development of computational models to filter, transform, merge and analyze data from many different areas is complex and challenging. This complexity increases when combining several climatic and agrometeorological variables and also when using climate and agriculture models together. In particular, the importance of the agricultural production for the Brazilian economy requires impact assessment studies of seasonal climate variations and climate changes.

In the past few years, the improvements in data acquisition technology decreased the time interval for gathering data, leading institutions to store huge amounts of data, such as remote sensing images and time series of climate data. Agrometeorologists have used climate data from ground-based meteorological stations (for instance, rainfall and temperature) for a long time. Recently, remote sensing data are being employed to improve traditional agrometeorological methods. Nowadays, data are more accessible and there is more appropriate technology (regarding both software and hardware) to receive, distribute and process long time series of satellite images.

The huge volume of environmental data (obtained mainly from ground stations or from computational models) associated to remote sensing images and geographical information are important motivations to the development of new data mining algorithms, as they provide important tools to identify relationships, patterns and correlations that are not previously known by the specialists. Extracting new knowledge from data involves several areas of computer science, such as statistics, databases, artificial intelligence, pattern recognition, machine learning and visualization.

Consider, for example, datasets integrating climate data and remote sensing images from some sugar cane fields. A feature selection algorithm can identify the most relevant attributes of the datasets, which represent the majority of the information related to the agricultural yield and the correlations among attributes. Moreover, it is interesting to know which attributes can better approximate the values of the others. In fact, the detection of correlated attributes, their importance and precedence in climate datasets can improve the agricultural monitoring of sugar cane in Brazil.

Additionally, in real climate applications, an impressive amount of time series is available, both generated by meteorological stations and interpolated over distributed grid points. As the data distribution in this application domain usually changes over time, climate time series can be seamless considered as evolving data streams. Therefore, tracking the behavior of evolving climate data can be very useful to agricultural monitoring, as for example, to monitor the precipitation, air temperature and soil water content.

Another issue related to time series of climatic data is that they are usually composed of continuous data, which brings an additional challenge when these data have to be properly analyzed in order to identify relevant patterns and associations. Thus, relevant problems to be investigated include answering:

- How to mine patterns in time series when events are continuous?
- How to quantize time series retaining the temporal meaning of the patterns?
- How to discover relevant patterns in datasets that combine/match/link time series of climate data and of remote sensing images?

In this chapter, we present three different approaches to discover and analyze patterns on time series of climatic data and remote sensing images to improve yield forecast of sugar cane fields in Brazil, based on agrometeorological models. The methods described are based on different techniques, such as from Fractal Theory and time series mining. The first technique is the *FD-ASE* algorithm, applied to identify sets of correlated attributes and to select relevant attributes to represent the meaningful features in the data.

We also explore the fractal dimension as a tool to support a framework for data stream monitoring in agrometeorological applications. The suitability of the fractal-based approach to monitor data streams is obtained by employing a statistical approach to compare the data in consecutive time periods, pointing out the attributes that are responsible for the trend changes and how they influence them.

Finally, we present the *CLIPSMiner* (CLImate PatternS Miner) algorithm, which is able to discover relevant patterns on time series of climatic data and remote sensing images. This new method works on multiple time series of continuous data, identifying all defined patterns or the relevant ones according to a relevance factor, which can be tuned by the user.

## BACKGROUND

Fractal concepts have been applied to several data mining tasks. In particular, the intrinsic dimension based on the fractal dimension has been employed as a useful tool for clustering analysis (Barbará & Chen, 2003), mining of temporal association rules (Barbará et al., 2004), attribute selection (Traina Jr. et al., 2000), time series forecasting (Chakrabarti & Faloutsos, 2002) and spatial data mining (Traina et al., 2001). Fundamental concepts and definitions are presented as follows. Table 1 lists the symbols used hereinafter.

**Definition 1.** Fractal: *A fractal is an object that presents roughly the same characteristics regardless of the scale where it is analyzed, i.e. a self-similar object. Thus, small scale details are similar to large scale characteristics (Traina et al., 2005).*

**Definition 2.** Embedding dimension *E: Given a finite dataset A, the embedding dimension* $E \in \mathbb{N}$ *is the number of attributes that define A, i.e., E is the dimensionality of the space in which the dataset is embedded.*

**Definition 3.** Intrinsic dimension *D: Given a finite dataset A, its intrinsic dimension* $D \in \mathbb{R}^+$ *is the dimensionality of the object represented by the data, regardless of the dimension of the space in which it is embedded.*

The intrinsic dimension (*D*) measures the amount of information that the dataset represents. For instance, the intrinsic dimension of a set of points distributed along a line is equal to one. If the set of points is embedded in a higher dimensional

*Table 1. Table of symbols*

| Symbol | Definition |
|---|---|
| $A = \{a_1, a_2, \ldots, a_E\}$ | Definition of dataset $A$, composed of attributes $a_i$ |
| R | Domain of real numbers |
| $E$ | Embedding dimension |
| $D$ | Intrinsic dimension |
| $D_2$ | Correlation fractal dimension |
| $r$ | Side size of a grid cell |
| $pD()$ | Partial intrinsic dimension |
| $iC()$ | Maximum individual contribution of an attribute |
| $C_{r,i}$ | Count ('occupancy') of points in the *i-th* grid cell of side size $r$ |
| $\xi$ | Strength threshold of correlations to be retrieved |
| $\xi C$ | Attribute set core |
| $\xi B_p$ | Correlation base of a correlation group $p$ |
| $\xi G_p$ | Correlation group $p$ |
| $M_j(C) \rightarrow a_j$ | Mapping of attributes $C \subset A$ restricting the values of $a_j$ |
| $S$ | Time series |
| $e_i$ | Events of type $(b_i; t_i)$ |
| $S_e$ | Event Sequence |
| $S_{ea}$ | Ascending event sequence |
| $S_{ed}$ | Descending event sequence |
| $S_{es}$ | Stable event sequence |
| $V$ | Pattern of type Valley (negative peak) |
| $M$ | Pattern of type Mountain (positive peak) |
| $P$ | Pattern of type Plateau (small variation) |
| $y$ | Time series amplitude |
| $\delta$ | Minimum variation between two consecutive events |
| $\rho$ | Relevance Factor |
| $\lambda$ | Plateau Length |
| $n$ | Number of elements in a time series |

space, the intrinsic dimensionality continues equal to one. Faloutsos and Kamel (1994) proposed using the intrinsic dimension as a tool to measure the non-uniform behavior of real datasets. Moreover, the authors presented empirical studies to show that real data usually have self-similar behavior, which is the fundamental characteristic of fractal objects. Therefore, the intrinsic dimension $D$ of a real dataset can be estimated by calculating its fractal dimension.

The fractal dimension of statistically self-similar datasets can be determined by the Correlation Fractal Dimension $D_2$. An efficient approach to measure the fractal dimension of datasets embedded in $E$-dimensional spaces is the BoxCounting method (Schroeder, 1991), which defines $D_2$ as presented in Equation 1. An efficient algorithm (linear cost on the number of elements in the dataset) to compute $D_2$ was proposed in (Traina Jr. et al., 2000).

**Definition 4.** Correlation Fractal Dimension $D_2$: *Given a dataset self-similar in the range of scales* $[r_1; r_2]$, *its Correlation Fractal Dimension* $D_2 \rightarrow R^+$ *is measured as*

$$D_2 \equiv \frac{\partial \log(\sum_i C^2_{r,i})}{\partial \log(r)} r \in [r_1, r_2] \quad (1)$$

where r is the side size of the cells in a (hyper) cubic grid that divides the address space of the dataset and $C_{r,i}$ is the count of points in the i-th cell.

In traditional database applications, the embedded dimension *E* (the number of attributes) determines the dataset address space, while its behavior can be described through its intrinsic dimension *D*, which effectively measures the amount of information the data represent (Faloutsos & Kamel, 1994). *D* is usually lower than *E*, because real data rarely present independence and uniformity properties. Therefore, data mining tasks can benefit from the behavior information provided by *D*.

Similarly, in data stream mining continuous measurements of *D* can indicate general changes in data distribution over time, spotting meaningful occurrences. As applications in remote sensing and climate areas have generated continuous sequences of data over long periods of time, these data can be seamlessly considered data streams. In this chapter, we consider a data stream as an ordered sequence of events (or items) $\{c_1, c_2, ..., c_n\}$ in which an event $c_j$ is defined by a set of *E* attributes $a_i$, such that each $c_j = (a_1, ..., a_E)$.

The general idea of using the intrinsic dimension as a tool to monitor evolving data streams is to continuously measure *D* over time in order to detect significant variations of successive values of *D* and, consequently, identify meaningful behavior changes. An approach to measure the intrinsic dimension of data streams was proposed in (Sousa et al., 2007a). The authors present the *SID-meter* algorithm, which considers a data stream as a sequence of events $\{c_1, c_2, ..., c_n\}$, each one represented by an array of *E* measurements. The events occurring within a time interval are seen as a dimensional dataset of dimension *E*. The fractal dimension $D_2$ is used to estimate the intrinsic dimension *D* of a event sequence. *SID-meter* applies an event-based sliding window divided into $n_c$ sequential periods, named counting periods. Each period can process events arriving during a given time or a predefined number of incoming events, i.e., $n_i$ events are processed in each counting period. When a counting period is complete, the events of the oldest one are discarded. Therefore, $n_c$ and $n_i$ respectively specify the length of the window and the movement step of the measuring window. The value of *D* is based on the count of events inside the whole window, following Equation 1.

Several authors have worked on analysis of the behavioral changes of evolving data (Aggarwal, 2003; Kifer et al., 2004; Papadimitriou et al., 2004), burst detection (Kleinberg, 2003; Zhu & Shasha, 2003), classification (Aggarwal et al., 2004b; Gama et al., 2005; Ferrer-Troyano et al., 2006; Aggarwal & Yu, 2008), clustering (Guha et al., 2003; Aggarwal et al., 2004a, Rodrigues et al. 2008a; Rodrigues et al. 2008b), frequent items identification, maintenance and processing data streams (Jin et al., 2003; Manjhi et al.,2005; Sakurai, 2007).

In the time series domain, a usual approach to analysis and knowledge discovery is the application of data mining techniques. Wu et al. (Wu et al., 2008) proposed the GEAM (Geographic Episode Association Pattern Mining) algorithm to find association patterns in abnormal event sequence. Harms and Deogun (2004) developed the MOWCATL (Minimal Occurrences with Constraints and Time Lags) algorithm to mine frequent association rules from sequential datasets. They presented an application to drought risk management. Both algorithms work over event sequences with discrete events. For instance, an event type is denoted as a tuple of the form <*attribute,level*>, where *attribute* is a variable such as rain or temperature, and *level* is the corresponding value of the variable, such as low, normal or high. Using discrete events, it is possible to discover interesting patterns, but the problem was simplified. For example, it can be found an association pattern such as "*If El Niño then low precipitation in region X*". In this case, data are quantized according to the intensity of rainfall only, disregarding the period of time, which is important to understand the reasons of the phenomenon. Indeed, researchers are more interested in finding out the peaks (with different amplitudes) and their respective periods of occurrence in the El Niño time series, in order to study the effects of this phenomenon and its relation to the climate change scenario.

Honda and Konishi (2001) proposed a framework to mine image time series. They applied the method to weather satellite cloud images taken by GMS-5. The proposed algorithm extracts features from the images and clusters them according to changes in the mass of cloud. Julea et al. (2006) presented an application of the SPADE algorithm (Zaki, 2001) to extract frequent evolutions that are observed on geographical zoning represented by pixels. Experimental studies were performed on Meteosat (Meteorological Satellite) images. The authors use feature vectors to represent satellite images or symbols associated to quantized intervals representing reflectance values of satellite channels. However, this solution does not use indexes that can be generated from the combination of channels. The proposed algorithms also do not work with continuous data. Moreover, they do not combine climate and remote sensing data, which could be an important source of meaningful knowledge.

## ANALYZING CLIMATE AND REMOTE SENSING DATA THROUGH DATA MINING TECHNIQUES

Researchers in Agriculture and Agrometeorology mainly employ statistical models, such as principal component analysis, frequency distribution, geostatistics, cluster analysis, Fourier transform, non-parametric statistics and so on, to analyze and find patterns in earth science. However, specific data mining techniques for climate databases are still lacking. Hence, this volume of climate data from ground stations and computer models associated to remote sensing data and geographical information are important motivations to develop new data mining algorithms. In this context, we have proposed new approaches to analyze, monitor and discover patterns on time series of climate data and of remote sensing images in order to improve researches on monitoring of sugar cane fields, specially applied to yield forecasting. In the following sections, we detail three techniques: 1) The *FDASE* algorithm to identify groups of correlated attributes; 2) A method to monitor evolving climate and remote sensing data; 3) The *CLIPSMiner* algorithm to identify relevant and extreme patterns in multiple time series of continuous climate data.

### Correlation Detection

The number of attributes in a dataset defines its embedded dimension $E$, but if there are correlated attributes, its intrinsic dimension $D$ is smaller than $E$. The intrinsic dimension estimated by the Correlation Fractal Dimension $D_2$ indicates the minimum number of attributes needed to represent a dataset. $D$ can also be used to discover how many and which attributes may be employed to reduce the data dimensionality. With this purpose, the *FD-ASE* (*Fractal Dimension Attribute Significance Estimator*) algorithm aims at identifying different types of correlations (Sousa et al., 2007b). This technique applies the forward attribute inclusion approach and uses the intrinsic dimension as a criterion to identify groups of correlated attributes and to select a relevant attribute subgroup to represent the essential data characteristics. The following definitions base the *FD-ASE* algorithm.

**Definition 5.** Partial Intrinsic Dimension $pD()$: *Given a finite dataset A with E attributes and a subset of attributes $C \subset A$, the Partial Intrinsic Dimension $pD(C)$ is the intrinsic dimension projecting the dataset on the subset C.*

**Definition 6.** Individual Contribution $iC()$: *Given a finite dataset A with E attributes, the Individual Contribution $iC()$ of an attribute $a_k \in A$ is the maximum potential contribution of $a_k$ to the intrinsic dimension of A, and it is measured as $iC(a_k) = pD(\{a_k\}) \rightarrow [0,1]$.*

Consider a dataset defined on $A = \{a_1, a_2,...,a_E\}$ and a subset of attributes $C \subset A$ with partial intrinsic dimension $pD(C)$. An attribute $a_i \in (A - C)$ increases the partial intrinsic dimension of $C$ by at most its individual contribution $iC(a_i)$, according to the level of correlation between $a_i$ and the attributes of $C$. If $a_i$ is completely uncorrelated to the every attribute in $C$, the partial intrinsic dimension will increase by the individual contribution $iC(a_i)$, i.e., $pD(C \cup \{a_i\}) - pD(C) \cong iC(a_i)$. On the other hand, if $a_i$ is strongly correlated to the attributes in $C$, the partial intrinsic dimension will increase by a value of almost zero, i.e., $pD(C \cup \{a_i\}) - pD(C) \cong 0$. Additionally, if $a_i$ is weakly correlated to the attributes in $C$, the partial intrinsic dimension will increase by an amount between zero and the individual contribution $iC(a_i)$, i.e., $0 \leq pD(C \cup \{a_i\}) - pD(C) \leq iC(a_i)$.

Correlations mean that the value of an attribute can be approximated from some other attributes. Sousa et al. (2007b) define the terms 'strong correlation' and 'weak correlation'. The first one is used when the value of one attribute can be closely deduced from a subset of other attributes, as in linear correlations. A weak correlation indicates that an attribute can be only approximated from other attributes, as in fractal correlations. In order to quantify the correlation among attributes, a threshold $\xi$ ranges from zero –meaning complete correlation – up to one, when the attributes are independent.

**Definition 7.** ξ-Correlation: *Given a dataset defined on $A = \{a_1, a_2,...,a_E\}$, a subset $B \subset A$ is said to be $\xi$-correlated to a subset $C \subset A, B \cap C = \emptyset$ if each attribute $a_i$ in B does not contribute more than $\xi * iC(a_i)$ to the partial intrinsic dimension of C.*

**Definition 8.** Attribute Set Core $\xi C$: *Given a dataset defined on $A = \{a_1, a_2,...,a_E\}$ with intrinsic dimension D, an Attribute Set Core $\xi C$ is a subset of attributes in A such that, $|pD(\xi C) - D| < \sum \xi * iC(a_i)$," $a_i \in (A - \xi C)$, and there is no attribute $a_k \in \xi C$ such that $|pD(\xi C) - pD(\xi C - \{a_k\})| < \xi * iC(a_k)$.*

**Definition 9.** Correlation base $\xi B_p$: *Given a dataset defined on $A = \{a_1, a_2,...,a_E\}$ and an Attribute Set Core $\xi C \subseteq A$, a Correlation Base $\xi B_p$ is a subset of attributes $\xi B_p \subseteq \xi C$ such that either $\exists a_k \in (A - \xi C) \mid \exists M_k, M_k(\xi B_p) \rightarrow a_k$ or there are no $\xi$-correlated attributes in the dataset and $\xi B_p = \xi C = A$, where $M_k$ is a mapping indicating that $a_k$ is $\xi$-correlated to all the attributes in $\xi B_p$.*

**Definition 10.** Correlation group $\xi G_p$: *Given a dataset defined on $A = \{a_1, a_2,...,a_E\}$ and a Correlation Base $\xi B_p \subseteq A$, a Correlation Group $\xi G_p$ is the subset of attributes $\xi G_p \subseteq A$, such that, $\xi G_p = \xi B_p \cup \{a_k \in (A - \xi B_p) \mid |pD(\xi G_p) - pD(\xi G_p - \{a_k\})| < \xi * iC(\{a_k\})$ and $\exists M_k(\xi B_p) \rightarrow a_k\}$, where $M_k$ is a mapping indicating that $a_k$ is $\xi$-correlated to all the attributes in $\xi B_p$.*

A correlation group $\xi G_p$ includes the correlation base $\xi B_p$ and every attribute $\xi$-correlated to all attributes in $\xi B_p$, but excludes the attributes not $\xi$–correlated to the full correlation base $\xi B_p$. In other words, attributes $\xi$–correlated to some, but not to all the attributes in $\xi B_p$ are not in $\xi G_p$.

For example, consider a dataset defined by five attributes $A = \{a_1, a_2, a_3, a_4, a_5\}$ as illustrated at Figure 1a. This dataset has the mappings $M_2(\{a_1\}) \rightarrow a_2$ and $M_5(\{a_1\}) \rightarrow a_5$ as shown in Figure 1b. Then $A$ has the correlation group: $\xi G_1 = \{a_1, a_2, a_5\}$, with correlation base $\xi B_1 = \{a_1\}$. Figure 1c shows the mapping $M_4(\{a_1, a_3\}) \rightarrow a_4$. Then $A$ has another correlation group: $\xi G_2 = \{a_1, a_3, a_4\}$, with correlation base $\xi B_2 = \{a_1, a_3\}$. The attribute set core $\xi C = \{a_1, a_3\}$ of $A$ is composed of $\xi B_1$ and $\xi B_2$ as it can be seen in Figure 1d.

Notice that if a function exists that $f_k(\xi B_p) \rightarrow a_k$, $\xi B_p \subset \xi G_p$, $a_k \in A$, then $a_k \in \xi G_p$ and $pD(\xi G_p)$

*Figure 1. Example of correlations groups of a dataset with five attributes*

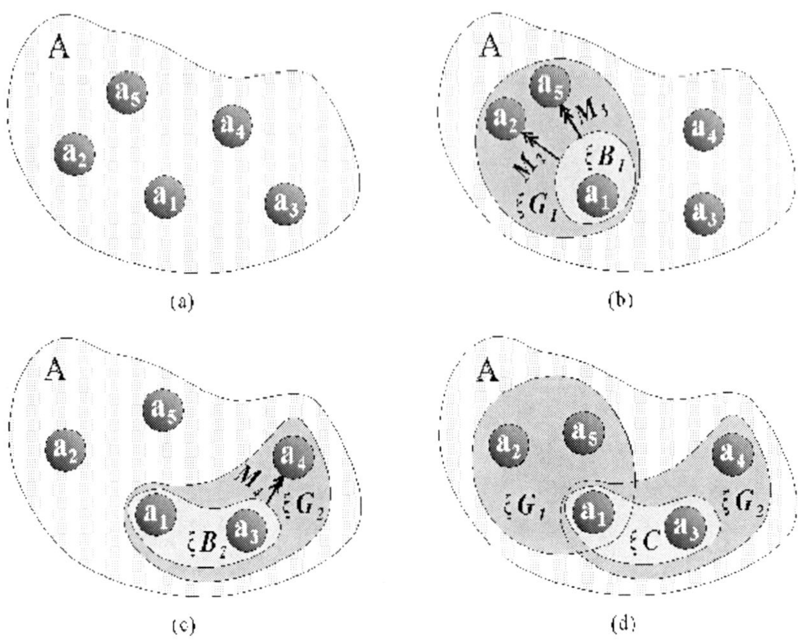


$= pD(\xi G_p - \{a_k\})$. However, correlations are not limited to functions. In fact, any mapping $M_k(\xi B_p) \rightarrow a_k$ such that $|pD(\xi B_p \cup \{a_k\}) - pD(\xi B_p)| < \xi * iC(a_k)$ is a $\xi$-correlation, so for every attribute $a_k \in \xi G_p$, $a_k \notin \xi B_p$ we have:

$$|pD(\xi G_p) - pD(\xi G_p - \{a_k\})| < \xi * iC(\{a_k\})$$

Therefore, an attribute $a_i$ in a correlation group but not in the correlation base does not increase the partial intrinsic dimension of the group by more than $\xi * iC(a_i)$, thus $|pD(\xi G_p) - pD(\xi B_p)| < \Sigma\, \xi * iC(a_i)$, ” $a_i \in (\xi G_p - \xi B_p)$.

This approach allows spotting the attributes that define others, and how strong are their correlations. Therefore, the analyst of the database can drop the attributes that are not meaningful and save memory space as well as time processing when managing and querying the data.

## Data Stream Monitoring

The information provided by the intrinsic dimension is explored by the *SID-meter* algorithm (Sousa et al., 2007a) to monitor evolving data streams, as described in the background section. By combining the *SID-meter* and a statistical method, one can compare the data in consecutive time periods, pointing out the attributes that are responsible for the trend changes and how they influence such changes (Romani et al., 2009a). Basically, *SID-meter* monitors a data stream by continuously measuring the intrinsic dimension and, when significant changes occur, the Data Analysis Module is triggered in order to analyze the data and to validate the variation of the intrinsic dimension, also identifying the changes that have occurred in the distribution of the attributes.

The Data Analysis Module employs a statistical test to identify which attributes have been responsible for the intrinsic dimension variation. Also, the variation is described in terms of mean and standard deviation of attribute values. As new events are received, *SID-meter* is executed to update the measures of $D$ for recent events. For explanation purposes, $D_p$ represents the intrinsic dimension calculated over the current sliding window and $D_{p-1}$ denotes the intrinsic dimension computed over the preceding window.

The values of $D_{p-1}$ and $D_p$ are continually monitored to detect if a significant difference between them occurs, as illustrated in Figure 2. We quantify the significance of the difference between $D_{p-1}$ and $D_p$ based on a user-defined parameter $\varepsilon$. A significant difference spotted between the current window $p$ and the preceding window $p-1$ occurred if and only if $|D_{p-1} - D_p| > \varepsilon$. The smaller the value of $\varepsilon$, the more sensitive the monitoring process. The monitoring process triggers the Data Analysis Module when $|D_{p-1} - D_p| > \varepsilon$, in order to analyze the data variation and to reveal which significant differences have occurred between the preceding $p-1$ and the current window $p$.

Consider that each event $c_j$ of a data stream $\{c_1, c_2, ..., c_n\}$ is defined by a set of $E$ measured values, one for each attribute $a$. The set of $a$ attribute values in the current window $p$ of m ($m \leq n$) events is given by $a_p = \{a1_p, a2_p, ..., an_p\}$. Similarly, the set of $a$ attribute values in the preceding window $p-1$ is given by $a_{p-1} = \{a1_{p-1}, a2_{p-1}, ..., an_{p-1ni}\}$.

The Data Analysis Module runs the *StARMiner* algorithm (Ribeiro et al, 2005) that statistically analyzes the data in the current window. For each attribute $a$, the Z-hypothesis test is employed to compare the mean $\mu$ between the preceding window $p-1$ and the current window $p$. Consider that the mean of $a$ values in the current and preceding windows are given respectively by $\mu(a_p)$ and $\mu(a_{p-1})$. The hypothesis $H_0$: $\mu(a_p) = \mu(a_{p-1})$ should be rejected with a significance $\alpha_{min}$ (the test's probability of incorrectly rejecting the null hypothesis), in favor of the hypothesis that the means $\mu(a_p)$ and $\mu(a_{p-1})$ are statistically different as shown in Figure 2.

If the hypothesis $H_0$ is rejected, the mean and standard deviations $\sigma$ of the $a$ values are collected for the current ($\mu(a_p)$, $\sigma(a_p)$) and preceding windows ($\mu(a_{p-1})$, $\sigma(a_{p-1})$) to describe the attribute changes in the data stream windows. Figure 2 presents a table with $\mu$ and $\sigma$ examples for the attribute $a$ in windows $p$ and $p-1$.

The proposed method shows that it is possible to gather in a single graph the behavior of several variables and parameters, which otherwise would

*Figure 2. Integration of SID-meter and data analysis module*

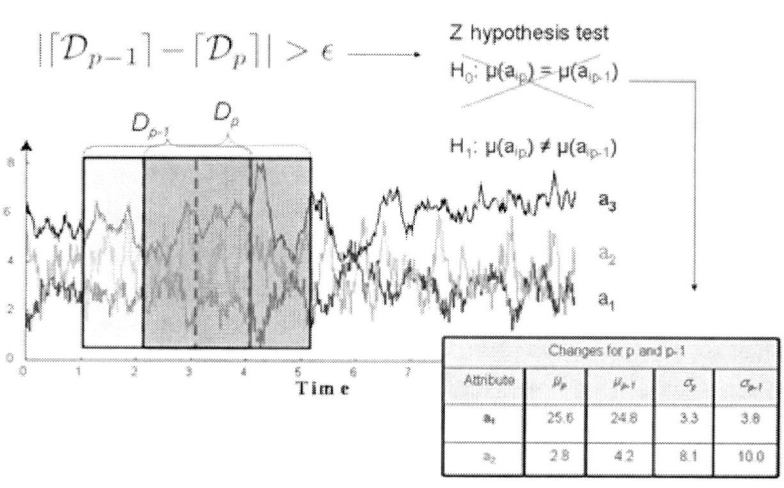

have to be analyzed separately, imposing to the specialist much more effort and time.

## Relevant Patterns Detection in Multiple Time Series

The *CLIPSMiner* (CLImate PatternS Miner) algorithm aims at discovering relevant patterns in climate and remote sensing time series (Romani et al., 2009b). The patterns represent relevant phenomena, such as periods with a small variation in data distribution or peaks. The peaks sizes may indicate normal episodes, such as season changes, or extreme phenomena, such as meaningful drops in temperature or heavy rains in a short period of time. Some definitions are fundamental to better understand the algorithm.

**Definition 11.** Time series *S*: *Formally, a time series is defined as a sequence of pairs* $(b_i, t_i)$ *with* $i = \{1, 2, ..., n\}$ *i.e.* $S = [(b_1,t_1),(b_2,t_2),\ldots,(b_i,t_i),\ldots,(b_n,t_n)]$ *with* $(t_1 < t_2 < \ldots < t_i < \ldots < t_n)$, *where each* $b_i$ *is a data value and each* $t_i$ *is a time value in which* $b_i$ *occurs.*

We call each pair $(b,t)$ an event $e_i$. A set of events contains *m* events of type $(b_i,t_i)$ for $i = \{1,2,\ldots,m\}$. Each $b_i$ is a continuous value. Each $t_i$ is a unit of time that, in our application domain, can be measured in days, months or years. All sets $S_i$ have the values $t_i$ measured in the same unit of time.

**Definition 12.** Event sequence $S_e$: *An event sequence* $S_e$ *is a set of consecutive events* $e_i$, *i.e.* $S_e = (e_i, e_{i+1}, \ldots, e_k)$, *where* $e_i = (b_i, t_i)$ *for* $i \geq 1$ *and* $k \leq n$, $k\text{-}i \geq q$, *where q is the user defined minimum number of events in an event sequence.*

Datasets of $S_e$ can generate some arrangements that are interesting in meteorological and agrometeorological studies. Then, we extract event sequences from a given sequence where the number of elements $e_i$ depends on the difference between the events given by $d_i = (b_{i+1} - b_i)$ and a given $\delta$ parameter. The value of $\delta$ is usually very small, tending to zero ($\delta \rightarrow 0$). Therefore, we define three exclusive types of event sequences, as follows.

**Definition 13.** Ascending event sequence $S_{ea}$: *An ascending event sequence is a set of consecutive events* $e_i$, *such that* $S_{ea} = (e_i, e_{i+1}, \ldots, e_k)$ *where* $\sum_{i}^{k}(d_i) > 0$, such that ” $d_i$, $d_i > 0$ *and* $|d_{k-i}| < \delta$ *to* $(k - i) \leq$ *user-defined parameter.*

**Definition 14.** Descending event sequence $S_{ed}$: *A descending event sequence is a set of consecutive events* $e_i$, *such that* $S_{ed} = (e_i, e_{i+1}, \ldots, e_k)$ *where* $\sum_{i}^{k}(d_i) < 0$, such that ” $d_i$, $d_i < 0$ *and* $|d_{k-i}| < \delta$ *to* $(k - i) \leq$ *user-defined parameter.*

**Definition 15.** Stable event sequence $S_{es}$: *A stable event sequence is a set of consecutive events* $e_i$, *such that* $S_{es} = (e_i, e_{i+1}, \ldots, e_k)$ *where* "$d_i$, $|d_i| < \delta$.

Combining different types of event sequences generates patterns that resemble peaks (negative and positive) and intervals with constant distribution.

For illustration purposes, let's consider that an event sequence with values equal to zero for precipitation may correspond to a drought period. Moreover, patterns can be an event sequence, where the starting and ending points have high values, and the middle $b_i$ has low values, corresponding to a negative peak. This type of pattern can represent for example a period of high temperatures with a sudden drop in the middle of the period. We define three types of patterns used to transform time series *S* into discrete values.

**Definition 16**. Valley patterns (V): *Valley patterns are defined as the concatenation of a descending event sequence and an ascending event sequence, i.e.* $V = S_{ed}S_{ea}$..

**Definition 17**. Mountain patterns (M): *Mountain patterns are defined as the concatenation of an ascending event sequence and a descending event sequence, i.e.* $V = S_{ea}S_{ed}$.

**Definition 18**. Plateau patterns (P): *Plateau patterns are defined as a stable event sequence, i.e.* $V = S_{es}$..

*Figure 3. Examples of V, P, M and relevant/extreme patterns*

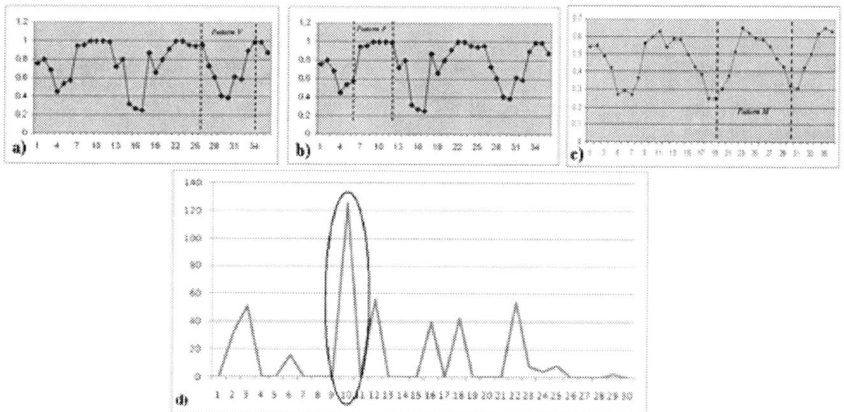


Figure 3a illustrates a pattern of type *V*. In real data, a pattern V can be observed for example when a sharp drop in the minimum temperature occurs.

Figure 3b and 3c illustrate respectively a pattern of type *P* and a pattern of type *M*. A pattern of type *P* can occur when there are many consecutive values with low variation in a time series. The latter is called a *Mountain* for resembling a real mountain peak. In a real dataset, pattern M occurs when there is an important variation in the amplitude, for example a very heavy rain.

The *CLIPSMiner* algorithm can retrieve all patterns (*V*), (*M*) and (*P*) in a time series, regardless of its scale or size. However, thresholds can be defined to search only for the relevant patterns. One of the thresholds is called relevance factor and can be changed by the user.

**Definition 17.** Amplitude $y$: *The amplitude is the difference between the maximum and the minimum values in the time series:*

$$y = (b_{max} - b_{min}).$$

**Definition 18.** Relevance factor $\rho$: *The Relevance factor is a percentage of the amplitude value.*

The relevance factor is a measure to identify whether an event sequence (peak) is an important/relevant pattern in a time series. For example, a positive peak of rainfall ranging from (0, 5, 0) is not representative because it has a very small variation (5 mm). However, a peak that ranges from (0, 120, 0) is an extreme phenomenon that may cause floods. In this case, the relevance factor indicates which event sequence will be considered a relevant or extreme pattern as showed in Figure 3d.

The default value for the relevance factor $\rho$ was defined empirically and corresponds to 40% of the amplitude, that is, $\rho = y*40/100$. For instance, let us consider $b_{max} = 0.65$ and $b_{min} = 0.23$ as the maximum and minimum values. The amplitude is $y = 0.42$ and the relevance factor is $\rho = 0.168$. Thus, only event sequences with amplitude equal or greater than $\rho$ are considered when searching for the relevant patterns.

Although *CLIPSMiner* employs a default value for $\rho$ (40% from the amplitude), specialists can change/tune it. As they know more about the distribution of the time series, they can properly tune the relevance factor to detect the most relevant patterns. Therefore, depending on the value of $\rho$, the algorithm finds more or less patterns. As another example, if a time series has $b_{max} = 1.0$ and $b_{min} = 0.25$, the relevance factor $\rho$ calculated by the algorithm is $\rho = 0.30$. However, this factor can be changed by the user to $\rho = 0.50$, setting the algorithm to find only more relevant or extreme patterns. The algorithm allows adjusting the method to find patterns according to the user's interest and can be changed in different analysis.

**Definition 19.** Plateau Length $\lambda$: *The plateau length is a value that defines the minimum length of a stable event sequence* $S_{es}$.

The default value for $\lambda$ is 4, allowing to discover plateaus composed of four consecutive events. This value was defined empirically.

*CLIPSMiner* tracks time series of continuous data and sets points of control as a method for quantization. However, the algorithm considers the occurrence instant of the events, organizing the quantized pieces in patterns that have their semantics related to weather events.

## Application Examples

In this section, we describe three case studies that exemplify the applicability of the approaches we presented in this chapter. The first case study applies the *FD-ASE* algorithm to analyze the *SugarCaneRegion* dataset, which is composed of rain, maximum and minimum temperature, NDVI (Normalized Difference Vegetation Index) and WRSI (Water Requirement Satisfaction Index) values taken from the 10 sugar cane productive areas from the State of Sao Paulo (Brazil) from 04/01/2001 to 03/31/2008. Sao Paulo is the major sugar cane producer in Brazil.

The NDVI index is closely correlated to the leaf area index, green biomass and productivity. NDVI is calculated using Equation 2.

$$NDVI = \frac{NIR - R}{NIR + R} \tag{2}$$

where $NIR$ = near-infrared (channel 2) and $R$ = red (channel 1). Channels 1 and 2 are acquired by the AVHRR sensor on board of the NOAA meteorological satellites. To obtain high quality images, we employed the Maximum Value Composite (MVC) technique, which is useful to minimize the effects of shadows, aerosols and water vapor present in atmosphere (Holben, 1986). Each image generates a month-long MVC, so there is one image per month.

The agroclimatic conditions through the period of analysis are described by the WRSI, which is the ratio between the real evapotranspiration and the maximum evapotranspiration. Evapotranspiration is the sum of evaporation and plant transpiration. The WRSI index varies from zero to one and it is the fraction of the total water that would be used by the plant to ensure a maximum productivity and what was really consumed.

The second case study employs the *SID-meter* and the Data Analysis Module to monitor data streams from the *ClimateCPS* dataset, which is composed of measures of daily rain, maximum and minimum temperature from Campinas city (SP, Brazil) collected from 01/01/1890 to 01/19/2009.

The last case study applies the *CLIPSMiner* algorithm to find patterns in the *ClimatePira* dataset, which is composed of measurements of daily rain, maximum and minimum temperature from Piracicaba city (SP, Brazil) collected from 01/01/1991 to 01/18/2009.

### Case Study 1: Correlation Detection (*SugarCaneRegion* Dataset)

We divided the *SugarCaneRegion* dataset into 10 subsets, one for each evaluated region. The dataset attributes are: $a_1$ (rainfall), $a_2$ (maximum temperature), $a_3$ (minimum temperature), $a_4$ (NDVI) and $a_5$ (WRSI). We first applied the *FD-ASE* algorithm to the dataset, and evaluated the threshold $\xi$ values between 0.4 and 0.7. The intrinsic dimension calculus presented $\lceil D \rceil = 3$ or $\lceil D \rceil = 4$, depending on the dataset. This value is lower than the embedded dimension $E = 5$. This means that there are 3 to 4 relevant attributes in the datasets that represent each of the 10 cities studied, indicating that at least one of the attributes is correlated to the others.

By analyzing the found correlations, we can observe some interesting relationships between regions. It can be noted that the groups of correlated attributes (Correlation Group), the relevant attributes in each group (Correlation Base) and the set of relevant attributes considering the whole dataset (Attribute Set Core) are similar for different regions. Table 2 presents the Attribute Set Core, Correlation Groups and their Bases generated for each region evaluated.

The groups of correlated variables at Table 2 are not equal for all regions, which ask for further analysis of what occurs in each area and the relationship of climate variables with NDVI for each region. For instance, *FD-ASE* found $\xi G_1 = \{a_4, a_1\}$ and $\xi B_1 = \{a_4\}$ for Jau city. This means that as the $\xi B_1$ base contains NDVI ($a_4$), we can affirm that rainfall ($a_1$) is correlated to NDVI. On the other hand, the algorithm generated $\xi G_1 = \{a_2, a_4, a_1, a_3\}$ and $\xi B_1 = \{a_2, a_4\}$ for Jaboticabal city. This means that as the base $\xi B_1$ contains maximum temperature ($a_2$) and NDVI ($a_4$), we can affirm that the rainfall ($a_1$) is correlated to the maximum temperature and to NDVI, and also that the minimum temperature ($a_3$) is correlated to the maximum temperature and to NDVI ($\xi B_1$). However, we cannot guarantee that the rainfall is correlated to the minimum temperature.

Almost all regions, except Pontal and Ribeirao Preto, keep NDVI and the maximum temperature in the Attribute Set Core (ξC), evidencing the importance of these variables in the datasets. Pontal is the region with the smallest planted sugar cane area ξC among all regions studied. Thus, this fact may be the main reason for non-selecting NDVI for the Pontal region.

Thereafter, by using the method of fractal correlation, we discovered the existence of correlations between NDVI and precipitation, which is not identified when employing the Pearson correlation (Pearson, 1896) technique. It is worth to mention that the Pearson correlation is the technique usually employed by agrometheorologists to find correlations among data. As the correlation found between NDVI and precipitation is not linear, it cannot be detected by Pearson. The *FD-ASE* method can also find correlation among more than two attributes, which is a big advantage when compared to other methods available in literature, such as the well-known Pearson correlation.

*Table 2. Results of FD-ASE execution*

| City | ξ >= 0.5 | |
|---|---|---|
| Araraquara | $\xi G_1 = \{a_4, a_3\}$ and $\xi B_1 = \{a_4\}$ | $\xi C = \{a_4, a_2, a_1, a_5\}$ |
| Araras | $\xi G_1 = \{a_4, a_3, a_1\}$ and $\xi B_1 = \{a_4\}$ | $\xi C = \{a_4, a_2, a_5\}$ |
| Jaboticabal | $\xi G_1 = \{a_2, a_4, a_1, a_3\}$ and $\xi B_1 = \{a_2, a_4\}$<br>$\xi G_2 = \{a_2, a_5\}$ and $\xi B_2 = \{a_2\}$ | $\xi C = \{a_2, a_4\}$ |
| Jardinopolis | $\xi G_1 = \{a_4, a_1, a_5, a_3\}$ and $\xi B_1 = \{a_4, a_1\}$ | $\xi C = \{a_2, a_4, a_1\}$ |
| Jau | $\xi G_1 = \{a_4, a_1\}$ and $\xi B_1 = \{a_4\}$<br>$\xi G_2 = \{a_2, a_4, a_3\}$ and $\xi B_2 = \{a_2, a_4\}$<br>$\xi G_3 = \{a_2, a_5\}$ and $\xi B_3 = \{a_2\}$ | $\xi C = \{a_2, a_4\}$ |
| Luis Antonio | $\xi G_1 = \{a_4, a_3\}$ and $\xi B_1 = \{a_4\}$ | $\xi C = \{a_4, a_2, a_1, a_5\}$ |
| Pitangueiras | $\xi G_1 = \{a_4, a_5\}$ and $\xi B_1 = \{a_4\}$<br>$\xi G_2 = \{a_1, a_3\}$ and $\xi B_2 = \{a_1\}$ | $\xi C = \{a_2, a_4, a_1\}$ |
| Pontal | $\xi G_1 = \{a_2, a_4, a_5\}$ and $\xi B_1 = \{a_2\}$<br>$\xi G_2 = \{a_1, a_3\}$ and $\xi B_2 = \{a_1\}$ | $\xi C = \{a_2, a_1\}$ |
| Ribeirao Preto | $\xi G_1 = \{a_4, a_2, a_1\}$ and $\xi B_1 = \{a_4\}$ | $\xi C = \{a_4, a_5, a_3\}$ |
| Sertaozinho | $\xi G_1 = \{a_2, a_4, a_1\}$ and $\xi B_1 = \{a_2, a_4\}$<br>$\xi G_2 = \{a_2, a_3\}$ and $\xi B_2 = \{a_2\}$ | $\xi C = \{a_2, a_4, a_5\}$ |

## Case Study 2: Data Stream Monitoring (*ClimateCps* Dataset)

The dataset *ClimateCps* has three attributes: the daily minimum ($t_{min}$) and maximum ($t_{max}$) temperatures (degrees Celsius), and precipitation (mm), measured for a period of 118 years in Campinas, an important region of Brazil. In this study, the dataset is interpreted as a data stream whose events are defined by these three attributes.

In order to calculate the intrinsic dimension of *ClimateCps* over time we applied the *SID-meter* algorithm with 3 counting periods ($n_c = 3$) and 365 events per period ($n_i = 365$), that is, $D$ is updated every 12 months in a three-year sliding window. The parameter to trigger the Data Analysis Module was empirically set as $\alpha = 0.1$, considering that we were interested in tracking small variations. The graph of Figure 4 shows the values of the intrinsic dimension over time for the climate data from Campinas.

The monitoring process triggered the Data Analysis Module for the following data windows: 17, 20, 25, 39, 43, 49, 57, 63, 69 and 87. First, the Data Analysis Module is triggered when the current window is $p = 17$. The statistical tests employed by the Data Analysis Module indicate that the attributes $t_{max}$ and rain had significant changes in their values in these consecutive windows. These changes are described in Table 3 in terms of mean and standard deviation of the values of the attributes in the current ($p = 17$) and in the precedent window ($p - 1 = 16$).

As shown in Table 3, rain is the attribute with the major significant variation. The mean value decreased 1.4 from period $p$-1 = 16 to period $p$ = 17. This result copes with the domain specialist's knowledge that the highest climate variation in the region was caused by rain variations along the years. As it can be seen in Figure 4, the window for $p = 63$ is the period with the highest variation in the intrinsic dimension. The Data Analysis Module triggered for this period indicates that the attributes $t_{min}$ and $t_{max}$ had significant changes in their values considering the preceding window starting at period $p$-1 = 62 and the current window starting at period $p$ = 63. The rain attribute was kept without significant variation in these windows.

*Figure 4. Graph of intrinsic dimension values for ClimateCps dataset*

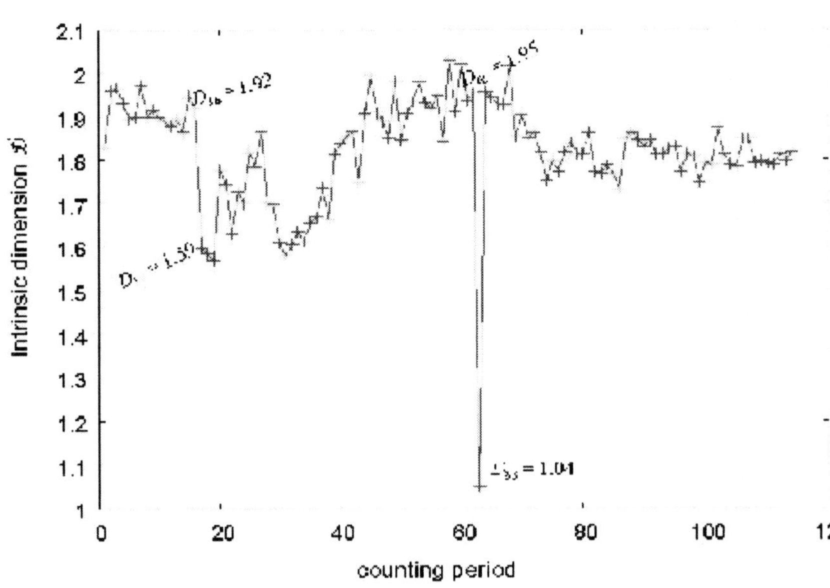

*Table 3. Attributes of ClimateCps dataset revealed by data analysis module*

| Attribute | $\mu_p$ | $\mu_{p-1}$ | $\delta_p$ | $\delta_{p-1}$ |
|---|---|---|---|---|
| $t_{max}$ (° Celsius) | 25.6 | 24.8 | 3.3 | 3.8 |
| rain (mm) | 2.8 | 4.2 | 8.1 | 10.0 |

The large difference between $D_{63}$ and $D_{62}$ associated with the rain attribute stable behavior indicates a meaningful change in climate conditions. This fact can be confirmed by observing the moving average of annual rain values where the lowest values coincide with the monitoring points indicated by *SID-meter*. As it can be seen in Figure 4, the most invariant interval of intrinsic dimension occurred between the periods $p = 90$ and $p = 93$. Just for test purposes, we also executed the Data Analysis Module for the window $p = 92$ and, as a result, no attribute was revealed by the Data Analysis Module as having a significant value change.

Additionally, Figure 4 indicates three different patterns in the data. From the period 17 (1906) to 45 (1933), there is a meaningful variation in the intrinsic dimension. In the next period (45 to 75), the values of the intrinsic dimension are close to 2. In the end of the stream, the intrinsic dimension varies less than in the other periods, showing certain stability in the data distribution. According to the meteorology researcher's team, these three patterns suggest a variation in the distribution of the rain and an increase in the minimum temperature in the last years.

## Case Study 3: Time Series Mining (*ClimatePira* Dataset)

The *ClimaPira* dataset has three attributes (daily rain (*rain*), maximum ($t_{max}$) and minimum ($t_{min}$) temperatures) measured for a period of 98 years at the Piracicaba region, an important production area of sugar cane in Brazil. For this study, the dataset is represented as multiple time series, one for each attribute.

Minimum, default and maximum values were defined to parameters $\rho$ and $\lambda$. Figure 5a, 5b and 5c show the number of patterns discovered for the three time series of the *ClimaPira* database when parameters were set to less ($\rho = y*20\%$ and $\lambda = 3$), average ($\rho = y*40\%$ and $\lambda = 7$) and more sensitive (($\rho = y*70\%$ and $\lambda = 10$) in finding the patterns, respectively.

Analyzing the results presented in Figure 5, we can observe a meaningful decrease of patterns when $\rho$ and $\lambda$ values increase, i.e. as they become more restrictive. Patterns of type *V* in time series $t_{min}$ changed from 1757 to 1. This pattern represents variations in the minimum temperature in different periods of time. The *V* patterns found in the time series $t_{min}$ using $\rho$ = default ($y$*40%) are presented as follows:

- 1: [16.0; 0.0; 15.0] [06/18/1918-06/22/1918]
- 2: [15.0; -1.8; 13.0] [06/22/1918-07/01/1918]
- 3: [18.6; 4.8; 18.0] [05/02/1931-05/15/1931]

*Figure 5. Discovered patterns for ClimatePira dataset*

| Attribute | a) less sensitive (subtle patterns found) | | | b) average sensitive | | | c) more sensitive (only extreme patterns found) | | |
|---|---|---|---|---|---|---|---|---|---|
| | M | V | P | M | V | P | M | V | P |
| tmin | 306 | 475 | 940 | 7 | 15 | 63 | 1 | 0 | 13 |
| tmax | 594 | 1208 | 1341 | 22 | 97 | 238 | 0 | 2 | 74 |
| rain | 899 | 143 | 1916 | 135 | 9 | 850 | 3 | 0 | 525 |

- 4: [15.6; -1.6; 14.0] [06/27/1931-07/06/1931]
- 5: [16.0; 2.1; 18.0] [09/12/1935-09/18/1935]
- 6: [16.7; -0.5; 12.0] [06/15/1942-06/25/1942]
- 7: [15.5; 0.7; 17.0] [09/12/1943-09/19/1943]
- 8: [17.1; 4.5; 16.9] [09/07/1944-09/15/1944]
- 9: [19.3; 2.1; 17.8] [09/29/1947-10/01/1947]
- 10: [15.4; 2.2; 14.9] [08/05/1948-08/12/1948]
- 11: [15.2; -2.6; 9.9] [07/27/1955-08/08/1955]
- 12: [13.6; 1.2; 15.2] [07/10/1965-07/15/1965]
- 13: [16.0; 2.2; 15.0] [08/06/1987-08/13/1987]
- 14: [15.8; -0.4; 13.0] [06/25/1994-07/05/1994]
- 15: [16.15; 3.5; 16.2] [08/30/2002-09/06/2002]

As it can be seen, *CLIPSMiner* discovered relevant *V* patterns in the time series $t_{min}$, such as a large decrease in the minimum temperature in 1918, 1931 and 1955. In those years, the minimum temperature reached negative values after a short time interval. The results also show meaningful changes in temperature in September, when spring begins.

*M* patterns found in time series rain using $\rho$ = 80% are presented as follows.

- 1:[1.4;113.2;0.0][15/02/1998-17/02/1998]
- 2:[0.0;126.9;0.0][03/10/1999-03/13/1999]
- 3:[0.0;139.1;0.0][05/23/2005-05/26/2005]

The results show a high increase in the total rainfall in a short period of time. This extreme phenomenon causes serious problems to both towns and crop fields. Therefore, researchers are interested in finding out when such phenomena occurred in the series of data and the intensity of rainfall that occurred in few days. Recently, these extreme events have occurred with greater frequency and specialists want to know if these phenomena are related to climate changes. According to our results, these phenomena have indeed occurred recently (from 1998), confirming the hypothesis that the distribution of the total rainfall has changed in the past few years.

Many *P* patterns were found in all time series, especially for *rain*. Furthermore, as this dataset is composed of daily values of temperature and rain, the algorithm can detect periods with low variation in temperature or days without rain, for example. By changing the values of parameters $\rho$ and $\lambda$, *CLIPSMiner* is also able to discover prolonged droughts, which is a pattern that has been widely studied by experts because of its consequences for agriculture, especially if it occurs in periods when it is not expected.

## FUTURE RESEARCH DIRECTIONS

The knowledge discovery in agricultural data has a wide variety of subjects still not explored, especially with regard to the analysis of climate data associated to remote sensor data. Techniques for correlation detection, missing data filling, association patterns and forecasting should be further developed. Some improvements have been accomplished using statistical models. However, the growing volume of data from climate change models, meteorological stations, radars, remote sensing equipment and the complexity of their analyses become an important opportunity to data mining researchers.

Climate change researchers have studied extreme events and their correlation to other data such as sea surface temperature. Methods for correlation detection between several types of variable are required. Future work on fractal-based correlation detection includes the development of techniques to find attribute correlations in datasets characterized by the presence of clusters. In real data, correlations may differ considering distinct clusters, resulting in different sets of relevant attributes. For instance, data from remote sensing images and from meteorological stations may contain clusters related to distinct regions in a big country like Brazil, and each region may have a different climate factor impacting its agricultural productivity. Identifying these varying correlations in a broader scenario is useful to the specialists.

Methods to analyze forecasting data aiming to discover possible model errors can be employed in order to calibrate the climate change forecasting models. Computational environments to model and simulate different scenarios of future climate changes can help researchers to verify the impacts of global warming in agriculture. These tools can use information visualization techniques to facilitate users' manipulation and exploration of time series.

Another question to be considered is related to the association of agricultural and environmental (mainly climate, and meteorological) data. Development of association rule mining techniques, specifically directed to agrometeorological analysis, can help the understanding of this relationship, including the generation of temporal association rules. Data mining techniques can be applied to forecast time series, as well.

## CONCLUSION

In this chapter we presented an approach to analyze, monitor and discover knowledge from remote sensing images associated to climate data. We described a set of techniques based on the fractal theory, data streams and time series mining. Three case studies were presented to exemplify the applicability of these techniques. The case studies were performed on datasets from the Sao Paulo State, due to its importance as the main sugar cane producer in Brazil and its strategic relevance to the national economy.

Knowing how the attributes extracted from the raw data are correlated helps the specialists during the analysis of the data gathered. Furthermore, since the amount of data acquired and provided by satellites and meteorological stations is very large and grows in a very fast pace, a tool that highlights where the specialists should pay more attention is a valuable asset. Therefore, the fractal-based monitoring process presented in this chapter is an important and well-suited tool to monitor evolving climate data. In other words, instead of spending hours analyzing many different graphs and charts, the specialists now can count on methods that spot the regions and the periods of interest, where they should pay more attention during the decision making process. For instance, the described methods may help farmers to improve the monitoring of agricultural crops and to avoid production damages along the years.

Finally, the *CLIPSMiner* algorithm can automatically detect patterns that are manually identified by the specialists. The case studies show the correctness and power of the algorithm. Additionally, patterns detected using the highest relevance factor coincide with extreme phenomena, as heavy rain or many days without rain. Climatic extremes are widely studied and may be responsible for natural disasters, such as floods, especially now, taking into account the predictions of the researches on climate changes.

## ACKNOWLEDGMENT

We thank Fapesp, CNPq, CAPES, Microsoft and Embrapa for financial support, Agritempo for climate data, and CEPAGRI/Unicamp for the remote sensing images.

## REFERENCES


Aggarwal, C. C. (2003). A framework for diagnosing changes in evolving data streams. In *Proceedings of the ACM SIGMOD*, 575-586.

Aggarwal, C. C., Han, J., Wang, J., & Yu, P. S. (2004). On demand classification of data streams. In *Proceedings of the ACM Knowledge Discovery & Data Mining Conference*, (pp. 503-508).

Aggarwal, C. C., Han, J., Wang, J., & Yu, P. S. (2004a). A framework for projected clustering of high dimensional data streams. In *Proceedings of the Very Large Data Base*, (pp. 852-863).

Aggarwal, C. C., & Yu, P. S. (2008). LOCUST: An Online Analytical Processing Framework for High Dimensional Classication of Data Streams. In *Proceedings of the International Conference on Data Engineering*, (pp. 426-435).

Barbará, D., & Chen, P. (2003). Using self-similarity to cluster large data sets. *Data Mining and Knowledge Discovery*, *7*(2), 123–152. doi:10.1023/A:1022493416690

Barbará, D., Chen, P., & Nazeri, Z. (2004). Self-similar mining of time association rules. *Proceedings of the Pacific-Asia Conference on Knowledge Discovery and Data Mining*, *3056*, 86–95.

Chakrabarti, D., & Faloutsos, C. (2002). F4: large-scale automated forecasting using fractals. In *Proceedings of the International Conference on Information and Knowledge Management*, McLean, VA - EUA, (Vol. 1, pp. 2-9).

Esquerdo, J. C. D. M., Antunes, J. F. G., Baldwin, D. G., Emery, W. J., & Zullo, J. Jr. (2006). An automatic system for AVHRR land surface product generation. *International Journal of Remote Sensing*, *27*(18), 3925–3942. doi:10.1080/01431160600763956

Faloutsos, C., & Kamel, I. (1994, 1994). Beyond Uniformity and Independence: Analysis of R-trees Using the Concept of Fractal Dimension. In *Proceedings of the ACM Symposium on Principles of Database Systems*, Minneapolis, MN, (pp. 4-13).

Ferrer-Troyano, F., Aguilar-Ruiz, J. S., & Riquelme, J. C. (2006). Data streams classification by incremental rule learning with parameterized generalization. In *Proceedings of the ACM Symposium of Applied Computing*, (pp. 657-661).

Gama, J., Medas, P., & Rodrigues, P. (2005). Learning decision trees from dynamic data streams. In *Proceedings of the ACM Symposium of Applied Computing*, (pp. 573-577).

Guha, S., Meyerson, A., Mishra, N., Motwani, R., & O'Callaghan, L. (2003). Clustering Data Streams: Theory and Practice. *IEEE Transactions on Knowledge and Data Engineering*, *15*(3), 515–528. doi:10.1109/TKDE.2003.1198387

Harms, S. K., & Deogun, J. S. (2004). Sequential association rule mining with time lags. *Journal of Intelligent Information Systems*, *22*(1), 7–22. doi:10.1023/A:1025824629047

Holben, B. N. (1986). Characteristics of maximum value composite images from temporal AVHRR data. *International Journal of Remote Sensing*, *7*, 1417–1435. doi:10.1080/01431168608948945

Honda, R., & Konishi, O. (2001). Temporal rule discovery for time-series satellite images and integration with RDB. In *Proceedings of the European Conference on Principles and Practice of Knowledge Discovery in Databases*, Freiburg, Germany, (pp. 204-215).

IBGE. *Instituto Brasileiro de Geografia e Estatística.* (2007). Retrieved April 02, 2007 from http://www.ibge.gov.br

IPCC. (2007). *Climate Change 2007: Synthesis Report*. Retrieved March 07, 2009, from http://www.ipcc.ch/pdf/assessment-report/ar4/syr/ar4_syr.pdf

Jin, C., Qian, W., Sha, C., Yu, J. X., & Zhou, A. (2003). Dynamically maintaining frequent items over a data stream. *Proceedings of the ACM Conference on Information and Knowledge Management*, 287-294.

Julea, A., Méger, N., & Trouvé, E. (2006). Sequential patterns extraction in multitemporal satellite images. In *Proceedings of the European Conference on Principles and Practice of Knowledge Discovery in Databases*, Berlin, Germany, (pp. 96-99).

Kifer, D., Ben-David, S., & Gehrke, J. (2004). Detecting change in data streams. In *Proceedings of the Very Large Data Base*, (pp. 180-191).

Kleinberg, J. M. (2003). Bursty and hierarchical structure in streams. *Data Mining and Knowledge Discovery*, *7*(4), 373–397. doi:10.1023/A:1024940629314

Manjhi, A., Shkapenyuk, V., Dhamdhere, K., & Olston, C. (2005). Finding (recently) frequent items in distributed data streams. In *Proceedings of the IEEE International Conference on Data Engineering*, (pp. 767-778).

Papadimitriou, S., Brockwell, A., & Faloutsos, C. (2004). Adaptive, unsupervised stream mining. *Very Large Data Base Journal*, *13*(3), 222–239.

Pearson, K. (1896). Mathematical contributions to the theory of evolution. III regression, heredity and panmixia. *Philos Trans Royal Society London Ser A*, *187*, 253–318. doi:10.1098/rsta.1896.0007

Ribeiro, M. X., Balan, A. G. R., Felipe, J. C., Traina, A. J. M., & Traina, C., Jr. (2005), Mining Statistical Association Rules to Select the Most Relevant Medical Image Features. In *First International Workshop on Mining Complex Data, in conjunction with ICDM'05*, Houston, TX, (pp. 91-98). Washington, DC: IEEE Computer Society.

Rodrigues, P., Gama, J., & Pedroso, J. P. (2008b). Hierarchical Clustering of Time-Series Data Streams. *IEEE Transactions on Knowledge and Data Engineering*, *20*(5), 615–627. doi:10.1109/TKDE.2007.190727

Rodrigues, P. P., Gama, J., & Lopes, L. M. B. (2008a). Clustering Distributed Sensor Data Streams. In *Proceedings of the European Conference on Principles and Practice of Knowledge Discovery in Databases*, (pp. 282-297).

Romani, L. A. S., Ávila, A. M. H. d., Zullo, J., Jr., Traina, C., Jr., & Traina, A. J. M. (2009). Mining Climate and Remote Sensing Time Series to Discover the Most Relevant Climate Patterns. In *Proceedings of the XXIV Simpósio Brasileiro de Banco de Dados*, Fortaleza, CE, Brasil.

Romani, L. A. S., Sousa, E. P., Ribeiro, M. X., Zullo, J., Jr., Traina, C., Jr., & Traina, A. J. M. (2009). Employing Fractal Dimension to Analyze Climate and Remote Sensing Data Streams. In *Proceedings of the SIAM Multimedia Data Mining Workshop (SDM)*, Sparks, Nevada, 1-15.

Rosseti, L. (2001). Zoneamento agrícola em aplicações de crédito e securidade rural no brasil. *RBAgro*, *9*(3), 386–399.

Sakurai, Y., Faloutsos, C., & Yamamuro, M. (2007). Stream Monitoring under the Time Warping Distance. In *Proceedings of the IEEE International Conference on Data Engineering*, (pp. 1046-1055).

Schroeder, M. (1991). *Fractals, Chaos, Power Laws* (6 ed.). New York: W. H. Freeman.

Sousa, E. P. M. d., Traina, A. J. M., Traina Jr., C., & Faloutsos, C. (2007a). Measuring Evolving Data Streams' behavior through their Intrinsic Dimension. *New Generation Computing Journal*, 25(Special Issue on Knowledge Discovery from Data Streams), 33-59.

Sousa, E. P. M. d., Traina, C. Jr, Traina, A. J. M., Wu, L., & Faloutsos, C. (2007b). A Fast and Effective Method to Find Correlations among Attributes in Databases. [DMKD]. *Data Mining and Knowledge Discovery*, *14*(3), 367–407. doi:10.1007/s10618-006-0056-4

Traina, A. J. M., Traina, C., Jr., Papadimitriou, S., & Faloutsos, C. (2001, August 26-29, 2001). Tri-Plots: Scalable Tools for Multidimensional Data Mining. In *Proceedings of the ACM International Conference on Knowledge Discovery and Data Mining*, San Francisco, CA, (pp. 184-193).

Traina, C. Jr, Sousa, E. P. M., & Traina, A. J. M. (2005). Using Fractals in Data Mining. In Kantardzic, M. M., & Zurada, J. (Eds.), *New Generation of Data Mining Applications*. New York: Wiley/IEEE Press.

Traina, C., Jr., Traina, A. J. M., Wu, L., & Faloutsos, C. (2000, october 2-4, 2000). Fast feature selection using fractal dimension. *Proceedings of the Brazilian Symposium on Databases*, João Pessoa, PB, 158-171.

Wu, T., Song, G., Ma, X., Xie, K., Gao, X., & Jin, X. (2008). Mining geographic episode association patterns of abnormal events in global earth science data. *Science in China*, *51*, 155–164. doi:10.1007/s11431-008-5008-3

Zaki, M. J. (2001). Spade: An efficient algorithm for mining frequent sequences. *Machine Learning*, *42*(1-2), 31–60. doi:10.1023/A:1007652502315

Zhu, Y., & Shasha, D. (2003). Efficient elastic burst detection in data streams. In *Proceedings of the ACM Knowledge Discovery & Data Mining*, (pp. 336-345).

## ADDITIONAL READING

Agrawal, R., Faloutsos, C., & Swami, A. (1993). Efficient similarity search in sequence databases. *Proceedings of the 4th Int'l. Conference on Foundations of Data Organization and Algorithms*, Chicago, IL, 69–84.

Aksoy, S., Koperski, K., Tusk, C., & Marchisio, G. (2004). Interactive Training of Advanced Classifiers for Mining Remote Sensing Image Archives. *Proceedings of the ACM Knowledge Discovery & Data Mining*, Seattle, Washington, USA.

Bettini, C., Wang, X. S., Jajodia, S., & Lin, J. L. (1998). Discovering frequent event patterns with multiple granularities in time sequences. *IEEE Transactions on Knowledge and Data Engineering*, *10*(2), 222–237. doi:10.1109/69.683754

Camastra, F., & Vinciarelli, A. (2002). Estimating the intrinsic dimension of data with a fractal-based method. *IEEE Transactions on Pattern Analysis and Machine Intelligence*, *24*(10), 1404–1407. doi:10.1109/TPAMI.2002.1039212

Cao, H., Mamoulis, N., & Cheung, D. W. (2005). Mining frequent spatio-temporal sequential patterns. *Proceedings of the Fifth IEEE International Conference on Data Mining*, Houston, Texas, USA.

Chaudhry, N., Shaw, K., & Abdelguerfi, M. (Eds.). (2005) *Stream data management. Advances in Database Systems*. New York, USA: Springer Science+Business Media.

Chen, K. S., Yen, S. K., & Tsay, D. W. (1997). Neural classification of spot imagery through integration of intensity and fractal information. *International Journal of Remote Sensing*, *18*(4), 763–783. doi:10.1080/014311697218746

Cihlar, J., Latifovic, R., Chen, J., Trishchenko, A., Du, Y., & Fedosejevs, G. (2004). Systematic corrections of AVHRR image composites for temporal studies. *Remote Sensing of Environment*, *89*, 217–233. doi:10.1016/j.rse.2002.06.007

Das, G., Lin, K., Mannila, H., Renganathan, G., & Smyth, P. (1998, Aug 27-31, 1998). Rule discovery from time series. *Proceedings of the 4th Int'l Conference on Knowledge Discovery and Data Mining*, New York, NY, 16-22.

Daschiel, H., & Datcu, M. (2005). Design and Evaluation of Human–Machine Communication for Image Information Mining. *IEEE Transactions on Multimedia*, *7*(6), 1036–1046. doi:10.1109/TMM.2005.858383

Daschiel, H., & Datcu, M. (2005). Information Mining in Remote Sensing Image Archives: System Evaluation. *IEEE Transactions on Geoscience and Remote Sensing*, *43*(1), 188–199. doi:10.1109/TGRS.2004.838374

Datcu, M., Daschiel, H., Pelizzari, A., Quartulli, M., Galoppo, A., & Colapicchioni, A. (2003). Information Mining in Remote Sensing Image Archives: System Concepts. *IEEE Transactions on Geoscience and Remote Sensing*, *41*(12), 2923–2936. doi:10.1109/TGRS.2003.817197

Datcu, M., & Seidel, K. (2005). Human-Centered Concepts for Exploration and Understanding of Earth Observation Images. *IEEE Transactions on Geoscience and Remote Sensing*, *43*(3), 601–609. doi:10.1109/TGRS.2005.843253

Ding, W., Eick, C., Wang, J., & Yuan, X. (2006). A framework for regional association rule mining in spatial datasets. *Proceedings of the 6th IEEE International Conference on Data Mining*, Hong Kong, China.

Ding, W., Stepinski, T., & Salazar, J. (2009). Discovery of geospatial discriminating patterns from remote sensing datasets. *Proceedings of the SIAM International Conference on Data Mining*, Nevada, USA.

Esquerdo, J. C. D. M., Antunes, J. F. G., Baldwin, D. G., Emery, W. J., & Jr, J. Z. (2006). An automatic system for AVHRR land surface product generation. *International Journal of Remote Sensing*, *27*(18), 3925–3942. doi:10.1080/01431160600763956

Gaber, M. M., Krishnaswamy, S., & Zaslavsky, A. (2005). Resource-aware Mining of Data Streams. *Journal of Universal Computer Science*, *11*(8), 1440–1453.

Gama, J., & Rodrigues, P. P. (2009). An Overview on Mining Data Streams. *Foundations of Computational*, *6*, 29–45.

Harms, S. K., Deogun, J., Saquer, J., & Tadesse, T. (2001). Discovering representative episodal association rules from event sequences using frequent closed episode sets and event constraints. *Proceedings of the International Conference on Data Mining*, San Jose, California, USA.

Harms, S. K., Goddard, S., Reichenbach, S. E., Waltman, W. J., & Tadesse, T. (2001). Data mining in a geospatial decision support system for drought risk management. *Proceedings of the National Conference on Digital Government Research*, Los Angeles, California, USA.

Hinke, T. H., Rushing, J., Ranganath, H., & Graves, S. J. (2000). Techniques and Experience in Mining Remotely Sensed Satellite Data. *Artificial Intelligence Review: Issues on the Application of Data Mining*, *14*, 503–531.

Klemettinen, M., Mannila, H., Ronkainen, P., Toivonen, H., & Verkano, A. I. (1994). Finding interesting rules from large sets of discovered association rules. *Proceedings of the Conference on Information and Knowledge Management*, Gaitherburg MD, USA.

Mandelbrot, B. (1983). *The fractal geometry of nature*. New York.

Qiu, H. a. (1999). Fractal characterization of hyper-spectral imagery. *Photogrammetric Engineering and Remote Sensing*, *65*(1), 63–71.

Silva, M. P. S., Câmara, G., Souza, R. C. M., Valeriano, D. M., & Escada, M. I. S. (2005). Mining Patterns of Change in Remote Sensing Image Databases. *Proceedings of the Fifth IEEE International Conference on Data Mining*.

Sun, J., Papadimitriou, S., & Faloutsos, C. (2006). Distributed Pattern Discovery in Multiple Streams. *Proceedings of the Pacific-Asia Conference on Knowledge Discovery and Data Mining*, Berlin Heidelberg.

Wang, J., Price, K. P., & Rich, P. M. (2003). Temporal responses of NDVI to precipitation and temperature in the Central Great Plains. *International Journal of Remote Sensing*, *24*(11), 2345–2364. doi:10.1080/01431160210154812

Yang, Y., Linb, H., Guob, Z., & Jiang, J. (2007). A data mining approach for heavy rainfall forecasting based on satellite image sequence analysis. *Computers & Geosciences*, *33*, 20–30. doi:10.1016/j.cageo.2006.05.010

## KEY TERMS AND DEFINITIONS

**Agrometeorology:** A subarea of Meteorology that studies the effect of weather and climate on agriculture.

**Climate Change:** A term used to indicate a change in the climate that can be identified by changes in the mean and/or the variability of its properties and that persists for a long period.

**Maximum Evapotranspiration:** Is the maximum value for the evapotranspiration measure.

**NOAA Satellites:** A set of meteorological satellites of low resolution with polar orbit, which is used in geosciences.

**Real Evapotranspiration:** Is the real value of evaporation and plant transpiration.

# Chapter 5
# Objective Sampling Estimation of Crop Area Based on Remote Sensing Images

**Alfredo José Barreto Luiz**
*Embrapa, Brazil*

**Antonio Roberto Formaggio**
*INPE, Brazil*

**José Carlos Neves Epiphanio**
*INPE, Brazil*

## ABSTRACT

*Having the ability to estimate crop areas is a necessity ever more pressing for all the stakeholders of productive chains. For many scientists involved in agricultural research it is also important to know the location of crops and the area they occupy. With this information, considered together with data on soil, climate, availability of infrastructure for storage and transportation, among others, it is possible, for example, to build scenarios and fit models to attend multiple demands. In this chapter the authors propose a simple method combining the techniques of statistical sampling with the characteristics of images obtained by remote sensing, to construct estimates of acreage in a micro-region scale. The objective to be achieved is to produce an estimate of crop area in a defined territory, with estimated statistical error associated.*

## INTRODUCTION

Area estimation is one of the most obvious applications of remote sensing because, among other reasons, it has a direct economic impact and very early found a user community with clearly defined accuracy specifications, at least for applications to agriculture statistics (Gallego, 2004). In his chapter, Gallego (2004) cited an abundant literature that made use of different ways to use satellite images for land cover area estimation. However, the linkage with statistical sampling techniques and field work to check or substitute image interpretation are not presented in a simple way yet.

DOI: 10.4018/978-1-61692-871-1.ch005

## Context and History

Proper utilization of statistics within any scope depends primarily on how well the nature of employed data is known, and on how clearly the goals are established. Remote sensing offers a rather particular set of data, which almost pre-determines the characteristics that must be taken into consideration when choosing the statistical methods to be used in analysis. Application of those data in agriculture, particularly when the aim is to quantify crop areas, will define quite specific goals that should influence the selection of statistical analysis techniques. An alliance between statistics and remote sensing, based on theory and used in proper ways to estimate crop areas, shall result in a step ahead in the efficient use of data from orbital sensors for agriculture purposes. In this direction, this document presents: 1) a method to prepare and use satellite images in agricultural surveys by sampling; 2) the way to calculate estimates over objective data of crop area and their respective variances; 3) a case study consisting in the estimates of soybean area in a municipality, using remote sensing as auxiliary data; and 4) how to use stratification to improve the quality of estimates.

The availability of reliable information about agricultural production is an increasingly fundamental demand in the decision making process. One of the main variables involved in the assessment of agricultural production is the sowed or planted acreage of important crops (Epiphanio et al, 2009). Methods of survey that take into consideration the increasing availability of remote sensing images, and the use of Global Positioning Systems (GPS) and Geographic Information Systems (GIS) may turn out to be the most practical way for a country produce basic data on major farming commodities (FAO, 1996).

The challenge that arises is the establishment of a method which allows associating the technology provided by orbital remote sensing to the procedures used in agricultural statistics surveys.

## Objectives

The purpose of this chapter is to present a simple and reliable method based on the use of remote sensing images and statistical sampling, which allows the quantification of areas occupied by a particular crop in micro-scale regional or municipal level. This method was developed by Luiz (2003).

To achieve this general goal the following specific objectives are established:

a. Provide a guide for the preparation of images from orbital remote sensing in order to allow their use in agricultural sample surveys;
b. Establish a procedure to extract a random sample of image's pixels from a set of pixels spatially delimited;
c. Present formulas for the calculation of estimated planted area by micro-region or municipality, as well as its variance from data obtained by sampling.

# THEORETICAL FUNDAMENTALS

## Agriculture Estimations and Relationships with Remote Sensing

Despite the decline in the proportional importance of agriculture in global economy, there is a growing need to monitor the agricultural complex (Ryerson et al., 1997). This interest is justified when we consider absolute figures are significant and reach US$ 171 billion annually in the U.S. alone (The World Factbook, 2009).

According to Pino (1999), more countries need to have information and agricultural forecasts that are effective and allow fast perception of change. For Ray et al. (1999), estimates of production in the micro-regional level are essential for management decisions related to the farming-sector economy of any country.

Since the cost of applying traditional techniques over large areas is a limiting factor, the use of remote sensing becomes a practical alternative for some stages of the statistics-obtaining process (Thenkabail et al., 1994). Gurgel et al. (2003) proved that it is possible to use remote sensing data to monitor spatial and temporal dynamics of land covering over large areas.

## Image-Acquiring Systems

Images obtained by remote sensing are indirect ways to capture spatial information. The images obtained by sensors placed in airborne or orbital platforms are stored as matrices. For each picture element (called a "pixel": contraction of "pix" for "picture", and "el" for "element") a value is attributed, which keeps a relationship with the radiance characteristics of an object located on the surface imaged. In digital imaging, each portion of the area imaged by the sensors corresponds to a matrix cell or pixel that must be geographically identified. The pixel is the smallest addressable screen element and each pixel has its own digital address. The address of a pixel corresponds to its coordinates. Pixels are regularly arranged in a 2-dimension grid, and are often represented using dots or squares.

The process of scanning a non-digital image (continuous image) corresponds to a discretization of the scene under observation through a hypothetical mesh overlay and the assignment of an integer value (as represented by digital numbers) to each point of that mesh (in a process called quantization). The spatial resolution of a sensor is defined by the size of its pixels. Sensors commonly used to monitor the characteristics of land surface have spatial resolution ranging from fractions of meters to tens of meters.

The scanned image has a finite number of bits to represent the radiance in each pixel. Radiance is the radiant flux that comes from a source (reflected or emitted) in a given direction, per unit of area. The continuous measurement of the radiance of a scene, represented by digital numbers in the image, is stored in a number of bits per pixel. In general, each object (e.g., a soybean field or a water reservoir) has a typical behavior in relation to the properties of reflectivity or emissivity along the electromagnetic spectrum.

Images obtained by remote sensing are available in the form of "scenes" that correspond to a certain fraction of land surface (Figure 1). Each scene of the Thematic Mapper (TM) sensor installed aboard the Landsat satellite, for example, is provided in the form of 6,000 rows and 6,000 columns (or 36 mega pixels), and corresponds to an approximate area of 180 km x 180 km of land. Thus, each pixel in the TM image corresponds, on the ground, to an area of 30 m x 30 m.

TM images are recorded in seven bands (wavelengths intervals) along the electromagnetic spectrum, numbered from 1 to 7. We can choose any combination of these bands to reproduce the imaged surface into a printed image, by inputting to the digital numbers of each band, one of three colors: red = R, green = G, and blue = B. In general, a composition that represents band 3 by the blue color, band 4 by red, and band 5 by green is named 4R5G3B.

To go deeper into concepts such as swath, geometric resolution, radiometric resolution, spectral resolution, and spectral behavior of objects of land surface, an excellent source of reference is the book of Jensen (2000).

## Crops as Target of Remote Sensing

From the perspective of remote sensing, agricultural activities have certain characteristics that interfere with their identification in digital images obtained by orbital sensors. Unlike natural targets, agricultural targets undergo intense and intentional anthropogenic influence, which helps to give them an aspect of regular geometric figures, with great internal homogeneity. According to Luiz (2003), this is due to some procedures that include:

- plowing and harrowing, which reduce irregularities of the terrain;

*Figure 1. Schematic representation of a remote sensing system, with the source (sun), the platform + sensor, the imaged scene and the pixel*

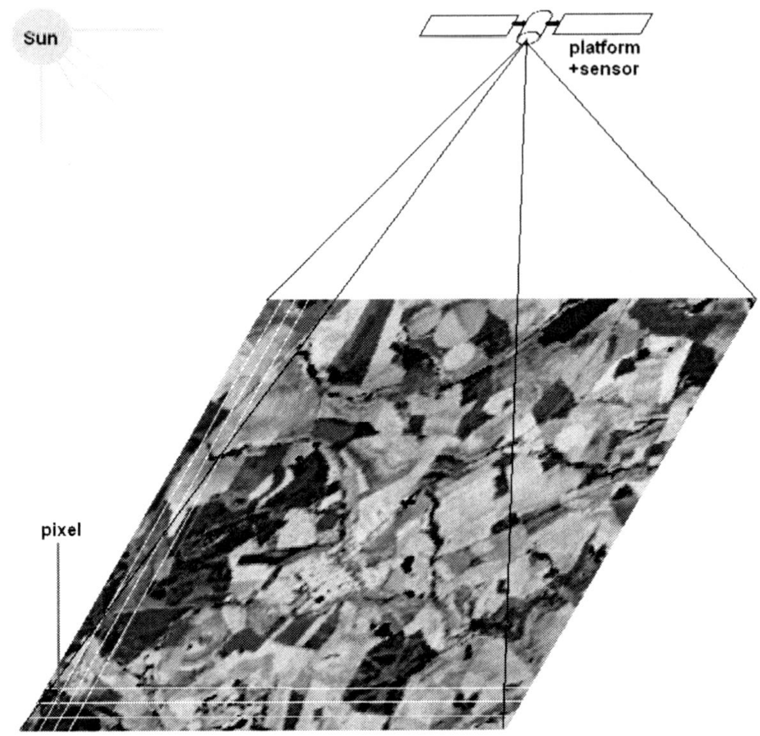


- liming and fertilization, which subdue differences in soil chemical components;
- mechanized sowing, which provides a geometric character to "plots";
- sowing in a short period of time, and the use of selected seeds, which confer regularity to the green cover across subsequent instants in time; and
- control of pests, diseases and weeds, which warrants regularity and purity for the plant population.

Other factors affecting the homogeneity of agricultural targets are the regional vocation and agricultural calendar, both of which join together to result in a stable and relatively limited number of crops that can occur in a specific image. These features help in the identification of limits of agricultural targets in images obtained by remote sensing satellites (Luiz, 2002).

Moreover, the responses obtained over time, for the same point on the ground occupied by a crop, changes dramatically according to both seasonal factors (climate) and anthropogenic factors (planting, harvesting, soil preparation, etc.).

Agriculture covers a vast number of activities and involves cultivation of several hundred plant species. However, due to environmental and economic constraints, and to agricultural tradition, in a given county only a few species predominate on the cultivated area and constitute a truly significant group for that region (Luiz & Epiphanio, 2001).

An example of this occurs with soybean and corn in Brazil. Speaking of municipalities, the 50 largest producers of these crops, representing 0.91% of the 5,564 Brazilian municipalities, concentrated 14.11% of the area under maize and 36.45% of the area under soybean in 1999 (Tsunechiro & Freitas, 2001).

The fact that the production of a certain crop is concentrated in a few counties, and in these counties it occupies a great proportion of the territory, can mean on the one hand a greater environmental or economic risk, but on the other hand it facilitates the use of remote sensing and statistical sampling.

## Image Preparation

A complete scene obtained by orbital sensors may include many municipalities. To facilitate image handling during fieldwork, it is recommended to extract from the whole scene only the smallest rectangle that will encompass the entire area of the survey's target region. Several types of GIS software can be used to register an image, and then overlap information layers to the limits of the study area, to create a minimal enclosing rectangle. In Brazil there is the SPRING software (Camara et al., 1996), available for free downloading at http://www.dpi.inpe.br/spring/, where the software's operation manual can also be obtained. Image processing procedures described in this chapter were made in SPRING software using Landsat-TM images.

Over the pre-registered image, official municipality boundaries is digitized, which are usually supplied in vector files or in the form of maps by governmental institutions.

When transferring digital material to printed form, an adequate color composition must be chosen to allow for interpretation of the most interesting targets. The R4G5B3 composition is indicated because although it doesn't presents immediate visual identity between colors of targets in real scene and those in the image, it allows a good discrimination between major kinds of land use found in agricultural areas.

To allow for easy handling of the printed image, while at the same time producing printed material in a scale that renders possible the identification of agricultural plots and the main physiognomic features of the terrain, we suggest the production of image modules in A4-size paper, scale 1:60,000. Prints in this size and scale are fully suitable to a regular clipboard. Each A4 sheet will cover an area of approximately 18,000 ha, which ensures a reasonable synoptic view, without losing the ability to distinguish the main agricultural targets - for example, a squared-shape plot with 9 ha would appear in the print as a square of 5x5 mm.

However, as it is needed several modules in that scale to cover one municipality, it is also necessary to produce a synthesis print in a less detailed scale. This synthesis should present all the modules together in a single A4 sheet, to allow for an overview of the total area of the municipality. Figure 2 shows an image synthesis module of the municipality of Ipuã, State of São Paulo, Brazil, in the summer of 2002.

The SCARTA application, which is part of the SPRING software, can be used to prepare and print A4 modules. The first procedure is the preparation of a printing template, which includes the definition of: paper size (A4), sheet position (landscape), print area (18.7 x 27.0 cm) and scale (1: 60,000). Next, the distribution of modules must be determined to cover the image. The first module should be positioned in the upper left corner of the study area, and the other modules located on the right horizontally and down vertically. It is recommended to leave a small overlap between the modules.

Aiming to not overload the image with texts and lines, and in order to allow overwriting notes during fieldworks, an auxiliary set of modules is printed containing only data such as geographical coordinates and the boundaries of the municipality. These modules are printed on transparent films and then overlaid upon their respective image modules and fixed with adhesive tape, ensuring geographic coincidence through common fiduciary marks that are made for this purpose. Image modules must be printed on high quality paper, specific for images or photographs, using a regular color printer; in other words, it is not necessary to use sophisticated resources such as plotters

*Figure 2. Synthesis module image of the municipality of Ipuã, State of São Paulo, Brazil, Landsat ETM +, WRS 220/74, 4R5G3B, Jan/05/2002*

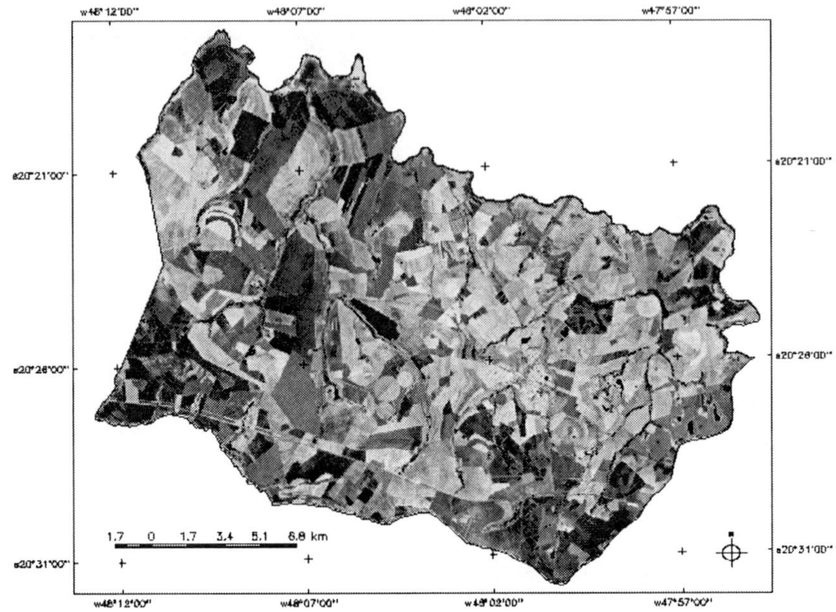


or photographic techniques in the production of fieldwork material (Luiz et al., 2002).

## Statistical Analysis: Option for Sampling

Obtaining information on planted areas and crop development should be a continuing, periodical, quick, georeferenced, transparent, auditable process with a known accuracy. Such requirements impose certain restrictions on the methods that can be adopted. In Brazil, for example, nationwide coverage in a country of more than 5,500 municipalities and 850,000,000 hectares can hardly be obtained by the census method and, therefore, sampling is more appropriate.

The need for information produced quickly enough to allow intervention still during that harvest cycle requires that time spent gathering, processing and analyzing data must as short as possible, which limits the size of the sample.

For agricultural production seasonality is an important factor, and an effective national program of crop monitoring can only exists through a continuous process of data collection. It requires that sample units be visited and analyzed several times throughout the year, which further restricts the number of sample units that can be observed at a reasonable cost.

Information collected in the sampling units should be georeferenced and stored in a database to allow data be audited at any time, which ensures transparency and the possibility of quality control in the process. This is facilitated by the combined use of remote sensing data in the form of images, global positioning devices and GIS software.

For the survey to produce objective estimates with known precision, it is necessary to define a sampling plan with a probabilistic method, and it is also necessary that data about planted areas be obtained by measurements, not by interviews or questionnaires. The use of images obtained by remote sensing allows the incorporation of the concept of image pixel as the sampling unit, which facilitates achieving an objective measurement (Luiz et al., 2002; Luiz & Gürtler, 2003). The possibility of establishing the accuracy of estimates is ensured by a study of the nature of variables being measured, and by designing an appropriate sampling plan (Epiphanio et al. 2001; Epiphanio et al., 2002; Epiphanio & Luiz, 2001).

## Simple Random Sampling

The Simple Random Sampling (SRS) is a method consisting of selecting a sample of $n$ elements from a population total of $N$, such that every possible sample has the same probability of being chosen. To draw a simple random sample of $n$ pixels in a digital image, the scene elements that make up the image must be numbered from 1 to $N$, and then using an algorithm, random numbers should be generated to select the $n$ elements of the sample.

Because the sampling unit is a pixel that has a known and constant area across the whole region under investigation, the proportion in which pixels with the target crop occur in the sample allows, through expansion, an estimation of the proportion of occurrence of that crop in the municipality. It is therefore necessary and sufficient to measure just one category variable per sampling unit. In other words, in an ordinary case of just one crop ($x_1$), a given element will fall under one of only two categories: $U$ if it belongs to a segment occupied by crop $x_1$; or $\bar{U}$ if it belongs to a segment not occupied by crop $x_1$. In order to facilitate exposition of the method, the following notation is adopted:

A number of $U$-class elements in the population ($a$ in the sample)

P $= \frac{A}{N}$ proportion of $U$-class elements in the population ($p = \frac{a}{n}$ in the sample)

Q $= 1 - P$ proportion of $\bar{U}$-class elements in the population ($q = 1 - p$ in the sample)

f $= \frac{n}{N}$ sample fraction (or $\frac{1}{f} = \frac{N}{n}$ = expansion factor)

Considering that the variable is categorical, the process described to obtain a sample, according to Johnson & Kotz (1969), is the classic situation that fits naturally to a discrete distribution of the hypergeometric type. When $N$ is big enough, this distribution is approximated to a binomial (Zwillinger & Kokoska, 2000). In this case, according to Cochran (1977), estimation of $P$ is given directly by $\hat{P} = p$, while the estimation of $A$ is obtained through the application of the expansion factor, namely, $\hat{A} = a \times (\frac{1}{f})$.

Admitting that for each element in the population there is a variable $y_i$ that will take value 1 if that element belongs to class $U$, or value 0 if that element belongs to class $\bar{U}$, it is clear that the total for the population ($Y$) is obtained through Equation 1:

$$Y = \sum_{i=1}^{N} y_i = A \tag{1}$$

and that the averages for population and sample can be calculated through Equations 2 and 3 respectively:

$$\bar{Y} = \frac{\sum_{i=1}^{N} y_i}{N} = \frac{A}{N} = P \tag{2}$$

$$\bar{y} = \frac{\sum_{i=1}^{n} y_i}{n} = \frac{a}{n} = p \tag{3}$$

Consequently, the problem of obtaining $\hat{A}$ and $\hat{P}$ may be reduced to a simple estimation of the total and the average of a population where every

$y_i$ is either 1 or 0. It is then possible to express population variance ($S^2$) and sample variance ($s^2$) as a function of $P$ and $p$, as follows:

$$S^2 = \frac{N}{N-1} PQ \tag{4}$$

$$s^2 = \frac{n}{(n-1)} pq \tag{5}$$

Under such conditions it is possible to affirm that the sample proportion $p = \frac{a}{n}$ is a non-biased estimation of the population proportion $P = \frac{A}{N}$. Also, in practice, in cases where the sample fraction does not exceed 5% ($f$< 0,05), it is possible to consider that the variance of sample average (that is, $p$ variance) can be estimated in a non-biased way through Equation 6:

$$\hat{s}^2_p = \frac{pq}{(n-1)} \tag{6}$$

The estimated total of $U$-class elements in the population is represented as $\hat{A} = Np$, and a non-biased estimation of its variance can be obtained through Equation 7:

$$\hat{s}^2_{Np} = \frac{N^2}{(n-1)} pq \tag{7}$$

With this we can have a good indication of the sample size that is adequate to identify events (agricultural crops, in the present case) with different probabilities of occurrence in the population. In other words, starting from the area of a given municipality and the dimensions of a scene element, we have the $N$ value, and then based on information from previous years, obtained in other surveys, we obtain a value for $p$. It will allow the construction of a table or graph representing the relation between sample size ($n$) and variance of the estimation ($s^2Np$) or its derivatives, such as the coefficient of variation $CV$% that can be calculated with the following expression:

$$CV_{Np} = \frac{100}{p} \times \sqrt{\frac{pq}{n-1}} \tag{8}$$

The graph in Figure 3 shows an example where values are the area of the municipality of Ipuã (467,058,750 m$^2$), the area of a scene element corresponding to 25 m × 25 m (625 m$^2$), and the value estimated by IBGE/PAM for the area of harvested soybean in that municipality in 2001, which was 162,000,000 m$^2$ (IBGE, 2003b). In this case, then, $N$ = 747,294 and $P$ = 0.34685144 ≅ 0.35.

## Building a Random Sample

Building a random sample is not a trivial task, and in fact the difficulty to obtain it in agriculture surveys has been a challenge for some decades now, to statisticians involved with the subject. When we consider the problem in a traditional way, basic sampling units consist in harvest plots, farms, units of production or some other type of segment, but always their limits are defined from the perspective of the predominant socioeconomic activity, that is to say, farming. It poses a serious problem for the definition of random samples, because if randomizing is to be warranted, it is necessary to guarantee that all components of a population will have a non-zero and known chance to be part of the sample. Thus, it is necessary to know all elements of a finite population, or the distribution of probabilities in the case of infinite populations.

The case of estimating a planted area characteristically belongs to the first groups, that of finite populations. Therefore, if the basic sampling unit chosen is a farm, for example, in order to

*Figure 3. Curve representing the relation between sample size (n) and the expected coefficient of variation (CV %), given a probability p = 0.35 of occurrence of a crop in the area*

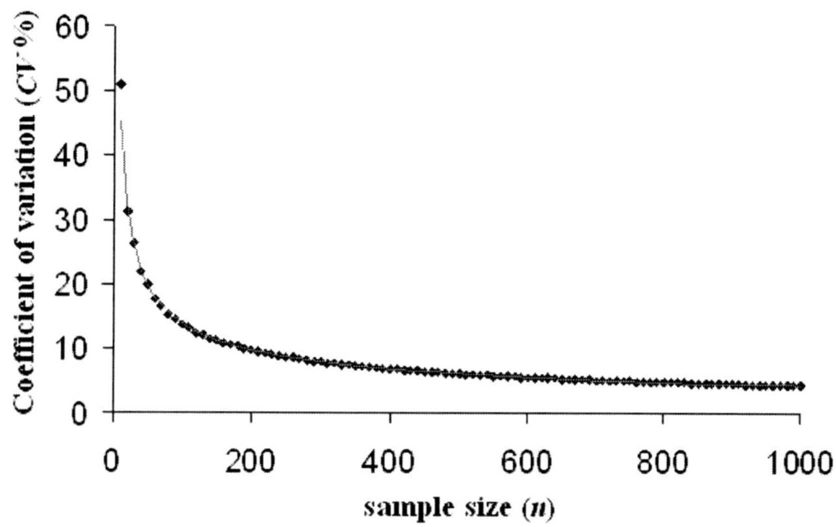


proceed to random sampling it is necessary to have a complete list of all farms existing in the considered population. Among several impediments to such a list, some are: a) the sum of farm areas is not equal to the population's total area, due to measuring errors or mistakes in declaration; b) estate/property registries, depending on when they've been made, are to be found in different bases – public notaries, counties, districts, municipalities, states or territories that may have been created before or after the property was registered; c) plain inexistence of any registry; d) vacant areas; e) areas under litigation – with two or more registries or owners; f) fast modifications in number, size or ownership of properties, due to owner's death, property sale, separation or dissolution of a partnership.

An alternative to bypass such obstacles is the use of sampling panels by areas. Although in theory the panels may contemplate elements with artificially-defined boundaries (Luiz, 2002), this was not the norm until recently, and segments were typically defined by physical boundaries of the terrain (Collares et al., 1993).

Although the advantages that remote sensing images could bring to agriculture sample surveys have been known for a long time, it was only towards the end of the 1990's that utilization of the pixel as a basic sampling unit began to be seen as possible in practice, for agricultural surveys (Luiz et al., 2002). It happened due to the availability of precise GPS equipment at low costs, as well as the uprising of friendly GIS programs – that sometimes are even free of charge, such as SPRING. Starting from that time, the joint utilization of remote sensing images, GIS and GPS devices has been increasing also in agricultural applications (Wilkinson, 1996; Pradhan, 2001; Gao, 2002).

Thus, today it is feasible to use remote sensing images which naturally provide elements with artificial boundaries – the pixels – to obtain random samples for the estimation of the planted area. It renders much easier both the sampling design and the statistical analysis of data collected. Nonetheless, it is necessary to guarantee that the practical procedures to obtain the sample will really result in random process, a fundamental condition for the application of all the remaining statistical concepts to be used in calculation of estimations.

In the past, the use of random sampling per points in statistical surveys to estimate a planted area was nearly impossible, due to the difficulty of point location and the inexistence of an enumerable list of such points. The availability of all three technologies: remote sensing images, GPS equipment, and geographic information systems (GIS) has changed the scenario. The benefit of the use of images when building sampling panels is that the image produces an imaginary division of land surface into scene elements (pixels) of equal dimensions that cover a land area completely and without overlapping. If the area of a municipality is the region of interest, there is a finite number N of pixels that represent in the image the real surface area, where N = (municipality area)/(pixel area). Therefore a numerable list of components of the population is obtained, from which random samples may be drawn. The difficulty of identifying on the actual terrain sample elements with imaginary limits (not real limits such as rivers, woods etc), as is the case with the pixel, is bypassed thanks to the precision and easy operation of GPS equipments that are currently available at low costs. Finally, to integrate benefits from both the use of images and the use of GPS devices, there are the GIS that facilitate the handling of georeferenced information (Luiz et al., 2002).

## Hypotheses

Considering that each digital image obtained by orbital sensors is represented by a $n \times m$ matrix where $n$ is the number of columns and $m$ is the number of lines, and knowing that each cell of the matrix is an element of scene (pixel) that can have its spatial position represented by plane coordinates $X$ and $Y$, it is possible to affirm that:

1. if, through a random number-generating process, $t$ ordered pairs $(x_i, y_j)$ are produced such that determine the positions of $t$ pixels in the image, those pixels shall constitute a Random Sample (RS) that is representative of the image.
2. if, after overlapping an irregular polygon (such as for example the boundaries of a municipality) upon the image, only the $t'$ pixels ($t' \leq k$) are conserved that the centers of which are located within the polygon, the new group of pixels shall continue to constitute a RS that is representative of the region of the image corresponding to the municipality area.

## Demonstration

If $X$ is a random variable (r.v.) with univariate discrete uniform distribution, it means that it may adopt $n$ different values $\{x_1, x_2, \ldots, x_n\}$ with equal probability. Among the properties of such a distribution, the probability density function (pdf) for $X$ is:

$$f_{(x)} = \begin{cases} \dfrac{1}{n} & \forall x_i \quad if \quad x_i \in \left\{x_1, x_2, \ldots, x_n\right\} \\ 0 & otherwise \end{cases} \tag{9}$$

Let $Y$ be another r.v., independent and identically distributed (i.i.d.) in relation to $X$, that can adopt $m$ different values $\{y_1, y_2, \ldots, y_m\}$. Then, if $Z$ is defined by the ordered pair $(X, Y)$ in such a way that $z_k = (x_i, y_j)$, where $1 \leq i \leq n$, $1 \leq j \leq m$, and $1 \leq k \leq (n \times m)$, and $i, j, k, n, m$ are integer positive numbers; then we can affirm that $Z$ has a discrete uniform distribution with the following pdf:

$$f_{(y)} = \begin{cases} \dfrac{1}{b-a} & se \quad a < y < b \\ 0 & caso\ contrário\ (c.c.) \end{cases} \tag{10}$$

This is true if and only if $X$ and $Y$ are i.i.d.

Now, if $W$ is a r.v. that can adopt $t'$ different values, and $t' < n \times m$, so that $w_r = z_k$, where $1 \leq r \leq n \times m$, $t'$ and $r$ being integer positive numbers; then it follows that $W$ also has a discrete uniform distribution with the following pdf:

$$f_{(y)} = \begin{cases} \dfrac{1}{b-a} & se \quad a < y < b \\ \\ 0 & caso\ contrário\ (c.c.) \end{cases} \tag{11}$$

In such case, to answer hypothesis a), if the $t$ ordered pairs taken from the population have been produced with any electronic spreadsheet where the cells of two columns, each with $t$ elements, have been filled with integer positive numbers through a random number-generating procedure based on a univariate uniform distribution – in one column numbers vary from 1 to $n$ (number of columns of the image), and in the other column numbers vary from 1 to $m$ (number of lines in the image) – when making the equivalence between those numbers and the pointers of lines and columns in a GIS, the $t$ pixels have been selected that shall constitute a RS representing the image.

In relation to the hypothesis b) exactly the same argumentation is valid, and it is applied twice. In other words, when municipality boundaries are overlapped over the image, a new population is created that is constituted by a subset of the pixels in the original population, in which all elements continue to have identical probabilities of being selected to make the sample; so in parallel, among the pixels selected to make a sample of the original population, the subset of pixels located within the municipality boundaries will still constitute a RS.

It is thus demonstrated that a sample obtained by selecting a set of pixels in an image, and then later taking the subset of pixels contained within the limits of a municipality, is a simple random sample.

## Randomization of Points in the Municipality

After defining a sampling method aided by a remote sensing image, and having selected an area to be studied, it is now necessary to apply the method and obtain estimations originated by the sampling.

The first step is to draw pixels to make the sample. In order to illustrate the method, a sample with size 200 has been drawn.

Considering the irregular characteristics of municipal boundaries, in order to distribute sample points randomly across the municipality, a first draw is made for the enclosing rectangle upon the image stored in SPRING, promoting the equivalence between point and pixel, and then discarding the points that have fallen outside municipal boundaries.

By working upon a matrix representation of the image, the position of each pixel has been defined by the binomial line and column. The name $x$ was given to numbers in the columns, and the name $y$ to line numbers in an image (note that $x$ and $y$ may be multiplied by the pixel dimension, thus transforming their unit in meters, or they may be converted to coordinates of latitude and longitude). The enclosing rectangle will be formed by $r$ lines and $s$ columns, and will be absolutely defined by the ordered pairs of just two scene elements: one in the bottom left corner $(x_1, y_1)$ and another in the upper right corner $(x_2, y_2)$. Observe that $r = (|x_2 - x_1| + 1)$, and $s = (|y_2 - y_1| + 1)$, and that the total number of pixels in the rectangle equals the product of $r$ and $s$, which will be called $T$. In order to allow a random selection of the $m$ candidate sample points among all $T$ pixels, it is necessary to generate $m$ ordered pairs $(x_i, y_j)$. Because the generator used returns two random $z$ numbers $\{0 \leq z < 1\}$ uniformly distributed, it is enough to apply $m$ times the transformations: $x_i = (z_1 \times r)$ and $y_j = (z_2 \times s)$.

These points are considered to be candidates. Since the rectangle is not entirely occupied by the municipality area, a fraction of the $m$ elements will be located outside the limits of the region of interest, and will not be sampled. Only $n$ drawn elements the locations of which are inside the defined limits shall be considered sample points. Because the selection of points is randomized, it is expected that on average the relation between $n$ and $m$ will be the same as exists between the area of the municipality ($A$) and the area of the enclosing rectangle ($R$). From that relation it will be possible to determine the best estimation of $m$ that will produce a certain value of $n$, which is calculated as follows: $m = \{1 + \text{INT}\ [n \times (R/A)]\}$, where INT means the integer part of the expression within brackets.

Considering that the area of the municipality of Ipuã is 46,837.4 ha (IBGE, 2003a), and the area of the enclosing rectangle is 86,265 ha; in order to be able to select a sample of size $n_1 = 200$ it is necessary to draw a total of: $m_2 = \{1 + \text{INT}[200 \times (86{,}265/46{,}837.4)]\} = 369$. That is, if 369 points are generated within the enclosing rectangle, the mathematical expectation is that 200 of them will be located within the municipality boundaries. In order to increase the probability of having the desired number of points inside the area of interest, and because the cost incurred to generate random numbers is computationally very low, we have generated twice that number of points, that is, 738 points to secure 200.

In order to transform the chosen ordered pairs $(x_i, y_j)$ in values compatible with the plane coordinates $(X_i, Y_j)$ of the enclosing rectangle defined in SPRING, in the case of an image with spatial resolution of 25 meters, the following transformation is used: $X_i = X_1 + 12{,}5 + [\text{INT}\ (x_i \times 25)]$ and $Y_j = Y_1 + 12{,}5 + [\text{INT}\ (y_j \times 25)]$, where the pair $(X_1, Y_1)$ defines the position of the southeastern corner of the enclosing rectangle, in the Southern Hemisphere. The procedure guarantees that each pixel in the image will have the same chance to be picked to make up the sample, and that the selection is made considering the center of the pixel.

The coordinates of points randomly generated as described are inputted into the SPRING GIS software through importation as points' files in the format ASCII-SPRING, which will generate an Information Layer (IL) for each set of points. To eliminate points that are not within the polygon corresponding to Ipuã municipality's boundaries, the IL are cut out using a mask with the boundaries. As the resulting IL contain a number of points that exceeds the desired number, IL are exported to be opened in an Excel spreadsheet. Points are then classified in crescent order (relative to their position in generation process), and from the lists a group is selected with the desired number of points for each file, that is, the 200 first points. The remaining points are deleted from the files that are then imported again into SPRING, finishing this stage. Figure 4 exemplifies the distribution of sample points across Ipuã's territory, for a sample of size 200.

After samples are generated, with IL of the selected points, image modules are printed for fieldwork, with the location of points printed only on plastic films that may be overlay the image, to enable localization of points in the field. Points' coordinates may also be exported directly to the GPS memory, which will further facilitate their localization. It is important to remember that when you can count on a relatively recent image of the area under study, several selected points will be easy to classify without the need of an actual visit, because they will present easily identifiable characteristics such as water bodies, urban areas, forests etc. After filling in the tables with cadastral data and exporting them to an Excel spreadsheet, data are worked on to originate estimations. The procedure consists in a simple count of points in the class of interest.

The area occupied by the class of interest in the municipality can be estimated based on the 200 point sample. You will calculate the area dividing the frequency observed for that class by the size of the sample, and then multiplying the result by the municipality's total area. The ratio between

*Figure 4. Example of distribution of the 200 points selected to make up a random sample in Ipuã, State of São Paulo, Brazil*

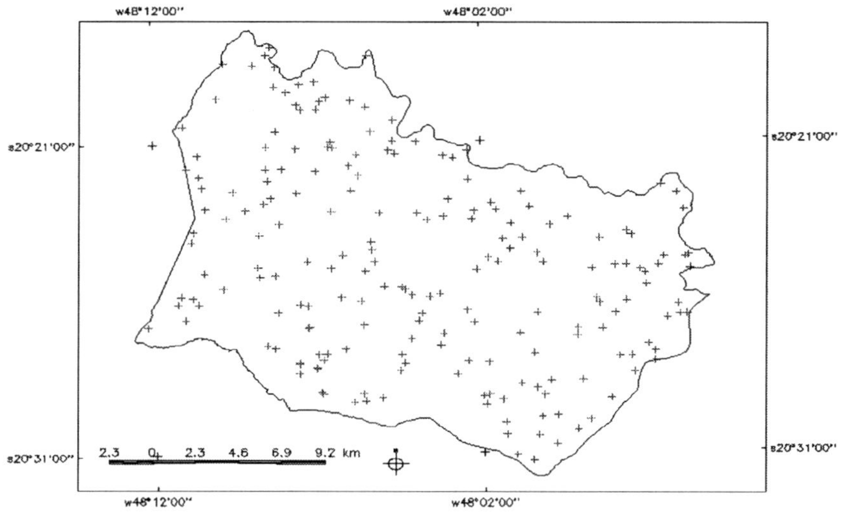


the observed frequency and the size of sample is the proportion $p$ of occurrence of that class in the sample, which is a non-biased estimator for the proportion $P$ of that class in the population. Therefore, it suffices to multiply the $p$ of each class by the total area of the municipality $AM$ to obtain the estimated area for that class, that is:

$$\hat{A} = p \times AM \tag{12}$$

In our example, among the 200 points selected and visited at the field, 68 were classified as soybean. Applying the indicated formulae, we have that the crop of interest occupies a proportion $p = 68/200$ of the territory in the municipality of Ipuã. When we multiply $p$ by the area of the municipality, we estimate as 15,925 ha the area planted with soybean in Ipuã in the Summer 2001/2002.

The procedure to calculate $CV$ is also very simple, and it suffices to apply Equation 8. Getting back to the example of Ipuã in the Summer 2001/2002, with $p = 68/200$ and $q = 1 - p$, we found that the $CV$ for the estimation of soybean area in the municipality is 9.85%.

## Stratification

Stratification is a regular sampling technique, and there are several reasons for that. Among those reasons we may highlight that when there is the desire to know the degree of data precision for certain subdivisions of the population, it is appropriate to treat each of these divisions as a separate "population", or in other words, each subpopulation is a stratum. Another strong reason would be the existence of administrative advantage; for example, the agency conducting data collection may have different offices, each of them with its own area of scope defined according to criteria that are often non-related to a specific survey. Another reason could be detected in the case of a sampling based on farmers working with a given crop, when one has only a list of large farms, which will be put in a separate stratum, while smaller unlisted properties will be divided across other strata where some type of area sampling could be applied (Cochran, 1977; FAO, 1996; FAO, 1998).

Separation in strata could also produce some gain in precision when estimating characteristics of the full population, since it is possible to divide a heterogeneous population into subpopulations that will each be internally homogeneous. This is suggested by the word "stratum", with its implication of a division by layers. If each stratum is homogeneous, so that measures will show little variation from one unit to another, a precise estimation of the average in each stratum can be obtained from a small sample in that stratum. Estimations from all strata can be combined into a single precise estimation of the whole population (Cochran, 1977).

The stratification can be used in two or more stages in a program of statistical surveys of crops. In a first moment, stratification will be made only based on information obtained from the remote sensing image, with the purpose of dividing the municipal territory into areas internally homogeneous, and different among them in relation to the predominant land use. In another way, a group of municipalities – belonging to a State, a Region or even the entire country – could be stratified in order to obtain homogeneous subpopulations as to the variable of interest, which in the present work is the planted area, but such stratification will not be focused in this work.

## Stratification within a Municipality

Stratification inside a municipality is essentially of the spatial type, and it is particularly adequate when the usage of soil varies greatly from one region to another within municipal boundaries. Several reasons may lead to such a situation Farming usage of soil is strongly affected by climate, type of soil and topography; for example, flooded rice in Brazilian conditions of crop handling occurs almost exclusively in hydromorphic plains; while large crops with intense mechanization occupy primarily terrains with little declivity and fertile soils; and perennial and semiperennial cultures may occur in areas of irregular terrain, as the selection of their localization is conditioned mainly by the climate. Another set of factors affecting soil usage are anthropomorphic factors, mainly the agrarian structure and the distribution of infrastructure for transportation and storage of farming products, which makes it relatively easy to understand why within the same municipality a region with difficult access – due to privation of roads of bridges – shall be neglected in comparison with another region that is better served in those aspects; crops appropriate for regions divided into small rural properties shall be different from those in regions occupied by great landed estates. It is important to observe that although the agrarian structures and the availability of infrastructure are highly correlated with environmental variables such as soil, geography and climate, this is not always valid.

Another factor determining the type of land use in each region in the municipality is the legal or juridical component. The appointment of laws on environment preservation may transform an entire area into a reserve and abolish every form of agricultural use; or else, laws establishing limits for a given farming activity may force a spatial rearrangement of production, as in the case of laws concerning the maximum declivity susceptible to mechanization, or the minimum distance from urban centers where sugar cane can be burnt in preparation for harvest. An excellent source of study to understand the impact that such types of restriction can have on the spatial distribution of crops is the work by Giannotti (2001). The paper deals with the impacts of the environmental law prohibiting the burning of sugar cane in the region of Piracicaba (State of São Paulo), where such prohibition fosters mechanization of crops, and environment variables condition the susceptibility of land to mechanization, which is modifying the spatial formation of the crops.

From the perspective of an estimation of the planted area, some or all the above mentioned factors may act to promote or discourage the spatial concentration of a given crop inside a municipality. It may also vary from one species to another, and there can be in the same territory an example of homogeneous distribution and another example of high spatial concentration. Then, the option for stratification will depend on the study's objective, as the method chosen should favor the estimation of the crop of greatest interest. In other words, depending on the crop of interest, one can choose to use or not use stratification in one municipality; and further still, even when there is an option for stratification, the optimal allocation of sample points across all strata shall also vary according to the crop of interest. For example, if a municipality is occupied mainly with sugar cane and soybean crops; and if spatial distribution of sugar cane crops is strongly influenced by their distance from refineries, thus causing a higher concentration of that culture in one region of the municipality, it could be the case of making a stratification in two areas, one with a higher occurrence of sugar cane, and the other with a higher incidence of soybean. Such stratification will result in a single segmentation of the municipality's territory, regardless whether the main interest is soybean or sugar-cane. But the optimal number of points in each stratum – keeping constant the total number of points – shall vary in accordance with the main interest of the study. In regard to the definition of limits for each stratum, considering the scale of that municipality and admitting that the person in charge of the survey is familiar with the territory under study, the manual method of segmentation is indicated, based on visual interpretation of the remote sensing image (King, 2002).

After the population (area of the municipality split into $N$ pixels) is divided into $l$ strata, and maintaining the notation used in the simple random sampling (SRS), we observe that in the stratified random sampling (StRS), strata have each $N_h$ units, where suffix $h$ defines the $h$-esimal stratum of the population; they do not overlap and constitute together the entire population, that is:

$$N = N_1 + N_2 + \ldots + N_h + \ldots + N_l = \sum_{h=1}^{l} N_h \quad (13)$$

In order to characterize a StRS it is necessary to take a SRS in an independent way in each stratum. Sample sizes inside strata are named $n_1$, $n_2$,..., $n_h$,..., $n_z$, respectively. The total size of the sample is still called $n$, and it is given by the sum of all $z$ values of $n_h$. To follow with the previous notation, we'll adopt the following:

$A_h$ = number of $U$-class elements in stratum $h$ ($a_h$ in the sample)
$P_h = A_h/N_h$ = proportion of $U$-class elements in stratum $h$ ($p_h = a_h/n_h$ in the sample)
$Q_h = 1-P_h$ = proportion of $\bar{U}$-class elements in stratum $h$ ($q_h = 1 - p_h$ in the sample)
$f_h = nh/N_h$ = sample fraction in stratum $h$ ($1/f_h = N_h/n_h$ = expansion factor)
$W_h = N_h/N$ = weight of stratum $h$

In stratified sampling, the number of sample points $n_h$ (or sample size) in each stratum is chosen by the person in charge of sampling. Several criteria may be adopted to determine that amount, and usually an objective is to reduce the estimated variance, keeping a fixed or reduced cost (Cochran, 1977). In a simple case, admitting a linear function of cost, and given the restriction meant by the total number of sample points, one will look for the best allocation of points in each stratum so as to obtain the least variance.

According to Cochran (1977), when StRS is applied to proportions, as is the present case, and one wants to estimate the proportion of units in the population that fall under $U$ class, the ideal stratification is achieved when one puts in the first stratum all units of such class, and in other strata all other classes. But as this usually is not possible in practice, especially when units and strata are spatially distributed, strata should be built in such a way that the proportion in $U$ class will vary as much as possible from one stratum to another. For this situation, the estimation of proportion $P$ that is adequate for StRS is named $\hat{P}_{st}$ and is given by Equation 14:

$$\hat{P}_{st} = \sum_{h=1}^{l} \frac{N_h \times p_h}{N} = \sum_{h=1}^{l} \left( W_h \times p_h \right) \tag{14}$$

And estimation of variance for $\hat{P}_{st}$ can be calculated with Equation 15:

$$\hat{s}^2_{\hat{P}_{st}} = \sum_{h=1}^{l} W_h^2 \frac{p_h \times q_h}{n_h} \tag{15}$$

## Applying Stratification in the Municipality

Aiming to evaluate the effect of stratification within a municipality upon the precision of estimations, Ipuã's territory has been divided into two strata. Stratum I was defined so that it covered approximately two thirds of the total area (27,858 ha), being located upon the center and east of the municipality, always north of the road SP 345 that links the municipality of Guaíra to Highway SP 330. Stratum II, in consequence, was located in the western side of the municipality, adjacent to Guaíra, and also included the entire area south of road SP 345, which represented one third of the total area (18,979 ha).

Stratification was made manually, based on visual interpretation of the image, facilitated by familiarity with the field that was acquired along several campaigns conducted in that municipality. We have also tried to use natural limits such as water streams, and anthropomorphic limits such as the road, to draw strata borderlines. The main visual characteristic influencing in the definition of both strata was the presence of sugar cane and soybean crops. Stratum I is predominantly occupied with soybean crops, while Stratum II has a higher frequency of sugar cane.

The same procedure for drawing points has been performed in this stage, and the number of points drawn for the municipality was 200. Two different distributions have been defined among both strata, one considering a greater interest in the estimation of soybean area, and the other admitting a greater interest in sugar cane. In both cases, 86 points were allocated in one stratum, and 114 in the other. These numbers resulted from the calculation of the expected variance that is given by Equation 5. Based on values for soybean and sugar cane in strata I and II, obtained from a mapping effort (Luiz, 2003), we found the value for the proportion $p$ of each crop in both strata. The optimal allocation of sample points is that produces the lowest sum of variances for a given culture in both strata. In the case under study, just as a coincidence, the optimal combination for soybean occurred with 114 points in stratum I and 86 in stratum II; for sugar cane it was just the opposite, that is, 86 points in stratum I and 114 in stratum II.

One example of drawing points with 114 points in stratum I and 86 in stratum II is shown in Figure 5. The distribution, in terms of precision, favors the estimation of the area planted with soybean, since the area with greater amount of points is precisely the one with most of the soybean planted in the municipality.

In the other case, opposed to the previous one, 114 points were distributed in stratum II, occupied primarily with sugar cane, and 86 points were distributed in stratum I, that has a higher frequency of soybean crops. This distribution favors the estimation of the area planted with sugar cane, because it intensifies the sampling in the area where cane is predominant.

The procedure employed to calculate the estimation of the planted area in both strata and the municipality, has been at first the same as for SRS. After obtaining the frequency of each class in each stratum, the area of culture estimated in the stratum was obtained simply through multiplication of the proportion found by the area of the stratum, $AH$, that is:

$$\hat{A}_h = p_h \times AH \tag{16}$$

*Figure 5. Example of distribution of 200 points chosen to make up a stratified random sample, with 114 points in stratum I and 86 points in stratum II, in Ipuã, State of São Paulo, Brazil*

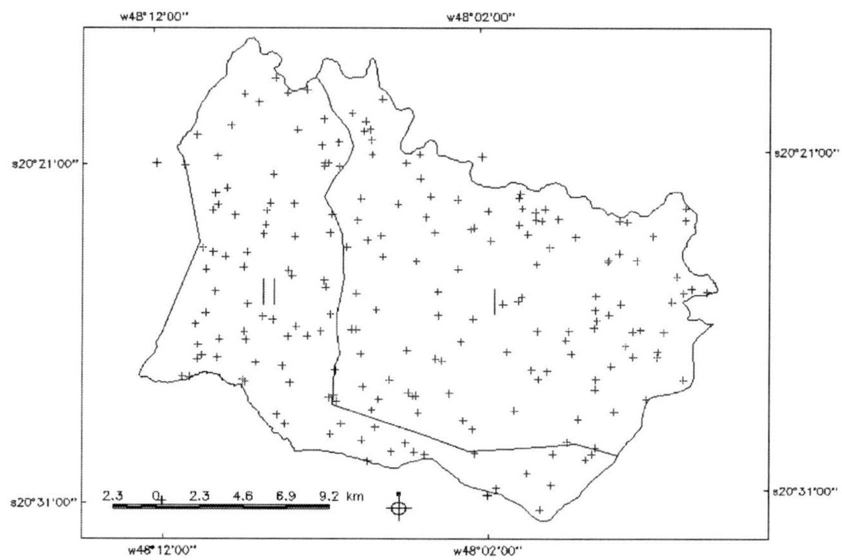


The procedure employed to calculate the coefficient of variation in each stratum, $CV_h$, has also been the same as in the case of SRS, that is:

$$CV_h = 100 \times \sqrt{\frac{q_h}{n_h p_h}} \tag{17}$$

The procedure has suffered a small change to calculate the estimation for the entire municipality, considering both strata. But even in this case it sufficed to add the areas estimated for each stratum, that is:

$$\hat{A} = \hat{A}_1 + \hat{A}_2 \tag{18}$$

But to calculate the $CV$ it was necessary to consider first the weight of each stratum, which has been calculated merely dividing the area of each stratum by the municipality area, that is:

$$W_h = \frac{AH}{AM} \tag{19}$$

Therefore, the $CV$ of the estimation for the municipality that was obtained through StRS, was calculated using the following equation:

$$CV_{\hat{A}} = 100 \times \frac{\sqrt{W_1^2 \frac{p_1 q_1}{n_1} + W_2^2 \frac{p_2 q_2}{n_2}}}{\hat{p}_{est}} \tag{20}$$

The municipality of Ipuã is relatively homogeneous regarding to its rural occupation, but in order to demonstrate the utility of a spatial stratification within a municipality, it has been split into two strata, one more dedicated to soybean, and another more dedicated to sugar cane. Then, just to verify the effect that a change of objective has upon the precision of estimates, two samplings were performed. In the first case, called case A, admitting that the interest of the survey relates to soybean estimates, a greater amount of points (114) were put in stratum I, where soybean has a higher presence, leaving the remaining 86 points (as the total sample has always been 200) for stratum II, where sugar cane is predominant. In the second case, called case B, the exact opposite actions were taken: admitting a greater interest in the estimation of sugar cane areas, 114 points were put in stratum II, and only 86 points into stratum I. In both cases, estimations of area and *CV* for both sugar cane and soybean were calculated as shown in Table 1.

We can observe in Table 1 that the *CV* for the estimation of soybean is lowest in Case A (8.1%), where sampling has been intentionally in favor of soybean culture, while the lowest *CV* for sugar cane (9.1%) occurred in Case B, for the same reason. It also worth observing that, although the *CV* of the estimation of soybean for the municipality has been lowest in Case A, inside stratum II there has been an increase of *CV* value, as compared with Case B. This is exactly how stratification works: if the interest is directed to a given variable, the method will try to make less mistake where that variable is more frequent (or where it occupies a larger area, in this case), even if it represents to err more where the variable is rare. The important thing is that in the final estimation there is a gain in precision, provided that the objective has been well defined, and stratification has been correctly done.

## FUTURE RESEARCH DIRECTIONS

The estimate of crop areas by objective sampling is of great potential, since it allows predictions provided with quantification of the error. It can be performed at different scales, from the smallest, as in the municipalities, to the larger, such as states, regions or countries.

In addition to its implementation, a verification of costs is recommended for the various phases involved, especially those related to data collection in the field. The costs involved are for equipment (e.g. GPS), daily payment and training for fieldwork data collectors, fuel, etc. One of the comparisons to be made is between these costs for the sampling method proposed here versus the costs for other methods. Also, it would be interesting to perform quantitative comparisons between the accuracy of the crop estimates obtained by this method and by other methods, and between the cost effectiveness of each.

New studies that explore the possibility of reducing the number of points to be visited in the fieldworks, through the use of current remote sensing images, so that points that fall into non-agricultural class can be discarded, should be encouraged. Efforts to improve the method should be directed towards the optimization and reduction of costs and laboriousness, while seeking to increase its optimization, speed and usefulness.

Another goal that can be sought in the near future is the increasing automation of the method, starting with the randomization of sampling points directly into the municipal area or region of interest with no need to go through the phase of the enclosing rectangle. The search and registration

*Table 1. Planted area and CV for sugar cane and soybean, in Ipuã, estimated through StRS*

| | | | Sugar cane | | Soybean | |
|---|---|---|---|---|---|---|
| | Stratum | *nh* | Area (ha) | *CV* (%) | Area (ha) | *CV* (%) |
| Case A | I | 114 | 3,912 | 23.2 | 15,158 | 8.6 |
| | II | 86 | 9,483 | 10.8 | 3,749 | 21.7 |
| | | | 13,395 | 10.4 | 18,907 | 8.1 |
| Case B | I | 86 | 4,861 | 23.5 | 11,991 | 12.4 |
| | II | 114 | 10,980 | 8.0 | 3,494 | 19.7 |
| | | | 15,841 | 9.1 | 15,485 | 10.6 |

of current images of the study area also may be automated so that at any time one wishes to carry out the fieldwork, the most recent image is available.

## CONCLUSION

The agricultural information agile, accurate, inexpensive and safe, with quantified statistical errors, constitutes material of increasing value in a high dynamic scenario within the agricultural national and international markets.

In this chapter we provided the theoretical and practical basis for a method of estimating crop areas, using statistical objective sampling and remote sensing data. The method can also be applied to other targets that occupy large portions of land surface in a delimited area, such as forest, water bodies, urban areas and so on.

The sampling based on the structure of digital images enables a quick survey with known accuracy of the area covered with predominant crops within a municipality or region.

The remote sensing images, for having a comprehensive coverage and are repeatedly obtained, may help to estimate the area occupied by crops in municipal or regional scale.

The stratification approach is an excellent contribution in reducing the sample size or increasing the precision of estimates, provided that there is a strong knowledge base on which we can determine the boundaries of the strata.

The use of objective sampling methods, by region, leads to increased knowledge on the part of local agricultural statistics staff, which will surely bring efficiencies and effectiveness to the process.

Databases, GPS and GIS are tools which increasing efficiency is essential to the objective sampling method. Year after year, the application of the methodology should allow the continued enrichment of the database, in addition to a growth in experience of technical people both in the fieldwork as in the office (those who prepare the material for the field work, receive and organize the data collected and generate final statistics), which will make the method more efficient.

The audits, which can improve the quality of data coming from the field, are of fundamental importance, since the quality of field data are crucial to the quality and accuracy of objective estimates sample.

## REFERENCES


Camara, G., Souza, R. C. M., Freitas, U. M., & Garrido, J. (1996). SPRING: Integrating remote sensing and GIS by object-oriented data modeling. *Computers & Graphics*, *20*(3), 395–403. doi:10.1016/0097-8493(96)00008-8

Cochran, W. G. (1977). *Sampling techniques*. (3ed.). New York: John Wiley & Sons.

Collares, J. E. R., Lauria, C. A., & Carrilho, M. M. (1993). Pesquisa de previsão e acompanhamento de safras baseada em painéis de amostras de áreas. In Instituto Nacional de Pesquisas Espaciais (Ed.), *VII Simpósio Brasileiro de Sensoriamento Remoto,* (Vol. 4, pp. 450-453). São José dos Campos, Brazil: INPE.

Epiphanio, J. C. N., Barros Neto, O. O., Luiz, A. J. B., & Formaggio, A. R. (2001). Sistema de amostragem em imagem como base para estimativa de áreas agrícolas no município de Ipuã-SP. In Instituto Nacional de Pesquisas Espaciais (Ed.), *X Simpósio Brasileiro de Sensoriamento Remoto*, (pp. 59-66). São José dos Campos, Brazil: INPE.

Epiphanio, J. C. N., Luiz, A. J. B., & Formaggio, A. R. (2002). Estimativa de áreas agrícolas municipais, utilizando sistema de amostragem simples sobre imagens de satélite. *Bragantia*, *61*(2), 187–197. doi:10.1590/S0006-87052002000200012

Epiphanio, R. D. V., Formaggio, A. R., Rudorff, B. F. T., Maeda, E. E., & Luiz, A. J. B. (2010). Estimating soybean crop areas using spectral-temporal surfaces derived from MODIS images in Mato Grosso, Brazil. *Pesquisa Agropecuaria Brasileira*, *45*(1), 72–80.

Food and Agriculture Organization of the United Nations – FAO. (1996). *Multiple frame agricultural surveys: volume 1 current survey based on area and list sampling methods*. Rome: FAO. (FAO Statistical Development Series, 7).

Food and Agriculture Organization of the United Nations – FAO. (1999). *Multiple frame agricultural surveys: volume 2 agricultural survey programmes based on area frame or dual frame (area and list) sample design*. Rome: FAO. (FAO Statistical Development Series, 10).

Gallego, F. J. (2004). Remote sensing and land cover area estimation. *International Journal of Remote Sensing*, *25*(15), 3019–3047. doi:10.1080/01431160310001619607

Gao, J. (2002). Integration of GPS with remote sensing and GIS: reality and prospect. *Photogrammetric Engineering and Remote Sensing*, *68*(5), 447–453.

Giannotti, M. A. (2001). *Geotecnologias na análise de impactos sócios-ambientais: o caso da queima da cana-de-açúcar na região de Piracicaba. Unpublished master's theses*. São José dos Campos, Brazil: Instituto Nacional de Pesquisas Espaciais.

Gurgel, H. C., Ferreira, N. J., & Luiz, A. J. B. (2003) Estudo da variabilidade do NDVI sobre o Brasil, utilizando-se a análise de agrupamentos. *Revista brasileira de engenharia agrícola e ambiental, 7*(1), 85-90. doi: 10.1590/S1415-43662003000100014.

Instituto Brasileiro de Geografia e Estatística – IBGE. (2003a). *CIDADES*. Retrieved May 20, 2003, from http://www.ibge.gov.br/cidadesat

Instituto Brasileiro de Geografia e Estatística – IBGE. (2003b). *SIDRA*. Retrieved April 22, 2003, from http://www.sidra.ibge.gov.br

Jensen, J. R. (2000). *Remote sensing of the environment: an earth resource perspective*. Upper Saddle River: Prentice Hall.

Johnson, N. L., & Kotz, S. (1969). *Discrete Distributions*. New York: John Wiley & Sons.

King, R. B. (2002). Land cover mapping principles: a return to interpretation fundamentals. *International Journal of Remote Sensing*, *23*(18), 3525–3545. doi:10.1080/01431160110109606

Luiz, A. J. B. (2002). *Sensoriamento remoto agrícola*. São José dos Campos, Brazil: INPE.

Luiz, A. J. B. (2003). *Estatísticas agrícolas por amostragem auxiliadas pelo sensoriamento remoto*. Unpublished doctoral dissertation, Instituto Nacional de Pesquisas Espaciais, São José dos Campos, Brazil.

Luiz, A. J. B., & Epiphanio, J. C. N. (2001). Amostragem por pontos em imagens de sensoriamento remoto para estimativa de área plantada por município. In Instituto Nacional de Pesquisas Espaciais (Ed.), *X Simpósio Brasileiro de Sensoriamento Remoto*, (pp. 111-118). São José dos Campos, Brazil: INPE.

Luiz, A. J. B., & Gürtler, S. (2003). Aleatorização de pontos no território de um município, usando o SPRING, para a estimativa de área agrícola por amostragem. In Sociedade Brasileira de Informática Aplicada à Agropecuária e Agroindústria (Ed.), *IV Congresso Brasileiro da Sociedade Brasileira de Informática Aplicada à Agropecuária e Agroindústria SBI-Agro*. (pp. 9-12). Lavras, Brazil: UFLa.

Luiz, A. J. B., Oliveira, J. C., Epiphanio, J. C. N., & Formaggio, A. R. (2002). Auxílio das imagens de satélite aos levantamentos por amostragem em agricultura. *Agricultura em São Paulo*, *49*(1), 41–54.

Pino, F. A. (1999). Estatísticas agrícolas para o século XXI. *Agricultura em São Paulo*, *46*(2), 71–105.

Pradhan, S. (2001). Crop area estimation using GIS, remote sensing and area frame sampling. *JAG: International Journal of Applied Earth Observation and Geoinformation*, *3*(1), 86–92. doi:10.1016/S0303-2434(01)85025-X

Ray, S. S., & Pokharna, S. S., & Ajai. (1999). Cotton yield estimation using agrometeorological model and satellite-derived spectral profile. *International Journal of Remote Sensing*, *20*(14), 2693–2702. doi:10.1080/014311699211741

Ryerson, R. A., Curran, P. J., & Stephens, P. R. (1997). Agriculture. In Philipson, W. R. (Ed.), *Manual of Photographic Interpretation* (pp. 285–397). Bethesda, MD: ASPRS.

*The World Factbook 2009*. (2009). Washington, DC: Central Intelligence Agency. Retrieved February 9, 2010, from https://www.cia.gov/library/publications/the-world-factbook/index.html

Thenkabail, P. S., Ward, A. D., Lyon, J. G., & Merry, C. J. (1994). Thematic mapper vegetation indices for determining soybean and corn growth parameters. *Photogrammetric Engineering and Remote Sensing*, *60*(4), 437–442.

Tsunechiro, A., & de Freitas, B. B. (2001). Os cinqüenta municípios brasileiros maiores produtores de milho e soja. *Informações Econômicas*, *31*(7), 53–58.

Wilkinson, G. G. (1996). A review of current issues in the integration of GIS and remote sensing data. *International Journal of Geographical Information Systems*, *10*(1), 85–101.

Zwillinger, D., & Kokoska, S. (2000). *CRC Standard probability and statistics tables and formulae*. Boca Raton, FL: Chapman & Hall/CRC.

## ADDITIONAL READING

Benedetti, R., Bee, M., Espa, G., & Piersimoni, F. (Eds.). (2010). *Agricultural survey methods*. New York: John Wiley & Sons. doi:10.1002/9780470665480

Bouma, J., Varallyay, G., & Batjes, N. H. (1998). Principal land use changes anticipated in Europe. *Agriculture Ecosystems & Environment*, *67*(2-3), 103–119. doi:10.1016/S0167-8809(97)00109-6

Brink, A. B., & Eva, H. D. (2009). Monitoring 25 years of land cover change dynamics in Africa: A sample based remote sensing approach. *Applied Geography (Sevenoaks, England)*, *29*(4), 501–512. doi:10.1016/j.apgeog.2008.10.004

Carfagna, E., & Gallego, F. J. (2005). Using remote sensing for agricultural statistics. *International Statistical Review*, *73*(3), 389–404. doi:10.1111/j.1751-5823.2005.tb00155.x

Carfagna, E., & Marzialetti, J. (2009). Sequential design in quality control and validation of land cover databases. *Applied Stochastic Models in Business and Industry*, *25*(2), 195–205. doi:10.1002/asmb.742

Eva, H. D., Carboni, S., Achard, F., Stach, N., Durieux, L., Faure, J.-F., & Mollicone, D. (2010). Monitoring forest areas from continental to territorial levels using a sample of medium spatial resolution satellite imagery. *ISPRS Journal of Photogrammetry and Remote Sensing*, *65*(2), 191–197. doi:10.1016/j.isprsjprs.2009.10.008

Fjellstad, W. J., & Dramstad, W. E. (1999). Patterns of change in two contrasting Norwegian agricultural landscapes. *Landscape and Urban Planning*, *45*(4), 177–191. doi:10.1016/S0169-2046(99)00055-9

Gonzalez-Alonso, F., Cuevas, J. M., Arbiol, R., & Baulies, X. (1997). Remote sensing and agricultural statistics: crop area estimation in north-eastern Spain through diachronic Landsat TM and ground sample data. *International Journal of Remote Sensing*, *18*(2), 467–470. doi:10.1080/014311697219213

Green, K. (1992). Spatial imagery and GIS: integrated data for natural resource management. *Journal of Forestry*, *90*(11), 32–36.

Griffith, D. A., & Amrhein, C. G. (1997). *Multivariate statistical analysis for geographers*. Upper Saddle River: Prentice Hall.

Hietala-Koivu, R. (1999). Agricultural landscape change: a case study in Yläne, southwest Finland. *Landscape and Urban Planning*, *46*(1-3), 103–108. doi:10.1016/S0169-2046(99)00051-1

Ippoliti-Ramilo, G. A., Epiphanio, J. C. N., & Shimabukuro, Y. E. (2003). Landsat-5 thematic mapper data for pre-planting crop area evaluation in tropical countries. *International Journal of Remote Sensing*, *24*(7), 1521–1534. doi:10.1080/01431160010007105

Klersy, R. (1992). The work and role of the Commission of the European Communities. *International Journal of Remote Sensing*, *13*(6-7), 1035–1058. doi:10.1080/01431169208904177

Lillesand, T. M., & Kiefer, R. W. (1994). *Remote sensing and image interpretation*. 3ed. New York: John Wiley & Sons. 750p.

MacDonald, R. B., & Hall, F. G. (1980, May). Global crop forecasting. *Science*, *208*(4445), 670–679. doi:10.1126/science.208.4445.670

Meyer-Roux, J., & King, C. (1992). Agriculture and Forestry. *International Journal of Remote Sensing*, *13*(6-7), 1329–1341. doi:10.1080/01431169208904194

Monitoring Agriculture with Remote Sensing Techniques (MARS). (1993). *Crop area estimation of annual crops through area frame sampling based on segments: results obtained in Europe in 1992*. Varese: Ispra. 21p.

Mulders, M. A., De Bruin, S., & Schuiling, B. P. (1992). Structured approach to land cover mapping of the Atlantic zone of Costa Rica using single date TM data. *International Journal of Remote Sensing*, *13*(16), 3017–3033. doi:10.1080/01431169208904099

Ortiz, M. J., Formaggio, A. R., & Epiphanio, J. C. N. (1997). Classification of croplands through integration of remote sensing, GIS, and historical database. *International Journal of Remote Sensing*, *18*(1), 95–105. doi:10.1080/014311697219295

Stehman, S. V. (2005). Comparing estimators of gross change derived from complete coverage mapping versus statistical sampling of remotely sensed data. *Remote Sensing of Environment*, *96*(3-4), 466–474. doi:10.1016/j.rse.2005.04.002

Stehman, S. V. (2009). Model-assisted estimation as a unifying framework for estimating the area of land cover and land-cover change from remote sensing. *Remote Sensing of Environment*, *113*(11), 2455–2462. doi:10.1016/j.rse.2009.07.006

Tsiligirides, T. A. (1998). Remote sensing as a tool for agricultural statistics: a case study of area frame sampling methodology in Hellas. *Computers and Electronics in Agriculture*, *20*(1), 45–77. doi:10.1016/S0168-1699(98)00011-8

Villalobos, A. G. (2006). Sample Survey Statistics Teaching: An almost worldwide problem on teaching Agricultural Survey Methods. In *Proceedings of International Conference on the Teaching of Statistics* (ICOTS 7). Salvador, Brazil: IASE - International Association for Statistical Education. http://www.stat.auckland.ac.nz/~iase/publications/17/4f3_gonz.pdf/.

Warner, T. A., Nellis, M. D., & Foody, G. M. (2009). *The SAGE handbook of remote sensing*. London: SAGE Publications.

Xiao, X., Boles, S., Frolking, S., Salas, W., Moore, B., & Li, C. (2002). Landscape-scale characterization of cropland in China using VEGETATION sensor data and Landsat TM imagery. *International Journal of Remote Sensing*, *23*(18), 3579–3594. doi:10.1080/01431160110106069

## KEY TERMS AND DEFINITIONS

**Class of Interest:** It's the kind of land cover that most interests the sample survey. For example, if an agricultural survey that seeks only to estimate the area planted with soybeans, this would be the class of interest and all other types of land use, including other crops, could be grouped in a class defined as non-soy

**Fieldwork:** The fieldwork may or may not be present in an agricultural survey, and is an activity performed at the site where the studied phenomenon occurs naturally and not in the office or lab. Involves all the steps necessary to collect and record data, features and information related to the phenomenon or object of study, including travel, preparation of material and handling equipment.

**Image Processing Procedures:** Image processing procedures are all the procedures used on the raw data of images obtained by remote sensing in order to present them the best possible way for their analysis and interpretation. Procedures can be divided into preprocessing, image enhancement, classification and information extraction.

**Information Layer (IL):** Information layers are digital files of data that can be represented spatially. Each layer can be seen like a map that brings one characteristic represented in space. The geographic information systems (GIS) allow overlapping information layers to cross characteristics and properties located in geographic space as appropriate to the purpose of each investigation.

**Objective Measurement:** Objective measurement is the act of obtaining a value associated with a characteristic of the object observed through the use of devices or instruments, and that does not depend on who makes the measure. It is the opposite of the subjective measure, which is dependent on the bias of who is measuring.

**Sampling Unit:** Each element of the population that can be identified and selected for the sample

**Simple Random Sample:** A simple random sample occurs when the sampling plan ensures that every element of the population has an equal and known probability of being sampled.

**Spatial Information:** Spatial information is any information about an object that is tied to its location in space and its spatial relationship with other objects of the same population.

**Stratified Sampling:** The stratified sampling uses a priori information to divide the target population into subgroups internally homogeneous, called strata. The strata can be defined based on various factors such as topography, political boundaries, roads, rivers, human characteristics, depending on the context of the problem. Once defined the strata, a sample will be extracted in each stratum. Generally, the objective of this procedure is to estimate the true average or the total for a variable in each stratum. If the stratification was correct (relatively homogeneous strata) the estimated average for the population is more accurate than could be obtained by a simple sampling.

# Chapter 6
# Study of Potential Climate for Sugarcane Production in the State of Rio Grande Do Sul, Brazil, by Means of Geoprocessing

**Fernando Ulhmann Soares**
*Federal Institute of Education, Science and Technology of Goias, Brazil*

**José Maria Filippini Alba**
*Embrapa Temperate Climate Research Center, Brazil*

**Elódio Sebem**
*Federal University of Santa Maria, Brazil*

**Marcos Silveira Wrege**
*Embrapa Forest Research Center, Brazil*

## ABSTRACT

*A potential climate study for sugarcane of a sector of Rio Grande do Sul State, Brazil is presented here. GIS technology was applied for evaluation of the risk of frost and for integration of spatial data. The problem was focused in regional scale and in local scale (Municipality of Jaguari). Results showed that cultivation can be programmed in order to obtain physiological maturity before the period of risk of frost, thus avoiding low production. Spatial analysis of the information allows rapid perspective for productivity of sugarcane in a specific territory. The Municipality of Jaguari has large potential for cultivation of sugarcane because of the absence of the risk of frost. Its productivity allows for expansion into suitable neighboring areas. Also, geoprocessing combined with the study of climate and soil appears as a significant tool for interpreting the areas with aptitude for production of sugarcane or for the industry of sugar and alcohol.*

DOI: 10.4018/978-1-61692-871-1.ch006

## INTRODUCTION

This chapter describes a potential climate study for sugarcane (*Sacharun ssp.*) using Geographic Information Systems – GIS. This is a representation of the real world using a computer where data are structured according to levels of information (ASSAD; SANO, 1998). Several municipalities of Rio Grande do Sul State were considered, with a specific history case in the Municipality of Jaguari, near the centre of Rio Grande do Sul State which has high production of said crop (Soares, 2008). The state has a significant production of sugarcane, from both a historical and geographic point of view, which could be used in the current context of agro-energy and sustainability.

The current expansion of production of sugarcane shows its permanent condition as an element of the social dynamics, due to the requirement of a new energetic matrix in accordance with less harmful practices to the environment and the substitution of fossil fuels. The sugarcane crop runs in tandem with the trend of the international market and the needs of conservation of natural resources, therefore, it constitutes an option for a scenario of more sustainable energy: agro energy.

Martin Afonso de Souza cultivated the first plants in Brazilian territory, in the village of São Vicente which was founded by emigrants from Portugal in 1532, a place known as the "Baixada Santista" today, in the State of São Paulo (União de Industrias da Cana de Açucar, 2007). However, the industry was developed in the states of Pernanbuco and Bahia, Northeastern Brazil, due to the tropical climate, the fertility of the soil and the presence of cheap labor (African slaves). The crop was fully oriented towards the production of sugar at first and it was then transformed into one of the most economic cycles in Brazil. After that, sugarcane became a significant crop in the national context, with increasing use in the production of fuel alcohol.

As mentioned before, the Northeastern region of Brazil took advantage of the planting of sugarcane and in turn, the crop spread throughout the Southeastern region, assuming a significant role in the national and international scenarios at the beginning of the 21st century. The old National Program for Alcohol was replaced by the National Plan for Agro energy for 2006 – 2011 (Brasil, 2006; 2007). This plan assembles strategic actions in order to reduce consumption of fossil fuels while at the same time, the federal government tries to improve the use of biofuels while conserving the environment, the preservation of the international market and the promotion of social inclusion. The plan foments the agro-ecological zoning for energetic crops in both traditional and new agricultural areas where the expansion of the agricultural frontiers is evident, in this way investment of the private sector is managed. Genetic improvement using traditional methods or biotechnology is also a target because productivity can be improved without a significant advance on unexplored lands.

In this way, several states in Brazil have incremented their plantations of sugarcane due to the improvements in the production and installation of plants for the sector of alcohol and sugar. The State of Rio Grande do Sul is inserted in that context when the investments of private and public sectors are oriented to the research of agricultural productivity through vegetal improvement, agronomic efficiency and zoning of climate and soil thereby enlarging the possibility of development of the crop in spite of the risk of frost and low temperatures derived from the subtropical climate.

Sugarcane was introduced in the territory of Rio Grande do Sul State in 1725, in the municipalities of Torres, Osório and Santo Antônio da Patrulha, but currently the regions of Missões, Middle and High Uruguai and Central constitute half of the production (Secretaria de Planejamento e Gestão - RS, 2007). Production is important for familiar agriculture from small and medium farms where the crop is used as food for cattle and as raw material for production of candies, brown sugar and spirits. The role of sugarcane as a commodity is increasing and the agro-climatic zonings are the instruments used for the selection of suitable areas for cultivation. The season of planting for Rio Grande do Sul State goes from October to December for the harvest of one year or from January to March for the harvest of one year and a half. The season for growing is October to April and the harvest begins in June or afterwards. Low temperatures (2 – 7 °C) and the short days of winter induce the deactivation of growing; promoting the maturity and the content of sucrose that is regulated until the point of harvest. Temperatures near 17 °C are advantageous for this process that can be disrupted or restricted by temperatures lower than 2.5 °C (Maluf, 2007).

Two kinds of data analysis were developed in this chapter: (1) Production data of sugarcane in Rio Grande do Sul State were associated with spatial territory, considering statistical tables (IBGE, 2007). (2) A study of potential climate for sugarcane production based on data from the minimum temperature of five weather stations, with daily records during 25 - 30 years. Preventing the risk of occurrence of frost events is a useful task for planning the production of sugarcane, as well as for guiding public and private policies for regional development. The organization of the territory through GIS strategies would be a complementary procedure.

## THE POTENTIAL OF THE STATE OF RIO GRANDE DO SUL FOR PRODUCTION OF SUGARCANE.

The theme agro-energy takes increasing importance, because of the concern of government and the pressure of society to optimize the impacts of current energy matrices, based on the burning of fossil fuels. The growing investment in structures of production and research come from a gradual expansion of bio-fuels due to their potential to generate sustainable energy.

According to Brunini (2007), sugarcane belongs to the family of *Poaceae*, genus *Saccharum* and traditional cultivation is between latitudes 35°N to 35°S and its capacity of adaptation ranges from temperatures of 34 – 35 °C. *Saccharum spp.* is an important inter-specific hybrid with high resistance to disease and good adaptation to several regional climates. Its condition of semi-perennial provides large representation among tropical species, but climate and seasonal variations must be considered for cultivation.

Brazil has favorable conditions due to its tropical climate, potential productive areas and tradition of production for sugarcane and derivatives. Similarly, sectors of the State of Rio Grande do Sul, with potential climate for cultivation, are improving research projects related to the production of alcohol. The State of Rio Grande do Sul is located at the southern point of the Tropic of Capricorn, in a region of temperate climate with features of subtropical climate, rains in uniform regime during the whole year and a hot summer (Brasil, 2007). However, there are strong occurrences of frost, especially during the months of winter, when the temperature is less than 16°C average. The temperature rises up to 30 °C in the hottest month.

A good production of sugarcane depends on alternation of periods of high and low temperature. Good vegetative development, with high production of sucrose, is reached in periods with high solar radiation and wetness of soil (Wrege, 2004). Moreover, accumulation of sucrose in the stalks, with reduced rates of growth is obtained because of restricted water and low temperatures. Cultivation in regions of sub-tropical climates is possible for mild winters, with good thermal conditions, but without early lethal frosts (Maluf, 2007). In this sense, low temperatures in Rio Grande do Sul State occur during the winter months, although sugarcane is cultivated from October to December for harvest of one year or from January to March for harvest of one year and a half.

According to recent data, the total production of sugarcane in Rio Grande do Sul is of 1,426,978 tons, while the cultivated area covers 35,767 hectares. Most municipalities have less than 500 hectares of planted area (Soares, 2008). Municipalities with 500 to 1000 hectares of planted area are distributed, usually, near the main water resources of the Northern Half of the State. Few municipalities reached planted areas greater than 1000 hectares, as in Roque Gonzales, Porto Xavier and Irai. Differences in soil types and temperature are essential in the spatial configuration of production. The majority of sugarcane production is in the northern sector of the State of Rio Grande do Sul. The southern sector is influenced by lower temperatures and significant risk of frost, thus, productivity is low. The growth of sugarcane is minimized when the temperature is below 18-20°C, depending on the existence or absence of irrigation respectively (WREGE, 2004). Some of the best levels of productivity are in the regions close to industrial areas, such as Coopercana (Porto Xavier), Limara (Jaguari) and Norobios (São Luiz Gonzaga). However, two municipalities, Salto Jacuí and Camargo, show higher production than the national average although they are located far away from industrial activities.

## STUDY OF POTENTIAL CLIMATE

Grodzki *et al.* (1996) indicated that the characterization of the regime of frost is extremely useful for orienting extension techniques, producers and governmental institutions as support to planning and management of the agricultural sector. Thus, temperatures lower than 3°C are considered as occurrences of frost, based on the comparison of data under cover. This allows us to define the periods of the year without frosts which is a vital parameter for the schedule of cultivation.

Several themes, such as climate, soil, water, flora, fauna and economy should be considered for the planning of plantations. The study of potential climate is important for conservation and environmental protection, because production and productivity are improved, as well as land use and the future risks related to weather events being minimized. The study of potential climate indicates the best areas for planting sugarcane in a region, taking into account the risk of weather events and the occurrences of frost. Therefore, the dates of the beginning and end of occurrences of frost are evaluated so that plants are cultivated with better chances of developing. At the same time, the study shows locations where production must be avoided because of low efficiency due to adverse weather conditions.

Thus, the data of temperature, altitude, latitude and longitude of five meteorological stations were taken into account and used to produce maps of risk of frost and maps of the ideal period for planting sugarcane.

In the region of Jaguari and in the territory of the municipality (Figure 1), The Municipality of Jaguari was chosen for a case study due to its high production of sugarcane and the location of micro-distilleries in the process of expansion.

Minimum temperatures near 3°C under cover represent about – 1°C in the field. This can cause heavy damage to sugarcane and other tropical crops (Wrege *et al.*, 2005). However, freezing of water is not necessary to cause damage, because a rapid drop in temperature is enough ("agronomic frost"). So, knowledge of the distribution of the risk of frost designates the regions and periods for cultivating, with the respective disposal of areas representing major risk, when cultivation can be programmed in such a way that physiological maturity occurs before the first frost.

## MODELING FROST RISK

Temperature data from five meteorological stations of the research network of the State of Rio Grande do Sul were processed by the Laboratory of Agro-meteorology of the Embrapa Temperate

*Figure 1. Location of the study area, the municipality of the case study and the municipalities with meteorological stations*

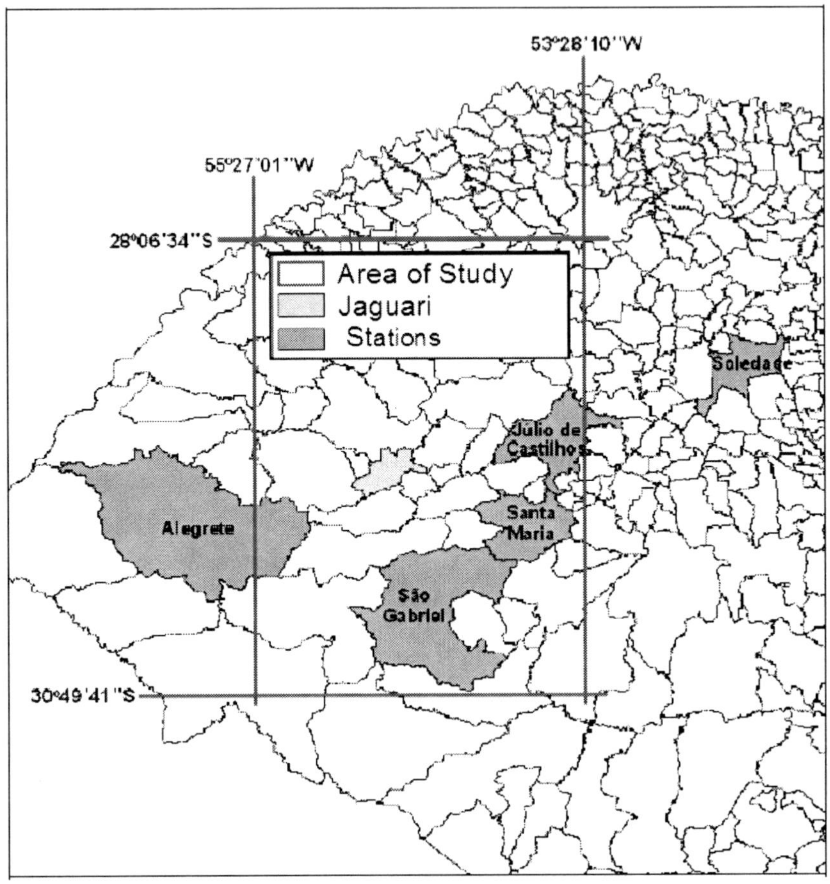


Climate, in cooperation with the State Foundation of Agricultural Research (Fepagro) and the National Institute of Meteorology (INMET). The risk of frost was evaluated using a series of daily data from the last 25 - 30 years. The occurrence of frost was considered when the temperature of the air was below 3 °C (Grodski *et al.*, 1996). The five meteorological stations (Figure 1) have different characteristics and heterogeneous patterns (Table 1). Soledade is out the area of study but, as its altitude is elevated (720 meters), the effect of that variable was adjusted from the resulting model. The other four stations are located at altitudes below 160 meters which might lead to undesirable results for the model of frost risk if they were studied on an isolated basis.

Data of temperature were classified in binary form, value one was assigned for temperatures lower than 3°C (agronomic frost); in other case value zero was considered. The days of frost per season, expressed in the Julian system (1 – 365/366), were correlated with altitude, latitude and longitude. Two equations were estimated for the occurrence of the first and the last occurrence of frost in the study area, according to a multiple linear regression model (Wrege, 2004). Regionalization of data was obtained through the regression model when latitude, longitude and altitude were transformed in raster files according to sectors of 30". The digital elevation model was extracted from the Shuttle Radar Topography Mission (Figure 2) with pixel of 90 meters and

*Table 1. Geographic location, altitude and period of measurement of data for the meteorological station considered in this work*

| Meteorological stations | Latitude | Longitude | Altitude | Period of measure |
|---|---|---|---|---|
| Alegrete | 29.7831° S | 55.7831° W | 96 m | 1966-1986 |
| Júlio de Castilhos | 29.1222° S | 56.5478° W | 78 m | 1956-1990 |
| Santa Maria | 29.6900° S | 53.8117° W | 153 m | 1963-1990 |
| São Gabriel | 30.3408° S | 54.3169° W | 109 m | 1963-1990 |
| Soledade | 28.8306° S | 52.5111° W | 720 m | 1967-1990 |

vertical precision of 30 meters (apud Weber; Hasenack; Ferreira, 2005). This process was performed in a GIS environment (ESRI, 1999) with geographic projection and South American Datum (SAD69).

The area of study covers part of the Central Depression where the interior altitudes are lower than the exterior (Magnoli, 2001). For instance, the Meridion Plateau with the highest altitude and the lowest temperatures and, the Mountains of the Southeast, a sequence of rounded hills. A small

*Figure 2. Digital elevation model of the area of study extracted from the shuttle radar topography mission (apud Weber et al., 2005)*

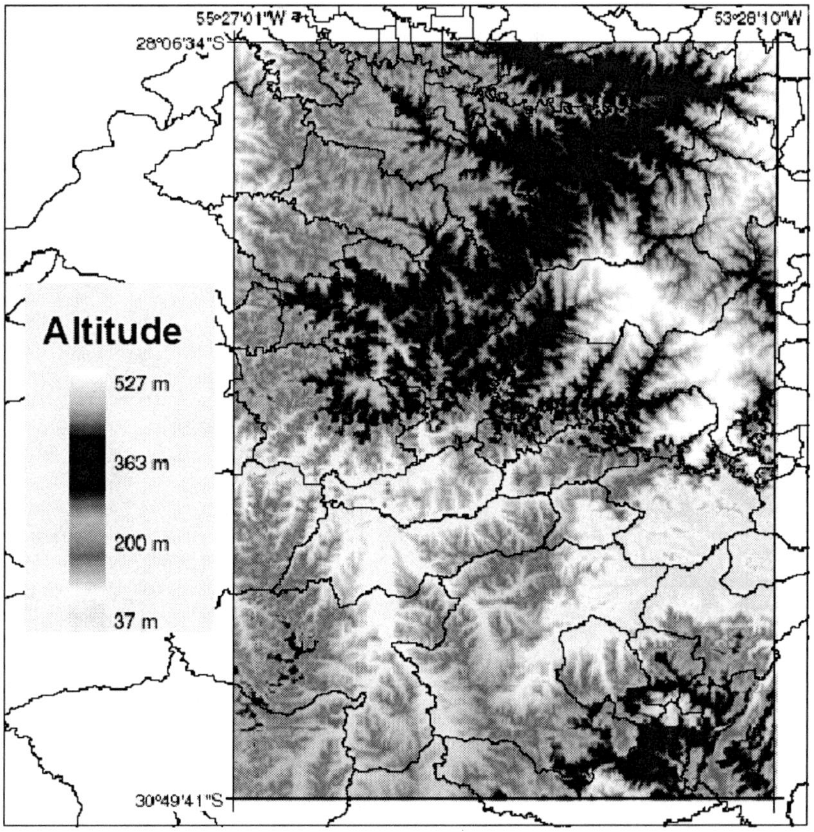

portion of the area of the Campaign with wavy relief also occurs in the area of study.

A second stage was developed in the municipality of Jaguari, a strong producer with extensively cultivated land and strategic location related to the meteorological stations and industry in expansion.

## THEMATIC MAPS FOR THE CULTIVATION AND HARVEST OF SUGARCANE

Temperature has high correlation with altitude, thus the colder regions are the higher ones. This is verified in the maps when the first frosts occur in April (Figure 3) and the last frosts in October (Figure 4), where the corresponding areas will have less possibility for developing sugarcane due to a longer period of frost risk.

The southern part of the area of study shows low risk of frost, but municipalities in this sector have low production of sugarcane, almost nil. Several possibilities can be considered: (1) Error of model due to absence of data in marginal regions; (2) Effect of extra variables, humidity for instance; (3) Deficit of water. The Municipality of Jaguari has the privileged condition of climatic aptitude with good possibilities for expanding cultivation of sugarcane, because the risk of frost occurs over a short period, so, more than a half of the year can be used (Figure 5 and 6).

## CONCLUSION

The map of the distribution of the risk of frost helps to identify regions for cultivation when the areas of major risk and minor risk are discriminated. In this sense, cultivation can be programmed in order to obtain physiological maturity before the period of risk of frost, thus avoiding low production. Spatial analysis of the information allows rapid perspective for productivity of sugarcane in a specific territory.

The Municipality of Jaguari has large potential for cultivation of sugarcane because of the absence of the risk of frost. Its productivity allows for expansion into suitable neighboring areas. Also, geoprocessing combined with the study of climate and soil appears as a significant tool for interpreting the areas with aptitude for production of sugarcane or for the industry of sugar and alcohol. The ability of GIS for processing and integrating information was tested, becoming a powerful instrument for environmental management and public policies.

*Figure 3. Estimation of the date of first frost for the area of study*

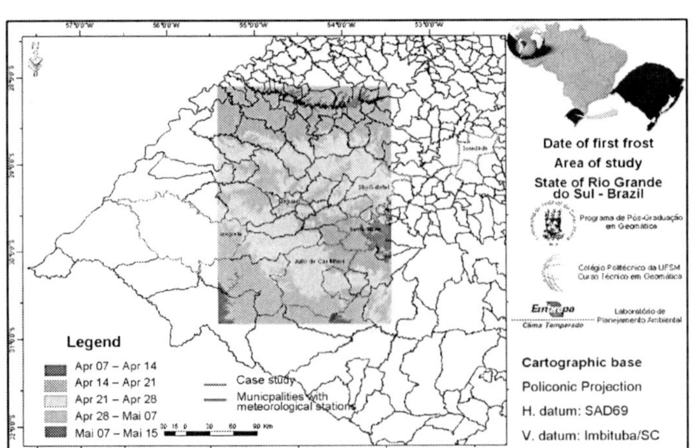

*Figure 4. Estimation of the date of the last frost for the area of study*

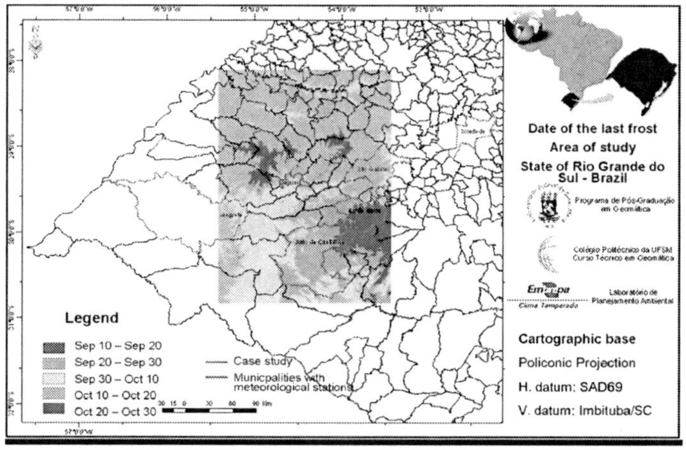


*Figure 5. Estimation of the date of the first frost for the municipality of Jaguari*

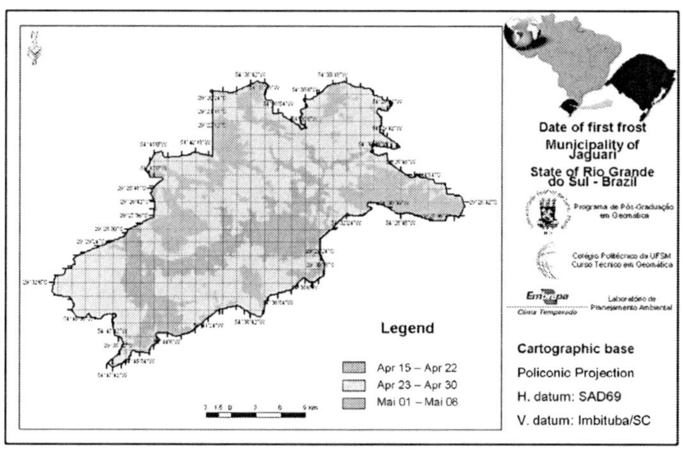


*Figure 6. Estimation of the date of the last frost for the municipality of Jaguari*

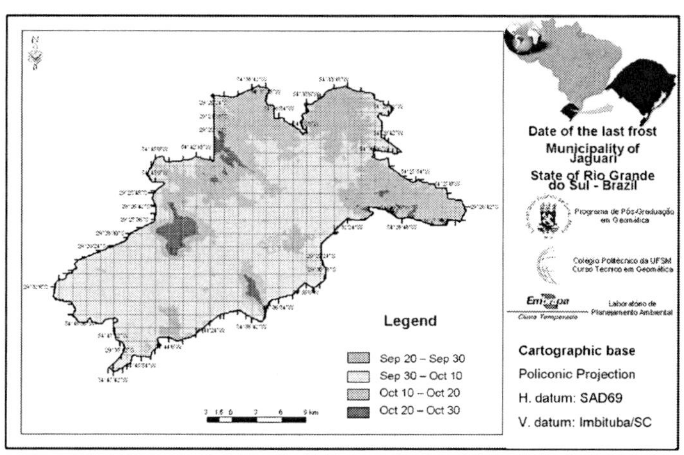

## REFERENCES

Assad, E. D., & Sano, E. E. (1998). *Sistemas de Informações Geográficas - Aplicações na Agricultura*. Brasília, DF: EMBRAPA.

Brasil. Ministério da Agricultura, Pecuária e Abastecimento. Secretaria de Produção e Agroenergia (2006). *Plano nacional de agroenergia 2006-2011*. (2. ed. rev. atual.). Brasília, DF: Embrapa Informação Tecnológica.

Brasil. Ministério da Agricultura, Pecuária e Abastecimento. Secretaria de Produção e Agroenergia. (2007). *Balanço nacional da cana-de-açúcar e de agroenergia 2007*. Brasília, DF: Secretaria de Política Agrícola.

Brunini, O. (2007). *Zoneamento de culturas bioenergéticas no Estado de São Paulo*. Retrived October 9, 2007, from http://www.infobibos.com/Artigos/2007_3/zoneamento/Index.htm.

Câmara, G., Davis, C., & Monteiro, A. M. V. (2001). *Introdução à ciência da geoinformação*. São José dos Campos, Brazil: INPE.

Costa Neto, P. L. de O. (2002). *Estatística*. São Paulo, Brazil: Editora Edgard Blücher Ltda.

da Cunha, G. R., & Assad, E. D. (2001). Uma visão geral do número especial da RBA sobre zoneamento agrícola no Brasil. *Revista Brasileira de Agrometeorologia*, *9*(3), 377–385.

ESRI. (1999). *ArcGis 9. ArcMap (software)*. Redlands, CA: ESRI.

Flores, C. A., Garrastazu, M. C., Hasenack, H., & Weber, E. (2006). *Zoneamento edáfico para as culturas da mamona, soja, girassol e canola na região Noroeste do Estado do Rio Grande do Sul*. Pelotas, Brazil: Embrapa Clima Temperado (Documentos, 174).

Grodski, L., Caramori, P. H., Bootsma, A., de Oliveira, D., & Gomes, J. (1996). Riscos de ocorrência de geada no Estado do Paraná. *Revista Brasileira de Agrometeorologia*, *4*(1), 93–99.

IBGE. (2007). *Cidades@*. Retrieved 18 June, 2007, from http://www.ibge.gov.br/cidadesat/default.php

Magnol, D., Oliveira, G., & Menegotto, R. (2001). *Cenário gaúcho: representações históricas e geográficas*. São Paulo, Brazil: Moderna.

Maluf, J. R. T., Westphalen, S. L., Matznauer, R., & Maluf, D. E. (2007). *Zoneamento agroclimático da cana-de-açúcar no Estado do Rio Grande do Sul, visando a produção de açúcar e álcool*. Paper presented at the XV Brasilian Congress of Agrometeorology, Aracajú, Brazil.

Secretaria de Planejamento e Gestão – RS. (2007). *Atlas socioeconômico do Rio Grande do Sul*. Retrieved July 1st, 2007, from http://www.scp.rs.gov.br/.

Soares, F. U. (2008). *Zoneamento agroclimático para cana-de-açúcar (Sacharum ssp.) no Rio Grande do Sul por meio de geoprocessamento: estudo de caso no município de Jaguari*. Unpublished master dissertation, Federal University of Santa Maria, Brazil.

União da Industria da Cana de Açucar. (n.d.). *Portal Única*. Retrieved in October 30, 2007 from http://www.portalunica.com.br/portalunica/

Weber, E., Hasenack, H., & Ferreira, C. J. S. (n.d.). *Adaptação do modelo digital de elevação do SRTM para o sistema de referência oficial brasileiro e recorte por unidade da federação*. Retrieved December 1st, 2005, from http://www.ecologia.ufrgs.br/labgeo.

Wrege, M. S. (2004). *Delimitação de regiões com potencial de cultivo econômico da cana-de-açúcar (Saccharum spp.), no estado do Paraná por meio da análise de riscos climáticos.* Unpublished Doctoral Dissertation, State University of Maringa, Maringa, Brazil.

Wrege, M. S., Caramori, P. H., Gonçalves, A. C. A., Bertonha, A., Ferreira, R. C., & Caviglione, J. H. (2005). Regiões potenciais para cultivo da cana-de-açúcar no Paraná, com base na análise do risco de geadas. *Revista Brasileira de Agrometeorologia, 13*(1), 113–122.

## ADDITIONAL READING

Alba, F. J.M. (Ed.) (2007). Trabalhos acadêmicos de geoprocessamento desenvolvidos no laboratório de planejamento ambiental em 2007. Pelotas, RS: Embrapa Clima Temperado (Documentos, 219).

Alba, F. J.M. (Ed.) (in press). Recuperação de areas degradadas. Pelotas, RS: Embrapa Clima Temperado.

Bonham-Carter, G. F. (1994). *Geographic information systems for geoscientists.* Ontario: Pergamon.

Falcade, I., & Mandelli, F. (Eds.). (1999). Vale do Vinhedos: Caracterização Geográfica da Região. Caxias do Sul, RS: EDUCS.

Filippini Alba, J. M., Berny, V. M., & da Costa, F. A. (2009). *Geoprocessamento no contexto de empresas de pesquisa.* Pelotas, RS: Embrapa Clima Temperado (Documentos, 254).

Filippini Alba, J. M., & de Souza Filho, C. R. (2006). *Geoprocessamento como instrumento de avaliação ambiental.* Pelotas, RS: Embrapa Clima Temperado (Boletim de Pesquisa e Desenvolvimento, 28).

Fischel, W. A. (1987). *A Property Rights Approach to American Land Use Controls.* London: The Johns Hopkins Univ. Press.

Flores, C. A., Filippini Alba, J. M., & Wrege, M. S. (Eds.). (2009). *Zoneamento agroclimático do eucalipto para o Estado do Rio Grande do Sul e edafoclimático na região do Corede Sul – RS.* Pelotas, RS: Embrapa Clima Temperado.

Flores, C. A., Garrastazu, M. C., & Filippini Alba, J. M. (2009). Metodologia de zoneamento edáfico de culturas para o Estado do Rio Grande do Sul. Pelotas, RS: Embrapa Clima Temperado (Documentos, 261).

Intergovernmental Panel on Climate Change. (2007). Climate in Peril: A Popular Guide to the Latest IPCC Reports presents the substance of the Climate Chang. Retrieved June 22, 2010 from http://www.grida.no/publications/climate-in-peril/

Jensen, J. R. (1996). *Introductory Digital Image Processing.* Upper Saddle River, NJ: Prentice Hall.

Molga, M. (1962). *Agricultural meteorology. Part II: Outline of agrometeorological problems.* Poland: National Science Foundation and the Department of Agriculture.

UN. (2004). Land degradation and Land Use/Cover Data Sources. Economic Retrieved June 22, 2010 from http://unstats.un.org/unsd/environment/envpdf/landdatafinal.pdf.

## KEY TERMS AND DEFINITIONS

**Agroenergy:** Biomass energy derived from agricultural practices.

**Agrometeorology:** Science investigating the meteorological, climatological and hydrological conditions that are significant to agriculture owing to their interaction with the objects and processes of of agriculture production (Molga, 1962).

**Geographic Information Systems (GIS):** Computer system for managing spatial data (Bonham-Carter, 1994).

**Edafology:** Soil science applied to agriculture.

**Land Use:** Natural resources classified according to human exploration.

**Pedology:** Soil science.

**Remote Sensing:** Acquisition of information of an object or phenomenon through sensing device(s) without physical or intimate contact with it.

**Zoning:** Division of territories in homogeneous zones since a Land Use perspective, considering criteria related to climate, soil, environment, economy, etc.

# Chapter 7
# Using Self-Organizing Maps for Rural Territorial Typology

**Marcos Aurélio Santos da Silva**
*Brazilian Agricultural Research Corporation, Embrapa Coastal Tablelands, Brazil*

**Edmar Ramos de Siqueira**
*Brazilian Agricultural Research Corporation, Embrapa Coastal Tablelands, Brazil*

**Olívio Alberto Teixeira**
*Federal University of Sergipe, Department of Economy, Cidade Universitária, Brazil*

**Maria Geovania Lima Manos**
*Brazilian Agricultural Research Corporation, Embrapa Coastal Tablelands, Brazil*

**Antônio Miguel Vieira Monteiro**
*National Institute for Space Research, Image Processing Division, Brazil*

## ABSTRACT

*This work assessed the capacity of the self-organizing map, an unsupervised artificial neural network, to aid the process of territorial design through visualization and clustering methods applied to a multivariate geospatial temporal dataset. The method was applied in the case study of Sergipe's institutional regional partition (Territories of Identity). Results have shown that the proposed method can improve the exploratory spatial-temporal analysis capacity of policy makers that are interested in territorial typology. A new partition for rural planning was elaborated and confirmed the coherence of the Territories of Identity.*

## INTRODUCTION

Rural territorial typology refers to the classification of commonalities found in contiguous rural areas that may be useful for a well definition of public policies and to discover identity elements. In an integrational perspective, all dimensions must be considered in a spatial analysis. For this study, it was preferred to adopt the definition of territory used by the Ministry of Agrarian Development (MDA, 2005, p.7).

DOI: 10.4018/978-1-61692-871-1.ch007

*... territories are defined as a physical space, geographically defined, generally continuous, encompassing urban and rural, characterized by multidimensional criteria, such as environment, economy, society, culture, politics and institutions, and a population with relatively distinct social groups that relate internally and externally through specific processes, which can distinguish one or more elements that indicate identity and social, cultural and territorial cohesion.*

Territories break down false boundaries (political divisions) among areas and facilitate the process of socio-cultural identity construction, discovery or investigation. The design of public policies also benefits from this strategy, in a way that watersheds, forests, ecosystems and other homogeneous zones can be analyzed altogether. Naturally, there are convergences and divergences about territorial definition, mainly concerning the focus of the study, for example, economical versus environmental analysis. A regional development study implies the use of an interdisciplinary approach, which reveals itself as a very difficult task due to the complexity and uncertainty of this field. Regional studies comprehend constraining factors, such as scales, physical amplitude, quantity and quality of data, diversity of applications, political and practical aspects (Sabourin & Teixeira, 2002).

The general definition of territory by the MDA (2005) argues that social-cultural-economical cohesion is the main factor to establish a regional aggregation into a territory. Although landscape homogeneity and distribution of natural resources are important aspects, they are not determinant. Therefore, multivariate socio-economical geospatial data integration and interpretation can be useful for policy makers that engage themselves in the process of territory design.

In Brazil, there are many initiatives towards the institutionalization of territories as a new strategy for public management (MDA, 2005; Bandeira, 2006). In this context of interdisciplinary studies and spatial integration, mainly for sustainable development, rural and urban issues are not viewed as different things, but treated as a systematic problem (Sabourin & Teixeira, 2002; Flores, 2004; MDA, 2005). Besides, the new trend for self-development by bottom-up as stated by Claval (2008) increases the demand for new and practical approaches to study and generate knowledge for a good territorial partition.

The absence of a territorial theory (Abramovay, 2006) and the rise of territorial policies, at least in Brazil, imply the demand for exploratory methods that could solve practical problems, such as exploratory spatial clustering and, at the same time, it gives us helpful insights for a territorial theory. There are many classical quantitative strategies to perform a multivariate exploratory analysis of geospatial data such as: colored maps, multivariate statistics coupled with these maps, spatial statistics, and geostatistics (Bailey, 1995). In this study, an Artificial Neural Network (ANN) was applied simultaneously for geospatial data visualization and clustering. The objective was to assess the capacity of the ANN, more precisely the Self-Organizing Map (SOM), to find hidden patterns in the geospatial dataset that could be useful for a territorial typology.

This ANN was applied in the analysis of Sergipe's institutional regional partition (Territories of Identity) created by Teixeira et al. (2007) using both subjective and analytical methods. On the other hand, the approach of this work used quantitative geospatial dataset aggregated by municipalities. Results were compared with this official Sergipe's clustering. These territories were chosen because a great effort for Sergipe's territorial partition was performed by academics and officials, so a huge amount of information was gathered and organized. Consequently, the know-how and knowledge about the way the study of regional commonalities is performed increased and, therefore, demanded for integrated quantitative analysis of multisource and multivariated geospatial data. In general terms, the main goal is to establish a semi-automatic territorial typology process to aid the rural planning by identification of homogeneous areas in Sergipe, Brazil.

## MATERIAL AND METHODS

### Self-Organizing Maps (SOM)

The Kohonen Self-Organizing Map (SOM) is a competitive artificial neural network structured in two layers (Kohonen, 2001). The first one represents the input vector of the dataset variables $x_k=[\xi_1, \xi_2, ..., \xi_d]^T$, $k=1,2,...n$, where $n$ is the number of input vectors and $d$ corresponds to data dimension, and the second one is a neuron's grid, usually bidimensional and fully connected (Figure 1a). Each neuron $j$ has one codevector $w_j=[w_{j1}, w_{j2}, ..., w_{jd}]^T$ associated to it. The main goal of this ANN is to order, by a learning algorithm, the input dataset into the neural grid, using the codevectors ($w_j$) as representations of the input data ($x_k$).

There are different topologies for the Self-Organizing Map architecture, while the most common is the two-dimensional topology. Figure 1 shows a one-dimensional SOM (b) and a toroidal SOM (c).

The learning process can be explained in three phases. In the first phase (competitive) each $x_k$ is presented to all neurons in search of the Best Match Unit (BMU) using Euclidean distance as a measurement reference of the distance feature. In the second phase (cooperative) a neighborhood relation between the BMU and the other neurons is defined for a smoothed update of each $w_j$. Finally, in the last phase (adaptive), all codevectors $w_j$ will be updated using some kind of adaptive rule, in general sequential or batch, see Equations 1 and 2 for the batch update rule used in this work (Vesanto, 1999).

$$s_i(t) = \sum_j^{n_{V_i}} x_i \tag{1}$$

$$w_i(t+1) = \frac{\sum_j^{m} h_{ji}(t) s_j(t)}{\sum_j^{m} n_{V_j} h_{ji}(t)} \tag{2}$$

Where $s_i$ represents an input pattern sum for the $i^{th}$-Voronoi region, $V_i$; and $n_{Vi}$ is the number of samples for the Voronoi dataset of the $i^{th}$-neuron. The neighborhood kernel function between neu-

*Figure 1. (a) A complete self-organizing NxM map architecture representation. (b) A one-dimensional SOM. (c) A toroidal SOM*

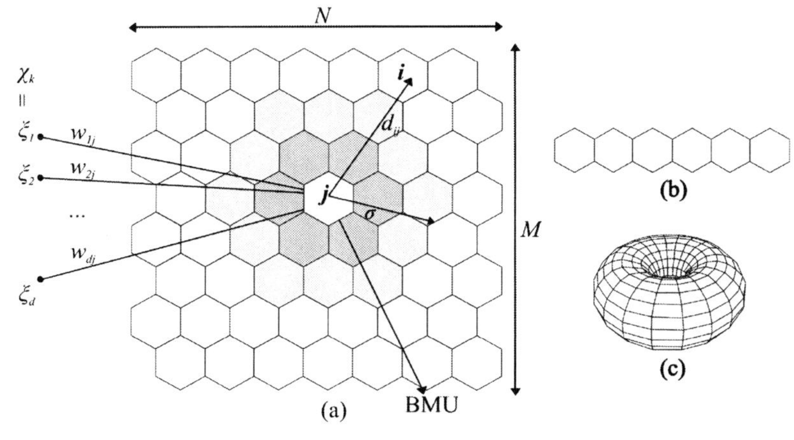

rons $i$ and $j$ at time $t$ is represented by $h_{ij}(t)=exp(-d_{ij}^2/2*\delta(t)^2)$, where $\delta(t)= \delta(0)*exp(-t/C)$, $d_{ij}$ is the distance between neurons $i$ and $j$ on the neural grid, $C$ is a constant, and $m$ is the number of Voronoi regions (number of neurons). See Equation 3 for the sequential update rule.

$$w_{ji}(t + 1) = w_{ji}(t) + \alpha(t)\, h_{ij}(t)\, [x_{ik}(t) - w_{ij}(t)] \quad (3)$$

Where $\alpha(t)$, $\alpha(0)<1$, represents the learning rate function. The learning rate function $\alpha(t)$, $h_{ij}(t)$ and $\delta(t)$ are monotonically decreasing functions.

After the learning process, codevectors should approximate, in a non-linear manner, the input dataset. In addition, SOM preserves the topological structure of the input data, and nearby patterns in the sample dataset are associated with nearby neurons in the SOM grid. For this work, a bi-dimensional SOM, the batch learning algorithm associated with hexagonal lattice, gaussian neighborhood kernel function, and linear initialization for visualization were used. For the automatic clustering it was used a uni-dimensional SOM.

The competitive process is the most time-consuming of the learning process. Usually this is a sequential search for the Best Match Unit (BMU). This process can be optimized by using some mechanism to minimize the heuristic search or through the parallelization of the learning code (Openshaw & Turton, 1996).

The parameters of learning are defined empirically, based on user experience and methods of trial and error. However, some techniques for automatically determining the parameters of learning have been proposed, either through genetic algorithms or numerical methods (Haese & Goodhill, 2001).

The dimensionality of self-organizing map and its size ($m$) depend on the type of problem and purpose. The literature shows that determining the size of the SOM is an empirical process (Flexer, 2001; Kohonen, 2001). In general, bi-dimensional SOM is used because of its ability to project the data of high dimension $d$ into a two-dimensional neural map grid.

## Assessment of Quality of SOM Learning Algorithm

There is a reasonable set of methods for assessing the quality of the generated map after the learning process. The most applied are the quantization error vector and the topological error (Kohonen, 2001).

The quantization error ($E_q$), Equation 4, is the average error of values of the difference between the feature vector $x_k$ and $w_{BMU}$ codevector, that is the winner in the competitive process for the vector $x_k$:

$$E_q = \frac{\sum_{k=1}^{n} \left\| x_k - w_{BMU} \right\|}{n} \quad (4)$$

The topological error ($E_t$), Equation 5, evaluates how the structure of the neural grid approximate vectors of the input space. Whereas each $x_k$ has a BMU as the first neuron in the order of competition in the neural grid, BMU2 correspond to second neuron at this scale. Thus, the error will correspond to the percentage of BMU and BMU2 whose are not neighbors in the grid for the same $x_k$:

$$E_t = \frac{1}{n}\sum_{k=1}^{n} u(x_k) \quad (5)$$

Where $u(x_k)$ is setted to one if the BMU and BMU2 are not neighbors, and zero otherwise.

## Data Visualization

After the learning process it is necessary to visually verify the result of topological ordering onto the SOM's grid. There are many ways of visual representation of the dataset using SOM (Vesanto, 1997; Vesanto, 1999). For this study, two of the most common methods were used. The first is the direct link between the artificial neuron unit and the $x_k$, in this case, the municipality.

The second form, the Component Planes, uses the values of each component of the codevector $w_j$ to color the Self-Organizing Map. This method allows the visual analysis of component correlation, outlier detection and spatial component distribution on SOM (Vesanto, 1999; Kohonen, 2000; Silva et al., 2004a).

Two periods were used and such temporal aspect was included in the SOM model by mounting these two datasets in only one, so each temporal set of values for each municipality became an entry $x_k$. This is the simplest strategy, but yet useful when the purpose in mind is a visual clustering and interpretation by component planes. According to Barreto & Araújo (2001), there is a plethora of SOM models for time processing, but the most common application of the SOM in sequence processing is the trajectory of BMU's in the latent map space (Varsta, 2002). Therefore, these two periods were compared visually by Component Planes following the trajectory of each municipality onto the neural map. To proceed this visual analysis, the last version of the SOM ToolBox was used (www.cis.hut.fi/somtoolbox/).

## Automatic Clustering

The SOM's codevectors can also be segmented by a wide range of strategies, including the visual ones (Vesanto & Alhoniemi, 2000). Costa (1999) partitioned the dataset throughout the segmentation of the image generated by the codevector's matrix of distance. However, in general, this is too complex and it does not exhibit clearly hidden patterns. Vesanto (1999) used SOM as a data compressor for a posterior statistical analysis, but this strategy needs a high level of human interaction. Costa & Andrade Netto (2003) used a conceptually simple and empiric approach (Costa-Netto algorithm) using only the information stored in codevectors after the learning process as the activity level of neurons, the quantization error and the neighborhood relation for an automatic clustering.

Considering the output layer as a non-oriented graph structure, Costa & Andrade Netto (2003) proposed a SOM clustering based on graph segmentation, in a way that the algorithm eliminates inconsistent connections among neurons, in attempt to find the ideal partition of the SOM's grid. The algorithm is based on the geometric distance information among neurons, quantization error and level of the neuron's activity. The strategy is to use heuristic rules to eliminate inconsistent connections between two neighbor neurons.

The algorithm:

1.1. Take all distances *d* between adjacent neurons *i* and *j* using the codevector $w_j$ as a measurement parameter, $d(w_i, w_j)$; and activity level for each neuron *i*, $H(i)$.
1.2. For each couple of adjacent neurons, *i* and *j*, the edge will be considered inconsistent if:
    1.2.1. the distance between codevectors exceed by 2 the median distance of the other adjacent neurons to i or j;
    1.2.2. two adjacent neurons have the level of activity ($H$) less than 50% of the allowed minimum ($H_{min}$), or some of them are inactive neurons ($H(i) = 0$); $H_{min} = \omega H_{med}$, $0.1 \leq \omega \leq 0.6$ and $H_{med} = n/m$;
    1.2.3. the distance between the centroids of the dataset associated to neurons *i* and *j* exceeds by 2 the distance between the codevectors of *i* and *j*, $d(w_i, w_j)$.
1.3. Remove all inconsistent edges.
1.4. Associate a unique code for each set of connected neurons.

At the end of the process, each group of connected neurons will represent one cluster of the codevectors. The algorithm uses some empirical values, defined by the user, but can segment the input data using only the SOM information after the learning process. To validate the clusters partition of the dataset, it was used the Compose Density between and within the clusters (CDbw) index (Halkidi & Vazirgiannis, 2002; Silva et al., 2004b; Wu & Chow, 2004). For this spatial analysis, the TerraView software (www.dpi.inpe.br/terraview) was used, as well as the TerraSOM pluging developed from the SOMCode library (www.cpatc.embrapa.br/somcode/) that implements some Self-Organizing Map related algorithms.

## Dataset

The State of Sergipe has an area of 21,910.34 km$^2$, accounting for 0.26% of national territory and 1.4% of the Northeast. Its absolute position is between the parallel 9° 31' and 11° 34' south latitude and the meridians 36° 25' and 38 ° 14' longitude West of Greenwich (Figure 2). Although it constitutes a state of small size and population, Sergipe occupies a privileged position in the economic and social development scenario of the Northeast, considering its recent entry into modern economic activities: off-shore oil exploitation, mining, tourism and ethanol production by sugar cane crops. The agriculture share in the GDP corresponds to approximately 3,0%, but it is very important in terms of employment and food supply.

Sergipe's economy is diversified. Nevertheless, it is important to highlight the mineral extractive-industry (based on non-metallic minerals) located in cities within a radius of up to 40 km from the capital, which intensely contributes to the concentration of the population, income and wealth in this area. Agricultural activities are distributed throughout the whole territory, occupying a significant area, as well as labor force, with a strong emphasis on family work. Sugar cane and orange stand out, as well as cattle and poultry. The tertiary sector has been showing changes with the addition of new functions, especially in the service sector (Teixeira et al., 2007).

*Figure 2. Municipalities of Sergipe's state*

For this study, 44 indicators (Table 1) were selected to represent demographic, social, economic, and agricultural production dimensions, as well as agricultural activities, such as land use and crop and animal production. Two periods (A and B) were considered in this study for a comparison between the 1990's and 2000's values. All variables were standardized according to their meaning and standard deviation. These variables were chosen according to their disponibility and proximity with those used by Teixeira et al. (2007).

The typology process used the institutional partition of the State of Sergipe held by the Department of Planning as a starting point, named Territory of Identity. This process takes into account quantitative indicators (Table 1), popular opinions gathered by workshops and multidisciplinary assessment of spatial aggregation (Teixeira et al., 2007). The main goal of this partition was to find out a well-stablished division of the State that could aid the process of general public policies design. Figure 3 shows the eight Territories of Identity: Central Wasteland, High Hinterland, Low San Francisco, South Center, Great Aracaju, East, Middle Hinterland and South.

*Figure 3. Territories of identity*

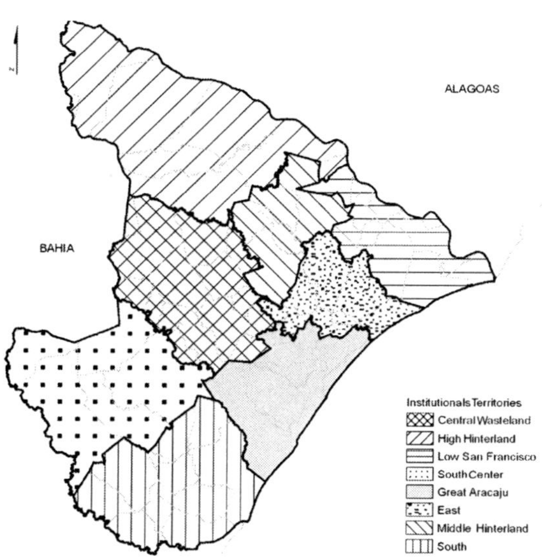


The **High Hinterland Territory** is formed by seven municipalities: Canindé do São Francisco (CSF), Grararu (GRR), Monte Alegre de Sergipe (MAS), Nossa Senhora da Glória (GLR), Nossa Senhora de Lourdes (LOU), Poço Redondo (PRD) and Porto da Folha (PFL). This territory represents 22.3% of the states's land area, covering 4908 $km^2$. In 2007, the High Hinterland Territory had a population of 137,926 inhabitants, with a density of 28.15 hab/$Km^2$, representing 7.11% of the total population of the state. In 2005, the GDP of this Territory represented 11.12% of the state's GDP, with a GDP per capita of R$ 10,730.00. The Index of Human Development in 2000 was 0.575.

The **Low San Francisco Territory** is formed by fourteen municipalities: Amparo do São Francisco (ASF), Brejo Grande (BRJ), Canhoba (CNH), Cedro de São João (CED), Ilha das Flores (IFL), Japoatã (JPT), Malhada dos Bois (MLH), Muribeca (MUR), Neópolis (NEO), Pacatuba (PAC), Propriá (PRO), Santana do São Francisco (SSF), São Francisco (SFR) and Telha (TEL). This territory corresponds to 9.0% of the state's land area, which represents 1986.3 $km^2$ of total area. In 2007, the Low San Francisco Territory had a population of 123,482 inhabitants, with a density of 63.45 hab/$Km^2$, representing 6.37% of the population of the state. In 2005, the GDP of this Territory represented 3.94% of the state's GDP, with a GDP per capita of R$ 4,176.00. The Index of Human Development in 2000 was 0.614.

The **Middle Hinterland Territory** is formed by six municipalities: Aquidabã (AQB), Cumbe (CUM), Feira Nova (FNV), Graccho Cardoso (GCR), Itabi (ITB) and Nossa Senhora das Dores (DOR). This territory represents 7.3% of the land area of the state, corresponding to 1612.6 $km^2$ of total area. In 2007, the Middle Hinterland Territory had a population of 62,644 inhabitants, with a density of 39.59 hab/$Km^2$, representing 3.23% of the population of the state. In 2005, the GDP of this Territory represented 1.57% of the state's GDP, with a GDP per capita of R$ 3,298.00. The Index of Human Development in 2000 was 0.621.

*Table 1. List of analyzed variables*

| | VARIABLE | DESCRIPTION | GROUP A | GROUP B |
|---|---|---|---|---|
| 1 | POPTOT | Total of resident population (people) | Pop. 1991[1] | Pop. 2000[2] |
| 2 | POPRUR | Total of rural resident population (people) | Pop. 1991 | Pop. 2000 |
| 3 | PRODMCOW | Milk cow production (1.000 l) | Census 1996[3] | Census 2006[4] |
| 4 | PRODMGOAT | Milk goat production (1.000 l) | Census 1996 | Census 2006 |
| 5 | PRODEGGS | Egg production (t) | Census 1996 | Census 2006 |
| 6 | TOTCOX | Cattle - Ox (unit) | Census 1996 | Census 2006 |
| 7 | TOTCS | Cattle - swine (unit) | Census 1996 | Census 2006 |
| 8 | TOTBIRDS | Birds (unit) | Census 1996 | Census 2006 |
| 9 | TRACTORS | Tractors (amount) | Census 1996 | Census 2006 |
| 10 | PAREAPERM | Permanent and temporary crops (ha) | Census 1996 | Census 2006 |
| 11 | PAREAPAST | Natural and artificial pastures (ha) | Census 1996 | Census 2006 |
| 12 | PAREAFOT | Cultivated and natural forests (ha) | Census 1996 | Census 2006 |
| 13 | PMCFSJ | Percentage of women heads of households with no spouse and with minor children (15 years old) | ATLAS 1991[5] | ATLAS 2000[5] |
| 14 | PPOOR | Percentage of poor people | ATLAS 1991 | ATLAS 2000 |
| 15 | PCI | Percentage of indigent children | ATLAS 1991 | ATLAS 2000 |
| 16 | PC714FE | Percentage of children out of school (range from 7 to 14 years old) | ATLAS 1991 | ATLAS 2000 |
| 17 | INPERCAPITA | Per capita income (R$) | ATLAS 1991 | ATLAS 2000 |
| 18 | PINGOV | Income's percentage related to governmental transferences | ATLAS 1991 | ATLAS 2000 |
| 19 | PINWORK | Income's percentage related to work's yield | ATLAS 1991 | ATLAS 2000 |
| 20 | RBRICSPOOR | Relation between 10% richest and 40% poorest | ATLAS 1991 | ATLAS 2000 |
| 21 | GINI | Gini's index | ATLAS 1991 | ATLAS 2000 |
| 22 | ITHEIL | Theil's index | ATLAS 1991 | ATLAS 2000 |
| 23 | INTIND | Indigence intensity | ATLAS 1991 | ATLAS 2000 |
| 24 | IDHMIN | Municipal Human Development Index - Income | ATLAS 1991 | ATLAS 2000 |
| 25 | IDHMLONG | Municipal Human Development Index - Longevity | ATLAS 1991 | ATLAS 2000 |
| 26 | ESPLIFE | Life expectancy at birth | ATLAS 1991 | ATLAS 2000 |
| 27 | IDHMEDUC | Municipal Human Development Index - Education | ATLAS 1991 | ATLAS 2000 |
| 28 | RILLI | Rate of illiteracy | ATLAS 1991 | ATLAS 2000 |
| 29 | GRSA | Gross rate of school attendance | ATLAS 1991 | ATLAS 2000 |
| 30 | RFECUN | Rate of fecundity | ATLAS 1991 | ATLAS 2000 |
| 31 | PPDOWN | Percentage of people living in their own and paid off house or land | ATLAS 1991 | ATLAS 2000 |
| 32 | P714A | Percentage of illiterate children (ranging from 7 to 14 years old) | ATLAS 1991 | ATLAS 2000 |
| 33 | PPWATER | Percentage of people living in houses with piped water | ATLAS 1991 | ATLAS 2000 |
| 34 | PPWC | Percentage of people living in houses with WC and piped water | ATLAS 1991 | ATLAS 2000 |
| 35 | PPGARB | Percentage of people living in houses with garbage service | ATLAS 1991 | ATLAS 2000 |
| 36 | PPENERGY | Percentage of people living in houses with electric energy | ATLAS 1991 | ATLAS 2000 |
| 37 | PPIBAGRI | Percentage of Agricultural GDP | IPEA 1996[6] | IPEA 2006[6] |

*continued on following page*

*Table 1. continued*

| | VARIABLE | DESCRIPTION | GROUP A | GROUP B |
|---|---|---|---|---|
| 38 | RTTotal | Constitutional transferences from National Treasure | MF 1996[7] | MF 2006[7] |
| 39 | MPRODMAIZE | Maize production (t) | Mean 90/96[8] | Mean 00/06[8] |
| 40 | MPRODCAS | Cassava production (t) | Mean 90/96 | Mean 00/06 |
| 41 | MPRODBEAN | Bean production (t) | Mean 90/96 | Mean 00/06 |
| 42 | MPRODSC | Sugar cane production (t) | Mean 90/96 | Mean 00/06 |
| 43 | MPRODORAN | Orange production (t) | Mean 90/96 | Mean 00/06 |
| 44 | MPRODCOCO | Coconut production (1.000 fruits) | Mean 90/96 | Mean 00/06 |

SOURCE: [1]Demographic census 1991 - IBGE. [2] Demographic census 2000 - IBGE. [3]Agricultural census 1995/1996 - IBGE. [4]Agricultural census 2006 - IBGE. [5]Atlas of Human Development - PNUD/Brasil. [6]IPEADATA (http://www.ipeadata.gov.br/). [7]Ministry of Finance - National Treasure (http://www.tesouro.fazenda.gov.br/estados_municipios/transferencias_constitucionais.asp). [8]Anual Municipal Agricultural Production - PMA/IBGE.

The **East Territory** is formed by nine municipalities: Capela (CPL), Carmópolis (CAR), Divina Pastora (DVN), General Maynard (GMN), Japaratuba (JTB), Pirambu (PIR), Rosário do Catete (RCT), Santa Rosa de Lima (SRL) and Siriri (SRR). This territory corresponds to 6.7% of the land area of the state, possessing 1474.1 km$^2$ of total area. In 2007, the territory of East Sergipe had a population of 90,452 inhabitants, with a density of 59.56 hab/Km$^2$, representing 4.66% of the population of the state. In 2005, the GDP of this Territory represented 6.47% of the state's GDP, with a GDP per capita of R$ 9,878.00. The Index of Human Development in 2000 was 0.643.

The **Central Wasteland Territory** is formed by fourteen municipalities: Areia Branca (ARB), Campo do Brito (CPB), Carira (CRR), Frei Paulo (FPA), Itabaiana (ITA), Macambira (MAC), Malhador (MLD), Moita Bonita (MOI), Nossa Senhora Aparecida (APA), Pedra Mole (PRM), Pinhão (PIN), Ribeirópolis (RIB), São Domingos (SDO) and São Miguel do Aleixo (SMA). This territory corresponds to 6.7% of the land area of the state, possessing 1474.1 km$^2$ of total area. In 2007, the Central Wasteland Territory Territory had a population of 222,197 inhabitants, with a density of 71.14 hab/Km$^2$, representing 11.46% of the population of the state. In 2005, the GDP of this Territory represented 6.70% of the state's GDP, with a GDP per capita of R$ 3,949.00. The Index of Human Development in 2000 was 0.646.

The **South Territory** is formed by eleven municipalities: Arauá (ARA), Boquim (BOQ), Cristinápolis (CRI), Estância (EST), Indiaroba (IND), Itabaianinha (ITN), Pedrinhas (PDR), Salgado (SAL), Santa Luzia do Itanhy (SLI), Tomar do Geru (TGE) and Umbaúba (UMB). This territory represents 14.5% of the land area of the state, with 3193.6 km$^2$ of total area. In 2007, the South Territory had a population of 241,192 inhabitants, with a density of 77.06 hab/Km$^2$, representing 12.44% of the population of the state. In 2005, the GDP of this Territory represented 8.40% of the state's GDP, with a GDP per capita of R$ 4,579.00. The Index of Human Development in 2000 was 0.616.

The **South Center Territory** is formed by five municipalities: Lagarto (LGT), Poço Verde (PVD), Riachão do Dantas (RDT), Simão Dias (SDI) and Tobias Barreto (TBA). This territory represents 16.1% of the land area of the state, possessing 3551.5 km$^2$ of total area. In 2007, the South Center Territory had a population of 213,492 inhabitants, with a density of 60.63 hab/Km$^2$, representing 16.07% of the population of the state. In 2005, the GDP of this Territory represented 5.67% of the state's GDP, with a GDP per capita of R$ 3,480.00. The Index of Human Development in 2000 was 0.599.

The **Great Aracaju Territory** is formed by nine municipalities: Aracaju (AJU), Barra dos Coqueiros (BCQ), Itaporanga d'Ajuda (IAJ), Laranjeiras (LRJ), Maruim (MRI), Nossa Senhora do Socorro (SOC), Riachuelo (RIA), Santo Amaro das Brotas (SAB) and São Cristóvão (SCR). This territory represents 9.9% of the land area of the state, with 2192 $km^2$ of total area. In 2007, the Great Aracaju Territory had a population of 847,941 inhabitants, with a density of 387.65 hab/$Km^2$, representing 43.72% of the population of the state. In 2005, the GDP of this Territory represented 56.12% of the state's GDP, with a GDP per capita of R$ 8,785.00. The Index of Human Development in 2000 was 0.750.

## GEOSPATIAL ANALYSIS BY SELF-ORGANIZING MAPS

The analysis of multivariate geospatial data could be pointed as one of the great scientific challenges to be solved in the present. Several institutions are providing amounts of geo-referenced data over the Internet, but there are many limitations to the effective use and interpretation of that type of dataset. In fact, the integration of all information depends on the ability to analyze the data and find relevant knowledge in such databases. As stated by Bação et al. (2005) "More than prediction tools, we need to develop exploratory tools based on classification and clustering, which would enable an improved understanding of the available data" (p. 155).

The Self-Organizing Map has been applied to the problem of geospatial analysis in many ways. Nevertheless, in general, the SOM is used for exploratory spatial analysis of census data aggregated by area. The way the analysis is performed varies, but the use of visual interpretation methods is registered in almost every application.

Winter & Hewitson (1994) applied the SOM associated with the Component Planes for the characterization of 21 districts of the Western Cape of South Africa using 10 census variables. From the Component Planes it was possible to characterize mainly the black population of the region. Kaski & Kohonen (1996) showed the feasibility of the visualization of multivariate data from the SOM. The dataset consisted of 39 indicators, which described different aspects of the welfare "states" of the countries around the world.

Koua (2003) analyzed 29 socioeconomic indicators related to municipalities. The author used the Component Planes and the matrix of distances among codevectors $w_j$ for a visual geospatial exploration and discovered that SOM improved the geographical analysis and offered support for the exploration of large geospatial datasets, increasing knowledge in this area. Silva (2004ab) performed a complete exploratory spatial analysis using eight socio-economic indicators related to census sectors. The author adjusted the algorithms related to the SOM neural network in a suite for outlier detection, spatial dependence analysis, automatic clustering and component analysis. The suite was applied in the problem of intra-urban analysis of social exclusion and was compared with statistical analysis (Genovez, 2002).

Aksoy (2006) applied the SOM to the problem of the regionalization of 923 districts in Turkey using 36 socio-economic indicators and compared the results with a statistical approach (k-means and Principal Component Analysis). Spielman & Thill (2008) studied the spatial relationship among 2,217 census tracts in New York City using a dataset of 79 attributes applying a simple method of geospatial data visualization linking the artificial neural network with a Geographical Information System (GIS).

There are some differences between the approach of this chapter and the others listed in this short review. First, in the present study, there is an explicit focus on rural aspects of territorial study while the majority of researches focus on urban problems; second, the chapter proposed a complete regional characterization (typology) using, at the same time, visual tools and an automatic clustering algorithm; finally, the work started from a previous qualitative territorial study, Territories of Identity, that makes a comparison of the results possible.

## RESULTS AND DICUSSION

### General Spatial Dependence by Visual Interpretation

After SOM's learning algorithm processing had been performed, the correspondence between each input vector (municipality-group) to its corresponding artificial neuron was realized, according to the Euclidean distance. Thus, it was possible to visualize the position in the neural map of each municipality-group and examine the proximity relations in the feature's space among them. Whereas the study has the institutional division of the state of Sergipe into Territories of Identity as its starting point, the neural map was partitioned having it as guide, which means looking for proximities among municipalities of the same Territory of Identity.

From the visual analysis of the distribution of the cities on the neural map it was found that there is spatial dependence for most of the input vectors. That is, neighboring municipalities in geophysical space are also deeply related in the space of attributes (neural map). From this observation it was possible to trace thresholds among the regions of the neural map that corresponded to most municipalities-group of each area previously defined (Territories of Identity). Therefore, Figure 4a illustrates how the Territories of Identity are distributed in the neural map. It was also observed that there is spatial dependence among these territories.

Figure 4b shows the distribution of some municipalities in the neural map so that it is possible to analyze two factors: discrepancies between the positions of each municipality according to data from two periods (A and B), and the inter-territorial shifting of municipalities in the space of attributes, so that it is possible to analyze the link between the territories constructed by the neural map and the Territories of Identity. It is important to note that when significant differences between the groups A and B for the same city are not present, it will be only represented by the symbol of the city. Otherwise, the municipal-

*Figure 4. (a) SOM after the visual partition. (b) Linking between each artificial neuron and the municipality associated to it. The number of municipalities represented onto the neural map was limited for legibility reasons*

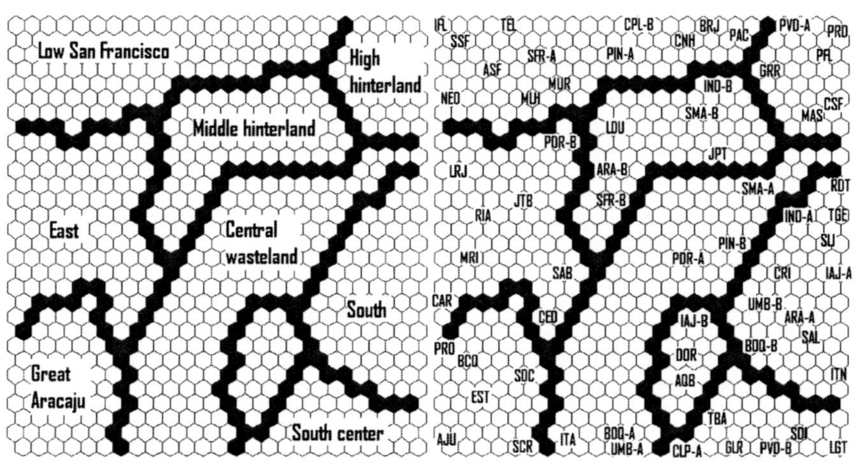

ity will be represented by its initials followed by a letter indicating the group (A or B).

From Figure 4b, it is possible to identify the local displacement between periods A and B. The municipality of Capela (CPL) migrated from the region linked to the South Central region to the Lower San Francisco Territory, approaching the municipalities of the East. The municipality of Poço Verde (PVD) migrated from the area linked to the High Hinterland to the area linked to the South Center. The municipality of São Miguel do Aleixo (SMA) migrated from the region associated with the Central Wasteland to the Middle Hinterland. The municipalities of Umbaúba (UMB) and Boquim (BOQ) migrated from the Central Wasteland to the South region. The municipality of Itaporanga d'Ajuda (IAJ) migrated from the South region to the region nearby the Central Wasteland, into the direction of the Great Aracaju. The municipalities of Indiaroba (IND), Pedrinhas (PDR) and Arauá (ARA) scattered from the area linked to the South, contrary to the behavior of BOQ and UMB.

The analysis of temporal shifts between groups A and B did not show significant changes for most municipalities. The Southern region showed greater variation between the two periods in the sense of closeness between municipalities (BOQ and UMB) and distance (IND, PDR, NIV). The municipalities of CPL and PVD had the highest displacements. The municipality of São Francisco (SFR) migrated from the Low San Francisco to the region associated to the Middle Hinterland.

Only half the regions (that represent the territories) formed by the neural map have at least one of its Centralities[1] not positioned in the region of concentration of the majority of the municipalities of the same Territory of Identity. The municipality of Estancia (EST), for example, belongs to the Southern Territory but is extremely close to the Great Aracaju Territory on the neural map. The municipalities of Aquidabã (AQB) and Nossa Senhora das Dores (DOR) are far away in the feature space of the other municipalities of the Middle Hinterland Territory. Canindé do São Francisco (CSF) is a centrality with characteristics similar to those of most municipalities in the High Hinterland Territory, which does not occur with the municipality of Nossa Senhora da Glória (GLR), that stands in the region of the South Central Territory. Another centrality that is positioned distantly from the other municipalities of the same Territory in the feature space is Propriá (PRO), which also has similar characteristics to the municipalities of the Great Aracaju. All the centralities are shown in Figure 3b.

The same occurs with other municipalities that are not Centralities. The municipality of Riachão do Dantas (RDT) belongs to the South Central Territory, but has features related to the South Territory. The municipality of Nossa Senhora de Lourdes (LOU) belongs to the High Hinterland Territory, but has characteristics more intimately related to the Middle Hinterland Territory. The municipalities of Santo Amaro das Brotas (SAB), Maruim (MRI), Riachuelo (RIA) and Laranjeiras (LRJ) are associated to the Great Aracaju Territory, in spite of presenting characteristics associated with the municipalities of the East Territory.

It was observed that there are some differences between the groups formed by the neural map and the Territories of Identity, although, in general, the neural approach confirms the cohesion intra-Territories in the feature space.

## Characterization by Component Planes

Through Components Planes (Figure 5) it was possible to characterize each group defined by neural map (that represents each territory) through the observation of the spatial distribution of each component intensity in the neural map. As each municipality is associated with only one artificial neuron it is possible to use the partition of the neural map to facilitate the characterization of the territories.

*Figure 5. Component planes for each variable*

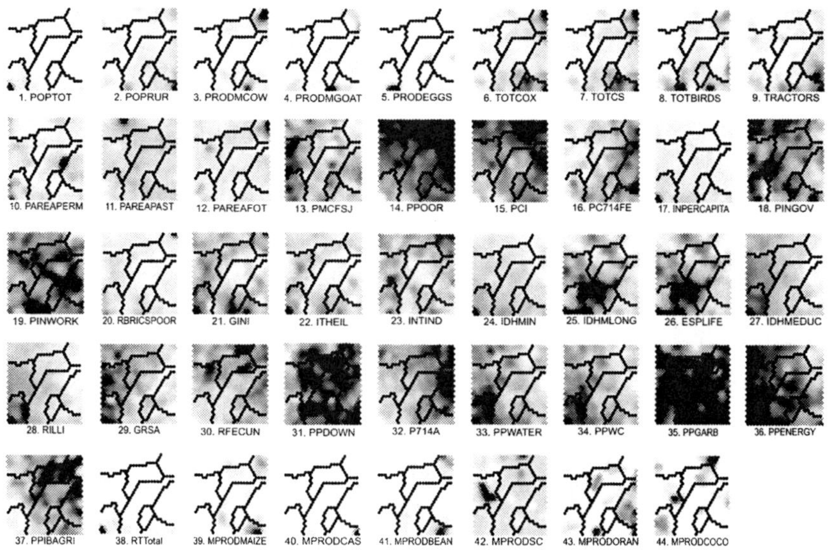


First, it was observed that there were outliers for some variables. Aracaju (AJU) was differentiated from the others with high values for the variables POPTOT, INPERCAPITA, ITHEIL, IDHMIN and RTTOTAL. The municipality of Lagarto (LGT) has high scores for the variables POPRUR and MPRODCAS. The High Hinterland and South Center territories presents high values for PRODMCOW, PRODMGOAT, TOTOCOX, TOTBIRDS, TOTCS, MPRODMAIZE and MPRODBEAN. Outliers were also detected for PRODEGGS, PAREAPAST, PAREAFOT, RBRICSPOOR variables. Correlation was also observed, same patterns for the Component Plane coloring, between variables PPOOR and PCI, IDHMLONG and ESPLIFE, IDHMEDUC and RILLI and PPWATER and PPWC.

The area which borders the High Hinterland region on the neural map was characterized by high values for the variables TOTCOX, PPOOR, PCI, INTIND, PPDOWN, P714A, PPGARB, PPIBAGRI and MPRODSC, average values for the variables PC714FE and GINI, and low values for the variables PAREAPERM, PMCFSJ, PINGOV, IDHMLONG, ESPLIFE, IDHMEDUC, RILLI, GRSA, PPWATER, PPWC, PPENERGY, MPRODORAN and MPRODCOCO.

The area which borders the Lower San Francisco region was characterized by high values for the variables PPOOR, PCI, PINGOV, RFECUN and PPENERGY, average values for the variables PMCFSJ, PINWORK, GINI, INTIND, IDHMEDUC, RILLI, GRSA, PPDOWN, P714A, PPWATER, PPWC and PPIBAGRI, and low values for the variables TOTCOX, TOTBIRDS, TRACTORS, PAREAPERM, PC714FE, IDHMLONG, ESPLIFE, PPGARB, MPRODSC, MPRODORAN and MPRODCOCO.

The area which borders the Middle hinterland region was characterized by high values for the variables PPOOR, PCI, PINWORK, PPDOWN, PPGARB and PPIBAGRI, average values for the variables PINGOV, GRSA, RFECUN, P714A and PPENERGY, and low values for the variables TOTCOX, TRACTORS, PAREAPERM, PMCFSJ, PC714FE, GINI, INTIND, IDHMLONG, ESPLIFE, IDHMEDUC, RILLI, PPWATER, PPWC, MPRODSC, MPRODORAN and MPRODCOCO.

The area which borders the East region on neural map was characterized by high values for the variables PMCFSJ, PPOOR, PCI, PPGARB, PPENERGY and MPRODSC, average values for the variables IDHMLONG, ESPLIFE, IDHMEDUC, RILLI, GRSA, RFECUN, PPDOWN, PPWATER and PPWC and low values for the variables TOTCOX, TRACTORS, PAREAPERM, PC714FE, P714A, PPIBAGRI, MPRODORAN and MPRODCOCO.

The area which borders the Central Wasteland region was characterized by high values for the variables IDHMLONG, ESPLIFE, PPDOWN, PPGARB and PPENERGY, average values for the variables PPOOR, PCI, PC714FE, PINGOV, PINWORK and PPIBAGRI, and low values for the variables TOTCOX, TRACTORS, PAREAPERM, PMCFSJ, GINI, INTIND, IDHMEDUC, RILLI, GRSA, RFECUN, P714A, PPWATER, PPWC, MPRODORAN and MPRODCOCO.

The area which borders the South region on the neural map was characterized by high values for the variables PPOOR, PCI, PC714FE, PINWORK, PPDOWN, P714A, PPGARB, PPIBAGRI, MPRODORAN, average values for the variables TRACTORS, PMCFSJ, GINI, RFECUN, and low values for the variables TOTCOX, PAREAPERM, PINGOV, INTIND, IDHMLONG, ESPLIFE, IDHMEDUC, RILLI, GRSA, PPWATER, PPWC, PPENERGY, and MPRODCOCO.

The area which borders the South Central region was characterized by high values for the variables TOTCOX, PPDOWN, and PPGARB, average values for the variables TRACTORS, PPOOR, PCI, PC714FE, PINGOV, PINWORK, GINI, INTIND, RFECUN, P714A, PPWATER, PPWC, PPENERGY and PPIBAGRI, and low values for the variables PAREAPERM, 13.IDHMLONG, ESPLIFE, IDHMEDUC, RILLI, GRSA, MPRODORAN and MPRODCOCO.

The area which borders the Great Aracaju region on neural map is characterized by high values for the variables PINGOV, GINI, IDHMEDUC, RILLI, GRSA, PPWATER, PPWC, PPGARB and PPENERGY, average values for the variables PAREAPERM, PMCFSJ, PC714FE, PINWORK, INTIND, IDHMLONG, ESPLIFE and MPRODCOCO and low values for variables TOTCOX, TOTBIRDS, TRACTORS, PPOOR, PCI, RFECUN, PPDOWN, P714A and PPIBAGRI.

It was observed that the municipalities of the High Hinterland region form a more homogeneous, cohesive and well defined region, while the Central Wasteland Territory, as well as the Low San Francisco, have some variation degree which could imply the existence of subdivisions. This suggests that policy makers should pay attention to internal territorial differences and particularities and decide according to the best solution of each subdivision. Other neural map territories have shown good degree of cohesion by the Component Planes analysis.

## Automatic Clustering by TerraSOM

From the distribution of the municipalities on the neural map and the variables analysis by the Component Planes it was possible to propose a new arrangement of the Territories of Identity. Although the neural map regionalization by visual interpretation was not conclusive, it was possible to proceed with a third clustering by the Costa-Netto algorithm to improve the amount of information about the aggregation of Sergipe's municipalities and validate the previous results.

Figure 6 shows the regionalization of the state of Sergipe by a one-dimensional SOM, with 10 neurons and linearly initiated. Several network sizes were tested, but basically the same groups were found, varying not more than in terms of aggregation or subdivision of each one. The automatic clustering algorithm found 6 groups, as shown in Figure 6. Important to note that there was no contiguity restriction on the clustering algorithm, so non-neighboring municipalities clusters were allowed.

The automatic regionalization confirms the previous results. The municipality of Riachão do Dantas belongs to the group 5, the same group of municipalities in the South Territory (Tomar do

*Figure 6. Automatic regionalization by the TerraSOMplugin without neighborhood constraints*

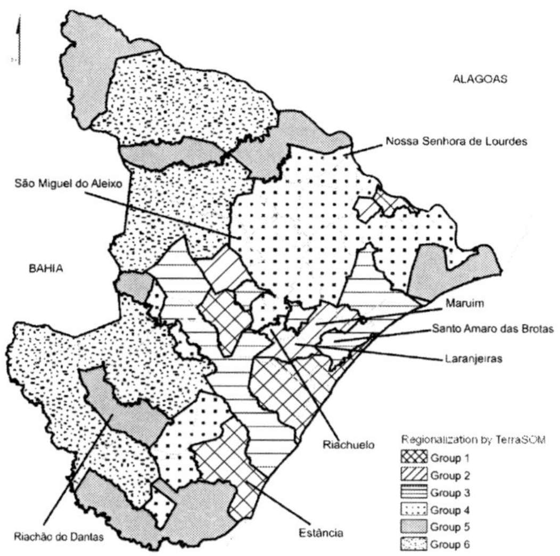


Geru, Indiaroba, Santa Luzia do Itanhy and Cristinápolis). The municipality of Estancia belongs to the group 1, the same group of the municipalities of the Great Aracaju Territory. Also note the distinction between these municipalities in the groups 2 and 3, which belong to the Grand Aracaju Territory. Nevertheless, according to the SOM network, they belong to the East Territory (Laranjeiras, Riachuelo, Maruim and Santo Amaro das Brotas).

It was also confirmed by the automatic regionalization that the municipalities of Nossa Senhora de Lourdes and São Miguel do Aleixo were closer in the feature space with the Middle Hinterland Territory. It was observed that the group 4 congregated municipalities from three Territories of Identity (Lower San Francisco, Central wasteland and Middle Hinterland), which demonstrates the difficulty of separating these municipalities.

## The Final Regional Partition

After the visual and automatic clustering and interpretation of the neural map a new regionalization of the state of Sergipe was proposed (Figure 7). It was taken into account the contiguity among municipalities during the process of construction of the new partition. The consistency of the results of the three forms of the SOM analysis improved the consistency of the proposal.

Figure 7 shows the final configuration of the state of Sergipe partition based on the forty-four variables. The municipality of Riachão do Dantas migrated from the South Central Territory to the South Territory, while Estância migrated from the Southern Territory to the Great Aracaju Territory. The municipalities of Laranjeiras, Maruim, Santo Amaro das Brotas and Riachuelo migrated from the Great Aracaju Territory to the Eastern Territory. The municipality of São Miguel do Aleixo migrated from the Central Wasteland Territory to the Middle Hinterland Territory. And finally, the city of Nossa Senhora de Lourdes migrated from the High Hinterland Territory to the Middle Hinterland Territory.

This final regionalization proposal is not intended to replace the official one, but rather demonstrates the ability to undertake the territorialization from an exploratory spatial analysis by a neural network. Teixeira et al. (2007) evidenced

*Figure 7. Proposed regionalization by SOM algorithm*

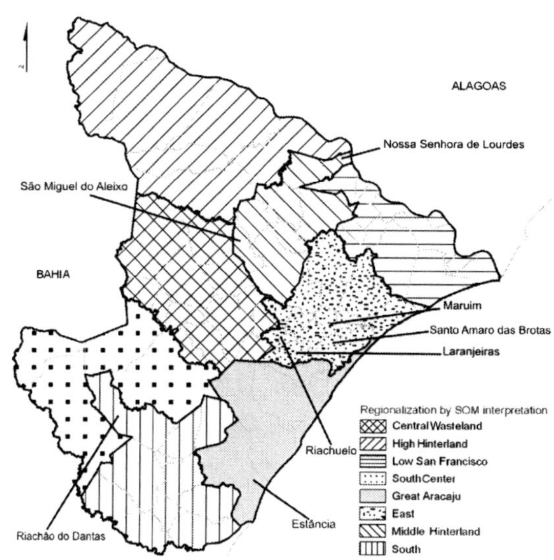

that several key criteria to describe the association to one or another territory is also intimately related to group and personal abstract values, e.g., Nossa Senhora de Lourdes (LOU) and São Miguel do Aleixo (SMA) are quantitatively within the reach of the Middle Hinterland, but the participative process resulted in a different association between these municipalities and the Territories of Identity.

Estância (EST), Riachão do Dantas (RDT) and Itaporanga D'Ajuda (IAJ) are vast municipalities, and the existence of concentrated activities could lead to misclassification. However, considering only the forty-four variables mentioned in this study, it was concluded that Estância is more related to the Great Aracaju Territory, once this municipality has a great industrial GDP and a low agricultural production. The geographical space of the RDT municipality goes from the Wasteland to the Hinterland, despite of the indication of a deeper relation to the South Territory, especially to the municipalities of the extreme south, as it can be observed in Figure 5. Maruim (MRI), Laranjeiras (LRJ), Santo Amaro das Brotas (SAB) and Riachuelo (RIA) present features nearer the East Territory than the Great Aracaju. The sugar cane activity, the governmental transferences (mainly the royaties), and the way of social organization may approximate this group to municipalities of the East Territory.

## FUTURE RESEARCH DIRECTIONS

The SOM acts as an extractor of features without reducing the size of the feature vector, i.e., without significant loss of statistical information on geospatial dataset. This allows the use of statistical methods associated with the neural network, especially in the post-processing of the codevectors $w$. Hierarchical clustering methods or algorithms based on vector quantization can be applied to the clustering process of the codevectors. Depending on the relationship between the number of input vectors and the size of the feature vector, the application of the Principal Component Analysis (PCA) could be necessary for a dimensionality reduction.

There is also the possibility to add new components in the input dataset so that new factors could be considered in the process of regionalization. It is possible, for example, to include an explicit component of spatial location of the municipalities into the input vector or into the learning process of the SOM algorithm for a spatial constraint. Numerical and categorical data fusion can be performed for a better representation of different dimensions, both environmental and economical. Additional temporal information can be added for a trend analysis using the SOM's projection algorithms.

As an interdisciplinary study a lot of work is necessary to join distinct concepts and actions from many disciplines, such as economy, sociology, geography and agronomy towards rural territorial typology for a sustainable development. However, this integrative and systemic approach is still a challenger for researchers and policy makers.

## CONCLUSION

The study of the sustainable development of rural areas depends on the growth of integrated and multidisciplinary studies. The regional approach has been one of the methods to design projects of development and regional research in order to consider social, economic, geophysical and cultural dimensions. Population growth, and hence the demand for quality public services and projects for regional infrastructure, require from public managers greater flexibility in decision-making, so that they observe the requirements of fairness, efficiency and effectiveness. Territorial typology certainly provides important feedback for the development of a collective solution to issues relating to rural development.

The large volume of data already available and their differences in origin and format imposes research challenges, requiring new methods of data analysis, especially those related to data mining. The possibility of visualization of multivariate data combined with the ability to use automatic clustering algorithms turned the neural network SOM into an important tool in the process of territorial typology.

Considering forty-four heterogeneous variables the Territory of Identity presented by the dataset distribution on artificial neural map intra and inter spatial dependence and low level of inter-territorial shifting of municipalities (only 12%), however, the analysis by Component Planes have shown some subdivision intra-territories showing that policy makers must be careful about simple and homogeneous solutions for each Territory. The automatic regionalization by the SOM algorithm integrated with the Costa-Netto partition strategy associated with the CDbw validity cluster index confirmed that there are significant differences intra-territories and that there are no-neighbor municipalities with similar characteristics (for example, Riachão do Dantas, Canindé do São Francisco and Brejo Grande). The automatic clustering has also shown that some Centralities (Itabaliana, Estância and Propriá) are closer to the Great Aracaju Territory and that some municipalities of the Great Aracaju Territory could be associated to another group or to the East Territory.

Based on the Territories of Identity and on the results of the SOM algorithm was constructed a regionalization for rural studies purposes. It can help the general understanding about the Sergipes's rural actual scenario and aid the process of regional planning. This is particularly important for regions where agriculture is a major employer of labor, family labor and, therefore, important source of income for the farming populations.

## REFERENCES

Abramovay, R. (2006). Para una teoría de los estudios territoriales. In Manzanal, M., Neiman, G., & Lattuada, M. (Eds.), *Desarrollo rural – Organizaciones, instituciones y territorios* (pp. 51–70). Buenos Aires, Argentina: Ediciones Ciccus.

Aksoy, E. (2006, October) *Clustering with GIS: an attempt to classify turkish district data*. Paper presented at the XXIII FIG Congress. Munich, Germany.

Bação, F., Lobo, V., & Painho, M. (2005). The self-organizing map. the Geo-SOM, and relevant variants for geosciences. *Computers & Geosciences, 31*, 155–163. doi:10.1016/j.cageo.2004.06.013

Bailey, T. C., & Gatrell, A. C. (1995). *Interactive Spatial Data Analysis*. Longman.

Bandeira, P. S. (2006). Institucionalização de regiões no Brasil. *Revista da Sociedade Brasileira para o Progresso da Ciência, 58*(1), 34-37.

Barreto, G. A., & Araújo, A. F. R. (2001). Time in self-organizing maps: An overview of models. *International Journal of Computer Research, 10*(2), 139–179.

Claval, P. (2008). Espaces et territoire: les bifurcations de la science régionale. *Géographie, Économie. Société, 10*, 157–184.

Costa, J. A. F. (1999) *Classificação automática e análise de dados por redes neurais auto-organizáveis*. Unpublished doctoral dissertation, University of Campinas, Campinas, São Paulo.

Costa, J. A. F., & Andrade Netto, M. L. (2003). *Segmentação do SOM Baseada em Particionamento de Grafos*. Paper presented at the VI Brazilian Congress on Neural Networks, São Paulo, (pp. 451-456).

Flexer, A. (2001). On the use of self-organizing maps for clustering and visualization. *Intelligent Data Analysis*, *5*, 373–384.

Flores, M. (2004). Desenvolvimento territorial rural: uma proposta de estudo para apoio à formulação de políticas públicas. In Lages, et al (Ed.), *Territórios em movimento: cultura e identidade como estratégia de inserção competitiva*. Brasília, DF: SEBRAE.

Genovez, P. (2002). *Território e Desigualdades: análise espacial intra-urbana no estudo da dinâmica de exclusão/inclusão social no espaço urbano em São José dos Campos-SP*. Unpublished master thesis, National Institute for Space Research, São José dos Campos, São Paulo, Brazil.

Haese, K., & Goodhill, G. J. (2001). Auto-SOM: Recursive Parameter Estimation for Guidance of Self-Organizing Feature Maps. *Neural Computation*, *13*, 595–619. doi:10.1162/089976601300014475

Halkidi, M., & Vazirgiannis, M. (2002). *Clustering validity assessment using multi representatives*. Paper presented at the SETN Conference, Thessaloniki, Greece.

Kaski, S., & Kohonen, T. (1996). Exploratory data analysis by the self-organizing map: structures of welfare and poverty in the world. In Refenes, A.-P. N., Abu-Mostafa, Y., Moody, J., & Weigend, A. (Ed.), *Third International Conference on Neural Networks in the Capital Markets* (pp. 498-507). Singapore: World Scientific.

Kohonen, T. (2001). *Self-organizing maps*. Berlin: Springer.

Koua, E. L. (2003). *Using self-organizing maps for information visualization and knowledge discovery in complex geospatial datasets*. Paper presented at the XXI International Cartographic Conference (ICC), (pp. 1694-1702).

Ministério do Desenvolvimento Agrário – MDA. (2005). *Referências para a gestão social de territórios rurais*. Brasília, DF: MDA. Série Documentos nº 3.

Openshaw, S., & Turton, I. (1996). A parallel Kohonen algorithm for the classification of large spatial datasets. *Computers & Geosciences*, *22*(9), 1019–1026. doi:10.1016/S0098-3004(96)00040-4

Sabourin, E., & Teixeira, O. A. (2002). *Planejamento e desenvolvimento dos territórios rurais*. Brasília: Embrapa Informação Tecnológica.

Silva, M. A. S., Monteiro, A. M. V., & de Medeiros, J. S. (2004a). *Visualization of geospatial data by component planes and u-matrix*. Paper presented at the VI *Brazilian Symposium on GeoInformatics*, Campos do Jordão, Brazil.

Silva, M. A. S., Monteiro, A. M. V., & de Medeiros, J. S. (2004b). *Semi-Automatic Geospatial Data Clustering by Self-Organizing Maps*. Paper presented at the VI Brazilian Symposium on Neural Networks, 2004, São Luis, MA.

Spielman, S. E., & Thill, J.-C. (2008). Social area analysis, data mining, and GIS. *Computers, Environment and Urban Systems*, *32*, 110–122. doi:10.1016/j.compenvurbsys.2007.11.004

Teixeira, O. A., Melo, R. L. de & França, V. L. A. (2007). *Estado de Sergipe: uma proposta de territorialização para o planejamento*. SEPLAN Report.

Varsta, M. (2002). *Self-organizing maps in sequence processing*. Unpublished doctoral dissertation, Helsinki University of Technology. Espoo, Finland.

Vesanto, J. (1997). *Data Mining Techniques Based on the Self-Organizing Map*. Unpublished master thesis, Helsinki University of Technology.

Vesanto, J. (1999). SOM based data visualization methods. *Intelligent Data Analysis*, *3*, 111–126. doi:10.1016/S1088-467X(99)00013-X

Vesanto, J., & Alhoniemi, E. (2000). Clustering of the Self-Organizing Map. *IEEE Transactions on Neural Networks*, *11*(3), 586–600. doi:10.1109/72.846731

Winter, K., & Hewitson, B. (1994). Selforganizing maps - applications to census data. In Hewitson, B., & Crane, R. (Eds.), *Neural nets: applications in geography*. Amsterdam: Kluwer.

Wu, S., & Chow, T. W. (2004). Clustering of the self-organizing map using a clustering validity index based on inter-cluster and intra-cluster density. *Pattern Recognition*, *37*, 175–188. doi:10.1016/S0031-3203(03)00237-1

## ADDITIONAL READING

Babu, G. P. (1997). Self-organizing neural networks for spatial data. *Pattern Recognition Letters*, *18*, 133–142. doi:10.1016/S0167-8655(97)00003-2

Bishop, C. (1995). *Neural Networks for Pattern Recognition*. Oxford University Press.

Cai, Y. (1995). Artificial neural-network method for soil-erosion forecasting. *Bodenkultur*, *46*(1), 19–24.

Cereghino, R., Giraudel, J., & Compin, A. (2001). Spatial analysis of stream invertebrates distribution in the Adour-Garonne drainage basin (France), using Kohonen self organizing maps. *Ecological Modelling*, *146*, 1–3, 167–180. doi:10.1016/S0304-3800(01)00304-0

Couclelis, H. (1998). Geocomputation in context. In Longley, P., Brooks, S., McDonnell, R., & Macmillan, B. (Eds.), *Geocomputation: a primer*. John Wiley and Sons.

Demartines, P., & Blayo, F. (1992). Kohonen self-organizing maps: is the normalization necessary? *Complex Systems*, *6*, 105–123.

Erwin, E., Obermayer, K. K., & Schulter, K. (1992). Self-Organizing maps: Stationary states, metastability and convergence rate. *Biological Cybernetics*, *67*, 35–45. doi:10.1007/BF00201800

Fausett, L. (1994). *Fundamentals Neural Networks: Architectures, algorithms, and applications*. Prentice Hall.

Fischer, M., & Getis, A. (1996). *Recent developments in spatial analysis*. Springer.

Foody, G. (1999). Applications of the self-organising feature map neural network in community data analysis. *Ecological Modelling*, *120*, 97–107. doi:10.1016/S0304-3800(99)00094-0

Gahegan, M. (1999). What is Geocomputation? *Transactions in GIS*, *3*, 203–206. doi:10.1111/1467-9671.00017

Gahegan, M., Takatsuka, M., Wheeler, M., & Hardisty, H. (2002). Introducing GeoVISTA Studio: an integrated suite of visualization and computational methods for exploration and knowledge construction in geography. *Computers, Environment and Urban Systems*, *26*, 267–292. doi:10.1016/S0198-9715(01)00046-1

Haese, K. (1998). Self-Organizing Feature Maps with Self-Adjusting Learning Parameters. *IEEE Transactions on Neural Networks*, *9*, 1270–1278. doi:10.1109/72.728376

Hewitson, B., & Crane, R. (1994). *Neural nets: applications in geography*. Kluwer.

Ji, C. Y. (2000). Land-use classification of remotely sensed data using self-organizing feature mapa neural networks. *Photogrammetric Engineering and Remote Sensing*, *66*(12), 1451–1460.

Kaski, S., Kangas, J., & Kohonen, T. (1998). Bibliography of Self-Organizing Map (SOM) Papers: 1981–1997. *Neural Computing Surveys*, *1*, 102–350.

Koga, D. (2003). *Medidas de Cidades: entre territórios de vida e territórios vividos*. Cortez.

Kohonen, T. (1990). The Self-organizing maps. *Proceedings of the IEEE*, *78*(9), 1464–1480. doi:10.1109/5.58325

Kropp, J. (1998). A neural network approach to the analysis of city systems. *Applied Geography (Sevenoaks, England)*, *18*(1), 83–96. doi:10.1016/S0143-6228(97)00048-9

Lo, Z., Yu, Y., & Bavarian, B. (1993). Analysis of the convergence properties of topology preserving neural networks. *IEEE Transactions on Neural Networks*, *4*, 207–220. doi:10.1109/72.207609

Muñoz, A., & Muruzábal, J. (1998). Self-organizing maps for outlier detection. *Neurocomputing*, *18*, 33–60. doi:10.1016/S0925-2312(97)00068-4

Oja, M., Kaski, S., & Kohonen, T. (2003). Bibliography of Self-Organizing Map (SOM) Papers: 1998-2001 Addendum. *Neural Computing Surveys*, *3*, 1–156.

Openshaw, S., & Abrahart, S. (2000). *Geocomputation*. Taylor & Francis.

Openshaw, S., & Openshaw, C. (1997). *Artificial intelligence in geography*. John Wiley.

Purvis, M., Zhou, Q., Cranefield, S., Ward, R., Raykov, R., & Jessberger, D. (2001). Spatial information modelling and analysis in a distributed environment. *Ecological Modelling & Software*, *16*, 439–445. doi:10.1016/S1364-8152(01)00014-7

## KEY TERMS AND DEFINITIONS

**Artificial Neural Network (ANN):** ANN is a computational model that simulates functional aspects of biological neural networks. It consists of an interconnected set of artificial neurons and processes information using a parallel approach to computation. ANN is usually used to find patterns in datasets.

**Best Match Unit (BMU):** BMU is the neuron j closer to some input $x_k$ than other neurons. It can also be called winner neuron.

**Codevectors:** Codevectors are the set of $w_j$, $j=1,2,\ldots,m$, $m$ is the number of neurons.

**Learning Algorithm:** Algorithm that allows computers to evolve behaviors based on empirical data, such as from sensor data or databases.

**Neural Topology:** Neural topology refers to the layout of connected neurons.

**Quantization:** Quantization is the process of mapping a large set of possible discrete values (or a continuous range of values) by a relatively small and finite set of discrete values which is also on that original interval.

**Voronoi Region:** Voronoi region is a kind of a metric space determined by Euclidean distances to a specified discrete set of objects (codevectors $w_j$). Each Voronoi region is formed by all input vectors $x_k$ closer to each reference object ($w_j$) than any other codevector.

## ENDNOTE

1 Municipalities that concentrate services and opportunities for the rest of the same Territory of Identity.

# Chapter 8
# Zoning Based on Climate and Soil for Planting Eucalyptus in Southern Region of Rio Grande do Sul State, Brazil

**José Maria Filippini Alba**
*Embrapa Temperate Climate Research Center, Brazil.*

**Marcos Silvera Wrege**
*Embrapa Forest Research Center, Brazil.*

**Carlos Alberto Flores**
*Embrapa Temperate Climate Research Center, Brazil.*

**Marilice Cordeiro Garrastazu**
*Embrapa Forest Research Center, Brazil.*

## ABSTRACT

*Zoning based on climate and soil characteristics does not represent a full ecological-economic zoning procedure, as the Brazilian law establishes, but it is a positive step forward toward a sustainable use of natural resources. The Embrapa Temperate Climate Research Center has developed, with the collaboration of several entities from Brazil and Uruguay, a zoning procedure based on climate and soil for planting eucalyptus in southern region of Rio Grande do Sul State covering a total area greater than 3,5 million hectares. Three eucalyptus species were considered: Eucalyptus grandis W. Hill ex Maiden, Eucalyptus dunnii Maiden and Eucalyptus globulus Labill. Data were processed and integrated through statistical procedures and by using a GIS. The potential surface area for forestry surpasses 900,000 hectares, but a homogeneous exploitation of that territory is limited by legal restrictions.*

DOI: 10.4018/978-1-61692-871-1.ch008

## INTRODUCTION

The Brazilian legislation that first established the country's official National Environmental Policy was Law 6938, signed into effect in 1981. It specifies environmental zoning as one of the instruments officially sanctioned to implement this national policy. Regulation of specific aspects of Law 6938 came to fruition 20 years later when ecological-economic zoning was detailed in 2002 by Ordinance Number 4297. This ordinance indicates the representation of the real world through a geographic information system (GIS), with minimum scale of 1:250000, including several information levels, as cartography, economics, geology, geomorphology, hydrology, meteorology, sociology, remote sensing and land use and performing analysis on these data by constructing models and simulating scenarios.

While zoning based on climate and soil characteristics does not represent a full ecological-economic zoning procedure, it is a positive step forward toward planning the sustainable use of natural resources. The Embrapa Temperate Climate Research Center has developed a zoning procedure based on climate and soil for planting eucalyptus in southern region of Rio Grande do Sul State covering a total area of 3,714,096 hectares (Figure 1). The region has a temperate to subtropical climate with elevation ranging from sea level to 400 meters. Prairies are the prevailing landscape, but rocky fields occur in the western sector, and wetlands together with sandy terrain appear in the eastern sector (coastal region).

Financial support for the project was provided by Foundation for Support of Research, State of Rio Grande do Sul. Brazilian and Uruguayan institutions provided meteorological data and productivity data for the three eucalyptus species (*Eucalyptus grandis* W. Hill ex Maiden, *Eucalyptus dunnii* Maiden and *Eucalyptus globulus* Labill) commonly planted in this region. The results of the project have been previously published (Flores, Filippini Alba & Wrege, 2009) and are summarized in this present chapter.

*Eucalyptus spp.* originated in Australia and on several islands of Oceania. More than 600 species and subspecies have been identified. One of the main characteristics of eucalypts is their ability to adapt to different environments. Some species occur in arid regions whereas others do well in cold and humid conditions, meaning eucalypts

*Figure 1. Location of the study region (red) within contexts of South America, Brazil and Rio Grande do Sul State (gray) and geographic details study area (yellow) including hydrographic network (blue lines), and municipal demarcations (gray lines on yellow background)*

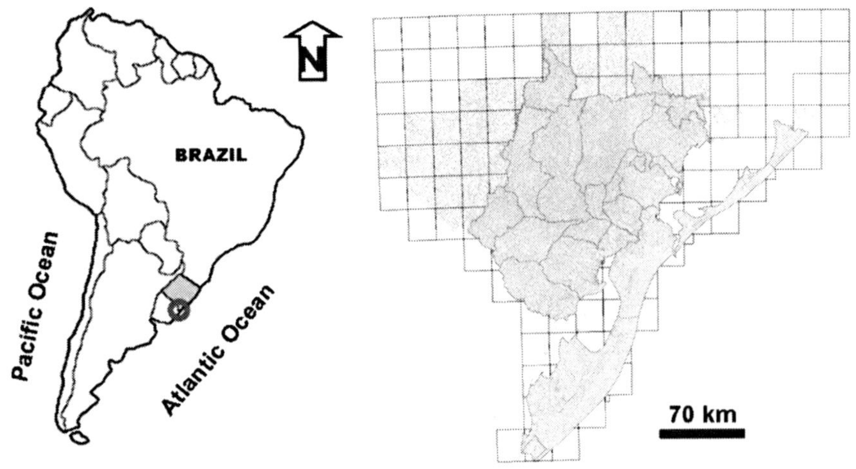

are included in all plant formations of Oceania. The name "eucalyptus" is derived from the conjunction of two Greek words and means "true coverage" due to the magnificent expansiveness of trees that may grow to a height of one hundred meters (Pryor, 1976). *Eucalyptus grandis* has a large volumetric annual increment when grown in favorable conditions of climate and soil, meaning geographic regions with low occurrence of frost and drought (Golfari et al., 1978; Poyton, 1979; Elridge et al., 1994). While *Eucalyptus dunnii* has lesser volumetric increment, it is more tolerant to frost. *Eucalyptus globules*, on the other hand, performs well in some regions for industrial purposes, but its geographic range is restricted because of intolerance to both high temperatures that facilitate disease and frost.

The territory subject to zoning was classified according to characteristics of soils, geomorphology and climate. Data were first processed and integrated through statistical procedures and by using a GIS (Davis, 1986; Bonham-Carter, 1994). Based on information from that initial procedure, the most suitable locations for development of *eucalyptus grandis*, *eucalyptus dunnii* and *ecualyptus globulus* were selected through multicriteria methods using Boolean operations.

## BACKGROUND: GIS AS ZONING STRATEGY

The term "Zoning" can be regarded under different points of view. For instance, in Mineralogy and Geology, when a crystal shows bands or rings of different colors or when characteristic zones occur in a mineral deposit, respectively (Augustithis, 1995). Since the perspective of this chapter, territories can be divided in different zones of land use, what allow a rational exploration of the natural resources. In that sense, the Ministry of Agriculture, Livestock and Supply (Brazil) has created the Program of Guarantee for Agriculture, which develops zoning for the climatic suitability of some crops, especially fruits and grains, according to technical-scientific criteria (Cunha; Assad, 2001). Lately, the Federal Government of Brazil has appointed the Ecological – Economic Zoning as an instrument of environmental planning and sustainability for the development of the States and the Macroregions of Brazil (Brasil, 2001).

Some years ago the zoning was done by superimposing analogic information (Mota et al., 1974). Nowadays, the integration of digital data can be performed by a geographic information system (GIS). Anyway, zoning and GIS are connected by two common aspects: spatial analysis and levels of information. GIS technology was developed in the last two decades and theory, software and application are really advanced nowadays (Burrough, 1986; Aronof, 1991; Bonham-Carter, 1994).

## MATERIAL AND METHODS

Various institutions provided data for this study which included the following subject areas: cartography, topography, hydrology, meteorology and pedology (Table 1). All were organized as levels of information within the GIS (Environmental Systems Research Institute, 1999) at the Laboratory of Environmental Planning, Embrapa Temperate Climate Research Center, Rio Grande do Sul State, Brazil. Data processing included regression models, interpolation, classification and multicriteria selection.

The GIS digital cartographic base was used in scale of 1:250000 for Rio Grande do Sul State (Weber & Hasenack, 2006) and 1:50000 for the region of study (Embrapa, 2006; Garrastazu et al., 2006), with SAD69 datum as reference in the both cases.

The soils map was produced by integrating the State's respective municipal maps (Cunha et al., 2006). The boundaries for soil unit were determined by overlaying the hydrographic network, digital elevation model and satellite imagery, with special attention given to territorial boundaries.

*Table 1. Description of data used for developing a zoning plan based on climate and soil conditions for cultivating eucalyptus in southern region of Rio Grande do Sul State. ANA = National Water Agency; FEPAGRO = The State Foundation for Agricultural Research (Rio Grande do Sul State, Brazil); INIA = National Institute for Agriculture and Livestock Research (Uruguay); INMET = National Institute of Meteorology; MAPA = Ministry of Agriculture, Livestock and Fisheries (Uruguay); SRTM = Shuttle Radar Topography Mission; UdelaR = University of de Republic (Uruguay); 1DL = First Division of Survey of the Army (Brazil)*

| Theme | Information level | Format | Source |
|---|---|---|---|
| Cartography | Hydrographic network | Lines | 1DL |
| | Urban area | Polygons | |
| | Roads, railroads and highways | Lines | |
| Forestry | Characteristics of eucalyptus | Textual | Embrapa Forests, INIA, MAPA, UdelaR |
| Meteorology | Temperature | Points | INMET, FEPAGRO, INIA |
| | Rainfall | Points | ANA |
| Pedology | Soils map | Polygons | Cunha et al. (2006) |
| Remote Sensing | Landsat imagery | Raster | NASA, INPE |
| Topography | Digital elevation model | Raster | SRTM (apud WEBER et al., 2005) |

This operation permitted the map legend to then be updated (Figure 2). This process necessarily fixed zoning scale at about 1:250000 because the climatic zoning procedure used a greater scale (1:1000000). This meets the minimum scale required for ecological-economic zoning according to the aforementioned Brazilian legislation.

*Figure 2. Example of soils map revision for the indicated sector (northeastern region of Hulha Negra Municipality, Rio Grande do Sul State, Brazil): (a) previous vector file; (b) modified vector file; (c) soil unit boundaries on Landsat imagery; (d) new revised map*

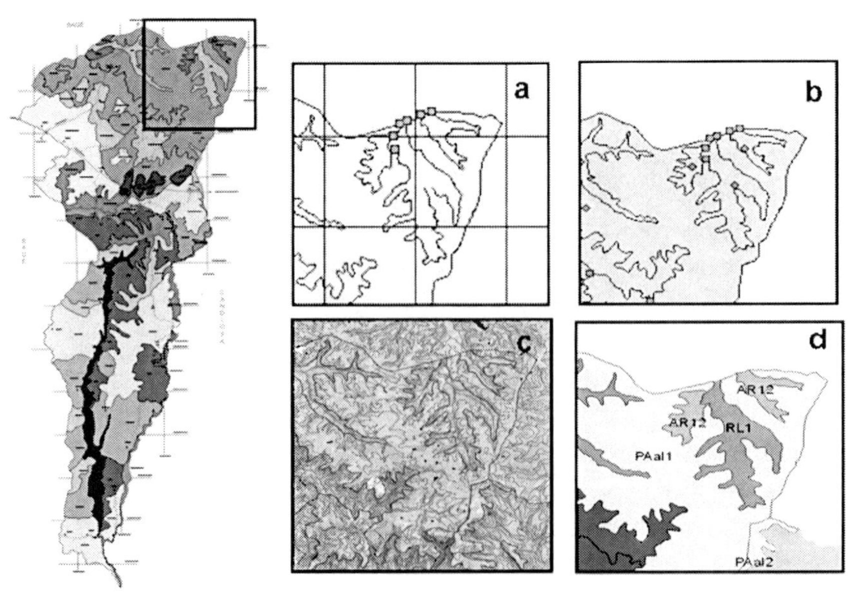

*Figure 3. Water deficit map for Rio Grande do Sul State also showing locations for hydrological stations belonging to the Brazilian national agency of water*

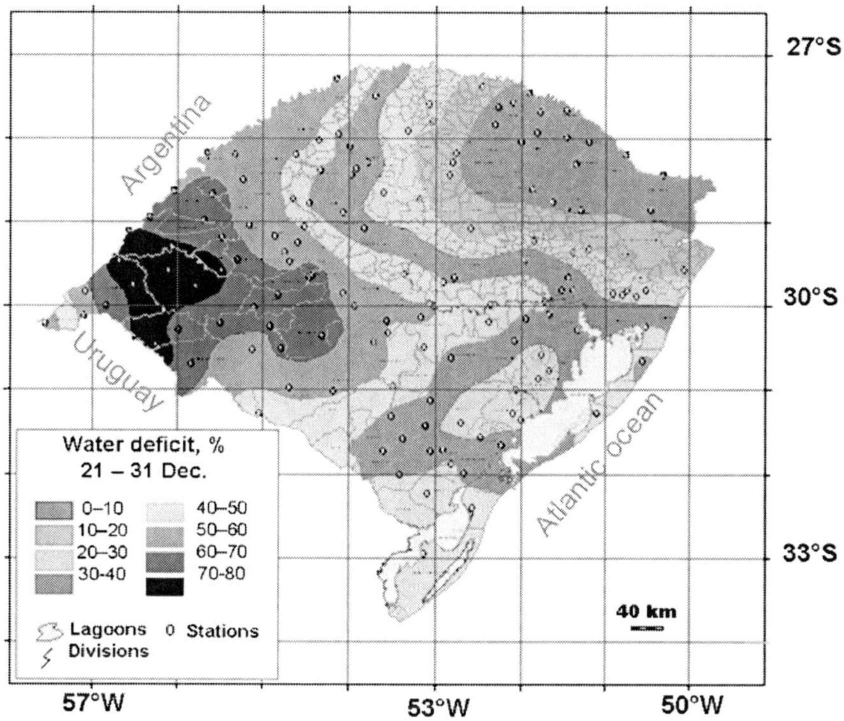


Water deficit was determined using data from the National Meteorological Station Network which has 160 stations throughout the State of Rio Grande do Sul (Figure 3). Temporal series with more than ten years were utilized. Using classes of soils defined by a state map at scale 1:750000 (Brasil, 1973), simulations were run for water storage capacity of 50 mm, 75 mm and 100 mm.

A special strategy had to be adopted for evaluation of water deficit due to the absence of overlapping for meteorological and hydrological stations. Measures of temperature were extrapolated and a continuous surface was obtained. Next, data of temperature were extracted from the extrapolated map for each hydrological station. Evapotranspiration was evaluated for each hydrological station as a function of temperature (Thornthwaite, 1948).

These data made it possible to calculate water balance for periods of deficit and for excess and storage of water in soil. Maps with probability of occurrence of water deficit greater than 20 mm in the 160 hydrological stations were simulated using GIS (Environmental Systems Research Institute, 1999). The largest deficit of water occurs in the West Frontier region, with a mean risk of 80% corresponding to water less than 20 mm during summer (Figure 3).

Risk of frost was estimated according to a regression model with temperature considered a function of location (X, Y) and elevation (Z), with subsequent regionalization obtained through raster data processed by GIS as illustrated in Figure 4. This model for the Southern Region of the State of Rio Grande do Sul was extracted from the model for the entire State by the GIS clipping function. This was necessary because the extrapolation

*Figure 4. Regionalization of temperature data derived from the meteorological stations and the description of the procedure used to construct the frost risk map. Parameters a, b, c and d are initially calculated through a regression model of temperature (T) considered as function of longitude, latitude and altitude (DEM). Following that, data are extrapolated by integration using GIS*

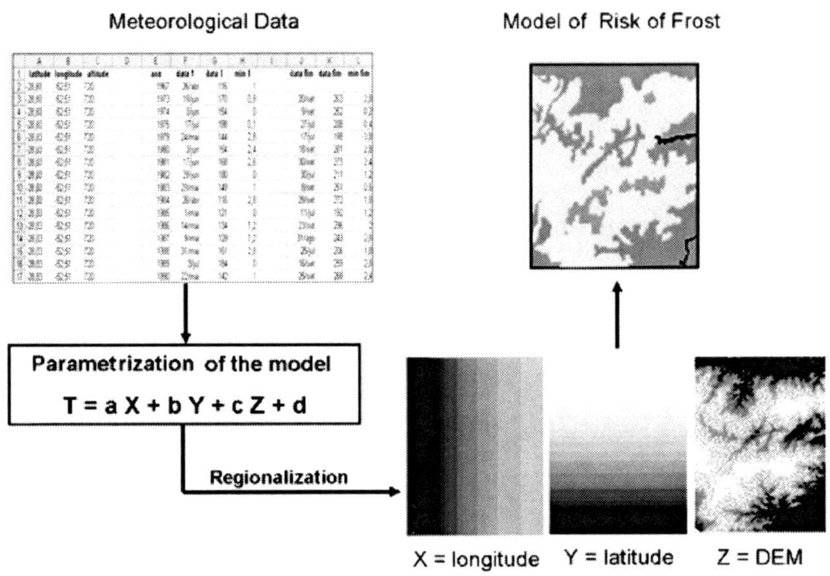


process is inadequate when a low number of points (meteorological stations) is evolved.

Each species of eucalyptus has its own special thermal requirements for proper development, including tolerance to frost (Table 2). Water balance and soil criteria were evaluated according to generic conditions. For example, *Eucalyptus grandis* is not tolerant to frost and minimum temperature must be greater than 3 °C and maximum temperature greater than 8 °C for good development. *Eucalyptus globulus* is more tolerant to frost but it does not adapt to temperatures greater than 23 °C. Another consideration is that a uniform sequence of rainfall during the year is desirable for any species of eucalyptus: 500 mm/year is necessary for survival and at least 1000 mm/year is optimal for production (Poyton, 1979; Pryor, 1991; Cromer et al., 1993; Darrow, 1994; Jovanovic & Both, 2002).

The climatic zoning was integrated with the soils zoning by overlay (common area intersection) according to criteria presented in Table 3. This method is very restrictive, but more reliable because the class with lower suitability controls the process. This means that the class will be "Highly recommended" when both variables, climate and soils, were "Highly recommended" themselves. Otherwise, the integrated suitability classification adopts the order of the lowest class and thereby proceeds.

## RESULTS AND DISCUSSION

### Climatic Zoning and Soils Zoning

The western frontier of Rio Grande do Sul State, a region where water deficits occur during the summer as mentioned, is very distant from the study region. For this reason, the climatic zoning was based on temperature. Low tolerance of *eucalyptus grandis* to frost is evident in climatic zoning for the State and especially for its southern region (Figure 5). This becomes especially

*Table 2. Criteria used to define the suitability classes for climate and soil zoning for planting eucalyptus in southern region of Rio Grande do Sul State. T = Temperature (°C); $T_m$ = Minimum temperature; $T_x$ = Maximum temperature. NR = Not recommended; RR = Rarely recommended; R = Recommended; HR = Highly recommended*

| Species | Variable (unit) | Suitability classes | | | |
|---|---|---|---|---|---|
| | | NR | RR | R | HR |
| *Grandis* | T (°C) | Disregarded | 20% probability that $T_m$<3 and $T_x$<8 | 10 -20% probability that $T_m$<3 and $T_x$<8 | < 10% probability that $T_m$<3 and $T_x$<8 |
| *Dunnii* | T (°C) | Disregarded | 20% probability that $T_m$<0 and $T_x$<8 | 10 -20% probability that $T_m$<0 and $T_x$<8 | < 10% probability that $T_m$ <0 and $T_x$ <8 |
| *Globulus* | T (°C) | Disregarded | 20% probability that $T_m$<0 and $T_x$<8 nor $T_x$> 23 | 10 -20% probability that $T_m$<0 and $T_x$<8 nor $T_x$>23 | < 10% probability that $T_m$<0 and $T_x$<8 nor $T_x$>23 |
| Generic | Water balance (mm/year) | < 500 | Disregarded | 500 - 1000 | > 1000 |
| Generic | Drainage | Very poorly suitable | Somewhat poorly or poorly suitable | Moderately well suitable | Excessively or well suitable |
| Generic | Effective depth | Very shallow | Shallow | Few deep | Deep |
| Generic | Texture | Organic | Clayey (2:1), silty or sandy | Very clayey (1:1) | Clayey (1:1) |
| Generic | Relief | Mountainous | Hilly | Irregular or escarpments | Flat or smooth |
| Generic | Fertility | Salinity | Very low | Low | High or medium |
| Generic | Stoniness | Abundant | Marked | Moderate | Absent or rare |

evident when compared to the range for more tolerant species such as *eucalyptus dunnii* (Figure 6). Nevertheless, altitude is a significant control factor, because the highlands are represented as being colder than regions of medium altitude or lowlands. Perhaps the prediction model for temperature would be more effective if water bodies and vegetation cover had been included.

*Table 3. Suitability for climate and soils was defined by overlay of soil suitability and climatic suitability. NR = Not recommended; RR = Rarely recommended; R = Recommended; HR = Highly recommended*

| Soils Suitability | Climatic Suitability | | |
|---|---|---|---|
| | HR (risk: 0-10%) | R (risk: 10-20%) | RR (risk: >20%) |
| HR | HR | R | NR |
| R | R | | |
| RR | RR | | |
| NR | NR | | |

The strong restriction for *eucalyptus globulus* plantations is due to intolerance for temperatures greater than 23°C. The southern region of the State could be transformed into a strong forestry producing area due to its unique conditions for plantations. This comparison is especially valid when one considers the full extension of Brazil's national territory (Figure 7).

As mentioned earlier, the soils map of the study region was adjusted for the zoning process (Figure 8 and Table 4), thereby allowing delineation of four suitability classes (Figure 9). The class identified as "Not recommended" corresponds to

*Figure 5. Climatic zoning of Eucalyptus grandis for Rio Grande do Sul State (left-superior) and in detail for southern region (clipping, see Material and Methods)*

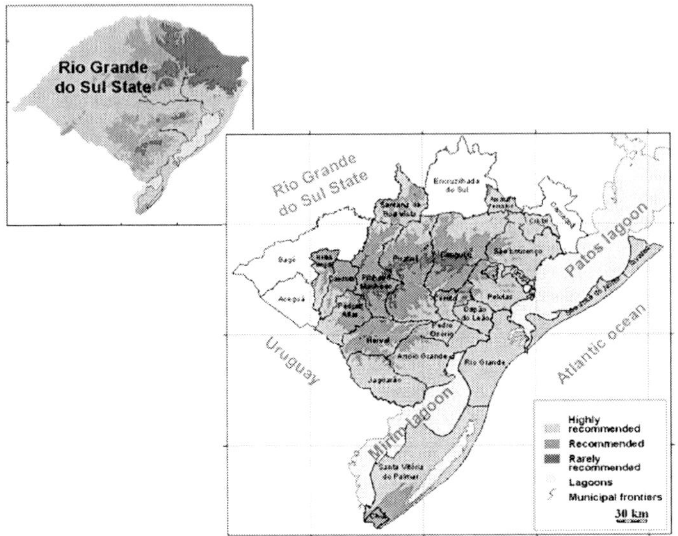


*Figure 6. Climatic zoning of Eucalyptus dunnii for the Rio Grande do Sul State (left-superior) and detail for southern region (clipping, see Material and Methods)*

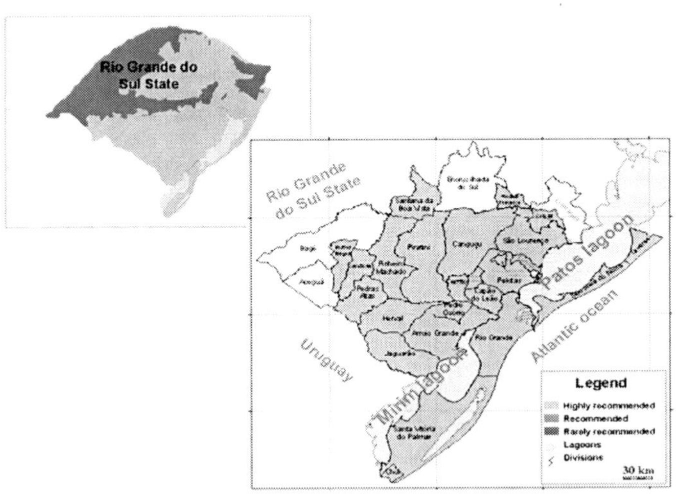


several different situations, including the following examples:

1. Soils with salinity problems, usually in coastal regions;
2. Shallow soils in which roots cannot properly develop;
3. Poorly drained soils such as in wetlands.

Classes "Highly recommended" and "Recommended" are located in regions of moderate altitude with predominantly good conditions for drainage, sufficient depth, suitable fertility and texture and little stoniness or rockiness. These classes cover more than 900,000 hectares of territory, constituting 25% of the study region (Table 5). Furthermore, 75% of the territory is classified

*Figure 7. Climatic zoning of Eucalyptus globulus for Rio Grande do Sul State (left-*perior) and detail for southern region of the State (clipping, see Material and Methods)*

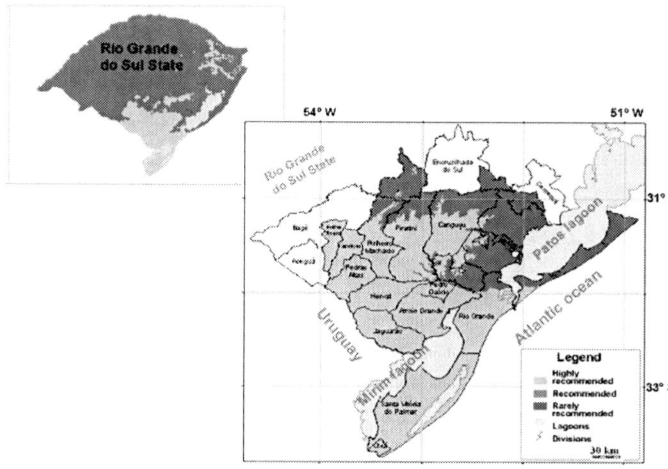


*Figure 8. Soils map for southern region of Rio Grande do Sul State. See Table 4 for legend description*

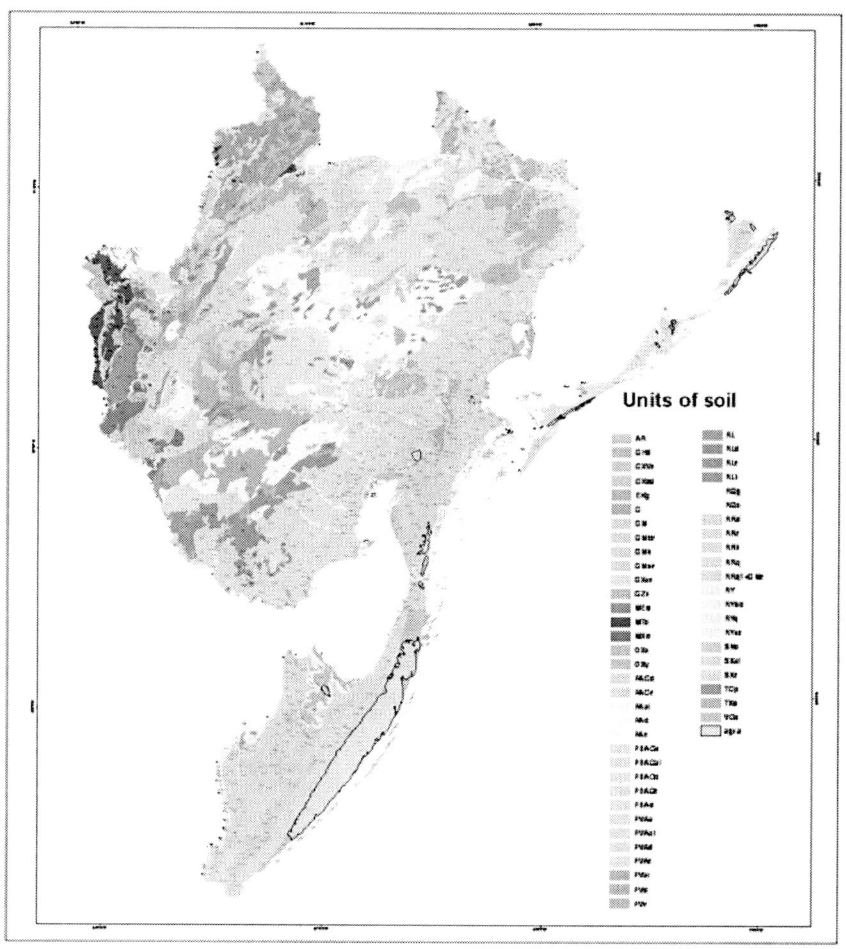

*Table 4. Soil mapping categories according to soil taxonomy (United States, 1999)*

| Identifier | Association |
|---|---|
| AR | Outcrops + Entisols + Inceptisols + Ultisols |
| CHd | Inceptisols + Entisols |
| CXVe, CXbd | Inceptisols + Ultisols + Entisols |
| EKg | Spodosols + Gleysols |
| G, GM, GMbe, GMk, GXve, GZ | Gleysols + Histosols + Alfisols + Entisols |
| MEo, MTb, MXo | Molisols + Vertisols + Entisols |
| OXg, Oxy | Histosols + Entisols |
| PACd, PVAe | Ultisols |
| Paal, PACe, Pad, PAe, PBACa, PBACd, PVAa, PVAal, PVe, | Ultisols + Entisols |
| PBACe | Ultisols + Molisols |
| PVa, PVAd | Ultisols + Entisols + Alfisols |
| RL, RLd | Entisols Lithic + Ultisols + Inceptisols |
| RLe | Entisols Lithic + Molisols + Outcrops |
| RLh | Entisols Lithic |
| RQg, RQo | Entisols Quartzipsamment + Ultisols |
| RRd | Entisols Psamment + Outcrops + Ultisols + Alfisols + Molisols + Vertisols |
| RRe | Entisols Psamment + Ultisols + Inceptisols + Molisols + Outcrops |
| RRh | Entisols Psamment + Ultisols |
| RY, RY, RYq, RYve | Entisols Fluvents |
| SNo, SXal, SXe | Alfisols + Entisols + Ultisols + Molisols |
| TCp, TXo | Alfisols + Entisols + Vertisols + Molisols |
| Water bodies | Lakes and lagoons |

*Figure 9. Soils zoning for planting eucalyptus in southern region of Rio Grande do Sul State*

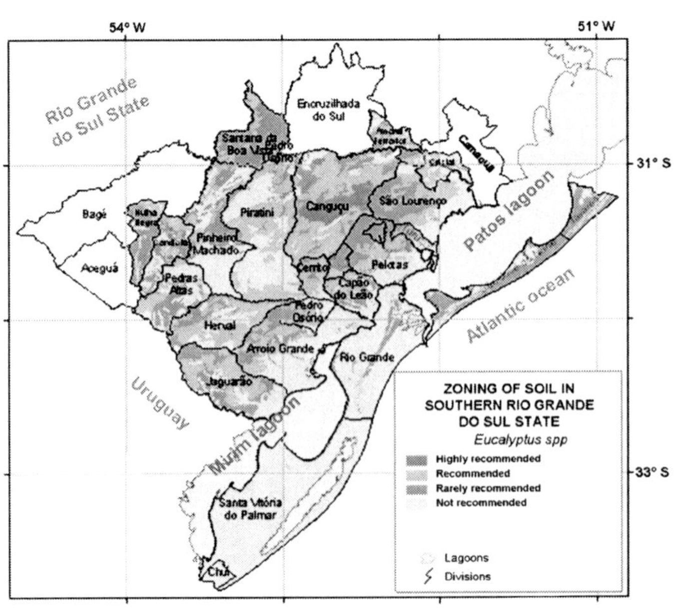

*Table 5. Areas occupied by each class of suitability based on the zoning of soils for planting eucalyptus in southern region of Rio Grande do Sul State*

| Suitability classes | Area (ha) | Area (%) |
|---|---|---|
| Water bodies | 100,650 | 3 |
| Highly recommended | 79,702 | 2 |
| Recommended | 846,700 | 23 |
| Rarely recommended | 994,380 | 27 |
| Not recommended | 1,692,664 | 45 |
| Total | 3,714,096 | 100 |

as "Rarely recommended" or "Not recommended" for planting of eucalyptus. Management methods can improve production in these cases, but this potential is restrained by market reasons.

The class "Not recommended" based on climate and climate zoning for planting eucalyptus was divided in three subclasses: (1) Restricted by soil conditions; (2) Restricted by frost risk; (3) Restrictions for both conditions (Figures 10, 11 and 12).

The greater susceptibility of *Eucalyptus grandis* to frost appears clearly in the region of study, indicated by colors red and brown (Figure 10). Litholic soils, near the central sector of the region of study, are usually shallow, thus productivity is restricted. Low productivity is also associated to the flat regions, where drainage is poor (the coastal region, for example). *Eucalyptus dunnii* is not restricted by frost or water requirements in the study region. Because of this, its zoning is only dependent on soil conditions (Figure 11). The susceptibility of *Eucalyptus globulus* to temperatures greater than 23°C appears at northern sector of the region of study in red color (Figure 12).

Descriptions of areas covered by the main suitability classes are summarized in Table 6.

*Figure 10. Zoning of climate and soils for planting Eucalyptus Grandis in southern region of Rio Grande do Sul State (cr = climate restriction; sr = soil restriction; csr = climate and soil restriction)*

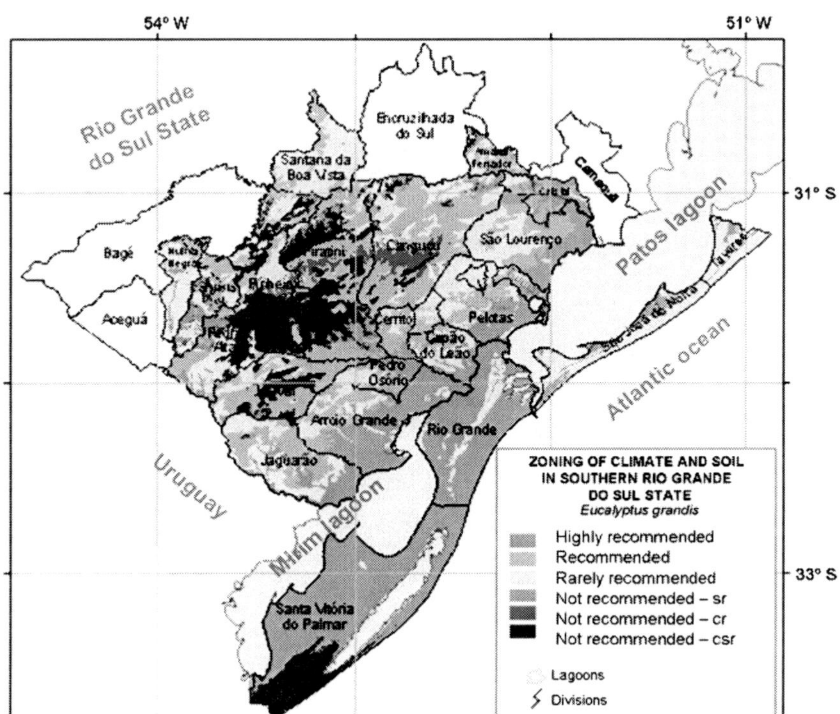

*Figure 11. Zoning of climate and soils for planting Eucalyptus dunnii in southern region of Rio Grande do Sul State (cr = climate restriction; sr = soil restriction; csr = climate and soil restriction)*

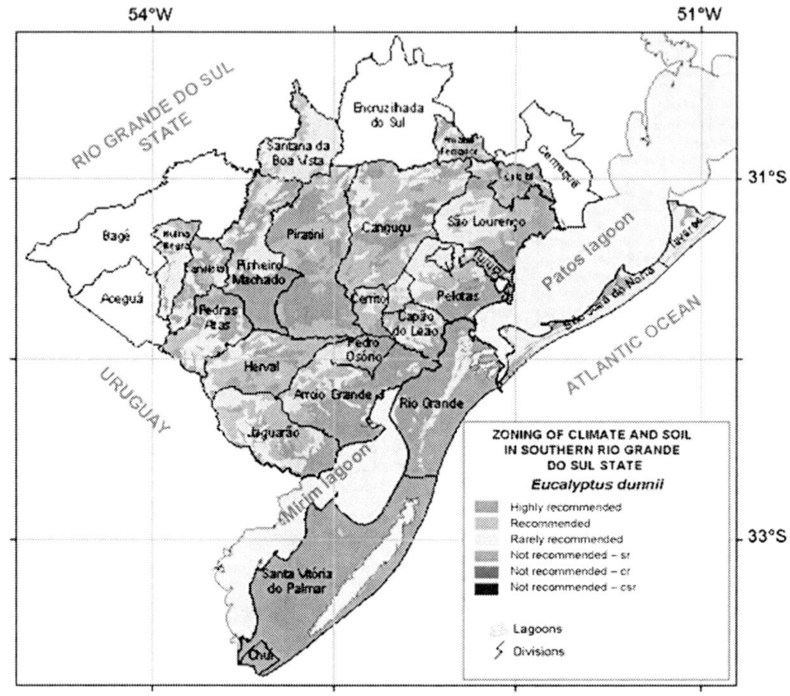


*Figure 12. Zoning of climate and soils for planting Eucalyptus globulus in southern region of Rio Grande do Sul State (cr = climate restriction; sr = soil restriction; csr = climate and soil restriction)*

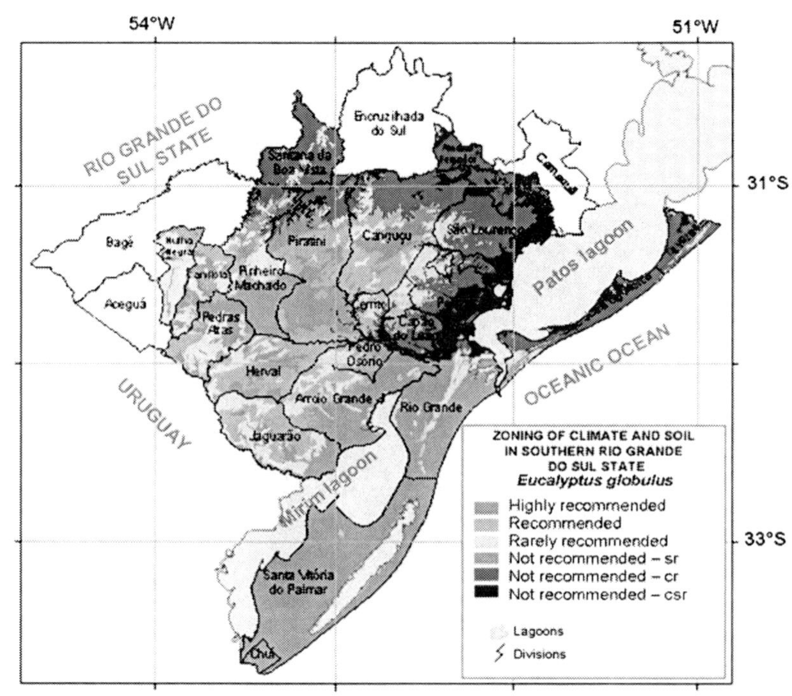

*Table 6. Surface area of the suitability classes for planting three species of eucalyptus in southern region of Rio Grande do Sul State, based on zoning of climate and soil*

| Suitability class | *Eucalyptus grandis* | | *Eucalyptus dunnii* | | *Eucalyptus globulus* | |
|---|---|---|---|---|---|---|
| | Area (ha) | % | Area (ha) | % | Area (ha) | % |
| Preferential | 5,462 | 0.2 | 79,702 | 2 | 75,600 | 2 |
| Recommended | 883,972 | 24 | 846,700 | 23 | 565,259 | 15 |
| Rarely recommended | 976,796 | 26 | 994,380 | 27 | 566,085 | 15 |
| Not recommended | 1,747,216 | 47 | 1,692,664 | 45 | 2,406,502 | 65 |
| Water bodies | 100,650 | 3 | 100,650 | 3 | 100,650 | 3 |
| **Total** | **3,714,096** | **100** | **3,714,096** | **100** | **3,714,096** | **100** |

As was suggested by the spatial distribution of the suitability classes, *eucalyptus globulus* is the more restricted specie from a production point of view. The sum of classes "Preferential" and "Recommended" occupies 17% of the study region, compared to 24.2% for *Eucalyptus grandis* and 25% for *Eucalyptus dunnii*. A similar situation occurs for the "Not recommended" class with a total area of 65% for *eucalyptus globules*. That value diminishes to 45 to 47% of territory for the other two species.

## CONCLUSION AND NEW PERSPECTIVES

Zoning based on climate and soil for optimizing eucalyptus production in southern region of Rio Grande do Sul State was performed using GIS technologies. The spatial distribution of suitability classes for three species was mapped and area coverage was calculated. The potential surface area for forestry surpasses 900,000 hectares, but a homogeneous exploitation of that territory is limited by legal restrictions.

Two questions need to be answered from a sustainability point of view:

1. How much of the total potential area is occupied by other land uses?
2. How much of this area must be reserved for environmental protection such as ecological reserves or permanent preservation areas?

Modeling of climate was controlled by the digital elevation model obtained from the Shuttle Radar Topography Mission with a 90-meter pixel which represents a scale of 1:500000 to 1:200000. This scale might be enhanced by using digitalized isolines from maps of scale 1:50000. A further advance for the climate model would be the correction of temperature as function of presence of water bodies and vegetation cover (XIAO et al., 2004).

The working scale of 1:250000 facilitated the ecological-economic zoning process largely due to integrating the soils map of approximately 1:200000 scale.

Geographic information systems (GIS) normally include several ways to overlay data such as using Boolean logic, fuzzy functions, classification schemes or other criteria. This study concentrated primarily on intersection which is a very restrictive method. Application of advanced methods, as mentioned before, could provide different results. As a consequence of such differences, it may be possible to evaluate the precision and the validity of the process used in the present study, a useful exercise from a production point of view.

## ACKNOWLEDGMENT

The authors would like to acknowledge the collaboration of the following persons and institutions: Dr. Bernadete Radin and Dr. Ronaldo Matzenauer, both from the Foundation for Agricultural Research of Rio Grande do Sul State, Brazil; Dr. Rosana Clara Victoria Higa and Dr. Antônio Rioyei Higa; Dr. Silvio Steinmetz and Dr. Carlos Reisser from Embrapa, Brazil, Dr. Flavio Herter retired from Embrapa, Brazil; M.Sc. Carlos Roberto Soares Severo from the National Institute of Colonization and Agrarian Reform, Brazil; M.Sc. Eliseu Weber and M.Sc. Heinrich Hasenack, both from the Federal University of Rio Grande do Sul, Brazil; Dr. Vilmar Luciano Mattei from the Federal University of Pelotas, Brazil; National Institute of Meteorology, Brazil; National Agency of Water, Brazil; Dra. Zohra Bennadji from the National Institute for Agriculture and Livestock Research, Uruguay; Dr. Carlos Mantero Álvares, Dr. Guillermo Moras Loyarte, Dr. Rafael Escudero, Dr. Rodolfo Pedocchi and Dr. Luis Silveira from the University of the Republic, Uruguay; Eng. Juan Carlos Sganga and Eng. Daniel San Román from the Ministry of Agriculture, Livestock and Fisheries, Uruguay. Financial support for the project of research was from the Foundation of Research Support of Rio Grande do Sul State, Brazil through the Councils for Regional Development, with the significant evolvement of Dr. Flávio Miguel Schneider (*in memoriam*) from the Federal University of Santa Maria. We give special thanks to Daiane Hellvnvig Zarnott for her assistance with graphic presentation. Finally, the authors are very grateful to Dr. James Jackson Griffith, Federal University of Viçosa, Brazil, for reviewing the content of the chapter.


## REFERENCES


Aronoff, S. (1991). *Geographic information systems: a management perspective*. Ottawa, Ontário, Canada: WDL.

Augustithis, S. P. (1995). *Atlas of textural patterns of ore minerals and metallogenic processes*. New York: de Gruyter.

Bonham-Carter, G. F. (1994). *Geographic information systems for geoscientists*. Ontario, Canada: Pergamon.

Brasil. Ministério do Meio Ambiente (2001). *Memória das Reuniões de Discussões sobre Zoneamento Ecológico-Econômico no Brasil*. Brasília: MMA. 1 CD-rom.

Burrough, P. A. (1986). *Principles of geographical information systems*. New York: Oxford. Brasil. Ministério da Agricultura. Divisão de Pesquisa Pedológica (1973). *Levantamento de reconhecimento dos solos do Estado do Rio Grande do Sul*. Recife, Brazil: MA/DPP – SA/DRNR (Boletim técnico, 30).

Cromer, R. N., Cameron, D. M., Rance, S. J., Ryan, P. A., & Brown, M. (1993). Response to nutrients in E. grandis. Nitrogen acumulation. *Forest Ecology and Management*, *62*(1-4), 231–243. doi:10.1016/0378-1127(93)90052-O

Cunha, G. R., & Assad, E. D. (2001). Uma visão geral do número especial da revista brasileira de agrometeorologia sobre zoneamento agrícola no brasil. *Revista Brasileira de Agrometeorologia*, *9*, 377–385.

Cunha, N.G., Silvira, R.J.C., Severo, C.R.S. (2006) *Solos e terras do planalto Sul-Rio-Grandense e planícies costeiras*. Pelotas: Embrapa Clima Temperado. (Circular técnica, 55).

da Mota, F. S., Beirsdorf, M. I. C., Acosta, M. J. C., Motta, W. A., & West, S. L. (1974). *Zoneamento agroclimático do Rio Grande do Sul e Santa Catarina*. Pelotas, Brazil: IPEAS (Circular, 50).

Darrow, W. K. (1994). The effect of drought en eucalypt species growing on shallow soil in South Africa: effect on mortality and growth. *ICFR Bulletin Series, 7*, 1–24.

Davis, J. (1986). *Statistics and data analysis in geology*. New York: Wiley.

Eldridge, K. G., Davidson, J., Hardwood, C. E., & van Wyk, G. (1994). *Eucalypt domestication and breeding*. Oxford, UK: Clarendon Press.

Environmental Systems Research Institute. (1999). *ArcGIS 9.1: handbook*. Redlands, CA: Environmental Systems Research Institute.

Flores, C. A., Filippini Alba, J. M., & Wrege, M. S. (Eds.). (2009). *Zoneamento agroclimático do eucalipto para o Estado do Rio Grande do Sul e edafoclimático na região do Corede Sul – RS*. Pelotas, Brazil: Embrapa Clima Temperado.

Garrastazú, M. C., da Silva, M. B., & de Siqueira, O. J. W. (2006). *Estruturação da base cartográfica digital para o extremo sul do RS*. Pelotas, Brazil: Embrapa Clima Temperado (Comunicado técnico, 142).

Golfari, L., Caser, R. L., & Moura, V. P. G. (1978). *Zoneamento ecológico esquemático para o reflorestamento no Brasil*. Belo Horizonte, Brazil: Centro de Pesquisa Florestal da Região do Cerrado (Série técnica, 11).

Jovanovic, T., & Booth, T. (2002). *Improved species climatic profiles*. Australia: Union Offset Printing.

Poyton, R. J. (1979). *Tree planting in Southern Africa: the eucalyptus*. Pretoria, South Africa: Department of Forestry.

Pryor, L. D. (1976). *Biology of eucalypts*. London: Edward Arnold.

Thornthwaite, C. W. (1948). An approach toward a rational classification of climate. *Geographical Review, 38*, 55–94. doi:10.2307/210739

United States. Department of Agriculture. Natural Resources Conservation Service. Soil Survey Staff. (1999). *Soil taxonomy: a basic system of soil classification for making and interpreting soil surveys*. Washington, DC: USDA.

Weber, E., Hasenack, H., & Ferreira, C. J. S. (2005) *Adaptação do modelo digital de elevação do SRTM para o sistema de referência oficial brasileiro e recorte por unidade da federação*. Porto Alegre, Brazil: UFRGS. Retrieved Dezember 01, 2005 from http://www.ecologia.ufrgs.br/labgeo

Xiao, X., Boles, S., Liu, J., Zhuang, D., Frolking, S., Li, Ch., & Salas, W., & MOORE, B. (2005). Mapping paddy rice agriculture in southern China using multi-temporal MODIS images. *Remote Sensing of Environment, 95*, 480–492. doi:10.1016/j.rse.2004.12.009

## ADDITIONAL READING

Alba, F. J.M. (Ed.) (2007). Trabalhos acadêmicos de geoprocessamento desenvolvidos no laboratório de planejamento ambiental em 2007. Pelotas, RS: Embrapa Clima Temperado (Documentos, 219).

Alba, F. J.M. (Ed.) (in press). Recuperação de areas degradadas. Pelotas, RS: Embrapa Clima Temperado.

Assad, S., & SANO, E.E. (1998). *Sistemas de Informação Geográfica: Aplicações na agricultura*. Brasília, DF: Embrapa.

Falcade, I., & Mandelli, F. (Eds.). (1999). Vale do Vinhedos: Caracterização Geográfica da Região. Caxias do Sul, RS: EDUCS.

Filippini Alba, J. M. (1998). Análise e integração de dados geoquímicso e de sensoriamento remoto em um setor do cristalino uruguaio. Unpublished doctoral dissertation, State University of Sao Paulo, Sao Paulo, SP, Brazil (http://www.cpact.embrapa.br/teses/tese_jmfa.pdf).

Filippini Alba, J. M., Berny, V. M., & da Costa, F. A. (2009). *Geoprocessamento no contexto de empresas de pesquisa.* Pelotas, RS: Embrapa Clima Temperado (Documentos, 254).

Filippini Alba, J. M., Crosta, A. P., & de Oliveira, S. M. B. (2001). Interpretation of surface geochemical data and integration with geological maps and Landsat-TM images for minral exploration from a portion of the Precambriam of Uruguay. *Revista Brasileira de Geociencias*, *31*(2), 123–130.

Filippini Alba, J. M., & de Souza Filho, C. R. (2006). *Geoprocessamento como instrumento de avaliação ambiental.* Pelotas, RS: Embrapa Clima Temperado (Boletim de Pesquisa e Desenvolvimento, 28).

Filippini Alba, J. M., & de Souza Filho, C. R. (2010). GIS-based environmental risk assessment in the Ribeira Valley, São Paulo, Brazil. *Environmental Earth Science*, *59*(5), 1139–1147. doi:10.1007/s12665-009-0104-7

Fischel, W. A. (1987). *A Property Rights Approach to American Land Use Controls.* London: The Johns Hopkins Univ. Press.

Flores, C. A., Garrastazu, M. C., & Filippini Alba, J. M. (2009). Metodologia de zoneamento edáfico de culturas para o Estado do Rio Grande do Sul. Pelotas, RS: Embrapa Clima Temperado (Documentos, 261).

Flores, C. A., Garrastazu, M. C., Hasenack, H., & Weber, E. (2006). *Zoneamento edáfico para as culturas da mamona, soja, girassol e canola na região Noroeste do Estado do Rio Grande do Sul.* Pelotas, RS: Embrapa Clima Temperado (Documentos, 174).

Intergovernmental Panel on Climate Change. (2007). Climate in Peril: A Popular Guide to the Latest IPCC Reports presents the substance of the Climate Chang. Retrieved June 22, 2010 from http://www.grida.no/publications/climate-in-peril/

Jensen, J. R. (1996). *Introductory Digital Image Processing.* Upper Saddle River, NJ: Prentice Hall.

Molga, M. (1962). *Agricultural meteorology. Part II: Outline of agrometeorological problems.* Poland: National Science Foundation and the Department of Agriculture.

UN. (2004). Land degradation and Land Use/Cover Data Sources. Economic Retrieved June 22, 2010 from http://unstats.un.org/unsd/environment/envpdf/landdatafinal.pdf.

Wrege, M. S. (2004). *Delimitação de regiões com potencial de cultivo econômico da cana-de-açúcar (Saccharum spp.), no Estado do Paraná por meio da análise de riscos climáticos.* Unpublished doctoral dissertation, State University of Maringá, Maringá, PR, Brazil.

## KEY TERMS AND DEFINITIONS

**Agrometeorology:** Science investigating the meteorological, climatological and hydrological conditions that are significant to agriculture owing to their interaction with the objects and processes of of agriculture production (Molga, 1962).

**Geographic Information Systems (GIS):** Computer system for managing spatial data (Bonham-Carter, 1994).

**Edafology:** Soil science applied to agriculture.

**Forestry:** Science of managing forests, tree plantations, and related natural resources.

**Land Use:** Natural resources classified according to human exploration.

**Pedology:** Soil science.

**Remote Sensing:** acquisition of information of an object or phenomenon through sensing device(s) without physical or intimate contact with it.

**Zoning:** Division of territories in homogeneous zones since a Land Use perspective, considering criteria related to climate, soil, environment, economy, etc.

# Chapter 9
# A Mixed Integer Programming Approach for Sugar Cane Cultivation and Harvest Planning

**Sanjay Dominik Jena**
*Pontifícia Universidade Católica do Rio de Janeiro, Brazil*

**Marcus Vinicius Soledade Poggi de Aragão**
*Pontifícia Universidade Católica do Rio de Janeiro, Brazil*

## ABSTRACT

*The planning of agricultural cultivation and harvesting is a complex task. However, this area of study is still relatively young. This chapter focuses on the tactical and operational planning for sugar cane cultivation and harvesting which determines the best moment to harvest the fields, maximizing the total profit given by the sugar content within the cane. It considers resources such as cutting and transport crews, processing capacities in sugar cane mills, the use of maturation products and the application of vinasse on harvested fields. The mixed integer programming model extends the classical Packing formulation, adding a network flow structure to represent the harvest scheduling. All experiments were performed with real-world instances provided by a Brazilian sugar cane producer. The suggested solutions are discussed in terms of quality and use in practice.*

## INTRODUCTION

Sugar cane is a sub-tropical and tropical genus of tall growing crop, counting 37 species plus a number of hybrid species. According to the Food and Agriculture Organization of the United Nations (FAO, 2008), sugar cane is one of the most important commodities in the world, commonly being further processed to sugar or agro fuel. With more than 420 billion tons of harvested sugar cane in the year 2005, Brazil is by far the largest producer of this grass worldwide, followed by India, China and Thailand. Among all agricultural commodities produced in Brazil, sugar cane is its most produced measured in biomass and its fourth most lucrative. Internationally, sugar cane is a highly competitive market. Recent international studies

DOI: 10.4018/978-1-61692-871-1.ch009

(Higgins, 2007; Gala et al., 2008; Boehlje et al., 2002) showed great opportunities to improve the value chain and reduce costs in the operational planning in order to remain competitive.

## Sugar Cane Harvest Process

The sugar cane harvest typically begins in May, sometimes April and prolongates to November, the time of the year when the sugar cane plants normally reach their maturation peaks. The maturation of sugar cane is measured in percentage of sucrose in the sugar cane, denoted by Pol and reduced sugar, denoted by AR. The maturation periods vary widely around the world from six to 24 months.

Manual and mechanical cutting crews cut the plants on the fields, chopping down the stems but leaving the roots to re-grow in time for the following harvest. The harvest is then immediately transported to the industrial sector, i.e. sugar cane mills, by trucks, rail wagons or by manual carriage (cart pulled by a bullock or a donkey). Figure 1 illustrates example routes for cutting and transporting. The cutting crews travel from one field to another harvesting the cane. The transportation crews commute between the fields and the mills.

In the mills, the sugar cane is crushed and the cane juice is extracted. The bagasse leftover, also referred to as fiber, is burned in boilers. The induced steam drives the turbines that generate the power for the mills. The sugar cane is further processed either to sugar or to ethanol. The sugar is also referred to as the total recoverable sugar (also referred to as ATR, originated in the Portuguese term *Açúcar Total Recuperável*, used in Brazil). For the sugar production, the sugar cane juice undergoes further processes such as heating, filtering and evaporation. The result is a syrup which is centrifuged to separate the sugar crystals from the molasses. To produce ethanol, the juice also undergoes processes as heating, filtering and evaporation. Afterwards, the juice is fermented in large vats, centrifuged and distillated to separate the ethanol.

A side effect of the alcohol distillation process is a residual industrial liquid called vinasse. Vinasse is a corrosive contaminant that contains high levels of organic matter, potassium, calcium and moderate amounts of nitrogen and phosphorus (Gómez and Rodríguez, 2000). However, vinasse is an efficient fertilizer, thus its application to harvested plantation fields has become common practice.

The use of maturation products is a common approach in agriculture to influence the natural maturation curve of plants. In the context of sugar cane, often used products are growth regulators which decrease the growth of the cane and therefore lead to an increase of the relative quantity of sugar within the plant. Growth regulators are commonly used to pre-pone a field's yield in order to provide raw material, i.e. sugar cane, for the mills. Thus, the use of such maturation products

*Figure 1. Example routes for cutting and transportation crews*

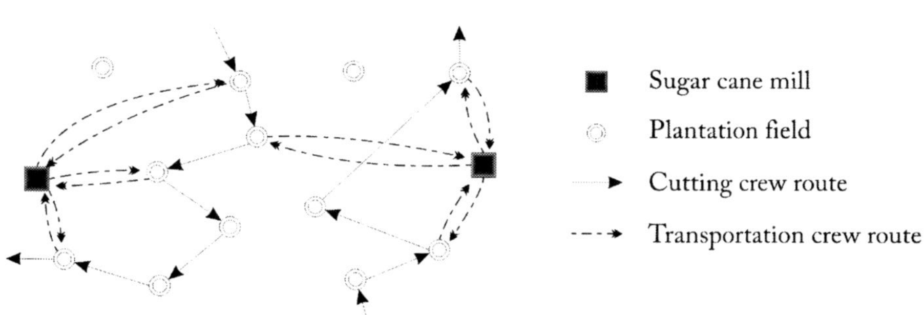

is directly related to the moment of sugar cane processing in the mills.

The cultivated areas can contain hundreds of lots with different varieties, each with distinct growth and maturation properties. One of the most difficult, but most important decisions is the determination of the ideal moment to cut each field and apply growth regulators in order to benefit best from the maturation peaks. The cutting crew's capacities and logistic factors are directly involved in such decisions, as they may constraint the harvest and therefore must be taken into account.

## Previous Work Review

### Decision Support Tools in Agriculture

Methods of Operations Research (OR) have been applied to the agricultural sector for more than five decades. Heady (1954) is frequently cited as one of the first applications of linear programming (LP) to agricultural planning. His model assigned farm land to various crops subject to operational constraints using a profit maximizing objective function. Audsley et al. (1978) compared different cultivation techniques for four types of crop: winter wheat, spring barley, sugar beets and potatoes. The linear programming model includes the allocation of resources such as land, labor and machinery. Although some variables in reality should be integers, the authors solve the problem as a linear problem and round the solution values. In the same year, McCarl (1978) presented a successful application of linear programming to grain crop production planning. Such crops include corn, soybeans, wheat and silage. The resulting program allowed the recommendation of production of one of the crop types. Its input data was provided by a 500 questions document answered by the farmers. The program has been used by the farmers for over 35 years in annual workshops.

OR was institutionally coupled to agricultural engineering already decades ago (Audsley, 1985) and has become common practice in the agricultural sector. Today, there is a considerable amount of research by the OR society, applied to almost all types of fruits, vegetables, crops, corns and so on. Masini (2003) applied mixed integer programming (MIP) to the supply chain planning optimization in fruit industry. Bixby et al. (2006) provide a complex system with several optimization models to support planning of operations and deliveries in meat industry. An example for the problems' diversity in agricultural planning is given by the problem of wine harvesting. Ferrer et al. (2008) present a model that both minimizes the operational costs and maximizes the wine quality. The model uses a quality loss function to relate these two goals to each other.

A first review of work applying decision support tools to agriculture was given by Glen (1987). Since his comprehensive survey, at least three recent surveys appeared. Lowe and Preckel (2004) highlight some of the important works based on linear programming, stochastic programming, risk programming, dynamic programming and simulation. Their publication also includes a call for future research, which was responded by the OR society. The review of Lucas and Chhajed (2004) focuses on location analysis applied to agriculture. Finally, Ahumada and Villalobos (2009) aim at completing the previous compilations. They list more than 40 works, where most of them have their scope on the tactical planning. In addition to the surveys, France and Thornley (2007) provide a rich introduction into mathematical models applied to agriculture.

### Decision Support Tools in Sugar Cane Industry

The efforts of the OR society to support sugar cane industry was mainly developed in the last ten years. They can be divided roughly into value chain optimization, harvest and crew scheduling and the prediction of sugar cane performance indicators. Efforts were mainly recognized from countries where sugar cane typically is planted, for example South America, South Africa (Gala et al., 2008), Australia and the United States of America.

Higgins strongly contributed to recent literature about optimization approaches in sugar cane industry, mostly based on experiences in the Australian sugar cane industry. His works include various planning aspects in that field. In one of his works Higgins and Laredo (2006) aim at better integration and optimization of the cane harvesting and transport sectors of the value chain. The formulation is based on the P-Median problem and spatial clustering. Further works to improve the value chain (Higgins et al., 2004; Higgins et al., 2007) include investigations of the capacities of cutting and transport crews as well as the mill capacities (Higgins et al., 2007). His works on harvest scheduling include an extension of the Generalized Assignment Problem with over 500,000 integer variables, subject to constraints of transportation and mill crushing capacities. The model is solved heuristically. Later exact approaches of Zhaorong et al. (2005) include a LP model and an integrated statistical and optimization approach that maximizes the sugar content in the crops for a harvest season. Higgins (2006) also separately tackled scheduling of cane transport, the simulation of transport capacity-planning (Higgins and Davies, 2005) and the optimization of harvester rosters (Higgins, 2002).

Alencar et al. (2006) use a genetic algorithm (GA) to maximize the produced sugar. Caliaria et al. (2004) use linear programming, based on the XA Callable library, to maximize the sugar production. Pacheco and Lima Neto (2008) seem to be the first ones to use the term *Sugar Cane Harvest problem* for the problem. They incorporated it as a 0/1-Knapsack problem and show experimental results with multi-objective evolutionary algorithms. The model considers mill and crew capacities. The chapter also proposes the incorporation of logistic data such as distances between sugar cane lots, though such logistic issues were not implemented into their model. Pacheco et al. (2007) and Pacheco et al. (2008) also presented many works using multi-objective Artificial Intelligence (AI) approaches. All these works incorporate the value of the sugar content (PCC, Apparent Percentage of sucrose in the sugar cane), the biomass per square meter TCH (Tons of sugar cane per hectare) and the total fiber in their objective functions. Pacheco et al. (2005) and Oliveira et al. (2006) present works that aim at predicting the performance indicators PCC, TCH and fiber using Artificial Neural Networks (ANN) and GAs. Finally, Díaz and Pérez (2000) focus on sugar cane transportation.

## Decision Support Tools for Other Harvest Types

The forest sector has also been an intensive user of Operations Research tools. The tactical and operational planning include planting, harvesting and transporting. Cutting crews possess cutting capacities and transportation crews are limited by its transportation capacities. Clients usually demand a certain timber quantity during several time periods.

Research work in the forest sector has been very versatile. Mitchell (2004) and Karlsson et al. (2003) focus on short-term operational planning in forest harvest optimization, including crew scheduling. As transport is a crucial issue in forest harvesting, many researchers such as Karlsson et al. (2004) and Henningsson et al. (2006) also focus on road planning to meet the demands for the timber transport. Further optimization approaches in silviculture include the works of Hao et al. (2004), Weintraub and Navon (1976), Equi et al. (1997) and Martella et al. (1998).

## THE SUGAR CANE CULTIVATION AND HARVEST PROBLEM

Throughout this chapter, the here presented problem will be denoted by the *Sugar Cane Cultivation and Harvest Problem (SCHP)*. One of its most important decisions is the determination of the optimal moment to harvest the plantation fields. Clearly, it is desirable to harvest each field at the peak of its sugar content, as the sugar indicated by the Pol and AR values vary as the cane grows. In the beginning of the planning, each field got a certain initial age. A field can only be cut within a given interval of its age defined in the input data. The complete problem is now described.

### Problem Description

#### Sugar Cane Mills

After harvesting a field, its sugar cane is immediately transported to one of the sugar cane mills to be crushed and further processed to sugar. The mills operation is one of the most important constraints as it must not interrupt sugar cane processing. Each mill contains minimum and maximum process capacities which must be respected by the planning solution. Plantation fields that have been selected for harvesting must be assigned to one of the available sugar cane mills. Furthermore, the processed sugar cane must contain a certain minimum quantity of fiber used to generate electricity to operate the mills.

#### Maturation Products

Some sugar cane varieties allow the use of maturation products to anticipate its harvest. In general, such products slow down the growing process of the cane mass, whereas the growth of the sucrose within the cane is not affected. Thus, the percentage of sugar in the cane increases. Maturation products can only be applied on fields with certain cane variety and when the cane reaches minimum age. Once applied, they directly influence the originally given interval in which the field can be harvested. For each combination of maturation products and sugar cane varieties, an interval is defined that determines the feasible period to harvest the field after the product was applied. The planning solution shall determine the type and exact application moment of maturation products in order to find the most lucrative harvesting schedule.

#### Cutting Crews

The sugar cane is harvested by cutting crews. A cutting crew can either be a manual one, that is a group of human workers, or mechanical. Each cutting crew may be eligible to cut only a certain subset of the fields. Cutting crews may not work every day and work a limited time at each day. Each cutting crew has its own properties. Its minimum and maximum cutting capacities, travel speed as well as cutting and travel costs are given in the input data and must be taken into account.

#### Transportation

Transportation crews carry the cut sugar cane from the fields to the sugar cane mills. Each crew possesses individual properties such as a certain transport capacity, speed and costs. The solution must suggest a transportation schedule that assigns exactly one transportation crew to each field selected to be harvested. Transportation crews' capacities, speed and costs must be considered in the solution of the problem.

#### Vinasse Application

After crushing the sugar cane within the mills, the waste dump vinasse remains. A common practice to remove this byproduct is its application at already harvested area. In order to allow its frequent application, a sufficiently large field area must be harvested in certain periods. The vinasse demands, i.e. the periods and the necessary field size, and the fields that are eligible for vinasse application are informed by the input data.

## Tactical and Operational Planning

The Pol and AR values for each cane variation are one of the most important performance indicators to select the best moment for a field's harvest. However, these values cannot be foreseen very well for a long period such as twelve months.

A detailed planning for such a broad time period would result in a huge problem size that could hardly be solved in adequate time. In many industrial planning problems, the division of the entire problem into smaller pieces is a natural approach. Many decisions have to be made only for a rough time period such as months or weeks, without determination of the exact detailed planning for shorter periods such as days or hours. Planning of such a broad time horizon is referred to as *tactical planning (TP)*. It is generally followed by an *operational planning (OP)*, where the problem is subdivided into smaller pieces and the resulting sub problems are solved in detail.

The here proposed model divides the whole planning into a *tactical planning* and an *operational planning*, also referred to as the SCHP-TP and SCHP-OP, respectively. The decisions of the TP directly influence the input data for the OP. First, the TP performs the planning for the whole planning horizon, i.e. up to a whole harvest season, suggesting decisions for each week. Afterwards, the total planning time is divided into smaller periods of up to 30 days. The OP is then performed for each of these sub periods, making the decisions for each day. Table 1 resumes the main differences between both planning activities.

The input data for each operational planning is based on the decisions of the tactical planning in the according time period. That is, the set of fields selected by the tactical planning for a certain time period forms the input data of an operational planning. In this operational planning, the exact days to harvest each of these fields are determined. In addition, the user may modify the input data for the operational planning.

Both the TP and the OP determine, amongst others, for each of their decision time periods:

- The plantation fields that should be harvested and the cutting crews to cut these fields.
- The sugar cane mills in which the cane should be processed. Each field is associated to exactly one mill.
- The types of vinasse that shall be applied on the fields after harvesting them and the according fields.

Both plannings maximize the total profit given by selling ATR, the quantity of theoretic total usable sugar. The ATR is computed against the quantity of reduced sugar (AR) and Pol, using coefficients for both values to normalize the relation between them:

$$ATR = coefPol \cdot tonsPol + coefAR \cdot tonsAR$$

During the maximization, the system must satisfy the following constraints:

- All plantation fields must be harvested.

*Table 1.*

| | TP | OP |
|---|---|---|
| Planning horizon | One season | 7 to 30 days |
| Objective | maximize profit | maximize profit |
| Decisions | weekly | daily |
| Cutting crews | yes | yes |
| Cutting crew harvest sequence | no | yes |
| Transportation crews | yes | no |
| Maturation products application | yes | no |
| Vinasse application | yes | yes |

- All cutting crews have a minimum and maximum cutting capacity of sugar cane per period.
- The sugar cane cut at a day shall be transported to and processed in the sugar cane mill within the same period. The sugar cane mills have minimum and maximum capacities of sugar cane that they can process. Consequently, the sugar cane quantity cut by the cutting crews shall satisfy the mills' demands.
- In all periods, the mills must produce a certain minimum quantity of fiber, given by a percentage within the processed sugar cane.
- The varieties of the different sugar cane species have a minimum and maximum age to be cut. The sugar cane can be cut only during these given intervals. These intervals may vary by week and field.
- In all periods, a minimum quantity of vinasse must be applied on the recently harvested plantation fields. In order to allow these applications, the number of hectares freed by field harvesting in each period must be sufficiently high.

## Tactical Planning

The TP supports the planning for a total planning horizon of up to twelve months, i.e. one harvest season. It may be applied to shorter periods such as a few months, usually in the replanning during a running season. In addition to the previously presented decision, the tactical planning also determines the following for each week:

- The transportation crews that carry the cut sugar cane to the mills. Each field is assigned to exactly one transportation crew.
- The maturation products that shall be applied and the fields at which the products shall be applied.

These decisions are subject to the following constraints:

- All transportation crews have a maximum capacity of sugar cane that can be carried.
- Maturation products can be applied only during a certain interval of the sugar cane's age. After applying a maturation product, the sugar cane must be harvested within a given number of weeks.

The complete harvest of any plantation field may not exceed one week, i.e. only plantation fields that can be harvested within one week are considered.

## Operational Planning

The OP realizes a detailed planning for a time horizon of up to 30 days. Its input data is based on the tactical planning performed for the chosen time period. The OP may redefine assignments between cutting crews, fields and mills. Valid assignments are informed in the input data. The application of maturation products is not determined by the OP. Instead, such decisions are already covered by the tactical planning and shall be informed in the input data for the operational planning. In practice, transportation crews are sufficiently available and can be hired on demand when necessary. Thus, assignments of transportation crews are also not considered within the OP. While the TP works with estimated maturation curves of the sugar cane, the OP is intended to work with updated recent values of the sugar cane's maturity, i.e. Pol, reduced sugar AR and fiber. These values result from the pre-analysis, where cane examples of a certain area are analyzed before the sugar cane is cut.

After finishing their work at the end of each day, mechanical cutting crews remain at the current field and start working in the beginning of the next day. Manual cutting crews return to a place where they are accommodated. In addition to the previously presented decisions, the OP should suggest the sequence in which each cutting crew cuts the fields. This includes the cutting sequence during the planning that considers the displacement costs and time from one field to another.

## Related Problems

The essential structures for the tactical and operational planning are very similar. Both try to maximize the profit by choosing the best moment to harvest the plantation fields, considering the minimum and maximum processing demands of the sugar cane mills and the limited capacities of the cutting crews. This effort can be seen as a *0-1 Multiple Knapsack problem*, a generalization of the *0-1 Knapsack problem* and identical to the Generalized Assignment Problem (GAP), one of the most studied combinatorial problems. Martello and Toth (1990) elaborately introduce into these problems and show that they are NP-hard.

In the 0-1 Multiple Knapsack problem, a number of objects must be packed into a given number of bins. Each object has a weight and a profit for each bin. Bins are limited in their capacity of weight. The objective of the problem consists in finding the configuration that maximizes the total profit, respecting the weight capacities of the bins. Pacheco and Lima Neto (2008) already modeled harvest scheduling as a 0-1 Multiple Knapsack Problem.

In the case of the tactical planning, the sugar cane mills can be modeled as bins, one for each week. All sugar cane mills own a maximum processing capacity, referring to the bins' capacities. The weights and profits to process a field's sugar cane at different weeks and mills may vary. The profit is obviously linked to the sugar content of the field at the processing week. Figure 2 (a) examples such an assignment. At each week, the sugar cane mill processes a different subset of fields. In the same way, other resources such as the cutting and transportation crews can be modeled. Figure 2 (b) examples this abstraction to the cutting crews: A cutting crew is assigned to harvest several fields within its cutting capacities in each week.

The operational planning is much more complex than the tactical, because it involves the cutting sequences for the cutting crews. A part of the operational planning can evidently be modeled in the same way. The sugar cane mills' and cutting

*Figure 2. Modeling the harvest problem as a knapsack problem*

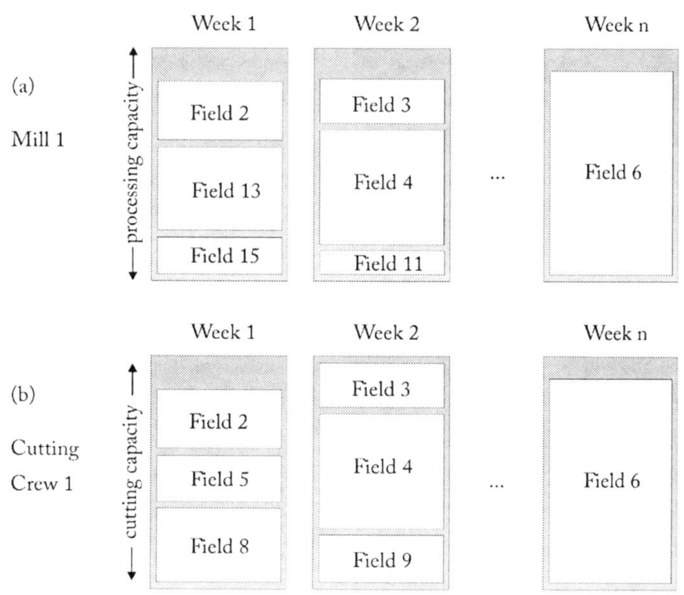

crews' capacities can be modeled based on a 0-1 Multiple Knapsack formulation, using a bin for each day. The main difference is the addition of the cutting sequences for the cutting crews which incorporate knowledge of the distances between one field to another. As such sequencing is often referred to as a harvest scheduling, the problem of determining the field harvest sequence naturally can be seen as a general type of scheduling as well.

The harvest sequence of the cutting crews also resembles the *Vehicle Routing Problem (VRP)*, a class of NP-hard problems. In the VRP, a fleet of vehicles supplies customers, starting from and returning to a depot where the vehicles are reloaded. Each vehicle has a certain capacity and each customer has a certain demand. The objective of these problems is to find optimal vehicle routes that satisfy the customer demands and either minimizes the total distance or the number of necessary vehicles. Rich (1999) broadly introduces into the VRP. For the SCHP, the vehicles can be represented by the cutting crews and the clients by the fields that have to be harvested. The VRP versions that partially resemble the SCHP-OP are the *Capacitated Vehicle Routing Problem (CVRP)*, the *Vehicle Routing Problem with Time Windows (VRPTW)* and the *Periodic Vehicle Routing Problem (PVRP)*. The CVRP limits the capacities for each vehicle. In the VRPTW, the customers have to be supplied within a certain time window. This equals the eligible days to harvest a plantation field. Finally, the PVRP allows that the deliveries may take more than one day. Though these problems are partially very similar to the SCHP-OP, they do not allow the incorporation of the most important aspects: the varying profit, i.e. the sugar content within the cane, as the sugar cane grows.

As the cutting crews' temporal resources may be a crucial factor in the harvest planning, it is desirable that they spend most of their time in cutting sugar cane and as less time as possible in moving from one field to another. This effort is directly linked to the distances and the length of the paths between the fields, leading to the attempt to minimize the total covered distance. From this point of view, the SCHP-OP also intersects with problems such as the *Traveling Salesman Problem (TSP)* and the problem of finding the shortest *Hamiltonian Path*. Latter one aims to find a shortest path between two given vertices that visits each vertex exactly once. In the problem of the *Hamiltonian Cycle*, the starting vertex is the same as the terminal vertex. The TSP is one of the most studied combinatorial problems. It consists in finding a path that covers all vertices so that the total distance is minimal. The *Multiple Traveling Salesman Problem (M-TSP)* is a generalization of the TSP, where more than one salesman can be used within the solution. The M-TSP can also be extended to a variety of vehicle routing problems. It strongly relates to the problem of finding the cutting sequences for the cutting crews. All these problems are known to be NP-hard.

### Complexity

Particularly the operational planning of the SCHP is a very versatile problem that intersects with many classical NP-hard problems known in Combinatorial Optimization. Thus, one may presume that the tactical and operational planning of the SCHP are also NP-hard. Indeed, this can be verified by observing that the corresponding formulations are generalizations of the Generalized Assignment Problem, the Knapsack and Packing problems. A complete discussion on this subject and proofs can be found in Jena (2009).

## MATHEMATICAL FORMULATION

It is now demonstrated how the Generalized Assignment Problem can be abstracted to model partial problems of the tactical and operational planning. Afterwards, the mathematical formulations for the SCHP-TP and SCHP-OP are given.

The GAP is used to assign fields to sugar cane mills in order to maximize the total sugar production profit. Such assignment can be applied to cutting crews and fields as well as transportation crews and fields in the same manner. A complete description and formulation for this problem can be found in Martello and Toth (1990).

## PROBLEM ABSTRACTION TO AN EXTENDED GAP

The bins within the GAP represent the different mills $P=\{p_1,\ldots,p_{nM}\}$ over the time where the sugar cane can be processed. The items represent the plantation fields $F=\{f_1,\ldots,f_{nF}\}$ where the sugar cane is harvested. The productivity of the plantation fields differ each week, as do the mills' processing capacities. In order to decide which plantation fields to harvest in which of the weeks $W=\{w_1,\ldots,w_{nW}\}$ so that the total profit is maximized, it is necessary to distinguish the input data within each week:

- $quantity^{w}_{f,p}$ : total quantity of sugar cane that will be processed at mill $p\in P$ when field $f\in F$ is cut within week $w\in W$.
- $gain^{w}_{f,p}$ : total profit of associating plantation field $f\in F$ to mill $p\in P$ within week $w\in W$. This value involves factors such as the sugar content within the cane and the ATR value at that day.
- $MaxProc^{w}_{p}$ : upper limit of sugar cane that can be processed in mill $p\in P$ within week $w\in W$.

**Observing the problem as a graph, there is one node $field_f$ for each field $f$, one node** $mill^{w}_{p}$ for each mill in each week and one edge $(f,(p,w))$ from each field $f$ to each mill in each week $mill^{w}_{p}$. The last inequality of the GAP is transformed into an equality to assure that all plantation fields are harvested exactly once. This extension results in the problem in Box 1.

**Consider** $MinProc^{w}_{p}$ as the lower limit of sugar cane that should be processed in mill $p$ within week $w$.

$$\sum_{f\in F} quantity^{w}_{f,p}\cdot x_{f,(p,w)} \geq MinProc^{w}_{p};$$
$$\forall p\in P,\ \forall w\in W$$

The input data for the linear program can then be computed as follows, considering the input data given in the mathematical formulation for the SCHP-TP. (see Box 2.)

*Box 1.*

$$MAX \quad \sum_{p\in P}\sum_{f\in F}\sum_{w\in W} gain^{w}_{f,p}\cdot x_{f,(p,w)}$$

subject to

$$\sum_{f\in F} quantity^{w}_{f,p}\cdot x_{f,(p,w)} \leq MaxProc^{w}_{p} \quad ;\forall p\in P,\ \forall w\in W$$

$$\sum_{p\in P}\sum_{w\in W} x_{f,(p,w)} = 1 \quad ;\forall f\in F$$

$$x_{f,(p,w)} \in \{0,1\} \quad ;\forall p\in P,\ \forall f\in F,\ \forall w\in W$$

*Box 2.*

- $quantity^{w}_{f,p} = Q^{w}_{f} \quad ;\forall f \in F, \forall p \in P, \forall w \in W$
- $gain^{w}_{f,p} = \left(ValueATR^{w}_{p} \cdot (CoefPol \cdot Pol^{w}_{f} + CoefAR \cdot AR^{w}_{f}) - ProcCost^{w}_{p} - TransportCost^{w}_{f,p}\right) \cdot Q^{w}_{f} \quad ;\forall f \in F, \forall p \in P, \forall w \in W$
- $MinProc^{w}_{p}$ and $MaxProc^{w}_{p}$ are the same as in the TP input data
- the Pol values are defined as the Pol values without the use of maturation products
- $TransportCost^{w}_{f,p}$ - is the cost to transport a ton of sugar cane from field f to mill p within week $w$

## MATHEMATICAL FORMULATION FOR THE TACTICAL PLANNING

### Input Data

For the mathematical formulation, consider the following input data separated into the following topics: sets, mills, plantation fields and maturation data, maturation products, cutting crews and transportation crews. Note that the following models for the TP and OP do not include third party suppliers. A further extension enables the TP model to harvest the same field more than once throughout the planning period. These two model extensions can be found in Jena (2009).

### Sets

- $W$- Set of all weeks considered during the planning.
- $F$- Set of all plantation fields.
- $P$- Set of all mills.
- $CC$- Set of all cutting crews
- $TC$- Set of all transportation crews.
- $M$- Set of maturation products.
- $V^{w}$- Set of plantation fields $f \in F$ that are eligible for the application of maturation products within week $w \in W$.

### Mills and Sugar Production Process

- $MinProc^{w}_{p}$: Minimum quantity of sugar cane (in tons) that has to be processed by mill $p \in P$ within week $w \in W$.
- $MaxProc^{w}_{p}$: Upper limit of sugar cane (in tons) that can be processed by mill $p \in P$ within week $w \in W$.
- $PFiberMin^{w}_{p}$: Minimum percentage of fiber that has to be processed by mill $p \in P$ within week $w \in W$.
- $ProcCost^{w}_{p}$: Processing Cost (per ton of sugar cane) in mill $p \in P$ within week $w \in W$.
- $ValueATR^{w}_{p}$: Value of one ton of ATR within week $w \in W$ for mill $p \in P$. This is the price at which the sugar is presumably sold.
- *CoefAR:* Coefficient of AR for the ATR calculation.
- *CoefPol:* Coefficient of Pol for the ATR calculation.

### Plantation Fields and Maturation Data

- $Pol^{w}_{f,m}$: Percentage of extracted sucrose from sugar cane in plantation field $f \in F$, that used the maturation product $m \in M$, within week $w \in W$.

- $AR_f^w$ - Percentage of reduced sugar within the sugar cane from plantation field $f \in F$ within week $w \in W$.
- $PFiber_f^w$ : Percentage of fiber within the sugar cane variety in plantation field $f \in F$ within week $w \in W$.
- $NumHA_f$: Size of plantation field $f \in F$ (in hectares).
- $Q_f^w$ : Productivity (of sugar cane in tons) of plantation field $f \in F$ within week $w \in W$.
- $DistFields_{f_1,f_2}$ : Distance (in km) between two plantation fields $f_1 \in F$ and $f_2 \in F$.
- $Dist_{f,p}$: Distance (in km) from plantation field $f \in F$ to mill $p \in P$.
- $InitialAge_f$: Initial age (in weeks) of the sugar cane in plantation field $f \in F$ at the beginning of the planning.
- $MinCutAge_{f,m}$: Minimum age (in weeks) of the sugar cane to be harvested at plantation field $f \in F$, treated by maturation product $m \in M$.
- $MaxCutAge_{f,m}$: Maximum age (in weeks) up to which the sugar cane can be harvested at plantation field $f \in F$, treated by the maturation product $m \in M$.

## Maturation Products

- $CoefReduc_m$: Reduction factor of sugar cane productivity by applying maturation product $m \in M$ on a field.
- $MinAgeAppl_{f,m}$: Minimum age (in weeks) of the sugar cane at plantation field $f \in F$ at which a maturation product $m \in M$ can be applied.
- $CostAppl_{f,m}$: Cost to apply maturation product $m \in M$ on plantation field $f \in F$.

## Cutting Crews

- $MinCut_c^w$ : Minimum quantity of sugar cane (in tons) that has to be cut by cutting crew $c \in CC$ within week $w \in W$.
- $MaxCut_c^w$ : Upper limit of the cutting capacity (in tons) of cutting crew $c \in CC$ within week $w \in W$.
- $CutCost_c^w$ : Cost to cut one ton of sugar cane by cutting crew $c \in CC$ within week $w \in W$.

## Transportation Crews

- $CostTransport_{f,t}^w$ : Transportation cost per km for one ton of sugar cane from plantation field $f \in F$ by transportation crew $t \in TC$ within week $w \in W$.
- $CostAlloc_t^w$ : Fixed allocation cost to use transportation crew $t \in TC$ within week $w \in W$.

## Vinasse

- $MinVin^w$: Minimum quantity of vinasse (in hectares) that should be applied to the plantation fields within week $w \in W$, i.e. the minimum quantity (in hectares) that have to be harvested within this week in order to provide sufficient area for vinasse application.

## Variables

The variables used in this model will now be listed. In contrast to the model of the operational planning, this model extensively uses continuous variables to facilitate the representation of harvested, transported and processed sugar cane quantities as well as the sugar cane reduction through the use of maturation products.

- $k^{w}_{f,m} \in \{0,1\}$: indicates whether the maturation product $m \in M$ is used at plantation field $f \in F$ within week $w \in W$.
- $x^{w}_{f,t,m} \in \{0,1\}$: indicates whether a plantation field $f \in F$, using the maturation product $m \in M$, is associated to the transportation crew $t \in TC$ within week $w \in W$. A value of m = 0 indicates that no maturation product is applied.
- $y^{w}_{f,c,m} \in \{0,1\}$: indicates whether plantation field $f \in F$, using the maturation product $m \in M$, is associated to the cutting crew $c \in CC$ within week $w \in W$. A value of m = 0 indicates that no maturation product is applied. The variables of this type are used only for the weeks in which the minimum and maximum cutting age of the fields are satisfied.
- $a^{w}_{f,m} \in \mathbb{R}_{+}$: quantity of cut sugar cane (in tons) within week $w \in W$ in plantation field $f \in F$, using the maturation product $m \in M$. This quantity does not consider the reduction of sugar cane caused by the application of the maturation product.
- $h^{w}_{f,t,m,p} \in \mathbb{R}_{+}$: quantity of sugar cane (in tons) produced by plantation field $f \in F$ using the maturation product $m \in M$ transported within week $w \in W$ by the transportation crew $t \in TC$ to mill $p \in P$.
- $s^{w}_{f,m,p} \in \mathbb{R}_{+}$: quantity of sugar cane (in tons) from plantation field $f \in F$ using the maturation product $m \in M$ processed by mill $p \in P$ within week $w \in W$.
- $z^{w}_{f,c,m} \in \mathbb{R}_{+}$: quantity of sugar cane (in tons) from plantation field $f \in F$, using the maturation product $m \in M$, cut by the cutting crew $c \in CC$ within week $w \in W$. This variable already considers the reduction of the sugar cane quantity caused by the application of the maturation product.

## Objective Function

The objective function for the tactical planning maximizes the total profit and is given by the equation in Box 3.

The profit is composed of the sugar production by processing harvested sugar cane. The first line already subtracts the processing costs. The following lines represent the cost of applying maturation products, transportation costs (considering the distance between the field and the mill), the cost to hire the transportation crews and the costs to harvest the fields.

*Box 3.*

$$\begin{aligned}
MAX \quad & \sum_{w \in W} \sum_{f \in F} \sum_{m \in M} \sum_{p \in P} \left( ValueATR^{w}_{p} \cdot \left( CoefPol \cdot Pol^{w}_{f,m} + CoefAR \cdot AR^{w}_{f} \right) - ProcCost^{w}_{p} \right) \cdot s^{w}_{f,m,p} \\
& - \sum_{w \in W} \sum_{f \in F} \sum_{m \in M} CostAppl_{f,m} \cdot k^{w}_{f,m} \\
& - \sum_{w \in W} \sum_{t \in TC} \sum_{f \in F} \sum_{m \in M} \sum_{p \in P} CostTransport^{w}_{f,t} \cdot Dist_{f,p} \cdot h^{w}_{f,t,m,p} \\
& - \sum_{w \in W} \sum_{t \in TC} \sum_{f \in F} \sum_{m \in M} \sum_{p \in P} CostAlloc^{w}_{t} \cdot x^{w}_{f,t,m} \\
& - \sum_{w \in W} \sum_{c \in CC} \sum_{f \in F} \sum_{m \in M} CutCost^{w}_{c} \cdot z^{w}_{f,c,m}
\end{aligned}$$

## Constraints

### Field Cutting Obligations

A plantation field $f \in F$ must be cut once within the complete planning period.

$$\sum_{c \in CC} \sum_{m \in M} \sum_{w \in W} y^{w}_{f,c,m} = 1 \quad ;\forall f \in F$$

### Cane Transportation within the Same Week

All sugar cane cut at a plantation field $f \in F$ within week $w \in W$ must be transported within the same week. This constraint passes the previous cutting obligation to the binary processing variables.

$$\sum_{t \in TC} \sum_{m \in M} x^{w}_{f,t,m} = \sum_{c \in CC} \sum_{m \in M} y^{w}_{f,c,m};$$
$$\forall w \in W, \forall f \in F$$

### Cutting Capacities of the Cutting Crews

A cutting crew $c \in CC$ must cut at least $MinCut^{w}_{c}$ tons and at most $MaxCut^{w}_{c}$ tons within week $w \in W$. These constraints match the capacity constraints for cutting crews within the GAP abstraction previously shown.

$$MinCut^{w}_{c} \le \sum_{f \in F} \sum_{m \in M} z^{w}_{f,c,m} \le MaxCut^{w}_{c};$$
$$\forall w \in W, \forall c \in CC$$

### Minimum and Maximum Processing Limits of a Mill

A mill $p \in P$ must process at least $MinProc^{w}_{p}$ tons and is limited to a maximum of $MaxProc^{w}_{p}$ processed tons within week $w \in W$. These constraints match the capacity constraints for mills within the GAP previously shown.

$$MinProc^{w}_{p} \le \sum_{f \in F} \sum_{m \in M} s^{w}_{f,m,p} \le MaxProc^{w}_{p};$$
$$\forall w \in W, \forall p \in P$$

### Sugar Cane Reduction Caused by Maturation Products

This constraint performs the reduction of the quantity of sugar cane, caused by the application of a maturation product.

$$\sum_{c \in CF} z^{w}_{f,c,m} = a^{w}_{f,m} \cdot CoefReduc_{m};$$
$$\forall w \in W, \forall f \in F, \forall m \in M$$

### Maximum Amount of Sugar Cane Cut from a Field

The quantity of sugar cane cut by a cutting crew $c \in CC$ from plantation field $f \in F$ within week $w \in W$ must be less than $Q_{w,f}$ if the crew c is associated with field $f$ within week $w$. Otherwise, this limit will be zero.

$$z^{w}_{f,c,m} \le Q^{w}_{f} \cdot y^{w}_{f,c,m};$$
$$\forall w \in W, \forall c \in CC, \forall f \in F, \forall m \in M$$

### Quantity of Sugar Cane Transported From a Field

The quantity of sugar cane from field $f \in F$ transported by crew $t \in TC$ within week $w \in W$, using the maturation product $m \in M$, must equal the cane quantity at plantation field $f$ within week $w$, using the maturation product $m$.

$$\sum_{t \in TC} \sum_{p \in P} h^{w}_{f,t,m,p} = a^{w}_{f,m} \cdot CoefReduc_{m};$$
$$\forall w \in W, \forall f \in F, \forall m \in M$$

## Maximum Amount of Sugar Cane Transported From a Field

The quantity of sugar cane from field $f \in F$ transported by crew $t \in TC$, within week $w \in W$, must be less than $Q^{w}_{f}$, if the crew $t$ is associated with the plantation field $t$ within week $w$. Otherwise, this limit will be zero.

$$\sum_{p \in P} h^{w}_{f,t,m,p} \leq Q^{w}_{f} \cdot x^{w}_{f,t,m};$$
$$\forall w \in W, \forall t \in TC, \forall f \in F, \forall m \in M$$

## Minimum Age for Application of a Maturation Product to a Field

A maturation product $m \in M$ can only be applied to a plantation field $f \in F$ within week $w \in W$ if this field already possesses the minimum age required for the application of the maturation product. (see Box 4)

## Quantity of Sugar Cane Cut from a Field Through The Use Of a Maturation Product

The quantity of cut sugar cane of from a plantation field $f \in F$, which used the maturation product $m \in M$, within week $w \in W$, must be non-zero only if field $f$, using the maturation product m, was harvested within week $w$.

$$a^{w}_{f,m} = Q^{w}_{f} \cdot \sum_{c \in CC} y^{w}_{f,c,m};$$
$$\forall w \in W, \forall f \in F, \forall m \in M$$

## Quantity of Sugar Cane Processed Per Week

The quantity of sugar cane processed within week $w \in W$ has to be equal to the quantity transported in that week.

$$s^{w}_{f,m,p} = \sum_{t \in TC} h^{w}_{f,t,m,p}$$
$$;\forall w \in W, \forall f \in F, \forall m \in M, \forall p \in P$$

## Minimum Percentage of Fiber

The average fiber percentage within the sugar cane processed at mill $p \in P$ must meet a certain minimum percentage demand within all weeks $w \in W$. (see Box 5)

*Box 4.*

$$\sum_{m \in M} \sum_{w'=w+1}^{w+MinAgeAppl_{f,m}} k^{w'}_{f,m} \leq 1 - \sum_{c \in CC} \sum_{m \in M} y^{w}_{f,c,m} \quad ;\forall f \in F, \forall w \in W$$

*Box 5.*

$$\frac{\sum_{f\in F}\sum_{m\in M} PFiber_f^w \cdot s_{f,m,p}^w}{\sum_{f\in F}\sum_{m\in M} s_{f,m,p}^w} \geq PFiberMin_p^w \quad ;\forall w\in W, \forall p\in P$$

## Relationship between Maturation Product Application and Field Cutting

If a maturation product $m\in M$ was applied to a plantation field $f\in F$, this plantation field must be harvested in one of the weeks within the planning period.

$$\sum_{w\in W}\sum_{c\in CC} y_{f,c,m}^w = \sum_{w\in W} k_{f,m}^w ;$$
$$\forall f\in F, \forall m\in M$$

## Prohibition of Cutting during Non-Valid Weeks When Using Maturation Products

If a maturation product $m\in M$ was applied to a plantation field $f\in F$ within week $w\in W$, then this plantation field cannot be cut during the weeks outside the valid cutting interval *[w+cuttingTimeMin, w+cuttingTimeMax]* defined for each maturation product. (see Box 6)

## Vinasse Application

In each week $w\in W$, there may be a minimum number of hectares that have to be harvested from a set of eligible fields $V_w$ in order to permit the vinasse application. (see Box 7)

## Mathematical Formulation for the Operational Planning

## Input Data

The input data for the OP is separated into the following topics: sets, mills, plantation fields, maturation data, maturation products and cutting crews.

*Box 6.*

$$\sum_{c\in CC}\sum_{w'=w+cuttingTimeMin+1}^{w+cuttingTimeMax-1} y_{f,c,m}^{w'} \leq 1-k_{f,m}^w \quad ;\forall w\in W, \forall f\in F, \forall m\in M$$

*Box 7.*

$$\sum_{c\in CC}\sum_{m\in M}\sum_{f\in V_w} NumHA_f \cdot y_{f,c,m}^w \geq MinVin^w \quad ;\forall w\in W$$

## Sets

- $D$: Set of days considered in this planning horizon.
- $I$: Set of instants at all days. $I^d$ is the set of all instants at day $d \in D$. $d_i$ is the day at instant $i$.
- $F$: Set of existing plantation fields.
- $V^d$: Set of all plantation fields on which vinasse can be applied within day $d$.
- $P$: Set of existing mills.
- $CC$: Set of existing cutting crews.

## Mills and Sugar Production Process

- *CoefAR* & *CoefPol*: Coefficient of AR and Pol for the ATR calculation.
- $ProCost_p$: Processing Cost (per ton) of mill $p \in P$ (equal at all days).
- $ValueATR_p^d$: Value of one ton of ATR within day $d \in D$ for mill $p \in P$.
- $MinProc_p^d$: Minimum quantity of sugar cane (in tons) that has to be processed by mill $p \in P$ within day $d \in D$.
- $MaxProc_p^d$: Upper limit of sugar cane (in tons) that can be processed by mill $p \in P$ within day $d \in D$.
- $PFiberMin_p^d$: Minimum percentage of fiber that has to be processed by mill $p \in P$ within day $d \in D$.

## Plantation Fields and Maturation Data

- $Q_f^d$: Productivity (of sugar cane in tons) of plantation field $f \in F$ within day $d \in D$.
- $Pol_f^d$: Percentage of sucrose extracted from sugar cane from field $f \in F$ at day $d \in D$.
- $AR_f^d$: Percentage of reduced sugar in sugar cane from field $f \in F$ at day $d \in D$.
- $PFiber_f^d$: Percentage of fiber within sugar cane from field $f \in F$ at day $d \in D$.
- $NumHA_f$: Size of plantation field $f \in F$ (in hectares).
- $Disst_{f,p}$: Distance (in km) from plantation field $f \in F$ to mill $p \in P$.
- $Dist_{f_1,f_2}$: Distance (in km) between two plantation fields $f_1 \in F$ and $f_2 \in F$.
- $flint_c$: The plantation field at which the cutting crew $c \in CC$ is located at the beginning of the planning.

## Maturation Products

- $MatGrain_f$: Factor by which the Pol value in the sugar cane (given in percentage) from plantation field $f \in F$ is multiplied (and thereby increased or decreased) if a maturation product has been applied.
- $MatReducFactor_f$: The reduction factor of the productivity of plantation field $f \in F$ if a maturation product has been applied.

## Cutting Crews

- $CuttingCost_c^d$: Cost to cut one ton of sugar cane by cutting crew $c \in CC$ at day $d \in D$.
- $TravelCostCF_c^d$: Displacement cost (per Km) of cutting crew $c \in CC$ at day $d \in D$.
- $MinCut_c^d$ & $MaxCut_c^d$: Minimum quantity and upper limit of sugar cane (in tons) that has to / can be cut by cutting crew $c \in CC$ at day $d \in D$.

## Others

- $MinVin^d$: Minimum quantity of vinasse (in hectares) that should be applied to the plantation fields within day $d \in D$.
- *TransportCost*: Average cost to transport one kg of sugar cane one km from the plantation fields to the mills.
- $\lambda_{c,f}^{i,d}$: Quantity (in tons) of sugar cane cut within day $d \in D$ by cutting crew $c \in CC$ at

plantation field $f \in F$, considering that the cutting crew began cutting at instant $i \in I$.

- $\delta^{i}_{c,f}$: Number of instants that cutting crew $c \in CC$ will need to harvest plantation field $f \in F$, starting at instant $i \in I$.
- $\theta^{i}_{c,f_1,f_2}$: Number of instants that cutting crew $c \in CC$ will need to travel from plantation field $f_1 \in F$ to $f_2 \in F$, starting the journey at instant $i \in I$.

The following lists the variables used within this model.

- $fh^{d}_{c,f} \in \{0,1\}$: Indicates whether cutting crew $c \in CC$ leaves accommodation at day $d \in D$ to start its work at plantation field $f \in F$ at the first available instant of the day.
- $fp_{c,f,p} \in \{0,1\}$: Indicates whether cutting crew $c \in CC$ leaves mill $p \in P$ and begins its activities at field $f \in F$.
- $n^{i}_{c,f} \in \{0,1\}$: Indicates whether cutting crew $c \in CC$ waits one instant at plantation field $f \in F$ (from instant $i$ to instant $i+1$).
- $o^{i_1,i_2}_{c,f_1,f_2} \in \{0,1\}$: Indicates whether cutting crew $c \in CC$ leaves plantation field $f_1 \in F$ at instant $i_1 \in I$ to arrive at plantation field $f_2 \in F$ at instant $i_2 \in I$. This notation is equal to $o^{i_1}_{c,f_1,f_2}$, where the instant at which the cutting crew arrives at plantation field $f_2$ is $i_1 + \theta^{i_1}_{c,f_1,f_2}$.
- $x_{f,p} \in \{0,1\}$: Indicates whether the sugar cane from plantation field $f \in F$ will be processed in mill $p \in P$.
- $y^{i_1,i_2}_{c,f} \in \{0,1\}$: Indicates whether cutting crew $c \in CC$ starts cutting the field $f \in F$ at instant $i_1 \in I$ and is available again at instant $i_2$. This notation is equal to $y^{i_1}_{c,f}$, where the cutting crew will be available again at instant $i_1 + \delta^{i_1}_{c,f}$
- $z^{d}_{f,p} \in \mathbb{R}_+$: Indicates the quantity of sugar cane from plantation field $f \in F$ that will be processed in mill $p \in P$ within day $d \in D$.

The objective function for the formulation aims at maximizing the total profit as shown in Box 8.

Just as in the TP, the profit is presented by the produced sugar by processing the sugar cane harvested in fields (line one). The calculation for the processing costs of harvested sugar cane involves the reduction factor for the total cane mass when a maturation product is applied. It also adds in the increase of the Pol value caused by the product. Line two covers the processing and transportation costs for harvested sugar cane. Line three and four represent harvesting and traveling

*Box 8.*

$$
\begin{aligned}
MAX \quad & \sum_{f \in F}\sum_{d \in D}\sum_{p \in P} MatReducFactor_f \cdot ValueATR^{d}_{p} \\
& \qquad \cdot \left(CoefPol \cdot \left(Pol^{d}_{f} + MatGain_f\right) + CoefAR \cdot AR^{d}_{f}\right) \cdot z^{d}_{f,p} \\
& - \sum_{f \in F}\sum_{d \in D}\sum_{p \in P} MatReducFactor_f \cdot \left(ProcCost_p + TransportCost^{d}_{f} \cdot Dist_{f,p}\right) \cdot z^{d}_{f,p} \\
& - \sum_{i \in I}\sum_{c \in CC}\sum_{f \in F} CuttingCost^{d_i}_{c} \cdot Q^{d_i}_{f} \cdot MatReducFactor_f \cdot y^{i}_{c,f} \\
& - \sum_{f_1 \in F}\sum_{f_2 \in F}\sum_{i \in I}\sum_{c \in CC} TravelCostCF^{d_i}_{c} \cdot Dist_{f_1,f_2} \cdot o^{i}_{c,f_1,f_2}
\end{aligned}
$$

costs for the cutting crews. The following presents the model's constraints.

## CONSTRAINTS

### Industrial and Resource Constraints

#### Plantation Field Cutting Obligation

All plantation fields have to be cut at least once during the planning period.

$$\sum_{c \in CC} \sum_{i \in I} y^{i}_{c,f} = 1 \quad ; \forall f \in F$$

#### Minimum and Maximum Processing Limits of a Mill

The quantity of sugar cane processed within day $d \in D$ by mill $p \in P$ must be at least $MinProc^{d}_{p}$ and no more than $MaxProc^{d}_{p}$.

$$MinProc^{d}_{p} \leq \sum_{f \in F} z^{d}_{f,p} \leq MaxProc^{d}_{p};$$
$$\forall d \in D, \forall p \in P$$

#### Minimum Percentage of Fiber

Each mill $p \in P$ demands a certain average percentage of fiber from the processed sugar cane within each day $d \in D$. The sugar cane cut on these days must meet these minimum values. (see Box 9)

#### Cutting Capacities of the Cutting Crews

The quantity of sugar cane cut by a cutting crew $c \in CC$ within a day $d \in D$ must be at least $MinCut^{d}_{c}$ and no more than $MaxCut^{d}_{c}$. (see Box 10)

#### Vinasse Application

In order to guarantee a sufficient area wherein to apply the vinasse, minimum quantities of cut fields (in hectares) may be defined for each day $d \in D$. The field cuts must finish by day $d$ and the fields $f \in F$ must be eligible for vinasse application, i.e. $f \in V^{d}$. (see Box 11)

*Box 9.*

$$\frac{\sum_{f \in F} PFiber^{d}_{f} \cdot z^{d}_{f,p}}{\sum_{f \in F} z^{d}_{f,p}} \geq PFiberMin^{d}_{p} \quad ; \forall d \in D, \forall p \in P$$

*Box 10.*

$$MinCut^{d}_{c} \leq \sum_{f \in F} \sum_{i \in I} y^{i}_{c,f} \cdot \lambda^{i,d}_{c,f} \leq MaxCut^{d}_{c} \quad ; \forall d \in D, \forall c \in CC$$

*Box 11.*

$$\sum_{f \in V^d} \sum_{c \in CC} \sum_{i_s \in I} \sum_{i_t \in I^d} NumHA_f \cdot y_{c,f}^{i_s,i_t} \geq MinVin^d \quad ;\forall d \in D$$

## Association between Field Cutting and Processing

If sugar cane from plantation field $f$ is processed in a mill, then the cutting variable must be non-zero for this field for any cutting crew.

$$\sum_{c \in CC} \sum_{i \in I} y_{c,f}^{i} = \sum_{p \in P} x_{f,p} \quad ;\forall f \in F$$

## Quantity of Cut Sugar Cane

This constraint sums up the quantity of sugar cane cut by all cutting crews within each day and field.

$$\sum_{p \in P} z_{f,p}^{d} \leq \sum_{c \in CC} \sum_{i \in I} \lambda_{c,f}^{i,d} \cdot y_{c,f}^{i};$$
$$\forall d \in D, \forall f \in F$$

Let $maxProd_f$ be the maximum productivity of field $f$ throughout all planning days. The following constraint allows the quantity of sugar cane cut from field $f$ and processed at mill $p$ to be non-zero only if the sugar cane is processed at this mill.

$$\sum_{d \in D} z_{f,p}^{d} = maxProd_f \cdot x_{f,p};$$
$$\forall f \in F, \forall p \in P$$

### Cutting Crews Network Flow Constraints

Each cutting crew possesses its own network that is independent of the networks of other cutting crews. Each network flows through time (i.e. all time instants) and the fields that the cutting crew may cut. The edges added to the model (representing harvesting, traveling and waiting of the crews) are based on the cutting rate, travel speed and daily time availability also informed within the input data. The network flows are handled by the following constraints:

### Initial Cutting Crew Position at a Mill

In the beginning of the planning, each cutting crew $c \in CC$ is located either at a plantation field or at a sugar cane mill. If the cutting crew is located at a mill, it must travel to one of its valid fields before the first instant of the planning. Let $CC_p$ be the set of cutting crews that are located at a mill. The mill at which $c$ is located is referred to as $p_c$. The flow must be inserted into one of the variables $fp$ for the mill $p_c$.

$$\sum_{f \in F} fp_{c,f,p_c} = 1 \quad ;\forall c \in CC_p$$

### Initial Cutting Crew Position at a Field and Passed Flow from a Mill

The flow for each cutting crew $c \in CC$ must pass to the node of the first available instant $iFirst_c$ of the cutting crew (RHS). The flow can then be used to cut a field, wait at the field or move to another field (LHS).

Let $loc_c$ be the location of $c$ at the beginning of the planning. (see Box 12)

*Box 12.*

$$y_{c,f}^{iFirst_c} + n_{c,f}^{iFirst_c} + \sum_{f_2 \in F} o_{c,f,f_2}^{iFirst_c} = \begin{cases} 1, & \text{if } f = fInit_c \text{ and } loc_c \text{ is a plantation field} \\ fp, & \text{if } loc_c \text{ is mill } p \\ 0, & \text{otherwise} \end{cases} \quad ;\forall c \in CC, \forall f \in F$$

## General Flow Conservation: Cutting, Waiting and Traveling

Once the flow entered the network, it must pass along time. The flow that enters a node at instant *i* must also leave it. Flow can enter from cutting variables for the field, from a waiting variable at the previous instant or by traveling from one of the other fields. If flow enters the node, it leaves it again either by cutting the field, waiting one instant or moving to another eligible field. In the flow conservation constraints, one must distinguish between mechanical and manual cutting crews. Mechanical cutting crews remain on the field during the night, whereas manual crews return to their accommodation where they spend the night. Let $IF_{cMan}$ be a set with all instants except the first available instant of each day, for each manual cutting crew *cMan*. Let $I_{cMec}$ be the set with all available instants of all days, available for each mechanical cutting crew.

Figure 3 (lower example) exemplifies a network flow for a mechanical cutting crew. The nodes in the grey area are not available for work. Hence, all flow variables skip the nodes of not available instants. (see Box 13)

Manual cutting crews are usually hosted at a place where they spend the night at the end of the day. The next day, they will be able to work at any field (eligible for these crews). The upper example in Figure 3 shows an example of a network for a manual cutting crew. Flow from cuttings and waiting that would enter into the first available instant of a day, now enters into the *accommodation node*. From the accommodation, the flow passes to any field that the cutting crew can cut.

The previous equation is valid for the nodes of manual cutting crews as well, except the ones valid for the first available instants of each day. The flow that would have entered in these first nodes of each day, now enter into the node of the crew's accommodation. This is guaranteed by the

*Figure 3. Example network flows for cutting crews: mechanical crews skip not available instants, whereas manual crews return to their accommodation*

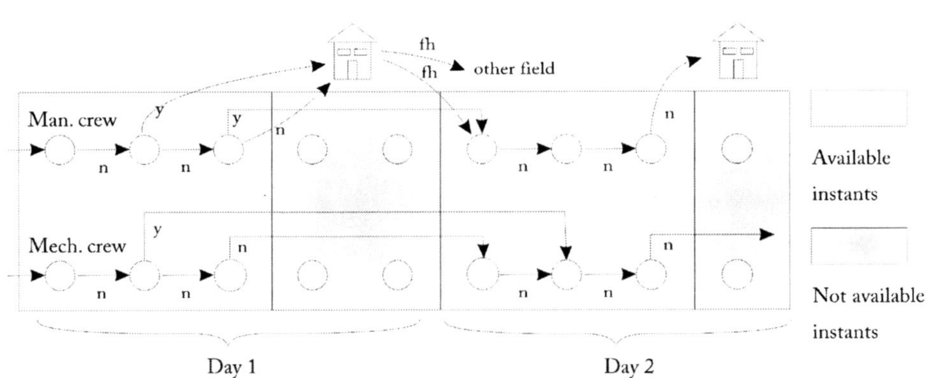

*Box 13.*

$$n_{c,f}^{i-1} + \sum_{i' \in I} y_{c,f}^{i',i} + \sum_{f' \in F} \sum_{i' \in I} o_{c,f',f}^{i',i} = n_{c,f}^{i} + y_{c,f}^{i} + \sum_{f' \in F} o_{c,f,f'}^{i} \quad ;\forall i \in IF_{cMan} \cup I_{cMec}^{\setminus \{i_0\}}, \forall f \in F, \forall c \in CC$$

following equation, where $I_c$ represents a set with the first available instant of each day of the cutting crew $c$. Note that this constraint is not generated for the node of the very first available instant of the entire planning. (see Box 14)

Finally, the following equation distributes the flow from the accommodation node to all plantation fields for each cutting crew.

$$fh_{c,f}^{d_i} = y_{c,f}^{i} + n_{c,f}^{i} + \sum_{f' \in F} o_{c,f,f'}^{i};$$
$$\forall i \in I_c^{\setminus \{i_0\}}, \forall f \in F, \forall c \in CC$$

## SOLUTION METHODS AND ANALYSIS

### Solution Techniques

Breakthrough advances in information technologies have provided new possibilities to the use Mathematical Programming. Such great developments in computers' capabilities make it possible to model large problems from practice as mixed integer programs and solve them effectively using mathematical solvers. Solvers such as ILOG CPLEX and XPRESS-MP underlie a continuous development process and had their efficiency improved by the years. For this reason they are now capable of solving even larger problems.

The use of decision support systems in industry is usually subject to a limited time available to find solutions. However, there is no guarantee that a MIP solver finds good quality solutions in a given time. As problems get bigger, i.e. the quantity of input data increases, using the MIP solver on the original problem does not always lead to a good solution quality in reasonable time.

Therefore, the Operations Research community strongly invests in finding techniques and methods to simplify mathematical models or support the actual MIP solver in order to solve the problems easier. The following techniques can typically be applied to the SCHP. Details regarding their implementation can be found in Jena and Poggi (2009) and Jena (2009).

### Preprocessing and Initial Solutions

In the case of the present problem, preprocessing aims at reducing the size of the input data in order to facilitate the solution of the problem. Typical preprocessing includes the elimination of redundant variables and rounding the values within the input data. A more specific technique is the clustering of geographically adjacent plantation fields with equal or similar decision properties such as sugar cane maturity, Pol, AR and fiber

*Box 14.*

$$\sum_{f \in F} \sum_{i' \in I} \left( y_{c,f}^{i',i} + n_{c,f}^{i-1} + \sum_{f' \in F} \sum_{i' \in I} o_{c,f',f}^{i',i} \right) = \sum_{f \in F} fh_{c,f}^{d_i} \quad ;\forall i \in I_c, \forall c \in CC$$

values. Furthermore, the set of direct connections between two plantation fields (used by the cutting crews to move from one field to another) can be reduced by applying certain filtering rules.

Initial solutions can be generated in order to provide starting points for the solver. Some MIP solvers make use of such solutions when applying heuristics or metaheuristics, for example evolutionary algorithms, in order to find new solutions. The structural diversity of the provided solutions as well as their quality are typically crucial to the success of such approaches.

### Valid Inequalities

Valid inequalities may be able to strengthen the MIP formulation and yield better bounds given through the optimal solution of the linear relaxation. In many cases, strong inequalities have a positive impact on the solution process through the solver. For many problems using the mathematical formulation for the operational planning presented in this chapter, the optimum solution of the linear relaxation turned out to be strongly fractional, especially for the traveling and harvesting variables. One set of valid inequalities aims at preventing strongly fractional results by limiting the use of travel variables where feasible.

### Alternative Solution Strategies and MIP Heuristics

Frameworks based on Branch-and-Bound for mathematical programming aim at solving a generic problem. However, there are several possibilities to involve semantic information of a problem to solve it. In the context of harvest scheduling in sugar cane industry, given by the SCHP-OP, there are several opportunities to apply such semantic information to solve the problem:

- The planning horizon can be split into smaller time periods, for example into weeks. Having information about the indicated week for each plantation field, the fields can be separated into each week so that the final problem is much smaller.
- The objective function can be simplified through disregarding certain variables. For example, travel or harvest costs only marginally influence the objective function and may therefore be disregarded from the objective function.
- The problem can be segregated into a separated planning for each cutting crew. After solving all subproblems, the solutions to these subproblems are composed to one solution feasible for the entire problem.

## Solution Analysis

The algorithms for the tactical and operational planning were tested with real-world instances, provided by the Brazilian sugarcane producer Grupo Virgolino de Oliveria (GVO). The instances include input data from four industrial units in the state of São Paulo. Monthly, about 1600 plantation fields are harvested, representing 1.3 million tons of sugar cane cut by almost 40 cutting crews. The mills possess a total processing capacity of 60,000 tons per day, equivalent to about 10 million tons per year. The manual cutting crews cut about 1,600 tons of sugar cane daily, whereas mechanical cutting crews cut 2,300 tons per day. In addition to these instances, artificial instances were created varying in the quantity of plantation fields, the availability of cutting crews and the processing capacities of the mills.

The MIP solver ILOG CPLEX version 11.2 was used to solve the problems. The tactical planning turned out to be easily solved to optimality for all available instances. In contrast to the tactical planning problem, experiments demonstrated that the basic use of CPLEX for the operational planning model was not sufficient to solve many of the instances in a reasonable time of up to 30 minutes. The previously presented pre-processing techniques were then applied in order to reduce the size of the problem. Distance filtering turned out to be a very important task as it effectively reduced the problem size. Though the limitation of distances used within the model results in a worse optimal solution for the problem, the quality of the final solutions obtained in time limited experiments improved significantly.

In the overall planning of the SCHP, the tactical planning has a crucial influence to the period in which a field will be harvested. This decision is essentially guided by the maturity of the cane. A simple example demonstrates the strong influence of these values. The example instance considers ten fields with a constant productivity of 150 tons (i.e. the productivity is the same in all weeks). The planning horizon contains eight weeks. The Pol values for all fields vary according to the values illustrated in Figure 4, whereas the AR values are constant. The maximum processing capacities at the sugar cane mill are limited to 10,000 tons per week, i.e. not all fields' sugar cane can be processed at one single day. It is supposed that the optimal solution cuts as many fields as possible within the week with the highest Pol value. The lower graph in Figure 4 confirms this assumption: the fields were cut within the weeks with the best Pol values.

Figure 5 presents the interval of the minimum and maximum processing capacities of the mills and the processed sugar cane tons (represented by the horizontal bar) at both mills suggested by an optimal solution. The minimum processing demands are respected in all weeks. An interesting observation is that as much sugar cane as possible was processed at the second mill. This in founded in the fact that the processing costs of the second mill are lower than at the first mill.

An analysis of the contribution of the different factors to the objective function, based on the real-world instances, demonstrated that the profit made by the sugar production is clearly the dominating variable within the objective function.

*Figure 4. Pol distribution per week for an example instance and the quantity of harvested sugar cane per week in the optimal solution for this instance*

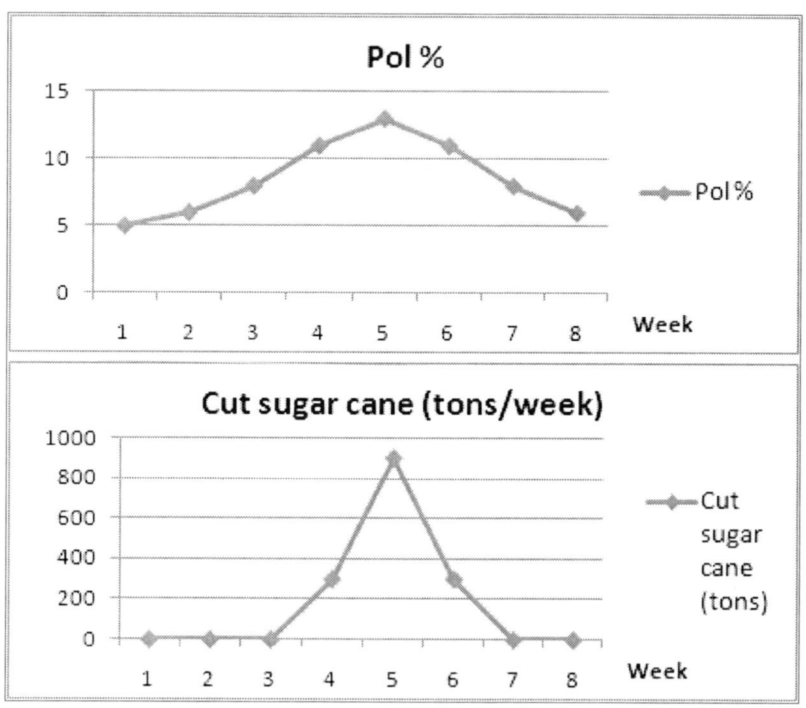

The profit made by sugar production is always much greater than the costs to harvest the field, transport the sugar cane and process it at a mill. Even with resource costs varying along the time, the solver still prefers to harvest the field close to its maturation peak as the profit increase is still much larger than the possible additional costs.

## Analysis of the Optimization Process

Whereas the MIP solver faced no problems to solve the instances for the tactical planning in short time, the operational planning problem turned out to be much more difficult to be solved. One big difference is the addition of the harvest sequence for each cutting crew, represented by a network flow. Instances become even harder to be solved when capacity bounds such as the cutting crews' minimum and maximum cutting capacities and the mills' minimum and maximum processing capacities are tight. Constant harvesting is often indispensable in order to satisfy such minimum cutting and processing demands. However, cutting crews spend time traveling from one plantation field to another. Throughout the optimization process, during constant improvement of the current solution, cutting crews tend to be associated to less travels, i.e. the solver prefers short connections between plantation fields or not traveling at all in order to have more time for harvesting.

## Analysis of the Planning Quality in Practice

The GVO was strongly involved in the specification of the industrial processes. This intense dialogue was held throughout the entire development of the model in order to identify the crucial decisions within the system and guarantee a realistic representation of the processes. Prior to the here presented algorithms, the GVO used a basic linear programming based solver as well as manual planning in order to generate the planning solutions, i.e. the harvest schedules including the allocation of resources. GVO members compared the solutions generated by the solver described in

*Figure 5. Processed sugar cane at the mills suggested by an optimal solution for a real-world instance*

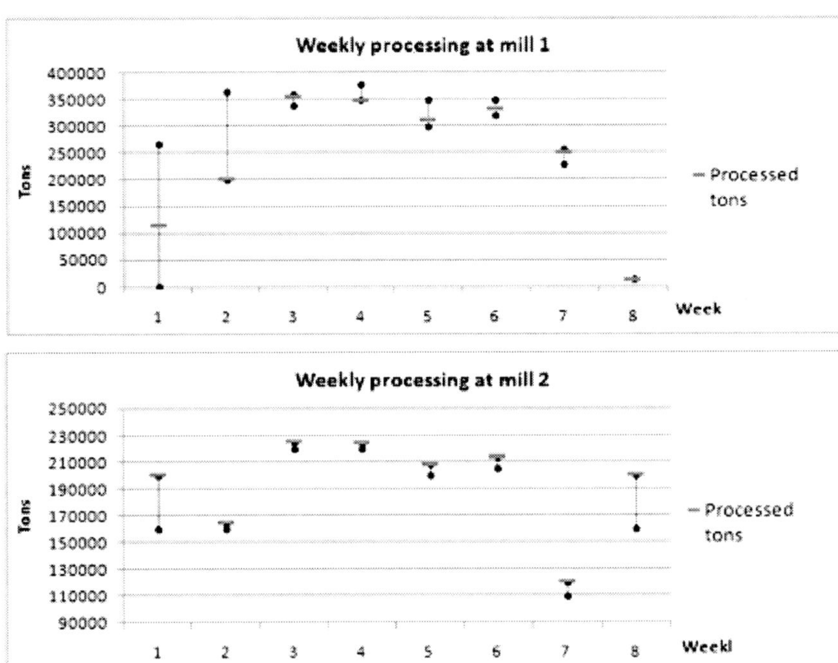

this work with their previous solutions by using an existing interface. This interface supports the analysis and comparison of the solutions proposed by the models, e.g. by visualization of the cutting crew routes and ample statistics about the use of transportation and cutting crews, harvested field areas, the total expected revenue and the compliance of capacity constraints and processing demands. This information enables the managers to evaluate the quality of the proposed planning as well as its execution in practice. The analysis performed on the solutions generated by the here proposed models and algorithms showed that it is possible to meet the industrial constraints better than before and significantly increase the total profit at the same time.

## FUTURE RESEARCH DIRECTIONS

This work aimed at providing a decision support tool for sugar cane harvest planning in practice. Due to constantly improving effectiveness of Operations Research tools in industrial problems, it is likely that their use in the agricultural sector will continuously increase, especially to support crucial decisions.

Experiences that will be made in the future through the use of the here presented tool may show that some of the model's functionalities or constraints still do not perfectly represent the requirements in practice. The *vinasse application* is one that might benefit from a model closer to the practical reality. In the models presented in this thesis, the demands for vinasse application have to be satisfied by fields whose harvesting terminates at the same week/day of the vinasse demand. For tight vinasse demands, i.e. demands where the required area is close to the total harvested area, the demands may be difficult to meet. However, in practice, vinasse may be applied on a field up to two months after it is harvested. The use of such constraints instead of the ones involved in the current models will facilitate meeting tight vinasse demands.

Finally, the models presented in this work hold potential to be used to simulate the competitiveness and feasibility of scenarios with different resource configurations. For example, the effect of using more mechanical cutting fronts or more mills can be tested. Such a simulation optimization is realized by introducing new variables that dimension the use of certain resources and may identify great potential to further reduce costs.

## CONCLUSION

This work investigated mixed integer formulations and algorithms for the planning of sugar cane cultivation and harvesting. The objective in such planning is the maximization of the total profit given by the quantity of sugar production. The quantity of sugar depends on the sucrose (Pol) and reduced sugar (AR) content within the cane as well as on the total quantity of cane. Since such values vary throughout the cane maturation, it is desirable to harvest a field at the maturation peak of the cane. In addition to the determination of the optimal moment to harvest the fields, the planning includes assignments of the cutting and transportation crews. It also suggests the application of vinasse in harvested field area and the application of maturation products in order to increase the sucrose percentage within the cane. The entire planning is subject to several industrial and resource constraints. These constraints include the availability of cutting and transportation crews as well as their cutting and transportation capacities, respectively. Sugar cane mills contain certain processing minimum demands and maximum processing capacities.

The system was divided into a tactical and an operational planning. Both problems were presented in detail. The former performs the planning for up to a whole harvest season, i.e.

approximately twelve months. Its decisions are made for each week. The latter performs a daily planning for up to 30 days and involves the determination of the harvest sequence for each cutting crew. A Brazilian sugar cane producer was involved in the development process in order to guarantee a realistic presentation of the processes. Real-world instances, based on a current harvest planning, were provided in order to evaluate the TP and the OP.

The experiences made within this work showed that the presented algorithms are very effective to solve even larger problems as they appear in practice in reasonable time. Some of the solution strategies that were suggested solved the instance benchmark only a few percents from optimality. The presented model considers numerous industrial and resource constraints that are crucial to the sugar cane cultivation and harvest planning. Due to this close representation of the real processes, it is likely that the obtained solutions are feasible in practice. Members of the sugar cane company performed evaluations of the generated solutions and confirmed the feasibility as well as the high quality of the obtained solutions. These results suggest that the use of the decision support system presented in this work is likely to make better use of the resources and the sugar cane itself and thus comprises a strong potential to increase the total profit to sugar cane companies and, consequently, reduce prices to the society. As sugar cane is also used to produce agro fuels, this approach also comprises potential to improve the efficiency of such production.

## ACKNOWLEDGMENT

We would like to thank the Grupo Virgolino de Oliveira for their constant support throughout the design, development and evaluation of this work. We also gratefully acknowledge the whole of the GAPSO company as a collaborator for the thesis research this work is based on.

## REFERENCES

Ahumada, O., & Villalobos, J. R. (2009). Application of planning models in the agri-food supply chain: A review. *European Journal of Operational Research*, *196*(1), 1–20. doi:10.1016/j.ejor.2008.02.014

Alencar, C. E. R., Corrêa, R. F., & de Lima Neto, F. B. (2006). *Ferramenta para suporte à decisão de frentes de corte de cana-de-açúcar usando algoritmos genéticos*. Universidade de Pernambuco.

Audsley, E. (1985). The development of operational research at the national institute of agricultural engineering. *The Journal of the Operational Research Society*, *36*(2), 117–123.

Audsley, E., Dumont, S., & Boyce, D. (1978). An economic comparison of methods of planting cereals, sugar beets and potatoes and their interactions with harvesting, timeliness and available labour by linear programming. *Journal of Agricultural Engineering Research*, *23*(3), 283–300. doi:10.1016/0021-8634(78)90102-6

Bixby, A., Downs, B., & Self, M. (2006). A scheduling and capable-topromise application for swift & company. *Interfaces*, *36*(1), 69–86. doi:10.1287/inte.1050.0184

Boehlje, M., Fulton, J. & AP Gray and T. N. (2002). *Strategic issues in the changing agricultural industry*. Purdue University Cooperative Extension Service.

Caliari, M., Júnior, M. S. S., & Serra, G. E. (2004). Programação matemática aplicada à colheita de cana-de-açúcar com enfase na recuperação máxima de açúcar. *Pesquisa Agropecuária Tropical.*, *35*, 1–8.

Díaz, J. A., & Pérez, I. G. (2000). Simulation and Optimization of sugar cane transportation in harvest season. *Proceedings of the Simulation Conference, Winter, 2*, 1114-1117.

Equi, L., Gallo, G., Marziale, S., & Weintraub, A. (1997). A combined transportation and scheduling problem. *European Journal of Operational Research, 97*(1), 94–104. doi:10.1016/S0377-2217(96)00085-9

FAO. (2008). Major Food and Agricultural Commodities and Producers, Food and Agriculture Organization of the United Nations, The Statistics Division, Economic and Social Department. Retrieved October, 2008 from http://www.fao.org/es/ess/top/country.html?lang=en.

Ferrer, J.-C., Cawley, A. M., Maturana, S., Toloza, S., & Vera, J. (2008). An optimization approach for scheduling wine grape harvest operations. *International Journal of Production Economics, 112*(2), 985–999. doi:10.1016/j.ijpe.2007.05.020

France, J., & Thornley, J. (2007). *Mathematical Models in Agriculture*. Wallingford, UK: Cabi Publishing.

Gala, P. Y. L., Lyneb, P., Meyerb, E., & Solerc, L. G. (2008). Impact of sugarcane supply scheduling on mill sugar production: A South African case study. *Agricultural Systems, 96*(1-3), 64–74. doi:10.1016/j.agsy.2007.05.006

Glen, J. J. (1987). Mathematical models in farm planning: a survey. *Operations Research, 35*(5), 641–666. doi:10.1287/opre.35.5.641

Gómez, J., & Rodríguez, O. (2000). Effects of vinasse on sugarcane (Saccharum officinarum) productivity (in spanish). [LUZ]. *Rev. Fac. Agron., 17*, 318–326.

Hao, Q., Meng, F., Zhou, Y., & Wang, J. (2004). Determining the optimal selective harvest strategy for mixed-species stands with a transition matrix growth model. *New Forests, 29*(3), 207–219. doi:10.1007/s11056-005-5656-0

Heady, E. O. (1954). Simplified presentation and logical aspects of linear programming technique. *Journal of Farm Economics*.

Henningsson, M., Karlsson, J., & Rönnqvist, M. (2006). *Optimization models for forest road upgrade planning*. International Series In Operations Research, Handbook of Operations Research In Natural Resources.

Higgins, A. (2002). Australian Sugar Mills Optimize Harvester Rosters to Improve Production. *Interfaces, 32*(3), 15–25. doi:10.1287/inte.32.3.15.41

Higgins, A. (2006). Scheduling of road vehicles in sugarcane transport: A case study at an Australian sugar mill. *European Journal of Operational Research, 170*(3), 987–1000. doi:10.1016/j.ejor.2004.07.055

Higgins, A., Antony, G., Sandell, G., Davies, I., Prestwidge, D., & Andrew, B. (2004). A framework for integrating a complex harvesting and transport system for sugar production. *Agricultural Systems, 82*(2), 99–115. doi:10.1016/j.agsy.2003.12.004

Higgins, A., & Davies, I. (2005). A simulation model for capacity planning in sugarcane transport. *Computers and Electronics in Agriculture, 47*(2), 85–102. doi:10.1016/j.compag.2004.10.006

Higgins, A. & Laredo, L. (2006). Improving harvesting and transport planning within a sugar value chain. *Journal of the Operational Research Society, 57* (4), 367–376 (10).

Higgins, A., Thorburn, P., Archer, A. & Jakku, E (2007). Opportunities for value chain research in sugar industries. *Agricultural Systems, 94*, 611-621 (3).

Jena, S. D. (2009). *A Mixed Integer Programming approach for sugar cane cultivation and harvest planning*. Master thesis, Pontifical Catholic University of Rio de Janeiro, Brazil.

Jena, S. D. & Poggi de Aragão, M. V. (2009). Sugar cane cultivation and harvesting: MIP approach and valid inequalities. *Anais do XLI Simpósio Brasileiro de Pesquisa Operacional.*, 969-980.

Karlsson, J., Rönnqvist, M., & Bergström, J. (2003). Short-term harvest planning including scheduling of harvest crews. *International Transactions in Operational Research, 10*(5), 413–431. doi:10.1111/1475-3995.00419

Karlsson, J., Rönnqvist, M., & Bergström, J. (2004). An optimization model for annual harvest planning. *Canadian Journal of Forest Research, 34*(8), 1747–1754. doi:10.1139/x04-043

Lowe, T. J., & Preckel, P. V. (2004). Decision technologies for agribusiness problems: A brief review of selected literature and a call for research. *Manufacturing and Service Operations Management, 6*(3), 201–208. doi:10.1287/msom.1040.0051

Lucas, M. T., & Chhajed, D. (2004). Applications of location analysis in agriculture: a survey. *The Journal of the Operational Research Society, 55*(6), 561–578. doi:10.1057/palgrave.jors.2601731

Martella, D. L., Eldon, A., & Weintraub, A. (1998). Forest management challenges for operational researchers. *European Journal of Operational Research, 104*, 1–17. doi:10.1016/S0377-2217(97)00329-9

Martello, S., & Toth, P. (1990). *Knapsack problems: Algorithms and Computer Implementations*. Chichester, UK: John Wiley and Sons.

Masini, G. (2003). *Supply chain planning optimization in the fruit industry*. CEPAC Working Paper.

McCarl, B. A., Candler, W. V., Doster, D. H. & Robbins, P., R. (1978). Experience with mass audience linear programming for farm planning. *Mathematical Programming Studies, 9*.

Mitchell, S. A. (2004). *Operational forest harvest scheduling optimisation: a mathematical model and solution strategy*. Doctoral dissertation, University of Auckland, New Zealand.

Oliveira, F. R., Pacheco, D. F., Leonel, A., & de Lima Neto, F. B. (2006). Intelligent Support Decision in Sugar Cane Harvest. *IV World Congress on Computers in Agriculture (WCCA 2006)*, Orlando, (pp. 456-462).

Pacheco, D. F., & Lima Neto, F. B. (2008). *A Fair Comparison of Representations, Operators and Algorithms for the Sugarcane Harvest Problem*. IEEE International Conference on Systems, Man, and Cybernetics.

Pacheco, D. F., Oliveira, F. R., & Lima Neto, F. B. (2008). *Including Multiobjective Abilities in the Hybrid Intelligent Suite for Decision Support*. International Joint Conference on Neural Networks-IJCNN (World Congress on Computational Intelligence-WCCI2008).

Pacheco, D. F., Pontes, T. D., & Lima Neto, F. B. (2007). *How to Obtain Fair Managerial Decisions in Sugar Cane Harvest Using NSGA II*. 7th International Conference on Hybrid Intelligent systems (HIS 07).

Pacheco, D. F., Regueira, F. S., & Lima Neto, F. B. (2005). Utilização de Redes Neurais Artificiais em colheitas de cana-de-açúcar para predição de PCC, TCH e fibra. *ALCOOLbrás, 90*, 60–63.

Rich, J. L. (1999). *A computational study of vehicle routing applications*. Doctoral dissertation, Houston, TX, USA.

Weintraub, A., & Navon, D. (1976). A forest management planning model integrating silvicultural and transportation activities. *Management Science, 22*, 1299–1309. doi:10.1287/mnsc.22.12.1299

Zhaorong, J., Higgins, A., & Prestwidge, D. B. (2005). An integrated statistical and optimisation approach to increasing sugar production within a mill region. *Computers and Electronics in Agriculture, 48*(2), 170–181. doi:10.1016/j.compag.2005.03.004

## ADDITIONAL READING


Aardal, K. I., Van Hoesel, S., Lenstra, J. K., & Stougie, L. (1997). A decade of combinatorial optimization. Technical Report UU-CS-1997-12, Department of Information and Computing Sciences, Utrecht University.

Andersson, G., Flisberg, P., Liden, B., & Rönnqvist, M. (2008). RuttOpt – A decision support system for routing of logging trucks. *Canadian Journal of Forest Research*, *38*, 1784–1796. doi:10.1139/X08-017

Blum, C., & Roli, A. (2003). Metaheuristics in combinatorial optimization: Overview and conceptual comparison. *ACM Computing Surveys*, *35*(3), 268–308. doi:10.1145/937503.937505

Bradley, S. P., Hax, A. C., & Magnati, T. L. (1977). *Applied Mathematical Programming*. Addison-Wesley.

Bredström, D., Lundgren, J. T., Rönnqvist, M., Carlsson, D., & Mason, A. (2004). Supply chain optimization in the pulp mill industry – IP models, column generation and novel constraint branches. *European Journal of Operational Research*, *156*, 2–22. doi:10.1016/j.ejor.2003.08.001

Carlsson, D., & Rönnqvist, M. (2007). Backhauling in forest transportation – models, methods and practical usage. *Canadian Journal of Forest Research*, *37*, 2612–2623. doi:10.1139/X07-106

Computational Infrastructure for Operations Research. C.-O. http://www.coin-or.org.

D'Amours, S., Rönnqvist, M., & Weintraub, A. (2008). Using Operational Research for supply chain planning in the forest product industry. *INFOR*, *46*(4), 47–64.

Dash Optimization. www.dashoptimization.com.

Dreyfus, S. (2002). Richard bellman on the birth of dynamic programming. *Operations Research*, *50*, 48–51. doi:10.1287/opre.50.1.48.17791

Flisberg, P., Liden, B., & Rönnqvist, M. (2009). A hybrid method based on linear programming and tabu search for routing of logging trucks. *Computers & Operations Research*, *36*, 1122–1144. doi:10.1016/j.cor.2007.12.012

Forsberg, M., Frisk, M., & Rönnqvist, M. (2005). FlowOpt – a decision support tool for strategic and tactical transportation planning in forestry. *International Journal of Forest Engineering*, *16*(2), 101–114.

Grötschel, M., Lovasz, L., & Schrivjer, A. (1988). Geometric Algorithms and Combinatorial Optimization. Springer, 1988.

Gunnarsson, H., Rönnqvist, M., & Carlsson, D. (2006). A combined terminal location and ship routing problem. *The Journal of the Operational Research Society*, *57*, 928–938. doi:10.1057/palgrave.jors.2602057

Handbook of Operations Research in Natural Resources. International Series in Operations Research & Management Science. Vol. 99. Springer US. ISSN0884-8289. 2007.

IBM ILOG CPLEX. www.ilog.com.

Institut für Informatik. Universität zu Köln. Abacus - a branch-and-cut system. http://www.informatik.uni-koeln.de/abacus/.

Linear Programming KitG. N. U. G. http://www.gnu.org/software/glpk/glpk.html.

Mixed Integer Optimizer. Coral. http://coral.ie.lehigh.edu/minto/.

Pochet, Y., & Wolsey, L. A. (2006). *Production Planning by Mixed Integer Programming*. Secaucus, NJ, USA: Springer Series in Operations Research and Bibliography Financial Engineering, Springer-Verlag New York, Inc.

Rönnqvist, M. (2003). Optimization in forestry, Mathematical Programming. *Ser. B*, *97*, 267–284.

SCIP. Solving Constraint Integer Programs. http://scip.zib.de.

Wolsey, L. A. (1998). *Integer Programming*. Wiley and Sons.

Wolsey, L. A., & Nemhauser, G. L. (1988). *Integer and Combinatorial Optimization*. Wiley-Interscience.

## KEY TERMS AND DEFINITIONS

**Cutting Crews:** A team of workers that cut sugar cane is referred to as a cutting crew. A cutting crew can either be manual or mechanical (i.e. machines).

**Decision Support System:** A decision support system (DSS) is a software that supports decision making, typically by analysis of existing data, derivation of new data and visualization of certain results. Usually, a DSS combines and filters several types of information in order to lift out trends and key results. Due to the complexity of information that can be difficult when done manually. Optimization methods are often included in order to derive results or solutions to a given problem.

**Harvest Scheduling:** This term refers to the schedule for harvesting sugar cane fields, i.e. it determines the exact moment when each of the fields will be harvested.

**Mixed-Integer Programming:** Mixed-Integer Programming (MIP) is a technique of mathematical optimization. A mixed-integer program represents a problem in terms of an objective function and a set of constraints. The variables used can be either real or integer variables.

**Operational Planning:** An operations planning determines how an actual planning will be performed for a short period. It commonly attributes resources and defines for traveling routes. In the context of sugar cane harvesting, the OP associates the cutting crews to fields and determines the harvesting sequence for each cutting crew.

**Operations Research:** Operations Research (OR) is an interdisciplinary branch of applied mathematics in order to derive good quality solutions to decision making problems. Operations Research combines several domains such as mathematical modeling, mathematical optimization, statistic analysis and simulation.

**Sugar Cane Harvesting:** Sugar cane harvesting refers to the process of harvesting and transporting the sugar cane on the fields. The sugar is cut by manual or mechanical cutting crews, and transported to a processing unit like a sugar cane mill. All decisions regarding the growth of sugar cane are included in the term of sugar cane cultivation.

**Tactical Planning:** A tactical planning (TP) aims at deciding for a mid-term period of around a year. It usually performs planning in terms of resources and locations. In the context of sugar cane harvesting, the most important decisions of the TP are the exact weeks in which the sugar cane fields should be harvested due to their maturation curve. It also may decide whether maturation products should be used in order to change the maturation curve.

**Transportation Crews:** A team of drivers and other specific operations persons together with trucks used to transport the sugar cane from a field to a processing unit is referred to as a transportation crew.

**Valid Inequalities:** An inequality is valid if it can be add to the model without changing the convex hull given by the feasible solutions of the model. Valid inequalities (VI) do not cut off feasible integer solutions, but aim at tightening the bound given by the linear relaxation.

# Chapter 10
# An Application of a Positive Mathematical Programming Model to Analyse the Impact of Agricultural Policy Measures in the Spanish Agricultural Sector[1]

**Lucinio Júdez Asensio**
*ETSIA/UPM, Spain*

**Rosario de Andrés Gómez de Barreda**
*CSIC, Spain*

**Miguel Angel Ibáñez Ruiz**
*ETSIA/UPM, Spain*

**José-Luis Miguel de Diego**
*COAG, Spain*

**Elvira Urzainqui Miqueleiz**
*CSIC, Spain*

## ABSTRACT

*In agricultural economics, one of the greatest weaknesses in mathematical programming models for the evaluation of agricultural processes is the calibration of the model in a base year. The reason for this is that it is extremely difficult, if not impossible, to introduce all the variables affecting farmers' decisions in the models and thus obtain reliable results. This chapter presents a method for calibrating mathematical programming models using limited information. From the mathematical programming properties, by using the dual form of the original model, this methodology allows the model results to reproduce the situation existing in a baseline situation of the unit (farm, region) modelled. This method, called Positive Mathematical Programming, is currently being used in a great number of analyses of new agricultural policies. In this chapter it is applied to analyse the impact of recent measures of the European Common Agricultural Policy in Spain.*

DOI: 10.4018/978-1-61692-871-1.ch010

## INTRODUCTION

In recent years Europe has seen important developments in the application of mathematical programming in the field of agricultural economics. This is due in part to the need for analytical tools to evaluate the impact of the Common Agricultural Policy (CAP) changes put in place since 1992 in the agricultural sector (the McSharry Reform), and also due to the appearance of positive mathematical programming (PMP) as devised by Howitt (1995). This technique, which can be applied when information is limited, gets round one of the greatest drawbacks of linear programming applications, i.e., the difficulty in obtaining realistic solutions when applying this technique in agricultural economics. PMP avoids this problem by way of the calibration of the model in a base year.

In Spain, the PROMAPA model (**Pro**gramación **Ma**temática para el **A**nálisis de **P**olíticas **A**grarias), a PMP model for analysing agricultural policy changes, was developed at the *Universidad Politécnica de Madrid* and the *Consejo Superior de Investigaciones Cientificas*..

On the basis of the PROMAPA model results, in this chapter the authors will present the analysis of the impact of the new CAP reform (Council Regulation (EC) 1782/2003) adopted in Spain in 2006 on Spanish agriculture. They will analyse as well the impact of some variants of the measures of this reform that could be or are going to be adopted in Spain at some future date.

Before presenting this analysis the authors will introduce first, a brief review of the applications and recent developments of PMP in the context of agricultural policy analysis; then, the mathematical basis of calibration in PMP and the method of calibration used in their application; and finally a description of PROMAPA, highlighting the formulation of the principal features of the CAP reform.

## POSITIVE MATHEMATICAL PROGRAMMING AND AGRICULTURAL POLICY

Mathematical programming and in particular linear programming has been and continues to be a widely used technique in the context of agricultural economics.

Despite this extensive use, however, considerable criticism has been levelled at linear programming – specifically, in obtaining solutions that accurately reflect reality, it is felt that certain– usually arbitrary – constraints must be included.

One way of avoiding this problem is to use PMP as devised by Howit (1995). Briefly, by estimating the coefficients of the objective function for a non-linear programming model, this technique can calibrate the model so that it reproduces the situation existing in a base year for the unit modelled (farm or region). The calibration method proposed by Howit was subsequently enhanced by including entropy maximization in the procedure (Paris and Howit, 1998).

The suitability of PMP for formulating and evaluating the Common Agricultural Policy (CAP) has been a driver for further development of this technique, as can be seen in the recent revisions by Heckelei and Britz (2005) and Henry de Frahan et al.(2007).

Variations designed to correct some of the shortcomings of the Howit and Paris calibration procedure have also been proposed. Such variations have been published, among others, by the following: Judez et al. (1998, 2001b) propose to perform calibration without running the first stage of PMP; Gohin and Chantreuil (1999) introduce a procedure for processing marginal activities; Helming et al. (2001) include supply elasticities obtained exogenously to calibrate the objective function coefficients; Röhm and Dabber (2003) propose a method for linking different variants of the same crop; and more recently Severini and Cortignani (2008) and Júdez et al (2008) suggest procedures for including activities in PMP that are not present in the base year. In addition to these proposals in which calibration is achieved for each unit modelled with the data for a single year, in Heckelei and Wolf (2003) and in Buysse et al. (2007) calibration is replaced with econometric estimation procedures using datasets.

In parallel with the theoretical developments around calibration, PMP has been recently applied in a fair number of cases to analyse the effects of agricultural policy (essentially Common Agricultural Policy measures) on the agricultural sector. In addition to the above papers, in which the authors illustrate their calibration proposals with applications, others have been published: Arfini and Paris (1995); Heckelei and Britz (1999); Barkaoui and Butault (1999, 2003); Röm and Dabbert (1999); CAPRI (2000); Paris et al. (2000) ; Osterburg et al. (2001); Arriaza and Gómez-Limón (2003); Júdez et al. (2003, 2001a) ; Buysse et al. (2004); Buysse and Van Huylenbroeck (2005); Offermann et al. (2005); Blanco and Iglesias (2005); Adenauer et al. (2006); and Kuepker and Klainhauss (2006).

Finally, for several years now a number of European teams have been developing PMP models at the farm level, using national and European FADN. Some of these models, which are often used by national and/or Community officials as a tool for analysing the impact of agricultural policy are: FARMIS (FAL-Germany), SEPALE (Ghent University, CAE Brussels, Catholic University of Louvain -Belgium) and CAPRI (Bonn University Germany)[2]. The PROMAPA model pursues the same line of research as the foregoing studies.

## CALIBRATION WITH PMP. PROBLEM AND SOLUTIONS

Let x be a vector with $n$ components, $x_j$, representing the levels of the farmholding's different activities, $b = [b_1, b_2, ..., b_i, ..., b_m]^T$ the vector representing the availability of resources and $A = \{a_{ij}\}$ the $m \times n$ matrix of technical coefficients in which $a_{ij}$ represents the needs for resource $i$ per unit of activity $j$.

Briefly, the *calibration problem* is to obtain a non-linear concave objective function $f(\mathrm{x})$ such that given a level of activities $\boldsymbol{x} = \bar{\boldsymbol{x}}$ for a base year, the following conditions are fulfilled:

$max\, f(\mathrm{x})$

subject to:

$A\mathrm{x} \leq b$

$\mathrm{x} > 0$

### Solution to the Calibration Problem

Let $\lambda^*$ be a positive n-dimensional vector such that:

$$\lambda_i^* > 0 \quad if \quad \sum_{j=1}^{n} a_{ij} \bar{x}_j - b_i = 0 \qquad (1)$$

$$\lambda_i = 0 \quad if \quad \sum_{j=1} a_{ij} \bar{x}_j - b_i < 0 \qquad (2)$$

The calibration problem is solved specifying any function $f(\mathrm{x})$ fulfilling the following conditions:

$$\left( \frac{\partial f(\boldsymbol{x})}{\partial x_j} \right)_{\boldsymbol{x}=\bar{\boldsymbol{x}}} = \sum_{i=1}^{m} a_{ij} \lambda_i^* \qquad j = 1, 2, 3, \ldots n$$

(3)

Proof of this result can be found in Júdez et al (1998)[3], where is proposed its application for the direct calibration of models, i.e., by-passing the so-called first phase of PMP (use of a linear program with calibration constraints to estimate $\lambda_i$). It is interesting to note in this regard that Heckelei and Wolff (2003) propose the Kuhn and

Tucker first order conditions as the basis for an alternative to PMP.

The application of the solution to the calibration problem set out above entails having i) a vector $\lambda$ that satisfies (1) and (2); and ii) specification of $f(x)$ as a function of certain parameters that can be estimated in a manner that satisfies (3).

## Determining λ*

The classical method to determine the values of the elements of $\lambda^*$ is the use of an auxiliary linear programming model (first phase of PMP) as proposed by Howitt (1995). References of other procedures van be found in the reviews above mentioned of Heckelei and Britz (2005) and Henry de Frahan et al.(2007) which highlights the various methods for by-passing the first phase of PMP.

## Specification of f(x) and Determination of its Parameters

Considering the cost function:

$$\mathrm{d} + \frac{1}{2}\boldsymbol{x}^T\mathbf{Q}\boldsymbol{x} = \sum_{j=1}^{n}\left[d_j + \sum_{k=1}^{n} q_{jk}x_k\right]x_j$$

where $Q$ is a positive simetric matrix of order n ($q_{jk}=q_{kj}$ for all $(k,j)$) and $d$ is a n-dimensional vector of elements $d_j$, the $f(x)$ most often used in PMP is:

$$\mathrm{f}(\boldsymbol{x}) = \sum_{j=1}^{n}\left[\mathrm{r}_j\text{-}(\mathrm{d}_j + \frac{1}{2}\sum_{\mathrm{k}=1}^{\mathrm{n}} \mathrm{q}_{\mathrm{jk}}\mathrm{x}_{\mathrm{k}})\right]\mathrm{x}_{\mathrm{j}}$$

where $r_j$ is the revenue per unit of the activity $j$. To estimate the n parameters $d_j$ and the n(n+1)/2 parameters $q_{jk}$ we have only n relationships derived from (3):

$$r_j - d_j - \sum_{k=1}^{n} q_{jk}\overline{x}_k = \sum_{i=1}^{m} a_{ij}\lambda_i^* \; j=1,2\ldots\ldots n \qquad (4)$$

So, an infinity of values of $d_j$ and $q_{jk}$ satisfies the conditions of calibration. To have a sole set of values it is necessary to add n(n+1)/2 relationships between the parameters. Several procedures have been proposed to solve this problem. A brief presentation of the most often used as well as its comparison based on simulated results when applied to several groups of farms in Germany is in Gocht(2005).

We will present here only the procedure used in the application Section which calibrate the model using exogenous supply elasticities (Helming et al., 2001 and Osterburg et al.,2001). In this procedure all off-diagonal elements of $Q$ are zero. If we let $q_{jj}=q_j$, Equation (4) that guaranty the calibration of the model become in this case:

$$r_j - d_j - q_j\overline{x}_j = \sum_{i=1}^{m} a_{ij}\lambda_i^* \qquad (5)$$

Letting $C(x_j)= (d_j+(1/2)q_jx_j)x_j$ Equation (5) is:

$$\left[\frac{\partial C(x_j)}{\partial x_j}\right]_{x_j=\overline{x}_j} = r_j - \sum_{i=1}^{m} a_{ij}\lambda_i^* \qquad (6)$$

As exogenous supply elasticity of activity $j$, $\varepsilon_j$ is:

$$\varepsilon_j = \frac{\partial x_j}{\partial r_j}\frac{r_j}{x_j}$$

The value of $q_j$, considering $\frac{\partial \lambda_i}{\partial x_j} = 0$, is[4]:

$$q_j = \frac{1}{\varepsilon_j}\frac{r_j}{\overline{x}_j}$$

The value of $d_j$ is obtained by replacing in (5) $q_j$ for its expression, so:

$$d_j = r_j\left(1 - \frac{1}{\varepsilon_j}\right) - \sum_{i=1}^{m} a_{ij}\lambda_j^*$$

In our application the elements of $\lambda^*$ are obtained by the auxiliary linear programming of the first phase of PMP and, with the exception of horticulture crops for which a elasticity of 0.1 was assumed[5], the exogenous supply elasticities adopted were the values used by FAL in the FARMIS model. Finally, we must notice that in the review of Heckelei and Britz (2005) the calibration incorporating exogenous elasticities is described when compared with other calibration methods as "a generally more convincing specification" even if they point out that considering $\frac{\partial \lambda_i}{\partial x_j} = 0$ "the current elasticities of the resulting model will deviate from $\varepsilon_j$".

## PROMAPA MODEL AND CAP REFORM

PROMAPA is a PMP representative farm model, designed to study the impact of change in agricultural policies on the Spanish agricultural sector. Model calibration can be performed by means of several procedures. Exogenous supply elasticities are used in the application of the next Section.

The activities covered by the model included some fifteen non-irrigated and around twenty five irrigated crops, as well as dairy cattle, rearing cattle, and dairy and non-dairy sheep. Livestock feeding is endogenous, whether produced on the farm or purchased to meet the energy and protein needs of different livestock categories; intake capacity was taken into account as well. The activities associated with the agricultural policy tools implemented in the model included the mandatory set-aside in irrigated and non-irrigated land requisite to receiving direct payments for COP (cereal, oilseed and protein) crops, several premiums for livestock (dairy and rearing cattle and sheep), direct payments for crops and livestock in the context of the Single Payment Scheme (SPS), and modulation of payments and crop and livestock quotas.

The primary source of the data for the model was the Spanish Farm Accountancy Data Network (FADN), although information provided by experts was also used, especially to determine unit costs for crops and different categories of livestock and to establish livestock feeding needs.

A detailed description of the model, without including the formulation of the SPS and the modulation, can be found in Júdez et al. (2005). The SPS and the modulation are aspects closely linked with the CAP reform. The SPS briefly consists of paying farmers the three-year average amount of payments (reference amount) they received from the EU in the reference period: 2000, 2001 and 2002. These payments, called direct payments or aid, which were coupled to different activities (area of crops, livestock numbers) before the reform, became decoupled from production with the reform. The decoupled payments are received as a payment per hectare called entitlement payment per hectare. This entitlement is the result of dividing the reference amount of the decoupled payments by the three-year average number of all hectares with right to direct payments in the reference period (we will call this three-year average area *reference area*). Not all of the member states of the EU strictly adopted the SPS. Spain, in particular, adopted a partial decoupling scheme of payments in 2006, so that one part of the direct payments received by the farmer was coupled to production and the other part was decoupled (single payments). In the next Section, we will see these payments in more detail.

The other characteristic of the CAP reform is the modulation of payments, which consists of the reduction of the total amount of direct payments (coupled and decoupled) exceeding a certain amount by a certain percentage.

In the following paragraphs we will introduce the formulation of the partially decoupled scheme and the modulation in PROMAPA[6]. This formulation has been included in PROMAPA to obtain the optimal results of farm types under the simulated scenario of the CAP reform after the model has been calibrated in a base year (2002 in the application).

The following variables are defined:

$x_{hj}$: area, in ha, of crop *j* on the land type *h* (*h*=1: non-irrigated, *h*=2: irrigated). $J_1$ is the subset of crops non eligible for single payment.
*XES*: area, in ha, eligible for the single payment under the simulated scenario.
*XE*: area, in ha, generating the single farm payment.
*XP1*: sum, in €, in the first payment interval exempt from modulation measures (regarded to be less than or equal to €5000).
*XP2*: sum, in €, in the second payment interval (more than €5000), subject to a modulation discount, assumed to be 5%.

The right-hand side elements and the coefficients of the objective function and constraints are likewise defined as follows:

$A_h$: area, in ha, of land type *h* on the farm
*AER*: reference area, in ha.
$s_{hj}$: coupled payment per ha, in €, for crop i on land type *h*.
*d*: payment entitlement per ha in €.
$r_{hj}$: revenue per ha, net of direct payments, of crop *j* on the land type *h*.
*dhj* and *qhj:* parameters of the crop j on the land type h estimated in the base year to calibrate the model.

The following system incorporates the specific characteristics of the partially decoupled scheme and the modulation:

$$\max : \sum_h \sum_j (r_{hj} - (d_{hj} + \frac{1}{2} q_{hj} x_{hj})) x_{hj} + XP1 + 0,95 * XP2 \quad (7)$$

$$\sum_j x_{hj} \leq A_h \; (h = 1,\, 2) \quad (8)$$

$$\sum_h \sum_{j \in J_1} x_{hj} + XES \leq A_1 + A_2 \quad (9)$$

$$XE \leq AER \quad (10)$$

$$XE - XES \leq 0 \quad (11)$$

$$-\sum_h \sum_j s_{hi} x_{hj} - d * XE + XP1 + XP2 \leq 0 \quad (12)$$

$$XP1 \leq 5000 \quad (13)$$

$$x_{hi}, XES, XE, XP1, > 0, XP2 \geq 0$$

Equation (7) is the target function. Equation (8) limits the cultivated area on land type h to the area of this land type on the holding. Equation (9), defines the eligible area for the single payment under the simulated scenario. Equations (10) and (11) define the area, XE, that serves as the basis for computing the single payment in the simulated scenario. This area is the minimum area between the reference area (AER) and the eligible area for direct payments in the simulated scenario (XES). Equation (12) defines the total sum of (coupled and decoupled) payments, XP1+XP2, in the simulated scenario. Finally, Equation (13) limits the sum of payments not subject to modulation measures.

The optimum solutions for the farm types are obtained with GAMS (General Algebraic Modelling System). Besides, a FORTRAN 95 software developed in close conjunction with PROMAPA is designed to interact with GAMS in such a way that: i) the model is provided with the necessary inputs for the various farm types, ii) the GAMS results are monitored, and iii) the optimal results for each farm type are captured and aggregated at several levels.

## APPLICATION

The application of PROMAPA in this section is a static comparative analysis between the situation of some farm types in the base year 2002 and the simulated situation (or simulation year) in 2013 (the last year before a new EU financial perspective) under the scenario of the partial decoupling scheme of the direct payments adopted in Spain and also under other scenarios such as: i) the continuation of the Agenda 2000 in place in 2002, ii) the full decoupling of directs payments, and iii) full decoupling in which the same payment entitlement per ha was applied to all farm types (regional flat rate model). In addition to a comparison of the findings for these different scenarios, this application provides an analysis of the consistency of the results obtained with the model.

Before proceeding to the analysis of the results of this application, the farm types, the detailed characteristics of the different policy scenarios and the price forecasts used are reviewed below.

## Farm Types, Policy Scenarios and Price Forecasts

### Farm Types Used

The farm types chosen for this application were the average farms listed in the Spanish FADN for 2002 for the type of farming (TF) most closely related to the CAP reform in each Comunidad Autonoma (Figure 1), with the exception of the Canary Islands (due to the highly distinctive farm types listed in the FADN for that region). A total of 86 farm types, covering 188,310 farms nationwide, were considered. The weighting factor used for a given farm type to compute the aggregate results was the number of farms covered by that farm type, according to the Spanish FADN.

### Scenarios

The measures considered in the partial decoupling scenario and in the base year are listed in Figure 2.

The measures used in the continuation of Agenda 2000 assumption (all payments are coupled) were the measures in place in the base year.

*Figure 1. Farm types considered in the application*

| | 01 Galicia | 02 Asturias | 03 Cantabria | 04 Basque Country | 05 Navarre | 06 Rioja | 07 Aragon | 08 Catalonia | 09 Balearic Isles | 10 Castile and Leon | 11 Madrid | 12 Castile-La Mancha | 13 Valencia | 14 Murcia | 15 Extrema-dure | 16 Andalusia |
|---|---|---|---|---|---|---|---|---|---|---|---|---|---|---|---|---|
| 1310 Specialist cereals (other than rice) oilseeds and protein crops | | | | X | X | | X | X | | X | X | X | | | X | X |
| 1320 Specialist rice | | | | | | | | | | | | | X | | | X |
| 1410 Specialist root crops | | | | | | X | | | X | X | | | | | | X |
| 1420 Cereals, and roots crops combined | | | | X | | X | | | | X | | | | | | |
| 1430 Specialist field vegetables | | | | | X | | | | | | | X | | X | X | |
| 1442 Specialist cotton | | | | | | | | | | | | | | | | X |
| 1443 Various field crops combined | | | | | X | X | X | X | | X | | X | | | X | X |
| 4100 Specialist dairying | X | X | X | X | X | | | X | X | X | | | | | | X |
| 4210 Specialist cattle-mainly rearing | X | X | X | X | X | | | | | X | X | | | | X | |
| 4300 Cattle-dairying, rearing and fattening combined | X | | | | | | | | | | | | | | | |
| 4410 Specialist sheep | | | | X | X | X | X | | | X | X | X | | X | X | X |
| 4450 Sheep, goat and cattle combined | | | | | X | | | | | | | X | | X | X | X |
| 6000 Mixed cropping | | | | | X | X | X | X | | X | X | X | X | | X | X |
| 7000 Mixed livestock holdings | | | | | | | X | | | | | | | | X | X |
| 8000 Mixed crops-livestock | | | | | X | | X | X | | X | | X | X | X | X | X |

*Figure 2. Agenda 2000 measures in place in 2002 and decoupling provisions considered in PROMAPA*

| | 2002 | CAP reform | |
|---|---|---|---|
| | | Coupled | Decoupled |
| COP (Cereal, oilseed, protein) crops | | | |
| Standard cereal payment (except maize) | €63 /t | €15.75 /t | €47.25 /t |
| Standard oilseed payment | €63 /t | €15.75 /t | €47.25 /t |
| Standard protein crops payment | €72.5 /t | €15.75 /t | €47.25 /t |
| Standard grain maize payment | €63 /t (€55.33 /t) | €15.75 /t (€13.83 /t) | €47.25 /t (41.50 /t) |
| Specific premium for protein crops | ------- | €55.57 /ha | ------- |
| Durum wheat supplementary payment | €344.50 /ha (€226.10 /ha) | €71.25 /ha (€46.76 /ha) | €213.75 /ha (€140.29 /ha) |
| Specific quality premium for durum wheat | ------- | €40 /ha | ------- |
| Standard set-aside payment | €63 /t | ------- | €63 /t |
| Rice | | | |
| Area payment for rice | €334.33 /ha (€224.40 /ha) | ------- | €647.70 /ha (€612.21 /ha) |
| Specific payment for rice | ------- | €476.25 /ha (€450.15 /ha) | ------- |
| Grain legumes | | | |
| Grain legumes payment (chick peas and lentils) | €181 /ha | ------- | €181 /ha |
| Grain legumes payment (vetches) | €181 /ha (€150.52 /ha) | ------- | €181 /ha (€150.52 /ha) |
| Cotton | | | |
| Deficiency payment paid to cotton production | €77.19 /100kg (€61.7/ 100kg) | ------- | ------- |
| Area payment | ------- | 1039 (*) | 1358 /ha (*) |
| Additional payment | | €151 /ha (*) | |
| Sugar beet | | | |
| Payment aid | ------- | ------- | €12.63/t (*) |
| Sheep | | | |
| Dairy ewe premium | €16.8/head | €8.4 /head | €8.4 /head |
| Non dairy ewe premium | €21 /head | €10.5 /head | €10.5 /head |
| Supplementary premium in less-favoured areas | €7 /head | €3.5 /head | €3.5 /head |
| Additional premium | €1 /head | ----- | €1 /head |
| Rearing cattle | | | |
| Suckler cow premium | €200/head | €186/head | ----- |
| Suckler cow additional payment | €24.15/head | €22.46/head | ----- |
| Extensification payment | €100/head | ------ | €93/head |
| New additional payment (2006) | | | |
| First 40 cows | ----- | €33.27/head | ----- |
| Between 41-70 cows | ----- | €22.18/head | ----- |
| Between 71-100 cows | ----- | €11.09/head | ----- |
| More than 100 cows | ----- | €0/head | ----- |
| Stocking density conditions | | | |
| General scheme | 1.9 LU/ha forage area | 1.9 LU/ha forage area (*) | |
| Extensification | 1.4 LU/ha forage area | ----- | |
| New additional payment | ----- | 1.5 LU/ha forage area | |
| Dairy cattle | | | |
| Additional payment for milk | ----- | €3.22/t | ----- |
| Cow milk premium | ----- | ------ | €29.07/t |

In brackets in 2002: payment actually received.

The decoupled payments in the full decoupling scenarios in the simulation year were defined to be the sum of the coupled and decoupled payments in Figure 2, with the exception of the following specific payments, which were considered to be coupled: i) specific premium for protein crops; ii) specific quality premium for durum wheat; iii) specific payment for rice; iv) coupled area payment for cotton.

In the full decoupling scenarios with a standard payment entitlement per ha (regional flat rate model), the amount used for all the farm types, €240.08/ha, was the entitlement per ha found for the full decoupling scenario in the nation-wide aggregated results.

Finally, the underlying assumptions for the partial and full decoupling scenarios were as follows:

1. Although modulation measures in the CAP reform are to be phased in, the simulations assumed them to be in the final phase, i.e., to consist of a 5% reduction in the total sum of direct payments, exempting only the first €5000.
2. The compulsory set-aside land considered in the simulation year was the same as in the base year.
3. The decoupled payment was determined for each farm type based on the area of crops and the number of head of livestock in the base year. In other words, the base year replaced the reference period (2000, 2001 and 2002) in the PROMAPA model.

## Price Forecasts

The price forecasts used in this application were taken from a communication from W. Kleinhanss and B. Kuepker, based on ESIM model predictions for the years 2002 to 2013. Such forecasts are given in Figure 3 for real prices (corresponding to nominal prices deflated by 1.5% per year).

As price variations were applied to both product sales and livestock feed purchases, the cost

*Figure 3. Real price change forecasts, in %, 2002 to 2013*

| | Agenda 2000 | Partial decoupling | Full decoupling |
|---|---|---|---|
| Wheat | -2.0 | 2.0 | 2.3 |
| Durum wheat | -6.1 | 3.5 | 3.7 |
| Barley | -11.9 | -6.1 | -5.7 |
| Maize | -21.0 | -15.3 | -14.8 |
| Rye | -14.0 | -14.0 | -14.0 |
| Other grains | -21.7 | -16.0 | -15.5 |
| Rice | -48.4 | -48.4 | -48.4 |
| Sugar | -27.6 | -25.0 | -24.9 |
| Potato | -43.9 | -38.0 | -37.7 |
| Rapeseed | -6.5 | -4.0 | -3.8 |
| Sunflower seed | -7.4 | -5.2 | -5.0 |
| Silage maize | -24.3 | -34.9 | -32.9 |
| Fodder | -24.7 | -48.0 | -46.2 |
| Grass | -28.0 | -44.8 | -42.5 |
| Milk | -19.0 | -22.8 | -22.4 |
| Beef | -18.4 | -8.8 | -4.6 |
| Sheep | 4.1 | 31.0 | 37.6 |

Source: Personal communication from W. Kleinhanss and B. Kuepker based on results of ESIM model.

of off-farm feed inputs declined, sometimes sharply, in several of the scenarios simulated in the present study.

It will be noted that Figure 3 shows no price forecast for certain products entered in the PROMAPA model (such as grain legumes, horticulture products, sheep's milk and veal and lamb). The price forecast for beef was adopted for veal and the forecast for mutton was used for lamb. As a general rule, the base year prices were used for all other products not shown in Figure 3.

## Analysis of Results

The optimal results for each farm type have been obtained with the GAMS SBB solver. The aggregated results for the different scenarios considered are presented in Figure 5. This table as well as similar tables obtained at the Comunidad Autonoma level and at farm type level are the basis of the analysis below.

### Results by Crop Groups

1. The decline in prices forecast for most of the crops considered under all the scenarios goes hand-in-hand with a decline in farming activity with respect to the base year. Quantified in terms of land use, this translates into a reduction of over 0.5% of the area used for farming, even assuming the continuation of the Agenda 2000 measures without decoupling. In decoupling scenarios the decline is logically more marked, reaching over 2% under partial and over 5% under full decoupling arrangements.
2. As Figure 4 shows, the results for the main crop groups are practically identical for two assumptions, full decoupling and the regional flat rate model. These similarities are likewise found when analyzing individual crops and livestock in all the regions of Spain. The findings for full decoupling and decoupling pursuant to the regional flat rate model simulated in this study only differ, in fact, in the economic results at the regional level, as discussed below.
3. According to the variations observed for crop groups in (full or partial) decoupling scenarios, the greater the degree of decoupling, the steeper the decline in area used for COP crops, grassland and fodder crops, all of which are impacted by that policy. No such trend is observed for "other crops" which, as

*Figure 4.*

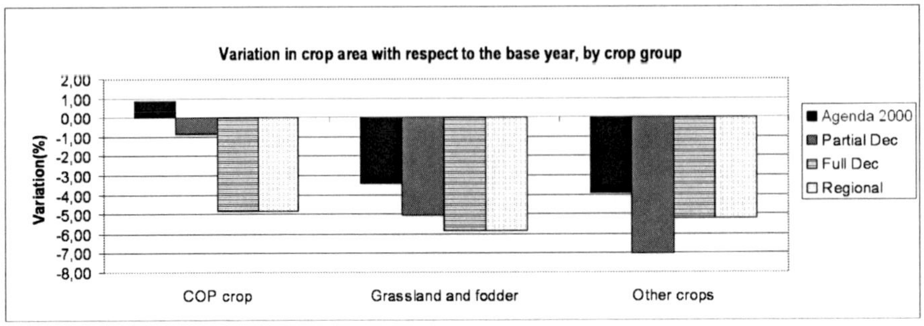


*Figure 5. Aggregated results of the 86 types of farms*

| | Base year 2002 | Agenda 2000 | | Partial decoupling | | Full decoupling | | Regional | |
|---|---|---|---|---|---|---|---|---|---|
| | | Value | Variation (%) | Value | Variation (%) | Value | Variation (%) | Value | Variation (%) |
| CROPS | | | | | | | | | |
| Non-Irrigated chickpea (thousand ha) | 25.37 | 28.57 | 12.62 | 26.59 | 4.81 | 29.2 | 15.12 | 29.2 | 15.12 |
| Irrigated rice (thousand ha) | 28.89 | 26.22 | -9.22 | 28.35 | -1.86 | 28.62 | -0.91 | 28.61 | -0.97 |
| Irrigated sugar beet (thousand ha) | 96.02 | 72.81 | -24.18 | 83.25 | -13.3 | 84.1 | -12.42 | 84.09 | -12.42 |
| Irrigated cotton (thousand ha) | 89.92 | 96.57 | 7.4 | 80.81 | -10.14 | 81.19 | -9.71 | 81.19 | -9.71 |
| Irrigated paprika pepper (thousand ha) | 2.32 | 2.34 | 0.59 | 2.34 | 0.58 | 2.34 | 0.59 | 2.34 | 0.7 |
| Irrigated early potato (thousand ha) | 8.32 | 7.03 | -15.44 | 7.35 | -11.63 | 7.36 | -11.52 | 7.36 | -11.52 |
| Irrigated medium season potato (thousand ha) | 12.29 | 10.67 | -13.19 | 11.18 | -9.07 | 11.2 | -8.88 | 11.2 | -8.88 |
| Irrigated late season potato (thousand ha) | 14.54 | 13.06 | -10.14 | 13.49 | -7.23 | 13.54 | -6.84 | 13.54 | -6.84 |
| Irrigated asparagus (thousand ha) | 0.54 | 0.55 | 1.57 | 0.55 | 1.42 | 0.55 | 1.47 | 0.55 | 1.8 |
| Irrigated melon (thousand ha) | 1.6 | 1.62 | 0.77 | 1.62 | 1.04 | 1.62 | 1.21 | 1.62 | 0.7 |
| Irrigated tomato (thousand ha) | 16.51 | 16.61 | 0.63 | 16.57 | 0.39 | 16.58 | 0.41 | 16.63 | 0.71 |
| Irrigated pepper (thousand ha) | 0.2 | 0.2 | 0.41 | 0.2 | 0.36 | 0.2 | 0.34 | 0.2 | 0.37 |
| Irrigated artichoke (thousand ha) | 3.61 | 3.64 | 0.79 | 3.63 | 0.49 | 3.63 | 0.48 | 3.63 | 0.53 |
| Irrigated cauliflower (thousand ha) | 0.53 | 0.53 | 0.7 | 0.53 | 0.57 | 0.53 | 0.55 | 0.53 | 0.63 |
| Irrigated garlic (thousand ha) | 10.95 | 10.97 | 0.24 | 10.99 | 0.37 | 10.99 | 0.39 | 10.97 | 0.22 |
| Irrigated onion (thousand ha) | 2.47 | 2.48 | 0.43 | 2.48 | 0.71 | 2.48 | 0.7 | 2.48 | 0.58 |
| Irrigated green bean (thousand ha) | 0.31 | 0.4 | 28.03 | 0.39 | 23.63 | 0.39 | 24.13 | 0.4 | 27.5 |
| Irrigated pea (thousand ha) | 1.91 | 2.16 | 12.98 | 2.03 | 6.42 | 1.99 | 4.12 | 1.99 | 4.12 |
| Non-Irrigated durum wheat (thousand ha) | 355.44 | 359.55 | 1.16 | 282.48 | -20.53 | 248.07 | -30.21 | 248.02 | -30.22 |
| Irrigated durum wheat (thousand ha) | 47.26 | 49.62 | 5 | 47.39 | 0.28 | 46.18 | -2.29 | 46.17 | -2.3 |
| Non-Irrigated soft wheat (thousand ha) | 897.36 | 943.14 | 5.1 | 944.45 | 5.25 | 904.46 | 0.79 | 904.45 | 0.79 |
| Irrigated soft wheat (thousand ha) | 120.86 | 141.61 | 17.17 | 149.13 | 23.39 | 148.28 | 22.69 | 148.28 | 22.69 |
| Non-Irrigated rye (thousand ha) | 50.72 | 49.66 | -2.08 | 47.67 | -6.01 | 45.2 | -10.88 | 45.2 | -10.88 |
| Irrigated rye (thousand ha) | 3.9 | 4.13 | 5.85 | 3.99 | 2.27 | 3.91 | 0.21 | 3.91 | 0.21 |
| Non-Irrigated barley (thousand ha) | 2882.61 | 2835.2 | -1.64 | 2861.11 | -0.75 | 2764.17 | -4.11 | 2764.18 | -4.11 |
| Irrigated barley (thousand ha) | 188.51 | 218.77 | 16.05 | 227.42 | 20.64 | 223.04 | 18.32 | 223.05 | 18.33 |
| Non-Irrigated oats (thousand ha) | 157.24 | 142.65 | -9.28 | 142.47 | -9.39 | 140.39 | -10.72 | 140.39 | -10.72 |
| Irrigated oats (thousand ha) | 0.55 | 0.56 | 1.1 | 0.58 | 5.84 | 0.57 | 2.93 | 0.57 | 2.93 |
| Non-Irrigated grain maize (thousand ha) | 2.73 | 2.87 | 5.21 | 2.87 | 5.15 | 2.62 | -3.87 | 2.62 | -3.87 |
| Irrigated grain maize (thousand ha) | 374.37 | 341.23 | -8.85 | 331.91 | -11.34 | 320.64 | -14.35 | 320.58 | -14.37 |
| Non-Irrigated sunflower (thousand ha) | 496.41 | 518.69 | 4.49 | 476.18 | -4.08 | 447.11 | -9.93 | 447.11 | -9.93 |
| Irrigated sunflower (thousand ha) | 103.82 | 123.67 | 19.12 | 115.75 | 11.49 | 111.88 | 7.76 | 111.86 | 7.74 |
| Non-Irrigated vetch (thousand ha) | 34.16 | 40.64 | 18.97 | 33.69 | -1.36 | 35.91 | 5.14 | 35.91 | 5.14 |
| Non-Irrigated alfalfa (thousand ha) | 158.21 | 137.82 | -12.89 | 117.78 | -25.56 | 126.04 | -20.34 | 126.04 | -20.34 |
| Irrigated alfalfa (thousand ha) | 138.22 | 116.86 | -15.46 | 88.26 | -36.15 | 94 | -32 | 93.99 | -32 |
| Non-Irrigated winter forage cereals (thousand ha) | 3.16 | 2.87 | -9.38 | 3.17 | 0.03 | 3.17 | 0.03 | 3.17 | 0.03 |
| Non-Irrigated forage maize (thousand ha) | 2.17 | 1.44 | -33.65 | 1.72 | -20.81 | 1.7 | -21.69 | 1.7 | -21.69 |
| Irrigated forage maize (thousand ha) | 8.03 | 7.16 | -10.94 | 7.73 | -3.74 | 7.86 | -2.16 | 7.86 | -2.16 |
| Non-Irrigated temporary grassland (thousand ha) | 247.03 | 236.6 | -4.22 | 240.03 | -2.83 | 245.16 | -0.76 | 245.16 | -0.76 |
| Irrigated temporary grassland (thousand ha) | 21.89 | 22.17 | 1.3 | 22.35 | 2.11 | 22.34 | 2.06 | 22.34 | 2.06 |
| Non-Irrigated permanent grassland (thousand ha) | 1817.11 | 1817.11 | 0 | 1817.11 | 0 | 1817.11 | 0 | 1817.11 | 0 |
| Irrigated permanent grassland (thousand ha) | 23.32 | 23.32 | 0 | 23.32 | 0 | 23.32 | 0 | 23.32 | 0 |
| LIVESTOCK | | | | | | | | | |
| Suckler cows ( thousand head) | 961.35 | 960.8 | -0.06 | 926.36 | -3.64 | 858.51 | -10.7 | 858.51 | -10.7 |
| Dairy cows (thousand head) | 999.59 | 943.3 | -5.63 | 969.24 | -3.04 | 961.19 | -3.84 | 961.19 | -3.84 |
| Dairy sheep (thousand head) | 5430.28 | 5467.69 | 0.69 | 5535.45 | 1.94 | 5430.7 | 0.01 | 5430.7 | 0.01 |
| Non dairy sheep (thousand head) | 8385.38 | 8411.54 | 0.31 | 8840.94 | 5.43 | 8431.03 | 0.54 | 8431.03 | 0.54 |
| LU (thousand LU) | 4348.58 | 4289.94 | -1.35 | 4357.07 | 0.2 | 4194.22 | -3.55 | 4194.22 | -3.55 |
| NON UTILIZED AREA | | | | | | | | | |
| Non irrigable non used area (thousand ha) | 0 | 41.25 | Inf | 157.33 | Inf | 380.33 | Inf | 380.38 | Inf |
| Irrigable non used area (thousand ha) | 0 | 0.19 | Inf | 38.09 | Inf | 54.38 | Inf | 54.43 | Inf |
| UTILISED AREA (SUMMARY) | | | | | | | | | |
| Non irrigated other crops area (thousand ha) | 59.52 | 69.21 | 16.26 | 60.28 | 1.27 | 65.11 | 9.39 | 65.11 | 9.39 |
| Non irrigated COP crops area (thousand ha) | 4842.51 | 4851.75 | 0.19 | 4757.23 | -1.76 | 4552.02 | -6 | 4551.96 | -6 |
| Non irrigated grassland and fodder crops area (thousand ha) | 2227.69 | 2166.24 | -2.76 | 2154.88 | -3.27 | 2132.26 | -4.28 | 2132.26 | -4.28 |
| Irrigated other crops (thousand ha) | 289.03 | 265.72 | -8.07 | 263.73 | -8.75 | 265.32 | -8.2 | 265.34 | -8.19 |
| Total irrigated COP crops (thousand ha) | 841.18 | 881.76 | 4.82 | 878.2 | 4.4 | 856.47 | 1.82 | 856.41 | 1.81 |
| Irrigated grassland and fodder crops area (thousand ha) | 191.47 | 169.5 | -11.47 | 141.67 | -26.01 | 145.5 | -24.01 | 145.49 | -24.01 |
| DUAL VALUES OF LAND | | | | | | | | | |
| Mean dual value of non irrigated land (€) | 254.51 | 143.25 | -43.71 | 78.52 | -69.15 | 52.84 | -79.24 | 52.85 | -79.24 |
| Mean dual value of irrigated land (€) | 568.57 | 346.54 | -39.05 | 189.11 | -66.74 | 150.8 | -73.48 | 150.82 | -73.47 |
| ECONOMIC RESULTS | | | | | | | | | |
| Objective function (million €) | 6887.45 | 6224.03 | -9.63 | 6949.3 | 0.9 | 7046.77 | 2.31 | 7046.5 | 2.31 |
| Coupled aid (million €) | 1977.75 | 2005.15 | 1.39 | 735.03 | -62.84 | 121.39 | -93.86 | 121.38 | -93.86 |
| Decoupled aid (million €) | 0 | 0 | 0 | 1551.27 | Inf | 2162.48 | Inf | 2162.5 | Inf |
| Total aid before modulation (million €) | 1977.75 | 2005.15 | 1.39 | 2286.3 | 15.6 | 2283.87 | 15.48 | 2283.88 | 15.48 |
| Modulation reduction (million €) | 0 | 0 | 0 | 50.7 | Inf | 50.68 | Inf | 50.98 | Inf |
| Total aid after modulation (million €) | 1977.75 | 2005.15 | 1.39 | 2235.59 | 13.04 | 2233.19 | 12.92 | 2232.9 | 12.9 |
| Gross margin after modulation (million €) | 6359.78 | 5580.3 | -12.26 | 6281.78 | -1.23 | 6298.65 | -0.96 | 6298.45 | -0.96 |
| Mean % of aid in margin | 31.1 | 35.93 | | 35.59 | | 35.45 | | 35.45 | |
| Average payment entitlement per ha (€) | 0 | 0 | | 172.23 | | 240.09 | | 240.09 | |

discussed below, is a heterogeneous group whose behaviour in these scenarios is less dependent on decoupling. For these crops, the decline is smaller under full decoupling because their potential replacements (essentially COP crops) have higher coupled revenues per ha under partial decoupling arrangements.

4. Assuming the continuation of the Agenda 2000 measures, crop variations are not affected by the decoupling of direct aid (non-existent in this scenario). The outcome is that the area devoted to COP crops, which is deeply influenced by decoupling policies, actually increases under the Agenda 2000 scenario. The reasons for the smaller decline in area devoted to grassland and fodder crops and "other crops" in the Agenda 2000 compared to the decoupling scenarios will be addressed below.

## COP Crop Variations

Figures 6 to 8 show the variations in the area used for the main irrigated and non-irrigated COP crops. The following remarks refer to these results.

1. Focusing on the partial and full decoupling scenarios only, the graphs show that:
   - Due to decoupling of direct crop payments, the greater the extent of decoupling, the slighter are the increases in cultivated area and the steeper the decreases.
   - The decline in coupled revenues in the different scenarios due to price forecasts and decoupling, as shown in Figure 9[7], are a key factor in the explanation of the irrigated and non-irrigated COP crop variations depicted in Figures 7 and 8.

*Figure 6.*

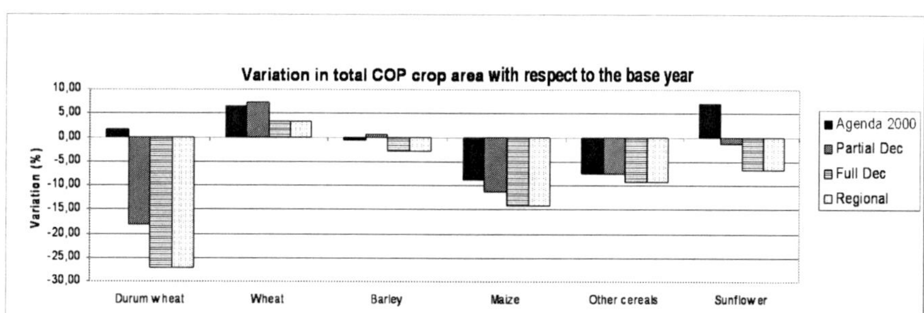


*Figure 7.*

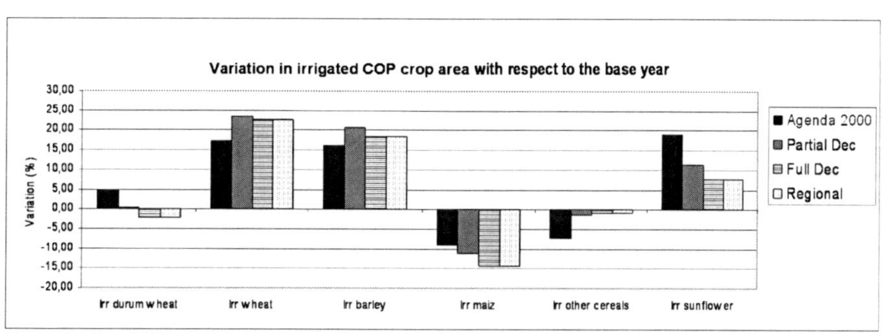

*Figure 8.*

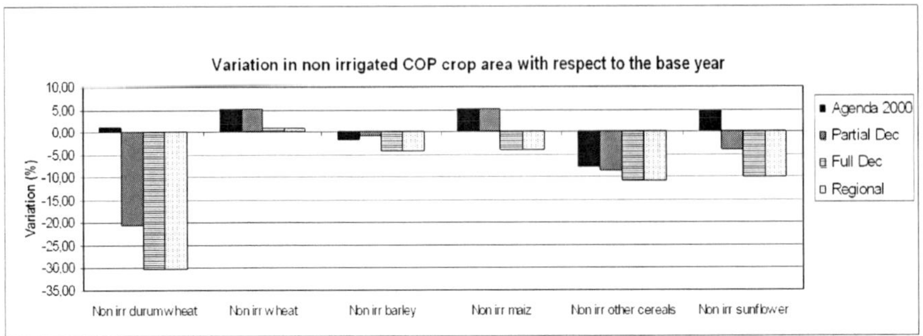


*Figure 9. Mean loss (in Euros) of coupled revenues per ha for COP crops in decoupling scenarios, with respect to the base year*

| | Partial decoupling | Total decoupling |
|---|---|---|
| Non irrigated wheat | 123.12 | 159.86 |
| Non irrigated barley | 147.39 | 184.06 |
| Non irrigated sunflower | 144.07 | 181.33 |
| Non irrigated durum wheat | 259.66 | 343.66 |
| Irrigated wheat | 204.76 | 262.82 |
| Irrigated barley | 242.57 | 300.72 |
| Irrigated sunflower | 312.60 | 395.11 |
| Irrigated durum wheat | 409.80 | 539.06 |
| Irrigated maize | 541.76 | 635.87 |

Although irrigated crop loss is greater than the loss for non-irrigated crops, greater increases are observed for COP crops area in the former category. This is because in irrigated farming they compete with other crops more highly penalized than they are in price (sugar beet, potatoes, alfalfa) or decoupling (such as cotton, whose coupled payment is just over half what was received in the base year under the deficiency payment scheme).

The only non-irrigated crop with an increase in area in all scenarios is wheat (the one least penalized). The case of non-irrigated maize, which also increases with partial decoupling, is in fact exceptional, inasmuch as the area involved is very small and concentrated in Galicia, where, in addition to this crop, there are temporary and permanent grasslands.

The area for the three irrigated crops least penalized by the combination of price and decoupling (wheat, barley and sunflower) grows at the expense of the area of other COP and non COP crops.

2. Assuming the continuation of the Agenda 2000 measures, the loss of coupled revenues per ha, due solely to the decline in prices, is shown in Figure 10[8].

Since, as in the preceding case, even if no exact inverse relationship can be drawn between mean loss and crop increase[9], we can see that the crops with the smaller losses are the ones that exhibit crop area increases, both in non-irrigated (wheat, durum wheat and sunflower) and irrigated (the above three plus barley) farming.

*Figure 10. Mean loss (in Euros) of coupled revenues per ha for COP crops in the Agenda 2000 scenario, with respect to the base year*

| | Non irrigated | Irrigated |
|---|---|---|
| Wheat | 7.08 | 11.89 |
| Durum Wheat | 17.17 | 29.70 |
| Sunflower | 19.73 | 35.68 |
| Barley | 33.53 | 50.56 |
| Maize | - | 276.49 |

## Grassland and Fodder Crop Variations

In the analysis of this group, the most important fodder crop, namely alfalfa, must be studied separately from all other forage, essentially comprising temporary and permanent grassland (but also forage maize and forage winter cereals, which account, however, for less than 1% of the total). While alfalfa may be used as fodder for a farm's own livestock or sold, the other crops in this group can only be used as feed.

Figure 11 shows the variation in the crop area for these two subgroups. Since most of the alfalfa is grown to be sold, the area used to grow this crop depends on price and the revenues per ha earned with competing crops, essentially COPs. In all the scenarios studied, alfalfa is less profitable than the other crops. At the same time the substantial decline in price in decoupling scenarios based on ESIM (over 45%) is nearly twice as large as the decrease under the Agenda 2000 assumption. Consequently, alfalfa crop area drops more when the former scenarios are assumed, with the largest decrease found for partial decoupling, where the COP crops generate larger direct revenues per ha than in the event of full decoupling.

The variation in the remaining forage crops and grassland is closely associated with cattle variations studied below.

## Other (Non-COP, Non- Fodder Crop) Crop Variations

Grain legumes (chick pea and vetch), cotton, rice, sugar beet and potatoes account for over 85% of the area of the non-COP, non-fodder crops considered in the model.

The variations in these crops under the different agricultural policies scenarios are shown in Figure 12.

All these crops, except the grain legumes, are irrigated. As the figure shows, their variations depend on the agricultural policy implemented.

Prices were assumed to be constant for grain legumes, giving them a clear advantage in the continuation of Agenda 2000 scenario over the

*Figure 11.*

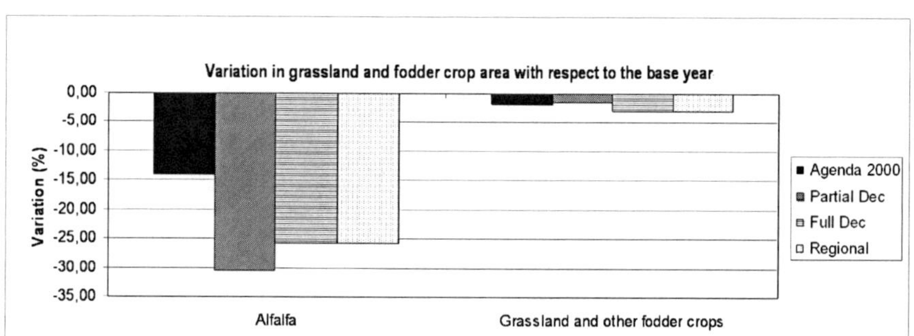

*Figure 12.*

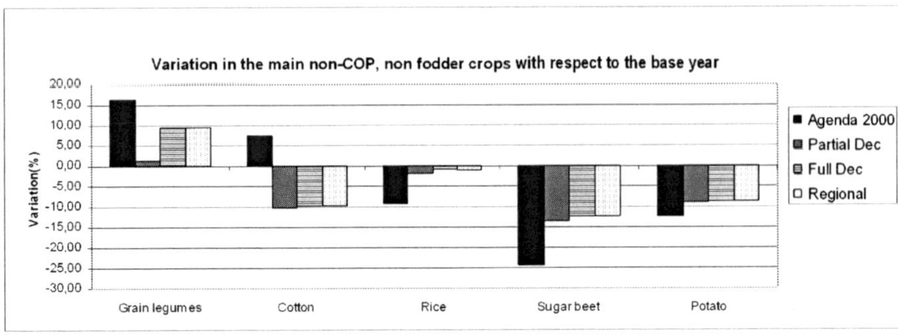


COP crops with which they compete, for prices decline in all of the latter. This is the reason for the substantial rise in these crops under the Agenda 2000 scenario. Since the constant price assumption makes these crops more profitable than some of the non-irrigated COP crops, in the decoupling scenarios too, their area rises here as well, although not as steeply as under the Agenda 2000 measures. Since grain legume decoupling is regarded to be the same under partial and full decoupling arrangements, the increase in the respective area is much greater in the latter.

This application assumes that the world price of cotton remains the same as in the base year. Under that assumption, its crop area grows, naturally, in the Agenda 2000 scenario. In the decoupling scenarios, the deficiency payment scheme existing in the base year is replaced by an area payment whose coupled portion accounts for around one half of the sum of the deficiency payment. As a result, cotton crop area declines under these scenarios.

The steep decline in prices assumed for rice, sugar beet and potatoes signifies a reduction in crop area in all the scenarios studied. The loss is more intense under Agenda 2000 arrangements due to the higher revenue per hectare earned with competing crops (essentially COP crops). Despite the substantial slide in rice prices, however, its area declines very little in both the partial and full decoupling scenarios because the direct aid per ha in such scenarios is nearly double the amount received in the base year.

All other crops in this group (paprika pepper, asparagus, melon, tomato, pepper, artichoke, cauliflower, garlic, onion and green bean) are to be found essentially in farm type 1430-Specialist field vegetable- in Navarre, Castile-La Mancha, Murcia and Extremadura. In such farm types these crops are preponderant and they replace irrigated COP and non-COP crops in the simulation year. The scale of the substitution is, however, rather small, as it is shown by the minor increase in the area devoted to these crops compared to the base year, which is no greater than 0.8% in any of the simulated scenarios.

## Livestock Results under Different Agricultural Policy Scenarios

Figure 13 shows the change with respect to the base year, under the various agricultural policy scenarios, for the different categories of livestock included in the model.

Due to the reduction in milk and beef prices, the number of cows declines in all scenarios, whereas the sheep herd size increases because of the upward trend in prices. This general observation is discussed in greater detail below.

### Suckler Cows

The pattern of variation in the number of head shown in the figure is not equally representative

*Figure 13.*

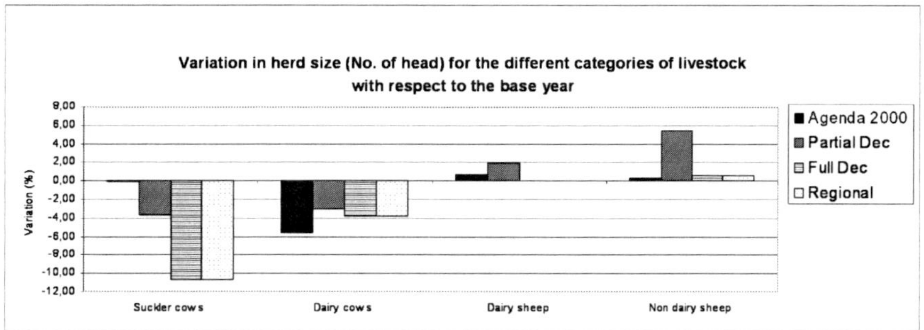


of all the regions of Spain, which differ substantially depending on parameters such as stocking density, meat prices and the proportion and price of off-farm feed.

The sum of the revenues per head from the sale of young animals[10] and coupled aid[11] depends on stocking density. Specifically:

- For high stocking densities that receive no aid, the highest revenues are found under the full decoupling scenario and the lowest under continuation of the Agenda 2000 arrangements.
- With densities between 1.4 and 1.9 LU/ha in the base year (and between 1.5 and 1.9 in the simulation year) which qualify farms for suckler cow premium excluding extensification premium, the highest revenues are obtained with partial and the lowest with full decoupling.
- With low stocking densities, which qualify farms for additional premiums, the highest revenues are found for the continuation of Agenda 2000 measures and the lowest for the full decoupling scenario.

Low livestock densities are generally found in central and southern Spain. These are the regions where variation follows the pattern shown in Figure 13, with different degrees of intensity.

In the regions of northern or so-called "wet Spain" the density per ha is much higher and many farm types opt for the general payment scheme, while others maintain a stocking density that disqualifies them for aid altogether.

Herd size varies very little in the farm types in the regions comprising "wet Spain" (Galicia, Asturias, Basque Country and Navarre), where pastures abound and very little off-farm feed is purchased (under 7% of the total dry matter consumed)[12].

The largest variations for this type of livestock are observed under decoupling policies in other Spanish regions where substantial portions of off-farm feed are purchased.

The decline in the price of feed with respect to the base year mitigates the decrease in herd size. This explains why herd size variation is small under Agenda 2000 arrangements (where there is no decoupling) in southern and central Spain, where large proportions of feed have to be purchased. In some exceptional cases this decline prompts an increase in the herd size in certain types of holdings in northern Spain where stocking densities are high[13]. Another aspect observed on some farms in this region is the increase in the number of suckler cows at the expense of the number of dairy cows in an attempt to maximize profits from grassland use.

**Dairy Cows**

The parameters affecting this type of livestock, in addition to milk prices, are the price of beef (since dairy farms also sell young animals) and the price and proportion of off-farm feed purchased. The coupled aid received by such farms under partial decoupling arrangements is too little to have any significant effect on livestock trends.

Referring to the revenue per head generated by milk and young animals sales in all types of farm holdings, better results (smaller declines with respect to base year revenues) are obtained under the Agenda 2000 than with the decoupling scenarios. The steeper decline in the price of purchased feed, which represents in all cases in this sector over 30% of the total dry matter consumed, leads in the decoupling scenarios, in many cases, to higher coupled revenue per head, corrected for the cost of purchased feed. This holds true particularly in the regions outside "wet Spain" where the proportion of off-farm feed inputs is larger. This would explain the pattern of variations for the different agricultural policy scenarios shown in Figure 13.

The pattern is not the same, however, in all the Spanish Comunidades Autonomas. In some regions of wet Spain, such as Navarre, with high milk yields per cow, and relatively small amounts of off-farm feed purchased, the decline in feed prices does not suffice to make dairy cows more profitable in decoupling scenarios than in Agenda 2000 scenario.

**Sheep**

Considering only the revenues per head from the sale of young animals and milk (in the case of dairy sheep) and coupled aid, the most favourable agricultural policy is the partial decoupling scheme, which preserves nearly 50% of the coupled payments and provides for a substantial increase in the sale of young animals. The second most favourable conditions are found under the Agenda 2000 scenario, where the increase in the price of young animals is much smaller, but the total aid is coupled.

The coupled revenues per head are higher in both above mentioned scenarios than in the base year; whereas under the full decoupling arrangements such revenues are lower than in that year, as the rise in the price of lamb does not wholly offset the total lack of coupled payments.

When the cost of purchased feed, which declines in all scenarios (although more under decoupling arrangements) is subtracted from the above direct revenues per head, the advantage of partial decoupling over the base year widens. This greater advantage translates into substantial increases in non-dairy livestock in regions with a high proportion of off-farm feed. The lesser impact on dairy sheep is due to the fact that, as mentioned above, the price of sheep's milk was assumed to be the same as in the base year.

When the cost of purchased feed is subtracted from the proceeds of the sale of young animals and milk, the revenues per head under full decoupling arrangements are higher than in the base year and close to the figures obtained for the Agenda 2000 scenario. In some regions revenues for non-dairy sheep are actually higher under full decoupling than with the Agenda 2000 measures, but in any event the variations observed for all types of sheep livestock under these two agricultural policies are small or even nil.

## Economic Results Under Different Agricultural Policy Scenarios

Figure 14 shows the variations in the target function, gross margin with and without aid and total aid (area payment and premiums) after modulation.

The objective function is the gross margin with costs represented by a quadratic function estimated in such a way that the model is calibrated with the base year results. It differs from what has been referred to here as gross margin in that the latter costs are linear and constitute part of the data used. Moreover, the gross margin includes all aid, coupled and decoupled; and although the latter is not strictly associated with each of the farm's activities, it is related to the farm's business *per se*, inasmuch as it depends on the eligible area devoted by the farm to receive such aid.

*Figure 14.*

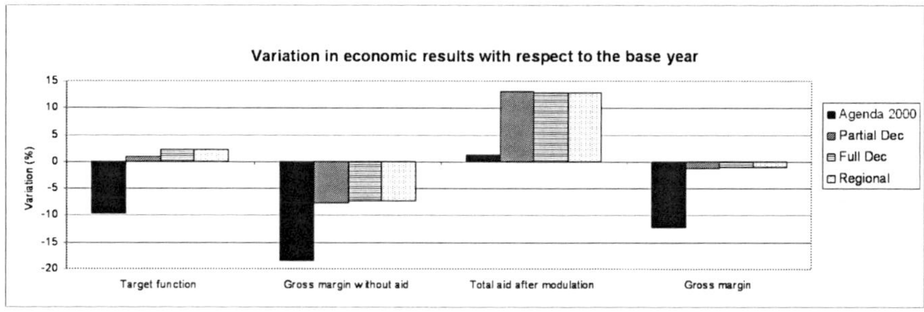


**Target Function**

Assuming constant prices and a constant sum of aid (coupled and decoupled) per production unit, this function, which is assumed to be maximized by farmers, should be larger at higher degrees of decoupling. As a result of the price and aid forecasts entered, however, this is not always the case in the different farm types. Nonetheless, as the figure shows for the aggregated results, this function rises slightly with the decoupling scenarios, more intensely under the full decoupling scheme.

**Results with the Regional Flat Rate Model Scenario**

As noted earlier, this scenario consists of the full decoupling of aid with the assumption that all farm types are paid the same entitlement payment per hectare.

Although the national aggregated results, given the total aid received and gross margin, are the same under this scenario as with full decoupling, the results for the various regions fluctuate widely. These variations are due to the fact that, although farming, which depends essentially on coupled payments per unit of activity, is the same as under full decoupling in the different farm holdings, the sum of decoupled aid differs. Such a sum is higher in regions where the entitlement payment per hectare under full decoupling is smaller than obtained for Spain as a whole (such as in Castile and Leon, Madrid and Castile-La Mancha). Everywhere else the payment is higher (see Figure 18[14]).

**Gross Margin Net Of Aid**

1. Under the Agenda 2000 scenario, only sheep prices rise slightly, while the price of all other products declines more or less sharply[15]. As a result, the gross margin net of aid declines substantially with respect to the base year. As Figure 15 shows, this downturn is visible to a greater or lesser extent in nearly all regions. The one exception is Murcia, where sheep husbandry is a predominant activity and off-farm feed purchases constitute a very high proportion of the total feed.
2. Similar variations in this economic variable are observed in the partial and full decoupling scenarios, both nation-wide and in the various regions. Compared to the Agenda 2000 scenario, the decoupling arrangements, excepting for fodder crops, show fewer declines in prices (with a slight rise in the price of wheat) and a substantial increase in the price of sheep products. This explains why the gross margin net of aid is generally higher in all regions under the decoupled aid scenarios[16].
3. As mentioned above, in the decoupling scenarios only wheat and sheep product prices are higher than in the base year. The outcome is that the gross margin net of aid is lower than in the base year in most regions. The margin is higher only in La Rioja, Aragon, Castile-La

*Figure 15. Gross margin net of aid (% variation)*

| | Agenda 2000 | Partial Decoupling | Full Decoupling | Regional |
|---|---|---|---|---|
| 01.Galicia | -22.35 | -17.09 | -16.23 | -16.23 |
| 02.Asturias | -19.93 | -17.76 | -16.35 | -16.35 |
| 03.Cantabria | -22.82 | -21.4 | -20.41 | -20.41 |
| 04.Basque Country | -18.28 | -15.09 | -13.56 | -13.56 |
| 05.Navarre | -13.53 | -3.21 | -1.06 | -1.06 |
| 06.La Rioja | -28.29 | 5.01 | 6.22 | 6.22 |
| 07.Aragón | -18.87 | 1.67 | 3.31 | 3.31 |
| 08.Catalonia | -31.38 | -33.86 | -35.13 | -35.13 |
| 09.Balearic Isles | -36.23 | -17.58 | -17.62 | -17.62 |
| 10.Castile-Leon | -18.22 | -6.47 | -7.12 | -7.12 |
| 11.Madrid | -13.47 | -1.74 | -2.51 | -2.51 |
| 12.Castile-La Mancha | -7.77 | 6.64 | 7.04 | 7.02 |
| 13.Valencian C. | -60.77 | -59.58 | -59.51 | -59.51 |
| 14.Murcia | 12.52 | 41.09 | 42.11 | 42.14 |
| 15.Extremadura | -10.01 | 10.76 | 13.48 | 13.53 |
| 16.Andalusia | -25.88 | -12.45 | -12.99 | -12.99 |
| 00.Spain | -18.41 | -7.66 | -7.22 | -7.22 |

Mancha, Murcia and Extremadura, regions where sheep husbandry is dominating.

## Payments and Modulation

1. As Figure 16 shows, under the Agenda 2000 scenario aid is approximately the same as in the base year in most regions (the result for the Balearic Isles is not significant in light of the small amounts involved: 30 euros in the base year and 60 under Agenda 2000), which would explain the small nation-wide variation. The most prominent upward variation is observed in Andalusia, due essentially to the increase in the area devoted to cotton.
2. The nation-wide increase in aid observed in the decoupling scenarios is largely due to the fact that of the products receiving coupled and/or decoupled aid, some (rice) obtain more than in the base year, while for others (milk and sugar beet) no aid existed in 2002. This also explains the substantial increase in aid observed in the Balearic Isles and Valencia. In the former, the only activity of any importance that is eligible for aid under the decoupled scenario is dairy farming, which, as noted, received no aid in the base year. In Valencia, the increase is due to the substantial rise in aid for rice. In addition to these regions, substantial increases in aid are recorded in the northern regions of Spain (Galicia, Asturias and Cantabria), where dairy production prevails. Finally, it should be noted, that aid declines in some regions, most visibly in Madrid, because the assumptions made in this study would lead to a substantial decrease in the number of suckler cows, and none of the products receiving higher payments is grown or raised in this region.
3. The reduction in aid due to modulation accounts for slightly more than 2% of total aid. This proportion differs from one region to another, however, ranging from 1% in Cantabria to a little over 3% in Andalusia, when we consider only the regions with an amount of aid large enough to be affected by aid reductions.

## Gross Margin

1. This economic indicator (the sum of gross margin net of aid, plus aid) shows that decoupling policies produce better results than continuation of the Agenda 2000 measures in all regions (Figure 17). The downward variation of this variable under the latter scenario is observed both nation-wide (Figure8) and

*Figure 16. Total aid after modulation (% variation)*

| | Agenda 2000 | Partial Decoupling | Full Decoupling | Regional |
|---|---|---|---|---|
| 01.Galicia | 0.27 | 294.99 | 294.02 | 278.79 |
| 02.Asturias | 0 | 183.03 | 183.08 | 150.68 |
| 03.Cantabria | 0 | 189.35 | 190.84 | 124.28 |
| 04.Basque Country | 0 | 50.48 | 50.76 | 39.09 |
| 05.Navarre | 0.88 | 8.65 | 8.69 | -3.33 |
| 06.La Rioja | 0.71 | 33.38 | 30.79 | -39.96 |
| 07.Aragón | 1.41 | 0.91 | -1.21 | -1.67 |
| 08.Catalonia | 2.06 | 16 | 17.98 | 14.52 |
| 09.Balearic Isles | 65.59 | 7292.95 | 7282.8 | 6711.34 |
| 10.Castile-Leon | 1.42 | 11.62 | 11.36 | 26.87 |
| 11.Madrid | -0.02 | -16.07 | -39.77 | -27.47 |
| 12.Castile-La Mancha | -0.05 | -1.59 | -1.78 | 24.3 |
| 13.Valencian C. | 0.37 | 344.37 | 344.45 | 195.08 |
| 14.Murcia | 0.36 | 4.67 | 0.18 | -86.53 |
| 15.Extremadura | 0.33 | -1.96 | 0.21 | -4.86 |
| 16.Andalusia | 2.97 | 9.66 | 10.56 | -11.35 |
| 00.Spain | 1.39 | 13.04 | 12.92 | 12.9 |

in all regions except Murcia, which benefits from the aforementioned increase in earnings from sheep.

2. Under decoupling policies, the variation in gross margin is likewise negative in most regions. The exceptions are the regions with upward variations in the gross margin net of aid (La Rioja, Aragon, Castile-La Mancha, Murcia and Extremadura) and Navarre, where aid offsets the tiny loss in the gross margin net of such support.
3. In most regions, full decoupling yields higher gross margins than partial decoupling. Where this is not the case, the move from partial to full decoupling entails often a loss of agricultural activity, such as in Madrid[17]. The loss in this region translates into a substantial decline in the number of suckler cows.

**Dual Values of Land and Entitlement Payment per Hectare**

The PROMAPA generates two land value-related indicators: the dual value (one for non-irrigated and the other for irrigated land) and the entitlement payment per hectare, which is qualified to receive decoupled payments. Both nation-wide and for all regions, the former indicator, associated with the revenues per hectare deriving from land use in farming, declines with increasing degrees of decoupling, while the latter, which is linked to decoupled payments, follows an upward trend[18].

*Figure 17. Gross margin (% variation)*

| | Agenda 2000 | Partial Decoupling | Full Decoupling | Regional |
|---|---|---|---|---|
| 01.Galicia | -21.47 | -4.94 | -4.15 | -4.74 |
| 02.Asturias | -18.9 | -7.37 | -6.03 | -7.7 |
| 03.Cantabria | -21.44 | -8.67 | -7.64 | -11.66 |
| 04.Basque Country | -15.52 | -5.19 | -3.85 | -5.62 |
| 05.Navarre | -9.56 | 0.06 | 1.62 | -1.68 |
| 06.La Rioja | -18.69 | 14.41 | 14.36 | -9.08 |
| 07.Aragón | -9.88 | 1.33 | 1.31 | 1.1 |
| 08.Catalonia | -22.93 | -21.26 | -21.71 | -22.59 |
| 09.Balearic Isles | -36.01 | -1.85 | -1.91 | -3.14 |
| 10.Castile-Leon | -11.89 | -0.64 | -1.17 | 3.83 |
| 11.Madrid | -8.6 | -6.93 | -16.01 | -11.55 |
| 12.Castile-La Mancha | -5.07 | 3.76 | 3.96 | 13.06 |
| 13.Valencian C. | -52.97 | -8.06 | -7.98 | -27.04 |
| 14.Murcia | 9.58 | 32.28 | 31.96 | 11.01 |
| 15.Extremadura | -5.82 | 5.61 | 8.11 | 6.09 |
| 16.Andalusia | -11.2 | -1.2 | -1 | -12.15 |
| 00.Spain | -12.26 | -1.23 | -0.96 | -0.96 |

*Figure 18. Average entitlement payment per ha (€)*

| | Agenda 2000 | Partial Decoupling | Full Decoupling | Regional |
|---|---|---|---|---|
| 01.Galicia | 0 | 181.99 | 249.74 | 240.09 |
| 02.Asturias | 0 | 163.86 | 271.12 | 240.09 |
| 03.Cantabria | 0 | 212.68 | 312.59 | 240.09 |
| 04.Basque Country | 0 | 173.83 | 261.09 | 240.09 |
| 05.Navarre | 0 | 185.85 | 270.72 | 240.09 |
| 06.La Rioja | 0 | 384.81 | 530.83 | 240.09 |
| 07.Aragón | 0 | 165.33 | 241.34 | 240.09 |
| 08.Catalonia | 0 | 195.34 | 248.34 | 240.09 |
| 09.Balearic Isles | 0 | 234.45 | 260.98 | 240.09 |
| 10.Castile-Leon | 0 | 152.18 | 210.01 | 240.09 |
| 11.Madrid | 0 | 145.19 | 198.16 | 240.09 |
| 12.Castile-La Mancha | 0 | 137.88 | 188.61 | 240.09 |
| 13.Valencian C. | 0 | 580.12 | 584.49 | 240.09 |
| 14.Murcia | 0 | 1052.64 | 1960.11 | 240.09 |
| 15.Extremadura | 0 | 134.22 | 253.42 | 240.09 |
| 16.Andalusia | 0 | 264.8 | 322.18 | 240.09 |
| 00.Spain | 0 | 172.23 | 240.09 | 240.09 |

## FINAL REMARKS

The preceding pages contain a detailed analysis, based on PROMAPA results, of the impact of different agricultural policy measures on Spanish agriculture.

One general conclusion is that while farming activity is lower under the partial and full decoupling assumptions than under the continuation of Agenda 2000 scenario, farmers' earnings are higher in the former. Nonetheless, a great part of this improvement is the outcome of more favourable price changes as well as of increased aid for some products under the decoupling scenarios and new forms of aid not provided for under Agenda 2000 arrangements for others.

Irrespective of the results, the positive mathematical programming model PROMAPA is presently a flexible model, relatively quick and easy to use for simulating nation-wide and regional results associated with different agricultural policy assumptions and price forecasts.

The model is essentially designed to be a tool for reflection about the expected consequences of implementing agricultural policy measures. An illustrative example of such use of the model is given in the preceding pages, in an attempt to explain how the results obtained are affected by the assumptions proposed.

## ACKNOWLEDGMENT

This work was supported by the Spanish MEC (Project SEJ2007-63642) and by the EU (GENEDEC project FP6-502184).

## REFERENCES

Adenáuer, M., Britz, W., Gocht, A., Gömann, H., Cristoiu, A., & Ratinger, T. (2006). *Modelling impacts of decoupled premiums: building-up a farm type layer within the EU-wide regionalised CAPRI model*. Selected paper presented at the 93rd EAAE Seminar. Prague, 22-23 September 2006.

Arfini, F., & Paris, Q. (1995). A positive mathematical programming model for regional analysis of agricultural policies. In E, Sotte, (Ed.), *The Regional Dimension in Agricultural Economics and Policies, EAAE, Proceedings of the 40 th Seminar, 26-28 June 1995, Ancona,* (pp. 17-35).

Arriaza, M., & Gómez-Limón, J. A. (2003). Comparative performance of selected mathematical programming models. *Agricultural Systems, 77,* 155–171. doi:10.1016/S0308-521X(02)00107-5

Barkaoui, A., & Butault, J. P. (1999). *Positive Mathematical Programming and Cereals and Oilseeds Supply within EU under Agenda 2000*. Paper presented at the 9th European Congress of Agricultural Economics, Warsaw, August 1999.

Barkaoui, A., & Butault, J. P. (2003). *Révision à mi-parcours de la PAC: simulations de l'effect du découplage sur l'offre dans les régions françaises. Document de travail*. INRA-ESR Nancy.

Blanco, A., & Iglesias, E. (2005). *Modelling New EU Agricultural Policies: Global Guidelines, Local Strategies*. Selected paper presented at the 89th EAAE Seminar. 3-5 February, Parma, Italy.

Buysse, J., Fernagut, B., Harmignie, O., Henry De Frahan, B., Lauwers, L., Polomé, P., et al. (*2004*). *Modelling the impact of sugar reform on Belgian Agriculture*. Selected paper presented at the International Conference on Policy Modelling, Paris, 30 June- 2 July, 2004.

Buysse, J., & Van Huylenbroeck, G. (2005). *Impact of alternative implementations of the Agenda 2000 Mid Term Review*. Selected Paper presented at the 11 th Congress of the EAAE. Copenhagen, Denmark, 24-27 August, 2005.

CAPRI. (2000). *Common Agricultural Policy Regional Impact (Fair3-CT96-1849). Final Consolidated Report, Institute for Agricultural Policy, University of Bonn, Department of Economics, University College Galway, Institut Agronomique Meditérranéen de Montpellier, Departamento di Economia, Sociologia y Politica Agraria, Universidad Politécnica de Valencia, Università degli Studi di Bologna, Dipartamento di Protezione e Valorizzazione Agro-Alimentare (DIPROVAL)*. Sezione Economía.

Gocht, A. (2005). *Assessment of simulation behaviour of different mathematical programming approaches*. Selected paper presented at the 89th EAAE symposium on Modelling agricultural policies: state of the art and new challenges, Parma, 3-5 February.

Gohin, A., & Chantreuil, F. (1999). La programmation mathématique positive dans les modèles d'exploitation agricole: principes et importance du calibrage. *Cahiers d'Économie et Sociologie Rurales.*, *52*, 59–78.

Heckelei, T., & Britz, W. (1999). *Maximum entropy specification of PMP.* CAPRI Working Paper 08199.

Heckelei, T., & Britz, W. (2005). *Models Based on Positive Mathematical Programming: State of the Art and Futher Extensions*. Plenary paper presented at the 89th EAAE Seminar, 3-5 February, Parma, (pp. 48-73).

Heckelei, T., & Wolff, H. (2003). Estimation of Constrained Optimisation Models for Agricultural Supply Analysis Based on Generalised Maximum Entropy. *European Review of Agriculture Economics*, *30*(1), 27–50. doi:10.1093/erae/30.1.27

Helming, J. F. M., Peeters, L., & Veendendaal, P. J. J. (2001). Assessing the Consequences of Environmental Policy Scenarios in Flemish Agriculture. In Heckelei, T., Witzke, H. P., & Henrichsmeyer, W. (Eds.), *Agricultural Sector Modelling and Policy Information Systems*. Kiel, Germany: Vank Verlag.

Henry de Frahan, B., Buysse, J., Polomé, P., Fernagut, B., Harmignie, O., & Lauwers, L. (2007). Positive mathematical programming for agricultural and environmental policy analysis: review and practice. In Weintraub, A., Bjorndal, T., Epstein, R., & Romero, C. (Eds.), *Management of Natural Resources: A Hanbook of Operations Research Models, Algorithms and Implementations*. Berlin: Springer, International Series in Operations Research and Management Science.

Howitt, R. E. (1995). Positive Mathematical Programming. *American Journal of Agricultural Economics*, *77*(2), 329–342. doi:10.2307/1243543

Ibáñez, J., & Pérez, C. (1999). Impactos de la reforma de la PAC de 1992 sobre el subsector agrícola español. *Estudios Agrosociales y Pesqueros*, *185*, 9–30.

Jayet, P. A., Marzochi, E., Hofstetter, A., & Donati, M. (2000). *Modular approach of agricultural supply modelling through linear programming.* EUROTOOLS (Proyect FAIR 5-CT97-3403), Working Paper Series n.26.

Júdez, L., Chaya, C., Martínez, S., & González, A. A. (2001). Effects of the Measures Envisaged in "Agenda 2000" on Arable Crop Producers and Beef and Veal Producers: An Application of Positive Mathematical Programming to Representative Farms of a Spanish Region. *Agricultural Systems*, *67*, 121–138. doi:10.1016/S0308-521X(00)00051-2

Júdez, L., de Andrés, R., Ibáñez, M., de Miguel, J. M., & Urzainqui, E. (2005). *The PROMAPA.G Model. Document of GENEDEC (Project FP6 502184)*. 31 March.

Júdez, L., de Andrés, R., Ibáñez, M., & Urzainqui, E. (2008). *A method for including in PMP models activities non –existent in the baseline situation.* Selected poster presented at the XII th Congress EAAE, Ghent, Belgium, 26-29 August.

Júdez, L., de Miguel, J. M., Mas, J., & Bru, R. (2002). Modeling Crop Regional Production Using Positive Mathematical Programming. *Mathematical and Computer Modelling*, *35*, 77–89. doi:10.1016/S0895-7177(01)00150-9

Júdez, L., de Miguel, J. M., Piniés, M., Legorburu, I. G., & Miguel, J. L. (2003). *Effects of Mid-Term review measures on some representative Spanish farm: an application of the PROMAPA model.* Selected paper presented 80 th EAAE Seminar, Ghent, 24-26 September 2004.

Júdez, L., Fuentes-Pila, J., de Miguel, J. M., & Martínez, S. (2001a). Programación Matemática Positiva: Un análisis de los efectos de la Agenda 2000 en algunas explotaciones tipo españolas. Agricultura en Sao Paulo. *Journal of Agricultural Economics*, *48*(1), 99–118.

Júdez, L., Ibáñez, M., de Andrés, R., & Urzainqui, E. (2006). *Single Farm Payment, Modulation and Dual Values of Land.* Document of GENEDEC (Project FP6 502184). Presented at Rethymno meeting 12-13 May.

Júdez, L., Martínez, S., & Fuentes-Pila, J. (1998): Positive Mathematical Programming Revisited. EUROTOOLS (Proyect FAIR 5-CT97-3403), Working Paper Series n.2.

Kuepker, B., & Kleinhanss, W. (2006). *Comparative Assessment of National and Alternative Decoupling Schemes in EU Member States.* Selected paper presented at the 93 rd EAAE Seminar, Prague, 22-23 September 2006.

Offermann, F., Kleinhanss, W., Huettel, S., & Kuepker, B. (2005). Assessing the 2003 CAP reform impacts on German Agriculture using the farm group model FARMIS. In F. Arfini, (ed.), *Modelling agricultural policies: state of the art and new challenges. Proceedings of the 89 th EAAE Seminar,* Parma, Italy.

Osterburg, B., Offermann, F., & Kleinhanss, W. (2001). A sector consistent farm group model for German agriculture. In Heckelei, T., Witzke, H.P. & Henrichsmeyer, W. (eds.), *Agricultural Sector Modelling and Policy Information Systems.* Kiel, Germay: Wissenschaftsverlag, Vauk.

Paris, Q., & Howitt, R. E. (1998). An Analysis of Ill-posed Production Problems Using Maximum Entropy. *American Journal of Agricultural Economics*, *80*(1), 124–138. doi:10.2307/3180275

Paris, Q., Montresor, E., Arfini, F., & Mazzocchi, M. (2000). An Integrated Multi-Phase Model for Evaluating Agricultural Policies Through Positive Information. In Heckelei, T., Witzke, H.P. & Henrichsmeyer, W. (eds.), *Agricultural Sector Modelling and Policy Information Systems. Proceedings of the 65 th EAAE Seminar*, March 29-31, 2000 at Bonn University, (pp. 100-110). Kiel, Germany: Vauk Verlag.

Polomé, P., Harmignie, O., Henry de Frahan, B., Buysse, J., Van Huylenbroeck, G., Van Meensel, J., et al. (2005). *Farm–level Multi-output Multi-input Cost Function: a Comparison of Least Squares and Entropy Estimators*. Selected paper presented at the 11 th EAAE Congress, Copenhagen, 23-27 August 2005.

Röhm, O., & Dabbert, S. (1999). Modelling regional production and Income Effects. In Huylenbroeck, G., & Witby, M. (Eds.), *Countryside Stewardship: Farmers, Policies and Markets*. Amsterdam: Pergamon.

Röhm, O., & Dabbert, S. (2003). Integrating Agri-Environmental Programs into Regional Production Models: An Extension of Positive Mathematical Programming. *American Journal of Agricultural Economics*, *85*(1), 254–265. doi:10.1111/1467-8276.00117

Severini, S., & Cortignani, R. (2008). *Introducing deficit irrigation crop techniques derived by crop growth models into a Positive Mathematical Programming model.* Selected paper presented at the XII th Congress EAAE, Ghent, Belgium, 26-29 August.

## KEY TERMS AND DEFINITIONS

**Calibrated Model:** Model capable of reproducing the real results of the unit modeled (farm or region) in a baseline situation (or base year).

**Common Agricultural Policy (CAP):** Is a set of measures of the European Union concerning agricultural market and subsidies and rural development..The CAP came into being in the late 1950s. Since then it has undergone several reforms. The most recent are that of 1999 (called Agenda 2000 reforms) and that of 2003 (called the Mid-Term-Review (MTR) reform).

**Decoupling and Single Payment Scheme (SPS):** Before the MTR reform, and in particular with Agenda 2000, all direct payments were coupled to the agricultural production sectors. The MTR reform introduced the single payment, decoupled from the agricultural production sectors and calculated on the basis of the payments received during a past reference period.

**Direct Payments:** European Union subsidies paid to the farmers to guarantee more stable incomes. These payments can be coupled and decoupled for one year with respect to the different agricultural production sectors. These payments are also referred to as couple aid and decouple aid respectively.

**Farm Accountancy Data Network (FADN):** Is a network which collects yearly accountancy data on farms from the different regions of the European Union. It was designed to be a quantitative database which would contribute to the analysis of the effects of agricultural policy measures and the changes in farm income.

**Positive Mathematical Programming (PMP):** In general, positive farm models are models which give an almost exact reproduction of the real results of the farms. These models are the opposite of normative models whose results are obtained to help farmers' decisions. Recently, the term PMP refers to mathematical programming models calibrated on the basis of the procedure formalized in Howitt (1995).

**PROMAPA Model:** Is a positive mathematical programming model, developed by the authors to simulate the impact of new agricultural policy measures in the Spanish agricultural sector using

Spanish FADN data. PROMAPA is the acronym of **PRo**gramación **M**atemática para el **A**nálisis de **P**olíticas **A**grarias.

## ENDNOTES

1 This study has been developed in the context of the Spanish MEC project SEJ2007-63642 and of the EU project GENEDEC (FP6-502184).

2 The AROPAJ model developed in France applying linear programming, which also uses information from the Farm Accountancy Data Networks, is used to analyse agricultural and agro-environmental policies (see Jayet et al., 2000)

3 This result was later set out in Júdez et al. (2001b) and proved in greater detail in Júdez et al. (2002).

4 In effect from (6):

$$\frac{\partial^2 C(x_j)}{\partial x_j^2} = \frac{\partial r_j}{\partial x_j} - \sum_{i=1}^{m} a_{ij} \frac{\partial \lambda_i}{\partial x_j} \text{ So:}$$

$$q_j = \frac{\partial r_j}{\partial x_j}$$

5 As estimated by Ibáñez and Pérez (1999).

6 To simplify the formulation we do not consider livestock.

7 The table was drawn up with data on the mean farm type for Spain as a whole.

8 Like Figure 5, this table was drawn up with data on the mean farm type for Spain as a whole.

9 This is because the figures in the tables of Figures 5 and 9 are means and in each of the 86 farm types studied in this chapter, the crops have different yields, prices and competing crops.

10 The greater with higher degrees of decoupling, since beef prices decline less.

11 The smaller at higher degrees of decoupling.

12 For all farm types in these regions taken as a whole, the variation with respect to the base year is; -0.07% under Agenda 2000; 0.15% with partial decoupling and -0.19% with full decoupling.

13 Such is the case in Navarre, for instance, where under full decoupling direct revenues less cost of purchased feed per suckler cow in TF 8000 are approximately 4% higher than the base year, prompting an increase of around 7% in the number of head.

14 The high figures for Murcia and Valencia stand out in the table. The sum for the former is due to the high sheep density per ha and for the latter to the large area devoted to rice.

15 With the exception of products for which no price information was available, which were regarded to be constant.

16 The sole exception is Catalonia, where the alfalfa sold plays a predominant role.

17 The same situation exists in La Rioja, Aragon, Murcia, Catalonia, Balearic Isles and Castile-Leon, but here the difference in the gross margins under the two scenarios is small.

18 For a theoretical study of these questions, see Júdez et al. (2006).

# Chapter 11
# Mathematical-Modelling Simulation Applied to Help in the Decision-Making Process on Environmental Impact Assessment of Agriculture

**Maria Conceição Peres Young Pessoa**
*Embrapa Environment, Jaguariúna/SP, Brazil*

**Elizabeth Nogueira Fernandes**
*Embrapa Dairy Cattle, Juiz de Fora/MG, Brazil*

**Sonia Cláudia do Nascimento de Queiroz**
*Embrapa Environment, Jaguariúna/SP, Brazil*

**Vera Lúcia Ferracini**
*Embrapa Environment, Jaguariúna/SP, Brazil*

**Marco Antonio Ferreira Gomes**
*Embrapa Environment, Jaguariúna/SP, Brazil*

**Manoel Dornelas de Souza**
*Embrapa Environment, Jaguariúna/SP, Brazil*

## ABSTRACT

*The present chapter provides a brief explanation on some aspects involved in the development of models and mathematical-modelling simulations, to show their benefits to the decision-making process in the environmental impact assessment of agriculture. Aspects concerning the agroecosystems were also presented toward the sustainability of Brazilian agricultural production systems. Some applications which have been developed in Brazil were pointed out, as well as a specific case study conducted at the Guarani aquifer recharge area located in Ribeirão Preto, São Paulo state, in order to show the influence of input data on the results provided by CMLS94 simulator.*

DOI: 10.4018/978-1-61692-871-1.ch011

## INTRODUCTION

It is known that the globalization of the markets and, consequently, the of easy ways to obtain a wide variety of product options have shown that the agility in decision-making is the main factor in the quest for efficiency and quality competitiveness. However, the decision-making process involves both the acquisition of knowledge on a given problem and the prediction of risks and uncertainties associated with it. Thus, it requires prior knowledge of the main factors that represent the problem to be solved, as well as their interactions with other factors which could also influence them either positively or negatively.

The advent of the internet and the facility to gain access in order to obtain and make information available almost instantly, increase the perception that people do not have enough knowledge about a particular problem, causing delays in the decision making process. The huge demand for details, the large amount of conflicting information available on web sites and the absence of a minimum well substantiated standard related to important aspects for the correct definition of a specific problem lead to errors in the model formulation, delays and decrease the choice of a most likely decision to succeed. Thus, it emerges the need for action planning for the appropriate management of activities, in order to face the need of knowledge of controllable factors and of risks.

Tashizawa & Rezende (2000; pg 16) presented the action planning as "*a method to prioritize activities in order to achieve a proposed objective and therefore a desired future.*" For this reason, planning requires the availability, the organization and the speeding up of the information retrieval, as well as the use of agile methods to promote the integration of knowledge. It allows the analysis of elements that can interfere significantly in the decision-making process, which are present in the method of activities sort, making sure that the activity flows aligned with the objectives, even if problems occur during its implementation.

When the focus of attention is on the background of the uncontrollable factors of the problem, i.e., on the "risk" of their occurrence, which is still doubtful, techniques of statistical inference and simulation systems become great allies. They encourage both the identification of the degree of perceived risk and the resulting evaluation of different possibilities to be investigated, although they must also be based on the known aspects.

Considering that a real problem to be solved can be represented in all sorts of possibilities, there are different ways to design a model of a same situation, which are presented according to the goal and the particular view of the modeller (perception, intuition, technical and cultural knowledge and ability to rescue information in advance) about the actual exposition of risk in relation to future possibilities of consequences. For this reason the model will be capable of success or failure in finding solutions, as your modeller may overestimate or underestimate common risks. Therefore, to provide computer tools capable of well representing a real problem, it is required consistency and reliability of its models related to actual object of research, as well as the acquisition of safe input data, based on information and method of realistic perspective for its representation.

Frequently, databases accessible through the internet make available the mean data values of simulator inputs such as soil characteristics and physicochemical properties of pesticides. It is important to notice that some data can vary in a wide range of values that may affect the mean data presented. In addition, some data can also vary according to local environmental factors. As a consequence, the use of mean data can produce simulation results quite different from those observed in actual experiments. Thus, it is important to properly understand each input of the simulator in order to know when it is possible, or not, to use mean values for data entry. It is also important to take into account that mathematical models and simulators need calibrations and validations, which are frequently not performed.

Several examples of mathematical-modelling simulation applications have been presented in the technical literature, including for the Brazilian agricultural research applications, which have been widely used in the investigation of strategies aimed at increasing the knowledge on the possible environmental impacts (positive or negative) resulting from agricultural practices. The systemic view provided by them has been helpful in identifying strategies to promote more accurate production systems of higher environmental quality. Among them it could be mentioned field monitoring, laboratory-oriented actions for both the insects rearing for biological control of agricultural pests and for the identification of pesticides less persistent in the environment. They have also been helpful in the decision-making processes in management practices that minimize seasonal emissions of greenhouse gases from agricultural practices, as well as applied to the conservation of soil and water, among others. The foregoing applications provide essential information for decision-making processes focused on good agricultural practices and sustainable management of the Brazilian environment. Below, some briefly aspects concerning agricultural environmental issues are presented in order to better understand the mathematical-modelling simulation used to promote the sustainability of Brazilian agricultural production systems.

## ENVIRONMENTAL AND MATHEMATICAL-MODELLING ASPECTS OF THE INTEREST AND SOME USES FOR ENVIRONMENTAL IMPACT ASSESSMENT OF AGRICULTURAL ACTIVITIES IN BRAZIL

The interest of humans in learning about issues related to the environment where they live has been reported since the beginning of mankind. In an attempt to better understand their relationship with other animals, plants and natural phenomena in the environment, which have directly influenced their survival, humans have tried to understand and use the resources of their environment, interfering and modifying it.

Over the centuries, technological advances of the modern era, not only drove humans to progress, comfort and economical growth, but also led to negative consequences for the environment.

The concern with environmental issues only came to be known in the 60´s after the United Nations Conference on the Human Environment, in Stockholm in 1972. Although the term ecology (from the Greek *oikos* that means "*house*" or "*place where we live*"(ODUM, 2004; pg. 4)) has been formally mentioned for the first time in 1869 by Ernst Haeckel, it was from this conference that it has been further disseminated to the general population. Defined by Odum (2004; pg. 4) as *the study of the relationships of organisms or groups of organisms with their environment, or the science of the interrelationships that bind its living organisms to their environment*, the ecology studies the structure, the relations and inter-relationships that link the living organisms to their environment and the workings of nature, considering humanity as part of it.

The rise of collective perception that human actions were causing damage to natural resources, some of which are irreversible, motivated actions arising from the needs to better understand the negative impacts of human activities in order to mitigate or prevent them. Thus, several environmental impact aspects associated with agricultural production systems were also identified. Among them we could mention the indiscriminate use of pesticides to agricultural plague control, the soil erosion due to inadequate methods of crop management and grazing, the soil salinization promoted by inadequate method of irrigation, the water and soil contaminations promoted by pesticide leaching and run-off, the inappropriate use of pesticide application technologies, the improper disposal of farm wastes.

It is also important to notice that the assessment of the environmental impacts of agriculture involves:

- the diagnosis of the actual state of the environment;
- the prevision and tendencies of the future state of the environment considering scenarios which maintain the present activities and others one which consider that some changes could be done, in order to investigate the possible consequences;
- to find and to spread procedures that allow minimizing or mitigating the negative environmental impacts identified on the local;
- to perform local monitoring activities in order to verify the space-time dynamic and the persistence of some important variables.

In order to better understand the process of occurrence of environmental impacts, the environment where agricultural production takes place can be considered as a part of the ecosystem named agroecosystems. According to Odum (2004), the ecosystem concept is broad and its main function highlight compulsory relationships related to interdependences and causal relationships. So, he brings *components to form functional units, or parts of inseparable whole, which therefore, makes the ecosystem level of biological organization appropriate to the application of techniques of systems analysis*. When considering the agricultural environment, the term agroecosystems emerges. It is a spatially and functionally coherent unit of agricultural land, which is managed by human activity, where frequently there is a simpler diversity of species (insects, plants), energy fluxes and nutrient balances when compared to the natural ecosystem, demanding for high quantities of nutrient inputs.

The agroecosystem relations and interrelations provoke modifications inside the system, as well as on its outside, causing changes also in the region where it is present, that interferes as its feedback. Therefore, it is important to well know the environmental aspects of the area where the agricultural activities have been done. Among them, we could mention weather, physicochemical attributes of soils, crops, hydrogeology, relief, human cultural aspects, natural vegetation, available technologies and economic aspects of the production systems, among others. However, it is also important to establish a terrestrial basic scale of work which makes possible to recognize these important aspects mentioned and their interrelation in order to assure the equilibrated research level of detail. As a result, the catchment scale, which is also known as micro-basin scale, has been adopted (BOTELHO, 1999). A number of models and simulators were developed to investigate or to better understand the relations and contributions of some variables of the agroecosystems in the catchment scale in order to promote the correct management of agricultural production systems.

In a general way, a model can be classified as: conceptual, mathematical (analytical, empirical, deterministic, mechanistic, screening, lumped, static, steady-state, dynamic, continuous, discrete, compartmental, numeric or normative (which uses operational research techniques such as linear programming, dynamic programming, multi-objective programming, among others), holistic, reductionist, and others), statistical, physical, biogeochemical, research, among others (PESSOA et al., 1997b; DOURADO NETO et al., 1998). The model scale (punctual, local, municipal, regional, global) depends on its objective, the technique used, the amount of information and the resources available. They have different characteristics that help classify them into one or more of the types described above.

Models which are described into mathematical language are called "mathematical-models" and the process of constructing this representation into mathematical-language is named "mathematical-modelling". In this sense, mathematical-modelling is a process of developing this way of knowledge representation, which can be made by means of logical operators, functions, equation systems, mathematical intervals, among others. In mathematical models, their components such as input, variables, parameters and outputs are represented, as well as functional relationships based on a greater or lesser degree of detail, always depending on their respective goals. They are a systemic representation.

The complexity of a mathematical modelling is associated with their objective and data availability to present it. This will orient the choice between the holistic or the reductionist ways. The holistic approach came from the holism theory (from the Greek *holos* that means "all" or "entire") where "*The whole is more than the sum of its parts*" as presented by the philosopher Aristotle. Each part has a different contribution to the significance of the whole and can be understood only in this sense. The opposite sense of holism is the reductionism, where a complex problem can be explained by a single reduction of it, presenting its minimum fundamental parts. This is a very simplistic way to describe a real problem to be modelled.

The foregoing allows the mathematical-modelling process to well identify the influencing factors of the investigated problem, considering its main characteristics, interactions with other processes, risk perceptions, as well as other aspects involved, such as socio-economic, physical and ecological.

When developing or using a computer environment, it will be possible to create different scenario options, stressing the correlation of the parameters and allowing to reach and to investigate a set of possible results, that case we are dealing with computational simulation, also called mathematical-modelling simulation.

Here, the mathematical models, interrelationships and user interfaces (inputs, outputs and scenarios) are developed in a specific computer language. It is important to notice that within a single mathematical-modelling simulator, one or more mathematical models may be present in order to deal with the problem to be investigated.

It is further argued that simulators need calibration and validation. Calibration adjusts the models parameters which are present within the simulator. It takes into account the results observed in the field or in the laboratory, in order to achieve greater precision of the tool before the beginning of the investigation of the scenarios. During the validation process, the resulting data from simulations being performed are compared with those resulted from experimental work for the same scenarios, in order to verify whether the built in tools has the expected reliability. It is necessary to point out that data used for validation must be different from those used during calibration.

A great number of simulators were developed using different mathematical modelling techniques, where some of them were also integrated with other computing features such as Geographic Information System (GIS), Database, Expert Systems, among others. Following, we presented some applications conducted in Brazilian environment.

The use of mathematical-models as a tool for decision-making planning, helping on the management of agricultural farm is not new. Initially they were presented as the "farm models", which mainly considered the conceptual and operational research methods for modelling. They have been available in Brazil since the 1970s.

In the beginning, they made possible an evaluation of the production systems at the local scale (property), progressing later to regional scales. As examples, one should mention the works of Praseres (1976), which presented the use of linear-programming to agricultural company activities located in Maranhão State, those developed by Azevedo Filho & Neves (1982), which made use of mathematic-programming to analyse the cattle investments, and also Rodrigues (1987) which presented the agricultural-planning using a non-determinist linear-programming. Peres (1990) also presented a decision-making model for confined cattle and Dossa (1994) proposed the use of a linear-programming model to the farm management. Regarding economic models, Ozeki (1998) presented a review for Brazilian farm management also considering an operational-research method called Revised-Linear Programming,, Veloso & Chaib Filho (1999) indicated the use of mathematical-modelling strategies to evaluate the impact of production systems on the Brazilian savannah (which is known in Brazil as "cerrado"), in terms of social, economic and ecological aspects, using techniques of forecasting prices, productivity of alternative crops, linear programming, integer programming, multi-criterions programming and Geographic Information System (GIS) techniques. Recently, Mendonça (2004) proposed a software application applied for beef cattle bioeconomic simulation. It promoted the comparison of various scenarios in order to identify a strategy to be adopted for the farm located in Brazilian Central region. Moreover, Veloso et al (2007) made use of mathematical programming to evaluate the economic viability of milk production on little scale, as well as Borges Junior et al (2008) made use of computational modelling tools to investigate alternatives for planning the irrigated agriculture activities at the farm level for Jaíba irrigation in the state of Minas Gerais.

Nowadays, there are a great number of methods considering environmental economy and social responsibility (SOUZA, 2007).

The pollution of water resources is an international concern. Exposed to different agricultural activities, which make use of this non-renewable natural resource or make pressure on it, it requires planning activities in order to ensure their conservation and preservation. The pesticide uses and soil erosion have been the main worries towards the conservation of catchments where agricultural activities have been done. Due to that, the number of mathematical-model simulations dealing with the assessment of the pesticide dynamics and their implications on the water quality is increasing.

Some mathematical-modelling simulators also enable to investigate the interactions among crop, soil, irrigation, weather, and pesticide application managements during all the crop development period. Thus, the use of models and simulators is raising in Brazil to predict crop growth and harvest, as well as to follow the insect-plant environment, especially for the expressive commercial cultures. These computational tools made possible a greater understanding of specific aspects of the culture, such as physiological, genetic, nutritional, and others. Moreover, they represent the process of operation of the environment and, consequently, their interactions enabling the identification of how a particular factor (abiotic, economic, social, etc.) interferes with the culture as a whole.

Most mathematical-models of agricultural cultures were the result of empirical observation, i.e., resulting from laboratory experiments also conducted under the observation of some variables of the real system under controlled conditions. Therefore, the subsequent relationship observed must be strictly explained with their objective and the pre-defined hypotheses. Most of these mathematical models were derived from linear and non-linear regression models fitted to the observed data. Thus, the subsequent investigation, that intends to use their results, must also maintain the same conditions under which the empirical relationships had been established. In addition, it is also needed to stay within the range for which they had been calibrated, otherwise the subsequent results obtained will be less reliable to reality and will little contribute for the decision-making to planning. It is worth remembering that any mathematical-model is a representation of knowledge about the object of model development, therefore, has limitations for its extrapolation to different situations.

Some mechanistic models incorporated one or more processes which occur in the growth development, such as photosynthesis, respiration, fructification, soil erosion, water flow, leaching of pesticides, microbial degradation, pesticide translocation, evapotranspiration, among others. Their computational representations and interactions have been promoting to follow the development of the plant, its productivity, insect-pest population control considering the strategies for biological and chemical controls, among other scenarios even before not evaluated in practice by means of simulation systems. Through its performances, many of these programs enable the assessment of crop growth for specific varieties, building the sensitivity analysis of variables that interfere with their development and productivity. These took into account the influence of the development of culture, such as climate, leaf area, nitrogen dynamics, soil type and its representative variables, such as organic matter content, bulk density, wilting point, saturation, field capacity, irrigation, evapotranspiration, among others.

Regarding the estimates of harvest, several mathematical models of culture have been incorporated into the simulators in order to evaluate the performance of production systems under different management alternatives for soil, water and plant, among others. These programs are also important for the environmental planning of the farm, since they allow decision-making to more appropriate management in the face of local characteristics and thus may lead to the use of strategies to promote the best yields in the context of agricultural sustainability (PESSOA et al., 2006b; CERDEIRA et al., 2007d).

It is important to notice that the Brazilian researchers have been looking for more representative models for tropical environment also working on the development of specific softwares.

Pessoa et al (1997b) and Pessoa & Scramin (2004) presented a literature review of the main mathematical-models used worldwide. Among them, they mention those that help a rapid assessment of the pesticide active-ingredient potential to be transported both dissolved in water and adsorbed in the soil, as well as the more complex simulators. Some of them are the fugacity models, which evaluate the tendency to leak substances for compartments such as soil, water, air and biota (LEWIS, 1901; PAUSNITZ, 1969; VAN NESS & ABBOTT, 1982; MACKAY, 1993), and screening models like those proposed by the *GOSS Method* (GOSS, 1992), the *GUS Index* -"Groundwater Ubiquity Score" (GUSTAFSON, 1989), the Screening Models and Screening level tier used by USA-Environmental Protection Agency (COHEN et al., 1995; COHEN et al., 1984), among others. Other important models to be considered are the Attenuation Factor (AF) or rate of emission mass of pesticides in soil (RAO et al., 1985) and Retardation Factor (RF) or delay index of pesticides in soil (RAO et al., 1985).

The potential for erosion was calculated making use of the Universal Soil Loss Equation (USLE) (WISCHMEIER & SMITH, 1978), which was afterwards presented as Revised Universal Soil Loss Equation (RUSLE) (RENARD et al., 1991) and Modified Universal Soil Loss Equation (MUSLE) (WILLIAMS, 1975).

More complex mathematical-modeling simulators have also been available for this analysis, like Pesticide Root Zone Model (PRZM) (CARSEL et al. 1985), Groundwater Loading Effects of Agricultural Management Systems (GLEAMS) (LEONARD et al. 1987), Chemical Movement in Layered Soils (CMLS-94) (NOFZIGER & HORNSBY, 1994), Soil and Water Assessment Tool (SWAT) (USDA/GSWRL), Agricultural Non-point Source Pollution Model (AGNPS) (YOUNG et al., 1986), Erosion Productivity Impact Calculator (EPIC) (SHARPLEY & WILLIANS, 1990), Water Erosion Prediction Project (WEPP) (LANE & NEARING, 1980), Leaching Estimation and Chemistry Model (LEACHM) (WAGENET & HUTSON, 1989), Decision Support System for Agrotechnology Transfer (DSSAT) (TSUJI et al., 1994), among others. They have been gradually improved to different crops and environmental conditions, as well as calibrated and validated for the Brazilian environmental conditions, which include their crop varieties. Afterwards, some of the simulators were integrated to features like Geographic Information Systems (GIS), Databases and Expert Systems to provide the analysis of regional productivity of the crop (LAL et al., 1993), the assessment of the global climate change impact on crops (CURRY et al. 1990; EARSTERLING et al., 1993), and the spatial distribution of pesticides or expansion of crops or plagues. Some of Brazilian applications were presented, as follows.

Liu (1989) used the *CERES-Maize* to evaluate the yield of a Brazilian hybrid corn variety, while Tojo-Soler et al (2001) applied the same model to identify the genetic traits and sowing dates of corn cultivars sowed at out of seasons. The physiological model of *CERES-Wheat* was used on data collected in an experimental field in order to estimate the yield of wheat under São Paulo conditions (ANUNCIAÇÃO & LIU, 1991). Medeiros et al (2000) proposed modifications on a specific subroutine of *Ceres-Sorghum* to evaluate the dry matter production.

*LEACHM-P* was calibrated and used by Giuliano (1995) to evaluate the diffusion movement of one organic solute inside a porous media located at São Paulo State, while *CMLS94* was applied for sugarcane cropped in the Guarani aquifer recharge area located in Ribeirão Preto Municipality, in the state of São Paulo (PESSOA et al., 1997a,1998, 1999, 2003ab; CERDEIRA et al., 2000, 2004, 2005abcde, 2007a). Pessoa et al (2003a) also presented an integration of CMLS94 with GIS, in order to identify areas with great potential to find pesticides applied in the recharge area of Guarani Aquifer, located in Ribeirao Preto/SP. Afterwards, Pessoa et al. (2005) used the same simulator to evaluate the dynamic of active principles applied in soy, corn and "out of season" corn (also called "safrinha" in Brazil) crops cultivated in Mineiros Municipality, Goiás State – Brazil, in the vicinity of "Chitolina" erosion gully (also called "voçoroca" in Brazil), which are present in another recharge area of Guarani Aquifer. These areas have been frequently exposed to pesticides, requiring a close follow up in order to access the risk of water contamination. CMLS94 simulator was also applied to evaluate the dynamic of pesticides applied on fruits like banana cropped at the semi-arid region (PESSOA et al.2003b), located in the states of Pernambuco and Bahia, and passion-flower cropped at the Igarapé-Açú Cumarú located in the eastern region of Amazon region, in the state of Pará (LIMA et al., 2007; FIGUEIREDO et al., 2007), where the Pós-Barreira Aquifer is present, considering their respectively pesticide uses and local environmental characteristics.

Paraíba & Spadotto (2002) developed the Temperature Leaching Potential Index model (TLPI) to evaluate the dynamic of pesticides considering the influence of the soil temperature on the dynamics of the products applied, which was mention as an important differential by the authors when evaluating tropical environments. Afterwards, the TLPI model was applied for cerrados region (PARAÍBA et al, 2003). Coutinho et al. (sd, accessed on 2009) recently presented a work that integrated the TLPI model with a GIS to evaluate the dynamic of pesticides on the recharge area of the Guarani Aquifer located in the state of Mato Grosso. DAPESTE model, developed by Paraíba & Pulino (2003), was also available to simulate pesticide leaching in soil taking into account the daily average soil temperature at different depths.

The SWAT model had some soil physic-hydro parameters calibrated by Machado et al, (2003) for a catchment located in São Paulo. In addition, the integration of SWAT with GIS for simulate streamflows within the watershed was pointed out.

Medeiros et al., (2001) studied the saturated hydraulic conductivity of some soil from Campinas municipality, São Paulo, under different levels of compactations in order to establish their relationship with the solute transport and the water balance. For this purpose, the EPIC model was used. Picini et al (2005) also used EPIC but in order to adjust its meteorological data simulators for different localities in the State of São Paulo.

Considering the applications to estimate erosion losses using USLE, integrated or not with other tools, we could mention Minoti (2006), Tomazoni & Guimarães (2005), Minoti & Crestana (2004), Fernandes et al. (2003), Fernandes (1997), Fernandes et al. (1995) and Resende & Almeida (1985).

Gonçalves (2008) presented the WEPP model validation in order to predict the water erosion, considering the edaphic climatic characteristics of Viçosa region, in Minas Gerais State.

Experiences on the use of fugacity model were conducted by Paraíba et al., (2007), Silva (2007) and Plese (2009). Paraíba et al (2007) applied the model to evaluate an insecticide used on rice crop in order to observe its distribution in the soil, water, plant and air compartments. Silva(2007) applied another fugacity model to evaluate the environmental fate of fungicides, herbicides and insecticides applied on rice crop and Plese et al (2009) also applied this type of model considering herbicides applied on cotton, coffee and citrus.

When focusing applications on Brazilian South region, AGNPS was used by Souto & Crestana (2000), who determined the potential sources of non-point pollution and evaluated the soil conservation practices in a catchment located in the state of Paraná. For Santa Catarina State, applications of AGNPS toward parameter calibrations and sensitivity analysis were done by Bacic et al. (2008). They also evaluated different management scenarios considering a catchment area of swine production that had been causing problems of manure accumulations and consequently water pollution. Dias (2003) applied a hydrologic model AGNPS2001 to evaluate a catchment located in the state of Rio Grande do Sul.

PRZM3 was applied for Santa Catarina State by Gebler (2000), who presented a case study of pesticide dynamics (atrazine, simazine and diazinon).

Siqueira (1991) made use of DSSAT for the evaluation of soil-plant-climate interactions while Asadi & Clemente (2003) compared differences between Ceres-maize and DSSAT for simulations of nitrate leaching, among others.

Siqueira & Berg (1991) presented strategies for the SOYGRO-Soy validation considering the Rio Grande do Sul State environment in order to provide this tool for the performance of the evaluation of culture conditions in the South of Brazil.

Silva & Bergamasco (2001) presented a survey of mathematical models for sugarcane crop. They mentioned models and simulators used to estimate crop growth and yield, as well as to promote the estimative of carbon balance of a plant community, in order to provide dry matter production of sugarcane, the transfer of heavy metals during the way (compost-soil-root-shoot) and the growth of leaf area index. Bergamasco et al. (2002), also considering this culture, pointed out the use of nitrogen balance model to predict the response of sugarcane to nitrogen fertilizer uses.

Nowadays it is very important to consider the bioconcentration factor potential of organic pollutants, which can be found in crops cultivated with sewage sludge. Thus, Paraíba et al (2006) presented an application of the Fruit Tree Model (FTM) for an hypothetical culture of orange orchard by using molar mass, vapour pressure, water solubility and the octanol-water partition coefficient (non ionic organic compounds) that can be found in sludge. Paraíba (2007) also reported the pesticide bioconcentration modelling for fruit trees.

When considered the screening mathematical-models, like Gus Index and Goss Method, the applications for water contamination potential were done for various Brazilian regions and crops. Ferracini et al (2001) pointed out the importance of both methods to evaluate the pesticides applied in mango and grape crops on sub-middle portion of San Francisco river basin. In order to facilitate the calculus for a great number of active principles, Pessoa et al.(2004b, 2007) proposed a computer program, which was called Agroscre, that promotes the automatic calculus required for the GUS index, GOSS methods and one of Cohen screening model used by EPA. Agroscre has been helped to spread the use of these models in Brazilian country (CRUZ, 2006). Among them we can mention Ferracini & Pessoa (2008), who studied the toxicology and environmental aspects of pesticides applied on melon crop, as well as Pessoa et al. (2007), who evaluated the water contamination potential for all the pesticides applied on Brazil. Ferracini et al. (2005) also evaluated the potential environmental contaminants to subside the production quality of citrus, in which also made use of the screening models evaluated by the Agroscre.

Another screening model called Leaching IndeX (LIX) was proposed by Spadotto (2002) in order to evaluate the leaching potential of pesticides. Spadotto (2002) also compared its uses with Gus Index and Goss Method. Afterwards, Lourencetti et al (2005) compared the use of EPA screening criteria, indices GUS, LIX, simple models RF, AF and TLPI considering literature data to evaluate the pesticide leaching potential and their potential for groundwater contaminations. Thus, it was considered the pesticides applied in Brazilian sugar cane crop.

In relation to applications of mathematical-modelling to population dynamic or predator-prey interactions for agricultural-pest control, it is necessary to mention the large use of the most traditional mathematical-models like those proposed by Malthus, Verhulst (also known as logistic), Gompertz, Lotka- Volterra, Nicholson-Bailey, Kolmogorov, among others (BASSANEZI & FERREIRA JUNIOR, 1988). These models have been applied for the Brazilian aspects of phytopatology as mentioned by Vilcarromero (1996), who cited applications for *Phytophthora parasita*, *Phytophthora infestans*, *Sclerotinia minor*, *Cochliobulus sativus*, *Erysiphe graminis*, among others.

Other applications for biological control were the mathematical-modelling simulation for the cotton-boll weevil (*Anthonomus grandis* Boheman) population dynamics in the context of Integrated Pest Management strategies developed by Pessoa (1994), which evaluated the influence of systemic pesticide and of a boll-weevil larval parasitoid on its control considering different scenarios. In the same year, Suji (1994) presented aspects of *Deois flavopictds* (Homoptera: Cercopidae) distribution patterns and ecophysiological responses, which are relevant for the design of a mathematical model of the population dynamics of this pasture spittlebug. The resulting simulation helped decision-making on control measures to be applied against this important pest of cultivated pastures in Central Brazil. In addition, the analysis of spatial distribution pattern was made using the Dispersion Index (Variation/Mean), the k value of the binomial distribution, the Morizita Index and the Morizita Standardized Coefficient. Ambrosano et al (1996) also presented a compartmental and deterministic model to evaluate the dynamic of sugarcane-borer *Diatraea saccharalis* considering the presence of its parasitoid *Trichogramma galloi*. Gazzoni et al. (1998) developed a mathematical-modelling simulator for soybeans caterpillar (*Anticarsia gematalis*). Suji et al. (2002) also developed a system to integrate a mathematical model, which was created to simulate occurrence of fungus *Nomuraea rileyi* on populations of velvetbean Caterpillar *Anticarsia gemmatalis* in soybean crops of Florida, USA model, with several other models (phenology of pest and crop, insect plant interactions, and population dynamics of the pest), in order to generate more accurate simulations on the pest management. The systems took into account the equations of differences that are processed as differential equations through the software STELLA version 5.0. Ternes et al (2005) developed the SIMCOBIM simulator, which made use of MATLAB language, to evaluate the biological control of citrus leafminer, *Phyllocnistis citrella,* by *Galeopsomyia fausta* and *Ageniaspis citricola*.

Recently, Pessoa et al (2008) also developed a mathematical-modelling simulator to follow the red-gum lerp psyllid populational dynamic, in order to identify its nymph-stages which are preferable for *Psyllaephagus bliteus* parasitoid attacks, taking into different levels of initial psyllid and parasitoid populations and dates of parasitoid releases. The red-gum lerp psyllid is a very serious plague of *Eucaliptus sp.* Many other applications can be found in literature.

The random generations of climatic factors, like those promoted by the use of Weather Generators (WGEN) (RICHARDSON & WRIGHT, 1984) and WMAKER (KELLER, 1982), are also important for mathematical modelling simulation. Silva (2004) presented the main agrimetheorological mathematical-modelling simulators which turn available to simulate the growth and development of cultures like potato, corn, rice, soya, sorgo, sugarcane, wheat and sunflower. There were also recordings of the use of mathematical models and simulators for estimating crops adapted to Brazilian conditions as those proposed by Silva (2004), which examined the effects of hydro-climatic factors in the variability of corn crop yield in non-tillage in the Brazilian savannah (known also as "cerrado"). Thus, it was made possible to evaluate the effect of climatic changes into the different stages of culture development, as well as to generate information for decision support on improving of climatic risk zoning for different crops. This work made use of the **S**imulateur mul**T**Idisciplinaire pour lê **C**ultures **S**tandard (STICS). Siqueira et al (1994) simulated the production of wheat, maize and soybean with CERES and SOYGRO for different sites in Brazil, under climate change scenarios generated by international known climatic models (GISS, GFDL and UKMO GCM). For the assessment of crop production in the occurrence of dry spells, Sousa (1999) provided the program VERANICO 1.0, which was used by Hernandez et al. (2000) to assess the climate effect on cultures like cotton, bean, banana, grape, citrus and corn in the region of Palmeira D'Oeste, São Paulo State. Acuña et al. (nd), mentioned by the Federal University of Viçosa (2005), also made available PROGRESSA, which allows us to analyze the relationship between climatic variables and the development of plant diseases, using data of severity, among others.

It is important to notice that some aspects of the agricultural production systems also influences the greenhouse gas emissions (GEE). Therefore it is important to know them and to evaluate the interactions between the management practices and the emissions. Some applications of mathematical-modelling simulation can be useful to help the understanding of the process involved with, as well as to create different options of scenarios with potential to minimize the GEE emissions. Despite the fact that there are simulators available for the assessment of the dynamic of Carbon and Nitrogen that permits to assess the influence of agriculture management on the gas emissions, the difficult activity to quantify the emitted gases on field has limited the validation process and the spread of their uses. Nevertheless, some Brazilian researchers still insist on that, because the data is necessary for the validation process of these tools, which is fundamental to precisely evaluate of the management influence on the emissions. On some Brazilian applications already available, the DeNitrification-DeComposition (DNDC) model has been used to estimate the methane emissions from paddy-rice in the state of São Paulo (PLEC et al., 2007). DNDC is a process-based model that simulates carbon and nitrogen biogeochemical cycles occurring in agricultural systems. CQESTR has also been applied to evaluate the dynamics of the soil organic matter on crop-livestock system (LEITE et al., 2007). CQESTR is a process-based model (USDA/ARS) that relates additions of crop residues and the management (crop and soil) to soil organic-matter. Streck & Alberto (2006) studied the climate change impacts on the water soil availability by the simulation of the fraction of transpiring soil water (FTSW) considering wheat, soy and corn agroecosystems in Santa Maria municipality, Rio Grande do Sul state. Alves & Pesquero (2009) presented simulations to predict the future climate change for South America using the "Earth Simulator", which is a supercomputer developed in a NEC Cooperation

Project, the AGCM model from MRI-JMA, and the emission scenarios proposed by IPCC (2001).

Mathematical-modelling can also be applied on genetic resources. Multivariate Analysis techniques have been applied to plant breeding making use of multivariate techniques (CRUZ, 1990) and biometric models (CRUZ & REGAZZI, 1994). Some examples can also be found for *E. camaldulensis*, where it was evaluated by predicting the genetic gains of the forest breeding making use of phenotypic and genotypic correlation models. De Paula et al (2005) also presented another interesting point of model applications in order to verify the importance of genetic and environmental factors that can affect the reproductive biology of *Acacia mearnisii.*

Programs and mathematical-modelling techniques applied to planning activities for cattle, pigs and poultry are also presented in the literature. Guahyaba (2004) pointed out the use of mathematical techniques as the main responsible for the development of poultry farming. This author strengthens the need for greater use of national human resources and improvements in the transfer of technology already available in our country. Among them, the author mentioned:

- the use of mathematical models to explain the production of antibodies (post vaccination), for different ages of birds, Gumboro and Newcastle diseases and infectious bronchitis (SALLE et al., 1998b; SOARES, 1995);
- use of models to correlate levels of aflatoxin and ochratoxin in food and viscera of broiler with production parameters of the lot (SALLE et al., 1998a);
- use of mathematical-modelling to understand parts of the production process, such as to estimate the production of eggs (FIALHO & LEDUR, 2000a; 2000b), on the assessment of development and growth of broilers(RONDON et al., 2002; Ivey, 1999), among others (SALLE et al., 2001).

To assist the decision making on production systems of ruminants at grazing, Medeiros (2003) evaluated several models, including the model STOCKPOL on Brazilian environmental conditions. This is a model of forage accumulation as a function of climatic variables and models to estimate consumption of forage in the production system of ruminants on grazing.

Veloso et al (2007) also pointed out the use of mathematic-programming toward the evaluation of the economic viability on bovine milk production considering a small scale in the South of the state of Goiás. Barioni et al (2007) presented application of deterministic and stochastic models in programs of embryo transfer in cattle, while Tedeschi et al (2005) presented results using mathematical models applied to ruminant nutrition. Recently, Verneque (2009) also presented the genetic and environmental aspects of the lactation curve of cows Guzera making use of a new version for Gama model (PAPAJCSIK & BODERO,1988) and a model proposed by Cobuci et al. (2000). According to the author, a non linear regression of milk production was used to the estimation of parameters of each model and the Restrict Maximum Likelihood (REML) for estimation of genetic parameters. A more detailed review evaluating the use of simulation models for cattle was presented by Fontoura Junior et al (2007; 2009).

Trindade Neto (2009) presented an evaluation of the biological flow of P in pigs fed diets based on corn, soybean meal, defatted rice bran and soybean oil using deterministic and compartmental mathematical-models.

For aquiculture, the work of Dias (2009) presented a critical appraisal on the use of the 'broken-line' model in the analysis of growth data derived from a dietary protein requirement assay with "tambaqui" (*Colossoma macropomun*). Results show that the 'broken-line model analysis', as performed by a linear regression procedure, is simple, fast and reliable, suitable for the analysis of fish growth data.

As noticed, the examples listed above are just a set of representative applications of mathematical-modelling in decision-making practices in the pursuit of less impacting to agroecosystems are enormous. An exhaustive analysis is not the main objective of this work

Following, a case study will be presented to show the influence of average data, which are available as default data, and experimental ones, which were calculated in laboratory from specific local sample, on CMLS94 evaluations.

## CASE STUDY: VERTICAL MOVEMENT OF HEXAZINONE HERBICIDE INTO TYPIC EUTHORTHOX SOIL OF ESPRAIADO CATCHMENT, RIBEIRÃO PRETO/ SP: BRAZIL, CULTIVATED WITH SUGARCANE CROP:

### Local and *Default* Data Scenarios Using Mathematical-Modelling Simulation

Studied area: The Espraiado catchment has 4.141 hectares and it is located at the North region of the São Paulo State, landmark of Ribeirão Preto, Cravinhos and Serrana municipalities. According to Koppen classification, the climate of the region is winter tropical of dry winter of savannah (AW). The average annual temperature varies from 21° to 22°C with pluviosity that varies from 1300 to 1500 mm.year$^{-1}$. The potential evapotranspiration reach 1000 mm.year$^{-1}$, according to Thorntwaite method. In relation to the hydrology the presence of Guarani Aquifer (also called Botucatu Aquifer due to the Botucatu and Pirambóia Formation) must be highlighted, with one of recharge areas occuring at Espraiado catchment (GOMES, 1995). In the area Oxisoil, Alfisol and a sandy, amongst other transition soil types are predominant (MIKLOS & GOMES, 1996). The sugarcane crop is prevalent in the studied area, although a little quantity of corn, bean, peanuts, pasture, eucalyptus and pines are also found. Due to the faster initial development of the type of sugarcane cultivated in the area, it is more susceptible for the occurrence of weeds, as sugarcane type plant (PARANHOS, 1987). Because of that, it is necessary to apply pre-emergent herbicides to control. The sugarcane intensive monoculture has demanded the constant use of pesticides, including pre-emergence herbicides. The high soil permeability found in this region associated with the application of pesticides with high mobility have been the main factors that raise risk potential to undesirable contaminations of the aquifer. This negative environmental impact risk increases when considered that the aquifer provides drinking water to the cities and rural properties located in the region (GOMES et al, 2006b; GOMES et al., 2002; GOMES et al., 2006a; MIKLÓS & GOMES, 1996; GOMES, 1995; IPT, 1994). Thus, it is necessary to monitor the depths and pesticide concentrations in the water table and groundwater in order to ensure the quality of the hydric resource (FERRACINI et al., 2007; CERDEIRA et al., 2007b; FERRACINI et al., 2006; GOMES et al., 2006a; GOMES et al., 2004, CERDEIRA et al., 2004; CERDEIRA et al., 2000). The behaviour of the active principles of the pesticides applied in the soil have also been studied (CERDEIRA et al. 2007a,c; CERDEIRA et al, 2006; CERDEIRA et al., 2005a,b,c,d,e; GOMES et al., 2002). Despite of, it is difficult to identify the factors that effectively contribute to the negative environmental quality risks because the interaction between them cannot be followed. Among the active ingredients currently used in the area of Espraiado´s catchment, the hexazinone herbicide is highlighted, dose of 0.40 kg.ha$^{-1}$ is often present in the formulation of the commercial product used in the area.

Hexazinone and its laboratorial considerations: Hexazinone (CAS number: 51235-04-2) is a systemic herbicide of the triazine group, toxicological class I in accordance with the *Environmental Protection Agency*-EPA. It presents, on average, from high to moderate persistence in soil with its half-lives ranging from 30 to 180 days (being the most representative value of the 90 days). It is also very soluble in water. For this reason it is probably moving in many soil types and has potential for contamination of groundwater (EXTOXNET/PIP, consulted on may/2008). Laboratorial studies to estimate Koc and soil half-life of hexazinone in typic euthorthox soil type of Espraiado catchment allowed to validate an hexazinone method of residue analysis (QUEIROZ et al., 2007, 2006), which was able to detect and analyze the presence of hexazinone residues. QUEIROZ et al (2007) have also provided information about the detection limit of the method, without which it is not possible to estimate the dose to which the device is able to detect the active ingredient, depending on the dose applied in the field. This information is essential to progress in knowledge about the influence of parameters Koc and soil half life (t½) in obtaining the results of trends in pesticide leaching by simulation for the herbicide hexazinone.

CMLS-94: The mathematical-modelling simulator *Chemical Movement in Layered Soils* (CMLS-94) (NOFZIGER & HORNSBY, 1994) was used in order to estimate the relative quantity of the hexazinone present in the soil at any instant of time within the period stipulated for the simulation. The simulator inputs were: a) daily maximum and minimum temperatures at Ribeirão Preto, São Paulo-Brazil, b) daily pluviosity at Ribeirão Preto, São Paulo-Brazil; c) organic-carbon coefficient (KOC) and soil half-life (t½) of hexazinone in the typic euthorthox soil type of Espraiado catchment; d) percentage of organic carbon, bulk density (mg.m $^{-3}$), field capacity, wilting point and saturation for typic euthorthox soil type of Espraiado catchment e) sugarcane crop cultural coefficients (Kc); f) hexazinone herbicide initial dose (kg.ha$^{-1}$), and time (months) and depth (m) of the application. Climatic data relating to a period of one year and four months were reorganized into a file type *.w, containing for each Julian day: maximum and minimum temperatures (°C) and rainfall (mm) provided to the Ribeirão Preto city by the Climatology Section of the Agronomy Institute of Campinas (APTA/IAC-SP). Data of organic carbon percentage, bulk density (mg.m$^{-3}$), volumetric water content (%), field capacity, wilting point and saturation, raised in the study area at the depths of 0-20 cm, 20-40 cm, 40-60 cm, 60-80 cm, 80-100 cm and 100-120cm to typic euthorthox soil type (LVd) were inserted into the CMLS-94 soil database, presented in Table 1 in units of the international system. The cultural crop coefficients (Kc) for different seven ages of culture (days) presented by Paranhos (1987) (option PLANALSUCAR) (Table 2), were inserted into the CMLS94.

*Table 1. Values of organic carbon (OC) (g.kg$^{-1}$), density (ds) (mg.m$^{-3}$) and the water volumetric content for field capacity (dm$^3$ dm$^{-3}$) (FC) and saturation (dm$^3$ dm$^{-3}$) (S), as well as, the gravimetric content for the permanent wilting point (dm$^3$ dm$^{-3}$) (PWP), at different depths of typic euthorthox soil type (LVd)*

| Dephts (cm) | OC (g kg$^{-1}$) | ds (mg m$^{-3}$) | FC (dm$^3$dm$^{-3}$) | PWP (dm$^3$dm$^{-3}$) | S (dm$^3$dm$^{-3}$) |
|---|---|---|---|---|---|
| 0-20 | 9.4 | 1.28 | 0.268 | 0.155 | 0.52 |
| 20-40 | 8.5 | 1.44 | 0.204 | 0.167 | 0.46 |
| 40-60 | 6.5 | 1.25 | 0.198 | 0.173 | 0.53 |
| 60-80 | 6.0 | 1.13 | 0.197 | 0.175 | 0.57 |
| 80-100 | 5.0 | 1.12 | 0.210 | 0.169 | 0.58 |
| 100-120 | 4.5 | 1.04 | 0.187 | 0.172 | 0.61 |

*Table 2. Cultural crop coefficients (Kc) for the sugarcane (PARANHOS, 1987)*

| Age of culture (days) | Growth stage | Values of (Kc) Eta/ Etb |
|---|---|---|
| 1 | Planting to 0.25 of closing | 0.50 |
| 60 | 0.25 to 0.5 of closing | 0.65 |
| 90 | 0.5 to 0.75 of closing | 0.75 |
| 120 | 0.75 until closing | 0.90 |
| 270 | Maximum development | 1.10 |
| 300 | Beginning of maturation | 0.70 |
| 360 | maturation | 0.60 |

Baseline scenario: The baseline simulation scenario showed the cut-off date of sugarcane on 20 August and the simulation period of one year and four months. The product was launched on 20 September on the first year. The herbicide application was considered as done on the soil surface in September, taken into account 0.40 $kg.ha^{-1}$ as initial dose. Successively reapplication of the product was not considered, since the simulator is not capable to evaluate its influence on trend analysis.

Scenarios evaluated: Two different scenarios were simulated. The first one, which is henceforth called **Scenario 1**, considered Koc and t½ values from performance determined at the Pesticide Residue Laboratory of Embrapa Environment, Jaguariúna – São Paulo, which were presented by Queiroz et al. (2007, 2006). The average values found at depth of 0-20cm was considered. These data are presented in Table 3. They were also inserted into the CMLS94 pesticide database. The second one, or **Scenario 2**, considered t½ and Koc default values for the herbicide hexazinone, which were already available in the CMLS94 database. In the end of the simulated period, the depth (m) reached in each scenario was compared both with the water table, which varies from 0 to 20m in the area, and with the saturated zone level (40m). The remaining quantities of hexazinone ($kg.ha^{-1}$) found in each scenario was also registered.

Hexazinone limit dose of detection by the residue analysis method: In order to present useful values resulted from the simulation analysis, the values capable to be recovered by the available method of residue analysis were considered. Thus, a limit dose of detection by the equipment considering the field dose used was considered. It was calculated by the method presented by Queiroz et al. (2007) which considered the density of the soil ($g\ cm^{-3}$), the initial dose of hexazinone applied ($kg\ ha^{-1}$) and the detection limit of the laboratorial method of residue analysis (QUEIROZ et al., 2007). Using the information of the detection limit of the hexazinone residue analyze method, which was determined by Queiroz et al (2007) as 0.016 mg $kg^{-}$, the detection limited dose was subsequently estimated ($kg\ ha^{-1}$), which considered the field conditions for the studied soil. This value was used as reference in the simulation analysis because the laboratory method of hexazinone analysis of residue does not testify values below it, despite the fact that the simulator can present them.

Results: The results obtained in the scenario simulations done are briefly shown in Table IV. The comparison between the values recorded in the simulations of hexazinone considering both the use of default and calculated values for t ½ and Koc throughout the simulated period (1 year and 4 months) are presented in Figures 1 and 2. The

*Table 3. Organic adsorption coefficient ($K_{OC}$) and soil half-life (t½) of hercide hexazinone*

| HEXAZINONE | Koc ($mL.g^{-1}$) | t½ (days) |
|---|---|---|
| Scenario 1 - Laboratory | 26.3[2] | 134.5[3] |
| Scenario 2 – Default | 54[1] | 90[1] |

where: [1]default values available at CMLS-94 database, which are default values of literature);[2]Laboratory Koc average value reached at 0- 20cm of depth; [3]Laboratory t½ soil average value reached at 0- 20cm of depth;

detection limit of the hexazinone residue detection method was calculated and determined as 0.06 kg $ha^{-1}$ of hexazinone. In general, hexazinone showed greater mobility in the vertical profile of typic euthorthox soil when considered t½ and Koc estimated at laboratory from the local soil. Significant variations in the curves of depth (Figure 1) were observed near the 63th day of application. It had become more disparate from the 95th day after application, when **Scenario 2** registered depth of 0.82 m while in the **Scenario 1** 1.27 m one. Since then, further significant changes were recorded at 100th, 135th and then to 150th and 162th days of application, where the greatest variations between the **Scenario 2** and **Scenario 1** depths were recorded. The most leaching of hexaninone presented at the **Scenario 1** (laboratory) can be explained by the lower Koc (Table 3) of typic euthorthox soil, when compared with the default, which gives to the product a major mobility. The depths recorded at the end of the period simulated in both scenarios indicated that hexazinone has no potential to reach the groundwater in the saturated zone (40 m), since in both scenarios the potential to reach at the most 3 m of depth was indicated. In accordance with IPT(1994), although variations are found at the water table levels presented at the studied area, which varies from 0 to up 20m, the fact that the typic euthorthox soil cultivated with sugarcane appears only at the top of the catchment was verified, comprising about 10% of its total area, where the depths of water range from 10-30 m. Thus, hexazinone has no potential for direct contamination of deepest groundwater in studied area. The remaining quantities of hexazinone reached by simulations done indicated that hexazinone does not compromise the water quality in the confined aquifer (located at 40 m) for the initial dose of 0.40 kg $ha^{-1}$ applied. Variations in the quantities of hexazinone observed at **Scenario 1** and **Scenario 2** remained in the same order of magnitude, although major changes were observed from the 75th day of the application and also after the 160th day of the application. Considering the hexazinone potential to reach surface waters, it was observed that the highest amounts of hexazinone were observed in the topsoil in both scenarios, which registered values very close during the first 2 months. It is probably justified by the larger organic matter content in soil layers of 0-80 cm which favoured a greater adsorption of the product, as pointed out by Levy et al (1989) as likely to occur for the chemical, despite the lower Koc values recorded for the **Scenario 1** (Table III). In addition, this scenario indicates that after the 46th day of application the product would present the potential to reach about 13 cm of deep retaining 82.5% of the original amount applied, while in **Scenario 2** would be the same depth reached two days after that observed at **Scenario 1** retaining 70% of the amount applied.

These trends corroborate with those presented by Roy et al (1989), who studied the behaviour of hexazinone in clayed soils and reported high concentrations of hexazinone in the first 15 cm of soil (about 98%). In this sense, areas susceptible to the run-off of soil surface, considering the weather conditions of the particular period studied, tend to contaminate the water surface, which should be further investigated by local monitoring followed by specifics laboratory residue analysis. The results also indicate that the future local monitoring for groundwater residue analyses for the studied area must be driven to different depths, depending on the date of the hexazinone application. Until the first 30 days after product application it was verified that hexazinone can more likely be found very near the soil surface layer, while after the 60th days until 30 cm and $90^{th}$ days around 1 meter deep. After this period, the product would be found in deeper layers of soil (see Table IV), but never at depths greater than 3 meters when considered the studied soil and the hexazinone quantities that can be detected by the residue method used in the laboratory. It must be remembered that CMLS94 does not evaluate the residual effect of successive reapplication of the product, which were not considered in this simu-

*Figure 1. Comparison between the depths reached by hexazinone in typic euthorthox soil during the simulated period considering both scenarios simulated*

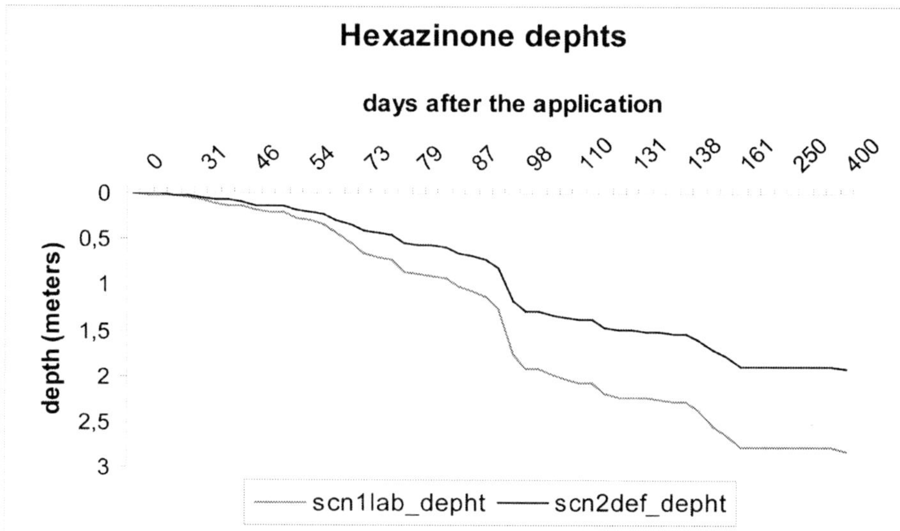


*Figure 2. Comparison between the remaining quantities reached by hexazinone in typic euthorthox soil during the simulated period considering both scenarios simulated*

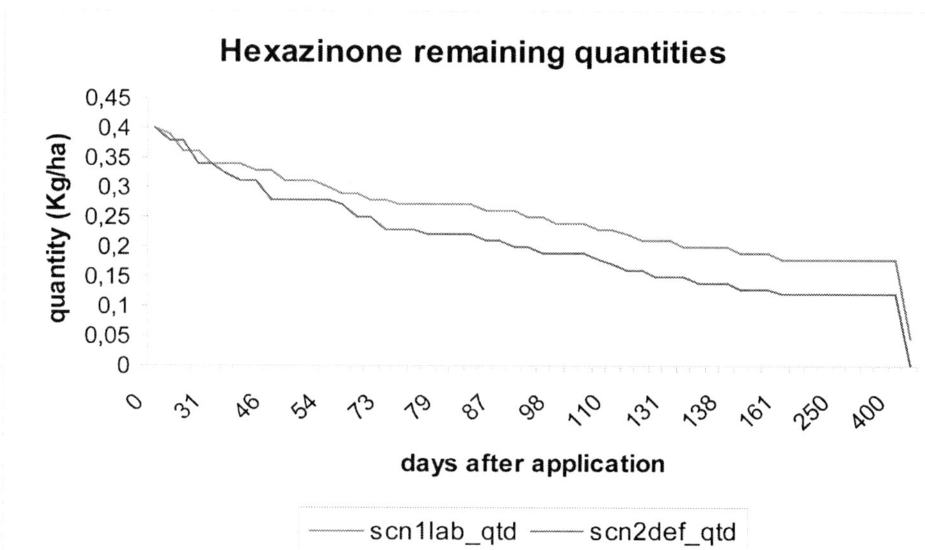


lation analysis. Regarding the differences found when evaluated both scenarios, it was verified that both simulated scenarios recorded depths and final amounts of hexazinone quantities in the same order of magnitude, showing no significant differences that would invalidate the simulation using default data.

*Table 4. Depth values (m) and hexazinone remaining quantities reached at the first six months after application of hexazinone on sugarcane crop planted at typic euthorthox (LVd) considering both scenarios simulated*

| **Hexazinone** | **- Scenario 1 - Koc and t ½ Laboratory** | | **- Scenario 2 - Koc and t ½ Default** | |
|---|---|---|---|---|
| **Days after application** | **DPT (m)** | **QT (Kg ha$^{-1}$)** | **DPT (m)** | **QT (Kg ha$^{-1}$)** |
| **0** | 0 | 0.40 | 0 | 0.40 |
| **30** | 0.031 | 0.36 | 0.019 | 0.34 |
| **60** | 0.279 | 0.30 | 0.185 | 0.28 |
| **90** | 1.121 | 0.25 | 0.722 | 0.20 |
| **120** | 2.021 | 0.23 | 1.354 | 0.17 |
| **150** | 2.544 | 0.19 | 1.721 | 0.13 |
| **180** | 2.782 | 0.18 | 1.889 | 0.12 |

where: DPT= depth (m); QT= product quantity (kg/ha); LVd= **typic euthorthox** soil type of Espraiado catchment.

## Solutions and Recommendations

The final depths found for hexazinone, using Koc and soil half-life default values and calculated ones were 1.89 m to 2.78 m, respectively, at the 422$^{th}$ day after the product application. These depths do not indicate potential to compromise the aquifer water in its saturated zone (40 m).

A tendency that the product reaches levels of depths ranging from 0 to 3 m in the studied soil was verified but these watertable depths are not found in the area where the studied soil is found at the catchment.

The hexaninone final concentrations found at simulations were 0.18 kg ha$^{-1}$ for the **Scenario 1** and 0.12 kg ha$^{-1}$ for **Scenario 2**. These were recorded at 422$^{th}$ day after initial application of the product. In both scenarios, trends to the presence of significant amounts of hexazinone, 70 and 82.5% of the initially applied, were still present after the 46$^{th}$ day of the initial application at 13 cm of depth, indicating a potential risk of contamination for surface water by runoff.

The values obtained by using t½ and Koc calculated at laboratory for the soil at the studied area showed quite differentiated when compared with those resulting from the default data available in CMLS-94 database (average from literature). Despite of that, both results remained within the same order of magnitude.

According to the results observed at the case studied, it is important to point out that not only the availability of databases is necessary for more precise mathematical-modelling simulation results or to spread applications of this research method. It is also important to have specific data collected at the studied area and some additional laboratory information.

Another important aspect presented was the necessity to define a limit dose of detection by the laboratory equipment considering the field dose used, which will enable the selection of the simulation results capable of being recovered by the available method of residue analysis. Due to that, this limit dose of detection must be used as reference in the simulation analysis because the laboratory method of pesticide residue analysis will not testify values below it, despite the fact that the simulator can present them.

As an ending remark on the case study presented, it is necessary to mention that the previous result also shows the importance to get a multidisciplinary and interdisciplinary team involved on the mathematical modelling simulation research from the concept of project proposal until the result analysis.

## FUTURE RESEARCH DIRECTIONS

The importance of computational methods applied to agricultural research can even be more prominent, as observed by Capra (2003; pg 22), who cited that "*The system's theory teaches that all living systems share the properties and principles of common organization. This means that the systemic thinking can be used for integrate disciplines and to discover similarities between different phenomena inside a wide range of living systems*".

On this context, mathematical-modelling and simulation systems play an important role as science. Considering that each person has his own impressions about the world, according to his own senses, it is possible to conclude that nothing is totally "real", even though most people's opinions on the same topic lead to a "common sense" about it. Due to that, science needs to be supported on concepts already accepted and based on data that have sense on the available theory, and not only on observations. Therefore, it is also important to keep in mind what was observed by Quirino (1990, pg.T05), who pointed out that *the scientific methodology is the knowledge that gives procedures and ways to be followed on scientific investigation. It permits to discover the ways of a existent phenomena, as well as, their connections in order to generalize and deepen the knowledge, test them rationally, to testify them on experimentation and on observation and through their technical applications.*

When taking into account the level of knowledge accuracy, it is possible to find: description (that presents the event), explanation (shows the reasons why the event happens), prediction (forecast what may happen) and control (change reality to avoid something wrong happens). Therefore, the importance of mathematical-modelling and simulation systems on the different level of knowledge's accuracy to the decision making on agricultural research is clear.

A great variety of methods, ranging from simple regression methods to complex modelling studies are available. Despite of that, the correct model formulating must take into account information required and the appropriate methods for the modelling objective, which also required the well understanding of the available knowledge level about the variables involved, gaps in information, degree of details on the expected results and level of accuracy. Thus, it is also important to spread the effectual computational methods applied to agricultural research to the scientific community, presenting their advances and applications in order to show different points of view on the same issue involved with the modelling, as well as their benefits for the decision-making process on environmental impact assessment of Brazilian agriculture.

Cheng (1990) presented some demands related to modelling research applied to pesticide behaviour and fate, which remain current today probably due to the necessity of background information related to the identification of critical processes, which remain unavailable on the adequated scale for modelling.

When considering the Brazilian environment, Trigueiro (2003) pointed out some challenges for the 21$^{st}$ century. Some of them, mentioned by Almeida (2003), was also related to decision making for sustainable agricultural practices like:

a. the assessment of the technology impacts (technological risks);
b. the increase in the eco-efficiency practices (renewable energy, recycling, reducing gas emissions, and others);
c. the introduction of the sustainable development concept on the public polices (best practices of management, hydric resources

sustainability, certified products, and others); and

d. the establishment practices of environmental valuation; and others.

Moreover, Almeida (2003; pg. 136) also pointed out that "*an important contribution of the sciences that study the natural systems to business management is the notion of resilience: the limits and capabilities of a system to resist to impacts*." Because of that, it will be necessary to well known the natural resources, the abiotic and biotic factors involved in the agricultural practices, the management, as well as their relationships. Therefore, mathematical modelling and simulation systems used as science could be explored on different ways, as pointed out by the review of applications presented on the present chapter. Thus, the presented references could be useful.

## CONCLUSION

The understanding of the cause-effect relationship involved in the environmental impact of agricultural systems is complex and made up of many multidisciplinary and interdisciplinary components and information.

Due to that, considerable attention has focused on the use of mathematical-modelling and simulation systems applied on environmental impact analysis in order to identify alternative strategies for the correct identification of indirect impacts and control of pollution and degradation sources. They have provided decision makers with the most precise descriptions of negative impacts and their respective analysis of alternatives which promotes the consequent choice justification (limitations, benefits and expected effects).

The present chapter, as a while, showed that there are many types of mathematical modelling simulation applications, as well as mathematical methods to deal with decision making in environmental impact assessment of agricultural systems. It was an important overview for researchers involved with some specific issue of the mathematical modelling in order to promote the correct understanding of this research topic at all. The main models and simulators mentioned on the literature, as well as their uses, were outlined on the present chapter focusing on the support of decision making in environmental impact assessment of agriculture and hence for agricultural research. Considerable practical applications for Brazilian environment have been carried out, and some of them were quoted on the present chapter showing the progress in the use of these techniques and mathematical methods in the Brazilian country.

A case study presented trends on a vertical movement of the hexazinone herbicide into the soil cropped with sugarcane at Espraiado Catchment - Ribeirão Preto /SP Brazil – using CMLS94 simulation. This study emphasized the importance of mathematical-modelling simulation process to assess the risk of surface and groundwater contamination located in vulnerable areas, as well as the response of the simulator according to the use of data calculated from located samples and literature default data inputs. It also pointed out that when mathematical model simulation is used for scientific research, not only the availability of databases for more precise mathematical-modelling simulation results are necessary, but also having information at the studied area and laboratory, as well as a multidisciplinary and interdisciplinary research team involved.

The chapter also presented emerging trends for mathematical-modelling simulation implementations and use on agricultural research.

## REFERENCES


Acuña, R. S., Cruz, C. D., Do Vale, F. X. R., & Zambolim, L. (2005). PROGRESSA. In *Informática na agricultura, agroindústria e meio ambiente*. Universidade Federal De Viçosa (sd. UFV). Consultado em janeiro/2005 em http://www.ufv.br/ctq/home.htm#progressa

Almeida, F. (2003). o mundo dos negócios e o meio ambiente no século 21. In Trigueiro, A. (Ed.), *meio ambiente no século 21: 21 especialistas falam da questão ambiental nas suas áreas de conhecimento*, (pp. 123-141).

Alves, L. M., & Pesquero, J. F. (2009). Previsões de mudanças climáticas para a América do Sul: resultados da aplicação do "Earth Simulator". Boletim do Grupo de Pesquisa em Mudanças Climáticas–GPMC, São José dos Campos: CCST/INPe, n.7, abril de 2009, semestral, pg.2-4.

Ambrosano, G. M. B., Stimac, J. L., Silveira-Neto, S., Igue, T., & Nagai, V. (1996). Modelo matemático para simulação do controle biológico da broca-da-cana com o parasitóide *Trichogramma galloi. I. modelos conceituais. Bragantia, Campinas*, *55*(2), 371–382.

Anunciação, Y. M. T da, Liu, W. T. (1991) Estimativa da produtividade do trigo em campo experimental, utilizando o modelo fisiológico CERES trigo em condições de campo experimental. In *Congresso Brasileiro de Agrometeorologia, 7, 1991,* Viçosa, MG. Resumos..., (pp.22-25). Viçosa, Brazil: UFV.

Asadi, M. E., & Clemente, R. S. (2003). Evaluation of CERES-MAIZE of DSSAT model to simulate nitrate leaching, yield and soil moisture content under tropical conditions. *Food. Agriculture & Environment*, *1*(3&4), 270–276.

Azevedo Filho, A. J. B. V., & Neves, E. M. (1982). O uso de programação matemática na análise de investimento na pecuária de corte: técnicas intensivas X mercado de capitais. In *Anais do I Congresso Latino-Americano de Pesquisa Operacional e Engenharia de Sistemas,* (pp. 41-55). Rio de Janeiro, Brazil: Sobrapo.

Bacic, I. L. Z., Rossiter, D. G., & Mannaerts, C. M. (2008). Applicability of a distributed watershed pollution model in a data-poor environment in Santa Catarina State, Brazil. *Rev. Bras. Ciênc. Solo. Viçosa*, *32*(4), 1699–1712.

Bassanezi, R. C., & Ferreira Junior, W. C. (1988). *Equações diferenciais com aplicações*. São Paulo, Brazil: Harbra Ltda. barioni, L. G., Beltrame, R. T., Quirino, C. R., Fernandes, D. R. (2007). Modelos determinista e estocástico em programas de transferência de embriões em bovinos. *Archivos Latinoamericanos de Producción Animal*, *15*, 107–113.

Bergamasco, A. F., Silva, F. C. D. A., & Rodrigues, L. H. A. Trivelin, P.C.O. (2002). Aplicabilidade do modelo de balanço de Nitrogênio para previsão de resposta da cana-de-açúcar ao fertilizante nitrogenado. *Campinas: Embrapa Informação Agropecuária, dezembro, 9p. (Comunicado Técnico, 28*).

Borges Junior, J. C. F., Ferreira, P. A., Andrade, C. L. T., & Hedden-Dunkhorst, B. (2008). Computational modeling for irrigated agriculture planning. Part I: general description and linear programming. *Eng. Agríc.*, *28*(3), 471–482.

Botelho, R. G. M. (1999). Planejamento ambiental em microbacias hidrográficas. In *Guerra, A.J.T., Silva, A.S., Botelho, R.G.M. (orgs.), Erosão e conservação do solo: conceitos, temas e aplicações* (pp. 269–300). Rio de Janeiro, Brazil: Bertrand Brasil.

Brooke, A., Kendrick, D., & Meeraus, A. (1992). *GAMS: a user's guide, release 2.25*. San Francisco: The Scientific Press.

Capra, F. (2003). Alfabetização ecológica: o desafio para a educação do século 21. In *Trigueiro, A. (coord) (2003). Meio ambiente no século 21: 21 especialistas falam da questão ambiental nas suas áreas de conhecimento* (pp. 19–33). Rio de Janeiro: Sextante.

Carsel, R. F., Mulkey, L. A., Lorber, M. N., & Baskin, L. B. (1985). The pesticide root zone model (PRZM), a procedure for evaluating pesticide leaching threats to ground water. *Ecological modelling. Amsterdan, 30*, 49–69.

Cerdeira, A. L., Gomes, M. A. F., Pessoa, M. C. P. Y., Bonato, P. S., & Lanchote, V. L. (Maggio. (2000). Tebuthiuron In Soil And Groudwater In Sugarcane Area In Brazil. *Bollettino Dei Chimici Igiesnisti, 51*(4), 51–57.

Cerdeira, A. L., Pessoa, M. C. P. Y., Bolonhezi, D., Gomes, M. A. F., Spadotto, C. A., Souza, M. D., et al. (2007b) Effects of agriculture on ground water in a recharge area of Guarany Aquifer in Brazil. In *37th Annual Mississipi Water Resources Conference,* April 24-25, Jackson, MS, (pp.120-125).

Cerdeira, A. L., Pessoa, M. C. P. Y., Bolonhezi, D., Gomes, M. A. F., Spadotto, C. A., Souza, M. D., et al. (2007c). Agriculture and ground water quality in a sugarcane area in São Paulo State, Brazil. In *Annual Mississipi Water Resources Conference, 37*, Jackson, MS, (p. 39).

Cerdeira, A.L., Pessoa, M. C. P. Y., Gomes, M. A. F., Bolonhezi, D., Souza, M. D., Farjani Neto, C. (2007d). Proposta de Boas Práticas Agrícolas para as Áreas de Afloramento do Aqüífero Guarani em Ribeirão Preto, SP. *Jaguariúna/SP: Embrapa Meio Ambiente, agosto, 65,* 86.

Cerdeira, A. L., Pessoa, M. C. P. Y., Santos, N. A. G., & Lanchote, V. L. (2005d). Lixiviação De Atrazina Em Solo Em Área De Recarga Do Aqüífero Guarani. *Revista Brasileira De Herbicidas, 2*, 92–101.

Cerdeira, A. L., Santos, N. A. G., Pessoa, M. C. P. Y., Gomes, M. A. F., & Lanchote, V. L. (2005b). Herbicide Leaching On A Recharge Area Of The Guarany Aquifer In Brazil. *Journal of Environmental Science and Health. Part. B, Pesticides, Food Contaminants, and Agricultural Wastes, 40*(1), 159–165. doi:10.1081/PFC-200034294

Cerdeira, A. L., Santos, N. A. G., Ueta, J., Shuhama, I. K., Pessoa, M. C. P. Y., Smith, S. Jr, & Lanchote, V. L. (2004, July). Atrazine In Water And Biodegradation In A Recharge Area Of Guarany Aquifer In Brazil. *Bulletin of Environmental Contamination and Toxicology, 73*(1), 117–124.. doi:10.1007/s00128-004-0402-4

Cerdeira, A. L., Souza, M. D., Bolonhezi, D., Queiroz, S. C. N., Ferracini, V. L., Ligo, M. A. V., et al. (2005a, April) Soil Tillage Systems And Herbicide Leaching In Brazil, In *35th Annual Mississipi Water Resources Conference,* April 26-27.

Cerdeira, A. L., Souza, M. D., Bolonhezi, D., Queiroz, S. C. N., Ferracini, V. L., & Ligo, M. A. V. (2005c). Effects Of Sugar Cane Mechanical Harvesting Followed By No-Tillage Crop Systems On Leaching Of Triazine Herbicides In Brazil. *Bulletin of Environmental Contamination and Toxicology, 75*(4), 805–812. doi:10.1007/s00128-005-0822-9

Cerdeira, A. L., Souza, M. D., Ferracini, V. L., Queiroz, S. C. N., Bolonhezi, D., Gomes, M. A. F., et al. (2006, April). Effect Of Sugarcane Coverage On The Behavior Of Tebuthiuron In Soil In Brazil. In *36th Annual Mississippi Water Resources Conference*, Jackson, Mississippi, 25 A 26/04/2006, (pp. 83-85).

Cerdeira, A. L., Souza, M. D., Pessoa, M. C. P. Y., Queiroz, S.C.N., Ferracini, V. L., Lanchote, V. L., Queiroz, R. H. C., Bolonhezi, D., Farjani Neto, C. (Outubro/2005e). Efeito De Sistemas De Cultivo Com Cana-De-Açúcar Sobre A Lixiviação Dos Herbicidas Do Grupo Das Triazinas Em Região De Recarga Do Aqüífero Guarani. *Jaguariúna/Sp.: Embrapa Meio Ambiente, Pp.1-25 (Boletim De Pesquisa E Desenvolvimento, N° 32).*

Cerdeira, A. L., Souza, M. D., Queiroz, S. C. N., Ferracini, V. L., Bolonhezi, D., & Gomes, M. A. F. (2007a). Leaching and half-life of the herbicide tebuthiuron on a recharge area of Guarany aquifer in sugarcane fields in Brazil. *Journal of Environmental Science and Health*, *42*, 635–639. doi:10.1080/03601230701465593

Cobuci, J. A., Euclydes, R. F., & Verneque, R. S. (2000). Curva De Lactação Na Raça Guzerá. *Revista Brasileira de Zootecnia*, *29*(5), 1332–1339. doi:10.1590/S1516-35982000000500011

Cohen, S. Z., Creeger, S. M., Carsel, R. F., & Enfield, C. G. (1984). Treatment And Disposal Of Pesticide Wastes. In Kruger, R. F., & Seiber, J. N. (Eds.), *Acs Symp. Ser. 259* (pp. 297–325). Washington, DC: Amer. Chem. Soc.

Cohen, S. Z., Wauchope, R. D., Klein, A. W., Eadsporth, C. V., & Grancy, R. (1995). Offsite Transport Of Pesticides In Water – Mathematical Models Of Pesticide Leaching And Runoff. *Pure and Applied Chemistry*, *67*, 2109–2148. doi:10.1351/pac199567122109

Coutinho, H. L. C., Leite, A.A.S., Paraiba, L.C., Cerdeira, A., Filizola, H., Silva, E.F., Gomes, M.A.F. (n.d.). *Análise Espacial Do Potencial De Lixiviação De Herbicidas Em Áreas De Recarga Do Aqüífero Guarani Na Alta Bacia Do Rio Taquari*. Retrieved from Http://Www.Riosvivos.Org.Br/Arquivos/1934241170.Pdf20p. (Accessed On September, 2009)

Cruz, C. D. (1990). *Aplicação De Algumas Técnicas Multivariadas No Melhoramento De Plantas. Piracicaba: Escola Superior De Agronomia Luiz De Queiroz/Usp*. Tese De Doutorado.

Cruz, C. D., & Regazzi, A. J. (1994). *Modelos Biométricos Aplicados Ao Melhoramento Genético*. Viçosa, Brazil: Ufv.

Cruz, F. M. (2006). *Avaliação Do Potencial De Contaminação Dos Recursos Hídricos Por Pesticidas Utilizando O Software Agroscre*. Belém: Ufpa/Ccnt, 2006. 86p. (Monografia, Engenharia Ambiental. Curry, R.B., Peart, R.M., Jones, J.W., Boote, K.J., Allen Jr., L.H. (1990). Simulation As A Tool For Analyzing Crop Response To Climate Change. *Transactions of the ASAE. American Society of Agricultural Engineers*, *33*, 981–990.

De Paula, K. R. (2005). *Fatores Genéticos E Ambientais Na Produção De Sementes De Acácia Negra*. Curitiba: Engenharia Florestal/Universidade Federal Do Paraná. 77p. (Dissertação De Mestrado).

De Siqueira, O. J. F. (1991). Exemplo Utilizando O Sistema De Suporte De Decisão Dssat-Nitrogênio X Trigo. In *Reunião Nacional De Pesquisa De Trigo, 16, Resumos... Dourados: Embrapa/Uepae*. Avanços Na Informática Aplicados Às Relações Solo-Planta-Clima.

De Siqueira, O. J. F. Berg, M. Van Den. (1991). Validação Do Modelo De Simulação Soygro-Soja – Passo Fundo, Rs. In *Soja: Resultados De Pesquisa 1990-1991*, (pp. 139-144). Passo Fundo: Embrapa/Cnpt.

Dias, C. T. S. (2009). *Regressão Segmentada Como Modelo Na Determinação De Exigências Nutricionais De Peixes*. Retrieved from http://Br.Monografias.Com/Trabalhos/Regressao-Segmentada-Exigencias-Nutricionais-Peixes/Regressao-Segmentada-Exigencias-Nutricionais-Peixes.Shtml (Accesed On 1st September, 2009).

Dias, J. R. (2003). *Aplicação Do Modelo Hidrologico Agnps2001 Utilizando Dados Observados Na Baica Do Arroio Vacacaí-Mirim*. Santa Maria,Rs/Ufsm-Ppge, Dissertação De Mestrado.

Dossa, D. (1994). Programação Linear Na Gestão Da Propriedade Rural: Um Enfoque Alternativo. *Teoria E Evidencia Econômica,* (4), 33-60.

Dourado-Neto, D., Teruel, D.A., Reichardt, K., Nielsen, D.R., Frizzone, J.A., Bacchi, O.O.S. (1998) Principles Of Crop Modeling And Simulation: I. Uses Of Mathematical Models In Agricultural Science. *Scientia Agricola, Piracicaba, 55*(Número Especial), 46-50.

Earsterling, W.E.Iii, Crosson, R.P., Rosenberg, N.J., Mc Kenney, M.S., Katz, L.A., Lemon, K.M. (1993). Agricultural Impacts And Responses To Climate Change In The Missouri_Iowa_Nebraska_Kansas (Mink) Region. *Climatic Change, 24*, 23–61. doi:10.1007/BF01091476

Extoxnet/Pip. (2008). *Extension Toxicology Network/ Pesticide Information Profiles*. http://Extoxnet.Orst.Edu/Pips/Hexazin.Htm (Accesed On 18th August, 2008).

Fernandes, E. N. (1997). *Sistema Inteligente De Apoio Ao Processo De Avaliação De Impactos Ambientais De Atividades Agropecuárias. Tese (Doutorado)*. Universidade Federal De Viçosa, Viçosa, Mg.

Fernandes, E. N., & Silva, C. A. B. (1995). Agroflo: Sistema Especialista Para Planejamento De Sistemas Agroflorestais. In *Agrosoft 95 - Feira E Congresso De Informática Aplicada À Agropecuária E Agroindústria, Juiz De Fora*. Anais... Disponível Em: http://Www.Agrosoft.Org.Br/Trabalhos/Ag95/Doc35.Htm>. Acesso Em: 12 Out. 2004.

Fernandes, E. N., Silva, E., Fernandes Filho, E. I., & Silva, C. A. B. (2003). Erosys: Sistema Integrado Para Avaliação De Impactos Ambientais De Atividades Agropecuárias. In João Fernando Marques, Ladislau Araujo Skorupa, Jose Maria Gusman Ferraz. (Org.), *Indicadores De Sustentabilidade Em Agroecossistemas,* (pp. 251-273). Jaguariuna: Embrapa Meio Ambiente.

Ferracini, V. L., & Pessoa, M. C. P. Y. (2008). *Aspectos Toxicológicos E Ambientais Dos Agrotóxicos Aplicados Na Cultura Do Melão Produção Integrada De Melão*, (pp. 237 - 247).

Ferracini, V. L., Pessoa, M. C. P. Y., Oliveira, F. C., & Silva, A. De S. (2005). *Potenciais Contaminantes Ambientais E Subsídios Para Produção Com Qualidade Na Citricultura Citros*. Campinas, Brazil: Instituto Agronômico E Fundag.

Ferracini, V.L., Pessoa, M.C.P.Y., Silva, A.S., Spadotto, C.A. (2001). Análise De Risco De Contaminação De Águas Subterrâneas E Superficiais Da Região De Petrolina (Pe) E Juazeiro (Ba). *Pesticidas: Revista De Ecotoxicologia E Meio Ambiente, 11*, 1-16.

Ferracini, V. L., Queiroz, S. C. N., Gomes, M. A. F., Cerdeira, A. L., Pereira, A. S., Souza, M. D., Santos, G. L. (2006, June). *Monitoramento Do Herbicida Tebutiuron Em Água Subterrânea Na Microbacia Do Córrego Espraiado, Região De Ribeirão Preto*, (pp. 1-10). Sp Jaguariúna/Sp: Embrapa Meio Ambiente (Documentos, Nº 54).

Fialho, F.B., Ledur, M.C. (Setembro, 2000a). *Modelo Matemático Para Curvas De Produção De Ovos*. Concórdia: Embrapa - Suínos E Aves, Comunicado Técnico N. 256, pp.1-3.

Fialho, F.B., & Ledur, M.C. (Setembro. (2000b). *Estimador Da Produção De Ovos*. Concórdia: Embrapa - Suínos E Aves. *Comunicado Técnico, 257*, 1–2.

Figueiredo, R. O., Watrin, O. S., Gerhard, P., Kato, O. R., Souza, E. L., Oliveira, F. A., et al. (2007). Watershed Studies In A Region Mainly Occupied By Small Holder Farms In The Eastern Amazon. In *Lba-Eco 11th Science Team Meeting, 2007*, Salvador, Book Of Abstracts.

Fontoura Junior, J. A. S., Menezes, L. M., Corrêa, M. N., & Dionello, N. J. L. (2007). Utilização De Modelos De Simulação Em Sistemas De Produção De Bovinos De Corte. *Veterina-ria e Zootecnia, 14*(1), 19–30.

Fontoura Junior, J. A. S., Siewerdt, F., Dionello, N. J. L., & Correa, M. N. (2009). Modelo De Simulação Do Desempenho Reprodutivo De Fêmeas Bovinas De Corte Com Base No Escore De Condição Corporal. *Revista Brasileira de Zootecnia, 38*(8), 1627–1635. doi:10.1590/S1516-35982009000800030

Fortaleza, E. A. T., & Brasil, B. D. N. D. Cap.20, Ferracini, V. L., Queiroz, S. C. N., Souza, M. D., Gomes, M. A. F., Cerdeira, A. L., Rampazzo, P., Balderrama, O. (May, 2007). Tebuthiuron Monitoring In Sandy Soils Of The Guarany Aquifer Recharge Zone In Southeastern Brazil. In *1st Latin American Pesticide Residue Workshop (Laprw 2007), Santa Maria, Rs, Brazil - May 6-11, 2007, Book Of Abstratcts*, (pp. 65).

Gazzoni, D. L., Pedroso Junior, M., Garagorry, F., & Moscardi, F. (1998). Mathematical Simulation Model Of The Velvetbean Caterpillar. *Pesquisa Agropecuária Brasileira, Brasília, 33*(4), 385–396.

Gebler, L. (2000). *Aplicabilidade Da Rotina Przm Do Modelo Matemático Przm3 Em Uma Situação Ambiental Do Oeste De Santa Catarina. Florianópolis: Universidade Federal De Santa Catarina.* Tese De Mestrado.

Giuliano, A.D. (Agosto 1995). *Verificação Experimental De Um Modelo De Simulação Do Movimento De Solutos Em Um Meio Poroso.* São Carlos: If/Usp, Dissertação De Mestrado, Departamento De Física E Ciência Dos Materiais.

Gomes, M. A. F., Queiroz, S. C. N., Ferracini, V. L., Pessoa, M. C. P. Y., Pereira, A. S., Filizola, H. F., Souza, M. D. (Junho, 2004). *Ocorrência Dos Herbicidas Tebutiuron E Hexazinone E Implicações Na Qualidade Da Água Subterrânea - Estudo De Caso Na Microbacia Do Córrego Espraiado.* Ribeirão Preto/Sp. In.: I Congresso De Meio Ambiente, Paulínia/Sp, 21 A 25/6/2004, Anais... 2004, Pp.150-151.

Gomes, M. A. F. (1995) Impacto Da Agricultura Intensiva De Cana-De-Açúcar Sobre As Propriedades Do Solo E Da Água Subterrânea - Estudo De Caso Em Ribeirão Preto, Sp. In *Encontro Nacional Da International Association For Impact Assessment - Iaia. 4*, (pp. 383-387). Belo Horizonte, Anais... Belo Horizonte: Gráfica E Editora Cultura.

Gomes, M. A. F., Filizola, H. F., & Spadotto, C. A. (2006a). Classificação Das Áreas De Recarga Do Sistema Aqüífero Guarani No Brasil Em Domínios Pedomorfoagroclimáticos - Subsídio Aos Estudos De Avaliação De Risco De Contaminação Das Águas Subterrâneas. *Revista Do Departamento De Geogragia (Usp), 18*, 67–74.

Gomes, M. A. F., Filizola, H. F., Spadotto, C. A., Pereira, A. S (Junho/2006b). *Caracterização Das Áreas De Afloramento Do Aqüífero Guarani No Brasil - Base Para Uma Proposta De Gestão Sustentável*, Jaguariúna/Sp.: Embrapa Meio Ambiente,. Pp.1-20 (Documentos, N° 52).

Gomes, M. A. F., Queiroz, S. C. N., Ferracini, V. L., Pereira, A. S., Santos, G. L. (Setembro/2006c). *Monitoramento Do Herbidicida Diuron Em Água Subterrânea Na Microbacia Do Córrego Espraiado*. Região De Ribeirão Preto/Sp, Jaguariúna/Sp: Embrapa Meio Ambiente, Pp 1-11(Documentos, Nº 55).

Gomes, M. A. F., Spadotto, C. A., & Pessoa, M. C. P. Y. (2002). Avaliação Da Vulnerabilidade Natural Do Solo Em Áreas Agrícolas: Subsídio À Avaliação Do Risco De Contaminação Do Lençol Freático Por Agroquímicos. *Pesticidas: Revista De Ecotoxicologia E Meio Ambiente, Centro De Pesquisa E Processamento De Alimentos, Curitiba, 12*(Jan./Dez.), 169-179.

Gonçalves, F. A. (2008). *Validação Do Modelo Wepp Na Predição De Erosão Hídrica Para Condição Edafoclimática Da Região De Viçosa, Mg. Viçosa, Mg/ Ufv-Ppgea*. Tese De Doutorado.

Goss, D. W. (1992). Screening Procedure For Soils And Pesticides For Potential Water Quality Impacts. *Weed Technology, 6*(3), 701–708.

Guahyaba, A. Da S. (2004). Modelagem Matemática Em Avicultura: Uma Questão De Sobrevivência. *Aveworld, 2(*9, Abril/Maio), 36-38.

Gustafson, D. I. (1989). Groundwater Ubiquity Score: A Simple Method For Assessing Pesticide Leachability. *Environmental Toxicology and Chemistry, 8*(4), 339–357. doi:10.1002/etc.5620080411

Hernandes, F. B. T., Sousa, S. A. V., Zocoler, J. L., & Frizone, J. A. (2000) Simulação E Efeito De Veranicos Em Culturas Desenvolvidas Na Região De Palmeira D`Oeste, Estado De São Paulo. In *Xxix Congresso Brasileiro De Engenharia Agrícola (Conbea), Fortaleza-Ceará, 4 A 7 De Julho De 2000*. Available At: Http://Www.Agr.Feis.Unesp.Br/Veranico.Htm (24/10/2009)

Intergovernmental Panel On Climate Change (Ipcc). (2001). *Climate Change 2001: The Scientific Basis: Contribution Of Working Group I To The Third Assessment Report Of The Intergovernmental Panel On Climate Change, (Edited By J. T. Houghton Et Al.)*. New York: Cambridge Univ. Press.

Ipt -,Instituto De Pesquisas Tecnológicas. (1994). Caracterização Do Potencial De Contaminação Das Águas Subterrâneas. Por Agroquímicos. *Hidrogeologia Da Microbacia, Relatório, 32*(605).

Ivey, F. J. (1999). Desenvolvimento E Aplicação De Modelos De Crescimento Para Frango De Corte. In *Simpósio Internacional Acav-Embrapa Sobre Nutrição De Aves, 1999, Concórdia, Santa Catarina, Brasil*, (pp. 22 –35).

Keller, A. A. (1982). *Development And Analysis Of An Irrigation Scheduling Program With Emphasis On Forecasting Consumptive Use*. Ms Thesis, Agricultural And Irrigation Engineering Department, Utah State University, Logan.

Krol, M. S., Jaeger, A., & Bronstert, A. (2003). Integrated Modeling Of Climate Changes Impact In Northearstern Brazil. In *Global Change And Regional Impacts, Water Availability, Vulnerability Of Ecosystems And Society In The Semi-Arid Northeast Of Brazil, Gaiser, Krol, Frischkorn, And Araújo*. Berlin: Springer. Disponivel Em Http://Documente.Nl/44592/1/2003_Krol_Et_Al_Waves_Book_Springer.Pdf

Lal, H., Hoogenboom, G., Calixte, J. P., Jones, J. W., & Beinroth, F. H. (1993). Using Crop Simulation Models And Gis For Regional Productivity Analysis. *Transactions of the ASAE. American Society of Agricultural Engineers, 36*(1).

Lane, L. J., & Nearing, M. A. (Eds.). (1980). *Usda Water Erosion Prediction Project: Hill Slope Profile Model Documentation*, (2ed.). West Lafayette: Usda: Sea, (Usda-Sea Research Report, 26).

Leite, L. F. C., Madari, B. E., Machado, P. L. O. A., & Doraiswamy, P. C. (2007) Modelagem Da Dinâmica Da Matéria Orgânica Do Solo Em Sistema Integração Lavoura-Pecuária Usando Cqestr. In *Anais Congresso Brasileiro De Ciência Do Solo. Gramado, Ufrgs*.

Leonard, R.A, Knisel, W.G, Still, D. A. (1987). Gleams: Groundwater Loading Effects Of Agricultural Management Systems. *Transactions Of Asae, St. Joseph, 30* (5), 1403-1418.

Levy, T. L., Mattice, J. D., & Kochenderfer, J. N. (1989). Hexazinone Persistence And Mobility Of A Steep Forested Watershed. *Journal of Environmental Quality*, *18*, 507–514. doi:10.2134/jeq1989.00472425001800040021x

Lewis, G. N. (1901). The Law Of Physico-Chemical Change. *Proceedings of the American Academy of Arts and Sciences*, *37*, 49–69.

Lima, L. M., Watrin, O. S., Pessoa, M. C. P. Y., Pereira, A. S., Campinas, D. S. N., Figueiredo, R. O., & Costa, F. R. (Abril, 2007). Simulação Da Movimentação De Agrotóxicos No Solo Com O Uso De Geotecnologias, Como Instrumento Para Avaliação Do Potencial De Risco De Contaminação De Águas Subterrâneas Na Amazônia Oriental. In *Anais Xiii Simpósio Brasileiro De Sensoriamento Remoto*, Florianópolis, Brasil, 21 A 26 Abril 2007, Inpe, (pp. 3397-3404).

Liu, W. T. (1989). Application Of Ceres-Maize Model To Yield Prediction Of A Brazilian Maixe Hybrid. *Agricultural and Forest Meteorology*, *45*, 299–312. doi:10.1016/0168-1923(89)90050-6

Lourencetti, C., Spadotto, C. A., Santiago-Silva, M., & Ribeiro, M. (2005) Avaliação Do Potencial De Águas Subterrâneas Por Pesticidas: Comparação Entre Métodos De Previsão Por Lixiviação. *Pesticidas: Revista De Ecotoxicologia E Meio Ambiente, Curitiba, 15*, 1-14.

Machado, R. E., & Vettorazzi, C. A. (2003). Simulação Da Produção De Sedimentos Para A Microbacia Hidrográfica Do Ribeirão Dos Marins (Sp). *Revista Brasileira de Ciencia do Solo*, *27*(4), 735–741. doi:10.1590/S0100-06832003000400018

Machado, R. E., Vettorazzi, C. A., & Cruciani, D. E. (2003, Março). Simulação De Escoamento Em Uma Microbacia Hidrográfica Utilizando Técnicas De Modelagem E Geoprocessamento. *Rev. Bras. Recursos Hídricos.*, *8*(1), 147–155.

Mackay, D. (1993). Toxic Substances Modeling: Fugacity Concepts. In Systems, C. E. O. E. (Ed.), *Young, P.C* (pp. 611–615). New York: Pergamon Press Ltd.

Medeiros, H. R. (2003). *Avaliação De Modelos Matemáticos Desenvolvidos Para Auxiliar A Tomada De Decisão Em Sistemas De Produção De Ruminantes Em Pastagens*. Piracicaba: Esalq/Usp 98p.

Medeiros, S. L. P., Gosse, G., Chartier, M., & Manfron, P. A. (2000). Ajuste Do Módulo De Produção De Matéria Seca No Modelo Ceres-Sorghum Para Simulação Da Cultura Do Sorgo Sacarino. *Revista Brasileira De Agrometeorologia. Santa Maria*, *8*(2), 223–229.

Mendonça, L. R. C. (2004) *Simulador De Cenários Bioeconômicos Para Suporte A Decisão No Gerenciamento De Fazendas Produtoras De Gado De Corte*. Goiânia: Universidade Federal De Goiás- Escola De Engenharia Elétrica E De Computação, 116p. (Dissertação De Mestrado).

Miklós, A. A. W., & Gomes, M. A F. (1996). *Levantamento Semidetalhado Dos Solos Da Bacia Hidrográfica Do Córrego Do Espraiado*, Ribeirão Preto- Sp. (Relatório De Consultoria).

Minoti, R.T. (2006). *Abordagens Qualitativa E Quantitativa Das Micro-Bacias Hidrográficas E Áreas Alagáveis De Um Compartimento Do Médio Mogi-Superior/Sp*. Tese Doutorado Em Ciências Da Engenharia Ambiental, São Carlos: Usp/Eesc.

Minoti, R. T. E. Crestana, S. (2004). Perda De Solo Como Indicador De Vulnerabilidade De Bacias Hidrográficas: Artigos Publicados Na Revista Brasileira De Ciência Do Solo E Em Congressos Brasileiros De Ciência Do Solo, Manejo E Conservação. In Espíndola, E.L.G., Schalch, V. (Orgs.), *Bacia Hidrográfica: Diversas Abordagens Em Pesquisa,* (pp. 47-54). São Carlos: Rima.

Nofziger, D. L., & Hornsby, A. G. (1994). [- *Chemical Movement In Layered Soils*. Gainesville, FL: University Of Florida.]. *Cellular and Molecular Life Sciences*, 94.

Odum, E. P. (Trad. Gomes, A.M.A.) (2004). *Fundamentos De Ecologia,* (7ª Ed.). Lisboa: Fundação Calouste Gulbenkian.

Oseki, H. (1998). *Economic Models For Brazilian Farm Environment*. Available At Http://Www.Jircas.Affrc.Go.Jp/Kankoubutsu/News/Newsletter/Nl1998/No.15/03ozeki.Htm (Accessed On August 2009).

Papajcsik, I. A., & Bodero, J. (1988). Modeling Lactation Curves Of Friesian Cow In A Subtropical Climate. *Animal Production*, *47*, 201–207. doi:10.1017/S0003356100003275

Paraíba, L. C. (2007). Pesticide Bioconcentration Modelling For Fruit Trees. *Chemosphere*, *66*, 1468–1475. doi:10.1016/j.chemosphere.2006.09.017

Paraiba, L. C., Boeira, R. C., Jonsson, C. M., & Carrasco, J. M. (2006). Fator De Bioconcentração De Poluentes Orgânicos De Lodo Em Frutos De Laranjeiras. *Pesticidas: Revista De Ecotoxicologia E Meio Ambiente, Curitiba, 16*(Jan./Dez), 125-134.

Paraíba, L. C., Cerdeira, A. L., Silva, E. F., Martins, J. S., & Coutinho, H. L. C. (2003). Evaluation Of Soil Temperature Effect On Herbicide Leaching Potential Into Groundwater In The Brazilian Cerrado. *Chemosphere*, *53*, 1087–1095. doi:10.1016/S0045-6535(03)00594-0

Paraíba, L. C., Plese, L. P. M., Foloni, L. L., & Carrasco, J. M. (2007). Simulation Of The Fate Of The Insecticide Carbofuran In A Rice Field Using A Level Iv Fugacity Model. *Spanish Journal of Agricultural Research*, *5*(1), 43–50.

Paraíba, L.C., Pulino, P. (2003). *Simulação Numérica Da Dispersão-Advecção De Pesticidas No Solo Sob Efeito Da Temperatura*. Jaguariúna: Embrapa Meio Ambiente (Embrapa Meio Ambiente. Documentos, 35).

Paraíba, L. C., & Spadotto, C. A. (2002). Soil Temperature Effect In Calculating Attenuation And Retardation Factors. *Chemosphere*, *48*, 905–912. doi:10.1016/S0045-6535(02)00181-9

Paranhos, S. B. (1987). *Cana-De-Açúcar - Cultivo E Utilização*. Campinas: Fundação Cargil.

Pausnitz, J. N. (1969). *Molecular Thermodynamics Of Fluid-Phase Equilibria*. Englewood Cliffs, NJ: Prentice-Hall.

Peres, F.C. (1990). *Confinamento De Gado De Corte: Um Modelo De Analise De Decisão*. Tese De Livre Docência Em Administração Rural, Piracicaba: Usp/Esalq.

Perkins, L. B., Yarborough, D. E., & Guthrie, K. E. (2006). Detection Of Hexazinone In Maine Groundwater- A Nine Year Study. *Acta Hort. (Ishs), 715*, 329-336. Accessed At: Http://Www.Actahort.Org/Books/715/715_49.Htm

Pessoa, M. C. P. Y. (1994). *Simulação E Inteligência Artificial Aplicadas Ao Estudo Da Dinâmica Populacional Do Bicudo Do Algodoeiro Na Região De Campinas/Sp*. Campinas: Densis-Fee/Unicamp, Tese De Doutorado.

Pessoa, M. C. P. Y., Chaim, A., Capalbo, D. M. F., Hamada, E., Tambasco, F. J., Ferraz, J. M. G., et al. (2006b). *Boas Práticas Agrícolas E Meio Ambiente - Elementos De Apoio Para As Boas Práticas Agrícolas E O Sistema Appcc/Pas Campo.* - 2.Ed. Rev., Atual. - Brasília, Df: Embrapa, (pp. 13-33) (Série Qualidade E Segurança Dos Alimentos).

Pessoa, M.C.P.Y., Chaim, A., Gomes, M.A.F., Silva, A. S., Soares, J. M. (2003b). Simulação Do Movimento De Aldicarb E Tebuthiuron Em Solos Sob Cultivos De Banana E Cana-De-Açúcar No Semi-Árido Brasileiro. *Rev. Bras. De Eng. Agrícola E Ambiental (Agriambi), 7* (2), 297-302.

Pessoa, M. C. P. Y., Ferracini, V. L., Chaim, A., & Scramin, S. (Dezembro/2004b). *Software Agroscre - Apoio À Avaliação De Tendências De Transporte De Princípios Ativos De Agrotóxicos.* Jaguariúna/Sp: Embrapa Meio Ambiente, (pp. 1-23). (Boletim De Pesquisa E Desenvolvimento, Nº 26).

Pessoa, M. C. P. Y., Filizola, H. F., & Gomes, M. A. F. (2005). *Simulação Da Movimentação Vertical De Produtos Aplicados Em Milho, Milho Safrinha E Soja Em Latossolos Vermelhos E Neossolos Quartzarênicos De Mineiros*, Jaguariúna: Embrapa Meio Ambiente, 27p. (Embrapa Meio Ambiente. Boletim De Pesquisa E Desenvolvimento, 31).

Pessoa, M. C. P. Y., Gomes, M. A. F., Neves, M. C., Cerdeira, A. L., & Souza, M. D. (2003a). Identificação De Áreas De Exposição Ao Risco De Contaminação De Águas Subterrâneas Pelos Herbicidas Atrazina. Diuron E Tebutiuron. *Pesticidas: Revista De Ecotoxicologia E Meio Ambiente, Curitiba, 13* (Jan./Dez.), 111-122.

Pessoa, M. C. P. Y., Gomes, M. A. F., Nicolella, G., Souza, M. D., Cerdeira, A. L., & Monticelli, A. (1997a). Simulação Do Movimento Dos Herbicidas Hexazinone, Diuron, Atrazina, Ametrina E Simazina Aplicados Na Cultura De Cana-De-Açúcar Em Solos Da Microbacia Do Córrego Do Espraiado. Ribeirão Preto,Sp. In *Congresso Brasileiro De Ciências De Solo. 26, Rio De Janeiro, Resumos Do Congresso Brasileiro De Ciências De Solo, 26, Cdrom.*

Pessoa, M.C.P.Y., Gomes, M.A.F., Scramin, S., Silva, A. De S., Gussakov, K. C. (2006a). Vulnerabilidade Natural Das Grandes Bacias Hidrográficas Brasileiras À Tendência De Contaminação De Águas Por Agrotóxicos Em Função Dos Tipos De Solos Predominantes. Pesticidas: Revista De *Ecotoxicologia E Meio Ambiente, 16*(Jan/Dez), 39-52.

Pessoa, M. C. P. Y., Gomes, M. A. F., Souza, M. D., Cerdeira, A. L., Nicolella, G., & Monitcelli, A. (1999). Simulação Do Movimento De Herbicidas Utilizados No Monocultivo De Cana-De-Açúcar Em Latossolos Da Área De Recarga Do Aquífero Botucatu (Guarani) Em Ribeirão Preto, Sp. *Revista Científica Rural. Bagé, 4*(1), 15–24.

Pessoa, M.C.P.Y., Luchiari Junior, A., Fernandes, E.N., Lima, M.A. De. (1997b). *Principais Modelos E Simuladores Usados Em Análise De Impacto Ambiental Da Agricultura.* Jaguariúna: Embrapa-Cnpma, 87p., (Documentos, N.8).

Pessoa, M. C. P. Y., Sá, L. A. N., Kodaira, J. Y., Wilcken, C. F., & Almeida, G. R. (2008). *Simulação Da Dinâmica Populacional Do Psilideo-De-Concha, Glycaspis Brimblecombei (Hemiptera: Psyllidae) E Identificação De Estratégias Para A Criação Laboratorial De Seu Parasitoide Psyllaephagus Bliteus (Hymenoptera: Encyrtidae).* Jaguariúna: Embrapa Meio Ambiente, 32p. (Embrapa Meio Ambiente. Boletim De Pesquisa E Desenvolvimento, 49).

Pessoa, M. C. P. Y., & Scramin, S. (2004a). Modelagem Matemática E Simulação Da Exposição Do Ambiente À Aplicação De Agrotóxicos: Apoio À Avaliação De Risco Ambiental. In: Silva, C.M.M.S, Fay, E. F. (Eds.), *Agrotóxicos E Ambiente. Embrapa Informação Tecnológica,* (pp. 319-364). Brasília/Df.

Pessoa, M.C.P.Y., Scramin, S., Chaim, A., Ferracini, V.L. (Junho/2007). *Avaliação Do Potencial De Transporte De Agrotóxicos Usados No Brasil Por Modelos Screening E Planilha Eletrônica.* Jaguariúna/Sp: Embrapa Meio Ambiente. Boletim De Pesquisa E Desenvolvimento, Nº 44.

Pessoa. M. C. P. Y., Gomes. M. A. F., Souza. M. D., Cerdeira. A. L., Nicolella. G., Monticelli, A. (1998). Simulação Do Movimento De Herbicidas Utilizados No Monocultivo De Cana-De-Açúcar, Em Areia Quartzosa Da Área De Recarga Do Aquífero Guarani (Antigo Botucatu) Em Ribeirão Preto,Sp. *Revista Científica Rural. Bagé/Rs, 3* (2), 11-19.

(1990). Pesticides. InCheng, H. H. (Ed.), *The Soil Environment: Process, Impacts, And Modeling.* Madison, WI: Soil Science Society Of America Inc.

Picini, A. G., Valeriano, M. De M., Lombardi Neto, F. And Zullo Junior, J. (2005) Ajuste Dos Simuladores De Dados Meteorológicos Do Modelo Epic Para Diferentes Locais Do Estado De São Paulo. *Revista Brasileira De Engenharia Agrícola E Ambiental., 9*(2), 239-246, 1415-4366.

Plec, D. F., Pessoa, M. C. P. Y., Lima, M. A., & Vilella, O. V. (Agosto, 2007). Avaliação Do Fluxo De Emissão De Gases Estufa Provenientes De Fontes Agrícolas Por Simulação De Sistemas: Estudo De Caso De Emissão De Metano Por Arroz Irrigado Por Inundação Em Pindamonhangaba, Sp. In *Anais... Congresso Interinstitucional De Iniciação Científica (Pibic/Cnpq)*, Campinas/ Sp: Instituto Agronômico De Campinas, 1 A 2 De Agosto De 2007. Cd-Rom.

Plese, L. P. M., Silva, C. L., & Foloni, L. L. (2009). Distribuição Nos Compartimentos Ambientais Dos Herbicidas Utilizados Nas Culturas De Algodão, Café E Citros. *Planta Daninha. Viçosa-Mg, 27*(1), 123–132.

Praseres, A. L. (1976). *Combinação De Atividades Pela Programação Linear Em Empresas Agrícolas Na Microrregião De Imperatriz, Maranhão.* Piracicaba: Usp/Esalq, 150p. (Dissertação De Mestrado).

Queiroz, S. C. N., Ferracini, V. L., Rosa, M. A., & Cerdeira, A. L. (Maio, 2006) Validação De Método Por Clae Para Determinação De Hexazinone E Diuron Em Solo. In *29ª Reunião Anual Da Sociedade Brasileira De Química. Águas De Lindóia/Sp, 19 A 22/05/2006.*

Queiroz, S.C.N., Ferracini, V.L., Rosa, M. A., Cerdeira, A.L. (2007, Dezembro). Método Para Determinação De Hexazinona E Diurom Em Solo. *Boletim De Pesquisa E Desenvolvimento,* (47).

Quirino, T. R. (1990). *Construção Do Conhecimento Científico.* Brasília: Embrapa Sede.

Rao, P. S. C., Hornsby, A. G., & Jessup, R. E. (1985). Indices For "Ranking" The Potential For Pesticide Contamination Of Groundwater. *Soil And Crop Science Society Of Florida, 44,* 1–8.

Renard, K. G., Foster, G. R., Weesies, G. A., & Mccool, D. K. (Eds.). (1991). *Predicting Soil Erosion By Water - A Guide To Conservation Planning With The Revised Universal Soil Loss Equation (Rusle).* Washington, DC: Usda-Ars.

Resende, M., & Almeida, J. A. (1985). Modelos De Predição De Perda De Solo: Uma Ferramenta Para Manejo E Conservação Do Solo. *Informe Agropecuário, Belo Horizonte, 11*(128), 38–54.

Richardson, C. W., & Wright, D. A. (1984). *Wgen: A Model For Generating Daily Weather Variables.* Washington, DC: Agricultural Research Service, Usda.

Rickman, R. W., Douglas, C. L. Jr, Albrecht, S. L., Bundy, L. G., & Berc, J. L. (2001). Cqestr: A Model To Estimate Carbon Sequestration In Agricultural Soils. *J. Soil Cons.*, *56*, 237–242.

Rodrigues, L. C. L. (1997). *Planejamento Agropecuário Através De Um Modelo De Programação Linear Não-Desterminista. Piracicaba: Usp/ Esalq*. Dissertação De Mestrado.

Rondon, E. O. O., Murakami, A. E., & Sakaguti, E. S. (2002). Modelagem Computacional Para Produção E Pesquisa Em Avicultura. *Revista Brasileira De Ciência Avícola, 4*(1), 9. Available At: Http://Www.Scielo.Br/Scielo.Php?Pid=S1516-635x2002000100001&Script=Sci_Arttext (25/10/2009)

Roy, D. N., Konar, S. K., Charles, D. A., Feng, J. C., Prasad, F., & Campbell, R. A. (1989). Determination Of Persistence. Movement And Degradation Of Hexazinone In Selected Canadian Boreal Forest Soils. *Journal of Agricultural and Food Chemistry*, *37*, 443–447. doi:10.1021/jf00086a039

Salle, C. T., Guahyba, A. S., Wald, V. B., Silva, A. B., Salle, F. O., & Fallavena, L. C. B. (2001). Uso De Redes Neurais Artificiais Para Estimar Parâmetros De Produção De Galinhas Reprodutoras Pesadas Em Recria. *Revista Brasileira de Ciência Avícola*, *3*, 257–264. doi:10.1590/S1516-635X2001000300008

Salle, C. T. P., Cé, M. C., Lorenzini, G., Sfoggia, M. V. B., Guahyba, A. S., Moraes, H. L. S., & Nascimento, V. P. (1998a). Correlation Between Aflatoxin And Ocratoxin Levels With Production Parameters In *A Poultry Company. Abstracts Of The Iv Asia-Pacific Poultry Health Conference. Melbourne - Australia,* (pp. 131-131).

Salle, C. T. P., Cé, M. C., Santos, C. H. C., Guahyba, A. S., & Nascimento, V. P. (1998b). Use Of Statistical Technique On The Interpretation Of Routine Serologic Data Produced By Poultry Industry. *Abstracts Of The Iv Asia-Pacific Poultry Health Conference. Melbourne - Australia*, (pp. 148-148).

Sharpley, A. N., & Willians, J. R. (Eds.). (1990). *Epic- Erosion Productivity Impact Calculator: I. Model Documentation.* Springfield, IL: Ntis, (Usda. Technicall Bulletin, 1768).

Silva, C. L., Foloni, L.L., Paraíba, L.C., Plese, L.P.M. (2007). Previsão Ambiental Da Distribuição Dos Pesticidas Aplicados Em Arroz. Pesticidas: R. Ecotoxicol. E Meio Ambiente, *17*(Jan./Dez.), 75-86.

Silva, F. A. M. (2004). Parametrização E Modelagem Do Balance Hídrico Em Sistemas De Plantio Direto No Cerrado Brasileiro. Campinas,Sp: Unicamp-Feagri, Tese Doutorado.

Silva, F. C., & Bergamasco, A. F. (2001). *Levantamento De Modelos Matemáticos Descritos Para A Cultura Da Cana-De-Açúcar*. Campinas: Embrapa Informática Agropecuária. Também Disponível Em Http://Www.Unitau.Br/Prppg/Publica/Biocienc/Downloads/Levantamento-modelos-N1-2001.Pdf Consulted In 12/11/2004

Siqueira, O. J. F., Farias, J. R. B., & Sans, L. M. A. (1994). Potential Effects Of Global Climate Change For Brazilian Agriculture And Adaptive Strategies For Wheat, Maize And Soybean. Revista *Brasileira De Agrometeorologia. Santa Maria*, *2*, 115–129.

Soares, R. C. B. (1995). *Avaliação Da Resposta Imunológica Em Reprodutoras De Perus (Meleagridis Gallopavo) Diante Do Vírus Vacinal Da Doença De Newcastle Através De Modelos Matemáticos*. Porto Alegre: Ufrgs/Faculdade De Veterinária, Dissertação De Mestrado.

Sousa, S. A. V. (1999). Programa Computacional Para Simulação Da Ocorrência De Veranicos E Queda De Produção. [P. Tese ] [Doutorado] [ – Escola Superior De Agricultura "Luiz De Queiroz" - Universidade De São Paulo.]. *Piracicaba, 1999*, 124.

Souto, A. R., Crestana, S. (2000) Identificação Das Áreas Potenciais De Produção De Sedimentos Com O Modelo Agnps E Técnicas De Sig Em Uma Microbacia Hidrográfica. *Rev. Bras. Eng. Agríc. Ambient., Campina Grande, 4*(3), Dez.

Souza, R. F. P. (2007). Economia Do Meio Ambiente E Responsabilidade Social: Os Métodos De Valoração Econômica E Controle Ambiental. In *Anais... Xlv Congresso Da Sociedade Brasileira De Economia, Administração E Sociologia Rural*, (pp.1-11). 22 A 25 De Julho 2007, Londrina: Uel.

Spadotto, C. A. (2002). Screening Method For Assessing Pesticide Leaching Potential. *Pesticidas: Revista De Ecotoxicologia E Meio Ambiente, Curitiba, 12*, 69-78.

Streck, N. A., & Alberto, C. M. (2006). Simulação Do Impacto Da Mudança Climática Sobre A Água Disponível Do Solo Em Agroecossistemas De Trigo, Soja E Milho Em Santa Maria, Rs. *Ciência Rural, 36*, 424–433.

Suji, E. R., Tigano, M. S., & Sosa-Gomes, D. (2002). Simulação Do Impacto Do Fungo *Nomuraea Rileyi* Em Populações Da Lagarta Da Soja. *Anticarsia Gemmatalis* Pesquisa. *Agropecuária Brasileira, Brasília, 37*(11), 1551–1558.

Suji. E. R. (1994). *Padrão De Distribuição Das Populações Anuais E Modelo Fenológico Para O Manejo Da Cigarrinha-Das-Pastagens, Deois Flavopicta (Homoptera: Cercopidae)*. Dissertação Mestrado. Campinas/Sp: Ib-Unicamp.

Tashizawa, T., & Rezende, W. (2000). *Estratégia Empresarial*. São Paulo: Makron Books.

Tedeschi, L. O., Fox, D. G., Sainz, R. D., Medeiros, S. R., Boin, C., & Barioni, L. G. (2005). Mathematical Models In Ruminant Nutrition. *Scientia Agricola. Piracicaba, Brasil, 62*(1), 76–91.

Ternes, S., Fernandes, J. F. R., & Yang, H. M. (2005). O Uso De Modelagem Matemática E Simulação Na Avaliação Do Controle Biológico Da Minadora Do Citros. *Revista Brasileira De Agroinformática, 7*(2), 1–16.

Tese De Doutorado. Medeiros, G.A, Perini, M., Lucarelli, J.R.F., Daniel, L.A. (2001) Influência Do Impacto Da Compactação Do Solo Sobre A Condutividade Hidráulica Saturada Para Um Solo Argiloso. *Revista Ecossistema, 26*(2), Agosto-Dezembro.

Tojo-Soler, C., Folegatti, M. V., & Faria, R. T. (2001). Uso Do Modelo Ceres-Maize Para Identificação De Características Genéticas Desejáveis E De Praticas Adequadas De Manejo De Milho "Safrinha". *Revista Brasileira De Agrometeorologia. Santa Maria, 9*(2), 339–346.

Tomazoni, J.C. & Guimarães, E. (2005). A Sistematização Dos Fatores Da Eups Em Sig Para Quantificação Da Erosão Laminar Na Bacia Do Rio Jirau. *Revista Brasileira De Cartografia, N°57*(03), 235-244.

Trigueiro, A. (Coord). (2003). *Meio Ambiente No Século 21: 21 Especialistas Falam Da Questão Ambiental Nas Suas Áreas De Conhecimento*. Rio De Janeiro: Sextante.

Trindade Neto, M. A. (n.d.). *Fluxo Biológico Do Fósforo No Metabolismo De Suínos Alimentados Com Dietas Contendo Fitase*. Http://Br.Monografias.Com/Trabalhos/Fluxo-Biologico-Fosforo-Metabolismo-Suinos/Fluxo-Biologico-Fosforo-Metabolismo-Suinos.Shtml. (Accessed On 01/09/2009).

Tsuji, G. Y., Uehara, G., & Balas, S. (Eds.). (1994). *Dssat: A Decision Support System For Agrotechnology Transfer. Version 3. V.1,2 And 3.* University Of Hawaii, Honolulu, HI.

Van Ness, H. C., & Abbott, M. M. (1982). *Classical Thermodinamics Of Non-Electrolyse Solutions*. New York: Mcgraw-Hill. Veloso, R. F., Chaib Filho, H. (1999). *Uma Proposta De Modelagem De Sistemas Agrícolas Nos Cerrados*. Juiz De Fora: Agrosoft. Retrieved from http://Agrosoft.Softex.Br/Trabalhos/Ag99/Artigo56.Htm

Veloso, R. F., Gomes, E. G., Barioni, L. G., Fernandes, F. B., & Chaib Filho, H., Oliveira. A. J. De, Silva, A.P. Da, Coloca, A. (2007). Programação Matemática Para Avaliação De Viabilidade Econômica Da Produção De Leite Bovino Em Pequena Escala No Sul De Goiás. *Engevista, 9*(1), 57–81.

Verneque, R. S. (2009). *Aspectos Genéticos E Ambientais Da Curva De Lactação De Vacas Da Raça Guzerá*. Retrieved from http://Www.Br.Monografias.Com/Trabalhos/Aspectos-Geneticos-Ambientais/Aspectos-Geneticos-Ambientais.Shtml (Accessed On 01/09/2009)

Vilcarromero, A. C. S. (1996). *Análise De Modelos Determinísticos Em Fitopatologia*. Dissertação De Mestrado, Imecc/Unicamp, Campinas/Sp.

Wagenet, R. J., & Hutson, J. L. (1989). *Leachm: Leaching Estimation And Chemistry Model.* Ithaca, NY: Cornell University Center For Environment Resource, V.2, Version 2.

Williams, J. R. (1975). Sediment Yield Prediction With Universal Equation Using Runoff Energy Factor. In *Present Prospective Technology For Predicting Sediment Yields And Source,* (pp. 244-252). Washington, DC: Usda: Ars.

Wischmeier, W. H., & Smith, D. D. (1978). *Predicting Rainfall Erosion And Losses: A Guide Of Conservation Planning*. Washington, DC: Usda.

Young, R., Onstad, C., Bosch, D., & Anderson, W. (1986). *Agricultural Nonpoint Source Pollution Model: A Watershed Analysis Tool, Model Documentation.* Morris, MN: Agricultural Research Service, U.S. Department Of Agriculture.

## ADDITIONAL READING

Barroso, L. C., Barroso, M. M. A., Dampos Filho, F. F., Carvalho, M. L. B., & Maia, M. L. (1992). *Calculo Numérico (Com Aplicações)*. São Paulo: Harbra.

Caixeta Filho, J. V. (2004) Pesquisa Operacional – Técnicas De Otimização Aplicadas A Sistemas Agroindustriais. 2.Ed., São Paulo: Atlas, 169p.

Castro Sobrinho, A. S. (2006). *Introdução Ao Método Dos Elementos Finitos*. Rio De Janeiro: Editora Ciência Moderna Ltda.

Chaim, A., Pessoa, M. C. P. Y., Neto, J. C., & Hermes, L. C. (2002, Abr.). Comparison Of Microscopic Method And Computational Program For Pesticide Deposition Evaluation Of Spraying. *Pesquisa Agropecuária Brasileira, Brasília, 37*(4), 493–496.

Filizola, H., Pessoa, M. C. P. Y., Gomes, M. A. F., & De Souza, M. D. (2002) Contaminação Dos Solos Em Áreas Agrícolas. In: Manzato, C.V.; Freitas Jr, E. De; Peres, J.R.R. (Eds). Uso Agrícola Dos Solos Brasileiros. Rio De Janeiro: Embrapa Solos, 174p. Capítulo 7, Pg. 79- 85.

Lachtermacher, G. (2002). *Pesquisa Operacional Na Tomada De Decisões*. Rio De Janeiro: Campus.

Lima, M. A., Cabral, O. M. R., & Miguez, J. D. G. (Eds.). (2001). Mudanças Climáticas Globais E A Agropecuária Brasileira. Jaguariúna: Embrapa Meio Ambiente, 397p.

Melo, I. S., Silva, C. M. M. S., Scramin, S., & Spessoto, A. (2001). Biodegradação. Jaguariúna: Embrapa Meio Ambiente, 440p.

Pessoa, M. C. P. Y. P., & Silva, A. De S.; Camargo, C. P. (2002) Qualidade E Certificação De Produtos Agropecuários. Brasília: Embrapa Informação Tecnológica, 188p. (Texto Para Discussão; 14).

Santos, J. W., & Gheyi, H. R. (2003) Estatística Experimental Aplicada. Campina Grande: Editora Grafica Marcone Ltda. 213p.

Seganfredo, M. A. (2007). *Gestão Ambiental Da Suinocultura*. Brasília: Embrapa Informação Tecnológica.

Souza, G. S. (1998). *Introdução Aos Modelos De Regressão Linear E Não-Linear*. Brasília: Embrapa-Spi/Embrapa-Sea.

Stoorvogel, J. J. (1995). Integration Of Computer-Based Models And Tools To Evaluate Alternative Land-Use Scenarios As Part Of An Agricultural Systems Analysis. *Agricultural Systems*, *49*, 353–367. doi:10.1016/0308-521X(95)00030-9

Trigueiro, A. (Coord). (2003). Meio Ambiente No Século 21: 21 Especialistas Falam Da Questão Ambiental Nas Suas Áreas De Conhecimento. Rio De Janeiro: Sextante,367p.

Tucci, C. E. M. (Org). (2001) Hidrologia – Ciência Aplicada. Porto Alegre: Ed Universidade/Ufrgs: Abrh, 2001). 943p. (Coleção Abrh De Recursos Hídricos; V.4).

Vieira, V. P. P. B. (1999) Avaliação Quantitativa De Riscos Econômicos E Ambientais. In: Galvão, C. O.; Mêuser, J. S. (Orgs.). Sistemas Inteligentes-Aplicações A Recursos Hídricos E Sistemas Ambientais. Porto Alegre: Abrh, P. 246. (Capítulo 5)

## KEY TERMS AND DEFINITIONS

**Agroecosystem:** Originally a natural-ecological system, which was transformed into an agrarian space and used for crop or livestock productions, according to different type and level of managements.

**Decision-Making Process:** Process which involves both the acquisition of knowledge on a given problem and the prediction of risks and uncertainties associated with it, requiring prior knowledge of the main factors that represent the problem to be solved, as well as their interactions with other factors, which could also influence them either positively or negatively.

**Environmental Impact Assessment (EIA):** Environmental policy tool, which consists of procedures to ensure the development of a systematic examination of the environmental impacts of a proposal and its alternatives. In fact, EIA is a method of analysis to identify, to predict, and to describe the possible impact of an agricultural occupation on the environment in which it operates, taking into account the environmental effects of a specific intervention in the short term (during its implantation phase), and in the long-term (during its operation and development phase).

**Environment:** A set of ecological, social, economic, and cultural factors which surrounds the individual, and with which it interacts, influencing and being influenced by them.

**Environmental Planning:** Proposal of a set of actions, which define the forms of use and occupancy of a determined area (property, county, territory, region, country), based on the previous analysis of its main factors (physic-natural, economic, social, and political) in order to assure the improvement of economic development taking into account necessities of preservation and conservation of the environment.

**Environmental Modeling:** Process of model development, which take into account the actual knowledge available and the capacity of understanding and to represent the behavior of the environmental system in focus, at defined scales, in order to well-comprehension of both its dynamic and factor interactions, as well as to allow observing its behavior considering different scenarios (predictions), to advance the science and improve decision making about its resources and environmental issues involved with.

**Holistic Vision:** Advocates an organic and systemic way of observing the things, i.e., to prioritize the behavior of the integrated whole to the detriment of the behavior of the parties isolated. It is based on the observation that when the universe is considered, the sum of the parts does not represent the whole.

# Chapter 12
# A Linear Optimization Approach for Increasing Sustainability in Vegetable Crop Production

**Lana Mara R. dos Santos**
*Universidade Federal de Viçosa, Brazil*

**Marcos N. Arenales**
*Universidade de São Paulo, Brazil*

**Alysson M. Costa**
*Universidade de São Paulo, Brazil*

**Ricardo H. S. Santos**
*Universidade Federal de Viçosa, Brazil*

## ABSTRACT

*This chapter is concerned with a set of optimization problems associated to crop rotation scheduling in the context of vegetable crop production according to some ecological criteria: no crop of the same botanic family is planted in sequence, green manure and fallow periods must be present in any schedule. A core mathematical model called the crop rotation scheduling model is proposed to represent these ecological criteria together with specific technical constraints associated to the growing of vegetable crops. Three optimization problems based on crop rotation schedules are written in detail in this chapter. For each problem, the authors present a general modeling framework and a solution methodology based on a technique known as column generation, which iteratively builds crop rotation plans for a number of plots. Some extensions are also presented, with the aim of incorporating additional characteristics found in production field conditions. This chapter ends with a brief discussion on a set of computational experiments and some suggestions for future research.*

DOI: 10.4018/978-1-61692-871-1.ch012

## INTRODUCTION

Large monoculture-based production systems have been present in Brazil since sugar cane was grown for sugar production in the 16th – 18th centuries. Furthermore, it became more important in the sixties and seventies. Although these systems are able to produce food at low prices, they are also associated with many social and environmental problems. Indeed, besides depending on extensive use of highly soluble mineral fertilizers, pesticides and fossil energy, large monocultures need a small number of employees and usually exclude small family growers due to the dependence on large amounts of capital. This has raised concerns about the economical, environmental and social sustainability of these food production system. In response to these concerns, alternative production models such as 'Natural', 'Organic', and 'Biodynamic' agricultures have attracted attention. Implementing these models has been boosted by private initiatives, such as the promotion of specific fairs and distribution strategies, as well as governmental incentives in the form of regulations and guidelines for organic production in the country.

The main cornerstone of more sustainable models in crop production is crop diversification and the introduction of biological processes in the system. Various agricultural and ecological studies demonstrate that the increase in crop species diversity leads to beneficial effects, such as better exploitation of productive resources, less pest damage, a higher control of spontaneous weeds and an increase in production stability (Altieri, 1995; Gliessman, 2000).

Biological processes are fundamental for maintaining ecological-based crop production. Among the important processes that should be stimulated, we can cite nutrient cycling, nitrogen biological fixation and the processes associated with the thriving of spontaneous vegetation at specific times and places. Crop diversification associated to biological processes in cropping systems allows for the conservation of productive resources, as well as less dependence on industrial inputs, such as fertilizers and pesticides. Moreover, they might reduce the production costs and result in social benefits, such as a decrease in grower's vulnerability to the market prices and a better distribution of grower's income and food for the population throughout the year. One of the most widespread strategies for enhancing crop diversification over time is crop rotation. In these production systems, a sequence of different crops is cultivated in the same field in a finite time interval and this production sequence is repeated indefinitely. Besides the mentioned potential benefits, crop rotation helps to protect soil from eroding and makes it possible to use the land in another period during the year.

Among the ecological criteria and biological methods that can be used to define rotation plans, one might list the following (Gliessman, 2000):

a. avoid the growth of a crop immediately after a crop from the same botanical family has been harvested as many herbivores and pathogens only attack crops from a specific botanical family;
b. include green manure crops (usually a leguminous) for biological nitrogen fixation; and
c. include fallow periods, in which the spontaneous vegetation is left to thrive for a given period of time, thus contributing to the biological control of insects and improving soil biological, chemical and physical characteristics.

In this chapter, we focus on the scheduling of vegetable crop rotation. In this context, one must consider that vegetable crops embed many different species and cultivars, they present very different cropping times and a high specific cropping season due to their sensibility to climatic conditions. Vegetable crops are also more susceptible to pest attacks. Moreover, it is common practice to cultivate many different crops in the same area throughout the year. These characteristics make the rotation planning using ecological criteria (a)–(c) a complex task, which is why the use of support software is very useful, if not essential. In addition, for commercial reasons, vegetable crop growers must ensure a constant supply of products, sometimes to meet known demands, which is a characteristic that further complicates production planning.

## BACKGROUND

The idea of obtaining crop rotation plans with the aid of mathematical tools is not new to operation researchers. Indeed, as early as 1939, Kantorovich had already mentioned this problem as one that could be dealt with by linear programming (Kantorovich, 1960 – translated from the Russian original, dated 1939). The author proposes a mathematical model which selects the optimal partition of land in order to maximize production. Hildreth and Reiter (1951), in turn, in one of the first conferences on applications of operations research, stated that crop rotation can be beneficial as it can increase crop yields (due to the fact that different rotations produce distinct effects on the soil). Moreover, the authors had already mentioned that crop diversification provides some sort of security against the failure of production of a given crop in a given year in addition to the fact that by dividing the available land among different rotations, one might reduce the need for resources (such as water, labor, etc) as different crops also have distinct needs throughout the year. The authors use a list of pre-determined crop rotations, which greatly simplifies the problem.

Since these two seminal papers, crop rotation schedules have been obtained with the aid of (integer) linear optimization models such as those proposed by El-Nazer and McCarl (1986), Clarke (1989), Haneveld and Stegeman (2005), Detlefsen and Jensen (2007), Santos et al. (2008; 2010) and Costa et al. (2009). Other manuscripts on crop rotation concentrate on the description of decision-aid tools which somehow evaluate the benefits of a given crop rotation plan (Jones et al., 2003; Stöckle et al., 2003, Bachinger and Zander, 2006). Among this list of contributions, we are briefly going to describe those which have made use of some optimization approach in order to obtain solutions which maximize or minimize a given goal.

El-Nazer and McCarl (1986) assume that the yield of a crop depends only on the crops grown on the same piece of land in the previous years. Using this simplification, they are able to eliminate the need for pre-determining all possible feasible crop rotation schedules. Haneveld and Stegeman (2005) use the idea of crop succession requirements, which are expressed in terms of an infeasible sequence of crops. Detlefsen and Jensen (2007), in turn, take into account the fact that crop rotation influences the needs for nitrogen and the yield of the fields. Their model considers that the amount of land to be planted with each crop is given, for each year, which enables the formulation of the crop rotation problem as a transportation problem (Appa, 1973). In this and in the previous mentioned articles, the crop production times are assumed to be equal to one year, the available cropping area is homogeneous and the authors explicitly forbid infeasible crop sequences. Clarke (1989) considers crops with different production times and planting seasons of the year. However, the proposed model does not allow constraints concerning the feasible sequence of crops.

Sustainability constraints of the type described in the introduction have been used by Santos et al. (2008; 2010) and by Costa et al. (2009), who have dealt with crop rotation problems focusing on the sustainable production of vegetable crops. Production planning of vegetable crops is particularly complex as it generally involves a large number of crop species having different specific limitations regarding the planting seasons and highly varied production times and productivity. Santos et al. (2008) have considered the scheduling of crop rotations in order to avoid the spatial propagation of pests: crops of the same botanical family are not allowed to grow simultaneously in adjacent plots. The goal was to maximize production but no demand aspects were considered. Santos et al. (2010) have introduced the idea of a "supply problem" with the inclusion of known demands. Costa et al. (2009) have extended this previous work to include uncertainty considerations and to study the effect of allowing products to be stocked. Due to the perishable nature of the products, stocking was only allowed for a limited number of periods and with associated losses.

## MAIN FOCUS OF THE CHAPTER

In this chapter, we present a general modeling and solution framework for the three optimization problems associated with vegetable crop rotation scheduling mentioned in the last paragraph:

1. The crop rotation scheduling problem with known demands (Santos et al., 2010);
2. The crop rotation scheduling problem with perishable stocks (Costa et al., 2009);
3. The crop rotation scheduling problem with adjacency constraints (Santos et al., 2008).

Our idea here is to give a general framework for these studies in order to facilitate the task of modeling new similar problems and developing solution methodologies. We concentrate on the deterministic case. The reader who is interested in uncertainty considerations is refereed to (Costa et al., 2009).

The remainder of this chapter is organized as follows. Next, we present the concept of crop rotation in more detail and develop a crop rotation scheduling model, which respects ecologically-based criteria (a)-(c) and specific growing technical constraints. We then discuss each one of the above problems (1)-(3) and show how the crop rotation scheduling model can be seen as core to the mathematical formulations developed in all of these cases (and, certainly, to many other similar variants). Using the same modeling framework allows us to develop a similar solution methodology for the three cases. This methodology, which is based on the iterative generation of crop rotation schedules respecting both ecological and technical constraints, will also be described in detail. This chapter ends with a brief discussion on some computational results that indicate the efficacy of the proposed strategy, as well as some conclusions and suggestions for future research.

## THE CROP ROTATION SCHEDULING MODEL

In an agricultural production system, a crop rotation can be defined as the sequence of crops that should be planted, one after another, in a given area during a time interval. This time interval is called the length of the rotation and the idea of *rotation* comes from the fact that the same sequence is repeated indefinitely. For instance, the 3 year crop rotation corn-potatoes-potatoes indicate that one should grow corn, then one should grow potatoes twice, after which the rotation should be restarted with the growing of corn and so on.

Most of the articles mentioned earlier deal with grain crops and/or high latitude environments and, in these cases, each crop has a specific growing season and production time. However, in the context of vegetable crop production, the definition of interesting crop rotation plans has some particularities. Vegetable crops embed many species and cultivars which have different cropping times and usually have specific cropping seasons, due to their sensibility to climatic conditions. Thus the vegetable crop schedule must consider the different growing and production times of each crop, which greatly complicates the problem. In the case of the vegetable crop rotation schedule problems presented, we first define the time interval in which the sequence of crops will be repeated, say T years, which is divided into M periods. A crop rotation schedule is a planting calendar for each crop in the rotation.

Moreover, ecologically-based production systems impose additional constraints on the crop rotations. Among many possible sustainability practices, we concentrate on the following constraints that result from the ecological criteria (a)-(c) mentioned in the previous section:

1. *Botanic family*: crops belonging to the same botanic family can not be planted in sequence.
2. *Green manure*: crops for green manuring must be planted in each crop rotation.
3. *Fallow*: a minimum number of fallow periods must be present in each rotation.

As an example, let us consider a one year- crop rotation schedule, divided into twelve one-month periods (M = 12). There are three crops, X, Y and Z with production times of 5, 4 and 2 months, respectively. Crops X and Y belong to the same botanic family while Z is a crop for green manuring. Crop X can be planted between January (month 1) and July (month 7), while crops Y and Z can be planted all year round. Finally, let us assume that one green manure and one month of fallow (F) are mandatory per schedule. According to (i)-(iii) constraints, a feasible crop rotation schedule is depicted in Figure 1.

In the crop rotation schedule presented in Figure 1, crop X is planted in month 3 and occupies the land until period 7 (as its production time is 5 months). In period 8, crop Z is planted, staying on the land until period 9, when it is harvested. In period 10, crop Y is planted and its final harvesting occurs in period 1 (of the next production year). In period 2, there is one month fallow (F). Each crop respects its planting times and crops X and Y (which belong to the same botanic family) are not planted in sequence. Moreover, as desired, in each crop rotation a green manure crop (Z) is cultivated and there is a period of fallow (F).

As the planting of a crop in a given period implies one or more subsequent harvesting periods, with known productivities, one might easily convert the crop rotation schedule into a crop production schedule, which is a harvesting calendar indicating the amount (in mass, volume, units or other appropriate measure) of each crop which is being produced in each period.

Consider the crop rotation schedule in Figure 1. Assume that crop X has its first harvest three months after planting and other harvests occur monthly, until the final harvest, five months after the time of planting. Suppose the productions in months 3, 4 and 5 after planting are 1, 2 and 1 kg/m$^2$. Crop Y has no partial harvestings and its production is estimated at 3 kg/m$^2$ with the harvesting occurring in the last period of its production time, i.e., four months after planting. Having these data, and assuming the planting calendar in Figure 1 has been used in an area of 2 m$^2$, we can express the harvesting calendar (or production schedule) in Figure 2. The numbers in italics represent the partial harvests of each crop: 2, 4 and 2 kg of X in months 5, 6 and 7, respectively, and 6 kg of Y in month 1. Crop Z is a green manure.

In the remainder of this section, we present a linear model to describe all feasible crop rotation

*Figure 1. A feasible crop rotation schedule*

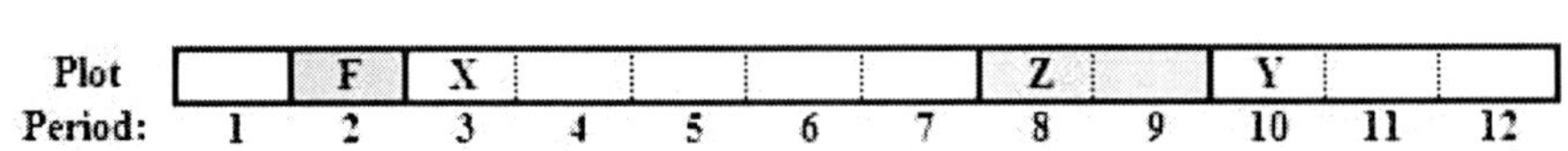

*Figure 2. Production schedule for a 2 m² cropping area*

| Plot | 6 | F | X | | 2 | 4 | 2 | Z | | Y | | |
|---|---|---|---|---|---|---|---|---|---|---|---|---|
| Period: | 1 | 2 | 3 | 4 | 5 | 6 | 7 | 8 | 9 | 10 | 11 | 12 |

schedules according to the ecological criteria (i)-(iii). In order to do this, we introduce the following notation:

M number of periods in each crop rotation (in a predefined unit of time);
C set of crops that can be selected for planting, except the ones associated to green manuring;
G set of crops that can be selected for green manuring;
N cardinality of $C \cup G$;
*NF* number of the botanic families;
*F(p)* set of crops of botanic family *p*, *p* = 1, …, *NF*;
$t_i$ production time of crop *i*, including the time estimated for preparing the soil and harvesting;
$I_i$ set of periods in which crop *i* can be planted;
N + 1 hypothetical crop for fallow. We adopt $I_{N+1} = \{1,\dots, M\}$.

We also define the decision variables:

$x_{ij} = 1$, if crop *i* is planted in period *j*, 0 otherwise, $i = 1,\dots, N + 1, j \in I_i$.

The model that describes all feasible crop rotation schedules (CRS model) can thus be written as:

$$\sum_{i=1}^{N+1} \sum_{r=0}^{t_i - 1} x_{i,j-r} \leq 1, \ j = 1,\dots, M \tag{1}$$

$$\sum_{i \in F(p)} \sum_{r=0}^{t_i} x_{i,j-r} \leq 1, \tag{2}$$
$$p = 1,\dots, NF, j = 1,\dots, M$$

$$\sum_{i \in G} \sum_{j \in I_i} x_{ij} = 1 \tag{3}$$

$$\sum_{j=1}^{M} x_{N+1,j} = 1 \tag{4}$$

$$x_{ij} \in \{0,1\}, \ i = 1,\dots, N + 1, j \in I_i \tag{5}$$

where, in constraints (1)–(2), $j - r$ is replaced by $j - r + M$, whenever it is less or equal to zero.

Constraints (1) ensure that at most one crop is cultivated at each period while constraints (2) forbid crops belonging to the same botanic family from being cultivated one immediately after the other. Constraints (3) and (4) force the planting of green manure and the existence of fallow. The model above allows for solutions in which a given period has no associated crop. This is equivalent to allowing more times of fallow in each schedule. Finally, constraints (5) define the used variables.

In model (1)–(5), the numbers of mandatory green manures and fallows have been set at one. Depending on the crop rotation length (T), these numbers might need to be increased. In this case, consider *g* the number of crops in G that must be planted per crop rotation, and *f*, the number of fallows per crop rotation. Constraints (3) and (4) can then be written as:

$$\sum_{i \in G} \sum_{j \in I_i} x_{ij} = g \tag{6}$$

$$\sum_{j=1}^{M} x_{N+1,j} = f \tag{7}$$

One might also need to ensure a minimum time delay between the planting of two green manures or two fallows. This can be done with the following additional constraints:

$$\sum_{i\in G}\sum_{r=0}^{tg} x_{i,j-r} \leq 1,\ j = 1,\ldots,M \quad (8)$$

$$\sum_{r=0}^{tf} x_{N+1,j-r} \leq 1,\ j = 1,\ldots,M \quad (9)$$

where *tg* and *tf* are the minimum time delays for green manures and fallows.

## SOME CROP ROTATION OPTIMIZATION PROBLEMS

Many optimization problems are associated with obtaining crop rotation. In this section, we describe three of them and show how, in each of the cases, constraints (1)–(5) appear as a core part of the model.

### The Crop Rotation Scheduling Problem with Known Demands (CRS-D Problem)

There is increasing interest in the study of vegetable crop production as it represents an important alternative income for family farmers. These farmers usually own small plots of land and, therefore, have small associated productions. These productions are usually discontinuous, mainly due to climatic conditions, which restrict the growth of some species in some regions or lead to significant reductions in the yields. These small and discontinuous productions do not justify investments in cleaning, packing and logistic structures necessary to make production available to customers.

In order to cope with this problem, an increasing number of farmers are gathering together in collective movements, such as cooperatives, so that the combined productions justify the investments in the necessary infrastructure. These cooperatives have the role of deciding the amount of land to be planted for each crop so that the production supplies the demand. In addition, the cooperatives must decide when such productions should occur, by taking into consideration differences between soil and climatic conditions for each one of the farmers' land, which might influence the crop yields and even forbid farmers to grow some crops.

Santos et al. (2010) deal with this problem with the aim of obtaining crop rotation plans that meet demands for vegetable crops, given a set of heterogeneous (in respect to soil and climatic conditions) pieces of land. Because of this, the set of growing crops and the associated productivity might differ from one cropping area to another. This problem is written as a linear optimization model, in which each variable is associated with a plot where a crop rotation schedule is given by constraints (1)–(5).

### Mathematical Model

Let $C$ be a set of crops with a positive demand and $k$=1,…,$L$ be the available cropping areas. Due to possible heterogeneities in climate and soil conditions, the set of ordered crops and the associated productivity might differ from one cropping area to another. Taking this into account, for each cropping area $k$, we have a specific crop rotation schedule model (1)–(5), based on the set $C_k$ of crops that can be planted in cropping area $k$ together with crops for green manure available to area $k$ ($G_k$).

In order to write a mathematical model for the crop rotation scheduling problems with known demands (CRS-D problem), consider the following parameters:

$d_{ij}$ the demand of crop $i$ in period $j$, given in appropriate units (kg, pack etc.);
$A_k$ the size of cropping area $k$, $k$=1,…,$L$;
$S_k$ the number of feasible crop schedules associated to cropping area $k$, i.e., the number of all possible solutions $(\hat{x}^s_{ijk})$ of (1)–(5), restrict to $i \in C_k \cup G_k$;
$r_{ks}$ income per m$^2$ when crop rotation schedule $s$ is used in area $k$.

A crop rotation schedule $(\hat{x}^s_{ijk})$ defines which and when crops are planted in area $k$. Once a planting schedule is fixed, one can determine the crop productions. For this, the following additional parameters are defined:

$o_i$ the number of periods between planting and the first harvesting of crop $i$;
$p_{irk}$ the production (units according to the demand) per m$^2$ of the $r^{th}$ harvesting of crop $i$ in the area $k$, for $r$=1,…,$t_i$-$o_i$-1.

Therefore, $p_{irk}\hat{x}^s_{ijk}$ is the outcome per m$^2$ of the $r^{th}$ harvesting of crop $i$ in area $k$ at period $j+o_i+r$ due to planting according to the crop rotation schedule $s$. Similarly, $p_{irk}\hat{x}^s_{i,j-o_i-r,k}$ is the outcome per m$^2$ of the $r^{th}$ harvesting of crop $i$ in area $k$ at period $j$. Despite the fact that we assume that $o_i$ and $t_i$ are independent of area $k$, one can easily extend the model for this case.

Defining variables:

$\lambda_{ks}$ size of the plot planted according to crop rotation schedule $s$ in area $k$.

We can write the CRS–D problem as:

$$\max \sum_{k=1}^{L}\sum_{S=1}^{S_k} r_{ks}\lambda_{ks} \tag{10}$$

subject to

$$\sum_{k=1}^{L}\sum_{s=1}^{S_k}\sum_{r=0}^{t_i-o_i-1} p_{irk}\hat{x}^s_{i,j-o_i-r,k}\lambda_{ks} \geq d_{ij},\ i \in C, j = 1,\dots, M, \tag{11}$$

$$\sum_{s=1}^{S_k} \lambda_{ks} \leq A_k,\ k = 1,\dots, L, \tag{12}$$

$$\lambda_{ks} \geq 0,\ s=1,\dots,S_k,\ k = 1,\dots, L. \tag{13}$$

where $j - o_i - r$ is replaced by $j - o_i - r + M$, whenever it is less or equal to zero.

The aim of model (10)–(13) is to maximize the return obtained by the crop rotation schedules for the $L$ areas. Given $c_{ij}$, the unit price (per kg, pack etc) of crop $i$ at period $j$, the income $r_{ks}$ can be expressed as:

$$r_{ks} = \sum_{i \in C_k}\sum_{j=1}^{M}\sum_{r=1}^{t_i-o_i-1} c_{ij}p_{irk}\hat{x}^s_{i,j-o_i-r,k} \tag{14}$$

Constraints (11) guarantee that the demands for the crops at all the periods are met while constraints (12) ensure that only the available lands are used.

In the section labeled *Solution Methodology*, we describe a procedure based on column generation to solve model (10)–(13).

## Example 1

Consider a one-year crop rotation schedule divided into twelve one month-periods ($M$ = 12) and 8 hypothetical crops, whose data are listed in Table 1. For each crop $i$, Table 1 presents the botanic family ($F_i$) to which it belongs, the set of possible planting periods ($I_i$), the production time ($t_i$), the number of periods before partial harvesting ($o_i$) and partial harvesting outcomes, $p_{ir}$. For example, crop 1 belongs to botanic family 1, it can be planted from month 1 to month 7, and it consumes 4 months from planting to final harvest. As $o_i = 1$ and $t_1$-$o_1$-1= 2, crop 1 has 1 period after the planting period and before the first harvesting period, and 2 periods of harvests. Table 1 also gives the demands for crops 1–6 at each period. Crops 7 and 8 are for green manuring and have no demand.

Consider 3 cropping areas with sizes $A_1 = 6$, $A_2 = 4$ and $A_3 = 5$ m$^2$ and assume that crop 1 cannot be planted in area 1, crops 2, 3 and 5 are not planted in area 2 while in area 3 one can not grow crops 2 and 4. In terms of our notation: $C_1 = \{2, \ldots, 8\}$, $C_2 = \{1, 4, 6, 7, 8\}$ and $C_3 = \{1, 3, 5, \ldots, 8\}$. Moreover, due to the different soil quality and climatic conditions, we assume that the productivity in areas 1, 2 and 3 are 110%, 100%, and 80% of the basis values presented in Table 1, respectively, in such a way that the productivities of crop 6, for example, are, in area 1 $p_{611} = 2.2$, $p_{621} = 1.1$, $p_{631} = 0$; in area 2, $p_{612} = 2$, $p_{622} = 1$, $p_{632} = 0$ and, in area 3, $p_{613} = 1.6$, $p_{623} = 0.8$ and $p_{633} = 0$. Of course, when dealing with real-word data, some pieces of land might be more productive for some crops and less productive for others.

Forcing one period of fallow (represented as crop 9) the solution presented in Figure 3 is the one that maximizes the total non-weighted production ($c_{ij}$=1, for all $i$ and $j$). Figure 3 uses the notation already introduced in Figure 2 where the italic figures correspond to the partial harvests.

In the optimal solution, the sizes of cropping areas 1 and 2 have been divided into two plots with different crop rotation schedules while the size of area 3 has been divided into three plots. The values of variables are: $\lambda_{11} = 1$, $\lambda_{12} = 5$, $\lambda_{21} = 1$, $\lambda_{22} = 3$, $\lambda_{31} = 2.5$, $\lambda_{32} = \lambda_{33} = 1.25$. Note that additional fallow periods may arise, as in the second schedule (plot of 5 m$^2$) for area 1 in period 3.

*Table 1. Problem data: crop data and demands for 8 hypothetical crops*

| Crop | Crop data | | | | | | | Demand per period | | | | | | | | | | | |
|---|---|---|---|---|---|---|---|---|---|---|---|---|---|---|---|---|---|---|---|
| | $F_i$ | $I_i$ | $t_i$ | $o_i$ | $p_{i1}$ | $p_{i2}$ | $p_{i3}$ | **1** | **2** | **3** | **4** | **5** | **6** | **7** | **8** | **9** | **10** | **11** | **12** |
| **1** | 1 | 1−7 | 4 | 1 | 1 | 1 | | | | 1 | 1 | | | | | | | | |
| **2** | 1 | 1−9 | 3 | 1 | 1 | | | | | | 1 | | | | | | | | |
| **3** | 2 | 3−11 | 3 | 0 | 1 | 1 | | 2 | | | 1 | 1 | | | 1 | 1 | | | |
| **4** | 2 | 2−11 | 4 | 0 | 1 | 2 | 1 | 4 | | | | 8 | 6 | 8 | | | 2 | 8 | 10 |
| **5** | 3 | 7−12; 1−5 | 3 | 0 | 1 | 1 | | | | 1 | 1 | | | | | | | | |
| **6** | 3 | 3−10 | 3 | 0 | 2 | 1 | | | | | 2 | 2 | 2 | 1 | 1 | 2 | 2 | 2 | 2 |
| **7** | 4 | 2−12 | 3 | 1 | 1 | | | | | | | | | | | | | | |
| **8** | 4 | 8−12; 1−6 | 2 | 0 | 1 | | | | | | | | | | | | | | |

*Figure 3. Crop production schedules for the three cropping areas*

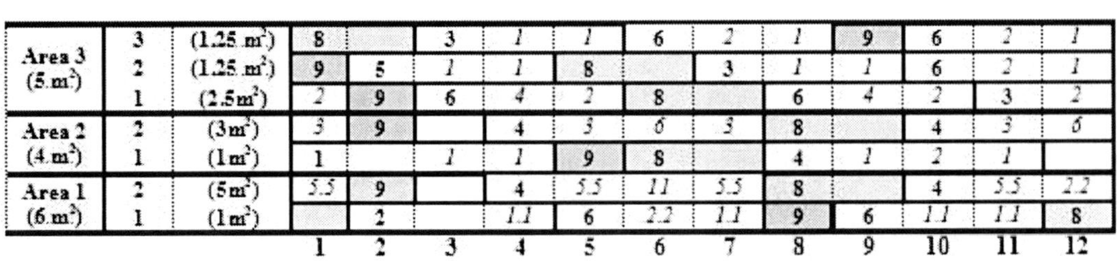

## Model Discussion and Extensions

In this subsection, we discuss some practical issues for the CRS–D problem and how they could be integrated into the model.

### Unmet Demand and Upper Bounds on Demands

The presence of ecological-based and technical constraints makes it harder to predict whether a feasible crop rotation schedule that supplies a given demand exists or not. In practice, however, one might be interested in solutions that minimize the unmet demand or obtain solutions presenting a trade-off between the gains associated with the harvests and the unmet demands. This multi-objective facet of the problem can be somehow incorporated into model (10)–(13) by changing the objective function and by using auxiliary variables $\theta_{ij}^{-}$ representing the unmet demand of crop $i$ at period $j$. To limit the amount of the unmet demand, one must bound the value of these variables. Model (15)-(18) formalizes these concepts:

$$\max \sum_{k=1}^{L} \sum_{s=1}^{S_k} r_{ks}\lambda_{ks} - \sum_{i\in C} \sum_{j=1}^{M} q_{ij}\theta_{ij}^{-} \tag{15}$$

subject to

$$\sum_{k=1}^{L} \sum_{s=1}^{S_k} \sum_{r=0}^{t_i-o_i-1} p_{irk}\hat{x}^{s}_{i,j-o_i-r,k}\lambda_{ks} \geq d_{ij} - \theta_{ij}^{-},\ i \in C, j = 1,\ldots, M, \tag{16}$$

$$\sum_{s=1}^{S_k} \lambda_{ks} \leq A_k,\ k = 1,\ldots, L, \tag{17}$$

$$\lambda_{ks} \geq 0,\ \theta_{ij}^{-} \geq 0,\ s=1,\ldots,S_k,\ k = 1,\ldots, L. \tag{18}$$

where $\theta_{ij}^{-}$ is the amount of unmet demand of crop $i$ at period $j$, and $q_{ij}$ is an associated penalty.

Besides allowing part of the demand to be unmet, it might be convenient to establish an upper bound on the exceeding production. The motivation for this would be to limit the production to levels that can be absorbed by the farmers′ clients. This can easily be done by imposing upper bounds to the slack variables of constraints (16). In the computational results section, we analyze some computational experiments based on real-world data for vegetable crops with the model (15)–(18).

### Considerations on the Number of Plots

Model (10)–(13) imposes no restriction on the number of plots that a cropping area might be divided into, allowing for solutions with a very large number of plots. In order to reduce the occurrence of these solutions, a fixed cost, say $z_k$, can be given for $\lambda_{ks} > 0$, i.e., whenever a new plot is considered. The original objective function is then replaced by:

$$\max \sum_{k=1}^{L} \sum_{s=1}^{S_k} (r_{ks}\lambda_{ks} - z_k\delta_{ks}) \tag{19}$$

and the following constraint is included:

$$\lambda_{ks} \leq A_k \delta k_{s,}\ \delta ks_{=\{}0,1\}, s=1,\ldots,Sk,\ k = 1,\ldots, L \tag{20}$$

The fixed cost $zk$ is selected according to the desired trade-off between the conflictive goals of minimizing the number of plots in the solution and maximizing income. Formulation (19), subject to constraints (11)–(13) and (20) is no longer a linear programming model but a more complex mixed-integer problem. The reader is referred to Santos et al. (2010) for a discussion on this subject.

### The Crop Rotation Scheduling Problem with Stocks (CRS-S Problem)

Vegetable crops are usually sent to packing houses after harvesting. In the packing houses, the vegetable crops are cleaned, packed and stored in cold chambers until delivery. Due to the high perishableness of many vegetable crops, however, they can only remain in stock for a limited amount of time and, even then, with losses. The possibility of stocking defines a more general and, therefore, more complex problem. We call this problem, which clearly is a general case of the CRS-D presented earlier, the crop rotation scheduling problem with stocks (CRS-S problem). In this new situation, the demand is no longer only supplied by fresh-harvested crops.

Formulation (10)–(13) and its variants do not model the possibility of keeping some of the harvested crops in stock in order to meet future demands. In order to overcome this limitation, we define the following variables:

- $I^w_{ij}$ stock of crop $i$ available at period $j$ which has been produced $w$ periods previously, $w=0,\dots,G_i$;
- $\eta^w_{ij}$ quantity of crop $i$ delivered at period $j$ using stocks $I^w_{ij}$, $w=0,\dots,G_i$;
- $\theta^+_{ij}$ quantity of crop $i$ delivered at period $j$ exceeding demand $d_{ij}$;
- $\theta^-_{ij}$ unmet demand $d_{ij}$.

In both sets of variables $I^w_{ij}$ and $\eta^w_{ij}$, index $w$ indicates the age of the product. When $w = 0$, it indicates a product that has just been harvested. In each period, the available stocks (of different ages) can either be used to meet the current demand or be held in stock for one additional period. To model the fact that each product can only be stocked for a limited amount of time, it suffices to limit index $w$ to appropriate values. For example, if a crop can be stocked for 3 periods, only the variables with indexes $w = 0,\dots,4$ should exist, index 4 being associated with a spoiled product (final losses). In the general case, parameter $G_i$ represents the number of periods a given crop $i$ can be held in stock.

The total harvested quantities of crop $i$, at period $j$, is modeled by variables $I^0_{ij}$ (Production balance). This quantity can either be used to supply the demand of period $j$, by means of variables $\eta^0_{ij}$ or be stocked, in which case it will appear (discounted the losses) as variable $I^1_{i,j+1}$ in the next period. The same is valid for values of $w$ other than zero (Stock balance). The quantities $\eta^w_{ij}$, for $w=0,\dots,G_i$, are used to supply the current demand (Demand supply). Figure 4 illustrates this general idea of the production balance.

Variables $\theta^+_{ij}$ and $\theta^-_{ij}$ model the unmet demand or the quantity over demand, respectively. With the following additional notation:

- C set of crops, not including those associated with green manure neither fallow;
- S the number of feasible crop rotation schedules for the cropping area;
- A the size of the cropping area;
- $c_{ij}$ unit income of selling ordered by demand $d_{ij}$ of the crop $i$ at period $j$;
- $c^+_{ij}$ unit income of selling non ordered by demand $d_{ij}$ of the crop $i$ at period $j$ ;
- $r_i$ fraction lost of the quantity already in stock of crop $i$, if it is held for one more period.

For the sake of presentation simplicity, we now consider a single cropping area. Taking this into account, index $k$ in parameters $p_{irk}$ and variables

*Figure 4. Relations among variables for crop i at period j*

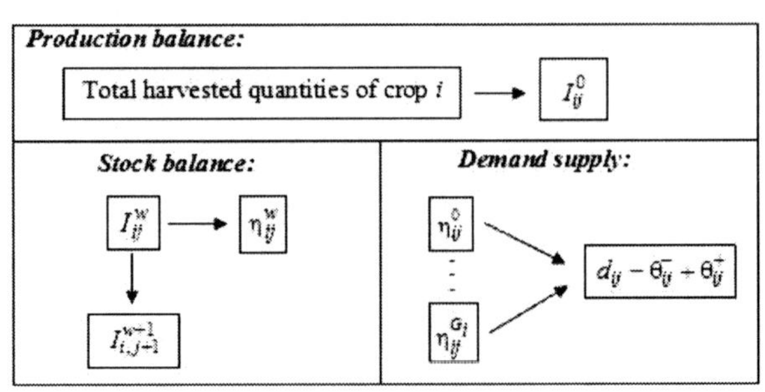

$\lambda_{sk}$ are no longer needed. The generalization for the case of many heterogeneous cropping areas is straightforward.

A model for the CRS-S problem can be written as:

$$\max \sum_{i \in C} \sum_{j=1}^{M} c_{ij}(d_{ij} - \theta_{ij}^-) + c_{ij}^+ \theta_{ij}^+ \quad (21)$$

subject to

$$\sum_{s=1}^{S} \sum_{r=0}^{t_i - o_i - 1} p_{ir} \hat{x}_{i,j-o_i-r}^s \lambda_s = I_{ij}^0, i \in C, j = 1,..., M, \quad (22)$$

$$\sum_{s=1}^{S} \lambda_s \le A, \quad (23)$$

$$(1 - r_i)(I_{ij}^w - \eta_{ij}^w) = I_{i,j+1}^{w+1}, \; i \in C \; w=0,...,G_i, j = 1,..., M, \quad (24)$$

$$\sum_{w=0}^{G_i} \eta_{ij}^w = d_{ij} - \theta_{ij}^- + \theta_{ij}^+, \; i \in C, j = 1,..., M, \quad (25)$$

$$\lambda_s \ge 0, \; s=1,...,S, \quad (26)$$

$$\theta_{ij}^+ \ge 0, \; \theta_{ij}^- \ge 0, \; \eta_{ij}^w \ge 0, \; I_{ij}^w \ge 0, \; I_{ij}^{G_i+1} \ge 0, \; i \in C, j = 1,..., M, w=0,...,G_i. \quad (27)$$

where, $j - o_i - r$ is replaced by $j - o_i - r + M$, whenever it is less or equal to zero.

The objective function (21) maximizes incomes, which can be obtained by selling products to meet the ordered demand or by selling the extra production (usually at a lower price). Constraints (22) define $I_{ij}^0$ as the harvested quantities of each crop $i$ in period $j$. Constraint (23) makes sure that only the available land is used.

Constraints (24) are the stock balance constraints and include a loss-factor for every additional period a product is kept in stock. These constraints can be read as follows: the quantity of stocked production of crop $i$ in period $j$ produced $w$ periods previously ($I_{ij}^w$) can either be used to meet the demand of period $j$ ($\eta_{ij}^w$) or remain in stock in the following period ($I_{i,j+1}^{w+1}$). For $w=G_i$, the variable $I_{ij}^{G_i+1}$ indicates losses (all remaining stock of crop $i$ after $G_i$ periods is considered spoiled). Constraints (25) control the demand supply: the demand of production of crop $i$ at period $j$, $d_{ij}$, can be met from fresh harvested products ($\eta_{ij}^0$) or from stocked products with $w \le G_i$ periods of age, ($\eta_{ij}^w$, $w \ge 1$).

In constraints (24) the fraction lost at the end of a period is independent of the period itself. This assumption can be easily generalized by considering the fraction of losses as a parameter dependent on the current period $w$, i.e., define $r_{iw}$ as the fraction lost in product $i$, at the end of period $w$, if the product is kept one period longer in stock.

## Example 2

In order to illustrate the impact of stocks, consider A = 200m$^2$ (total cropping area) and 5 hypothetical crops (crop 5 is used as green manure). Again we use a 1 year crop rotation scheduling, divided into 12 months and one fallow period (denoted by crop 6) per schedule. Table 2 presents the technical data and known demands for each one of the crops, including the number of periods in which crop $i$ can be held in stock ($G_i$) and the percentage loss ($r_i$) incurring in crop $i$ if it is stocked for one more period. For example, in the row associated with crop 1, we can read that it can be stocked for at most three periods ($G_i$= 3) but for each period in stock, 20% of the production stocked will be lost ($r_1$= 0.2).

We consider two cases: (a) no stock is allowed, and (b) stocking (under the conditions described in Table 2) is allowed. In both cases, we assumed $c_{ij}$=1 and $c_{ij}^+ = 0$, for all $i$ and $j$.

*Table 2. Problem data: crop data and demands for 5 hypothetical crops*

| crop | Crop data | | | | | | | | | Demand per period | | | | | | | | | | | |
|---|---|---|---|---|---|---|---|---|---|---|---|---|---|---|---|---|---|---|---|---|---|
| | $F_i$ | $I_i$ | $t_i$ | $o_i$ | $p_{i1}$ | $p_{i2}$ | $p_{i3}$ | $r_i$ | $G_i$ | 1 | 2 | 3 | 4 | 5 | 6 | 7 | 8 | 9 | 10 | 11 | 12 |
| 1 | 1 | 11-12;1−7 | 4 | 1 | 1 | 1 | | 0.2 | 3 | | | | | 10 | 10 | 10 | 10 | | | | |
| 2 | 1 | 1−9 | 3 | 1 | 1 | | | 0.1 | 2 | | | | | 20 | 20 | 20 | 40 | 40 | 40 | | |
| 3 | 2 | 4−12 | 3 | 0 | 1 | 1 | | 0.2 | 1 | 10 | 10 | 20 | 20 | 20 | | | | | | | |
| 4 | 2 | 1−11 | 4 | 0 | 1 | 2 | 1 | 0.4 | 2 | 10 | | | | 10 | 10 | 10 | | | | 10 | 10 |
| 5 | 3 | 1−12 | 2 | 0 | 1 | | | | | | | | | | | | | | | | |

In case (a), 8 plots are in the optimal solution with sizes $\lambda_1 = \lambda_2 = 10$, $\lambda_5 = \lambda_6 = \lambda_8 = 20$, $\lambda_3 = \lambda_4 = \lambda_7 = 40m^2$. The percentage of unmet demand is 1.9% and production over demand is 5.7%. In case (b), 5 plots are in the optimal solution, with sizes $\lambda_1 = 10$, $\lambda_2 = 35$, $\lambda_3 = \lambda_5 = 40$ and $\lambda_4 = 46.9\ m^2$, and corresponding crop rotation schedules are depicted in Figure5. The percentage of unmet demand is 0.95% and the extra production is 6.1%.

The fewer amount of unmet demands and the increase in extra production shows the greater flexibility introduced by the presence of stocks. Once product holdings are allowed, part of the production in some periods is stocked to meet the demand of the following periods. An illustrative example is the case of crop 1 in the solution shown in Figure 5. In fact, crop 1 is grown in plot 2 (of 35 $m^2$) in period 3, and its harvestings are forecasted to be 35 kg in period 5 and 35 kg in period 6. The demand of crop 1, as shown in Table 2, is of 10 kg for each one of periods 5 to 8.

To meet demands, one could proceed as follows: from the 35 kg harvested in period 5, 10 kg are immediately used to supply the demand of the period, 21.1 kg are sold as extra demand and 3.90 kg are kept in stock. Due to losses, only 80% of this amount is available at period 6 (3.125 kg), which associated with the harvest of the crop in the period, gives a total availability of 38.125 kg (35 kg with age zero and 3.125 kg one-period old). From this amount, 10 kg are used to supply the demand of the period (priority is given to the older product) and 28.125 kg remain in stock yielding an availability of 22.5 kg of one-period old product at period 7. The demand of 10 kg is therefore supplied and 12.5 kg of product is held in stock and appears as 10 kg of a two-period old product that is used to supply the demand of period 8.

**The Crop Rotation Scheduling Problem with Adjacency Constraints (CRS-A Problem)**

Consider a cropping area divided beforehand in plots. Since many herbivores and pathogens attack crops of the same botanic family, the first ecological criterion (a) listed in the Introduction Section can be seen in a spatial manner and one might therefore be interested in avoiding the simultaneous growth of crops belonging to the

*Figure 5. Optimal crop rotation schedules for case (b):stock allowed*

| | | | | | | | | | | | | | |
|---|---|---|---|---|---|---|---|---|---|---|---|---|---|
| 5 | (40m²) | | | 6 | | | 2 | | | | 5 | | |
| 4 | (46.9m²) | | | 2 | | | | 2 | | | | 6 | 5 |
| 3 | (40m²) | 2 | | | 3 | | | 6 | 2 | | | 5 | |
| 2 | (35m²) | | | 1 | | | | 6 | 5 | | | | 3 |
| 1 | (10m²) | | 6 | | 4 | | | | 5 | | 4 | | |
| | Period: | 1 | 2 | 3 | 4 | 5 | 6 | 7 | 8 | 9 | 10 | 11 | 12 |

same botanic family in adjacent plots. This gives rise to the following constraint:

*Spatial neighborhood:* crops belonging to the same botanic family are not cultivated simultaneously in adjacent plots.

In CRS–D and CRS–S problems, adjacencies among plots are not taken into account, since they are not known *a priori* but rather defined during the optimization process. For these problems, decisions are made so that the yields meet the expected demand. In the crop rotation scheduling problem with adjacency constraints (CRS-A problem), the goal is to maximize land occupation while respecting ecological criteria (i)–(iv). In this case, it is necessary to know the neighborhood of each plot in the cropping area.

A cropping area divided into $L$ plots can be represented by a planar graph G(V, A), in which a node in V corresponds to a plot and an edge ($u$, $v$) belongs to A if and only if plots $u$ and $v$ are adjacent. Figure 6(a) depicts a cropping area divided into 5 plots and Figure 6(b) shows the associated graph G(V, A).

**The Mathematical Model**

In order to present a mathematical model for the CRS–A problem, consider in addition to the parameters previously defined for the crop rotation schedule for one plot, the connected planar graph

*Figure 6.(a): Cropping area divided into five plots, and (b): the associated planar graph G(V, A)*

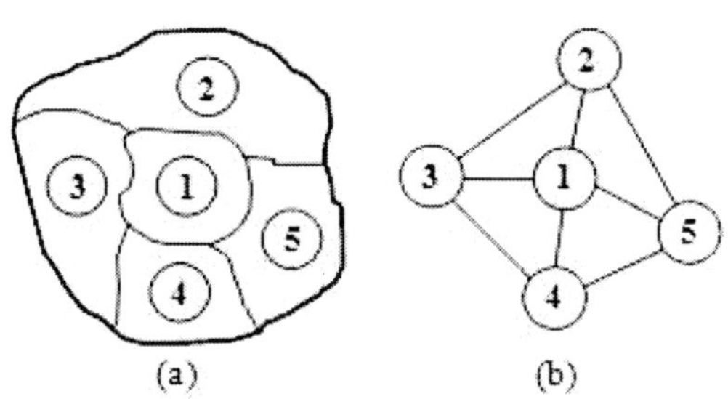


G(V, A) associated to the cropping area and the following decision variables:

$x_{ijk}$= 1, if crop $i$ is planted at period $j$ in plot $k$, 0 otherwise, $i$ = 1,...,$N$ + 1, $j \in I_i$, $k$ = 1,..., $L$.

For each plot $k$, the decisions of planting crop $i$ in period $j$ are restricted to constraints (i)-(iii), that is, the decisions are restricted to the crop rotation scheduling problem (1)–(5). Moreover, the crop rotation schedules for adjacent plots need in addition to ensure ecological constraint (iv) for adjacent plots.

A model for the CRS–A problem can be thus written as:

$$\max \sum_{k=1}^{L} \sum_{i \in C} \sum_{j \in I_i} t_i x_{ijk} \tag{28}$$

subject to

$$\sum_{i=1}^{N+1} \sum_{r=0}^{t_i - 1} x_{i,j-r,k} \leq 1, \quad j = 1, ..., M, k = 1, ..., L \tag{29}$$

$$\sum_{i \in F(p)} \sum_{r=0}^{t_i} x_{i,j-r,k} \leq 1, \quad p = 1..NF, j = 1, ..., M, k = 1, ..., L \tag{30}$$

$$\sum_{i \in G} \sum_{j \in I_i} x_{ijk} = 1, k = 1, ..., L \tag{31}$$

$$\sum_{j=1}^{M} x_{njk} = 1, k = 1, ..., L \tag{32}$$

$$\sum_{i \in F(p)} \sum_{r=0}^{t_i - 1} (x_{i,j-r,u} + x_{i,j-r,v}) \leq 1, \quad p = 1, ..., NF, j = 1, ..., M, (u, v) \in A \tag{33}$$

$$x_{ijk} \in \{0,1\},\ i = 1,\ldots, N+1, j \in I_i, k=1,\ldots,L. \quad (34)$$

where, as usual, in (29), (30) and (33), $j - r$ is replaced by $j - r + M$, whenever it is less or equal to zero.

The objective function (28) gives the total occupation of the plots in each crop rotation schedule. The variables related to green manuring crops and fallow periods were excluded from the objective function. Constraints (29)–(32) are analogous to constraints (1)–(5), which define a feasible crop rotation schedule, now for a plot $k$ known a priori: constraints (29) ensure that two crops cannot be grown simultaneously, constraints (30) prevent crops from the same botanic family being planted in sequence, and constraints (31)-(32) force the planting of a green manure crop and the presence of one fallow in each crop rotation.. Constraints (33) prohibit crops of the same family being grown at the same time in adjacent plots.

## Example 3

To illustrate the CRS-A problem, we consider a cropping area divided into 5 plots with a spatial configuration as illustrated in Figure 6. Each rotation is two years long divided into ten days (M = 72 periods). In this case, we used real-world data for the set of 28 crops from 10 botanical families obtained in a non-commercial organic vegetable garden in Viçosa - Brazil. The data for the 28 crops can be found in Table A.1 in the Appendix. A fallow period of 3 periods (1 month), represented as crop 29, was established for each crop rotation schedule. Figure 7 shows an optimal crop rotation schedule for the 5 plots over 24 months. In the scheduling for each plot, the bold number represents the crop number (for example, crop 1 is lettuce) and indicates its planting date.

It can be seen in Figure 7 that constraints (i)–(iv) are fulfilled: for each plot, the corresponding crop rotation schedule associated with the core problem (1)–(5) is respected. Moreover, no crops of the same botanic family are simultaneously cultivated in adjacent plots.

### Cropping Area with Special Spatial Configuration

For the CRS–A problem, there are some characteristics of the spatial configuration of the cropping area that can be explored in order to obtain feasible or optimal solutions more easily. Consider G(V, A) the connected planar graph of the cropping area divided into L plots. Suppose V is partitioned into $n$ independent subsets (i.e., there is no edge linking vertices in an independent subset). For example, set V of graph G given in Figure 6(b) can be partitioned into three independent sets: {1}, {2, 4} and {3, 5}.

If a feasible solution for the CRS–A problem exists for a complete graph $K_n$ (a graph with $n$ vertices and one edge for all pair of nodes), then a feasible solution for G can be constructed from this solution as follows. As any two vertices of an independent subset of V are not adjacent to each other, the same schedule can be assigned to all plots of this subset without violating adjacency constraints (33). We obtain a feasible solution

*Figure 7. Optimal crop rotation schedule for 5 plots*

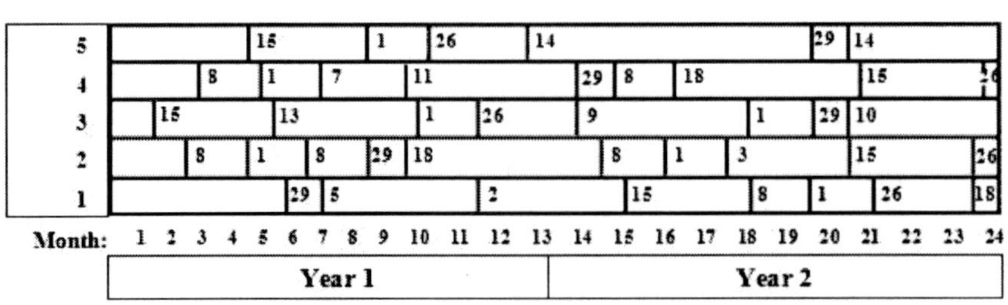

to graph G, assigning a schedule of one of the vertices of $K_n$ to all the plots of each independent subset of V so that no schedule is attributed to more than one subset.

For instance, consider any feasible crop rotation scheduling to graph $K_3$ and graph G in Figure 6(b). We can assign the schedule of plot 1 in $K_3$ to plot 1 in G; the schedule of plot 2 in $K_3$ to plot 2 and 4 of G; and finally, the schedule of plot 3 in $K_3$ to plot 3 and 5 of G. However, even though we have the optimal scheduling for $K_3$, this solution may not be the optimal to G (Santos, 2009). Moreover, the non-existence of a solution for $K_3$ does not imply that there is no solution for graph G. Nevertheless, in the next section, we present some situations where this strategy produces an optimal solution to the original problem.

**Cropping Areas Divided Into Parallel Plots or Checkerboard Patterns**

A simple and common situation found in practice is where the cropping area is divided into parallel plots or checkerboard patterns (see Figure 8(a)). In these cases, the plots can be divided into just two independent subsets. Santos et al. (2008) prove it is then possible to construct an optimal solution from the solution for only two adjacent plots. Consider, for example, a cropping area with a checkerboard pattern, as shown in Figure 8, and an optimal solution for two adjacent plots (that is, for graph $K_2$). An optimal crop rotation scheduling for the 9 plots is obtained by assigning one of the schedules of $K_2$ to plots 1, 3, 5, 7 and 9 (colored in Figure 8(b)), and the other schedule of $K_2$ to the plots 2, 4, 6 and 8.

*Figure 8. The cropping area divided into L = 9 plots*

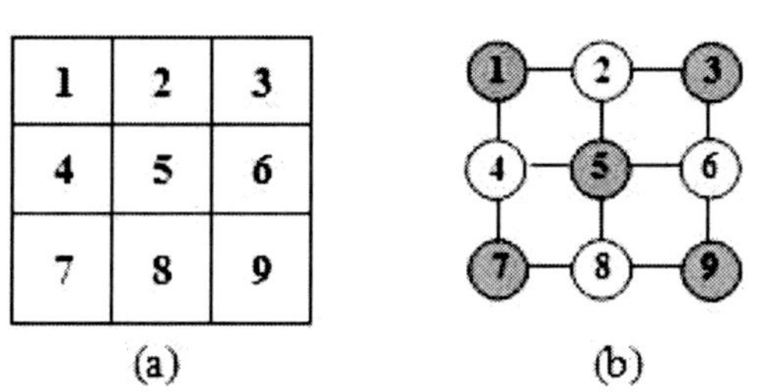


**Model Reformulation**

The mathematical formulation (28)–(34) for the CRS-A problem has, in practice, a very large number of constraints and variables which limit the use of exact solution methods. However, the special structure found in (29)–(33) can be exploited. Constraints (29) to (32) are associated with independent plots and are only linked by constraints (33), suggesting the use of decomposition methods.

The problem of finding a crop rotation schedule for $L$ plots can also be seen as one of choosing, among all feasible schedules for a plot, $L$ schedules satisfying plot adjacency constraints (33). Using this decomposition, Santos et al. (2008) reformulated model (28)–(34) as follows, and devised exact and heuristic methods to solve it.

Consider all feasible crop rotation schedules for one plot, $(\hat{x}_{ij}^{s})$, $s = 1,\ldots, S$, defined by constraints (1)–(5), and let the binary decision variables be:

$\lambda_{sk} = 1$, if, plot $k$ is planted according to $s^{\text{th}}$ feasible crop rotation, 0 otherwise, for $s = 1,\ldots, S$.

Consequently, any feasible crop rotation schedule to plot $k$, $(x_{ijk})$, can be written as:

$$x_{ijk} = \sum_{s=1}^{S} \hat{x}_{ij}^{s}\lambda_{sk}\,,\ \sum_{s=1}^{S}\lambda_{sk} = 1\,,\ \lambda_{sk} \in \{0,1\} \qquad (35)$$

Therefore, important combinatorial aspects of the CRS-A problem (to obtain crop rotation schedules for a plot) are supposed to be already solved and the decision now consists of which schedule should be used for each plot, taking into account adjacency constraints among the plots. Therefore, model (28)-(34) is rewritten by substituting (35) by (28)-(33), which leads to its reformulation called the *master problem*, given by:

$$\max \sum_{k=1}^{L} \sum_{i \in C} \sum_{j \in I} \sum_{s=1}^{S} t_i \hat{x}_{ij}^{s} \lambda_{sk} \tag{36}$$

*subject to*:

$$\sum_{i \in F(p)} \sum_{\substack{r=0 \\ j-r \in I_i}}^{t_i - 1} \sum_{s=1}^{S} (\hat{x}_{i,j-r}^{s} \lambda_{su} + \hat{x}_{i,j-r}^{s} \lambda_{sv}) \leq 1,$$
$$p = 1, ..., NF, \; j = 1, ..., M, \; (u, v) \in A \tag{37}$$

$$\sum_{s=1}^{S} \lambda_{sk} = 1, \; k = 1, ..., L \tag{38}$$

$$\lambda_{sk} \in \{0,1\}, s = 1, ..., S, k = 1, \ldots, L. \tag{39}$$

As a final comment to this section, we observe that uncertainty can be of importance when dealing, for example, with the yields of the crops and selling prices. For a discussion on this issue and an extension of model (21)-(27) to incorporate uncertainty, the reader is refereed to Costa et al. (2009).

## SOLUTION METHODOLOGY

In the previous sections, we have described and presented mathematical models for three crop rotation problems and extensions focusing on vegetable crop production. Now, we discuss some methodological approaches, both exact and heuristic, that have been used to solve them. First, we present a brief review of some of the available methodologies, which is followed by a discussion of the specific strategies that have been proposed for each particular case.

The problems discussed have been modeled as linear programs (CRS–D and CRS–S problems) or as linear programs with integer variables (CRS-A problem). This enables the use of a vast arsenal of methods that have been proposed over the years since the development of the simplex algorithm by Dantzig (1947 [Nemhauser and Wolsey, 1988)]). The simplex algorithm exploits the property that in a linear program, an optimal solution, if it exists, can always be found in a vertex of the polyhedron defined by the constraints of the problem. The method works by moving between neighbor solutions (vertices) until the one with the best objective function value has been found. Other efficient strategies do exist for solving linear programs. In particular, interior point methods (Vanderbei, 2008) have received a great deal of attention in the literature. Nevertheless, the simplex algorithm is still the most popular approach in part because of its good performance in practice, in spite of the fact that it has been shown that for some instances it visits $2^n$ vertices before finding the optimal solution, where $n$ is the number of variables in the problem (Vanderbei, 2008).

One of the advantages of the simplex algorithm is its ability to deal with some of the variables in an implicit manner. Indeed, one can start the method with only a few variables (columns) and iteratively include promising variables generated by an auxiliary problem. This methodology, known as column generation, was firstly proposed in the 60's and it has been successfully applied since then in many different optimization linear problems (Lübbecke and Desrosiers, 2005). Procedures based on column generation were used to solve the three problems presented in this chapter. In all of them, columns are generated by the auxiliary problem (also called subproblem or pricing problem), which is the crop rotation scheduling problem (1)-(5).

In the case where some variables need to be integers, the simplex algorithm can only supply a bound on the value of the objective function. This bound is associated with the solution obtained when the integrality constraints are relaxed. There are many techniques to round the solution obtained by the simplex algorithm to its closest integer, but there is usually no guarantee that a good (or even feasible) integer solution will be obtained. For this reason, more elaborate methods have been proposed, two of which emerged almost simultaneously in the sixties: the branch and bound algorithm (Land and Doig, 1960) and the cutting planes method (Gomory, 1958). Both of these methods rely on a number of calls to the simplex algorithm but with different strategies: while the branch and bound approaches do an implicit tree enumeration of all solutions (using the bound given by the simplex algorithm to cut non-promising branches), the cutting plane methods alternate between the solution of the simplex algorithm and the generation (and subsequent addition to the problem) of a valid inequality which cuts the current simplex optimal solution without eliminating any feasible integer solution, until the solution given by the simplex is, itself, integer. Nowadays, probably one of the most used strategies is a combination of these two methods, called branch-and-cut, which implicitly enumerates solutions as in the branch and bound, but also uses the idea of the valid inequalities to improve the value of the bound obtained at each call of the simplex algorithm (Johnson et al., 2000). An extension of the branch-and-bound algorithm, which is able to cope with a large number of variables in an implicit manner (as was done by the column generation algorithm in the case of pure linear program), also exists and is known in the literature as the branch-and-price (Barnhart et al., 1998). In branch-and-cut-price algorithms, cutting planes are added to accelerate the branch-and-price.

Commercial mixed-integer LP solver packages, such as CPLEX and XPRESS, and non-commercial, such as lp_solve and SIMPHONY, usually rely on sophisticated versions of the branch-and-cut algorithms, both because of its efficiency and because of its generality (Linderoth and Ralphs, 2005; Atamtürk and Savelsbergh, 2005). In spite of the ability of the algorithms implemented in these packages to obtain solutions to many practical problems, they can fail to converge or even to obtain a single feasible solution to many others. The alternative is then to rely on heuristic methodologies that do not ensure optimality, but that can be rather efficient in obtaining – usually quickly – good quality solutions. These methods can vary from very intuitive strategies based on a sequential (and often myopic) construction of the solution, to local descent methodologies which try to find the best solution in a given neighborhood of the search space, to the so-called metaheuristics which are approaches that rely on different mechanisms to try to avoid premature convergence to low quality local optima. Examples of metaheuristics are abundant in the literature. Among the most used we find tabu search, grasp, simulated annealing and genetic algorithms, to mention a few (Blum and Roli, 2003).

A more recent trend that can be found in the literature is the combination of exact and heuristic algorithms yielding the so-called hybrid approaches. Numerous strategies have been developed and can be roughly classified into two groups. In the first group, exact and heuristic algorithms are applied sequentially, each one of them finding part of the solution, often in an iterative process. In a second group of algorithms, there is a more hierarchical structure with the exact approach making successive calls to the heuristic method or vice-versa (Raidl and Puchinger, 2008).

The key technique behind the solution methods for the three crop rotation optimization problems presented earlier is the column generation approach. The column-generation methodology is devised to solve a master problem containing a large number of columns, which cannot be explicitly determined within reasonable computational times. These columns are obtained from solutions of a core model, which is, for the three presented optimization problems, the crop rotation scheduling model (1)–(5). In the context of this chapter, this model is solved by a conventional branch-and-bound algorithm. The column generation technique works iteratively with a sequence of restricted master programs, which contain only a small subset of variables (or columns) of the master problem. At each iteration of the algorithm, a new restricted master problem is obtained by adding new columns in order to improve the current solution. When no improving column can be found, the optimal solution of the current restricted master program is also an optimal solution of the original master problem (Lübbecke and Desrosiers, 2005).

We briefly discuss in the following section how the column generation technique can be applied to each one of the presented problems. For the reader who is interested in a more thorough presentation, please refer to Santos et al (2010), Costa et al. (2009) and Santos (2009). For more details on the column generation technique, see Desaulniers et al. (2005).

## Solving the CRS–D Problem by the Column Generation Technique

In model (10)–(13) for the CRS–D problem, columns are built from crop rotation schedules for the $L$ cropping areas. Let ( $\hat{x}^s_{ijk}$ ) be the $s^{\text{th}}$ feasible crop rotation schedule for cropping area $k$. The coefficient of the variable $\lambda_{ks}$ in constraint (11) for a given $i$ and $j$ is:

$$a^s_{ijk} = \sum_{r=0}^{t_i - o_i - 1} p_{irk} \hat{x}^s_{i,j-o_i-r,k}$$

indicating the production of the $r^{\text{th}}$ harvesting of crop $i$ at period $j$ per m$^2$ cultivated according to the $s^{\text{th}}$ crop rotation schedule for area $k$.

Therefore, column $\mathbf{a}_{sk}$ in the CRS-D model can be written as:

$$\mathbf{a}_{sk} = \begin{bmatrix} (a^s_{ijk}) \\ \mathbf{e}_k \end{bmatrix} \begin{matrix} \rightarrow \text{constraints (11)} \\ \rightarrow \text{constraints (12)} \end{matrix}$$

where $\mathbf{e}_k$ is an $L$–dimensional vector with $e_{kk} = 1$ and $e_{kj} = 0$ for $j \neq k$.

By associating dual variables $\pi_{ij}$ to constraints (11) and dual variables $\alpha_k$ to constraints (12) the column that gives the largest reduced cost for area $k$ can be obtained by solving the following pricing problem $P_k$:

$$\max \sum_{i \in C_k} \sum_{j=1}^{M} \sum_{r=1}^{t_i - o_i - 1} (c_{ij} - \pi_{ij}) p_{irk} x_{i,j-o_i-r,k} - \alpha_k$$

subject to variables ($x_{ijk}$) that satisfy (1)–(5) where only crops in $C_k \cup G_k$ are considered.

Based on the above we present a column generation algorithm for the CRS-D problem:

In order to solve model (15)–(18) for the case where unmet demands are allowed, other types of columns associated to variables $\theta^-_{ij}$ are considered. In this case, the initial columns for the restricted master problem can be assumed to be those associated with these variables. If no initial

*Algorithm 1. Column generation for the CRS-D problem*

| |
|---|
| 1. Choose an initial set of columns for the restricted master problem. |
| 2. Solve the restricted master problem and obtain $(\pi,\alpha)$, the dual variables associated to constraints (11) and (12), respectively. |
| 3. Solve the sub problem $P_k$, for $k = 1,..., L$, and obtain the reduced costs $\hat{c}_{ks}$ and column $\mathbf{a}_{sk}$. |
| 4. If $\hat{c}_{ks} = 0$, for $k = 1,..., L$, then STOP. The current solution is optimal. Otherwise, add column $\mathbf{a}_{sk}$ with relative costs $\hat{c}_{ks} > 0$ in the restricted master problem and go to step 2. |

set of columns is given for problem (10)–(13), these variables can serve as artificial variables, where $q_{ij}$ is a large penalty.

## Solving the CRS–S Problem by the Column Generation Technique

In model (21)–(27) for the CRS-S problem, we consider just one cropping area, allowing index $k$ to be omitted. Let $(\hat{x}_{ij}^s)$ be the $s^{\text{th}}$ feasible crop rotation schedule. Then the coefficient of the variable $\lambda_s$ in constraint (22) for given crop $i$ at period $j$ is $a_{ij}^s = \sum_{r=0}^{t_i - o_i - 1} p_{ir} \hat{x}_{i,j-o_i-r}^s$ and column $\mathbf{a}_s$ in the CRS-S model can be written as:

$$\mathbf{a}^s = \begin{bmatrix} (a_{ij}^s) \\ 1 \\ \mathbf{0} \end{bmatrix} \begin{matrix} \rightarrow \text{constraints (22)} \\ \rightarrow \text{constraint (23)} \\ \rightarrow \text{constraints (24)-(25)} \end{matrix}$$

This model also contains a number of additional columns associated to the different existing sets of variables: $I_{ij}^w$, $\eta_{ij}^w$, $\theta_{ij}^+$, $\theta_{ij}^-$. Each one of these columns has its own particular form. For instance, the column associated to variable $I_{ij}^0$ is given by:

$$\begin{bmatrix} -e_{ij} \\ 0 \\ (1-r_i)e_{ij}^w \\ \mathbf{0} \end{bmatrix} \begin{matrix} \rightarrow \text{constraint (22)} \\ \rightarrow \text{constraint (23)} \\ \rightarrow \text{constraint (24)} \\ \rightarrow \text{constraints (25)} \end{matrix}$$

where $e_{ij}$ and $e_{ij}^w$ are appropriate columns of identity matrices.

In order to solve model (21)–(27), the initial restricted master problem can be formed with the columns associated to variables $I_{ij}^w$, $\eta_{ij}^w$, $\theta_{ij}^+$ and $\theta_{ij}^-$ and then, promising columns $\mathbf{a}^s$ are iteratively added into the restricted master problem until optimality is achieved. By associating dual variables $\pi_{ij}$ with constraints (22) and $\alpha$ with constraint (23), the column with the largest reduced cost can be obtained by solving the following pricing problem $P$:

$$\max \sum_{i \in C} \sum_{j=1}^{M} \sum_{r=1}^{t_i - o_i - 1} -\pi_{ij} p_{ir} x_{i,j-o_i-r} - \alpha$$

subject to that $(x_{ij})$ satisfies (1)–(5).

Algorithm 2 summarizes this column generation procedure:

## Solving the CRS-A Problem

Earlier in this chapter, we present a 0-1 linear optimization model (28)–(34) for the CRS-A problem. Traditional branch-and-bound techniques are only able to solve small instances of this model and, therefore, the development of new strategies is in order. With the goal of reducing the problem's dimension, Santos (2009) proposed a decomposition strategy, in which several sub problems, each one associated with one plot, are solved in sequence, in such a way that adjacency constraints (33) are

*Algorithm 2. Column generation for the CRS-S problem*

| |
|---|
| 1. Consider the initial restricted master problem (21)–(27) consisting of columns associated to variables $I_{ij}^w$, $\eta_{ij}^w$, $\theta_{ij}^+$ and $\theta_{ij}^-$. |
| 2. Solve the restricted master problem and obtain the dual variables $(\pi,\alpha)$ associated to constraints (22)–(23). |
| 3. Solve the pricing problem ($P$) and obtain the reduced cost $\hat{c}_S$ and column $\mathbf{a}^s$.<br>4. If $\hat{c}_S = 0$, then STOP. The current solution is optimal. Otherwise, add column $\mathbf{a}^s$ to the restricted master problem (21)–(27) and go to step 2. |

fulfilled. Analogously to what was done for the CRS-D and CRS-S problems, the sub problems are solved by branch-and-bound algorithms. In this section, we present three algorithms that use this strategy (Santos, 2009). The first one is a constructive heuristic (*plot-to-plot*) which can be used to quickly provide feasible solutions for the other two algorithms which are based on the column generation technique.

## A *Plot-To-Plot* Constructive Heuristic

A simple decomposition strategy can be devised in order to solve the CRS-A problem by first determining a feasible schedule for one plot and then finding a feasible schedule for a neighboring plot while forbidding the growth of crops which violate adjacency constraints (33). This is done by setting appropriate variables at zero which would violate the adjacency constraints. The strategy is repeated iteratively until a schedule has been assigned to all plots. Algorithm 3 formalizes the *plot-to-plot* constructive heuristic.

The performance of this heuristic strongly depends on the sequence in which the plots are treated. In our implementations, it was found to be useful to travel from the plot with the greatest number of neighbors to the one with the least number of them.

## A Column Generation Procedure with *Plot-To-Plot* Heuristic (CGH)

Many exact and heuristic methods based on the column generation technique can be devised to solve the master problem (36)-(39). In the column generation procedure CGH, besides the columns with the largest reduced costs, other columns are also added to the restricted master problem in order to find a feasible solution to the CRS-A problem (Santos et al.; 2008; Santos, 2009). In model (36)–(39), each column is associated with the periods of growth of the crops determined by a crop rotation schedule.

Given the $s^{\text{th}}$ schedule ( $\hat{x}^s_{ijk}$ ) for plot $k$, let $q^s_{pjk}$ be the parameter which has value 1, if a crop $i$ belonging to botanic family $p$ is growing in period $j$ in cropping area $k$, and has value 0, otherwise. Therefore:

$$q^s_{pjk} = \sum_{i \in F(p)} \sum_{r=0}^{t_i - 1} \hat{x}^s_{i,j-r,k} \,.$$

Consider now the vector $q_{uk}$ where its components are $(q^s_{ijk})$, if $(u, k) \in A$, and $q_{uk} = \mathbf{0}$, otherwise. The column $\mathbf{q}^s_k$ associated to variable $\lambda_{ks}$ in model (36) – (39) can be defined as:

$$\mathbf{q}^s_k = \begin{bmatrix} (q_{uk}) \\ \mathbf{e}_k \end{bmatrix} \begin{matrix} \rightarrow \text{ constraints (37)} \\ \rightarrow \text{ constraints (38)} \end{matrix}$$

*Algorithm 3. The plot-to-plot constructive heuristic*

| |
|---|
| 1. Find feasible crop rotation schedules for the $L$ plots by considering only green manure crops and fallow mandatory periods. If there are not $L$ feasible schedules, then STOP. The CRS–A problem has no solution. Otherwise, set $k = 0$ and sort the $L$ plots. |
| 2. If $k = L$, then STOP. An overall crop rotation schedule for the $L$ plots has been obtained. Otherwise, for plot $k + 1$, with some variables already fixed in step 1 for green manure and fallow period, and taking into account the crops already fixed to plots $1,\dots, k$, set to zero appropriate variables in order to ensure: |
| (a) constraints (29) – at most one crop is cultivated in each period, in the same plot,<br>(b) constraints (30) – crops of the same botanic family are not cultivated one immediately after the other in the same plot, and<br>(c) constraints (33) – crops of the same botanic family are not cultivated simultaneously in adjacent plots. |
| 3. Find a feasible (or optimal) crop rotation schedule for plot $k + 1$. Set $k = k + 1$ and go to step 2. |

where vector $\mathbf{e}_k$ is an $L$–dimensional vector with $e_{kk} = 1$ and $e_{kj} = 0$ for $j \neq K$.

Model (36)-(39) is an integer optimization problem. In order to solve the linear relaxation of this problem by column generation methodology, consider the dual variables $\pi_{pj}^{ku}$ referring to constraints (37) for adjacent plots $k$ and $u$, and dual variables $\alpha_k$ referring to constraints (38) for plot $k$. The column $\mathbf{q}_k^s$ associated with plot $k$ that has the largest reduced cost is determined by solving the following pricing problem $P_k$:

$$\max \sum_{i=1}^{N+1} \sum_{j=1}^{M} t_i x_{ijk} - \sum_{j=1}^{M} \sum_{p=1}^{NF} \sum_{i \in F(p)} \sum_{r=0}^{t_i - 1} \sum_{(u,k) \in A} \pi_{p,j+r}^{ku} x_{ijk} - \alpha_k$$

subject to that $(x_{ijk})$ satisfies (1)–(5).

The *plot-to-plot* heuristic is used in order to insert new columns during the iterations of the CGH procedure. All columns obtained in this way with positive reduced costs are added to the linear relaxation of the master problem, otherwise the above pricing problem is solved. The aim of this strategy is to find feasible solutions for the CRS-A problem while solving the linear relaxation of the master problem (36)–(39). The columns obtained by a conventional column generation procedure might not be enough to provide a feasible integer solution for the problem, because the pricing problems that generate them do not take explicitly into account the adjacency constraints.

Algorithm 4 briefly presents this procedure.

Adding new columns by the *plot-to-plot* heuristic allows finding feasible solutions to the CRS–A (lower bounds). This strategy also accelerates the convergence in solving the linear relaxation of the master problem (upper bound). For a computational study of the impact of adding these new columns see Santos (2009).

*Algorithm 4. A column generation procedure with the plot-to-plot heuristic (CGH)*

| |
|---|
| 1. Sort the $L$ plots and apply the *plot-to-plot* heuristic in order to find an initial set of columns for the first restricted master problem (RMP). Keep the solution obtained by the *plot-to-plot* heuristic as incumbent. |
| 2. Solve the RMP and obtain the dual variables $\pi_{pj}^{ku}$ associated with constraints (37), and the dual variable $\alpha_k$ associated with the convexity constraints (38).<br>3. Set $col = 1$. Solve the pricing problem $P_1$ and obtain $\hat{x}_1 = (\hat{x}_{ij1})$ and the reduced cost $\hat{c}_1$.<br>If $\hat{c}_1 > 0$ then add the column associated to $\hat{x}_1$ in RPM and set $col = 2$. |
| 4. For $k = 2, \ldots, L$, do:<br>a) If $col = k$, then<br>i. solve $P_k$ but fixing to zero appropriate variables that would violate adjacency constraints (33), taking into account the crop rotation scheduling $\hat{x}_1, \ldots, \hat{x}_k$ already determined to plots $1, \ldots, k-1$ (strategy analogous to the *plot-to-plot* heuristic).<br>ii. If $P_k$ (fixing variables) is feasible then $col = col + 1$.<br>Else, solve $P_k$ without fixing variables.<br>b) Let and $\hat{c}_k$ be the obtained solution from $P_k$ and its reduced cost. If $\hat{c}_k > 0$ then add the column associated to $\hat{x}_k$ into RMP and go to step 5.<br>5. If $col = L$, a feasible solution $(\hat{x}_1, \ldots, \hat{x}_L)$ of CRS–A problem is obtained. If this solution is better than the incumbent one, keep the new solution as incumbent. |
| 6. If $col = 0$ (i.e., no column with relative cost is found) then STOP. The linear relaxation of the master problem solved. Return the incumbent solution. Otherwise, go to step 2. |

### Branch-And-Price with CGH Procedure

The CGH procedure always finds a feasible solution to the CRS–A problem. In the worst case, it provides the solution obtained by the *plot–to–plot* heuristic in step 1. However, there is no guarantee that the obtained solution is optimal. If proven optimality is needed then a branch–and–price algorithm can be devised to solve the (integer) master problem.

In both branch-and-price and branch-and-cut-price algorithms the linear programs which arise in the nodes of the tree of enumeration are solved by the column generation technique.

The adopted branching rules and the proper use of valid inequalities are crucial issues for successful implementation of branch-and-price algorithms (Barnhart et al., 1998). When the master problem originates from a reformulation of an original 0-1 linear program (such as the master problem (36)–(39) which is a reformulation of problem (28)–(34)), the branching constraints can be built on the original variables and by fixing these variables (to zero or one) in the pricing problems. Santos (2009) solves model (36)–(39) by a branch-and-price algorithm using branching in the original variables and the CGH procedure to solve the linear relaxation of the master problem in each node of the enumeration tree. The performance of the branch-and-price algorithm with the CGH procedure is much better than the performance obtained by the column generation procedure without the *plot-to-plot* heuristic. This can be explained by the fact that the CGH procedure produces an upper bound more quickly in addition to the fact that it can obtain good quality lower bounds (Santos, 2009).

In the following section, we present some computational results in order to assess the efficiency of the proposed methods.

## COMPUTATIONAL RESULTS

In Solution Methodology Section, we presented some procedures based on the column generation technique to solve the CRS–D, CRS–S and CRS–A problems. In this section, we briefly discuss computational experiments run in order to analyze the performance of algorithms (For more details, see Santos et al., 2010, Costa et al., 2009 and Santos, 2009). The algorithms have been implemented in C++ and the linear (integer) models solved by the commercial package CPLEX 10.1 (ILOG, 2006). The experiments were carried out on a computer with 2.33 GHz and 4 Gb of RAM, under Linux. For all the problems, we used 2-year crop rotation lengths, and 1-month fallow in the scheduling of each crop rotation.

For the CRS–D and CRS–S problems, the 2-year rotation length was divided into weeks. Data in Appendix B comes from an ecologically based production unit situated in the city of Barbacena, Brazil (21° 13' 30" S and 43° 46' 25" W, 1165 meters above sea level). The demands for the crops were randomly generated taking into account planting and harvesting periods of each crop.

For the CRS–A problem, we consider 10-day periods. Data in Appendix A was obtained in an organic vegetable garden of the Federal University of Viçosa, Brazil (20° 45' 14" S and 42° 52' 53" W, 648 meters above sea level).

### Computational Experiments for the CRS-D Problem

In this section, we briefly present and discuss a number of computational experiments run in order to analyze the performance of the column-generation algorithm devised for the model (15)–(18) for the CRS–D problem. Different parameters were varied such as the number of crops ($N = 12, 16, 20, 24$), the number of cropping areas ($L = 1, 3, 5$) and their sizes (the total area was kept constant at 10,000 $m^2$), We consider the case without upper bounds on extra production ($d\infty$) and the case where there is a limit on the extra production equal to 200% of the given demand (*d*200). In order to mimic the heterogeneity of lands found in practice, we defined, for each cropping area, a set of crops that could be grown and set their productivities by making random changes to the original data (with a maximum variation of 30% regarding the data in Table B.2). The combination of the different characteristics presented above gives rise to 24 different instance classes.

Table 3 presents a summary of the obtained results. For each class of instances, the table provides the algorithm's running times in seconds (*time*), the number of plots in the optimal solution (*#plots*), and the percentage of residual unmet demand (*%resid*). Equal costs were used to penalize the unmet demands of the different crops. The algorithm was allowed to run for 1 hour for each instance.

As shown in Table 3, the algorithm was able to obtain solutions within the allowed time limit for most of the instances. The average running times reached 1h only for the class with $N = 24$ and $d\infty$, indicating that the algorithm was usually unable to prove the optimality of the obtained solutions for these instances in the allowed time limit. The running times increased with the number of crops but showed no clear dependency on the number of cropping areas. Moreover, they are usually larger for the cases where no limit is imposed on the unmet demand. This is particularly true for the classes with fewer crops. However, the average percentage of unmet demand is smaller. The number of cropping areas has no clear effect on the number of plots in the optimal solutions, but these numbers are heavily dependent on the number of crops, reaching around 600 different plots for the instances with 24 crops. Santos et al. (2010) and Santos (2009) present some heuristic procedures to reduce the number of plots.

## Computational Experiments for the CRS-S Problem

The computational experiments presented in this section try to analyze the performance of the column-generation algorithm in solving the CRS–S problem and the impact of allowing stocks (under the conditions described in Table B.2 in Appendix). In these tests, the instances were divided into classes according to the total cropping area (A = 500, 1000, 2000 and 4000 m$^2$) and the total number of crops ($N$ = 10, 15, 20). For the sake of simplicity, a single cropping area is considered. The combination of these parameters yields 12 classes, each one containing 10 instances. All instances were solved for both the situations with stocks allowed and with stocks

*Table 3. Average results for the 24 classes of instances*

| | | $L = 1$ | | | $L = 3$ | | | $L = 5$ | | |
|---|---|---|---|---|---|---|---|---|---|---|
| *crop* | *demand* | *time* | *#plots* | *%resid* | *time* | *#plots* | *%resid* | *time* | *#plots* | *%resid* |
| **12** | *d200* | 250.20 | 224.30 | 0.73 | 257.37 | 266.10 | 0.51 | 129.83 | 249.60 | 0.68 |
| | *d∞* | 757.25 | 119.80 | 0.00 | 646.62 | 116.89 | 0.00 | 708.97 | 135.60 | 0.00 |
| **16** | *d200* | 910.35 | 395.40 | 0.47 | 811.76 | 335.40 | 0.63 | 766.70 | 394.50 | 0.48 |
| | *d∞* | 2331.47 | 237.00 | 0.00 | 1283.94 | 244.50 | 0.00 | 2722.03 | 241.50 | 0.00 |
| **20** | *d200* | 3283.76 | 494.70 | 0.50 | 2630.80 | 497.90 | 0.49 | 2062.26 | 502.40 | 0.67 |
| | *d∞* | 3075.58 | 426.30 | 0.00 | 3430.26 | 457.56 | 0.00 | 2947.01 | 424.80 | 0.00 |
| **24** | *d200* | 3230.28 | 549.40 | 0.60 | 3230.28 | 549.40 | 0.60 | 3168.13 | 546.67 | 0.60 |
| | *d∞* | 3601.36 | 528.22 | 0.00 | 3596.96 | 594.50 | 0.00 | 3612.39 | 635.89 | 0.00 |

not allowed. For all the experiments, we fixed $c_{ij}$ = 1 and $c_{ij}^{+}$ = 0, for all *i* and *j*.

Table 4 presents the average results obtained, in terms of the algorithm running times in seconds (*time*), the number of plots in the optimal solution (*#plots*), the percentage of unmet demand (*%resid*), and the percentage of production exceeding nominal demand (*%extra*). Run times were limited on 1 hour for each instance.

The computational experiments show that the proposed algorithm is able to find optimal solutions within quite reasonable running times for both the situations with and without stocks. In general terms, the instances become harder as the planting areas and the number of crops increase. From Table 4, one can see that by enabling stocks extra production increases and unmet demand is reduced. In average, a reduction of 5.35% was obtained for the unmet demand with an increase of 8.97% in extra production.

Table 4 also shows that solution allowing stocks present a smaller numbers of plots, up to 50% when A = 4000 m$^2$. This is a particularly desired characteristic since solutions with a very high number of plots can be infeasible in real-life situations (Santos et al., 2010). Considering all area sizes, the average reduction was 21.20%.

## Computational Experiments for the CRS-A Problem

When dealing with the Crop rotation scheduling problem with adjacency constraints, we presented two integer optimization models for the CRS–A problem and some of the solution procedures which were proposed by Santos et al. (2008) and Santos (2009). We will now discuss some results obtained with a heuristic based on column generation (CGH) and branch-and-price (B&P) algorithms.

As mentioned in the Solution Methodology Section, most optimization linear models with integer variables belong to the NP class, which means that no efficient algorithms are known that can obtain provable optimal solutions. This is the case of the CRS-A problem. In these situations, it is common to rely on heuristic methods, which provide feasible solutions. Evaluating the quality of solutions obtained is an important issue to

*Table 4. Average results for the 12 classes of instances when no stock is allowed and when stock is allowed according to the conditions presented in Table B.2*

| | | *no stock is allowed* | | | | *stock is allowed* | | | |
|---|---|---|---|---|---|---|---|---|---|
| *A* | *crop* | *time* | *#plots* | *%resid* | *%extra* | *time* | *#plots* | *%resid* | *%extr* |
| **500** | **10** | 53.40 | 62.40 | 18.16 | 69.51 | 52.6 | 52.6 | 17.9 | 77.5 |
| | **15** | 96.30 | 93.80 | 43.08 | 94.86 | 99.7 | 90.4 | 43.0 | 97.3 |
| | **20** | 404.10 | 112.50 | 66.63 | 96.39 | 406.8 | 110.4 | 66.6 | 97.6 |
| **1000** | **10** | 15.30 | 69.80 | 3.57 | 56.29 | 11.6 | 38.7 | 3.5 | 67.3 |
| | **15** | 184.40 | 200.70 | 16.10 | 89.85 | 272.4 | 184.7 | 15.9 | 93.2 |
| | **20** | 181.89 | 121.00 | 46.55 | 99.21 | 220.7 | 120.2 | 46.5 | 99.9 |
| **2000** | **10** | 2.70 | 40.00 | 0.13 | 39.11 | 2.5 | 21.4 | 0.1 | 49.4 |
| | **15** | 94.40 | 189.20 | 0.88 | 72.21 | 154.4 | 155.4 | 0.6 | 75.4 |
| | **20** | 392.70 | 349.10 | 3.39 | 91.14 | 918.3 | 313.4 | 2.8 | 93.5 |
| **4000** | **10** | 1.10 | 25.60 | 0.00 | 25.61 | 1.1 | 15.8 | 0.0 | 40.8 |
| | **15** | 19.90 | 109.50 | 0.00 | 55.14 | 9.0 | 49.8 | 0.0 | 61.9 |
| | **20** | 79.70 | 218.40 | 0.00 | 68.66 | 61.1 | 105.4 | 0.0 | 72.8 |

be addressed. The objective value of a feasible solution provides a lower bound (LB) to the optimal value for maximizing problems. If we have an upper bound (UB) to the objective function, then we have an interval that the optimal value belongs to. Therefore, we can define a gap value that gives a maximum deviation of the feasible solution value to the optimal value, as follows.

$$gap = 100 \frac{\text{UB} - \text{LB}}{|\text{LB}| + 10^{-10}}$$

Typically, UB are obtained by relaxing the master problem. A particular relaxation of (36)-(39) consists of relaxing the integrality conditions and solving it by the column generation method described in the Solution Methodology Section.

For the computational tests, the instances were divided into classes according to the total number of plots in the cropping area ($L$ = 4, 8, 12, 16, 20) and the total number of crops ($N$ = 6, 16, 26). Each class consists of 10 crops randomly chosen from the 28 crops in Appendix A. Spatial configurations for cropping areas were also randomly generated. By combining these different parameters, 15 classes were obtained.

Table 5 shows results comparing CGH and B&P algorithms. For each procedure, we present the average results for the running times in seconds (*time*), the percentage of instances solved to optimality (*%opt*), and the gap (*gap*). Run times were limited to 30 minutes for each instance.

The CGH algorithm was able to prove optimality of about 55% of the instances. The average running time was 45.14 seconds and the average gap was 1.9%. The algorithm B&P was able to prove optimality for all instances with $N$ = 16 and $N$ = 26, in less than 70 seconds. In general, the run time slightly increases with the number of plots. However, the performance of the algorithm seems to be very dependent on the number of crops. For the classes with $N$ = 6 crops, it proves optimality within 30 minutes for only 6% of the instances, and a gap of 3.59%. The fact that the algorithm behaves better for instances with a

*Table 5. Average results for 15 classes of instances*

| | | CGH | | | B&P | | |
|---|---|---|---|---|---|---|---|
| *L* | *N* | *time* | *%opt* | *gap* | *time* | *%opt* | *gap* |
| **4** | **6** | 214.92 | 0.00 | 3.62 | 1482.30 | 20.00 | 1.97 |
| | **16** | 3.08 | 80.00 | 0.17 | 3.65 | 100.00 | 0.00 |
| | **26** | 2.61 | 100.00 | 0.00 | 2.52 | 100.00 | 0.00 |
| **8** | **6** | 121.83 | 0.00 | 4.58 | 1817.93 | 0.00 | 3.12 |
| | **16** | 4.87 | 60.00 | 0.33 | 8.91 | 100.00 | 0.00 |
| | **26** | 5.48 | 100.00 | 0.00 | 5.44 | 100.00 | 0.00 |
| **12** | **6** | 59.89 | 0.00 | 5.07 | 1718.15 | 10.00 | 3.37 |
| | **16** | 7.13 | 80.00 | 0.06 | 12.23 | 100.00 | 0.00 |
| | **26** | 8.79 | 90.00 | 0.07 | 9.96 | 100.00 | 0.00 |
| **16** | **6** | 76.94 | 0.00 | 3.82 | 1834.29 | 0.00 | 3.09 |
| | **16** | 12.24 | 70.00 | 0.08 | 29.02 | 100.00 | 0.00 |
| | **26** | 14.17 | 80.00 | 0.07 | 15.75 | 100.00 | 0.00 |
| **20** | **6** | 110.54 | 0.00 | 10.31 | 1884.43 | 0.00 | 6.37 |
| | **16** | 16.99 | 40.00 | 0.21 | 68.50 | 100.00 | 0.00 |
| | **26** | 17.56 | 80.00 | 0.17 | 19.40 | 100.00 | 0.00 |

larger number of crops is interesting since this situation is more common in practice.

As a final note, we mentioned that the default branch-and-cut algorithm in CPLEX package performs poorly. Indeed, it was only able to prove optimality within 30 minutes for 40% of the instances with $L = 4$. For the other classes, convergence was not obtained within the time limit and, more importantly, the solver had trouble even finding feasible solutions. Indeed, no feasible solution was obtained for 33.7% of the instances with $L = 16$ and for 93.6% of the instances with $L = 20$. For the whole set of instances, the average gap of the solutions obtained was 13.26% and optimality was proved for only 8% of instances.

## CONCLUSION AND FUTURE RESEARCH

In this chapter, we presented three problems associated to crop rotation scheduling in the context of vegetable crop production according to ecologically-based criteria. The idea of gathering these problems was to somehow develop a framework that could be used to tackle similar situations. Indeed, the same core structure has been used to model each one of the problems as linear or integer linear programs and similar solution methodology strategies have been devised. A brief analysis of the results suggests that the proposed methodologies are adequate and can be extended to deal with other variants of the problems.

Probably the main assumption to develop the framework proposed in this chapter was related to the definition of a feasible crop rotation schedule. Our choice has been to use general guidelines imposing, for each crop rotation, some desirable features such as the presence of fallow and green manure crops, as well as the prohibition of growing crops of the same botanic family one immediately after another. Although this is a very particular choice, we believe that the proposed models are quite general and can be adapted to include other criteria. One important feature of the proposed methodology is the generality with which it copes with the crop rotation constraints, without the need to explicitly list feasible or infeasible crop sequences.

We foresee many extensions and future perspectives for the study presented here. We can cite the possibility of considering the economical aspect of production in a more explicit manner in our models, the inclusion of additional constraints specific for some growers (such as the climatic limitations to some crops or the options for few specific crops in certain seasons of the year) and the development of visual interfaces that can help the adoption of the methods in cooperatives, to mention a few. Moreover, one interesting research area is the adoption of a more flexible model by occasionally relaxing some of the ecologically-based constraints. We hope that this chapter will encourage research efforts and prove to be a helpful source of ideas and references.

## REFERENCES

Altieri, M. A. (1995). *Agroecology: the science of sustainable agriculture*. Boulder, CO: Westview press.

Appa, G. M. (1973). The transportation problem and its variants. *Operational Research Quarterly*, *24*, 79–99.

Atamtürk, A., & Savelsbergh, M. (2005). Integer-programming software systems. *Annals of Operations Research*, *140*, 67–124. doi:10.1007/s10479-005-3968-2

Bachinger, J., & Zander, P. (2006). ROTOR, a tool for generating and evaluating crop rotations for organic farming systems. *European Journal of Agronomy*, *26*, 130–143. doi:10.1016/j.eja.2006.09.002

Barnhart, C., Johnson, E. L., Nemhauser, G. L., Savelsbergh, M. W. P., & Vance, P. H. (1998). Branch and price: Column generation for solving huge integer programs. *Operations Research*, *46*(3), 316–329. doi:10.1287/opre.46.3.316

Blum, C., & Roli, A. (2003). Metaheuristics in combinatorial optimization: Overview and conceptual comparison. *ACM Computing Surveys*, *35*(3), 268–308. doi:10.1145/937503.937505

Campanhola, C., & Valarini, P. J. (2001). A agricultura orgânica e seu potencial para o pequeno agricultor. *Cadernos de Ciência & Tecnologia*, *18*(3), 69–101.

Castellazzi, M. S., Wood, G. A., Burgess, P. J., Morris, J., Conrad, K. F., & Perry, J. N. (2008). A systematic representation of crop rotations. *Agricultural Systems*, *97*, 26–33. doi:10.1016/j.agsy.2007.10.006

Clarke, H. R. (1989). Combinatorial aspects of cropping pattern selection in agriculture. *European Journal of Operational Research*, *40*, 70–77. doi:10.1016/0377-2217(89)90273-7

Costa, A. M., Santos, L. M. R., Alem, D. J., & Santos, R. H. S. (2009, July). Sustainable Vegetable Crop Supply Problem with Perishable Stocks. Paper presented at the EURO Summer Institute, Lleida, Spain.

Dantzig, G. B., & Wolfe, P. (1960). Decomposition principle for linear programs. *Operations Research*, *8*, 101–111. doi:10.1287/opre.8.1.101

Desaulniers, G., Desrosiers, J., & Solomon, M. M. (Eds.). (2005). *Column generation*. Berlin: Springer. doi:10.1007/b135457

Detlefsen, N., & Jensen, A. L. (2007). Modelling optimal crop sequences using network flows. *Agricultural Systems*, *94*, 566–572. doi:10.1016/j.agsy.2007.02.002

Dogliotti, S., Rossing, W. A. H., & Van Ittersum, M. K. (2003). ROTAT, a tool for systematically generating crop rotations. *European Journal of Agronomy*, *19*, 239–250. doi:10.1016/S1161-0301(02)00047-3

El-Nazer, T., & McCarl, B. A. (1986). The choice of crop rotation: A modeling approach and case study. *American Journal of Agricultural Economics*, *68*(1), 127–136. doi:10.2307/1241657

Gliessman, S. R. (2000). *Agroecology: ecological processes in sustainable agriculture*. Chelsea, UK: Ann Arbor.

Gomory, R. E. (1958). Outline of an algorithm for integer solutions to linear programs. *Bulletin of the American Mathematical Society*, *64*, 275–278. doi:10.1090/S0002-9904-1958-10224-4

Haneveld, W. K., & Stegeman, A. W. (2005). Crop succession requirements in agricultural production planning. *European Journal of Operational Research*, *166*, 406–429. doi:10.1016/j.ejor.2004.03.009

Hildreth, C. G., & Reiter, S. (1951). On the choice of a crop rotation plan. In Activity Analysis of Production and Allocation, T.C. Koopmans (Ed.), *Proceedings of the Conference on Linear Programming*, (pp. 177-188).

ILOG CPLEX 10.1 Reference Manual. 10.1. Copyright by ILOG, France, (2006).

Johnson, E. L., Nemhauser, G. L., & Savelsbergh, M. W. P. (2000). Progress in Linear Programming-Based Algorithms for Integer Programming: An Exposition. *INFORMS Journal on Computing*, *12*(1), 2–23. doi:10.1287/ijoc.12.1.2.11900

Jones, J., Hoogenboom, G., Porter, C., Boote, K., Batchelor, W., & Hunt, L. (2003). The DSSAT cropping system model. *European Journal of Agronomy*, *18*(3), 235–265. doi:10.1016/S1161-0301(02)00107-7

Kantorovich, L. (1960). Mathematical methods of organizing and planning production (translated from a report in Russian, dated 1939). *Management Science*, *6*, 366–422. doi:10.1287/mnsc.6.4.366

Land, A. H., & Doig, A. G. (1960). An Automatic Method for Solving Discrete Programming Problems. *Econometrica*, *28*, 497–520. doi:10.2307/1910129

Linderoth, J. T., & Ralphs, T. K. (2005). Noncommercial Software for Mixed-Integer Linear Programming. In *Integer Programming: Theory and Practice* (pp. 253-303). Boca Raton, FL: CRC Press Operations Research Series.

Lübbecke, M. E., & Desrosiers, J. (2005). Selected topics in column generation. *Operations Research*, *53*(6), 1007–1023. doi:10.1287/opre.1050.0234

Lyon, S. (2006). Evaluating fair trade consumption: politics, defetishization and producer participation. *International Journal of Consumer Studies*, *30*, 452–464. doi:10.1111/j.1470-6431.2006.00530.x

Makatouni, A. (2002). What motivates consumers to buy organic food in the UK? Results from a qualitative study. *British Food Journal*, *104*, 345–352. doi:10.1108/00070700210425769

Nemhauser, G. L., & Wolsey, L. A. (1988). *Integer and Combinatorial Optimization*. New York: Wiley-Interscience Publication.

Raidl, J. R., & Puchinger, J. (2008). Combining (Integer) Linear Programming Techniques and Metaheuristics for Combinatorial Optimization. [SCI]. *Studies in Computational Intelligence*, *114*, 31–62. doi:10.1007/978-3-540-78295-7_2

Santos, L. M. R. (2009). *Programação de Rotação de culturas – modelos e métodos de solução. Doctoral dissertation*, ICMC/USP - University of São Paulo, Brazil.

Santos, L. M. R., Costa, A. M., Arenales, M. N., & Santos, R. H. S. (2010). Sustainable vegetable crop supply problem. *European Journal of Operational Research*, *204*, 639–647. doi:10.1016/j.ejor.2009.11.026

Santos, L. M. R., Michelon, P., Arenales, M. N., & Santos, R. H. S. (2008). Crop rotation scheduling with adjacency constraints. *Annals of Operations Research*. doi:.doi:10.1007/s10479-008-0478-z

Seyfang, G. (2006). Ecological citizenship and sustainable consumption: Examining local organic food networks. *Journal of Rural Studies*, *22*, 383–395. doi:10.1016/j.jrurstud.2006.01.003

Stöckle, C. O., Donatelli, M., & Nelson, R. (2003). CropSyst, a cropping systems simulation model. *European Journal of Operational Research*, *18*, 289–307.

Tillman, D., Cassman, K. G., Matson, P. A., Naylor, R., & Polask, S. (2002). Agricultural sustainability and intensive production practices. *Nature*, *418*, 671–677. doi:10.1038/nature01014

Vanderbei, R. J. (2008). *Linear programming: foundations and extensions* (3rd ed.). New York: Springer.

## APPENDIX A

For the computational experiments of the crop rotation scheduling problem with adjacency constraints (CRS-A problem), 28 crops which are cultivated at an organic farming unit in Viçosa, Brazil, were selected. Crops 25 to 28 are used for green manuring. The production time (in number of periods of 10 days) for each crop includes the estimated planting and harvesting periods.

*Table A.1. Data for the 28 crops: name, botanic family, planting periods and production time*

| CROP | | BOTANIC FAMILY | | PLANTING MONTH | | PRODUCTION TIME (*10 days*) |
|---|---|---|---|---|---|---|
| | | | | *beginning* | *end* | |
| 1 | Lettuce | 1 | *Asteraceae* | all year round | | 4 |
| 2 | Watercress | 2 | *Brassicaceae* | all year round | | 6 |
| 3 | Chinese cabbage | 2 | *Brassicaceae* | March | June | 28 |
| 4 | Broccoli | 2 | *Brassicaceae* | February | June | 11 |
| 5 | Cabbage | 2 | *Brassicaceae* | February | July | 11 |
| 6 | Cauliflower | 2 | *Brassicaceae* | March | June | 10 |
| 7 | Beet | 3 | *Chenopodiaceae* | March | July | 7 |
| 8 | Spinach | 3 | *Chenopodiaceae* | February | July | 11 |
| 9 | Zucchini | 4 | *Cucurbitaceae* | August | March | 9 |
| 10 | Pumpkin | 4 | *Cucurbitaceae* | September | January | 15 |
| 11 | Cucumber | 4 | *Cucurbitaceae* | all year round | | 8 |
| 12 | Garlic | 5 | *Alliaceae* | March | April | 15 |
| 13 | Onion | 5 | *Alliaceae* | March | June | 12 |
| 14 | Okra | 6 | *Malvaceae* | August | March | 23 |
| 15 | Corn | 7 | *Gramineae* | August | April | 10 |
| 16 | Tomato | 8 | *Solanaceae* | all year round | | 18 |
| 17 | Sweet pepper | 8 | *Solanaceae* | all year round | | 15 |
| 18 | Eggplant | 8 | *Solanaceae* | September | February | 16 |
| 19 | Carrot | 9 | *Apiaceae* | March | July | 19 |
| 20 | Parsley | 9 | *Apiaceae* | September | March | 14 |
| 21 | Peruvian carrot | 9 | *Apiaceae* | March | August | 19 |
| 22 | Snap bean | 10 | *Leguminosae* | August | April | 11 |
| 23 | Pea | 10 | *Leguminosae* | March | April | 9 |
| 24 | Bean | 10 | *Leguminosae* | August | September | 9 |
| 25 | Sun hemp | 10 | *Leguminosae* | September | December | 9 |
| 26 | Jack bean | 10 | *Leguminosae* | September | December | 10 |
| 27 | Black velvet bean | 10 | *Leguminosae* | September | January | 11 |
| 28 | Vetch | 10 | *Leguminosae* | March | June | 14 |

## APPENDIX B

For the computational experiments of the crop rotation scheduling problem with known demands (CRS-D problem) and the crop rotation scheduling problem with perishable stocks (CRS-S problem) we select 23 crops cultivated at an organic farming unit in Barbacena, Brazil. Crops 20 to 23 are for green manuring. The production time (in number of periods of one week) for each crop includes the estimated planting and harvesting periods.

*Table B.1. Data for the 23 crops: name, botanic family, planting periods and production time*

| CROP | | BOTANIC FAMILY | | PLANTING PERIOD | | PRODUCTION TIME (weeks) |
|---|---|---|---|---|---|---|
| | | | | *beginning* | *end* | |
| 1 | Crisp head lettuce | 1 | *Asteraceae* | all year round | | 7 |
| 2 | Loose leaf lettuce | 1 | *Asteraceae* | all year round | | 7 |
| 3 | Butter head lettuce | 1 | *Asteraceae* | all year round | | 7 |
| 4 | Endive | 1 | *Asteraceae* | all year round | | 9 |
| 5 | Collards | 2 | *Brassicaceae* | February | September | 32 |
| 6 | Broccoli | 2 | *Brassicaceae* | February | October | 20 |
| 7 | Cauliflower | 2 | *Brassicaceae* | March | October | 18 |
| 8 | Beet | 3 | *Chenopodiaceae* | February | September | 11 |
| 9 | Spinach | 3 | *Chenopodiaceae* | February | September | 20 |
| 10 | Zucchini | 4 | *Cucurbitaceae* | October | February | 14 |
| 11 | Pumpkin | 4 | *Cucurbitaceae* | November | January | 19 |
| 12 | Cucumber | 4 | *Cucurbitaceae* | September | March | 13 |
| 13 | Garlic | 5 | *Alliaceae* | March | April | 24 |
| 14 | Onion | 5 | *Alliaceae* | March | July | 24 |
| 15 | Leek | 5 | *Alliaceae* | April | April | 12 |
| 16 | Okra | 6 | *Malvaceae* | November | January | 27 |
| 17 | Tomato | 7 | *Solanaceae* | all year round | | 24 |
| 18 | Carrot | 8 | *Apiaceae* | all year round | | 16 |
| 19 | Parsley | 8 | *Apiaceae* | October | February | 21 |
| 20 | Black velvet bean | 9 | *Leguminosae* | October | January | 16 |
| 21 | Jack bean | 9 | *Leguminosae* | October | February | 12 |
| 22 | Lupine | 9 | *Leguminosae* | March | July | 18 |
| 23 | Hairy vetch | 9 | *Leguminosae* | March | July | 20 |

*Table B.2. Data for the 19 crops with known demand: name, measuring unity, fraction lost ($r_i$) of the quantity already in stock of crop i, if it is held for one more period, number of periods(weeks) the crop can be held in stock ($G_i$), number of periods between planting and first partial harvesting ($o_i$), and expected harvests per square meter*

| CROP | | measuring unity | STOCK | | HARVESTING | | | | | | | | | | | | | | | | | | | | |
|---|---|---|---|---|---|---|---|---|---|---|---|---|---|---|---|---|---|---|---|---|---|---|---|---|---|
| | | | $r_i$ | $G_i$ | $o_i$ | 1 | 2 | 3 | 4 | 5 | 6 | 7 | 8 | 9 | 10 | 11 | 12 | 13 | 14 | 15 | 16 | 17 | 18 | 19 | 20 |
| 1 | Crisp head lettuce | *unities* | 0.1 | 1 | 5 | 9 | 3 | | | | | | | | | | | | | | | | | | |
| 2 | Loose leaf lettuce | *unities* | 0.1 | 1 | 5 | 9 | 3 | | | | | | | | | | | | | | | | | | |
| 3 | Butter head lettuce | *unities* | 0.1 | 1 | 5 | 9 | 3 | | | | | | | | | | | | | | | | | | |
| 4 | Endives | *unities* | 0.1 | 1 | 7 | 9 | 3 | | | | | | | | | | | | | | | | | | |
| 5 | Coolards | *pack* | 0.1 | 1 | 12 | 1 | 2 | 2 | 2 | 2 | 2 | 2 | 2 | 2 | 2 | 2 | 2 | 2 | 2 | 2 | 2 | 2 | 2 | 1 | 1 |
| 6 | Broccoli | *pack* | 0.2 | 1 | 10 | 1 | 1 | 2 | 2 | 3 | 3 | 3 | 2 | 2 | 1 | | | | | | | | | | |
| 7 | Cauliflower | *unities* | 0.1 | 2 | 16 | 1 | 3 | | | | | | | | | | | | | | | | | | |
| 8 | Beet | *kg* | 0.15 | 3 | 8 | 1 | 2 | 1 | | | | | | | | | | | | | | | | | |
| 9 | Spinach | *pack* | 0.1 | 1 | 5 | 3 | 4 | 4 | 4 | 4 | 4 | 4 | 4 | 4 | 4 | 4 | 4 | 3 | 3 | 2 | | | | | |
| 10 | Zucchini | *kg* | 0.15 | 2 | 9 | 0.2 | 0.3 | 0.3 | 0.3 | 0.2 | | | | | | | | | | | | | | | |
| 11 | Pumpkin | *kg* | 0.05 | 9 | 16 | 0.5 | 0.5 | 0.5 | | | | | | | | | | | | | | | | | |
| 12 | Cucumber | *kg* | 0.9 | 4 | 8 | 1 | 2 | 2 | 1 | 1 | | | | | | | | | | | | | | | |
| 13 | Garlic | *kg* | 0.02 | 24 | 23 | 0.3 | | | | | | | | | | | | | | | | | | | |
| 14 | Onion | *kg* | 0.05 | 12 | 23 | 3 | | | | | | | | | | | | | | | | | | | |
| 15 | Leek | *kg* | 0.15 | 2 | 8 | 3 | 3 | 3 | 3 | | | | | | | | | | | | | | | | |
| 16 | Okra | *kg* | 0.1 | 2 | 14 | 0.2 | 0.3 | 0.3 | 0.4 | 0.4 | 0.4 | 0.4 | 0.4 | 0.4 | 0.4 | 0.2 | 0.2 | 0.2 | | | | | | | |
| 17 | Tomato | *kg* | 0.1 | 2 | 15 | 0.8 | 0.8 | 1 | 1 | 1.2 | 1.2 | 1 | 0.8 | 0.2 | | | | | | | | | | | |
| 18 | Carrot | *kg* | 0.1 | 2 | 13 | 1.5 | 2 | 1.5 | | | | | | | | | | | | | | | | | |
| 19 | Parsley | *pack* | 0.1 | 1 | 8 | 14 | 16 | 16 | 16 | 16 | 16 | 16 | 16 | 16 | 16 | 16 | 14 | 10 | | | | | | | |

# Chapter 13
# Pesticide Leaching Models in a Brazilian Agricultural Field Scenario

**Rômulo Penna Scorza Júnior**
*Embrapa Western Region Agriculture, Brazil*

**Renê Luís de Oliveira Rigitano**
*Federal University of Lavras (UFLA), Brazil*

**Jos J. T. I. Boesten**
*Wageningen University and Research Centre (WUR), The Netherlands*

## ABSTRACT

*The use of Pesticide Leaching Models (PLM) for risk assessment may be an efficient and attractive way of assessing solutions to some agricultural and environmental problems. Many countries of the European Union and the USA have been using PLM for risk assessment already for a few decades. This chapter has the aim to present a successful application of two PLM (i.e. MACRO and PEARL) in a Brazilian very intensive agricultural area to simulate moisture profiles and the leaching of a water flow tracer (i.e. bromide) and the pesticides cyproconazole and thiamethoxam. Also attempts to summarize the available knowledge about the processes governing pesticide behavior in soil, types and classifications of PLM, the use of PLM for risk assessment at European Union, a theoretical description of PEARL and MACRO models and their testing in a Brazilian agricultural field scenario.*

## INTRODUCTION

Pesticides used in agriculture can pose contamination risks to water resources. Therefore, using pesticides while maintaining water resources quality is an important challenge. Groundwater contamination by pesticide residues has been a worldwide problem (Laabs et al., 2002; Bortoluzzi et al., 2006; Terry et al., 2008; Steele et al., 2009; Vryzas et al., 2009). Leaching through the soil profile has been identified as the major cause for the occurrence of pesticides in groundwater. Losses of pesticides by leaching to groundwater can be as high as 5% of the applied amount, but are typically less than 1% (Flury, 1996).

DOI: 10.4018/978-1-61692-871-1.ch013

There is an urgent need to assess the risks of groundwater contamination by pesticides as part of national pesticide registration procedures. This assessment at field scale is time- and money consuming considering all combinations of pesticides, soil and climatic conditions. However, to overcome this limitation, mathematical models have been developed to simulate pesticide leaching at field scale and considering all diversity of pesticides, soils and climate. The first attempts to simulate pesticide field behavior started around 1970 (Leistra, 1971; Walker, 1974). At that time, computational resource was considered a limitation for simulations. Also models were mostly used to optimise pesticide efficiency. After the introduction of personal computers in 1985, complex models were created that considered the interplay of factors affecting pesticide field behavior and more emphasis was given to environmental risk assessment. The use of Pesticide Leaching Models (PLM) for risk assessment may be an efficient and attractive way of assessing solutions to some agricultural and environmental problems. Moreover, PLM are useful for guiding experimentation and integrating, testing and improving the understanding of pesticide field behavior at process-level (Jarvis et al., 1995).

Before the use of any PLM for risk assessment, it is important to evaluate their performance by testing them against field measurements. Vanclooster et al. (2000) states that the validation status of PLM is considered low although several studies aimed to validate these models under various conditions (Scorza Júnior & Boesten, 2005; Castro et al., 2006; Bouraoui, 2007; Scorza Júnior et al., 2007; Bayless et al., 2008). PEARL (Leistra et al., 2002; Tiktak et al., 2002) and MACRO (Larsbo & Jarvis, 2003) are two PLM that have been used within European Union (EU) since 2000 to assess leaching risks as part of the procedure of pesticide registration (see http://focus.jrc.ec.europa.eu/). Although not all countries use PLM for risk assessment of water resources contamination by pesticides, its use has potential for growth by official agencies of various governments.

The main objective of this chapter is to present a successful application of two PLM (i.e. PEARL and MACRO) in a Brazilian very intensive agricultural area to simulate the leaching of a water flow tracer and two pesticides. We first present some background information about definitions, types and classifications of PLM and their use within EU for risk assessment. Thereafter, we describe the theory of the PEARL and MACRO models and their application in a Brazilian agricultural area including aspects such as input parameter estimation and calibration. Finally, some conclusions are drawn for future use of PLM in Brazil.

## PESTICIDE LEACHING MODELS: DEFINITION, CLASSIFICATION AND TYPES

De Wit (1993) classified simulation models in two broad categories: (i) descriptive and (ii) explanatory models. Descriptive models aim to show the existence of relationships among the elements of the system without any explanation. On the other hand, explanatory models aim to explain processes and are formed by different levels of organization or knowledge, which are distinguished themselves by the level of integration at which the process occur. These different integration levels may be classified according to the size of the system, such as molecules, soil particles, peds, soil profile, pedon, field, catchment area and region. The lower integration level is the explanatory level and the upper level is the level to be explained. Following this idea, we can attempt to describe pesticide behavior at the field scale based on laboratory data in combination with weather, soil and crop data collected at the field (Figure 1). This approach implies that no explanation of laboratory behavior of pesticide is attempted and that explanatory models attempt to generalize the knowledge of pesticide behavior under field conditions using the most important pesticide/soil properties. This generalization can be achieved via the modeling research cycle as proposed by Boesten (2000) and shown in Figure 2. In the first step of the modeling research cycle, it is necessary to define the problem for which a model should be developed. After that, a conceptual model is defined and is followed by the mathematical model and a computer code that solves the model equations via a stable and reliable numerical procedure. In the sequence, input parameters are obtained based on laboratory studies. The next important step is to conduct field experiments and obtain measurement that will be compared with the simulations using the laboratory data as input in the model. In the final step, it is necessary to decide whether the PLM test is acceptable or not. If not, it is necessary to reconsider all steps again starting from the conceptual model. Figure 2 implies that experimental methods in the laboratory and in the field are important steps in the modeling research cycle. Hence, inappropriate experimental methods can be a source of error in the model test.

PLM have been classified in several ways. Addiscott & Wagenet (1985) presented a hierarchical scheme for classification of PLM. The first distinction is made between deterministic and stochastic models. A deterministic model assumes that a single set of driving variables and input parameters leads to a single unique outcome. In contrast, a stochastic model assumes that the natural system is variable in space and time. Because of this, a stochastic model considers the model output to be uncertain and is structured to account for this uncertainty. For example, soil hydraulic properties are known to be spatially variable and thus can cause spatial variations in pesticide leaching.

Usually, input values for stochastic models are specified as statistical distributions that will lead to an ensemble of outputs characterized by an average and its statistical limits. Another distinction is made between mechanistic and functional PLM. A mechanistic model incorporates

*Figure 1.Scheme of an explanatory model to describe pesticide behavior at the field scale based on knowledge obtained from laboratory experiments (Adapted from Boesten, 2000)*

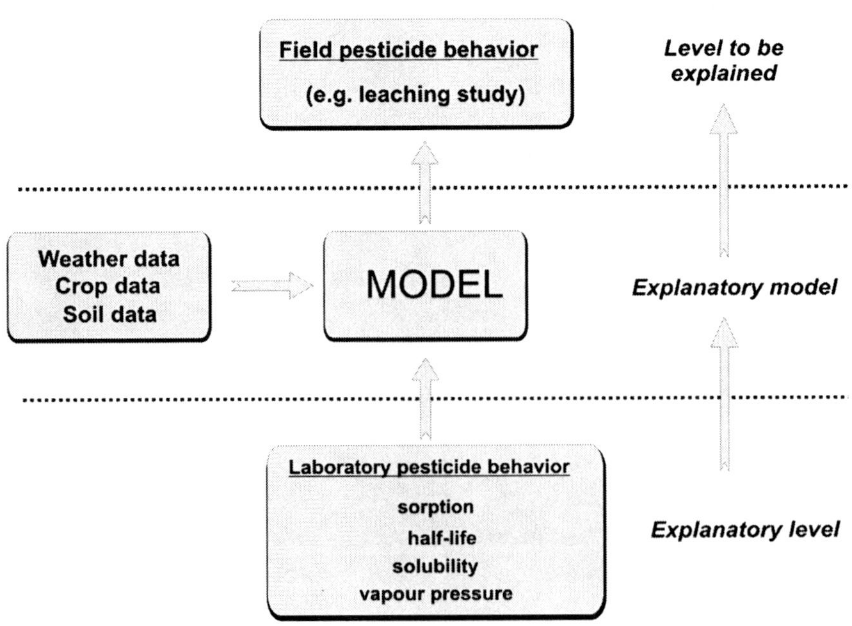

*Figure 2. Scheme of the research cycle for developing and testing models for pesticide behavior at the field scale (Adapted from Boesten, 2000)*

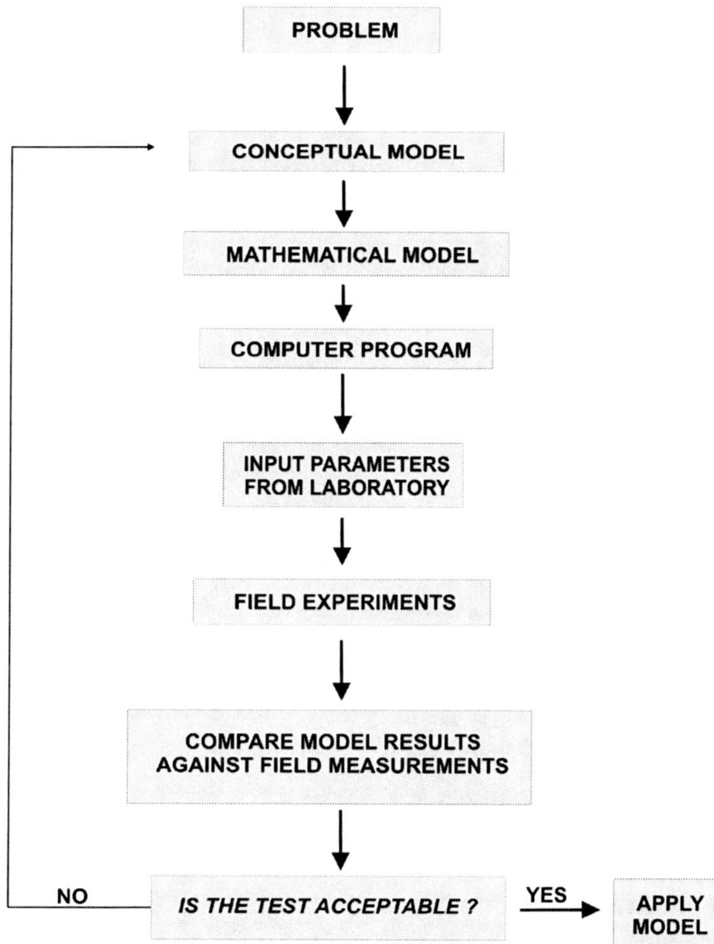


the most fundamental description of the important processes as presently understood. For example, to describe the pesticide leaching flux it uses an equation that combines the mechanisms of mass flow, diffusion and dispersion. A functional model incorporates simplified treatments of some or all processes to describe pesticide leaching. Because of that, functional models require less input data. Jarvis et al. (1995) point out that a clear distinction between mechanistic and functional PLM is difficult to make because PLM can be a mixture of mechanistic and functional submodels. It is important to understand that a mechanistic treatment of a process, in some cases, may not be possible because fundamental knowledge is lacking or because PLM that incorporate description of the current knowledge of a process may be too difficult to parameterize. Finally, there is also a distinction according to the use of PLM. These can be classified as research or management models. Research models are intended to improve the understanding of the processes governing pesticide leaching and help to identify gaps in knowledge. Moreover, research models are also developed to aid the testing of hypotheses. In contrast, management models are mainly used to guide the management of natural and agricultural areas.

PLM can be classified into two broad types: (i) chromatographic flow models (CF) and (ii) preferential flow models (PF). CF models are based on the convection/dispersion equation and thus presume a more or less uniform flow of solute and water through soil. Usually, CF models can describe the bulk leaching reasonably well for soils with uniform flow conditions. However, uniform flow is very unlikely to occur at field scale and there are many experimental evidences that preferential flow can be the rule rather than exception (Flury et al., 1994). Preferential flow is the rapid transport of water in the unsaturated zone of the soil profile via macropores (i.e. shrinkage cracks, earthworm and root channels). This macropore flow results when rates of lateral equilibration of water pressures and/or solute concentrations with the soil matrix are slow in relation to the vertical flow rates in macropores (Jarvis, 2007). Especially in structured and aggregated soils, macropores may dominate the soil hydrology as they function as high-conductivity flow pathways that bypass most of the soil matrix (Beven & Germann, 1982). Based on the advances of macropore flow understanding, PF models have been developed. The approach used in many PF models is to divide the soil into at least two flow domains with different flow rates. Because computer power is rapidly increasing, more complex PF models have been developed based on a more explicit and physically realistic representation of soil structure (Vogel et al., 2006).

## PESTICIDE LEACHING MODELS FOR RISK ASSESSMENT

The most important application of PLM is in the assessment of pesticide leaching to groundwater for pesticide registration. The PLM have been extensively used within EU and USA for risk assessment and there are potential of increasing adoption in other regions and countries. To illustrate how PLM have been used for risk assessment, we take the example in the Netherlands. The leaching assessment in the Netherlands is based on the requirement that the concentration of the pesticide or its relevant metabolites in the groundwater at 10 m depth does not exceed 0.1 $\mu g\ L^{-1}$. A stepwise approach is used for this decision-making regarding pesticide leaching in the Netherlands. This stepwise approach has the following sequence as stated by Boesten (2000): (i) measurements in laboratory studies of the most relevant pesticide properties such as transformation rates, sorption to soils, vapour pressure etc; (ii) assess the pesticide field behavior via modeling (i.e. using PEARL model) based on the properties measured in the laboratory and using simple scenarios; (iii) decide whether more information is needed, based on step (ii), about the pesticide behavior; (iv) if yes, consider more complex scenarios (e.g. based on Geographical Information Systems) and if necessary field or lysimeter experiments; (v) if yes, carry out field or lysimeter experiments etc. One important characteristic of this stepwise approach is to be pragmatic as required in a decision procedure. The first decision in this stepwise approach is based on the use of a PLM. A simple scenario is used for this simulation and the aim is to identify pesticides that will never cause leaching problems. In the Netherlands, the first scenario was developed by Leistra & Dekkers (1976) and used until 1988 when was replaced by a more vulnerable scenario described by Boesten & Van der Linden (1991). This scenario was constructed to be a realistic worst case scenario (i.e. 75% wet year in combination with a soil ranking approximately 80% in vulnerability) and a safety factor is applied to account for uncertainties in the scenario, in the model and in the model input to avoid unwarranted approval of pesticides. It is important to keep in mind that there are some limitations in the performance of PLM for the whole range of relevant scenarios (i.e. combinations of soil types, climatic conditions and pesticides). Because of that, it is important to use a stepwise approach in the risk assessment in which PLM are used to trigger requests for further information and not to provide a definitive judgment (Boesten, 2000).

A simplified proposal, for country level, of tiered approach for risk assessment of pesticide leaching, that uses Brazil as example, is shown in Figure 3. For the first tier, simple calculations using a PLM for realistic worst-case scenarios are required. It is important to consider in these calculations relevant agricultural scenarios for the different regions, if there are some in the country, (in Brazilian case there are, at least, five: South, Southeast, Central-West, Northeast and North) and should be based on pesticide data obtained in laboratory studies (e.g. transformation and sorption studies). If necessary, more than one relevant scenario can be used within the same region for the first tier. Based on the simulations of the first tier, more information may be necessary about the pesticide fate. In this case, we move to the second tier that has the aim to evaluate whether indeed a pesticide has some potential to leach and this conclusion should be based on field experiments.

*Figure 3. Proposal for a Brazilian tiered approach for risk assessment of pesticide leaching*

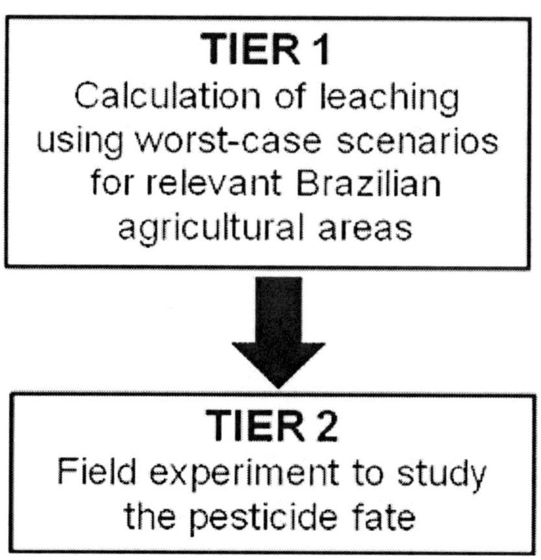


## THEORETICAL DESCRIPTION OF PEARL AND MACRO MODELS

### PEARL Model Description

The PEARL model (Leistra et al., 2002; Tiktak et al., 2002) was developed based on the concepts used in PESTLA (Van den Berg & Boesten, 1998; Boesten & Van der Linden, 1991) and in PESTRAS (Tiktak et al., 1994; Freijer et al., 1996). This model has been used in Dutch pesticide registration since about 1999 and is now being used to support authorities in decisions concerning the approval of pesticide registration in the European Union (FOCUS, 2000).

PEARL uses the SWAP model (Van Dam et al., 2008) as a sub-model for the simulation of water flow and soil temperature. SWAP describes water flow using the Darcy equation for one-dimensional, vertical and transient flow. The combination of the Darcy equation with the conservation equation results in the well-known Richards equation:

$$C\frac{\partial h}{\partial t} = \frac{\partial}{\partial z}\left[K\left(\frac{\partial h}{\partial z}+1\right)\right] - S$$

where $C$ is the water capacity $(d\theta/dh)$ (cm$^{-1}$), h is the soil water pressure head (cm), K is the hydraulic conductivity (cm day$^{-1}$), t is time (day), z is depth in soil (cm) and S is the rate of water uptake by plant roots (day$^{-1}$). To describe the hydraulic functions θ *(h)* and K*(h),* the following analytical functions are used (Van Genuchten, 1980):

$$\theta(h) = \theta_r + \frac{\theta_s - \theta_r}{\left(1+|\alpha h|^n\right)^{1-1/n}}$$

and

$$K\left(h\right)=K_s\frac{\left(\left(1+\left|\alpha h\right|^n\right)^{1-1/n}-\left|\alpha h\right|^{n-1}\right)^2}{\left(1+\left|\alpha h\right|^n\right)^{(1-1/n)(\lambda+2)}}$$

where $\theta$ is the volume fraction of liquid phase (cm$^3$ cm$^{-3}$), $\theta_r$ is the residual volume fraction of liquid phase (cm$^3$ cm$^{-3}$), $\theta_s$ is the saturated volume fraction of liquid phase (cm$^3$ cm$^{-3}$), $K_s$ is the saturated hydraulic conductivity (cm day$^{-1}$), and $\alpha$ (cm$^{-1}$), $\lambda$ (-) and $n$ (-) are fitting parameters.

The rate of water uptake, $S$, is calculated as suggested by Feddes et al. (1978):

$$S=\alpha_{red}S_p$$

in which $\alpha_{red}$ is a reduction factor (-) and $S_p$ is the potential rate of water uptake (day$^{-1}$), which can be obtained from:

$$S_p=\frac{T_p}{d_r}$$

where $T_p$ is the potential transpiration rate (cm day$^{-1}$) and $d_r$ is the root layer thickness (cm). The calculation of potential evapotranspiration rate, $ET_p$ (cm day$^{-1}$),is necessary to quantify the top boundary condition and the water uptake by roots. $ET_p$ is calculated using the Penman-Monteith method (Monteith, 1981; Kroes & Van Dam, 2003). The rate $T_p$ is thereafter obtained using the equation:

$$T_p=(1-f_w)ET_p-E_p$$

in which $f_w$ is the fraction of the day that canopy is wet (-) (i.e. no transpiration) and $E_p$ is the potential soil evaporation rate (cm day$^{-1}$) given by (Belmans et al., 1983):

$$E_p=e^{-kLAI}ET_p$$

in which $\kappa$ is the extinction coefficient for global solar radiation (-) and $LAI$ is the leaf area index (-). For calculation of the actual soil evaporation rate, $E$ (cm day$^{-1}$), $E_p$ is first limited to the maximum flux at soil surface based on the Darcy equation and $E$ is calculated using (Boesten & Stroosnijder, 1986):

$$\sum E=\beta\sqrt{\sum E_p}$$

where $\beta$ is a parameter (cm$^{1/2}$), $\sum E_p$ is the cumulative $E_p$ during a drying cycle and $\sum E$ is the cumulative $E$ during a drying cycle.

Soil temperature, $T$(°C), is simulated using the combination of one-dimensional soil heat flow and the conservation equation for heat in soil given by (Kroes & Van Dam, 2003):

$$C_{heat}\frac{\partial T}{\partial t}=\frac{\partial\left(\lambda_{heat}\frac{\partial T}{\partial z}\right)}{\partial z}$$

where $C_{heat}$ is the volumetric soil heat capacity (J cm$^{-3}$ °C$^{-1}$) and $\lambda_{heat}$ is the thermal conductivity (J cm$^{-1}$ °C$^{-1}$ day$^{-1}$). Both $C_{heat}$ and $\lambda_{heat}$ depend on physical properties of the soil. Daily average air temperature is used as upper boundary condition for simulation of soil temperature.

In PEARL, the total mass concentration of pesticide in the soil system, $C^*$ (g cm$^{-3}$), is defined as the sum of the mass concentrations in the liquid, solid and gas phases:

$$C^*=\theta CL_+\rho X+\varepsilon CG$$

where CL $_i$s the mass concentration of pesticide in the liquid phase (g cm-3$^)$, ρ *is* the dry soil bulk density (g cm-3$^)$, X *is* the sorbed pesticide content (g g-1$^)$, ε *is* the volume fraction of the gas phase (cm3 $^c$m-3$^)$ and CG $_i$s the mass concentration of pesticide in the gas phase (g cm-3$^)$. PEARL assumes that two types of sorption site exist in soil:

equilibrium sites and non-equilibrium sites. The sorption at the equilibrium sites is described by the Freundlich sorption isotherms:

$$X = K_F\, C_{L,REF} \left( \frac{C_L}{C_{L,REF}} \right)^N$$

in which $K_F$ is the Freundlich sorption coefficient ($cm^3\ g^{-1}$), $C_{L,REF}$ is a reference value of $C_L$ (usually set as 1 mg $L^{-1}$) and $N$ is the Freundlich exponent. The parameter $C_{L,REF}$ is introduced in Equation (11) to ensure dimensional consistency and thus $K_F$ is not dependent of the $N$ value (Boesten & Van der Linden, 1991). The sorption to the non-equilibrium sites is not considered in our study (see Leistra et al., 2002, for details of the description of the sorption on the non-equilibrium sites).

The mass flux of pesticide in the soil, $J$ (g $cm^{-2}$ $day^{-1}$), is described by considering convection, dispersion and diffusion in the liquid phase and only diffusion in the gas phase:

$$J = qC_L - D_L \frac{\partial C_L}{\partial z} - D_G \frac{\partial C_G}{\partial z}$$

in which $q$ is the volume flux of water in soil (cm $day^{-1}$), $D_L$ is the hydrodynamic dispersion coefficient ($cm^2\ day^{-1}$) and $D_G$ is the gas diffusion coefficient ($cm^2\ day^{-1}$). $D_L$ and $D_G$ are calculated from:

$$D_L = L_{dis}|q| + \zeta_L D_W$$

and

$$D_G = \zeta G_D A$$

in which $L_{dis}$ is the dispersion length (cm), $D_W$ is the pesticide diffusion coefficient in water ($cm^2$ $day^{-1}$), $D_A$ is the coefficient of pesticide diffusion in air ($cm^2\ day^{-1}$) and $\zeta_L$ and $\zeta_G$ are the tortuosity factors (-) for diffusion in the liquid and gas phase, respectively (Millington & Quirk, 1961).

The transformation rate of pesticide in soil, $R_t$ (g $cm^{-3}$ $day^{-1}$), is described using a first-order equation. If sorption to non-equilibrium sorption sites is not considered, the rate equation is given by:

$$R_t = k(\theta C^*)$$

where $k$ is the first-order transformation rate coefficient ($day^{-1}$). The effect of environmental factors in the pesticide transformation is calculated from:

$$k = f_T f_\theta f_z \frac{\ln(2)}{H_{REF}}$$

in which $f_T$, $f_\theta$ and $f_z$ are reduction factors (-) to account for the influence of temperature, soil moisture content and depth. $H_{REF}$ is the pesticide half-life (day) at reference conditions (i.e. soil collected from the top layer, at 20°C and soil water content corresponding with a pressure head of 100 cm of water). The reduction factor $f_T$ is calculated using Arrhenius equation:

$$f_T = \exp\left[ -\frac{U}{G} \left( \frac{1}{T} - \frac{1}{T_{REF}} \right) \right]$$

where $U$ is the Arrhenius activation energy (kJ $mol^{-1}$), $G$ is the gas constant (kJ $mol^{-1}$ $K^{-1}$), $T$ is soil absolute temperature (K), $T_{REF}$ is $T$ at reference temperature (i.e. 20 ºC). The moisture factor is given by (Walker, 1974):

$$f_\theta = \min \left[ 1, \left( \frac{\theta}{\theta_{REF}} \right)^B \right]$$

in which "min" stands for the minimum value of the two operands, $\theta_{REF}$ is the soil water content at a pressure head of 100 cm of water and $B$ is a constant. Equation (18) assumes that there is no influence of moisture content on $k$ when the soil is wetter than $\theta_{REF}$. The factor $f_z$ is not defined as an analytical function but its value can be specified for each soil horizon.

Pesticide uptake by plants, $R_U$ (g cm$^{-3}$ day$^{-1}$), is described as a passive process that is controlled by the rate of water uptake:

$$R_u = S f_u C_L$$

where $f_u$ is the transpiration stream concentration factor (TSCF), which can be related to the octanol/water partitioning (Briggs et al., 1982).

By combining all these equations, PEARL is based on the following mass conservation equation:

$$\frac{\partial C^*}{\partial t} = \frac{\partial}{\partial z}\left( q C_L - D_L \frac{\partial C_L}{\partial z} - D_G \frac{\partial C_G}{\partial z} \right) - R_t - R_u$$

## MACRO Model Description

MACRO model (Larsbo & Jarvis, 2003) has also been used to support authorities in decisions concerning the approval of pesticide registration in the European Union (FOCUS, 2000). This PF model has been selected because it calculates macropore flow with a physics-based approach that is relatively simple and parsimonious with respect to data requirements. Within the European Union procedures, MACRO is used to predict pesticide leaching to drains for a number of representative worst-case scenarios, some of which are precalibrated with respect to water flow and tracer transport (FOCUS, 2001).

MACRO is a physically based, one-dimensional, numerical model that simulates transient water flow and solute transport at the plot scale. In MACRO, the total porosity of each soil layer is partitioned into micropores and macropores. Each of these flow domains is characterized by a degree of saturation, conductivity and flux. The Richards equation is used to calculate the vertical movement of water in the micropores:

$$C \frac{\partial h_{mi}}{\partial t} = -\frac{\partial}{\partial z}\left[ K_{mi} \left( \frac{\partial h_{mi}}{\partial z} + 1 \right) \right] - \sum S_i$$

where $C$ is the water capacity ($d\theta/dhm_{ij}$ (cm$^{-1)}$, $hm_i$ is the soil water pressure head in the micropores (cm), t is time (day), z is depth (cm), $Km_i$ is the hydraulic conductivity in the micropores (cm day$^{-1)}$ and $Si$ are sink/source terms to account for water exchange with macropores and root water uptake (day$^{-1)}$. Soil water retention in the micropores is described by a modified form of the Van Genuchten equation (Van Genuchten, 1980), while hydraulic conductivity is calculated with the Mualem model (Mualem, 1976). These functions are given by:

$$S_{mi} = \frac{\theta_{mi} - \theta_r}{\theta_b - \theta_r} = \left( 1 + \left| \alpha h \right|^n \right)^{-m}$$

and

$$K_{mi} = K_b \left( \frac{S_{mi}}{S_{mi(\theta_b)}} \right)^l \left[ \frac{\left( 1 - \left( 1 - S_{mi}^{1/m} \right)^m \right)}{\left( 1 - \left( 1 - S_{mi(\theta_b)}^{1/m} \right)^m \right)} \right]^2$$

where $S_{mi}$ is the effective saturation in the micropores (cm$^3$ cm$^{-3}$), $\theta_{mi}$ is the volume fraction of liquid phase in the micropores (cm$^3$ cm$^{-3}$), $\theta_r$ is the residual volume fraction of liquid phase (cm$^3$ cm$^{-3}$), $\theta_b$ is the volume fraction of liquid phase at micropore/macropore boundary (cm$^3$ cm$^{-3}$), $h$ is the soil water pressure head (cm), $\alpha$ (cm$^{-1}$), $n$ (-), and $m$ (-) are fitting parameters (where $m$ is equal to 1-1/$n$), $K_{mi}$ is the hydraulic conductivity in the micropores (cm day$^{-1}$), $K_b$ is the saturated hydraulic conductivity in the micropores (cm day$^{-1}$), $l$ is the tortuosity factor in the micropores (-) and $S_{mi(\theta_b)}$ is the effective saturation (cm$^3$ cm$^{-3}$) for the soil water pressure head at the boundary between micropores and macropores ($h_b$). $S_{mi(\theta_b)}$ is given by:

$$S_{mi(\theta_b)} = \left[1 + \left(\left|\alpha\, h_b\right|\right)^n\right]^{-m}$$

A simplified approach is used to describe the water flow in the macropores, assuming a non-capillary, gravity-driven process (i.e. $\partial h/\partial z=0$). Therefore, description of the soil water pressure in the macropores is not required. The hydraulic conductivity function in the macropores, $K_{ma}$ (cm day$^{-1}$), is given as a simple power law function:

$$K_{ma} = \left(K_s - K_b\right) S_{ma}^{\,n^*}$$

and

$$S_{ma} = \frac{\theta_{ma}}{\theta_s - \theta_b}$$

where $K_s$ is the saturated hydraulic conductivity (cm day$^{-1}$), $S_{ma}$ is the effective saturation in the macropores (cm$^3$ cm$^{-3}$), $n^*$ is a kinematic exponent related to the macropore size distribution (-), $\theta_{ma}$ is the volume fraction of liquid phase in the macropores (cm$^3$ cm$^{-3}$) and $\theta_s$ is the saturated volume fraction of liquid phase (cm$^3$ cm$^{-3}$). Water exchange rates between micropores and macropores, $S_W$ (day$^{-1}$), are calculated as a function of an effective aggregate half-width, $d$ (cm), using an approximate physically based first-order function, which neglects the influence of gravity and assumes a rectangular slab geometry for the aggregates (Van Genuchten & Dalton, 1986; Booltink et al., 1993):

$$S_W = \left(\frac{G_f\, D_W\, \gamma_W}{d^2}\right)\left(\theta_b - \theta_{mi}\right)$$

where $G_f$ is a dimensionless geometry factor (set at a value of 3 in MACRO for a rectangular slab geometry as defined by Gerke & Van Genuchten, 1996), $D_W$ is an effective water diffusivity (cm$^2$ day$^{-1}$) and $\gamma_W$ is a dimensionless scaling factor to match the approximate and exact solutions to the diffusion problem (Gerke & Van Genuchten, 1993). In MACRO, $\gamma_W$ is set at an average value of 0.8. $D_W$ is given by:

$$D_W = \left(\frac{D_{\theta_{s(mi)}} + D_{\theta_{mi}}}{2}\right)$$

where $D_{\theta_{s(mi)}}$ is the water diffusivity at the saturated volume fraction of liquid phase in the micropores (cm$^2$ day$^{-1}$) and $D_{\theta_{mi}}$ is the water diffusivity at the current volume fraction of liquid phase in the micropores (cm$^2$ day$^{-1}$). $D_{\theta_{mi}}$ is given by (Van Genuchten, 1980):

$$D_{\theta_{mi}} = \left[\frac{(1-m)K_S^*}{\alpha\, m\left(\theta_b - \theta_r\right)}\right] S_{mi}^{\,l-1/m}\left[\left(1 - S_{mi}^{\,1/m}\right)^{-m} + \left(1 - S_{mi}^{\,1/m}\right)^{m} - 2\right]$$

where $K_S^*$ is a "fictitious" saturated hydraulic conductivity (cm day$^{-1}$) obtained by extrapolating the micropore conductivity to soil water pressure head to zero (Equation 23) and is given by:

$$K_S^* = K_b\left(\frac{1}{S_{mi(\theta_b)}}\right)^l\left[1 - \left(1 - S_{mi(\theta_b)}^{\,1/m}\right)^m\right]^{-2}$$

$D_{\theta_{s(mi)}}$ in Equation (28) is given by setting $S_{mi}$ in Equation (30) to $S_{mi(\theta_b)}$,which is calculated using Equation (24).

The rate of root water uptake, $S_U$ (day$^{-1}$), is calculated using a simple empirical function as described by Jarvis (1989):

$$S_U = \left(\frac{E}{\Delta z_i}\right)\left(\frac{r_i\,\omega_i}{\omega^*}\right)$$

where $E$ is the actual transpiration from the crop (cm day$^{-1}$), $\Delta z_i$ is the thickness of layer $i$ (cm), $r_i$ is the proportion of total root length assumed to be distributed logarithmically with depth (-), $\omega_i$ is the water stress reduction factor in layer $i$ (-) and $\omega^*$ is a water stress index (-). The root water uptake can occur in both regions although it is preferentially extracted from the macropores. A threshold type of response is assumed for $\omega_i$ that accounts for soil conditions that are too wet or too dry. $S_U$ is assumed to be zero if soil is saturated or if it is at the wilting point (Feddes et al., 1976). $\omega^*$ is given by:

$$\omega^* = \sum_{i=1}^{n} r_i\,\omega_i$$

where $n$ is the number of soil layers in the profile containing roots. MACRO needs daily or hourly rainfall and meteorological data (i.e. daily maximum and minimum air temperatures, daily wind speed, daily vapour pressure and daily global radiation), used in the calculation of potential evapotranspiration by the Penman-Monteith method. Based on this, potential transpiration rate from cropped surfaces and potential evaporation rate from the soil are calculated as a function of the crop surface resistance and aerodynamic resistances from soil to canopy and from canopy to air (Monteith, 1981). Thereafter, the actual soil evaporation rate is calculated by comparing the potential evaporation rate from the soil with the maximum possible rate of water supply to the soil surface.

The top boundary condition for water flow in PF models is very critical because it determines the partitioning of net precipitation and/or irrigation between micropores and macropores (Beven & Germann, 1981; Bronswijk, 1988). Basically, three different situations of flow can occur at the soil surface in MACRO. In the first situation, the precipitation rate is smaller than the infiltration capacity of the micropores. Then all rainfall will infiltrate into the micropores. In the second situation, precipitation is greater than the infiltration capacity of the soil matrix but smaller than infiltration capacity of the macropores. Then both micropores and macropores receive water. Finally, a third situation can occur in which the precipitation rate exceeds the total infiltration capacity of the soil. Then the excess of water at the soil surface leads to overland flow.

Soil temperatures are calculated from the heat conductivity equation as given by Equation 9. Daily average air temperature is used as upper boundary condition for simulation of soil temperature.

Solute transport in the micropores is calculated using the convection-dispersion equation:

$$\frac{\partial\left(\theta_{mi}\,C_{L,mi} + (1-f)\,X\,\rho\right)}{\partial t} = \frac{\partial}{\partial z}\left(D\,\theta_{mi}\,\frac{\partial C_{L,mi}}{\partial z} - qC_{L,mi}\right) - \sum R_i$$

where $C_{L,mi}$ is the concentration of solute in the liquid phase in the micropores (g cm$^{-3}$), $f$ is the fraction of sorption sites attributed to the macropores, $X$ is the content of solute sorbed onto the solid phase (g g$^{-1}$), $\rho$ is the dry soil bulk density (g cm$^{-3}$), $D$ is the dispersion coefficient (cm$^2$ day$^{-1}$), $q$ is the volume flux of water in soil (cm day$^{-1}$) and $\sum R_i$ represents the source/sink terms regarding mass exchange between flow domains, solute uptake by plant roots and solute transformation. Coefficient $D$ is calculated from:

$$D = L_{dis}\,v_{mi} + D_0\zeta L$$

in which $L_{dis}$ is the dispersion length (cm), $v_{mi}$ is the pore water velocity in the micropores (cm day$^{-1}$) given by $q/\theta m_{i,}$ $D0$ is the coefficient of solute diffusion in the water (cm2 day-$^{1)}$ and $\zeta L$ is the tortuosity factor (-) for diffusion in the liquid phase (Millington & Quirk, 1961). In macropores, solute dispersion is neglected because solute transport is assumed to be dominated by convection. The partitioning of reactive solutes between liquid and solid phases in either flow domain is described using the Freundlich isotherm:

$$X = K_F C_{REF} \left( \frac{C_{L,mi/ma}}{C_{REF}} \right)^N$$

where $K_F$ is the Freundlich sorption coefficient ($cm^3\ g^{-1}$), $C_{L,mi/ma}$ is the concentration of solute in the liquid phase in either micropores or macropores ($g\ cm^{-3}$), $N$ is the Freundlich exponent and $C_{REF}$ is a reference concentration ($g\ cm^{-3}$) introduced to make the units of $K_F$ independent of $N$ (Boesten, 1994).

The rate of solute exchange between micropores and macropores, $R_e$ ($g\ cm^{-3}\ day^{-1}$), is given by a combination of diffusion and convection (Van Genuchten & Dalton, 1986; Valocchi, 1990):

$$R_e = \left( \frac{G_f D_e \theta_{mi}}{d^2} \right) \left( C_{L,ma} - C_{L,mi} \right) + S_W C_{L,mi/ma}$$

where $D_e$ is an effective diffusion coefficient ($cm^2\ day^{-1}$), $C_{L,ma}$ and $C_{L,mi}$ are mass concentrations of solute in the liquid phase ($g\ cm^{-3}$) in macropores and micropores, respectively, and $C_{L,mi/ma}$ indicates either $C_{L,ma}$ or $C_{L,mi}$ depending on the direction of the water flow. Coefficient $D_e$ is approximated by:

$$D_e = D_0 \zeta L_S m_a$$

The rate of solute uptake by plant roots, $R_U$ ($g\ cm^{-3}\ day^{-1}$), is described as a passive process given by:

$$R_U = f_u S_u C_{L,mi/ma}$$

where $f_U$ is the transpiration stream concentration factor (-) (Boesten & Van der Linden, 1991).

Pesticide transformation in each soil phase and flow domain is described with first-order kinetics. The actual transformation rate in the field, $k$ ($day^{-1}$), is predicted from a laboratory-measured reference value, $k_{REF}$ ($day^{-1}$), using reduction factors to account for the influence of environmental conditions (Boesten & Van der Linden, 1991):

$$K = k_{REF} f_T f_\theta$$

where $f_T$ and $f_\theta$ are reduction factors (-) to account for the influence of temperature and soil moisture content respectively. Factor $f_T$ is given by a numerical approximation of the Arrhenius equation (Boesten, 1994) modified for low temperatures:

$$f_T = e^{\alpha(T - T_{REF})} \quad \text{for } T > 5°C$$

$$f_T = \left( \frac{T}{5} \right) e^{\alpha(5 - T_{REF})} \quad \text{for } 0 \leq T \leq 5°C$$

$$f_T = 0 \quad \text{for } T \leq 0°C$$

where $T$ is soil temperature (°C), $T_{REF}$ is $T$ at reference conditions (°C) and $\alpha$ is a composite parameter ($°C^{-1}$) dependent on $T$, $T_{REF}$, molar activation energy and the gas constant. The reduction factor $f_\theta$ is calculated from:

$$f_\theta = 1 \quad \text{for } \theta > \theta_{100}$$

$$f_\theta = \left[ \left( \theta - \frac{\theta_W}{2} \middle/ \theta_{100} - \frac{\theta_W}{2} \right) \right]^B \quad \text{for } \frac{\theta_W}{2} < \theta < \theta_{100}$$

$$f_\theta = 0 \quad \text{for } \theta < \frac{\theta_W}{2}$$

where $\theta$ is the volume fraction of liquid phase ($cm^3\ cm^{-3}$), $\theta_{100}$ is the volume fraction of liquid phase at a pressure head of 100 cm of water ($cm^3\ cm^{-3}$), $\theta_W$ is the volume fraction of liquid phase at the wilting point ($cm^3\ cm^{-3}$) and $B$ is a constant.

The water routed into the macropores is characterized by a concentration $C_{ma}$ (g cm$^{-3}$), which is calculated assuming complete mixing with solute stored in the soil liquid phase of a shallow mixing depth $z_d$ (cm) at the soil surface (Steenhuis & Walter, 1980):

$$C_{ma} = \frac{Q_S + \left(P\,C_P\right)}{P + \left(z_d\left(\theta_{mi,sur} + \left(1-f\right)\,\rho_{sur}\,K_F\right)\right)}$$

where $Q_S$ is the amount of solute stored in $z_d$ (g cm$^{-2}$), $C_P$ is the solute concentration in the precipitation (g cm$^{-3}$), $\theta_{mi,sur}$ is the volumetric fraction of liquid phase in the mixing depth $z_d$ (cm$^3$ cm$^{-3}$) and $\rho_{sur}$ is the dry soil bulk density in the mixing depth $z_d$ (g cm$^{-3}$).

## TESTING OF PEARL AND MACRO MODELS FOR PESTICIDE LEACHING IN A BRAZILIAN AGRICULTURAL FIELD SCENARIO

### Field Experiment

The field experiment was carried out in an area of 0.09 ha (30 x 30 m) located at *Embrapa Western Region Agriculture* in Dourados, Mato Grosso do Sul State, Brazil (22° 16' 26" S, 54° 48' 50" W, altitude of 408 m above sea level). The experimental field was divided into four plots containing 25 subplots of 3 x 3 m. Meteorological data were obtained from an automatic station located in an experimental area of Embrapa, a Brazilian governmental agriculture research institution, which is situated about 500 m from the experimental field. The soil is classified as Oxisol (Embrapa, 1997) and the basic properties are shown in Table 1. The clay content increases with depth from 630 g kg$^{-1}$ at soil surface to 713 g kg$^{-1}$ at 100 cm depth. These high clay contents are typical of heavy clay soils in Brazil. These soils have aggregates that have been formed as a result of flocculation of clays. The average macroporosity within the whole soil profile was about 14%. The organic matter content decreases with depth from 33.4 g kg$^{-1}$ at soil surface to 11.3 g kg$^{-1}$ at 100 cm depth. Undisturbed soil cores (three replicates) were taken at five different depths (i.e. 0-10, 10-30, 30-50, 50-70, 70-100 cm) in only one pit at the experimental field for the determination of soil water retention curves. In-situ measurements of saturated hydraulic conductivities were also carried out in this pit at five different depths using a permeameter of Guelph-type (Vieira, 1998). Soybean was sown at the experimental area on 21 November 2007 and harvested on 7 April 2008. Before soybean cultivation, all previous crop residues and straw at soil surface were removed from the experimental area.

All chemicals (cyproconazole, thiamethoxam and bromide) were applied separately from each other on 4 December 2007, using a tractor-mounted boom sprayer and using 280 L of water ha$^{-1}$. The bromide applied dose was 24 kg ha$^{-1}$ of KBr. Pesticides doses were 282 g a.i. ha$^{-1}$ of thiamethoxam as Engeo Pleno® and 240 g a.i. ha$^{-1}$ of cyproconazole as Alto 100®. Soil profiles (0-100

*Table 1. Basic soil properties at the experimental field in Dourados, Mato Grosso do Sul State, Brazil. OMC = Organic matter content; BD = dry soil bulk density*

| Properties | Layer (cm) | | | | |
|---|---|---|---|---|---|
| | 0-10 | 10-30 | 30-50 | 50-70 | 70-100 |
| OMC (g kg$^{-1}$) | 33.4 | 29.6 | 22.7 | 16.9 | 11.3 |
| BD (g cm$^{-3}$) | 1.45 | 1.36 | 1.12 | 1.09 | 1.07 |
| Clay (g kg$^{-1}$) | 630 | 663 | 697 | 713 | 713 |
| Sand (g kg$^{-1}$) | 245 | 229 | 212 | 195 | 195 |
| pH-KCl | 4.8 | 4.4 | 4.5 | 4.6 | 4.6 |

cm depth) were sampled at four times (0, 15, 73, and 156 days after application) during the experimental period. For each sampling date, three subplots were randomly selected and pits excavated till 100 cm depth. About 200 g of soil for each sampling depth were put in plastic bags and stored at -20°C until analysis. The 15 soil samples (5 depths and 3 subplots) on each sampling date were individually analyzed for moisture, bromide and pesticide residues.

## Laboratory Experiments

The degradation rates of cyproconazole and thiamethoxam were measured in soil samples taken from 0-30 and 50-70 cm depth. These soil samples were collected within the experimental field before its installation and application of pesticides. Analysis of blank samples from 0-30 and 50-70 cm depth at the experimental field did not show any remaining residues of these two pesticides in the soil. Before incubation, these soil samples were shadow-dried for two days, crumbled, mixed and sieved in a 1 mm mesh size. Thereafter, portions of 50 g of dry soil were added to glass cups and distilled water was added in each sample to reach 80% of field capacity (i.e. soil moisture at a pressure head of 100 cm of water). A total of 12 samples were prepared for each depth that corresponded to two replicates and six sampling dates. To each sample 1 mL of a solution containing analytical standards of thiamethoxam and cyproconazole was added, giving a total mass of 11.75 and 10 μg, respectively. This 1 mL was mixed through the soil. These amounts were based on the doses applied in the field.

All glass cups were capped with aluminum foil in which holes were made to ensure aerobic conditions. Thereafter, all samples were incubated at 25 ± 2°C in darkness. Soil moisture contents during incubation were adjusted every two days by adding distilled water until it reached 80% of field capacity using an electronic balance. At increasing time intervals (0, 14, 28, 56, 112 and 224 days after application), thiamethoxam and cyproconazole were extracted from the soil and data about remaining concentration as a function of time were fitted to first-order equation (Equation 15).

All degradation rate measurements for both pesticides and depths could be described well with the first-order kinetics (Figure 4). Determination coefficients ($R^2$) for the fitting were, for both pesticides, 0.7 ($P<0.05$) at 0-30 cm and 0.8 ($P<0.05$) at 50-70 cm depth. The corresponding half-lives assuming first-order kinetics were 142 and 113 days for cyproconazole at 0-30 and 50-70 cm depth, respectively, and 501 and 613 days for thiamethoxam at 0-30 and 50-70 cm depth, respectively. Coefficients of variation of the estimated half-life values for both pesticides and depths were between 13 and 23%. In general, Figure 4 shows that cyproconazole degraded faster than thiamethoxam for all depths. Thiamethoxam degradation rate decreased with increasing soil depth. This is probably the result of a decrease in the density of microbial population with depth. However, the opposite was observed with cyproconazole (Figure 4). The degradation rate of cyproconazole at 0-30 cm was slower than at 50-70 cm depth. Possibly, this was the result of a higher organic matter content at 0-30 cm than at 50-70 cm (Table 1) that lead to a higher sorption of cyproconazole at 0-30 cm. Because of that, less cyproconazole was available in the soil solution for degradation at 0-30 cm than at 50-70 cm. Urzedo et al. (2006) found half-life values for thiamethoxam in incubation studies for a Brazilian soil at 25°C between 117 and 301 days. These shorter half-life values found by Urzedo et al. (2006) compared to our study may be explained by the presence of light as thiamethoxam can be photodegraded (Maienfsch et al., 2001).

Sorption of cyproconazole and thiamethoxam in the soil was measured using a batch-equilibration approach as described by Lavorenti et al. (2003). The same soil samples collected at 0-30 and 50-70 cm depth for the degradation studies

*Figure 4. Remaining mass of cyproconazole and thiamethoxam as a function of time as measured in laboratory soil incubation studies at 25 ± 2°C and for 0-30 and 50-70 cm depth. The points are the measurements and the lines are fits assuming first-order kinetics. Note that the scales of horizontal axes are different*

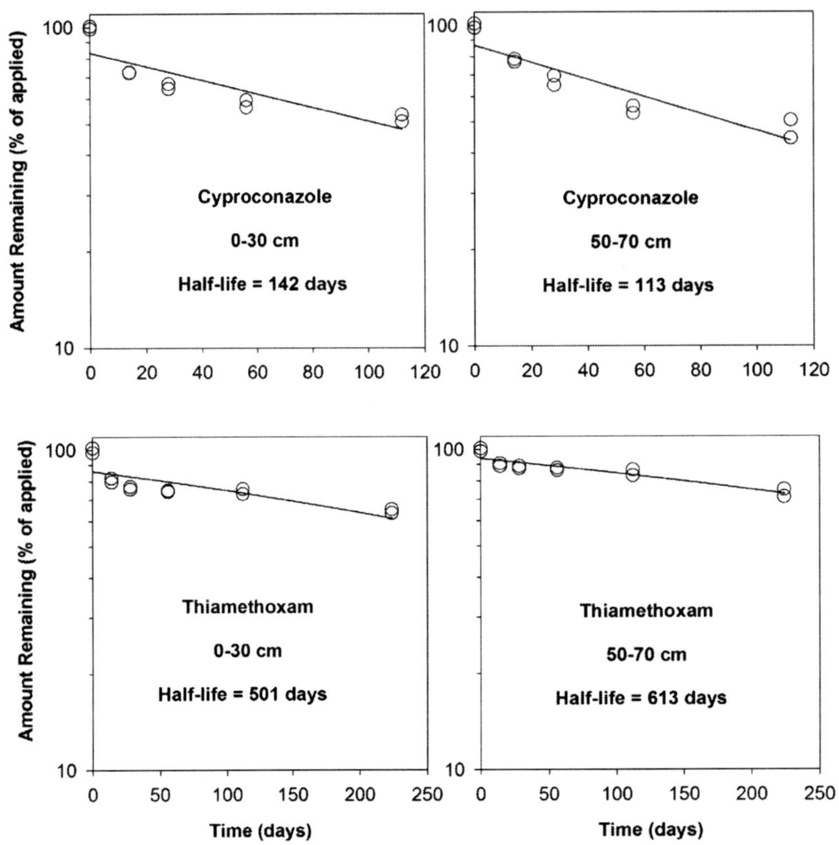


were used for sorption measurements. These soil samples were also shadow-dried for two days, crumbled, mixed and sieved in a 1 mm mesh size. Then, cyproconazole and thiamethoxam solutions with concentrations of 0.1; 0.5; 1; 5; and 10 µg $mL^{-1}$ were prepared in distilled water containing 0.01 M of $CaCl_2$. A mass of 2 g of soil for 0-30 and 50-70 cm depth was added to cap centrifuge tubes and 5 mL of the five different concentrations for each pesticide were added to each centrifuge tube. Two replicates were used for each concentration and depth. Thereafter, all tubes were rotated at 25°C during four hours to ensure sorption equilibrium. After this procedure, tubes were centrifuged for 10 min at 25°C and using a frequency of 10,510 g. An aliquot of 2 mL of the supernatant was collected and cyproconazole and thiamethoxam concentrations were measured. The measurements of cyproconazole and thiamethoxam sorption were described with the Freundlich equation.

Cyproconazole and thiamethoxam sorption at 0-30 and 50-70 cm depth could be described well with the Freundlich equation (Figure 5). Determination coefficients ($R^2$) for the fitting were, for both pesticides and depths, 0.9 ($P<0.05$). Fitted Freundlich sorption coefficients ($K_F$ in Equations 11 and 35) were 3.93 $cm^3\ g^{-1}$ at 0-30 cm and 1.46

cm$^3$ g$^{-1}$ at 50-70 cm for cyproconazole and 1.18 cm$^3$ g$^{-1}$ at 0-30 cm and 0.47 cm$^3$ g$^{-1}$ at 50-70 cm for thiamethoxam. The Freundlich exponents were: 0.85 at 0-30 cm and 0.74 at 50-70 cm for cyproconazole and 0.84 at 0-30 cm and 0.74 at 50-70 cm for thiamethoxam. The greater $K_F$ values for both pesticides at 0-30 cm depth than at 50-70 cm depth suggest stronger sorption at this depth. This can be explained by the higher organic matter content at 0-30 cm than at 50-70 cm depth (Table 1) as the soil organic matter is the most important component affecting sorption of thiamethoxam (Urzedo et al.,2006) and cyproconazole.

## PEARL and MACRO Simulations

### Estimation of Input Parameters for Soil Water Flow and Solute Behavior in PEARL

Parameters for the calculation of the potential evapotranspiration rate (i.e. wind velocity, relative humidity for air, hours of sunlight in the day and maximum and minimum daily air temperature) during the experimental period were obtained from the meteorological station. Rainfall rates were also measured and recorded on daily basis at the station. The bottom boundary condition used in the simulations was set to free drainage. This was done because no measurements of groundwater

*Figure 5. Sorption isotherms as measured for cyproconazole and thiamethoxam at 25 ± 2°C and for 0-30 and 50-70 cm depth. Points are the measurements and lines are fits to Freundlich isotherms*

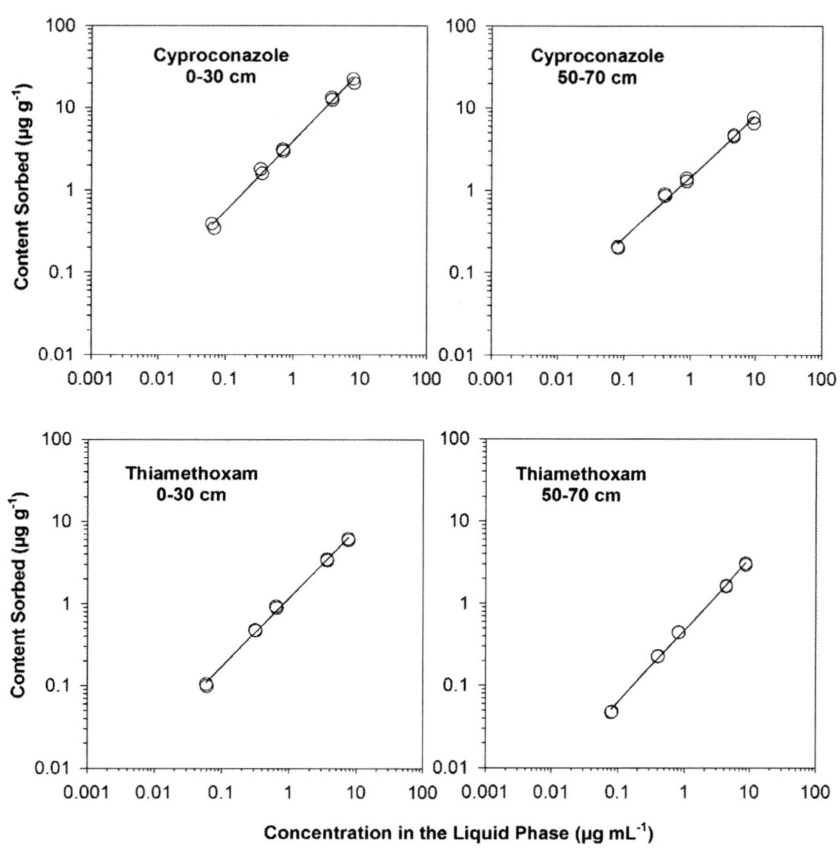

levels were available during the experimental period and also because previous measurements at the experimental area did show very deep groundwater levels (always deeper than 1 m below soil surface). Crop characteristics were obtained from measurements and also from literature. One soybean crop was grown within the simulation period. This was sown on 21 November 2007 and harvested on 07 April 2008. *LAI* (Equation 7) was obtained from Vissotto Júnior (2003) and the following values were used in the simulations: 0.5 on 26 November and 17 December 2007, 1.5 on 16 January 2008, 3 on 20 February 2008, 2.5 on 4 April 2008 and 2 on 7 April 2008. PEARL interpolates linearly between the specified *LAI* values for other dates. The estimation of $d_r$ was based on visual observations at different sampling dates. It was set to 10 cm on 26 November 2007 and increased to 30 and 60 cm on 17 December 2007 and on 16 January 2008, respectively, remaining constant to 80 cm until crop harvest. The factors $\alpha_{red}$ (Equation 4) as a function of soil water pressure head were obtained from Wesseling (1991). No water uptake by crop ($\alpha_{red}$ = 0) was possible for soil water pressure heads ≥ 0 and ≤ -16 000 cm. The maximum water uptake ($\alpha_{red}$ = 1) occurred between soil water pressure heads of -1 and -500 cm and a linear decrease/increase of $\alpha_{red}$ occurred in the ranges between -1 and 0 cm and between -16000 and -500 cm. The parameter $\beta$ (Equation 8) was set to 0.63 $cm^{1/2}$, which is in the range of 0.54 and 0.96 $cm^{1/2}$ proposed by Boesten (1986).

The soil profile was divided into six layers (0-10, 10-30, 30-50, 50-70, 70-100 and 100-200 cm depth) and the Van Genuchten parameters (Table 2) were derived from site-specific laboratory measurements using the RETC package (Van Genuchten et al., 1991). For $\theta_s$ it was assumed that it equals to soil moisture content at a pressure head of 60 cm of water (Claessen, 1997).

The application rate of bromide used in the simulations was 12.3 kg $ha^{-1}$ (the average amount recovered from soil just after application). This rate corresponds to 79% of the initially calculated dosage. The dispersion length, $L_{DIS}$ in Equation 13, was set to 5 cm, which is within the range of 3 to 10 cm proposed by Van Ommen et al. (1989). The reference diffusion coefficient of bromide in water, $D_W$ in Equation 13, was set to 1.3 $cm^2$ $day^{-1}$ (Weast, 1974). We assumed no transformation, no sorption and no anion exclusion of bromide in the simulations.

The application rate of cyproconazole used in the simulations was 0.20 kg $ha^{-1}$ (the average amount recovered from soil just after application). This rate corresponds to 87% of the initially calculated dosage. Pesticide/soil input parameters were derived from laboratory studies as shown in section "Laboratory experiments". The Freundlich sorption coefficient and the Freundlich exponent, $K_F$ and $N$ in Equation 11, were set to 3.93 $cm^3$ $g^{-1}$ and 0.85, respectively, at 0-30 cm soil layer. Below 30 cm, we considered that $K_F$ was reduced by a factor of 0.37 as shown by the laboratory experiments using 50-70 cm soil layer that gave $K_F$ equal to 1.46 $cm^3$ $g^{-1}$. Cyproconazole half-life ($H_{REF}$ in Equation 16) at topsoil (0-30 cm) was set to 142 days, which was based on laboratory measurements (see section "Laboratory experiments"). The Arrhenius activation energy, $U$ in Equation 17, was set to 68.1 kJ $mol^{-1}$ that is the

*Table 2. Van Genuchten parameters used as input values for simulations in PEARL. For definitions of the parameters see Equations 2 and 3*

| Input | Layer (cm) | | | | |
|---|---|---|---|---|---|
| | 0-10 | 10-30 | 30-50 | 50-70 | 70-100 |
| $\theta_S$ ($cm^3$ $cm^{-3}$) | 0.58 | 0.54 | 0.45 | 0.45 | 0.44 |
| $\theta_r$ ($cm^3$ $cm^{-3}$) | 0.20 | 0.27 | 0.14 | 0.13 | 0.06 |
| $\alpha$ ($cm^{-1}$) | 0.0186 | 0.0294 | 0.0097 | 0.0084 | 0.0052 |
| $n$ (-) | 1.955 | 1.957 | 1.437 | 1.446 | 1.487 |
| λ (-) | 0.5 | 0.5 | 0.5 | 0.5 | 0.5 |
| $K_s$ (cm $day^{-1}$) | 91 | 98 | 149 | 72 | 171 |

range found after reviewing 50 experiments with different soils and pesticides (FOCUS, 1997). The values of $f_Z$ in Equation 16 were set to 1.0 for the 0-30 cm layer and to 1.26 for the layers below 30 cm depth. This value of 1.26 was calculated by dividing $H_{REF}$ at 0-30 cm by the one at 50-70 cm depth. The plant uptake parameter, $f_U$ in Equation 19, was set to 0.47 based on the model proposed by Briggs et al. (1982) and using an octanol/water partitioning coefficient (i.e. log $K_{OW}$) equal to 2.91. The default value of 0.7 (Boesten & Van der Linden, 1991) was used for the parameter $B$ in Equation 18.

The application rate of thiamethoxam used in the simulations was 0.21 kg ha$^{-1}$ (the average amount recovered from soil just after application). This rate corresponds to 76% of the initially calculated dosage. Pesticide/soil input parameters were derived from laboratory studies as shown in section "Laboratory experiments". The Freundlich sorption coefficient and the Freundlich exponent, $K_F$ and $N$ in Equation 11, were set to 1.18 cm$^3$ g$^{-1}$ and 0.84, respectively, at 0-30 cm soil layer. Below 30 cm, we considered that $K_F$ was reduced by a factor of 0.40 as shown by the laboratory experiments using 50-70 cm soil layer that gave $K_F$ equal to 0.47 cm$^3$ g$^{-1}$. Thiamethoxam half-life ($H_{REF}$ in Equation 16) at topsoil (0-30 cm) was set to 501 days, which was based on laboratory measurements (see section "Laboratory experiments"). The Arrhenius activation energy, $U$ in Equation 17, was set to 68.1 kJ mol$^{-1}$ that is the range found after reviewing 50 experiments with different soils and pesticides (FOCUS, 1997). The values of $f_Z$ in Equation 16 were set to 1.0 for the 0-30 cm layer and to 0.82 for the layers below 30 cm depth. This value of 0.82 was calculated by dividing $H_{REF}$ at 0-30 cm by the one at 50-70 cm depth. The plant uptake parameter, $f_U$ in Equation 19, was set to 0.18 based on the model proposed by Briggs et al. (1982) and using an octanol/water partitioning coefficient (i.e. log $K_{OW}$) equal to -0.13. The default value of 0.7 (Boesten & Van der Linden, 1991) was used for the parameter $B$ in Equation 18.

## Estimation of Input Parameters for Soil Water Flow and Solute Behavior in MACRO

To provide adequate numerical accuracy, we considered a soil profile of 2 m that was divided into 60 numerical compartments with thicknesses varying between 0.3 (topsoil) and 4.85 cm (subsoil). Six different soil horizons (i.e. 0-10, 10-30, 30-50, 50-70, 70-100, and 100-200) were identified. Initial values of soil temperature and water content were fixed, respectively, at 15°C and 0.25 cm$^3$ cm$^{-3}$. Because of that, simulations were set to start two months before pesticide application to allow system stabilization (see section "Testing strategy and evaluation of model performance"). Parameters for the calculation of the potential evapotranspiration rate (i.e. wind velocity, relative humidity for air, hours of sunlight in the day and maximum and minimum daily air temperature) during the experimental period were obtained from the meteorological station. Rainfall rates were also measured and recorded on hourly basis at the weather station. The bottom boundary condition used in the simulations was set to free drainage. This was done because no measurements of groundwater levels were available during the experimental period and also because previous measurements at the experimental area did show very deep groundwater levels (the level was always deeper than 1 m below soil surface). Van Genuchten parameters to describe hydraulic functions in the micropores (Table 3) were derived from site-specific laboratory measurements using the RETC package (Van Genuchten et al., 1991). The assumed soil water pressure head at the micropore/macropore boundary was 10 cm for all layers.

Crop characteristics were obtained from measurements and also from literature. One soybean crop cycle did occur within the simulation period that was sown on 21 November 2007 and har-

*Table 3. Hydraulic parameters used as input values in MACRO. For definitions of the parameters see Equations 22, 23 and 27*

| Input | Layer (cm) | | | | |
|---|---|---|---|---|---|
| | 0-10 | 10-30 | 30-50 | 50-70 | 70-100 |
| $\theta_S$ (cm$^3$ cm$^{-3}$) | 0.59 | 0.50 | 0.55 | 0.58 | 0.59 |
| $\theta_r$ (cm$^3$ cm$^{-3}$) | 0 | 0 | 0 | 0 | 0 |
| $\theta_b$ (cm$^3$ cm$^{-3}$) | 0.37 | 0.43 | 0.47 | 0.42 | 0.45 |
| $h_b$ (cm) | 10 | 10 | 10 | 10 | 10 |
| $n$ (-) | 1.955 | 1.957 | 1.437 | 1.446 | 1.487 |
| $n^*$ (-) | 3 | 3 | 3 | 3 | 3 |
| $l$ (-) | 0.5 | 0.5 | 0.5 | 0.5 | 0.5 |
| $\alpha$ (cm$^{-1}$) | 0.0186 | 0.0294 | 0.0097 | 0.0084 | 0.0052 |
| $K_S$ (cm day$^{-1}$) | 91 | 98 | 149 | 72 | 171 |
| $K_b$ (cm day$^{-1}$) | 0.22 | 0.22 | 0.22 | 0.22 | 0.22 |
| $d$ (mm) | 450 | 450 | 450 | 450 | 450 |

vested on 07 April 2008. Leaf area indexes at different times were obtained from Vissotto Júnior (2003). The maximum root depth was based on field observations and set to 80 cm. It was assumed that 60% of the total root length was present in the top 20 cm of the soil profile.

The application rate of bromide used in the simulations was 12.3 kg ha$^{-1}$ (the average amount recovered from soil just after application). This rate corresponds to 79% of the initially calculated dosage. The dispersion length, $L_{DIS}$ in Equation 34, was set to 5 cm, which is within the range of 3 to 10 cm proposed by Van Ommen et al. (1989). The reference diffusion coefficient of bromide in water, $D_0$ in Equation 34, was set to 1.3 cm$^2$ day$^{-1}$ (Weast, 1974). It was assumed for bromide that the excluded soil water content due to anion exclusion was 5% of the micropore volume fraction and that no transformation and no sorption did occur in the simulations. The mixing depth, $z_d$ in Equation 42, was set at 0.1 cm for all simulations.

The application rate of cyproconazole used in the simulations was 0.20 kg ha$^{-1}$ (the average amount recovered from soil just after application). This rate corresponds to 87% of the initially calculated dosage. Pesticide/soil input parameters were derived from laboratory studies as shown in section "Laboratory experiments". The Freundlich sorption coefficient, $K_F$ in Equation 35, was set to 3.93 cm$^3$ g$^{-1}$ at 0-30 cm and to 1.46 cm$^3$ g$^{-1}$ at 50-70 cm soil layer. They were measured values as shown in section "Laboratory experiments". Between 30-50 cm and below 50 cm, we considered that $K_F$ was proportional to soil organic matter content, resulting in $K_F$ values of 2.67 and 0.98 cm$^3$ g$^{-1}$, respectively. The Freundlich exponent, $N$ in Equation 35, was set to 0.85 at 0-50 cm and to 0.74 below 50 cm depth. The transformation rates for cyproconazole at reference conditions were 0.0060 day$^{-1}$ for the 0-30 cm layer, 0.0041 day$^{-1}$ for the 30-50 cm layer, 0.0065 day$^{-1}$ for the 50-70 cm layer and 0.0044 day$^{-1}$ for the 70-100 cm layer. These transformation rates were based on laboratory experiments and converted to the reference conditions. Moreover, it was assumed that these transformation rates were the same in both phases and in both flow domains within each soil layer. The exponent $B$ in Equation 41 was set at 0.7 (Boesten & Van der Linden, 1991) and the parameter $\alpha$ in Equation 40 was calculated to be 0.097 K$^{-1}$. The plant uptake parameter, $f_U$ in Equation 38, was set to 0.47 based on the model proposed by Briggs et al. (1982) and using an octanol/water partitioning coefficient (i.e. log $K_{OW}$) equal to 2.91. The coefficient of cyproconazole diffusion in the water, $D_0$ in Equation 34, was set at 4.9 x 10$^{-10}$ cm$^2$ day$^{-1}$.

The application rate of thiamethoxam used in the simulations was 0.21 kg ha$^{-1}$ (the average amount recovered from soil just after application). This rate corresponds to 76% of the initially calculated dosage. Pesticide/soil input parameters were derived from laboratory studies as shown in section "Laboratory experiments". The Freundlich sorption coefficient, $K_F$ in Equation 35, was set to 1.18 $cm^3$ $g^{-1}$ at 0-30 cm and to 0.47 $cm^3$ $g^{-1}$ at 50-70 cm soil layer. They were measured values as shown in section "Laboratory experiments". Between 30-50 cm and below 50 cm, we considered that $K_F$ was proportional to soil organic matter content, resulting in $K_F$ values of 0.8 and 0.32 $cm^3$ $g^{-1}$, respectively. The Freundlich exponent, $N$ in Equation 35, was set to 0.86 for the whole soil profile. The transformation rates for thiamethoxam at reference conditions were 0.0023 $day^{-1}$ for the 0-30 cm layer, 0.0016 $day^{-1}$ for the 30-50 cm layer, 0.0012 $day^{-1}$ for the 50-70 cm layer and 0.0008 $day^{-1}$ for the 70-100 cm layer. These transformation rates were based on laboratory experiments and converted to the reference conditions. Moreover, it was assumed that these transformation rates were the same in both phases and in both flow domains within each soil layer. The exponent $B$ in Equation 41 was set at 0.7 (Boesten & Van der Linden, 1991) and the parameter $\alpha$ in Equation 40 was calculated to be 0.097 $K^{-1}$. The plant uptake parameter, $f_U$ in Equation 38, was set to 0.18 based on the model proposed by Briggs et al. (1982) and using an octanol/water partitioning coefficient (i.e. $\log K_{OW}$) equal to -0.13. The coefficient of thiamethoxam diffusion in the water, $D_0$ in Equation 34, was set at 4.9 x $10^{-10}$ $cm^2$ $day^{-1}$.

The fraction of sorption sites attributed to the macropores for both pesticides, $f$ in Equations 33 and 42, was fixed at the default value in the model of 0.02. Dubus et al. (2000) reported that $f$ values of between 0.01 and 0.04 are considered realistic for many soils.

## Testing Strategy and Evaluation of Model Performance

Simulations were carried out for the period from 1 October 2007 to 9 May 2008 and the outputs were compared with soil moisture profiles, bromide concentrations profiles and the two pesticides concentration profiles at different sampling dates. Important to point out that simulation for both PLM started two months before the beginning of the field experiment in order to allow system stabilization. By doing this, we minimized the influence of initial conditions on simulation outputs for the first sampling date. The performance of both tested PLM (i.e. PEARL and MACRO) was assessed visually by graphical display. For this, moisture and concentration profiles of simulated and measured values are presented.

A stepwise approach was used for the testing of both PLM (Armstrong et al., 1996; Vanclooster et al., 2000). In this approach, PLM test is a sequential process, which evaluates each submodel separately and sequentially. As a first step, the submodel for water flow for each PLM is tested and calibrated using moisture profiles measured in the field. Thereafter, PLM is tested against the bromide concentration profiles (i.e. the water flow tracer) using the calibrated water flow. Then, PLM is calibrated against the bromide concentration profiles. After this calibration, PLM was tested against field measurements of cyproconazole and thiamethoxam leaching.

## Uncalibrated and Calibrated PEARL and MACRO Simulations

Cumulative precipitation during the entire experimental period (i.e. 4 December 2007 until 8 May 2008) was 823 mm and the cumulative potential evapotranspiration was 587 mm. The greatest variations in soil moisture profiles for all sampling dates were observed between 0-30 cm depth due to high demand from atmosphere and crop uptake (Figure 6). Below 10 cm depth, the soil moisture content did not decrease below 0.3 $cm^3$ $cm^{-3}$ in all sampling dates. Uncalibrated simulations of water flow using PEARL resulted in moisture profiles that were much drier than measured between 10 and 100 cm depth (Figure 6). Differently from PEARL, uncalibrated simulations of moisture profiles using MACRO were reasonably good with the exception on 19 Dec. 2007 between 30 and 100 cm depth when there was clear tendency to overestimate the moisture values. These discrepancies suggested that the soil water retention curves measured in the laboratory were not representative for the experimental field. This lack of representativeness may have been caused by inadequate handling of spatial variability inasmuch as samples were taken from only one pit in the 0.09 experimental field that may not be representative for the entire area. The next step was to calibrate the water flow submodel using the measured moisture profiles. After calibration of soil water retention curves for all depths, PEARL and MACRO gave good simulations of soil moisture profiles for all sampling dates (Figure 6). Bromide was already found in the whole soil profile until 100 cm depth after 15 days of application (Figure 6). The cumulative precipitation after 15 days of application was 49 mm. Considering the piston flow approach and the average of soil moisture content equal to 0.35 cm$^3$ cm$^{-3}$, it would be expected that bromide had leached to 14 cm depth. This can be considered a strong evidence of preferential transport in this Brazilian Oxisol. PEARL tended to overestimate bromide leaching even after calibration of dispersion length ($L_{dis}$ in Equation 13) and of the saturated hydraulic conductivities ($K_s$ in Equation 3) for all depths. MACRO overestimated bromide concentrations between 0 and 50 cm depth on 15 February and 8 May 2008 (Figure 6). However, after calibration of $K_{mi}$, $d$ and $L_{dis}$ there was a good improvement in MACRO simulations of bromide leaching for all sampling dates.

Both pesticides did not leach below 50 cm depth for an accumulated precipitation of 823 mm (Figure 7). It would be expected more leaching for the more mobile pesticide thiamethoxam. It might be that the bulk of thiamethoxam leaching was protected against preferential flow of water in the macropores. In other words, macropore flow may have reduced thiamethoxam leaching in this clay soil as also observed by Larsson &

*Figure 6. Measured and simulated soil moisture and bromide concentration profiles for three sampling dates at the experimental field in Dourados, Mato Grosso do Sul State, Brazil. Solid black line is measured average and the areas within the dashed black lines are measured values plus and minus two times the standard deviations. Red dotted and dashed-dot-dot lines are uncalibrated and calibrated PEARL simulations, respectively, and blue solid and dashed lines are uncalibrated and calibrated MACRO simulations*

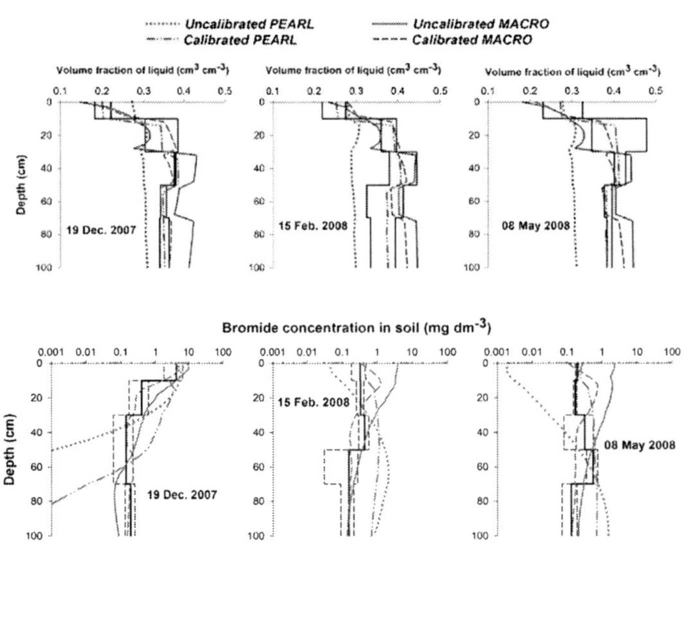

Jarvis (1999). Usually, macropore flow is considered to increase the risks of groundwater contamination by pesticides but these results suggest that this process reduced the risks of thiamethoxam leaching to groundwater.

There was a tendency for PEARL and MACRO to overestimate concentrations of both pesticides in the soil profile for all sampling dates (Figure 7). It is a clear indication that persistence estimated in the laboratory was not representative to describe field conditions. Then, the next step was to calibrate half-lives using the total amount of both pesticides in the soil profile. However, this procedure resulted in a decrease of half-lives for both pesticides that was sufficient to describe the fast decline in concentrations just after application (i.e. between application and first sampling date on 19 December 2007). Because of that, we had to reduce the amount applied used as input value for PEARL and MACRO. Probably this fast decline of concentrations in the field just after application was due to volatilization considering there was no precipitation after application and soil temperatures may have reached about 40°C or even higher. After calibration of persistence and the reduction in the amount applied used as input values, PEARL gave good simulation for leaching of both pesticides between 0 and 10 cm depth on 19 December 2007 but tended to overestimate it for the following two sampling dates (Figure 7). MACRO tended to overestimate the leaching of both pesticides for the 10-50 cm depth on 19 December 2007 but gave good simulations for all depths on the following two sampling dates.

## FUTURE RESEARCH DIRECTIONS

Brazilian government still does not use PLM for risk assessment. If there is an intention for using them by Brazilian authorities in the near future,

*Figure 7. Measured and simulated cyproconazole and thiamethoxam concentration profiles for three sampling dates at the experimental field in Dourados, Mato Grosso do Sul State, Brazil. Solid black line is measured average and the areas within the dashed black lines are measured values plus and minus two times the standard deviations. Red dotted and dashed-dot-dot lines are uncalibrated and calibrated PEARL simulations, respectively, and blue solid and dashed lines are uncalibrated and calibrated MACRO simulations*

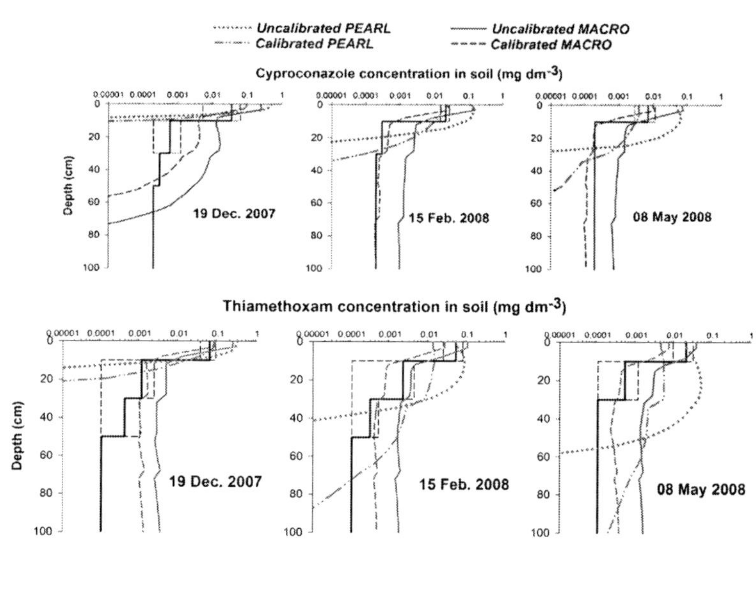

it is desirable to increase confidence in these models by performing in parallel more research on testing PLM for some important Brazilian agricultural areas. To do so, high quality datasets (i.e. detailed measurements of pesticide behavior in the field and in the laboratory) under Brazilian agricultural scenarios will be necessary. Some few studies carried out under Brazilian conditions have shown that important dissipation and transport processes of pesticides in the field deserve special attention, for example, volatilization and preferential transport. Pesticides that seem to be almost non-volatile in temperate regions can be very volatile under tropical conditions as soil surface temperatures can be very high. Modeling pesticide volatilization from soil surface before the first significant precipitation is still a challenge because the knowledge of the rate-determining processes is insufficient. Although there are some good models for simulation of preferential transport, more research is necessary to define measurement procedures of some important input parameters in these models. Another point that deserves special attention for research is the definition of criteria for the pesticide leaching assessment in Brazil. For example, definition of field scenarios (i.e. soil and climate types) to be used during the evaluation procedures, the toxicological threshold values for pesticide concentration for registration and also which PLM will be used for this. In conclusion, much research has to be done in order to test and use PLM in Brazil to support a scientifically sound pesticide leaching assessment in Brazil in future.

## CONCLUSION

Simulation of environmental fate of pesticides has become an important tool for risk assessment worldwide during the last decades. Knowledge of the main processes affecting pesticide leaching in soils has developed at a rapid pace and it has allowed the development and testing of pesticide leaching models (PLM). Moreover, these PLM have proved useful in guiding experimentation and in integrating and testing our understanding about the complex soil-plant-pesticide system. Giving the fact that PLM have to predict accurately very small amounts of the applied dose (i.e. 0.1% in some situations), it implies that these tools have to be reliably validated. However, modeling pesticide leaching at such low levels requires a very precise understanding of all relevant processes concerning pesticide transport in soil. A complication with regard to pesticide transport in soil is the occurrence of preferential transport that seems to be a rule more than an exception in many Brazilian agricultural soils. Modeling and simulating preferential transport of pesticides in soils is data demanding and calibration of some sensitive and difficult to measure parameters are necessary. However it is always important to keep in mind that there should be a balance between calibration and predictive ability of the model. Both PEARL and MACRO testing for simulation of moisture profiles, water flow tracer and two pesticides in a Brazilian agricultural area was successful after some calibration. Especially for pesticide concentration in soil, good simulations were obtained. This suggests that, after more testing studies in Brazil, PLM may be promising and useful tools to help authorities and government to take decisions about pesticide risks resulting from leaching to groundwater.

## REFERENCES

Addiscott, T. M. A., & Wagenet, R. J. (1985). Concepts of solute leaching in soils: a review of modeling approaches. *Journal of Soil Science*, *36*, 411–424. doi:10.1111/j.1365-2389.1985.tb00347.x

Armstrong, A. C., Portwood, A. M., Leeds-Harrison, P. B., Harris, G. L., & Catt, J. A. (1996). The validation of pesticide leaching models. *Pesticide Science*, *48*, 47–55. doi:10.1002/(SICI)1096-9063(199609)48:1<47::AID-PS434>3.0.CO;2-F

Bayless, E. R., Capel, P. D., Barbash, J. E., Webb, R. M. T., Hancock, T. C., & Lampe, D. C. (2008). Simulated fate and transport of metolachlor in the unsaturated zone, Maryland, USA. *Journal of Environmental Quality*, *37*, 1064–1072. doi:10.2134/jeq2006.0562

Belmans, C., Wesseling, J. G., & Feddes, R. A. (1983). Simulation of the water balance of a cropped soil: SWATRE. *Journal of Hydrology (Amsterdam)*, *63*, 271–286. doi:10.1016/0022-1694(83)90045-8

Beven, K., & Germann, P. (1981). Water flow in soil macropores. II. A combined flow model. *Journal of Soil Science*, *32*, 15–29. doi:10.1111/j.1365-2389.1981.tb01682.x

Beven, K., & Germann, P. (1982). Macropores and water flow in soils. *Water Resources Research*, *18*, 1311–1325. doi:10.1029/WR018i005p01311

Boesten, J. J. T. I. (1986). *Behaviour of herbicides in soil: simulation and experimental assessment*. Ph.D. thesis, Wageningen University and Research Centre, Wageningen.

Boesten, J. J. T. I. (1994). Simulation of bentazon leaching in sandy loam soil from Mellby (Sweden) with the PESTLA model. *Journal of Environmental Science and Health Part A*, *29*, 1231–1253. doi:10.1080/10934529409376105

Boesten, J. J. T. I. (2000). From laboratory to field: uses and limitations of pesticide behaviour models for the soil/plant system. *Weed Research*, *40*, 123–138. doi:10.1046/j.1365-3180.2000.00158.x

Boesten, J. J. T. I., & Stroosnijder, L. (1986). Simple model for daily evaporation from fallow tilled soil under spring conditions in a temperate climate. *Netherlands Journal of Agricultural Science*, *34*, 75–90.

Boesten, J. J. T. I., & Van der Linden, A. M. A. (1991). Modeling the influence of sorption and transformation on pesticide leaching and persistence. *Journal of Environmental Quality*, *20*, 425–435. doi:10.2134/jeq1991.00472425002000020015x

Booltink, H. W. G., Hatano, R., & Bouma, J. (1993). Measurement and simulation of bypass flow in a structured clay soil: a physico-morphological approach. *Journal of Hydrology (Amsterdam)*, *148*, 149–168. doi:10.1016/0022-1694(93)90257-A

Bortoluzzi, E. C., Rheinheimer, D. S., Gonçalves, C. S., Pellegrini, J. B. R., Zanella, R., & Copetti, A. C. C. (2006). Contaminação de águas superficiais por agrotóxicos em função do uso do solo numa microbacia hidrográfica de Agudo, RS. *Revista Brasileira de Engenharia Agrícola e Ambiental*, *10*(4), 881–887.

Bouraoui, F. (2007). Testing the PEARL model in the Netherlands and Sweden. *Environmental Modelling & Software*, *22*, 937–950. doi:10.1016/j.envsoft.2006.06.004

Briggs, G. G., Bromilow, R. H., & Evans, A. A. (1982). Relationships between lipophilicity and root uptake and translocation of non-ionised chemicals by barley. *Pesticide Science*, *13*, 495–504. doi:10.1002/ps.2780130506

Bronswijk, J. J. B. (1988). Modeling of water balance, cracking and subsidence of clay soils. *Journal of Hydrology (Amsterdam)*, *97*, 199–212. doi:10.1016/0022-1694(88)90115-1

Castro, N. R. A., Rigitano, R. L. O., Lima, J. M., & Scorza Júnior, R. P. (2006). Avaliação do simulador MACRO (versão 5.0) para predição da lixiviação do inseticida tiametoxam em duas classes de solos. *Pesticidas: Revista Brasileira de Ecotoxicologia e Meio Ambiente*, *16*, 1–12.

Claessen, M. E. C. (1997). *Manual de métodos de análise de solo*. Rio de Janeiro, Brazil: Centro Nacional de Pesquisa de Solos.

De Wit, C. T. (1993). Fundamentals of dynamic simulation: philosophy and terminology. In Leffelaar, P. A. (Ed.), *On systems analysis and simulation of ecological processes: with examples in CSMP and FORTRAN* (pp. 3–9). Dordrecht: Kluwer.

Dubus, I. G., Beulke, S., & Brown, C. D. (2000). *A briefing document on the application of inverse modeling techniques to pesticide leaching models*. Silsoe, UK: Cranfield University.

Embrapa. (1997). *Sistema brasileiro de classificação de solos*. Rio de Janeiro, Brazil: Centro Nacional de Pesquisa de Solos.

Feddes, R. A., Kowalik, P. J., Kolinska-Malinka, K., & Zaradny, H. (1976). Simulation of field water uptake by plants using a soil water dependent root extraction function. *Journal of Hydrology (Amsterdam)*, *31*, 13–26. doi:10.1016/0022-1694(76)90017-2

Feddes, R. A., Kowalik, P. J., & Zaradny, H. (1978). *Simulation of field water use and crop yield*. Wageningen, The Netherlands: PUDOC.

Flury, M. (1996). Experimental evidence of transport of pesticides through field soils – a review. *Journal of Environmental Quality*, *25*, 25–45. doi:10.2134/jeq1996.00472425002500010005x

Flury, M., Flüher, H., Jury, W. A., & Leuenberger, J. (1994). Susceptibility of soils to preferential flow of water: a field study. *Water Resources Research*, *30*, 1945–1954. doi:10.1029/94WR00871

FOCUS. (1997). *Surface water models and EU registration of plant protection products: Final report of the work of the Regulatory Modelling Working Group on Surface Water*. EC document reference 6476/VI/96.

FOCUS. (2000). *FOCUS groundwater scenarios in the EU review of active substances*: *Report of the FOCUS groundwater scenarios workgroup*. EC document reference Sanco/312/2000 rev.2.

FOCUS. (2001). *FOCUS surface water scenarios in the EU evaluation process under 91/414/EEC: Report of the FOCUS Working Group on Surface Water Scenarios*. EC Document Reference SANCO/4802/2001-rev.2.

Freijer, J. I., Tiktak, A., Hassanizadeh, S. M., & Van der Linden, A. M. A. (1996). PESTRAS: *Vol. 3. 1.: a one dimensional model for assessing leaching, accumulation and volatilization of pesticides in soil*. Bilthoven, The Netherlands: RIVM.

Gerke, H. H., & Van Genuchten, M. Th. (1993). Evaluation of a first-order water transfer term for variably-saturated dual-porosity flow models. *Water Resources Research*, *29*, 1225–1238. doi:10.1029/92WR02467

Gerke, H. H., & Van Genuchten, M. Th. (1996). Macroscopic representation of structural geometry for simulating water and solute movement in dual-porosity media. *Advances in Water Resources*, *19*, 343–357. doi:10.1016/0309-1708(96)00012-7

Jarvis, N. J. (1989). A simple empirical model for root water uptake. *Journal of Hydrology (Amsterdam)*, *107*, 57–72. doi:10.1016/0022-1694(89)90050-4

Jarvis, N. J. (2007). A review of non-equilibrium water flow and solute transport in soil macropores: principles, controlling factors and consequences for water quality. *European Journal of Soil Science*, *58*, 523–546. doi:10.1111/j.1365-2389.2007.00915.x

Jarvis, N. J., Bergström, L. F., & Brown, C. D. (1995). Pesticide leaching models and their use for management purposes. In Roberts, T. R., & Kearney, P. C. (Eds.), *Environmental behaviour of agrochemicals* (pp. 185–220). New York: John Wiley.

Kroes, J. G., & Van Dam, J. C. (Eds.). (2003). *Reference manual SWAP version 3.0.3*. Wageningen, The Netherlands: Alterra.

Laabs, V., Amelung, W., Pinto, A., Wantzen, M., Silva, C. J., & Zech, W. (2002). Pesticides in surface water, sediment, and rainfall of the northeastern Pantanal basin, Brazil. *Journal of Environmental Quality*, *31*, 1636–1648. doi:10.2134/jeq2002.1636

Larsbo, M., & Jarvis, N. J. (2003). *MACRO 5.0. A model of water flow and solute transport in macroporous soil: technical description*. Uppsala, Sweden: Swedish University of Agricultural Sciences.

Larsson, M. H., & Jarvis, N. J. (1999). Evaluation of a dual-porosity model to predict field-scale solute transport in a macroporous soil. *Journal of Hydrology (Amsterdam)*, *215*, 153–171. doi:10.1016/S0022-1694(98)00267-4

Lavorenti, A., Prata, F., & Regitano, J. B. (2003). Comportamento de pesticidas em solos – fundamentos. *Tópicos em Ciência do Solo*, *3*, 291–334.

Leistra, M. (1971). Diffusion of 1,3-dichloropropene from a plane source in soil. *Pesticide Science*, *2*, 75–79. doi:10.1002/ps.2780020206

Leistra, M., & Dekkers, W. A. (1976). Computed leaching of pesticides from soil under field conditions. *Water, Air, and Soil Pollution*, *5*, 491–500. doi:10.1007/BF00280848

Leistra, M., Van der Linden, A. M. A., Boesten, J. J. T. I., Tiktak, A., & Van den Berg, F. (2002). *PEARL model for pesticide behaviour and emissions in soil-plant systems: description of the processes in FOCUS PEARL v. 1.1.1*. Wageningen, The Netherlands: Alterra.

Maienfisch, P., Angst, M., Brandl, F., Fischer, W., Hofer, D., & Kayser, H. (2001). Chemistry and biology of thiamethoxam: a second generation neonicotinoid. *Pest Management Science*, *57*, 906–913. doi:10.1002/ps.365

Millington, R. J., & Quirk, J. P. (1961). Permeability of porous solids. *Transactions of the Faraday Society*, *57*, 1200–1207. doi:10.1039/tf9615701200

Monteith, J. L. (1981). Evaporation and surface temperature. *Quarterly Journal of the Royal Society*, *107*, 1–27. doi:10.1256/smsqj.45101

Mualem, Y. (1976). A new model for predicting the hydraulic conductivity of unsaturated porous media. *Water Resources Research*, *12*, 513–522. doi:10.1029/WR012i003p00513

Scorza Júnior, R. P., & Boesten, J. J. T. I. (2005). Simulation of pesticide leaching in a cracking clay soil with the PEARL model. *Pest Management Science*, *61*, 432–448. doi:10.1002/ps.1004

Scorza Júnior, R. P., Jarvis, N. J., Boesten, J. J. T. I., Van der Zee, S. E. A. T. M., & Roulier, S. (2007). Testing MACRO (version 5.1) for pesticide leaching in a Dutch clay soil. *Pest Management Science*, *63*, 1011–1025. doi:10.1002/ps.1434

Steele, G. V., Johnson, H. M., Sandstrom, M. W., Capel, P. D., & Barbash, J. E. (2009). Occurrence and fate of pesticides in four contrasting agricultural settings in the United States. *Journal of Environmental Quality*, *37*, 1116–1132. doi:10.2134/jeq2007.0166

Steenhuis, T. S., & Walter, M. F. (1980). Closed form solution for pesticide loss in runoff water. *Transactions of the ASAE. American Society of Agricultural Engineers*, *23*, 615–620.

Terry, A. S., Carter, A. D., Humphrey, R. L., Capri, E., Grua, B., & Panagopoulous, A. C. (2008). A monitoring programme for 1,3-dichloropropene and metabolites in groundwater in five EU countries. *Pest Management Science*, *64*, 923–932. doi:10.1002/ps.1585

Tiktak, A., Van den Berg, F., Boesten, J. J. T. I., Van Kraalingen, D., Leistra, M., & Van der Linden, A. M. A. (2002). *Manual of FOCUS PEARL version 1.1.1*. Bilthoven, The Netherlands: RIVM/Alterra.

Tiktak, A., Van der Linden, A. M. A., & Swartjes, F. A. (1994). *PESTRAS: A one dimensional model for assessing leaching and accumulation of pesticides in soil*. Bilthoven, The Netherlands: RIVM.

Urzedo, A. P. F. M., Rigitano, R. L. O., Guerreiro, M. C., & Castro, N. R. A. (2006). Dissipação do inseticida tiametoxam em solos da região de Lavras-MG. *Pesticidas: Revista de Ecotoxicologia e Meio Ambiente*, *16*, 31–38.

Valocchi, A. J. (1990). Use of temporal moment analysis to study reactive transport in aggregated porous media. *Geoderma*, *46*, 233–247. doi:10.1016/0016-7061(90)90017-4

Van Dam, J. C., Groenendijk, P., Hendriks, R. F. A., & Kroes, J. G. (2008). Advances of modeling water flow in variably saturated soils with SWAP. *Vadose Zone Journal*, *7*(2), 640–653. doi:10.2136/vzj2007.0060

Van den Berg, F., & Boesten, J. J. T. I. (1998). *Pesticide leaching and accumulation model (PESTLA) version 3.3. Description and user's guide*. Wageningen, The Netherlands: SC-DLO.

Van Genuchten, M. Th. (1980). A closed form equation for predicting the hydraulic conductivity of unsaturated soils. *Soil Science Society of America Journal*, *44*, 892–898. doi:10.2136/sssaj1980.03615995004400050002x

Van Genuchten, M. Th., & Dalton, F. N. (1986). Models for simulating salt movement in aggregated field soils. *Geoderma*, *38*, 165–183. doi:10.1016/0016-7061(86)90013-3

Van Genuchten, M. Th., Leij, F. J., & Yates, S. R. (1991). *The RETC code for quantifying the hydraulic functions of unsaturated soils*. Riverside, CA: US Salinity Laboratory.

Van Ommen, H. C., Van Genuchten, M. Th., Van der Molen, W. H., Dijksma, R., & Hulshof, J. (1989). Experimental and theoretical analysis of solute transport from a diffuse source of pollution. *Journal of Hydrology (Amsterdam)*, *105*, 225–251. doi:10.1016/0022-1694(89)90106-6

Vanclooster, M., Boesten, J. J. T. I., Trevisan, M., Brown, C. D., Capri, E., & Eklo, O. M. (2000). A European test of pesticide-leaching models: methodology and major recommendations. *Agricultural Water Management*, *44*, 1–19. doi:10.1016/S0378-3774(99)00081-5

Vieira, S.R. (1998). Permeâmetro: novo aliado na avaliação de manejo do solo. *O Agronômico, 47/50*, 32-33.

Vissotto Júnior, D. (2003). *Modelagem numérica para interações superfície-atmosfera no Estado do Paraná*. Master dissertation, Federal University of Paraná, Curitiba, PR.

Vogel, H. J., Cousin, I., Ippisch, O., & Bastian, P. (2006). The dominant role of structure for solute transport in soil: experimental evidence and modeling of structure and transport in a field experiment. *Hydrology and Earth System Sciences*, *10*, 495–506. doi:10.5194/hess-10-495-2006

Vryzas, Z., Vassiliou, G., Alexoudius, C., & Papadopoulou-Mourkidou, E. (2009). Spatial and temporal distribution of pesticide residues in surface waters in northeastern Greece. *Water Research*, *43*(1), 1–10. doi:10.1016/j.watres.2008.09.021

Walker, A. (1974). A simulation model for prediction of herbicide persistence. *Journal of Environmental Quality*, *3*, 396–401. doi:10.2134/jeq1974.00472425000300040021x

Weast, R. C. (1974). *Handbook of chemistry and physics*. Cleveland, OH: CRC Press.

Wesseling, J. G. (1991). *Meerjarige simulaties van grondwateronttrekking voor verschillende bodemprofielen, grondwatertrappen en gewassen met het model SWATRE*. Wageningen, The Netherlands: SC-DLO.

## ADDITIONAL READING

Allaire, S. E., Roulier, S., & Cessna, A. J. (2009). Quantifying preferential flow in soils: a review of different techniques. *Journal of Hydrology (Amsterdam)*, *378*, 179–204. doi:10.1016/j.jhydrol.2009.08.013

Arias-Estévez, M., López-Periago, E., Martínez-Carballo, E., Simal-Gándara, J., Mejuto, J. C., & Garcia-Río, L. (2008). The mobility and degradation of pesticides in soils and the pollution of groundwater resources. *Agriculture Ecosystems & Environment*, *123*, 247–260. doi:10.1016/j.agee.2007.07.011

Beulke, S., Brown, C. D., & Jarvis, N. J. (2001). MACRO: a preferential flow model to simulate pesticide leaching and movement to drains. In Linders, J. B. H. J. (Ed.), *Modelling of environmental chemical exposure and risk* (pp. 117–132). Amsterdam: Kluwer Academic Publishers.

Boesten, J. J. T. I. (2000). Modeller subjectivity in estimating pesticide parameters for leaching models using the same laboratory data set. *Agricultural Water Management*, *44*, 389–409. doi:10.1016/S0378-3774(99)00102-X

Boesten, J. J. T. I. (2007). Simulation of pesticide leaching in the field and in zero-tension lysimeters. *Vadose Zone Journal*, *6*(4), 793–804. doi:10.2136/vzj2007.0067

Boesten, J. J. T. I., & Van der Pas, L. J. T. (2000). Movement of water, bromide and the pesticides ethoprophos and bentazone in a sandy soils: the Vredepeel data set. *Agricultural Water Management*, *44*, 21–42. doi:10.1016/S0378-3774(99)00082-7

Bouraoui, F. (2007). Testing the PEARL model in the Netherlands and Sweden. *Environmental Modelling & Software*, *22*, 937–950. doi:10.1016/j.envsoft.2006.06.004

Celia, M. A. (1998). Fundamental concepts for numerical simulation of contaminant transport and biodegradation. In S.K. Sikdar, & R.L. Irvine (Eds.), *Bioremediation: Principles and Practice. Volume 1: Fundamentals and Applications* (pp. 59-88). Lancaster: Technomic Publishing.

Cheng, H. H. (Ed.). (1990). *Pesticides in the soil environment: processes, impacts, and modeling*. Madison, WI: SSSA.

Coppola, A., Comegna, V., Basile, A., Lamaddalena, N., & Severino, G. (2009). Darcian preferential water flow and solute transport through bimodal porous systems: experiments and modelling. *Journal of Contaminant Hydrology*, *104*, 74–83. doi:10.1016/j.jconhyd.2008.10.004

Dubus, I., & Brown, C. D. (2002). Sensitivity and first-step uncertainty analyses for the preferential flow model MACRO. *Journal of Environmental Quality*, *31*, 227–240. doi:10.2134/jeq2002.0227

Dusek, J., Sanda, M., Loo, B., & Ray, C. (2010). Field leaching of pesticides at five test sites in Hawaii: study description and results. *Pest Management Science*, *66*, 596–611.

Furman, A. (2008). Modeling coupled surface-subsurface flow processes: a review. *Vadose Zone Journal*, *7*, 741–756. doi:10.2136/vzj2007.0065

Hancock, T. C., Sandstrom, M. W., Vogel, J. R., Webb, R. M. T., Bayless, E. R., & Barbash, J. E. (2008). Pesticide fate and transport throughout unsaturated zones in five agricultural settings, USA. *Journal of Environmental Quality*, *37*, 1086–1100. doi:10.2134/jeq2007.0024

Heuvelink, G. B. M., Burgers, S. L. G. E., Tiktak, A., & Van den Berg, F. (2010). Uncertainty and stochastic sensitivity analysis of the GeoPEARL pesticide leaching model. *Geoderma*, *155*, 186–192. doi:10.1016/j.geoderma.2009.07.004

Jarvis, N., Larsbo, M., Roulier, S., Lindahl, A., & Persson, L. (2006). The role of soil properties in regulating non-equilibrium macropore flow and solute transport in agricultural topsoils. *European Journal of Soil Science*, *58*(1), 282–292. doi:10.1111/j.1365-2389.2006.00837.x

Jarvis, N. J. (2007). A review of non-equilibrium water flow and solute transport in soil macropores: principles, controlling factors and consequences for water quality. *European Journal of Soil Science*, *58*(3), 523–546. doi:10.1111/j.1365-2389.2007.00915.x

Jarvis, N. J., Brown, C. D., & Granitza, E. (2000). Sources of error in model predictions of pesticide leaching: a case study using the MACRO model. *Agricultural Water Management*, *44*, 247–262. doi:10.1016/S0378-3774(99)00094-3

Köhne, J. M., Köhne, S., & Simunek, J. (2009). A review of model applications for structured soils: b) pesticide transport. *Journal of Contaminant Hydrology*, *104*, 36–60. doi:10.1016/j.jconhyd.2008.10.003

Larsbo, M., Stenström, J., Etana, A., Börjesson, E., & Jarvis, N. J. (2009). Herbicide sorption, degradation, and leaching in three Swedish soils under long-term conventional and reduced tillage. *Soil & Tillage Research*, *105*, 200–208. doi:10.1016/j.still.2009.08.003

Larsson, M. H., & Jarvis, N. J. (2000). Quantifying interactions between compound properties and macropore flow effects on pesticide leaching. *Pest Management Science*, *56*, 133–141. doi:10.1002/1526-4998(200002)56:2<133::AID-PS103>3.0.CO;2-N

Lindahl, A. M. L., Söderström, M., & Jarvis, N. (2008). Influence of input uncertainty on prediction of within-field pesticide leaching risks. *Journal of Contaminant Hydrology*, *98*, 106–114. doi:10.1016/j.jconhyd.2008.03.006

Ma, L., & Selim, H. M. (1997). Physical nonequilibrium modeling approaches to solute transport in soils. *Advances in Agronomy*, *58*, 95–150. doi:10.1016/S0065-2113(08)60254-0

Nolan, B. T., Dubus, I. G., & Surdyk, N. (2009). A refined lack-of-fit statistic to calibrate pesticide fate models for responsive systems. *Pest Management Science*, *65*, 1367–1377. doi:10.1002/ps.1825

Roulier, S., & Jarvis, N. (2003). Modeling macropore flow effects on pesticide leaching: inverse parameter estimation using microlysimeters. *Journal of Environmental Quality*, *32*, 2341–2353. doi:10.2134/jeq2003.2341

Roulier, S., & Jarvis, N. (2003). Analysis of inverse procedures for estimating parameters controlling macropore flow and solute transport in the dual-permeability model MACRO. *Vadose Zone Journal*, *2*, 349–357.

Selim, H. M., & Ma, L. (Eds.), *Physical nonequilibrium in soils: modeling and application*. Chelsea, MI: Ann Arbor Press.

Tiktak, A., Boesten, J. J. T. I., Van der Linden, A. M. A., & Vanclooster, M. (2006). Mapping ground water vulnerability to pesticide leaching with a process-based metamodel of EuroPEARL. *Journal of Environmental Quality*, *35*, 1213–1226. doi:10.2134/jeq2005.0377

Tiktak, A., de Nie, D. S., Garcet, J. D. P., Jones, A., & Vanclooster, M. (2004). Assessment of the pesticide leaching risk at the Pan-European level. The EuroPEARL approach. *Journal of Hydrology (Amsterdam)*, *289*, 222–238. doi:10.1016/j.jhydrol.2003.11.030

Van der Linden, A. M. A., Tiktal, A., Boesten, J. J. T. I., & Leijnse, A. (2009). Influence of pH-dependent sorption and transformation on simulated pesticide leaching. *The Science of the Total Environment*, *407*, 3415–3420. doi:10.1016/j.scitotenv.2009.01.059

Van der Zee, S. E. A. T. M., & Van Riemsdijk, W. H. (1994). Transport of reactive solutes in soil. In Corapcioglu, M. Y. (Ed.), *Advances in Porous Media* (*Vol. 2*, pp. 1–105). Amsterdam: Elsevier.

Vanclooster, M., Boesten, J. J. T. I., Trevisan, M., Brown, C. D., Capri, E., & Eklo, O. M. (2000). A European test of pesticide-leaching models: methodology and major recommendations. *Agricultural Water Management*, *44*, 1–19. doi:10.1016/S0378-3774(99)00081-5

Webb, R. M. T., Wieczorek, M. E., Nolan, B. T., Hancock, T. C., Sandstrom, M. W., & Barbash, J. E. (2008). Variations in pesticide leaching related to land use, pesticide properties, and unsaturated zone thickness. *Journal of Environmental Quality*, *37*, 1145–1157. doi:10.2134/jeq2007.0245

## KEY TERMS AND DEFINITIONS

**Chromatography Flow:** Movement of water and solutes (i.e. pesticides) in a homogeneous soil profile.

**Degradation:** A transformation process that results when soil microorganisms or nonbiological reactions either partially or completely metabolize a pesticide.

**Leaching:** Vertical movement of pesticides within soil profile during percolation of water. This movement is the result of convective and diffusive transport in the liquid phase.

**Pesticide Leaching Models (PLM):** Computational tool to simulate pesticide leaching in soils after it has been applied to agricultural field. PLM are formed by many mathematical models that describe a variety of physical, chemical and biological processes that influence pesticide behavior in soil.

**Pesticide:** Synthetic organic chemicals that are manufactured for agricultural use to protect crops against pests, diseases, or to eliminate competition from aggressive weeds.

**Preferential Flow:** Rapid movement of water and solutes (i.e. pesticides) via macropores in the soil profile and that bypass a large portion of the soil matrix.

**Sorption:** Is a transfer process by which pesticides accumulate at soil-water and/or soil-air interfaces. It is an important process in regulating the concentration of pesticides in soil water.

# Chapter 14
# A Computational Agent Model of Flood Management Strategies

**Lisa Brouwers**
*The Royal Institute of Technology, School of ICT, Sweden*

**Magnus Boman**
*The Royal Institute of Technology, School of ICT, Sweden & The Swedish Institute of Computer Science (SICS), Sweden*

## ABSTRACT

*A geographically explicit flood simulation model was designed and implemented as a tool for policy making support, illustrated here with two simple flood management strategies pertaining to the Upper Tisza area in Hungary. The model integrates aspects of the geographical, hydrological, economical, land use, and social context. The perspectives of different stakeholders are represented as agents that make decisions on whether or not to buy flood insurance. The authors demonstrate that agent-based models can be important for policy issues in general, and for sustainable development policy issues in particular, by aiding stakeholder communication and learning, thereby increasing the chances of reaching robust decisions. The agent-based approach enables the highlighting and communication of distributional effects of policy changes at the micro-level, as illustrated by several graphical representations of outputs from the model.*

## INTRODUCTION

In the period 1988-1997, the cost for major natural disasters was approximately USD 700 billion, according to 1998 figures from Munich Re, with floods causing around 50 per cent of economic losses (Abramovitz, 2001). As a result, policy makers at the national level are facing increasing costs for compensation and mitigation of floods.

DOI: 10.4018/978-1-61692-871-1.ch014

Most countries have a disaster compensation policy based on solidarity, in which the government compensates for losses that individual property owners suffer. A higher degree of private responsibility is a distinguishing feature for compensation policies in only a few countries, such as the UK and Australia.

Our simulation model estimates the economical consequences of floods, with respect to different flood risk management strategies. Part of the Upper Tisza river in the relatively poor

Szabolcs-Szatmár-Bereg region of Hungary was investigated, with a focus on financial mitigation measures and insurance. For such a complex model—solving the kind of problem that Weaver (1948) famously dubbed a *problem of organized complexity*—to be useful, it must integrate data from different systems. In our case, this translates to geographical, hydrological, economical, land use, and social systems.

We seek to provide evidence for the usefulness of spatially explicit agent-based models in the domain of disaster policy management. In our model, the government, insurer, and property owner stakeholders are represented. Property owner agents are spatially explicit agents in the model, in that their properties have specific locations, affecting agent actions. In each simulated year, for example, the agent makes a decision on whether to buy insurance or not, for a particular property. In a simulation, the model is run to cover a long period of time, normally 50 years. Floods of different magnitude occur stochastically and cause damages to properties in the area. A flood causes either a levee failure or seepage. Different flood management strategies are tested and compared, and a strategy can implement a policy, e.g., subsidy insurance premiums for all poor households and set the compensation level from government to 50 per cent of property loss. The financial consequences for all stakeholders are then estimated at the end of each simulation round.

In the following section, we frame our work inside the field of related research. The third section describes our model, the fourth presents two policy scenarios and their analysis, and the final section gives our conclusions.

## BACKGROUND

While the agent area started off with artificial intelligence assistants, service robots, and many other forms of one-agent systems (Agre & Rosenschein, 1996), the potential for societal applications naturally reside with multi-agent systems (O'Sullivan & Haklay, 2000; Gimblett, 2002). In geography, physical components of a complex system—vegetation, fauna, soils, and climate/hydrology—are usually separated from socio-economic ones, like demography, culture, economy, and policy (Reenberg, 2001). An agricultural management strategy then advices and implements a policy by carefully considering those two sets of components separately. An agent-based model like our own, by contrast, lets the stakeholders be represented by agents taking actions in various states of information and in various geographical spaces, in situations where agents are informed of all components to very different extents. Heterogeneity in the agent population is thus an important feature (Wooldridge, 2002).

In microsimulation, the population is usually homogeneous. Early models (Orcutt 1957; Orcutt *et al.*, 1961; Hägerstrand 1975) purported to address the shortcomings of the macro models then reigning in economics, demography, and geography (Merz, 1991), a promise delivered on as computing power became available with which to execute models in scenario analysis fashion (Caldwell & Keister, 1996). The development in computer hardware and software has naturally also benefited agent-based simulations, making them a widely used complement to mathematical theorizing (Axtell, 2000). When to go for agent heterogeneity and when the simpler microsimulation models suffice is a question part of a methodological scientific development still in rapid progress (Boman & Holm, 2004), but agent-based social simulations have proved increasingly important to policy making (Polhill, Gotts & Law, 2001).

The computational model reported on here has in part been discussed in academic theses (Brouwers 2002; Brouwers 2005; Hansson 2007), with the decision-theoretic aspects of the model described in more detail elsewhere (Brouwers *et al.*, 2004; Hansson, Danielson & Ekenberg, 2008).

## A SCENARIO SIMULATION MODEL

The model was developed in an iterative process. Initially, an executable prototype model was implemented, using synthetic data. This prototype was based on an existing cyclone model, developed by Ermolieva (1997) at the International Institute of Applied Systems Analysis (IIASA), the host institution for the project. Hydrological inundation and flow models of the area of study in the Upper Tisza river basin were delivered by consultants linked to the project, and these models were included in simulation models developed at IIASA (Ermolieva *et al.*, 2003), where stakeholder interview studies were also carried out. The model described in this chapter consists of five modules, each briefly described in its own subsection below.

### The Monte Carlo Module

At the beginning of each simulation round, three stochastic variables are used to simulate flood events of two types: levee failure and seepage. The variables are assigned random numbers in the interval [0,1] from a uniform distribution. The term *flood* is used to describe that the water level in the river is higher than normal. In an unprotected river, all floods would overflow the embankments. However, the parts of the river investigated are protected by levees designed to hold back most floods. The hydrological model gives the probabilities for floods of different magnitudes to occur. The different magnitudes used in the simulations are 100-year floods (minor), 150-year (medium), and 1 000-year (extreme). A 100-year flood denotes a flood that in a long time perspective, the average time between floods of that magnitude or greater, is 100 years. The probability concerns a single year and says nothing of the accumulated risk. The probability distribution for floods is:

- no flood: 0.9823
- minor flood (100-year): 0.0100
- medium flood (150-year): 0.0667
- extreme flood (1 000-year): 0.0010

If the value of the variable *Magnitude* is less than or equal to the threshold for the three floods (0.0100, 0.0167, and 0.0177), the corresponding flood occurs. If a flood occurs, three things might happen:

- levee failure: a levee fails, either by being overtopped or getting burst
- seepage: water finds its way under a levee
- no flood event: the levee holds back the water

The first two events can be combined. The hydrological model gives the failure- and seepage risks for three specific locations along the river. If there is a flood, the values of the variables *Failure* and *Seepage* show whether or not the flood inundates the neighboring land. The variable *Failure* determines whether or not the flood will cause a levee failure, in one out of three possible locations (see Table 1). A careful estimate made by hydrologists in our project indicated that seepage occurs twice as often as levee failures, and so

*Table 1. Levee failure and seepage probabilities. Data from the 1999 World Bank report on Hungary's flood control development and rehabilitation project*

| Location | Magnitude | Probability of Levee Failure | Probability of Seepage |
|---|---|---|---|
| 1 | 100-year flood | 12% | 24% |
| 2 | 100-year flood | 20% | 40% |
| 3 | 100-year flood | 28% | 56% |
| 1 | 150-year flood | 18% | 36% |
| 2 | 150-year flood | 22% | 44% |
| 3 | 150-year flood | 40% | 80% |
| 1 | 1000-year flood | 19% | 38% |
| 2 | 1000-year flood | 33% | 66% |
| 3 | 1000-year flood | 45% | 90% |

this approximation was used in our model. The probability for a levee failure at location 3 to occur from a 1 000-year flood is thus 0.001 · 0.45.

## The Catastrophe Module

When a flood event has occurred, the effects of the failure and seepage events are calculated, considering, the elevation of the location, how long the inundation lasted and the magnitude of the flood. Only direct damages to private property are considered. The size of the losses further depends on soil type and building materials. The model uses 18 pre-compiled flood event damage scenarios: nine levee failure events and nine seepage events, with the corresponding losses for all properties. When a property is damaged both by a failing levee and by a seepage, only the largest loss is used.

## The Spatial Module

The river basin studied is represented explicitly, in the form of a grid with 1 551 · 1 551 cells, each measuring $100m^2$. For each cell, rich data is available, including elevation, current land use, asset value, soil type, and vegetation. Each cell belongs to one of eleven municipalities. Example values for the variables are: swamp (soil type), area for sport/recreation (land use), and broad-leaved forest (vegetation).

## The Agents Module

The stakeholders in the flood-risk management problem are represented in our model as agents: insurers, property owners, and government. Other agents could have been included, but these were considered by far the most important to financial flood management policy.

The insurer agent is represented by wealth and insurance contracts. The insurer agent is not situated, but the insurance contracts pertain to specific locations. The initial wealth of the insurer agent is HUF one million (corresponding roughly to ten per cent of the expected damage for all properties in the basin caused by levee failures, just to give an indication). An insurance contract is specified by coverage, premium size, and a deductible. The terms of the contracts are decided in the policy strategy currently used in the model. The insurance agent does not offer insurance against seepage events, reflecting the real situation in Hungary, a country in which approximately two per cent of the total claims from private property insurance come from flood damages.

The property owner agents have the following attributes: location, wealth, income, expenditures, and insurance. The location is the cell where the (mid-point of the) property is situated. Each property agent is assigned an initial random wealth, using a uniform distribution in the range [0,10000] HUF. The 2 580 property owner agents in our model have a yearly net income, which is a random value in the range [0,100000] HUF, taken from a normal distribution with the mean 36 900. This mean value was the average net earnings per month as of 1998, in the Szabolcs-Szatmár-Bereg county, pertaining to employed persons working in enterprises with more than 20 employees, or in the public sector (according to data from the Hungarian Statistics Central Office). The expenditures are assumed to be high, 90 per cent of the income. The area studied has a high degree of unemployment. In Hungary, 60 per cent of the property owners have property insurance with flood protection included. Property owner agents make annual insurance decisions, i.e., whether or not to buy insurance. Their decisions depend on their risk-willingness and their current wealth. A random value from a uniform distribution is compared with an *InsuranceRate* variable. When the property agents consider buying the more expensive top-insurance, with a risk-based premium, their financial situation determines if they can afford it or not.

Only the part of the governmental budget that concerns flood management is taken into consideration in the model, and only the financial load is considered; i.e., no incomes are included in the model. This is mainly because of the small size of the pilot basin and also due to lack of data. The government agent has a compensation level attribute, and a yearly maintenance cost attribute. The size of the compensation level is set in accordance with the current policy strategy. Maintenance costs for structural mitigation are also considered as a policy parameter, as the maintenance of levees is costly and the degree to which they are attended to vary with the financial climate.

## The Consequence Module

At the end of each simulation round, the Consequence module calculates the changes in wealth, for all agents, whereafter their economy is updated. The wealth transformation functions for the different agents are described in detail elsewhere (see, e.g., Hansson, Danielson & Ekenberg, 2008), and are described only informally below.

The policy vector $X$ contains possible policy strategies. An instance $x$ of the vector describes a specific policy strategy. A policy strategy consists of a number of policy parameters, each of which can take on different values. Policy parameters include level of governmental compensation, premium size, and size of the deductible. An example policy strategy could then correspond to, e.g., governmental compensation set to 40 per cent, premium size one per cent of property value, and size of deductibles set to HUF 100 000. The uncertainty is treated explicitly in the model, i.e., the vector $X$ contains all stochastic variables. The modeled system can be in different states, and a state can be thought of as a snapshot of the model with all the values of all variables displayed. The values of some of the variables are impossible to predict with certainty, and these stochastic variables include wind speed, precipitation, water level, and discharge. In the experiments reported on here, the vector $X$ contains only the three variables described in the Monte Carlo module subsection above: *Magnitude*, *Failure*, and *Seepage*.

Let $W^P$ be the wealth of the property agent $P$. The wealth is transformed over time as a function of the compensation received from the government, with the size depending on the severity of the flood, and the degree of compensation which is decided by the policy parameter $x$. Cost for flood damages on the property are deducted, and the income is added to the wealth, with expenditures deducted. The wealth $W^P$ is finally updated according to the collected claims, and the paid premiums. Naturally, more than one insurance company may be involved. How much insurance compensation the property owner receives from each insurer depends on two things:

- the current policy strategy $x$, which decides the size of the deductible, and coverage, i.e., the proportion of the property insured, and
- the severity of the flood.

A 70 per cent coverage of a building worth HUF one million means that the building is only insured up to 700 000. Such a building might have an insurance contract with the following features: coverage 70 per cent, deductible 10 per cent (with the first ten per cent of the losses not covered by the insurance, this being an example of a policy parameter). If a flood occurs that destroys 50 per cent of the property, the property owner would receive HUF 315 000 from the insurance company (losses: 500 000, deductibles: 50 000). Compensation from the government is added to the wealth, according to the policy parameter $x$, with the size depending on the severity of the flood. The cost for insurance premiums is then deducted. The premium is considered a policy parameter, which is dependent on the size of the coverage and can vary among the insurer agents. For example, if the premium is set to five per cent of the property value, and coverage to 50 per cent, then a HUF one million building would cost HUF 25 000 (per year) in insurance premiums.

In a similar fashion, the wealth $W^{Gov}$ of the government agent *Gov* can be calculated. It is reduced by flood compensation *G* paid to the property agents $P_p,...,P_n$, where *x* is the policy parameter, as explained above. The government agent also suffers direct costs for flood mitigation, as well as costs for maintenance of structural flood mitigation measures, represented together in the below equation as *M* for the flood ω:

$$W^{Gov}(x,\omega) = W - \sum_{P=1}^{n} G_P(x,\omega) - M(x)$$

The wealth of an insurance agent $W^I$ increases with the premiums received from the property agents. Coverage is determined by the policy strategy *x*, and *x* also determines how much compensation must be deducted from the wealth of $W^I$, given the severity of the flood. Risk-based premiums are calculated per location. Each levee failure scenario has its damage costs multiplied by the probability of the failure, and the expected damages for all scenarios are calculated. A safety load accounting for margins of profit and administrative costs is then added to the premium by $W^I$.

## TWO POLICY SCENARIOS

Policy strategies for coping with losses vary from country to country, and several grounds for choosing a policy can be identified. The capacity of the insurance industry in a country may affect the degree to which a local government becomes involved. In Hungary, an increased use of private catastrophe insurance is still a debated matter on the political agenda. The history and the tradition of how the government acts have had a significant influence on the prevailing flood management policies. At the outset of the research project, Vári (1999) carried out interviews with the stakeholders in the Upper Tisza region, and most respondents agreed to measures protecting the high-risk settlements, at all costs. Only a minority of respondents would support even partial relocation of the local population. Regarding strategies for sharing economical responsibility, most respondents thought that the government should compensate the victims. However, many would accept a policy combining government compensation with private or community self-insurance schemes. As many as 46 per cent of the respondents were of the opinion that taxes should be used as a means to compensate victims.

Linnerooth-Bayer (2001) subsequently designed three possible flood risk management policy scenarios for Hungary, based on interviews and surveys of the views of the stakeholders. A crucial problem was to find a balance between social solidarity, private responsibility and community responsibility. Two of the three scenarios were implemented and tested within our simulation experiments.

Scenario 1 represents the current flood management strategy in Hungary, which is why we refer to this scenario as *Business as usual*. The government continues to compensate the property owners from floods caused by levee failures, as this is considered the responsibility of the government. It also compensates for seepage, but the compensation is smaller than for levee failures. The government finances the costs for mitigation and compensation from its catastrophe budget, meaning that all taxpayers contribute. The insurance is a bundled property insurance, where the all-hazard insurance part accounts for two per cent of the total premiums (in accordance with assumptions recommended by insurance experts employed by the project). The size of the premium is based on property value alone, which implies a cross-subsidization of premiums from low-risk locations to high-risk locations.

In Scenario 2, which we named *Market Model*, part of the responsibility is shifted from the government to the individual property owner, which is why the compensation from the government is reduced. The seepage events are still compensated for, but only to a low extent. The property owner agents also receive less compensation than in Scenario 1 from their normal insurance, the intention behind this being to reduce the level of cross-subsidization. The gap in insurance coverage can be filled by buying a risk-based insurance, with its premiums based on the local flood-risks. The expected damage for a property is the base for calculating the size of the premium, and in our experiments, 545 properties were affected by levee failures. The expected damages are multiplied by a safety load, as was explained earlier. In Figure 1, the expected damages for the properties are presented.

## Experiments

The following settings were used for simulation experiments for the Business as Usual scenario.

- Maintenance cost: HUF 100 000 per year (for the levees in the basin)
- Compensation from government (levee failure): 100% of damages
- Compensation from government (seepage): 50% of damages
- Deductible: the first 5% of the damages
- Coverage: 100%
- Insurance rate: 60% of the households have property insurance
- Insurance premium: cross-subsidization, 0.2% of property value per year

It might seem strange that property owners are compensated from the government even if they hold private insurance, but the rationale for this policy of double compensation as implemented in Hungary was to encourage house owners to buy insurance. If insurance holders were punished by not being compensated by the government, the incentive to buy private insurance would be reduced.

*Figure 1. Expected damages for properties, given levee failures. The properties on the x axis are sorted according to the size of the expected damages*

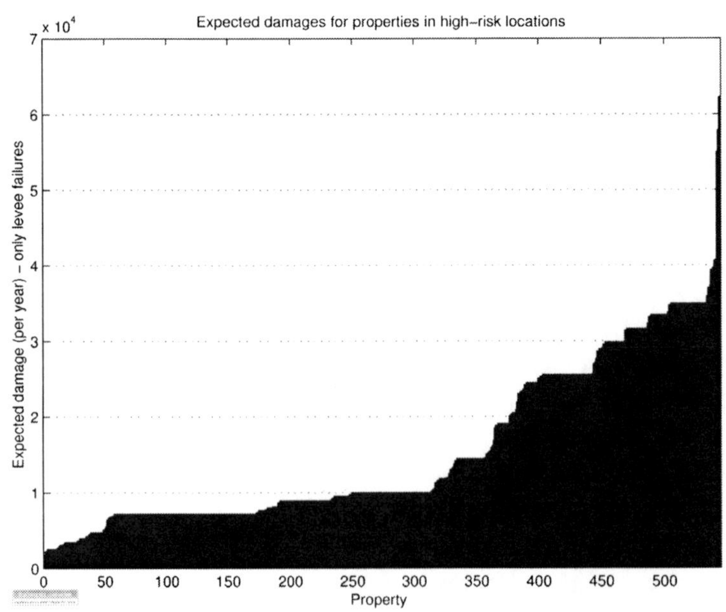

The following settings were used for the simulations of the scenario Market Model.

- Maintenance cost: HUF 100 000 per year (for the levees in the basin)
- Compensation from government (levee failure): 75% of damages
- Compensation from government (seepage): 30% of damages
- Deductible for cross-subsidization: the first 5% of the damages
- Coverage for cross-subsidization: 60%
- Insurance rate for cross-subsidization: 60%
- Premiums for cross-subsidization: 0.1% of property value/year
- Coverage for risk-based top-insurance: 100%
- Insurance premium for top-insurance: risk-based (as explained above)
- Insurance rate for top-insurance: approximately 10%; first decided by financial situation, then by risk-willingness
- Deductible for risk-based premium: 0%

The same distribution of flood events and baseline insurance were used for the two scenarios. Each scenario was first simulated only once, for a time-period of 50 years, as a pilot. The goal of the pilot was to investigate how the consequences for individual property owners may vary. The figures presented are merely included to show the variety in consequences among the agents. This serves here as an illustration of how our output can support policy considerations. The wealth of the property agents grows steadily during the simulated 50-year period, as the expenditures are smaller than the income. During this time-period in the pilot, four flood events occurred: two failures (in years 8 and 41) and two seepage events (in years 32 and 34). Looking carefully at the dynamic wealth trajectories (Figure 2), the effects of the double compensation are visible at the years in which the failures occurred. No property owners show a negative balance during the time period. When the seepage events occurred, the changes in wealth are too small to show, but by inspecting the wealth of a single property agent we can clearly see how its wealth is affected (Figure 3). This again serves to illustrate the usefulness of an

*Figure 2. Dynamic wealth of property agents in the pilot run for the business as usual scenario*

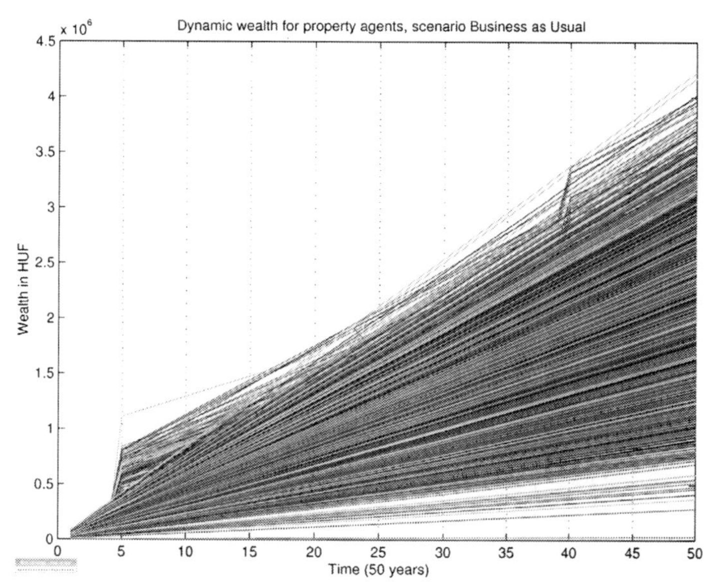

agent-based model: any agent can be monitored more carefully, as needed, e.g., for the purpose of fairness studies. Model heterogeneity is also displayed here, in that not all property owners are affected by all flood events; this example property owner was not affected by the failure event that occurred in year 41.

A second set of simulations was run after the pilot experiment, in which the number of simulations was increased to 1 000. The time period was still 50 years and both scenarios used the same distributions of flood events and cross-subsidized insurance contracts. These experiments showed that the difference between the two policy scenarios mainly affected the property owners. The property owners who experienced seepage failures suffered severe losses, since the government only compensated 30 per cent. The property owners in high-risk areas, who had much insurance, gained economically from levee failures. The majority of the property owners who most needed the top-insurance could not afford one, reflecting the real situation in the area. The experiments further indicate that the reduction in governmental compensation (as in the Market Model scenario, see Figure 4) was too small to bring any significant changes to the governmental flood management budget. If the compensation is reduced further, a market approach is likely to be very efficient for the government, since the responsibility of the loss lies more with the individual. This strategy also forces both industry and individuals to reduce risk, e.g., through constructing measures of their own. Governments carry a large responsibility for the infrastructure in the corresponding countries and they may wish to reduce catastrophic losses by diversifying with insurance, or other risk transfer instruments. On the other hand, the pressure of people living in risk prone areas might be high since they may not afford insurance. To illustrate how the government stakeholder perspective can be presented, we can simply turn to the wealth trajectory of the government agent (Figure 5).

The deductible can be used as a policy tool for reducing the concentration of people and properties in high-risk areas. A high deductible, or an upper limit on the insurance contract, puts a larger financial responsibility on the individuals. The fairness of subsidization of insurance premi-

*Figure 3. Dynamic wealth of a single property agent in the pilot run for the business as usual scenario*

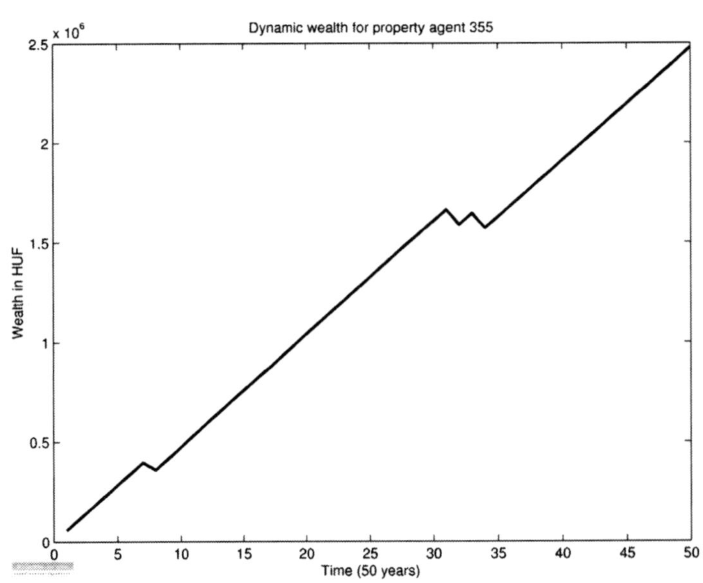

*Figure 4. Dynamic wealth of property agents in the pilot run for the market model scenario*

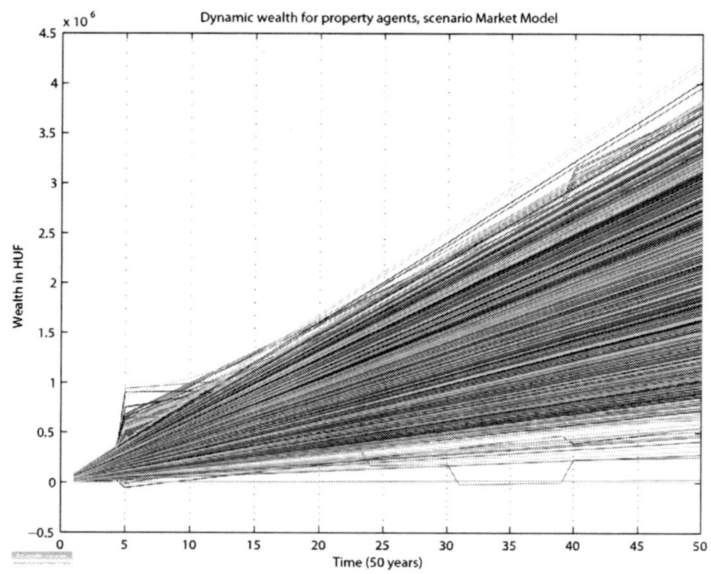


*Figure 5. Dynamic wealth of the government agent in one particular run of the business as usual scenario, in which the government went bankrupt after 37 years. The y axis shows wealth in HUF, and the x axis shows years*

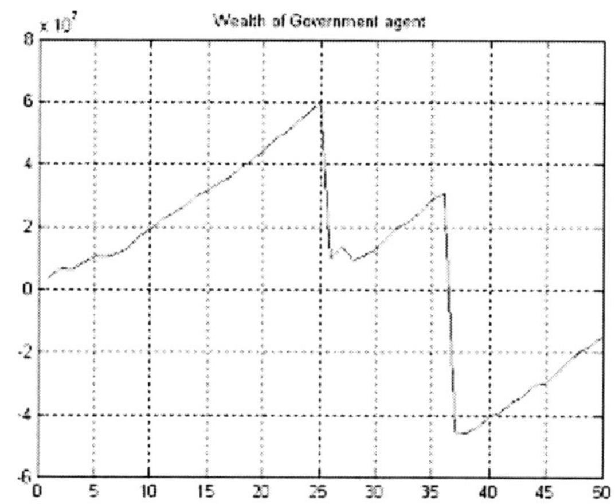


ums is also debatable. It is not self-evident that insurance should be cross-subsidized, for instance. In Figure 6, we illustrate how the fairness issue can be made clear by noting a number of property agents gone bankrupt.

## CONCLUSION

A flood-risk management model like ours is potentially useful as a basis for discussions of alternative policy strategies, in settings where policy

*Figure 6. Dynamic wealth of property agent in a set of runs of the market model scenario, in which several agents went bankrupt almost immediately, later followed by numerous agents in a less spectacular fashion. The y axis shows wealth in HUF, and the x axis shows years*

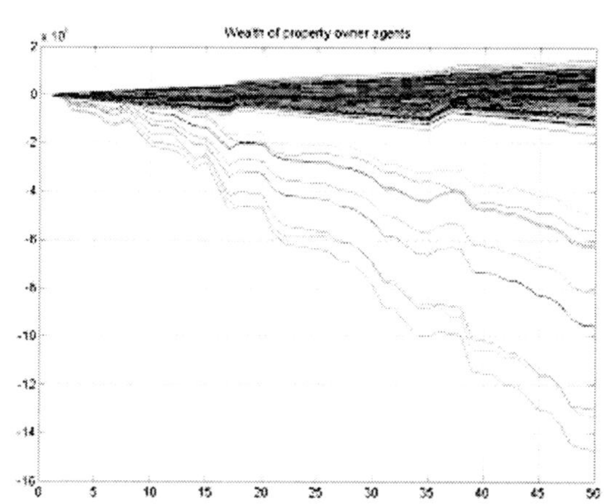

makers and stakeholders work together. As a case in point, one may show that a flood management strategy that seems fair at the aggregate level may hide large inequalities apparent only when considering the outcomes for specific agents. For a model like ours to be useful, it is important that the results are presented to the individuals also in more aggregate ways. The model can then contribute through visualization at stakeholder meetings on catastrophe management questions, on both mitigation strategies and issues of fairness.

Since the Tisza project was completed, we have pursued agent-based modeling as a way of supporting policy. Executable and spatially explicit models for several applications were made, chiefly within the area of epidemiology. Whatever the application area is, visualizing results down to the individual level has proven to be a powerful way of providing policy background material, and grounds for debate. With the advent of faster computers and improved methodologies for large-scale simulation, we expect even more interest for executable agent-based models in the future.

## ACKNOWLEDGMENT

We would like to thank Karin Hansson for her work on an earlier version of this chapter, and Maria Cheadle for comments on the final draft. Love Ekenberg, Joanne Linnerooth-Bayer, Tatiana Ermolieva, Yuri Ermoliev and Istvan Galambos shared information and contributed extensively in various ways to the parts of the project reported on here. Finally we would like to acknowledge Mats Danielson and Love Ekenberg for comments on relevant academic theses.

## REFERENCES

Abramovitz, J. N. (2001). *Unnatural Disasters*. Worldwatch Paper 158.

Agre, P. E., & Rosenschein, S. J. (Eds.). (1996). *Computational theories of interaction and agency*. Cambridge, MA: MIT Press.

Axtell, R. (2000). *Why agents? On the varied motivations for agent computing in the social sciences*. Working Paper 17, Center on Social and Economic Dynamics, Brookings Institution.

Boman, M., & Holm, E. (2004). Multi-agent systems, time geography, and microsimulations. In Olsson, M.-O., & Sjöstedt, G. (Eds.), *Systems Approaches and their Application* (pp. 95–118). Amsterdam: Kluwer Academic.

Brouwers, L. (2002). *Flood risk management policy from an insurance perspective*. Licentiate Dissertation, Stockholm University, Department of Computer & Systems Sciences, Report No. 02-008.

Brouwers, L. (2005). *Microsimulation models for disaster policy making*. Doctoral Dissertation, Stockholm University, Department of Computer & Systems Sciences, Report No. 05-012.

Brouwers, L., Danielson, M., Ekenberg, L., & Hansson, K. (2004). Multi-criteria decision-making of policy strategies with public-private re-insurance systems. *Risk Decision and Policy*, *9*(1), 23–45. doi:10.1080/14664530490429535

Caldwell, S., & Keister, L. A. (1996). Wealth in America: Family stock ownership and accumulation 1960-95. In Clark, G. P. (Ed.), *Microsimulation for urban and regional policy analysis*. Pion.

Ermolieva, T. (1997). *The design of optimal insurance decisions in the presence of catastrophic risks*. Interim Report IR-97-068. IIASA, Laxenburg, Austria.

Ermolieva, T., Ermoliev, Y., Fischer, G., & Galambos, I. (2003). The role of financial instruments in integrated catastrophic flood management. *Multinational Finance*, *7*(3&4), 207–230.

Gimblett, H. R. (Ed.). (2002). *Integrating geographic information systems and agent-based modeling techniques for simulating social and ecological processes*. Oxford, UK: Oxford University Press.

Hägerstrand, T. (1975). Survival and arena: On the life-history of individuals in relation to their geographical environment. *Monadnock*, *49*, 9–29.

Hansson, K. (2007). *A framework for evaluation of flood management strategies*. Doctoral Dissertation, Stockholm University, Department of Computer & Systems Sciences, Report No 07-005.

Hansson, K., Danielson, M., & Ekenberg, L. (2008). A framework for evaluation of flood management strategies. *Environmental Management*, *86*, 465–480. doi:10.1016/j.jenvman.2006.12.037

Linnerooth-Bayer, J. (2001). *An insurance system for Hungary: Three policy scenarios*. Working paper, IIASA, Laxenburg, Austria.

Merz, J. (1991). Microsimulation – A survey of principles, developments and applications. *Forecasting*, *7*, 77–104. doi:10.1016/0169-2070(91)90035-T

O'Sullivan, D., & Haklay, M. (2000). Agent-based models and individualism – Is the world agent-based? *Environment & Planning A*, *32*, 1409–1425. doi:10.1068/a32140

Orcutt, G. H. (1957). A new type of socio-economic system. *The Review of Economics and Statistics*, *58*, 773–794.

Orcutt, G. H., & Greenberger, M. Korbel, J. & Rivlin, A. (1961). *Microanalysis of socioeconomic systems: A simulation study*. New York: Harper & Row.

Polhill, J. G., Gotts, N. M., & Law, A. N. R. (2001). Imitative versus nonimitative strategies in a land-use simulation. *Cybernetics and Systems*, *32*, 285–307. doi:10.1080/019697201300001885

Reenberg, A. (2001). Agricultural land use pattern dynamics in the Sudan–Sahel: towards an event-driven framework. *Land Use Policy*, *18*(4), 309–319. doi:10.1016/S0264-8377(01)00020-5

Vári, A. (1999). *Floor control development in Hungary: Willingness-to-pay study*. Unpublished report, IIASA, Laxenburg, Austria.

Weaver, W. (1948). Science and complexity. *American Scientist*, *36*, 536–544.

Wooldridge, M. (2002). *An introduction to multiagent systems*. Chichester, UK: John Wiley & Sons.

# Chapter 15
# Computational Techniques for Biologic Species Distribution Modeling

**Pedro Luiz Pizzigatti Corrêa**
*Agricultural Automation Laboratory, Polytechnic School of the University of Sao Paulo, EPUSP, Brazil*

**Mariana Aparecida Carvalhaes**
*Brazilian Agricultural Research Corporation, EMBRAPA Middle-North, Brazil*

**Antonio Mauro Saraiva**
*Agricultural Automation Laboratory, Polytechnic School of the University of Sao Paulo, EPUSP, Brazil*

**Fabrício Augusto Rodrigues**
*Agricultural Automation Laboratory, Polytechnic School of the University of Sao Paulo, EPUSP, Brazil*

**Elisângela Silva da Cunha Rodrigues**
*Agricultural Automation Laboratory, Polytechnic School of the University of Sao Paulo, EPUSP, Brazil*

**Ricardo Luis de Azevedo da Rocha**
*Laboratory of Languages and Adaptive Techniques, Polytechnic School of the University of Sao Paulo, EPUSP, Brazil*

## ABSTRACT

*Computational modeling techniques for species geographic distribution are critical to support the task of identifying areas with high risk of loss of Biodiversity. These tools can assist in the conservation of Biodiversity, in planning the use of non-inhabited regions, in estimating the risk of invasive species, in the proposed reintroduction programs for species and even in planning the protecting endangered species. Furthermore, such techniques can help to understand the effects of climate change and other changes in the geographical distribution of species. This chapter presents concepts related to the species distribution modeling and algorithms based on Neural Networks and Maximum Entropy as alternatives*

DOI: 10.4018/978-1-61692-871-1.ch015

*for modeling of species distribution. The algorithms were integrated into the open source tool called openModeller used by biologists and other researchers in this area. A case study of modeling the distribution of babaçu (Orbignya phalerata) in the Piauí State – Brazil is presented, evaluating the potential distribution of this species used to produce bioenergy. Fifty models were generated and merged the ten models with best accuracy for each algorithm. The results show that the models obtained by both are consistent. The models obtained with Maximum Entropy seem to reflect best the reality, considering the occurrence pattern of babaçu as a secondary species.*

## INTRODUCTION

Brazil and other regions of the world face pressing environmental problems and Biodiversity loss associated with urbanization, industrialization, agriculture, use of resources and other human activities. These problems cut across national borders, scientific disciplines and observing systems. A key scientific and technical challenge, therefore, is to improve access to existing and emerging sources of environmental, biological, and socioeconomic data and to develop better ways to integrate these data to support a variety of disciplinary and interdisciplinary research and application efforts, and related policy-making initiatives (Canhos et al., 2005).

According to Myers et al. (2000), Brazil is believed to be the country with the richest flora in the planet, about one-sixth of the total. The Brazilian flora richness is distributed into one aquatic biome and six terrestrial ones: Amazon, Savannah, Atlantic Rainforest, Caatinga, Pampa and Pantanal. The Brazilian Biodiversity study can contribute economically and is very important for social and cultural life. Thus, the modeling of biological species can improve conservation and sustainable use of resources.

Advances in communication and information technology have provided an international environment for collaborative work, involving the sharing of knowledge. This fact, together with a clear demand from international conventions and treaties concerning environmental issues are contributing to a change in paradigm concerning free and open access to data, information, and software developments. These new demands are catalyzing the development of a new knowledge area, the Biodiversity Informatics (Canhos, 2003, Corrêa et al., 2006). Information is the basis for defining best practices and strategies for conservation and sustainable use of our natural resources. However, there are enormous gaps of data and knowledge about species occurrence and distribution in Brazil, and the use of species distribution modeling tools is a powerful means of overcoming them.

Computational modeling tools are essential to support the task of identifying regions that have high risk of Biodiversity loss. Such tools are important for directing the efforts to preserve the environment and the region's most threatened species. According to Pereira and Peterson (2001), the use of tools for modeling the distribution of species can assist in the conservation of Biodiversity in planning the use of non-inhabited regions, prediction of invasion of species, the design of programs for the reintroduction of species and even in projects for protecting endangered species. Furthermore, such tools can help understand the effects of climate change and other changes in the distribution of species (Santana et al., 2008).

The modeling of species distribution is mainly based on the concept of a species ecological niche. The term ecological niche was defined by Hutchinson (1957) as an n-dimensional area where each dimension represents the range of environmental conditions or resources necessary for survival and reproduction of the species, such as temperature, humidity, food resources, light intensity, population density, among others.

Such models are built up grouping and classifying the collected environmental data in order to create a set of well-connected data points: the ecological niche. Neural networks are fairly conventional computational techniques appropriate for both types of problem in which the modeling of species distribution can be inserted: clustering and classification (Beale & Jackson, 1990).

The process of species modeling becomes complex, among other factors, due to the variety of the data nature and mainly because it requires the treatment of lack of data needed to build a reliable model. This kind of problem considers a restricted optimization problem. To solve this issue, the concept of entropy could be applied, because it defines a kind of measure in the space of probability distributions, so that distributions with higher entropy can be favored over others (Tavares, 2004).

This chapter approaches a modeling method of biological species distribution and discusses the use of modeling techniques based neural network and entropy. The application of modeling method and these techniques can be clearer with the case study on the modeling of biologic species geographical distribution of babaçu. These two techniques were implemented in the open source tool called OpenModeller (Canhos et al., 2005). It provides a robust and flexible framework for modeling to facilitate the complex work of scientists in the predictive modeling of species distribution and for developing new algorithms.

## MODELING OF BIOLOGIC SPECIES GEOGRAPHICAL DISTRIBUTION

The destruction of Biodiversity in the world is a problem that has been quickly increasing in the last decades. This problem causes the decrease or even the extinction of animals and vegetables species in the planet. The effects of biodiversity loss are irreversible. Therefore, environmental conservation is one of the great current challenges of humanity. According to Myers et al. (2000), is very difficult to preserve all the endangered species, because the number of species to be protected exceeds the available resources for conservation. Thus, one of the ways to preserve the greater number of species with the smaller possible cost is hotspots identification that is, regions the biodiversity of which is significantly threatened with destruction.

Modeling of biologic species geographical distribution can contribute for Biodiversity conservation mainly in the decision-making process. Modeling can help in the identification of environmental alterations, environmental quality and identification of threatened species. Besides that, the modeling can be applied in the conservation of rare species, reintroduction of species, environmental risk identification, to evaluate the potential of invasive alien species, to consider possible routes of infectious diseases spread, planning of priority area for conservation and to improve our understanding of species habitat relationships in space and in time. However, the modeling process, in general, is very complex and demands a significant amount of time.

A modeling method consists in predicting the environmental quality for the species regarding the given environmental variables. A niche-based model represents an approximation of the species fundamental ecological niche in the studied area and in the considered environmental dimensions. In general, modeling does not represent the fundamental niche of the species (set of all conditions that allow the survival for a long period of time), but represents the observed niche (subset of the fundamental niche that can be smaller due to human influence or environmental barriers). The bigger the studied region, the closer to the fundamental niche will be the modeling (Phillips et al., 2006).

The modeling process involves data collecting and pre-processing, selecting an algorithm that best represents the problem, analysis and interpretation of results. The data available is a set of environmental variables, such as temperature, topography, rain precipitation, etc. Moreover, there is a list of occurrences at georeferenced localities (latitude and longitude). This list consists of geographic points where the species has been observed. Those data are called presence points. Sometimes, there is a set of points available where the species is known not to occur, called absence points. However, frequently, only presence points are available, indicating the occurrence of the species (Phillips et al., 2004; Fielding & Bell, 1997). The absence points are very difficult to obtain because the non-observation of a species in a region is not sufficient to guarantee the total absence of it.

The modeling algorithms are used to produce an approximation of the species fundamental niche, that is, they are applied with the aim of identifying environmental conditions where the species could exist. The algorithms receive niche points as input data. Niche points are values that the environmental variables assume in each occurrence point, transforming the geographical points into points in the environmental space. The output of an algorithm is a niche model, an occurrence probability function of a given species with domain in the space of the environmental variables. This model is projected in a geographic region of interest, producing a georeferenced map containing on each point the species occurrence probability (Phillips et al., 2006).

So as to model the distribution of species many methods that use only presence points have been applied, such as, Bioclim (Busby, 1986; Nix, 1986) and Domain (Carpenter et al. 1993). In addition, there are other statistical methods that use data sets containing presence and absence data used to model the distribution of species, such as linear models - GLMs (Generalized Linear Models) (Austin et al., 1994) and additive models - GAMs (Generalized additive Models) (Yee & Mitchell, 1991), but they are being applied in cases in which there are only presence points (Ferrier & Watson, 1997; Ferrier et al., 2002).

An algorithm very used in the environmental modeling community is GARP (Genetic Algorithm for Rule-set Production) (Stockwell & Peters, 1999; Stockwell & Noble, 1991; Stockwell, 1992; Persona et al., 2003). Besides GARP, other machine learning algorithms have been applied to modeling of biologic species geographical distribution, such as: SVMs (Support Vector Machines) (Vapinik, 1995; Lorena et al., 2008), Maximum Entropy (Jaynes, 1957a; Jaynes, 1957b; Phillips et al, 2006; Phillips et al., 2004) and Neural Networks (Haykin, 2001, Rodrigues et al., 2009). The AquaMaps (Kaschner et al., 2008) is an algorithm specifically developed for marine species distribution modeling and the Envelope Score (Nix, 1986; Piñero et al., 2007) is a variation of the BIOCLIM algorithm.

## The Process of Species Distribution Modeling

The process of creating a species distribution model is rather complex and involves a series of steps, concepts and techniques; it also demands obtaining and dealing with a large amount of data and using different software packages for model generation and for different types of processing and analyses, among other functionalities. Santana et al. (2008) developed a reference process with the aim of serving as a guide to that complex modeling activity. It is represented by a logical sequence of steps, each of them having one or more activities (tasks) that must be performed. These steps are briefly presented to clarify the process.

- Step 1: problem definition: consists of defining the modeling experiment itself, which includes the specification of a set of questions the model should help answer regarding the species and the area to be studied, among others.
- Step 2: species point's treatment: consists of preparing the species data points for modeling. It may involve indentifying and

correcting digitization errors, lack of geo-referenced data, etc.;

- Step 3: environmental data treatment: consists of identification, acquisition and conversion (if necessary) of the environmental data that is required to generate a model;
- Step 4: data viability analysis: refers to the analysis of the data obtained in the previous steps to decide if they are enough to answer the scientific questions;
- Step 5: algorithms choice: consists of selecting the algorithm(s) to generate the models, the data available;
- Step 6: parameter definitions: means to define the required parameters of the algorithm selected;
- Step 7: model(s) generation: consists of running the software tools using the input data and the selected algorithm to generate a model;
- Step 8: automatic post analysis: it consists of the evaluation of the model to assess its adequacy. Statistical measures may be calculated in order to indicate the performance (accuracy);
- Step 9: researcher (expert) validation: it involves analyzing the results obtained (model and post analysis data) to decide if the model is acceptable for that species and conditions.

## Maximum Entropy Applied to Modeling of Biologic Species Geographical Distribution

The Maximum Entropy (MaxEnt) methods are well known and largely used in other areas (Jaynes, 1982), but only recently has it been developed and used for modeling biologic species geographical distribution (Phillips et al., 2006). The MaxEnt showed good results for modeling in comparison with other algorithms (Elith et al., 2006).

The entropy notion can be used for modeling biologic species geographical distribution because it defines a type of measure in the space of probability distributions, so that distributions having higher entropy are favored in relation to others.

The amount of information of an event is related to its probability of occurrence. An unexpected event has a greater amount of information than an expected one. In particular, a certain event may have a null amount of information (Tavares, 2004).

Assume that a system the set of possible configurations of which is $X = \{x_k \mid k = 1, 2,..., N\}$ and the symbol $P$ is its probability distribution, so that each one of the configurations $X = x_k$ occurs with probability $p_k = P(X = x_k)$. The sum of the probabilities over all of these possible values is 1. The average amount of information associated with the possible configurations is called the entropy of $X$. The entropy can be defined formally as (Haykin, 1998):

$$H(X) = -\sum_{k=1}^{N} p_k \log p_k$$

When, for some configuration $x_k$, $p_k = 1, p_i = 0$ is required for all $i \neq k$, showing that there are no "surprises". In this situation, no "information" is transmitted, because the system configuration is completely known. However, if several configurations can occur with different probabilities and probability $p_k$ is small, then there is much more "surprise", and therefore "information", when $X$ assumes the value $x_k$ instead of any other value with higher probability. Thus, before the occurrence of a configuration, there is an amount of uncertainty. When the configuration occurs, there is an amount of surprise. After the occurrence of the configuration, there is an increase in the amount of information. These three amounts are obviously the same and the amount of information is related with the inverse of the occurrence probability (Haykin, 1998).

When the configurations are equally probable, that is they share the same probability, there is maximum entropy (Tavares, 2004). The entropy is maximum in the state of maximum clutter. In this case, the lack of organization is total and the choice is completely free.

In general, the principle of maximum entropy is applied when a system has a set of known states, but with unknown probabilities. In addition, some constraints about the probability distribution for the states must also be known. In most cases, the number of models that satisfy the constraints is infinite. Therefore, the problem consists in choosing a probabilistic model that achieves higher standards under some directions, given the previous knowledge of the model (Haykin, 1998).

The main advantage of the MaxEnt as an approach for modeling biologic species geographical distribution is its capacity of working only with presence points. In the modeling task, the environmental variables are called "features" and the constraints are "the expected value of each feature that should match its empirical average" (Phillips et al., 2006). The features can be the environmental variables or functions thereof.

Suppose that $X$ is a set of pixels representing the area of interest and $x_1,..., x_m$ is a set of points in $X$, representing the observed and recorded localities where the species occurs. The aim is to construct a distribution $\pi^*$ that approximates $\pi$. To construct $\pi^*$, a set of features $f_1,...,f_n$ is necessary, so that $f_j: X \rightarrow$ R. In (Ratnaparkhi, 1997) a proof can be found that distribution $\pi^*$ with maximum entropy must be equivalent to the form:

$$\pi^* = \lambda \prod_{j=1}^{k} \alpha_j^{f_j(x)}, 0 < \alpha_j < \infty$$

where $\lambda$ is a normalization constant and $\alpha_j$'s are the model parameters. Each feature $f_j$ has a corresponding parameter $\alpha_j$, which can be viewed as a "weight" for the feature.

There are many methods to estimate the maximum entropy parameters. The modeling algorithm available in the openModeller tool, based on the maximum entropy method, is the same sequential algorithm used in the MaxEnt software (Phillips et al., 2004; Phillips et al., 2006). This algorithm chooses a feature in each iteration and adjusts its parameter. This procedure is executed until either the convergence or the number of iterations predefined is reached.

The Maximum Entropy algorithm is used for presence-only modeling of biologic species geographical distribution. Some methods use background pixels in place of absence records. These pixels are taken randomly from the study area and also are called pseudo-absences. However, Maximum Entropy method does not require this approach, that is, background pixels being treated as absences. Instead it uses the presence data and the environmental layers over all the study area to generate a probability distribution under the region. Nevertheless, it is possible to define the number of random background pixels to be used with the aim of increase the processing time (Phillips et al., 2006).

The method explained above requires that the expected value of each feature match its empirical average, denoted by $\tilde{\pi}[f_j]$. However, this constraint can be relaxed using a regularization parameter $\beta_j$. The aim is to avoid over fitting in the training. Thus, the constraint $\pi^*[f_j] = \tilde{\pi}[f_j]$ is replaced by $\left|\pi^*[f_j] - \tilde{\pi}[f_j]\right| \leq \beta_j$, where $\beta_j = \beta\tilde{\sigma}[f_j] / \sqrt{m}$, $\tilde{\sigma}[f_j]$ is the standard deviation of $f_j$, $m$ is the number of samples and $\beta$ is a constant according to the number of samples (Phillips et al., 2004).

## Neural Networks Applied to Modeling of Biologic Species Geographical Distribution

Neural Networks are techniques inspired in intelligent biological mechanisms, more specifically the human brain. The human brain processes information more efficiently than conventional computers, except in relation to the speed with which mathematical calculations are carried through, and this characteristic is the main motivation for the research with Neural Networks (Beale & Jackson, 1990).

Modeling biological species geographical distribution can be seen as a prediction problem, because the aim is to determine the presence probability of a given biological species at each point of a georeferenced map. Among several applications of Neural Networks are prediction problems.

The prediction process in the modeling of biological species geographical distribution is based on environmental data and species occurrence data. A Neural Networks-based algorithm in the openModeller software uses the neural model MLP (Multilayer Perceptron), with Backpropagation training algorithm (Rumelhart et al., 1986). This is a classic model, well tested and adjusted to the prediction task. Moreover, the data classes that form the training and test sets are known and divided in to presence and absence points, making the supervised training suitable, such as Backpropagation. Even when only presence points, a strategy of pseudo-absences generation is used, allowing the supervised training.

The Neural Networks-based algorithm is executed until the stop criterion is satisfied. A normally criterion used is the minimum error, that is, when the error calculated in the output layer reaches a smaller value that a certain threshold determined at the beginning of the training, then the weight adjustment algorithm finishes. However, the Backpropagation training algorithm does not guarantee that the error will reach the minimum expected, since that the training can stop in local minimums. A possible strategy is to associate it with the training for minimum expected error an alternative stop criterion. Generally, the assistant criterion used is the number of training epochs. Every time the total training set is presented to the neural network, an epoch is completed. When the number of epochs previously established at the beginning of the training is reached, the weights adjustment algorithm finishes, even though the minimum error has not been reached.

A basic parameter that must be established in the definition of the neural network is the number of neurons in each layer. In a modeling task, the neural network input layer receives the values from the environmental layers. Therefore, the number of neurons in this layer corresponds to the number of environmental layers used in the modeling. The neural network output layer is formed by one neuron, which is responsible for presenting the network output value, that is, the species presence probability in each point of the georeferenced map. Only one neural network hidden layer is used in the current algorithm implementation. The definition of the number of neurons in the hidden layer is a neural networks classic problem that is normally empirically solved, although there are some approaches that aim to solve it as the validation techniques.

Other basic parameters in the learning process are the learning rate ($\eta$) and the momentum term. The learning rate controls the variations of the synaptic weights, in each iteration of the Backpropagation algorithm. The smaller its value, the smaller the variation in the weight adjustments and the slower will the learning process be. Conversely, the bigger the learning rate value, the bigger the variation in the weight adjustments; however, the neural network will be more instable during the learning process.

The momentum term is defined as $\alpha\Delta w_{ij}(n\text{-}1)$, where $\alpha$ is the momentum constant. This term is used to prevent the instability of the neural network when a high learning rate is used (Haykin, 1998). It is important to highlight that the inclusion of the momentum term is a modification in the original rule of Backpropagation weight adjustments. Furthermore, the use of this term implies the need to adjust an additional parameter during the training process, that is, to adjust the momentum constant $\alpha$.

## Software Tools for Species Distribution Modeling

There are some free software tools available for modeling biologic species geographical distributions. DesktopGarp is a software package that provides a desktop version of the GARP algorithm (Pereira, 2002). The MaxEnt is a program for modeling of species geographical distribution with an algorithm based on maximum entropy (Phillips et al., 2009).

A very interesting modeling software tool is the openModeller framework because it provides several functionalities that support the modeling process based on ecological niche (Muñoz et al., 2009; Sutton et al., 2007). Among the functionalities provided there are different modeling algorithms, data search and pre-analysis features, such as a resample algorithm called Jackknife (Quenouille, 1956; Efron, 1979, Rodrigues et al, 2008). A number of fundamental niche modeling algorithms are provided as plug-ins, including GARP, Neural Networks, SVM, Maximum Entropy, Minimum Distance, Climate Space Model, Bioclimatic Envelopes, and others.

The openModeller framework provides an interface that aims at a flexible, user friendly, and cross-platform environment in which the entire process of conducting a fundamental niche modeling experiment can be carried out. The software includes facilities for reading species occurrence and environmental data, selection of environmental layers on which the model should be based; creating a fundamental niche model and projecting the model into an environmental scenario (Figure 1).

Ecological niche modeling is typically a two-stage process. In the first stage, a model is constructed. In order to derive the ecological niche model, openModeller requires a set of points of species occurrence (latitude/longitude) as input for a given region. A second input required is a collection of one or more environmental layers (in the form of raster surfaces). Typically these raster layers contain climatic variables such as temperature and rainfall. Other abiotic raster layers can also be used – for example representing topography or soil properties. The environmental layers are sampled at each site in which an occurrence for the species under study has been observed. These environmental samples are then passed to an algorithm (for example Bioclim) that uses the samples to construct a model that describes the environmental preference or 'niche' of the species. In the second stage, this model is 'projected' into any geographic region containing a compatible set of environmental variables. The output from this modeling process modeling is a new raster layer in which each cell in the raster is assigned a value that represents the probability of the species occurring there. The output generated by openModeller is a gray-scale TIFF or Erdas Imagine '.img' file. The probabilities are expressed either as scaled integer values in the range [0, 255] or as floating point numbers in the range [0, 1].

Thus, the following items describe the openModeller architecture (Figure 2):

*Figure 1. The steps to build the probability distribution*

*Figure 2. Architecture of openModeller*

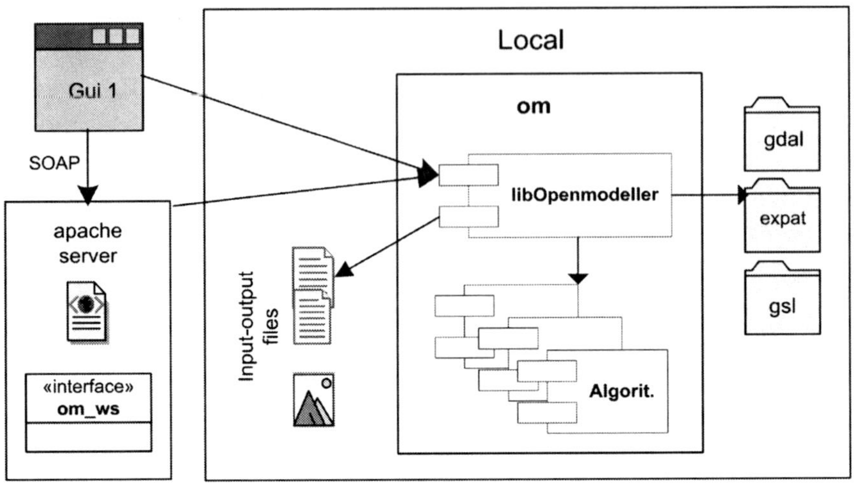


- A controller application: this can be a Graphical User Interface (GUI), a command line application or similar application that invokes operations in the openModeller library (libopenModeller). The controller application can access the libopenModeller locally or remotely via web services (om_ws);
- openModeller Library: handles input / output of model data, sampling of environmental values at species occurrence points and delegates the task of constructing the model to the algorithm plug-in;
- Algorithms: implement a standard Application Program Interface (API) and contain logic for constructing an ecological niche model and returning the probability of occurrence given a set of environmental conditions;
- Auxiliary libraries:
  - GDAL: Provides low level input / output support for raster layers;
  - EXPAT: An XML parser for reading configuration files;
  - GSL: Mathematical library used the Climate Space Model algorithm.

The neural network and maximum entropy algorithms, such as the others, are plugins of the openModeller library and they were written in C++. The algorithms generate and store models in a XML format file. The neural network algorithm was based on Paras Chopra algorithm (Chopra, 2002). The openModeller neural network implementation includes several improvements in relation to Paras Chopra neural network version, for example the momentum term inclusion. Besides that, several modifications were needed for the algorithm integration to openModeller library. For the neural network algorithm improvement, from new experiments some adjustments can be needed.

The maximum entropy algorithm was implemented in openModeller using the same algorithm used in software MaxEnt (Phillips et al., 2004). Currently, only the raw environmental variables are used as features in the models construction. The regularization parameter is calculated by a specific method based on the number of samples.

There are modeling problems involving a high number of layers that the algorithm takes a long time to execute (up to days) in a conventional desktop computer. An implementation of openModeller was developed that more deeply explores the possibilities of Service Oriented Computing (SOA) (Saraiva et al., 2009), including an enterprise service bus (Murakami et al., 2009) and of high performance computing (Araujo et al., 2007). The GUI module accesses an infrastructure for high performance computing services provided by a computer cluster with algorithms with different computational strategies, such as maximum entropy, neural networks, Parallel GARP (GARP using Parallel Algorithmic) and Adapt GARP (GARP using Adaptive strategies) (Bravo et al., 2007).

## CASE STUDY: BABAÇU DISTRIBUTION MODELING

### Modeling Aim

From a computational point of view, the case study aim was to exemplify the use of two computational techniques for modeling biological species geographical distribution. Resample techniques could be applied if the goal was to measure the algorithms performance. Modeling steps taken in the case study are similar to those adopted by Santana et al. (2008).

From a biological point of view, the importance of palms (*Palmae, Arecaceae*) in the Neotropical region is highlighted by several ethnobotanical studies related to aspects of food, medicines or extraction of palms, consisting of one more plant resources used by man.

The babaçu (*Orbignya* sp.) occurs in various regions of Brazil; in the Piauí State it has an important role in terms of social, economic and environmental aspects. This palm marks the physionomy in the transition area between the Amazon rainforests, *cerrado* and semi-arid, often forming almost monospecific forest. This is an important product of the extraction, supplying raw material for various products in Non-Timber Forest Products (NTFP), with full use of its fruits, stems, leaves, meristems and stem. In addition, the babaçu is an important source of renewable energy, as the various parts of the fruits have potential for production of ethanol, biodiesel, coal and gas; traditionally almond oil is extracted, which is one of the main products of exploitation this in the tropics (Mitja & Ferraz, 2001).

From the social point of view it is an important plant resource for small farmers and low-income populations (Mitja & Ferraz, 2001), and provides the opportunity to explore a plant for multiple uses, which grows in abandoned agriculture and / or livestock areas, even without cultural practices (Mitja et al., 2007). Data from the beginning of this decade estimate that about 300 thousand families survive from the extraction of babaçu in Brazil (Wagner et al., 2001).

This study aims to contribute to upgrading the stock of babaçu in areas of native vegetation of Piaui, since there are no current data on the size of exploitable babaçu forest areas. The size of the babaçu areas is an essential data for the realistic estimate of the stock available for its sustainable ecological management and for socio-economic development.

### Species Characterization

Babaçu is the generic name given to some palm trees species belonging to *Orbignya* and *Attalea* genera. *Orbignya phalerata* is a babaçu species with greater geographical distribution, morphological variation and socioeconomic importance, mainly for biodiesel production (Lorenzi et al., 2004). This palm is native to the transition zone between the cerrado and open forests of the southern Brazilian Amazon, areas of high biological complexity. Due to human actions, especially deforestation with successive forest fires, nowadays, the babaçu geographical distribution is concentrated in the Brazilian States of Maranhão, Piauí and Tocantins, where they constitute oligarchic forests.

It has a great economic importance for the State of Piauí, since it is an important extraction product. The almond, from which oil is extracted, is the main Brazilian NTFP with a stake of about 20%

of the total production of non-timber extraction in the country in 2005, totaling R$ 508.6 million (IBGE, 2005). From the socio-economic point of view, it is important to highlight the fact that its extraction is carried out by farmers with low income in the period between harvests of the major regional crops, working to supplement income of these families (May et al., 2000).

Figure 3 presents the climatic conditions of Brazilian regions and reference places where babaçu occurs.

## Occurrences Data Collection

The points of occurrence used in this study were georeferenced along paved roads and land in the Piauí State. The points represent only babaçu forest - homogeneous or not - but never isolated individuals. It was used 100 occurrence points obtained by Brazilian Agricultural Research Corporation – EMBRAPA Middle-North.

## Environmental Data Collection

The environmental data used in the babaçu distribution modeling were selected from the Worldclim (Hijmans et al., 2005) repository. Maximum temperature, minimum temperature, precipitation and altitude were the layers used. These variables were selected according to Table 1, mainly the variables related to temperature and precipitation. For maximum temperature, layers regarding the six hottest months were used, that is, from October to March. Similarly, for minimum temperature layers concerning the six coldest months were used, that is, from April to September. Besides that, layers related to monthly precipitation for the whole year and one layer concerning altitude were used. Since the models generated were projected in the Piauí State and considering the dimensions of this state, the resolution adopted for environmental variables was about 5 Km. According to related works (Santana at al., 2008), this resolution seems to be appropriate to the studied case.

*Figure 3. Climatic conditions of babaçu occurrence in the Brazilian northeast region (EMBRAPA, 1984)*

| Climate type | Average Temperature (°C) | Average Precipitation (mm) | Hydric Deficit (mm) | Hydric Index (mm) | Relative Humidity (%) | Reference Points |
|---|---|---|---|---|---|---|
| Semi-arid, tropical or subtropical | 21-28 | 500 to 1,000 | 500 to 1,000 | -33 to 70 | 60 to 70 | Chapada do Araripe and cities in south of Teresina (Floriano - PI region) |
| Sub-humid, dry or subtropical | 20-27 | 700 to 1000 | 200 to 600 | 0 to -33 | 65 to 76 | Caxias (MA), Teresina (PI) |
| Sub-humid, humid or subtropical | 20-27 | 1000 to 1700 | 50 to 300 | 0 to 120 | 70 to 80 | Imperatriz (MA), Sao Luiz (MA) |
| Humid, tropical or subtropical | 20-27 | 1500 to 2350 | 0 to 100 | 20 to 100 | 75 to 90 | Northwest of MA, Serras da Ibiapaba and Baturite |

## Algorithms Utilization

As mentioned in the previous section, the input parameters for the neural network are learning rate, number of neurons in the hidden layer, momentum constant, number of training epochs and minimum error. All the experiments were carried with the same values as the parameters. Normally, the definition of these parameters to the neural network is empirical. Some tests with the openModeller neural network algorithm showed that appropriate values for modeling are: learning rate = 0.3, number of neurons in the hidden layer = 8, momentum constant = 0.05, number of training epochs = 500,000 and minimum error = 0.01. It is important to observe that a small value for the minimum error do not guarantee a good generalization. One used strategy is to define the minimum error equal to 0.01 for the first training and adjust it according to the results. Thus, these values were used in the experiments with babaçu except the minimum error, which was defined 0.2. This value was chosen because the results were satisfactory and the aim was only to exemplify the modeling process. However, additional experiments are required to define the best value of minimum error parameter.

Two criterions are always used to stop the training: the minimum error and the number of training epochs, which occurs first. An additional input parameter is the training type that just determines how the training progress is shown for the user, that is, the percentage of training epochs or a percentage that represents how much the actual minimum error is close from the expected minimum error.

There are three input parameters in the maximum entropy algorithm: number of iterations, number of background points and convergence. Likewise in the experiments with neural network algorithm, the parameter values for the maximum entropy algorithm were the same in all executions. The parameter values defined empirically are: number of iterations = 500, number of background points = 10,000 and convergence = $10^{-5}$. The features used in the experiments were the raw environmental variables.

## Carried Through Experiments

Fifty babaçu distribution models were generated and merged the ten models with the best accuracy from each algorithm: neural network and maximum entropy algorithms. This approach had been applied because each algorithm execution can generate models with different accuracy and it is necessary to have many models to select those with the best accuracy (Siqueira, 2005). Two statistic measures were calculated for each distribution model generated, accuracy and AUC (Area Under the ROC Curve). ROC Curve (Receiver Operating Characteristic) is a measure used to characterize the model performance according to all thresholds (Phillips et al., 2006). Accuracy represents the percentage rate of examples correctly classified and AUC (Fielding & Bell, 1997) and gives a performance measure of the model, and whose values range from 0 to 1. The closer to 1 is the AUC value, the closer the generated model will be to the ideal model. Figure 4 shows the best models obtained from each algorithm, based on AUC.

The core area of natural occurrence of babaçu is represented in two models below (Figure 4). However in Figure 4.b it can be noted that the area is larger, which may best reflect the reality. This is because being a secondary species, babaçu must occupy larger areas due to human action in the growing region. One of the next steps of this experiment includes the validation of models in the field, checking the current area of occurrence validating the generated models.

The models generated by the neural network algorithm and by the maximum entropy algorithm are a little different. The neural network algorithm generates a value between 0 and 1 for each pixel, representing the probability of presence on that locality. For the carried out experiments the neural

*Figure 4. Best ten models with Neural Networks (a) and Maximum Entropy (b). Darker regions indicate more suitable conditions for babaçu occurrence, according to the used algorithms and based on the selected environmental layers*

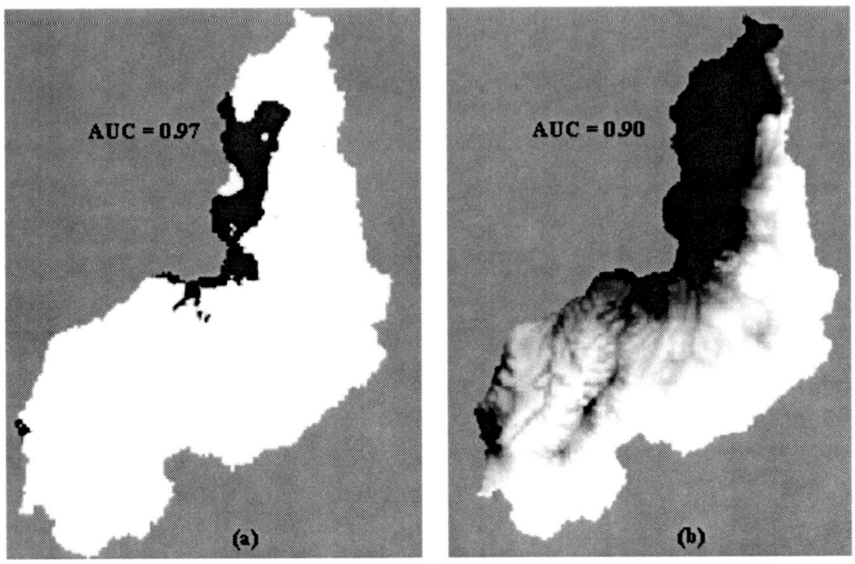


network generated values very close to 0 or very close to 1. Consequently, the generated distribution map is black and white. The maximum entropy algorithm also generates a value between 0 and 1 for each pixel, but the sum of all pixels is 1. Thus, the maximum entropy algorithm produces a probability distribution representing the environmental suitability of each locality for the species. Small values represent poor environmental conditions for the species and values closest to 1 represent good environmental conditions for the species. The difference between the generated maps does not invalidate its comparison by the biologist. Furthermore, the calculated statistics measurements can be used to compare the results.

## Post Analysis: Model Evaluation

The AUC values for the neural network algorithm ranged from 0.49 to 0.97. The AUC average was 0.92. The average training accuracy for the fifty models generated by neural networks was 90.36%. These results showed that the algorithm is suitable for modeling biologic species geographical distribution.

The Maximum Entropy algorithm is deterministic. Thus, the training accuracy was always the same, 98.21%. The AUC is not the same in all executions because of the background pixels generation. In each run the generated pixels can be different. Therefore, the AUC values for Maximum Entropy algorithm ranged from 0.88 to 0.90. Since Maximum Entropy algorithm is a method which uses species presence data, the maximum AUC value will always be less than one (Phillips et al., 2006). Figures 5 and 6 show the AUC frequency distribution for Neural Networks and Maximum Entropy algorithms, respectively.

The models obtained show possible distribution patterns of babaçu. The models were validated by experts based on the actual and potential occurrence of Babaçu in Piauí State. The models confirm the actual occurrence of babaçu in the northwest state and other regions of babaçu occurrence prediction are factible, considering some optimal abiotic conditions (Embrapa, 1984), as

*Figure 5. AUC distribution frequency for neural networks algorithm for the fifty models generated*

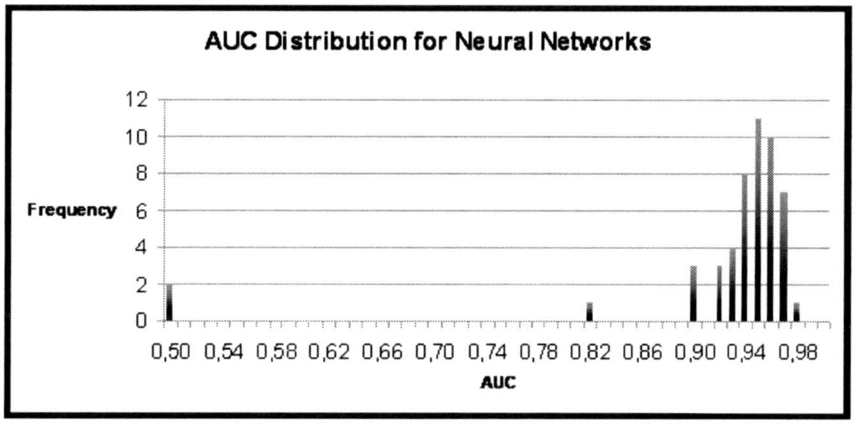


*Figure 6. AUC distribution frequency for maximum entropy algorithm for the fifty models generated*

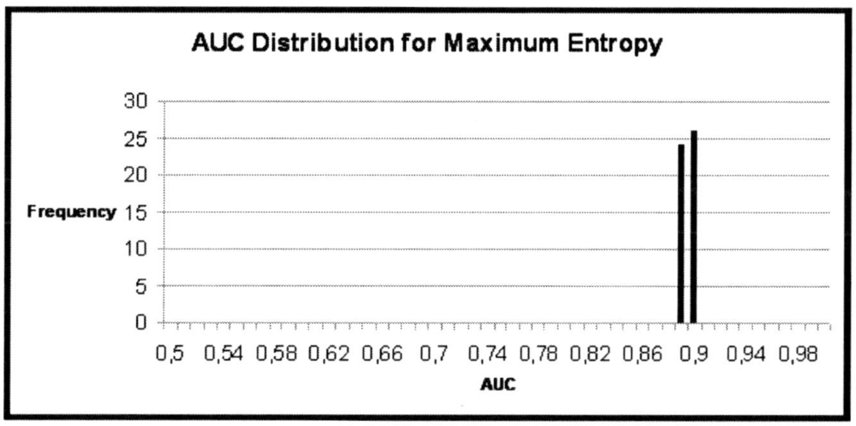


annual rainfall of not less than 1700 mm and insolation minimum of 2400 hours per year.

It is known that babaçu is an important NTFP since before the European colonization. But in this time these palms were diluted in the landscape of broad biological complexity. Nowadays it was changed by the human activity, and the babaçu natural distribution is concentrated in the Brazilian States of Maranhão, Piauí and Tocantins, where they constitute almost monoespecific stands. Therefore the babaçu distribution is determinated by their ecological requirements, it is also influenced by human actions. It is evident by its colonization in non-core areas and its dominance in core areas.

Based on the actual babaçu distribution it is possible to say that the model obtained from the Maximum Entropy algorithm reflects best the distribution of babaçu. By their unique characteristics it is necessary to develop more mapping and checking in natural babaçu areas to provide the model and make it more accurate.

## FUTURE RESEARCH DIRECTIONS

Nowadays, researches in modeling of biologic species geographical distribution present important results and development. However, there are several improvements and new research fields that can contribute with modeling. Some suggestions of future researches are:

- Pre-analysis tools that can assist the user to choose the suitable environmental layers according to the species, regions and modeling aim;
- Pre-analysis methods to support the identification of the best number of occurrence localities needed for each method to produce accurate models;
- Post analysis methods to help in the identification of optimal models;
- Development of new modeling algorithms using intelligent computational techniques not used yet for this task, considering not only the concept of Fundamental Niche, but species interactions too.
- Explore Adaptive techniques, such as in the Adapt GARP algorithm case, to enhance the performance of the already implemented algorithms;
- Development of methods to avoid over-fitting when the number of occurrence localities is very small;
- Hardware and software architecture for distribution models persistence, to allow use and sharing of these models.

## CONCLUSION

Computational techniques for biologic species distribution are important to preserve, protect and sustainably use Biodiversity, which is a relevant topic in Biodiversity Informatics research area. Among the software tools available for modeling, openModeller framework is the most comprehensive open source framework available with the possibility to include an experimental algorithm. Two new algorithms have just been added to it: one based on Neural Networks, and one based on Maximum Entropy. They have been used to develop geographical distribution models of babaçu, an important palm species in the middle-North of Brazil, in which it is used by low income communities as a means to increase their budget. The models obtained show possible distribution patterns of babaçu and an expert opinion is necessary to compare and validate them. Based on such heuristics it is possible to say that the model obtained from the Maximum Entropy algorithm reflects best the distribution of babaçu.

## REFERENCES

Araujo, M. J., Corrêa, P. L. P., Sato, L. M., & Saraiva, A. M. (2007). A framework for species distribution modelling - A performance evaluantion approach. In *6th International Information and Telecommunication Technologies Symposium, 2007*, Brasilia, (I2TS'2007).

Austin, M. P., Nicholls, A. O., Doherty, M. D., & Meyers, J. A. (1994). Determining species response functions to an environmental gradient by means of a beta-function. *Journal of Vegetation Science*, *5*, 215–228. doi:10.2307/3236154

Beale, R., & Jackson, T. (1990). *Neural Computing: An Introduction*. Bristol, UK: Institute of Physics Publishing.

Bravo, C., Neto, J. J., Santana, F. S., & Saraiva, A. M. (2007). Towards an Adaptive Implementation of Genetic Algorithms. In *Latin American Conference on Informatics, Proceedings Latin American Conference on Informatics*, San Jose.

Busby, J. R. (1986). A biogeographical analysis of Nothofagus cunninghamii (Hook.) Oerst. in southeastern Australia. *Australian Journal of Ecology*, *11*, 1–7. doi:10.1111/j.1442-9993.1986.tb00912.x

Canhos, V. (2005). *A framework for species distribution modeling. Thematic Research Project – Supported by FAPESP*. São Paulo, Brazil: CRIA/EPUSP/INPE.

Canhos, V. P. (2003). Biodiversity Informatics: patterns, protocols and tools. *Science and Culture, 55*(2), 45-47. Retrieved August 27, 2009, from http://cienciaecultura.bvs.br/pdf/cic/v55n2/15530.pdf. (in Portuguese).

Carpenter, G., Gillison, A. N., & Winter, J. (1993). DOMAIN: a flexible modeling procedure for mapping potential distributions of plants, animals. *Biodiversity and Conservation, 2*, 667–680. doi:10.1007/BF00051966

Chopra, P. (2002). *A very simple example of Neural Networks using back propagation*. Retrieved July 15, 2008, from http://paraschopra.com

Corrêa, P. L. P., Saraiva, A. M., Cugnasca, C. E., & Hirakawa, A. R. (2006). An architecture of systems for integration of Biodiversity distributed databases. *Journal of the Brazilian Society of Agroinformatics, 6*(1), 80–100.

Efron, B. (1979). Bootstrap Methods: Another Look at the Jackknife. *Annals of Statistics, 7*(1), 1–26. doi:10.1214/aos/1176344552

Elith, J., Graham, C. H., Anderson, R. P., Dudík, M., Ferrier, S., & Guisan, A. (2006). Novel methods improve prediction of species' distributions from occurrence data. *Ecography, 29*, 129–151. doi:10.1111/j.2006.0906-7590.04596.x

EMBRAPA. (1984). *Babaçu – Research National Program*. Brasília: EMBRAPA.

Ferrier, S., & Watson, G. (1997). *An evaluation of the effectiveness of environmental surrogates and modelling techniques in predicting the distribution of biological diversity*. Canberra, Australia: NSW National Parks and Wildlife Service. Retrieved August 27, 2009, from http://www.environment.gov.au/Biodiversity/publications/technical/surrogates/pubs/surrogates.pdf

Ferrier, S., Watson, G., Pearce, J., & Drielsma, M. (2002). Extended statistical approaches to modelling spatial pattern in Biodiversity in northeast New South Wales. 1. Species-level modeling. *Biodiversity and Conservation, 11*, 2275–2307. doi:10.1023/A:1021302930424

Fielding, A. H., & Bell, J. F. (1997). A review of methods for the assessment of prediction errors in conservation presence/absence models. *Environmental Conservation, 24*, 38–49. doi:10.1017/S0376892997000088

Haykin, S. (1998). *Neural Networks: A Comprehensive Foundation* (2nd ed.). New York: Prentice Hall.

Hijmans, R. J., Cameron, S. E., Parra, J. L., Jones, P. G., & Jarvis, A. (2005). Very high resolution interpolated climate surfaces for global land areas. *International Journal of Climatology, 25*, 1965–1978. doi:10.1002/joc.1276

IBGE. (2005). Geography and Statistic Brazilian Institute. *Production of vegetal extraction and forestry. v.20* – Brazil. (in Portuguese).

Jaynes, E. T. (1957a). Information theory and statistical mechanics. *Physical Review, 106*, 620–630. doi:10.1103/PhysRev.106.620

Jaynes, E. T. (1957b). Information theory and statistical mechanics II. *Physical Review, 108*, 171–190. doi:10.1103/PhysRev.108.171

Jaynes, E. T. (1982). On the rationale of maximum entropy methods. *Proceedings of the IEEE, 70*, 939–952. doi:10.1109/PROC.1982.12425

Kaschner, K., Ready, J. S., Agbayani, E., Rius, J., Kesner-Reyes, K., Eastwood, P. D., et al. (2008). *AquaMaps: Predicted range maps for aquatic species*. Retrieved August 27, 2009, from http://www.aquamaps.org. Version 10/2008.

Lorena, A. C., Siqueira, M. F., Giovanni, R., Carvalho, A. C. P. L. F., & Prati, R. C. (2008). Potential Distribution Modelling Using Machine Learning Classifiers. In *The Twenty First International Conference on Industrial, Engineering & Other Applications of Applied Intelligent Systems*, 2008, Wroclaw, *Proceedings The Twenty First International Conference on Industrial, Engineering & Other Applications of Applied Intelligent Systems* (LNAI 5027, pp. 255-264).

Lorenzi, H., Souza, H. M., Costa, J. T. M. C., & Ferreira, E. (2004). *Brazilian end exotic palm cultivated.* Editora Plantarum. (in Portuguese)

May, P., Veiga Neto, F. C., & Cheves Pozo, O. V. (2000). *Biodiversity economic valuation*. Brasília: Ministério do Meio Ambiente. (in Portuguese)

Mitja, D., & Ferraz, I. (2001). Establishment of babaçu in pastures in Pará, Brazil. *Palms*, *45*(3), 138–147.

Mitja, D., Martins, E. D., Silva, J. C. S., & Miranda, I. S. (2007). Biodiversité: les Palmiers, Babaçu, exploitation d'une invasion. *Science Studies*, *38*, 10.

Muñoz, M. E. S., Giovanni, R., Siqueira, M. F., Sutton, T., Brewer, P., & Pereira, R. S. (2009). openModeller: a generic approach to species' potential distribution modeling. *GeoInformatica*. doi:.doi:10.1007/s10707-009-0090-7

Murakami, E., Santana, F. S., Stange, R. L., & Saraiva, A. M. (2009). The integration of open standards for Enterprise Service Bus in a solution for agribusiness and environmental applications development. In *Proceedings of the Joint International Agricultural Conference*, Wageningen. Holand.

Myers, N., Mittermeier, R. A., Mittermeier, C. G., Fonseca, G. A. B., & Kent, J. (2000). Biodiversity hotspots for conservation priorities. *Nature*, *403*, 853–858. doi:10.1038/35002501

Nix, H. (1986). A biogeographic analysis of Australian elapid snakes. In *Atlas of Elapid Snakes of Australia* (pp. 4–15). Canberra, Australia: Australian Government Publishing Service.

Pereira, R. S. (2002). Retrieved August 27, 2009, from http://www.nhm.ku.edu/desktopgarp/index.html

Pereira, R. S., & Peterson, A. T. (2001). The use of modeling in the definition of strategies for conservation of Biodiversity. *Biodiversity - economic and social value*. (in Portuguese). Retrieved August 27, 2009, from http://www.comciencia.br/reportagens/biodiversidade/bio18.htm

Persona, L., Corrêa, P. L. P., & Saraiva, A. M. (2003). Environmental Niche Modeling in Biodiversity with Genetic Algorithms. (in Portuguese). In *Proceedings of the IEEE - I2TS'2003 2nd International Information and 68 Telecommunication Technologies Symposium*, Florianópolis.

Phillips, S. J., Anderson, R. P., & Schapire, R. E. (2006). Maximum entropy modeling of species geographic distributions. *Ecological Modelling*, *190*, 231–259. doi:10.1016/j.ecolmodel.2005.03.026

Phillips, S. J., Dudík, M., & Scaphire, R. E. (2009). Retrieved August 27, 2009, from http://www.cs.princeton.edu/~schapire/maxent

Phillips, S. J., Dudík, M., & Schapire, R. E. (2004). A Maximum Entropy Approach to Species Distribution Modeling. In *Proceedings of the 21st International Conference on Machine Learning*, Banff, Canada.

Piñeiro, R., Aguilar, J. F., Munt, D. D., & Feliner, G. N. (2007). Ecology matters: Atlantic-Mediterranean disjunction in the sand-dune shrub Armeria pungens (Plumbaginaceae). *Molecular Ecology*, *16*, 2155–2171. doi:10.1111/j.1365-294X.2007.03280.x

Quenouille, M. H. (1956). Notes on Bias in Estimation. *Biometrika*, *43*(3/4), 353–360. doi:10.2307/2332914

Ratnaparkhi, A. (1997). *A Simple Introduction to Maximum Entropy Models for Natural Language Processing.* Institute for Research in Cognitive Science - IRCS Report 97–08 at University of Pennsylvania.

Rodrigues, F. A., Avilla, A. O., Rodrigues, E. S. C., Corrêa, P. L. P., Saraiva, A. M., & Rocha, R. L. A. (2009). Species distribution modeling with neural networks. In *e-Biosphere 2009*, 1-3 June 2009, London-UK. Retrieved August 27, 2009, from http://www.e-biosphere09.org/assets/files/e-Biosphere%20Abstracts%20Volume%20-%20 FINAL.pdf

Rodrigues, F. A., Rodrigues, E. S. C., Sato, L. M., Midorikawa, E. T., Corrêa, P. L. P., & Saraiva, A. M. (2008). Parallelization of the Jackknife Algorithm Applied to a Biodiversity Modeling System. In *7th International Information and Telecommunication Technologies Symposium, 2008, Foz do Iguaçu. Proceedings of the 7th International Information and Telecommunication Technologies Symposium*, (pp. 58–65).

Rumelhart, D. E., Hinton, G. E., & Williams, R. J. (1986). Learning representations by back-propagating errors. *Nature*, *323*, 533–536. doi:10.1038/323533a0

Santana, F. S., Siqueira, M. F., Saraiva, A. M., & Corrêa, P. L. P. (2008). A reference business process for ecological niche modelling. [INSERT FIGURE 001]. *Ecological Informatics*, *3*, 75–86. doi:10.1016/j.ecoinf.2007.12.003

Saraiva, A. M., Correa, P. L. P., Sato, L. M., Rodrigues, F. A., Santana, F. S., Rodrigues, E. S. C., et al. (2009) A Service-Based Framework for Species Distribution Modeling. In *Conference Abstracts Of E-Biosphere Conference,* London, 1-3, june, 2009.

Siqueira, M. F. (2005). *Use of Fundamental Niche Modeling in the Pattern Evaluation of Vegetal Species Geographic Distribution.* PhD Thesis. Department of Ambient Engineering of University of Sao Carlos. Sao Carlos, Sao Paulo, Brazil. (In Portuguese).

Stockwell, D., & Peters, D. (1999). The GARP modelling system: problems and solutions to automated spatial prediction. *International Journal of Geographical Information Science*, *13*(2), 143–158. doi:10.1080/136588199241391

Stockwell, D. R. B. (1992). *Machine learning and the problem of prediction and explanation in ecological modelling.* Doctoral Thesis, Australian National University, Australia.

Stockwell, D. R. B., & Noble, I. R. (1991). Induction of sets of rules from animal distribution data: a robust and informative method of data analysis. *Mathematics and Computers in Simulation*, *32*, 249–254. doi:10.1016/0378-4754(90)90247-G

Sutton, T., Giovanni, R., & Siqueira, M. F. (2007). Introducing openModeller – a fundamental niche modeling framework. *OSGEO Journal*, *1*. Retrieved August 27, 2009, from http://www.osgeo.org/ojs/index.php/journal/article/viewFile/109/90

Tavares, J. N. (2004). *Free Course on Theory of the Field - Entropy. Measures of Gibbs. Mathematic Center of University of Porto*. Physics Center of Porto. (in Portuguese)

Vapinik, V. (1995). *The Nature of Statistical Learning Theory*. Berlin: Springer-Verlag.

Wagner, A. B. A. Shiraishi. J. N. & Alvino, B. M. (2001). *Babaçu economy, preliminary data survey*. MIQCB, Balaios Typographia, Sao Luis-MA, 294p. (in Portuguese)

Yee, T. W., & Mitchell, N. D. (1991). Generalized additive models in plant ecology. *Journal of Vegetation Science*, *2*, 587–602. doi:10.2307/3236170

# Chapter 16
# Seasonal Precipitation Forecast Based on Artificial Neural Networks

**Adriano Rolim da Paz**
*Federal University of Paraíba, Brazil*

**Cíntia Bertacchi Uvo**
*Lund University, Sweden*

**Juan Martín Bravo**
*Federal University of Rio Grande do Sul, Brazil*

**Walter Collischonn**
*Federal University of Rio Grande do Sul, Brazil*

**Humberto Ribeiro da Rocha**
*University of São Paulo, Brazil*

## ABSTRACT

*Agriculture is vulnerable to the interannual climate variability and to its unpredictability, in such a way that most agricultural decisions taken within the time horizon of several months are made in a conservative manner, supposing a near-pessimist scenario. The improvement of climate prediction may help the strategic view, mitigating unwanted impacts and taking advantage of favorable conditions. This chapter presents the development of an Artificial Neural Network (ANN) model for seasonal precipitation forecast based on climate indices, focusing on the practical aspects of selecting the best predictors, defining ANN architecture, data handling and ANN training and validation. The study case is the Pardo/Mogi-Guaçu rivers watershed in Brazil, which is characterized by intense sugarcane plantation for both ethanol and sugar industries. The results demonstrate how the methodology for seasonal precipitation forecast based on ANN can be particularly helpful, with the use of available time series of climate indices.*

DOI: 10.4018/978-1-61692-871-1.ch016

## INTRODUCTION

Several human activities are strongly dependent on the climate conditions and its variability, especially those related to water use. The seasonal precipitation controls the need for irrigation in agriculture, river navigation conditions, operation of multiple purposes river reservoirs, and the overall water resources for many other uses. The prediction of precipitation for weeks or months ahead may improve decision making, optimization of water resources management and have benefits for economy (Chiew et al., 2003).

Particularly, agriculture is vulnerable to the interannual climate variability and to its unpredictability (Hansen, 2002; Ogallo et al., 2000). The time of seeding, soil preparation, irrigation schedule and management, variety selection, productivity estimates and harvesting are some examples of agricultural practices which depend on the climate variability.

Owing to climatic uncertainty, most agricultural decisions taken with time horizon of several months are made in a conservative manner, which prescribes a near-pessimist scenario (Hansen, 2002; Jones et al., 2000). The improvement of climate prediction may help the strategic view, the mitigation of unwanted impacts and to take advantage of the most favorable conditions. Seasonal precipitation forecast in the early wet season has special importance to agriculture. Most crops depend on the first rains for seeding and emergence, rain amount and persistence for the development of the plants, and the occurrence of short dry spells for the harvesting. The prediction of a low rainfall in the early wet season may delay the timing of soil preparation and seeding, or even to provide irrigation and supply the crop water needs. On the other hand, if a pronounced wet season is expected, the decision may be to expand the crop area and/or reduce irrigation.

Global and regional atmospheric models have been widely applied for seasonal precipitation forecast. However, despite their increasing forecast performance, they still lack ability to produce reliable precipitation forecast for local specific areas (Koutsoyiannis et al., 2008). Additionally, running atmospheric models require large computational effort and model-specific knowledge. In turn, data driven models such as Artificial Neural Networks (ANN) are an alternative approach for local seasonal precipitation forecasting (Zwiers and von Storch, 2004; Nasrallah et al., 2001).

ANN may be concerned as a relatively recent solution to the problem of predicting the output of complex systems, used on a wide range of fields, such as environmental science, pattern recognition, remote sensing and financial management (ASCE, 2000; Dawson and Wilby, 2001). ANN models have also been used for a number of hydrologic problems, including rainfall-runoff modeling (Minns and Hall, 1996; Tokar and Johnson, 1999), streamflow forecasting (Dawson and Wilby 2001; Maier and Dandy 2000; Jeong and Kim, 2005) and short- and long-term precipitation forecasting (French et al., 1992; Luk et al., 2001; Silverman and Dracup, 2000).

Methods such as ANN applied for seasonal precipitation forecasting may be easily integrated to an expert agricultural decision support system, which helps improving irrigation management and strategic decisions (McIntosh et al. 2007; Hansen, 2002; Ogallo et al., 2000).

This chapter presents the main concepts concerning precipitation forecasting and how it may be used for agricultural purposes. The main focus of the chapter is how to develop an ANN model for seasonal precipitation forecast. The steps of defining the problem and the variable to be forecasted, choice of predictors, definition of ANN architecture, data handling, ANN training and skill assessment are discussed. To illustrate the methodologies presented, the case study is the Pardo/Mogi-Guaçu rivers watershed in Brazil, which is part of the contributing area of the Grande River, a region characterized by intense hydropower generation and sugarcane plantation for both ethanol and sugar industries, besides other seasonal crops are cultivated in a lesser extent.

## BACKGROUND

### Precipitation Forecast: Overview and Main Concepts

In simple words, forecast of precipitation means the prediction of the amount or intensity of precipitation for a given time period ahead. The time period may range from minutes or hours to an entire year or decade. Precipitation forecast may be as absolute values (amount of precipitation), or depicted through distinct levels of the wetness (or dryness) of the time period. The former is a quantitative precipitation forecast and the predictand is continuous, i.e. it may assume any value (within a reasonable range). In turn, the prediction of categories is a qualitative precipitation forecast, in which there is a discrete predictand that may assume one and only one of a finite set of possible values (Wilks, 2006). For example, it may be predicted a total precipitation of 250 mm for the next month (quantitative forecast), or the prediction may be that the precipitation of the next month will be above normal or very wet (qualitative-categorical forecast). The choice of quantitative or categorical forecast depends on several factors, including the desired time horizon, the purposes of the forecast, the data available, and the prediction method used.

For both quantitative and categorical forecast types, the forecast may be expressed as an areal average or as a spatial regional field, the choice being dependent on the forecast technique and the available data. It is also possible to forecast other variables related to precipitation, such as the number of rainy days (Tantanee et al., 2005), the onset and the end of the wet season or of the drought (e.g. Liebmann et al., 2007; Nguyen et al., 2007a), the distribution of wet/dry spells (Nguyen et al., 2007b) and the probability of persistence of an extreme drought or rainy period (Ogallo et al., 2000).

The time lag between the forecast calculation and the forecast event is called its time horizon, which broadly classifies it as short- or long-range forecast. Short-range forecasts deploy time horizons from hours to some days, generally called weather forecast, and primarily used for flood prediction, warning and control (Moore et al., 2005), operation of river reservoir (Yeh, 1982; Bravo et al., 2009a), natural disaster mitigation (Fall et al., 2007) and prediction of adverse conditions for agriculture such as frost and hailstorms (Sivakumar et al., 2005).

Long-range precipitation forecast prescribe time horizons from several weeks, months to years and are called climate forecast, mostly used for optimization of water resources management (Olsson et al., 2004; Nilsson et al., 2006), planning purposes of agriculture (Ogallo et al., 2000) and several other socioeconomic activities such as tourism, transport and navigation. Seasonal precipitation forecast is a subtype of climate prediction focused on the precipitation patterns for the seasonal rainy period, which usually comprise 1 to 6 months.

### Use of Precipitation Forecast in Agriculture

Agriculture is one of the socioeconomic activities with major potential to take advantage of precipitation forecasts, and this potential increases for developing countries where agriculture is more rain dependent and technological adaptations are low (Ogallo et al., 2000). In a broadly sense, these forecasts may minimize costs, reduce losses in droughts, improve risk management and decision-making capability (Ash et al., 2007; Everingham et al., 2002; Jones et al., 2000).

The agricultural decisions subject to precipitation forecasts vary from the farm scale to industry, regional and larger scales. In the farm scale, the prediction of a coming wetter or dryer season may be used to choose the most suitable crops and varieties, to decide on planting and harvesting times, to estimate and schedule irrigation water supplies and even to manage fertilizer and herbicides inputs, as precipitation influences the development and spread of insects, pests and fungal diseases (Everingham et al., 2002; Ash et al., 2007; Meinke and Stone, 2005; Jones et al., 2000). At the industry scale, decisions concerning grain storage and transport, livestock stocking rate, harvesting and milling operations and even marketing and selling strategies may be adjusted according to precipitation forecasts (Ash et al., 2007). At regional and larger scales, seasonal precipitation forecast may influence anticipation or delay of selling, prices variations and products availability for consumers.

Despite of the large range of potential applications of seasonal precipitation forecast for agriculture, it has no value unless it generates better decisions (Hammer et al., 2001). While for a forecaster the goodness of a prediction is based on the quality of forecast, generally given by the capability of reproducing observed conditions, the forecast user valuables the forecast only if it improves its decision-making process in some aspect (Murphy, 1993). To obtain a valuable forecast, first of all, the forecast must have some accuracy and convey the associated uncertainty (Meinke and Stone, 2005). Second, as the benefits provided by the forecasts are strongly dependent on the consequent responses by the decision-makers and farmers (McIntosh et al., 2005), they must be aware of the vulnerability to climate variability, motivated to improve management of that risk and have sufficient knowledge to use forecast information to improve their decisions (Hansen, 2002). They must also have developed a long-term trust and adoption of forecasts (Ash et al., 2007), which is conditioned by receiving skilful and reliable predictions along time. The way and effectiveness that predictions are delivered and communicated plays a major role in this sense, as forecast skills are imperfect and need to be well understood for the optimal use of the information (Hammer et al., 2001).

Additionally, to improve the benefits the forecast must be site-specific and focused on certain agricultural decisions, as each decision requires particular forecasts characteristics, such as time of realization, accuracy, scale and lead-time (Everingham et al., 2002; Hansen, 2002).

Finally, the value of precipitation forecasts for agriculture may be improved if developed in conjunction with farm programs and government interventions (Cabrera et al., 2007) and if coupled with crop yield simulation models (Moeller et al., 2008; Everingham et al., 2002). These models can convert climate predictions into more functional information at the field scale (Meinke and Stone, 2005).

## Methods for Seasonal Precipitation Forecasting

The oldest method used to forecasting precipitation is based on environmental indicators and traditional knowledge, such as the observation of atmospheric and astronomic conditions (wind, sea, clouds, moon), animal behavior and plants phenology (flowering, ripening and early rotting of fruits, development of buds, etc). This method is still used nowadays, however with lack of underpinning scientific background (for instance, Galacgac and Balisacan, 2009). The methods most used currently may be divided into empirical or data driven methods and dynamical or physical methods.

The empirical methods seek to identify relationships between past values of input variables ($X_i$, i=1,n) and the forecast variables ($Y_j$, j=1,m) in order to be able to predict future values of Y according with the availability of new values of X. The input variables are referred to as predictor variables, while Y are the forecast or predictand variables. The relationships are built using a statistical approach and may serve as a prediction model but not to explain or describe the functioning of the system. Regression analysis, multivariate analyses (e.g. Bretherton et al., 1992; Uvo et al. 2001) and Artificial Neural Networks (Bishop, 1995; Olsson et al., 2004) are examples of empirical methods used for precipitation forecast.

The dynamical precipitation forecast is realized using general circulation models (GCM), which uses mathematical equations related to the governing physical processes to simulate coupled ocean and atmospheric behavior, in the case of Coupled General Circulation Models (CGCM), or only the atmospheric conditions (Atmospheric General Circulation Models - AGCM). The GCM outputs a full range of meteorological variables besides precipitation fields, such as cloudiness, humidity, heat fluxes and winds. Due to the chaotic characteristic of the atmosphere, long-range predictions using GCMs are probabilistic and based on an ensemble of runs of the same model, varying the initial conditions, or even on an ensemble of distinct models. The main shortcomings of using GCMs for precipitation forecasting are the required large computational and modeling efforts, as running a GCM is not yet an user friendly and low cost task, as well as the limitation on forecasting site-specific values, as the model output represents areal averages over grid points usually covering tenths or hundreds of square kilometers.

On their turn, empirical (data driven) models have low computational and modeling costs when compared to full general circulation models, and may be developed for site-specific predictions. However, their disadvantages are to lack physical meaning and to be dependent on historical available data. A statistical relationship can be adjusted between predictor and forecast variables without the accurate definition of a physical relationship between them, and the predictability capacity of the relationship is constrained to the range of the existing data set.

Downscaling techniques may also be used to increase spatial resolution of GCM outputs and hence provide precipitation forecasts more valuable in local areas. In this sense one promising approach is the combination of GCM outputs and empirical models such as ANN (Ash et al., 2007; Goddard et al., 2001; Nilsson et al. 2008; Olsson et al., 2004; Ramirez et al., 2005).

## Forecast Verification

Forecast verification is the process of assessing the quality of the forecast. Knowing about performance of a forecast model is useful to monitoring quality, to improve quality and to compare quality of different forecast models (Brown et al., 2009).

Traditionally, forecast verification has emphasized two aspects of the forecast quality: accuracy and skill (Murphy, 1993). Accuracy refers to the average correspondence between pairs of forecast and observations, where the difference between the forecast and the observation values is named error. Skill refers to the relative accuracy of the forecast of interest with respect to a forecast produced by a standard control such as climatological average or persistence forecast (Wilks, 2006).

Several performance measures can be used in assessing accuracy and/or skill of forecasts (Legates and McCabe, 1999; Wilks, 2006; Casati et al. 2008; Brown et al., 2009), the choice being dependent, basically, on the type of the forecast. In this sense, we discuss about performance measures related to nonprobabilistic quantitative forecast. This type of forecast is characterized by not exhibiting the inherent uncertainty. However, it is one of the most used types of forecast in several areas of the hydrologic sciences.

The performance measures commonly used to assess accuracy on nonprobabilistic quantitative forecast are the mean-squared error (MSE – Equation 1), the mean-absolute error (MAE – Equation 2), the mean-relative error (MRE – Equation 3) and the Nash–Sutcliffe efficiency index (NS – Equation 4). Also, the typical form of a performance measure to assess forecast skill, usually named skill score (SS), is presented in Equation 5.

$$MSE = \frac{1}{N}\sum_{i=1}^{N}\left(F_i - O_i\right)^2 \quad (1)$$

$$MAE = \frac{1}{N}\sum_{i=1}^{N}\left|F_i - O_i\right| \quad (2)$$

$$MRE = \frac{1}{N}\sum_{i=1}^{N}\left|\frac{F_i - O_i}{O_i}\right| \quad (3)$$

$$NS = 1 - \frac{\sum_{i=1}^{N}\left(F_i - O_i\right)^2}{\sum_{i=1}^{N}\left(O_i - \overline{O}\right)^2} \quad (4)$$

$$SS = \frac{PMA - PMA_{ref}}{PMA_{perf} - PMA_{ref}} \quad (5)$$

where $F_i$ and $O_i$ are the forecasted and observed values of the predictand on sample i, respectively; $\overline{O}$ is the average observed predictand value; N is the total number of samples; PMA, $PMA_{ref}$ and $PMA_{perf}$ are, respectively, the value of an accuracy performance measure of the forecast, of the reference forecast and of the perfect forecast (hypothetical forecast without errors, where $F_i$ is equal to $O_i$).

As can be seen on Equation 1-4, performance measures of accuracy are function of the error (difference $F_i$ - $O_i$). Thus, lower the error, higher is the accuracy of the model. In this sense, MSE, MAE and MRE are equal to zero when perfect forecasts are used, and their value increase with larger prediction errors. In turn, NS has value equal to one when perfect forecasts are used. A model always predicting the climatological average has NS equal to zero, while models having better (worst) results than using climatological average present NS higher (lower) than zero (and lower than one in all cases).

## SEASONAL PRECIPITATION FORECASTING

### Developing a Data Driven Model

#### ANN Overview

Artificial neural networks models are one of the most used techniques of Artificial Intelligence (AI). AI is an area of computer science focused in the development of models that can emulate some intelligent entities or intelligent behavior. In this sense, ANN models are computational approaches based on analogies of the brain structure and brain functioning, having shown good performance in several practical problems relating input and output variables of complex systems. This had led to its application in such diverse areas as financial management, manufacturing, control systems, design, environmental science and pattern recognition (ASCE, 2000).

Since the early 1990's, ANN models have been successfully used in hydrometeorological-related areas such as rainfall-runoff modeling, streamflow forecasting, ground-water modeling, water quality, water management policy, precipitation forecasting, hydrologic time series, and reservoir operations (Dawson and Wilby, 2001). Nevertheless, few studies applying ANN models specifically to seasonal precipitation forecasting have been reported (Chattopadhyay, 2007; Karamouz et al., 2008; Silverman and Dracup, 2000).

ANN models have a set of intrinsic elements called artificial neurons, the fundamental processing units. Artificial neurons are arranged in processing layers and interconnected together by links. The strength of the connection between each two linked neurons is measured by a weight, and the weights of all connections constitute the parameters to be defined through an optimization process termed training. Each single neuron accumulates through a weighted sum all the information sent by the interconnected neurons and applies an

*Figure 1. Artificial Neural Network model and neuron functioning details*

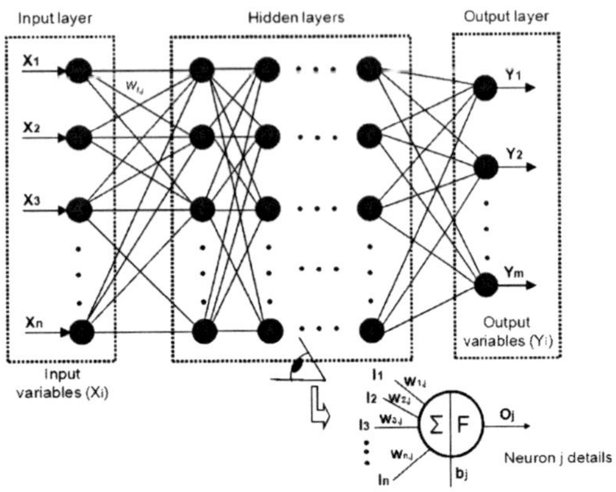


activation function to obtain a single respond or output (Equation 6 and Figure 1).

$$O_j = F\left(\sum_{i=1}^{nc} w_{i,j} \cdot I_i + b_j\right) \qquad (6)$$

where $O_j$ is the output of neuron j; F is the activation function; $w_{i,j}$ are the weights of the links between neuron j and its interconnected neurons (nc); $I_i$ are the inputs and $b_j$ is the bias.

Typically, an ANN model has one input layer, one or more hidden layers and one output layer (Figure 1). In most of ANN models, the input layer receives the input variables, e.g., the predictor variables, providing the model's information base. Through the hidden layer this information is processed and passed to the output layer, where the ANN final outputs, e.g., the predictand values, are generated.

The steps in developing a forecasting ANN model can be summarized as:

1. Definition of the forecast variables (predictands);
2. Identification and selection of the input variables (predictors);
3. Definition of the ANN architecture;
4. Data handling;
5. Training (definition of ANN weights and bias);
6. Validation or performance assessment.

These steps are discussed in the following sections.

## Definition of Forecast Variable

When building a data driven forecast model, the first issue relates to defining the predictand or forecast variable. This definition is intimately related to the aim of the forecast, but it is also dependent on the available data, as data driven relationships must be built based on historical data sets. In terms of seasonal precipitation forecasting, the usual forecast variables are the total precipitation along the entire wet season or the cumulated precipitation during a specific time period of that season, e.g. the monthly precipitation at the beginning of the season or the cumulated precipitation in a given trimester. In this sense, it is important to analyze the climate characteristics of the study region, such as precipitation seasonality and monthly average precipitation.

For providing specific precipitation forecast at a certain region (e.g. a catchment or an area intensively used for agriculture), the available precipitation records of gauging stations located within and around this area may be combined into one single variable representing the areal average precipitation, which will be considered as the forecast variable. Owing to the large spatial variability of precipitation, the averaging approach may lead to a variable that better represents the precipitation regime over the region of interest in comparison with using precipitation records of a particular gauging station alone, which may contain local particularities in data.

Some methods, such as ANN, may be designed to forecast more than one variable at a time, if necessary. In this case, the output layer of the ANN model would have two or more neurons. For example, an ANN model could be developed to predict monthly precipitation in August, September, November and December simultaneously. However, when a unique ANN model is developed to predict several variables at the same time, the representation of the individual underlying relationship between the predictors and each specific predictand may become weakened in favor of representing the overall relationship between predictors and all predictands simultaneously.

The most efficient procedure is to develop a specific ANN model to forecast each of the variables of interest. For the previous example, four distinct ANN models would be designed (model 1 to predict August precipitation, model 2 to predict September precipitation and so on). In this case, each model would have its own input variables and then the ANN could probably obtain the best representation of the individual relationship between predictors and the single predictand.

If the forecast model is developed for agricultural purposes, the forecast variable may also be chosen according with its usefulness to field decisions related to specific crops and regions, taking into account the climate characteristics of the region and its relation with crop practices.

Finally, the result of this step may be composed by a set of potential forecast variables, from which the final predictand is further chosen depending on the prediction power found for the selected predictor variables as well as according with the developed models accuracy.

## Identification and Selection of Predictor Variables

The identification and selection of predictor variables is a decisive step towards developing an ANN forecasting model. This selection concerns deciding which information in the past and/or the present to use to predict the future value of a given variable. The predictors represent the input variables to the model, i.e. the information from which the network produce the forecasts. This concept may give rise to the erroneous idea of including as many predictor variables as possible aiming at providing the maximum information for the ANN. One general restriction for the number of input variables of an ANN model is related to the amount of available data records (Dawson and Wilby, 2001). Using many predictor variables and few data records may lead to ANN overfitting, as the ANN will reproduce well the outputs which the model was developed with but will not have the capacity of generalization, i.e., will not be able to well represent an independent data set. Additionally, only variables with possible and reasonable physical relation with the predictand variable must be selected as potential predictors. Despite of ANN modeling not requiring the existence of such relation, it is important to limit the potential predictor variables to those related with the underlying physical phenomenon. In fact, this is recommended for any statistical forecasting method (Wilks, 2006).

For seasonal precipitation forecasting, both information collected at the forecast area and information related to remote areas may be used as predictors. The former comprises antecedent meteorological conditions at the forecast area, such as temperature and precipitation. For example, accumulated precipitation on a given month may be used as input to predict precipitation on the following month.

The second type of information used as predictors is based on the assumption of existing climate teleconnections, i.e. that a climatic anomaly in one part of the Earth can lead to another climatic anomaly in another distant region (Silverman and Dracup, 2000). The variables related with the teleconnections commonly used as predictors for seasonal precipitation forecasting comprise sea surface temperature, sea-ice extent, variables related with atmospheric circulation patterns such as the 700-mb height anomalies, and climate indices (Goddard et al., 2001; Wilby et al., 2004; Morid et al., 2007).

The use of climate indices is preferable in relation to using direct measures, as climate indices summarize the intensities of known modes of climate variability, which are relatively well described and understood, so that it becomes more plausible to explain the physical relation between the predictors and the precipitation at the interest area.

Even if only physically sound variables in relation to the precipitation conditions of the study region are selected, several potential predictors may be identified. Particularly, it is important to identify the most appropriate lead time for each predictor variable.

There is a wide range of techniques to accomplish the task of defining the best predictor variables. One traditional procedure used for any statistical forecasting method is the screening regression, through techniques such as forward selection, stepwise selection or backward elimination (Wilks, 2006; Wilby et al., 2004). Principal component analysis and singular spectrum analysis may also be used (Stokelj et al., 2002), or even more complex methods such as a combination of partial mutual information and self-organizing map techniques (Bowden et al., 2005; Frenzel and Pompe, 2007), which may detect nonlinear dependence between predictor and predictand variables.

A simple and practical approach to select predictor variables is the use of correlation testing, which may identify the strongest causal linear relationships from the potential predictors (Dawson and Wilby, 2001; Maier and Dandy, 1997). Despite the limitation of only assessing linear relationships, correlation analysis has been widely used for selecting input variables to ANN models with good results. Analyzing the correlation between the predictand and each predictor variable individually and considering varying lead times may roughly indicate appropriate predictor variables and their lead time (Dawson and Wilby, 1998; Bravo et al., 2009b). For example, given that a monthly climatic index was selected as potential predictor, the correlation analysis is useful to estimate the predictive power of each specific monthly index and to define which one to use.

An additional useful procedure for specifying the usefulness of a predictor candidate is the separation of records of the forecast variable into subsets according with intensity of the potential predictor. The available data records of each potential predictor may be classified into groups of low, normal and high intensity. These groups may be established in terms of standard deviation (above one standard deviation and below minus one standard deviation) or certain percentiles. The predictand records are then split into subsets corresponding to each predictor's group. These subsets are statistically compared, e.g. in terms of average values. If the predictor has large potential power, significant differences are expected to occur among the average values of the predictand subsets.

## ANN Architecture and Data Handling

Once ANN input and output variables are known, the next step is the ANN architecture setup. The network architecture is defined through selection of the ANN type, the number of hidden layers, the number of neurons per layer, and the activation functions of hidden and output layers.

Regarding to ANN type, multilayer feed-forward artificial neural network model (MFM) is currently one of the most widely used in forecasting hydro-meteorological variables (Maier and Dandy, 2000; Dawson et al., 2002). The MFM has one input layer, one or more hidden layers, and one output layer. In an MFM, each single artificial neuron may only receive input values from external inputs or from neurons in the preceding layer (Figure 2a). Other ANN type currently used in the prediction of hydro-meteorological variables is the recurrent neural network (RNN), as presented by Karamouz et al. (2008), Muluye and Coulibaly (2007), and Coulibaly et al. (2000). Although RNN models have also several layers, the main difference with respect to MFM models

*Figure 2. Types of three layer artificial neural network models with three input and one output variable: (a) Multilayer feed-forward ANN (MFM); (b) Recurrent neural network (RNN)*

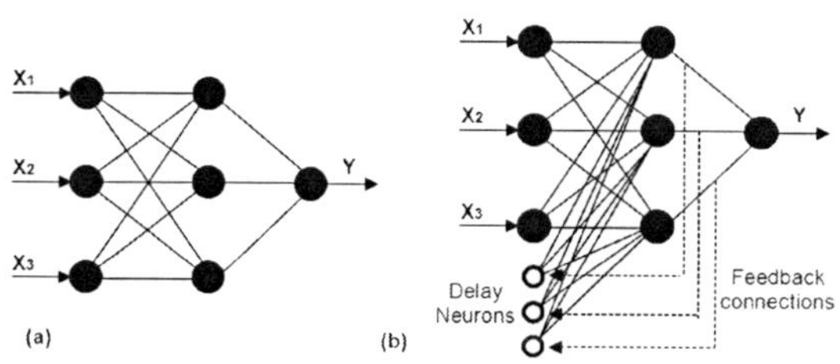


is the presence of feedback connections between neurons, through which input values to a particular neuron could be generated by neurons in following layers (Figure 2b).

In ANN models the input (output) layer has as many neurons as input (output) variables. Thus, each neuron of the input layer is associated to one input variable and each neuron of the output layer is associated to one output variable. However, the number of hidden layers and the number of neurons in each one may be specified by the modeler. Typically, they are defined by a trial and error process (Maier and Dandy, 2000), however specific methods based on either pruning or constructive processes can also be used (Dawson and Wilby, 2001).

In a pruning (constructive) process, the ANN architecture is set to a large (small) number of neurons in the hidden layer. One neuron or connection is removed (added) successively after each training and performance assessment (Bebis and Georgiopoulos, 1994). This procedure continues until a configuration with satisfactory performance is achieved with the smallest number of neurons and connections in the hidden layers, i.e. the most parsimonious ANN model architecture.

Finally, the ANN architecture is completed with the selection of activation functions of the neurons in the hidden and output layers. A neuron in the hidden or output layers processes, through an activation function, a weighted sum of its inputs.

The same activation function is usually adopted for all neurons in a given layer, the selection being dependent on the type of ANN and training algorithm. The functions more widely used are logistic sigmoid (Equation 7), linear (Equation 8) and hyperbolic tangent (Dawson and Wilby, 2001).

$$\log sig(n) = \frac{1}{1+e^{-n}} \tag{7}$$

$$linear(n)=n \tag{8}$$

For achieving the best performance of an ANN model, predictor and predictand data sets should be organized and standardized. This is the data handling process. Typically, the data sets to be used at the ANN should be divided into two independent parts: the training set and the validation set. The training data set is used during ANN model calibration to adjust the weights and the validation set is used to assess the extent to which the ANN model can generalize, i.e., the model skill. Ideally, the data must be split in three sets: training, testing and validation. This division is made in order to use cross-validation or early stopping, a common and practical method to determine the best moment for stopping the training process thus, overfitting is avoided and a proper generalization ensured (Bishop, 1995).

The division of the predictor and predictand data sets should be made in such way that each part is still representative of the whole population (Maier and Dandy, 2000). This may be tested by verifying if some statistics, as the mean and the standard deviation, are similar in the training and validation data sets.

Finally, before starting the training and validation procedures, the input data must be standardized, as for example, using Equation 9.

$$x_{new} = 0.1 + \left( \frac{x_{orig} - x_{\min}}{x_{\max} - x_{\min}} \right) \cdot 0.8 \tag{9}$$

where $x_{new}$ is the standardized value of the input variable x in the range [0.10, 0.90]; $x_{orig}$ is the original value of the input variable x in its appropriate unit and $x_{min}$ and $x_{max}$ are the minimum and maximum original values of the input variable x, respectively.

The range for which the input variables are scaled must be related to the asymptotic limits of the activation functions. For example, the outputs of the logistic sigmoid function are between 0 and 1, thus the data are generally scaled in the range [0.1, 0.9]. Through data standardization, input data measured in different scale units will receive the same consideration during the training process. It prevents a particular input variable to dominate the process and is also useful to improve the efficiency and efficacy of the training procedure, which can be affected when input variables have large values (Dawson and Wilby, 2001). In hydrometeorology-related areas, another advantage of adopting a scaled range inside the asymptotic limits of the activation function is related to the ability of the ANN model for representing extreme events even more severe than those existing in the training set (Hsu et al., 1995).

## ANN Training and Validation

ANN model weights are defined through a calibration procedure called training. During this process the weights are adjusted with the objective of minimizing/maximizing some performance measure based on the errors between the estimated and observed output, where the mean square error is the most widely used (Maier and Dandy, 2000; Dawson and Wilby, 2001).

If the data is split in only two sets (training and validation sets), the weights are adjusted based on the error in the training set and the process stops when a minimum/maximum of the performance measure is reached based only on the training data set. On the other hand, if the data is split in three sets (training, testing and validation sets), the weights are adjusted based on the error in the training set, but the process only stops when the performance measure calculated on the testing data set reaches a minimum/maximum (Bishop, 1995). In both cases, overfitting must be avoided. Overfitting occurs when ANN models reproduce existing noises in the training set, losing its ability to generalize when other data sets are used.

Finding the weights correspondent to global optima of the performance measure is the goal of ANN training. However, given the complicated mathematical shape typically arisen from the functions defining the performance measures, the training method can be trapped in a local optima. The local optima could be a good solution, however, always poorer than the global optima.

ANN training is usually based on local search methods, which can be broadly divided into two classes: first-order and second-order. The first-order methods are based on linear models (gradient descent), while the second-order methods are based on quadratic models, such as the Newton method. In both cases, several iterative techniques can be used during ANN training to minimize/maximize a performance measure based on the error between the observed and the calculated values of the output variables. The ANN weights are adjusted, at each iteration, according to a general equation of this type (Maier and Dandy, 2000):

$$w_{n+1} = w_n + \gamma n Dn \tag{10}$$

where $w_{n+1}$ is the weights vector at iteration n+1; $w_n$ is the weights vector at iteration n; $\gamma_n$ is the step size at iteration n and $D_n$ is a vector defining a search direction at iteration n.

The major difference between the distinct methods of ANN training is the choice of $D_n$ (Maier and Dandy, 2000). The mostly used method in hydrometeorology-related areas has been the backpropagation algorithm. In summary, this is a method characterized by $D_n$ being equal to the negative value of the error function gradient and by minimizing a given performance measure (Rumelhart et al., 1986). In this algorithm, the step size is known as learning rate and is usually adopted as a fixed value, although some versions of the algorithm have been adapted to allow its dynamic variation (Dawson and Wilby, 2001).

The learning rate is a very important parameter of the backpropagation algorithm. When this parameter has a small value, the process can be slow and it has a greater chance of being trapped in local optima. In turn, a too large value may cause instability in ANN training.

The second-order local search methods use the inverse of the Hessian matrix for defining $D_n$ at each iteration (Maier and Dandy, 2000), which tends to require a greater computational effort (storage and processing), in comparison to the first-order methods. However, the second-order techniques have a greater ability to escape of local optima relatively to first-order methods. Several second-order methods have been used for ANN training, such as the Levenberg-Marquardt method (Golden, 1996) and the scaled conjugate gradient method (SCGM) developed by Moller (1993).

It is important to emphasize that the use of a local search method in ANN training process does not guarantee finding the global optima of the optimization problem. This issue may be achieved repeating the ANN training several times, with different initial conditions. If the same weights are obtained, it can be sufficient evidence to consider having found the true global optima.

In none of the cases the validation set is used in the training process, but just for evaluating the skill of the ANN model, i.e., the extent to which the model can generalize. After ANN training, the adjusted weights are fixed and then the ANN model is run using the validation data set as input. This procedure is called ANN validation and is a simulation process. Several performance measures, as those defined by Equation 1 to Equation 4 are used in assessing the accuracy of the ANN model based on the validation data set.

## Study Case: The Mogi-Guaçu/Pardo Watershed

### The Study Area and Problem Definition

The study area is the Pardo/Mogi-Guaçu watershed located at the Southeast Brazil (Figure 3), encompassing 29,510 km² in the Northeast part of the federal state of São Paulo and 6,383 km² in the Southwest part of Minas Gerais state. This region is predominantly used for sugar cane production for both sugar and ethanol industries, but pasture, fruit crop production and forestry are also currently carried out at a less extent. Indeed, the Northeast part of São Paulo concentrates the major production of sugar cane in Brazil, which is the world largest producer of both sugar and bio-ethanol.

*Figure 3. Location of Pardo/Mogi-Guaçu watershed and of the available and selected rainfall gauging stations*

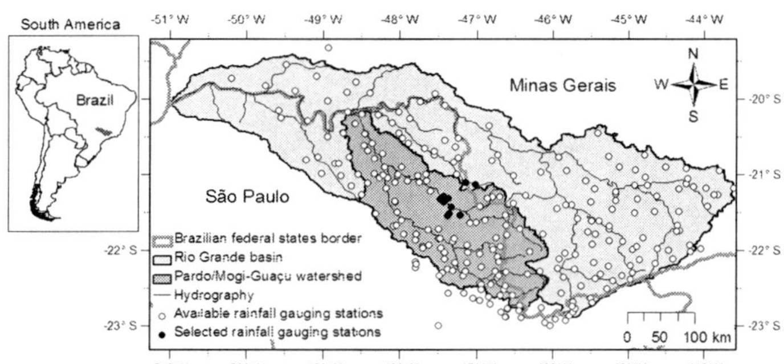

Additionally, the Pardo/Mogi-Guaçu watershed is part of the Rio Grande basin, in which there is an installed capacity of hydropower generation of about 7,722 MW, roughly 11.7% of the Brazilian total (ANEEL, 2005). Therefore, seasonal precipitation forecasting for this region may both benefit agriculture-related decisions and improve management of hydropower reservoirs.

For this study, daily precipitation data were available from 274 rainfall gauges from the Brazilian National Water Agency (ANA) database (http://hidroweb.ana.gov.br/) and from the São Paulo Water Resources Management Information System (http://www.sigrh.sp.gov.br/), spatially distributed over the Rio Grande basin, ranging from 1950 to 2003. Aiming to develop a forecast model focused specifically for a small area, a subset of ten rainfall gauges located in the central portion of the Pardo/Mogi-Guaçu watershed was selected (Figure 3). The daily precipitation in this area was determined as the average of daily precipitation in the ten gauges.

The precipitation pattern in the study area is markedly seasonal, the wet season beginning in September and extending up to March (Figure 4). The total annual rainfall is around 1500 mm, and almost 82% of this total occurs from October to March. Although the average precipitation on September is less than 5% of annual total, as this month represents the onset of the wet season it becomes a useful index for planning purposes, such as agriculture irrigation and reservoir operation. For this reason, this study has focused on developing a model to forecast monthly precipitation on September ($P_{set}$).

## The South American Monsoon and Its Influence on Precipitation Regime of the Study Area

The South American monsoon system (SAMS) refers to the austral summer season features of deep convective activity and large-scale circulation (Jones and Carvalho, 2002). Monsoon circulation systems, as SAMS, are a major component of continental warm season precipitation regimes. In South America, more than 50% of the total annual precipitation over tropical and sub-tropical regions occurs during the austral summer (DJF).

*Figure 4. Mean, minimum and maximum monthly precipitation over the study area, considering the average precipitation of the ten selected rainfall gauges (the thin vertical bars indicate ± 1 standard deviation)*

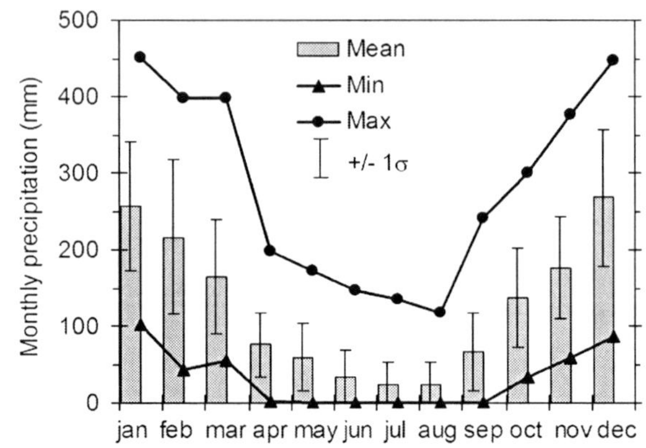


SAMS develops in response to seasonal changes in the thermal contrast between the continent and adjacent oceanic regions (Vera et al., 2006). In upper levels and during summer (Figure 5), SAMS is characterized by an anticyclonic circulation centered over Bolivia (named Bolivia high) and a trough over Northeast Brazil (named Nordeste trough). Besides, at low levels (Figure 6), the easterly flow from the Atlantic Ocean is channeled southward by the Andes mountains into the Chaco Low, located at Northern Argentina and Paraguay (Nogués-Paegle et al., 2002). This is made by the so called South American Low-level jet (SALLJ), characterized by a narrow stream that channels the near-surface flow between the tropics and mid-latitudes east of the Andes range, and related to moisture transport from Amazon region into Southern Brazil and Northern Argentina (Marengo et al., 2004).

Several studies, as presented by Marengo et al. (2004), have suggested an active role of the

*Figure 5. Mean January vector wind at 200 mb based on NCEP/NCAR Reanalysis (1949-2008)– location of Rio Grande basin is also indicated (Kalnay et al., 1996; Image provided by the NOAA/ESRL Physical Sciences Division, Boulder Colorado from their Web site at http://www.esrl.noaa.gov/psd/)*

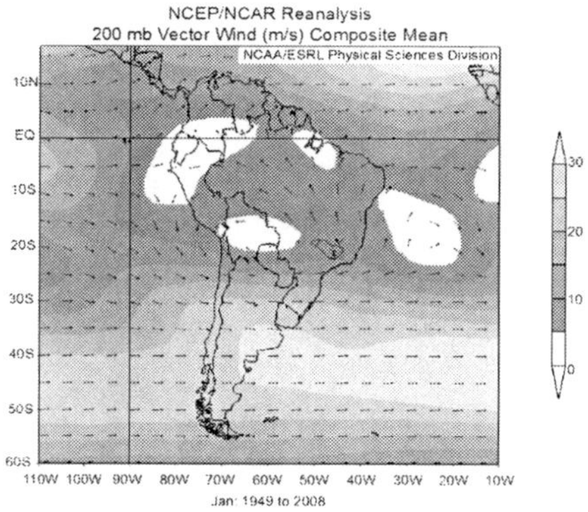


*Figure 6. Mean January vector wind at 850 mb based on NCEP/NCAR Reanalysis (1949-2008) - location of Rio Grande basin is also indicated (Kalnay et al., 1996; Image provided by the NOAA/ESRL Physical Sciences Division, Boulder Colorado from their Web site at http://www.esrl.noaa.gov/psd/)*

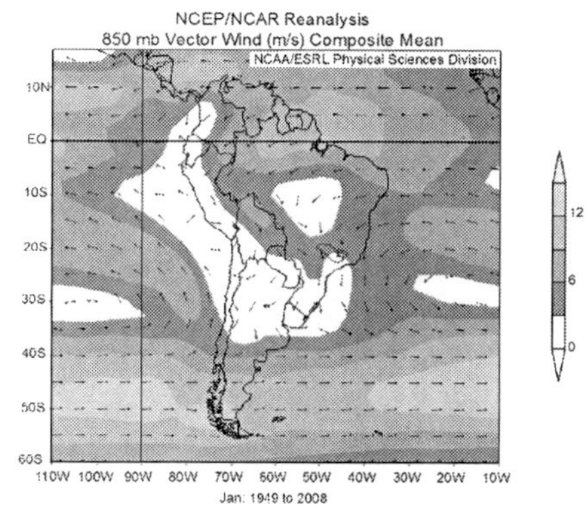


SALLJ in the positioning and intensity of the South Atlantic convergence zone (SACZ) and the rainfall and convection at the exit region of the jet in south-eastern South America. SACZ is a dominant summertime cloudiness feature of subtropical South America and the Western South Atlantic Ocean that appears to either emanate from or merge with the intense convection of the Amazon (Liebmann et al., 1999). Another tropical convergence zone influencing SAMS precipitation is the Inter-Tropical Convergence Zone (ITCZ), a low pressure zone, near the Equator, where Northeast and Southeast trade winds converge (Hastenrath, 1991).

Position and strength of the tropical convergence zones are one of the factors influencing inter-annual variability of SAMS precipitation. Other related factors that also influence SAMS precipitation are: SST anomalies, land surface conditions, water vapor transport and large-scale circulations (Vera et al., 2006).

In this sense, SAMS is influenced by several climate modes of variability, through teleconnections that can be defined as a strong statistical relationship between the weather in different parts of the globe.

## Main Climatic Modes of Variability Affecting the SAMS

Climate variability is the result of natural atmospheric and oceanic oscillations, differently from climate change, which is associated to human intervention, related to the emission of greenhouse gases. Interannual variations in precipitation and temperature are essentially related to climate variability, which is manifested through teleconnection.

Some of the most known atmospheric modes of variability are the North Atlantic Oscillation (NAO), the Southern Annular Mode (SAM) and the Northern Annular Mode (NAM). Climatic phenomena such as El Niño – Southern Oscillation (ENSO) and the Pacific Decadal Oscillation (PDO) are integrated oceanic-atmospheric phenomena. They all play an important role to climate variability worldwide through teleconnections and span different time frequencies ranging from intra-seasonal to multi-decadal.

- ENSO is characterized by an increase or decrease in the central Tropical Pacific sea surface temperature (SST) coupled to a decrease or increase of the speed of the easterly winds over the Tropical Pacific (Rasmusson and Carpenter, 1982).
- NAO is a redistribution of atmospheric mass between the North Atlantic subtropical high (Azores High) and Arctic low (Iceland Low). It swings between positive and negative phases, producing large changes in the mean wind speed and direction over Atlantic, in the heat and moisture transport between the Atlantic and the neighboring continents, and in the intensity and number of storm, their paths and their weather (Hurrel et al., 2003).
- SAM is also called Antarctic Oscillation (AAO) and refers to a large scale alternation of atmospheric mass between the mid-latitude and high latitudes surface pressure (Gong and Wang, 1999). This oscillation is also characterized by a zonally symmetric structure with geopotential height perturbations of opposing signs in Antarctica and in the surrounding zonal ring centered near 45° latitude (Reboita et al., 2009; Thompson and Wallace, 2000).
- PDO is an oceanic-atmospheric phenomenon, characterized by SST anomalies over the central North Pacific and North Pacific/North American coast and by SLP (sea level pressure) anomalies over the North Pacific (Mantua et al., 1997).

The intensity of these atmospheric modes and climate phenomena is expressed in terms of indices. Some of them are referred here. For further detailed information, the reader is referred to http://www.cpc.ncep.noaa.gov/products/precip/CWlink/MJO/climwx.shtml.

ENSO intensity can be measured by the intensity of the SST anomalies in central tropical Pacific or by its atmospheric counter-part, the normalized difference between the intensity of the sea level pressure (SLP) at Darwin, Australia and at Tahiti. When based on SST anomalies this index is named NINO, and the atmospheric index is the Southern Oscillation Index (SOI).

A common index used for expressing the NAO intensity, the NAOI, is calculated as the normalized difference between the SLP at Stykkisholmur, Iceland, which is representative of the intensity of the Iceland Low, and Ponta Delgada, Azores, which is representative of the intensity of the Azores High (Hurrell, 1995).

Two indices have been typically used to individually measure SAM intensities. The SAMI (southern annular mode index), given by Nan and Li (2003), is defined as the normalized difference between the intensity of the zonal mean SLP between 40°S and 70°S. Besides, the AOI (Antarctic oscillation index), given by Gong and Wang (1999), is also a normalized difference of SLP intensities, however between 40°S and 65°S.

PDO intensity can be measured both by the intensity of the SST anomalies in central North Pacific or by the intensity of the mean zonal SLP anomalies over the North Pacific, as presented by Trenberth (1990), Trenberth and Hurrell (1994), Zhang et al. (1997) and Mantua et al. (1997).

Atmospheric modes and climate phenomena are fundamental elements for understanding climate variability and important information when seasonal forecast is sought, thus, the climate indices mentioned above will be tested as predictor for the ANN model to be develop below.

## Definition of Predictor Variables

According to the known intensities of atmospheric circulation patterns and teleconnections, four climate indices were identified as possibly related to precipitation over the Pardo/Mogi-Guaçu watershed and thus with potential prediction power:

the North Atlantic Oscillation index (NAOI), the Southern Annular Mode index (SAMI), the Pacific Decadal Oscillation index (PDOI), and the NINO3.4 index. Monthly time series of SAM index are from Nan and Li (2003) for the period 1951-2002. The monthly time series of NAO, PDO and NINO3.4 indices were collected from the Climate Prediction Center site (http://www.cpc.noaa.gov/) for the following periods: 1937-2002 (NAOI), 1937-2002 (PDOI) and 1950-2002 (NINO3.4).

A simple approach for selecting the predictor variables was carried out based on correlation analysis and separation of precipitation records into categories according to the climate events phases (strongly positive - hereafter referred to as "high", strongly negative - hereafter referred to as "low" or weak - hereafter referred to as "normal").

The Pearson and Spearman correlation coefficient were calculated between possible predictors and predictand. This choice was made as Spearman correlation coefficient measures the linear relationship between variables and is also able to capture non-linear correlation between predictand and potential predictors. This characteristic is important considering the non-linear character of ANN and of the relations between precipitation and other climate phenomena.

Pearson (r) and Spearman's rank ($r_k$) correlation coefficients were calculated between the predictand, i.e., accumulated precipitation on September ($P_{set}$) and monthly values of the four indices with varying lead-times (Figure 7). Additionally, taking specifically each climate index at each month, the $P_{set}$ time series was split into three subsets according with a low, normal or high phase of each index. The average values of $P_{set}$ for each subset are shown in Figure 8.

Regarding the NAO index, weak correlations ($|r| < 0.10$ and $|r_k| < 0.19$) predominantly positive were found between $P_{set}$ and NAOI from January to June and August, but a single stronger negative correlation was obtained between the predictand and $NAOI_{jul}$ ($r = -0.28$ and $r_k = -0.35$, both significant at a 0.05 confidence level). Indeed, this negative correlation between $NAOI_{jul}$ and $P_{set}$ also appeared when analyzing the average precipitation for the $NAOI_{jul}$ different phases, as $P_{set}$ averages 71 mm for normal $NAOI_{jul}$ values and 39 mm for high $NAOI_{jul}$ (Figure 8-a).

For the PDO index, negative correlations appeared between $P_{set}$ and PDOI in January and February months and positive correlations between $P_{set}$ and PDOI from March to August. A peak of positive correlation of about 0.33 for Pearson coefficient and of 0.36 for Spearman coefficient was obtained between $P_{set}$ and $PDOI_{may}$, both be-

*Figure 7. (a) Pearson and (b) Spearman correlation between monthly climate indices and precipitation on September (the points outside the area delimited by the dashed straight lines are significant at the 0.05 confidence level)*

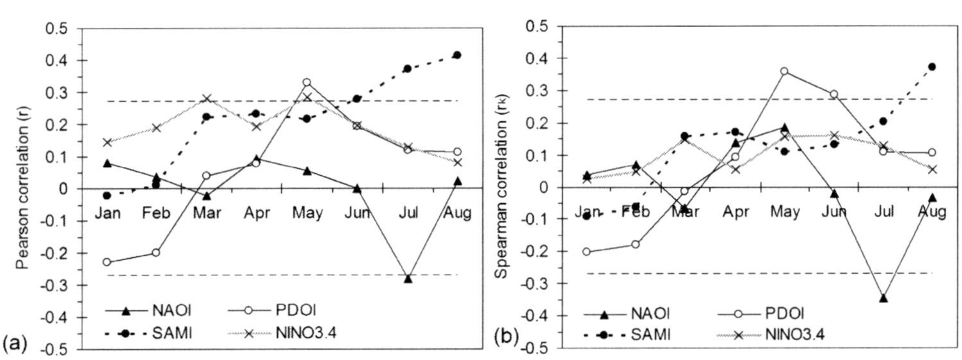

*Figure 8. Average precipitation on September at the study area, according with low, normal or high phases of monthly NAO, SAM, PDO and NINO3.4 climate indices (thin vertical bars indicate ± 1 standard deviation). Average $P_{set}$ values were significantly different (at 0.05 confidence level) between normal and high phases of $NAOI_{jul}$ (normal vs. high); low vs. high for $PDOI_{may}$; low vs. normal for $PDOI_{aug}$; normal vs. high for $SAMI_{may}$; normal vs. high and low vs. high for $SAMI_{aug}$; low vs. normal for $NINO3.4_{jul}$; and low vs. normal for $NINO3.4_{aug}$.*

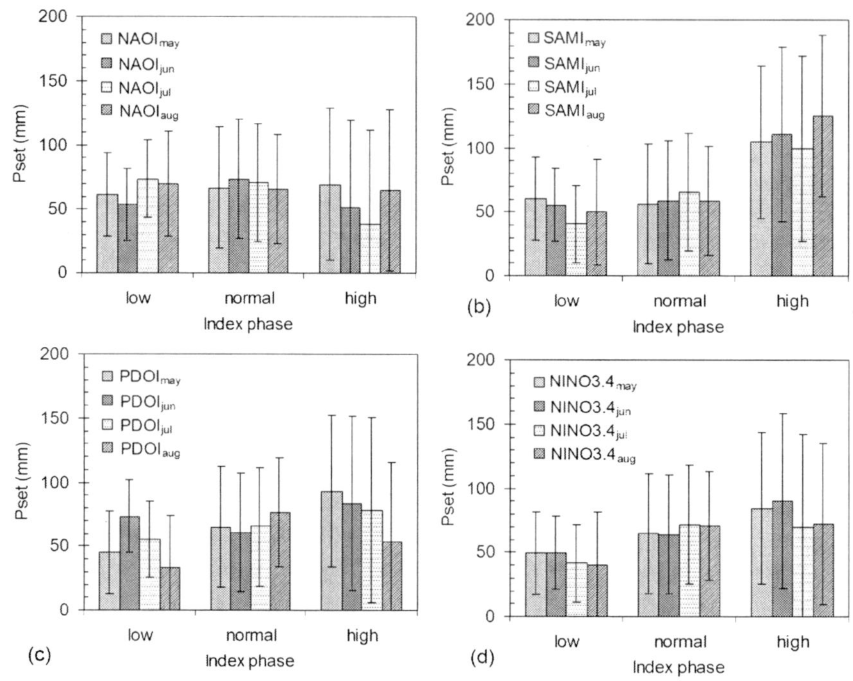


ing significant at the 0.05 confidence level. When $PDOI_{may}$ was low, normal or high the average $P_{set}$ was 45, 65 or 93 mm, respectively, corroborating the positive correlation between them (Figure 8-c).

A positive correlation was obtained between $P_{set}$ and SAMI from March to August considering both Pearson and Spearman coefficients. A maximum correlation was found between $P_{set}$ and $SAMI_{aug}$, reaching 0.41 for r and 0.37 for $r_k$ (both significant at 0.05 level). The analysis of SAM phases has also showed such strong positive correlation between the predictand and SAMI monthly values from May to August (Figure 8-b).

For NINO3.4, positive correlations were found between NINO3.4 from January to August and $P_{set}$, with peak values of r = 0.28 between $P_{set}$ and $NINO3.4_{mar}$ and $NINO3.4_{may}$ and of $r_k = 0.16$ between $P_{set}$ and $NINO3.4_{may}$ and $NINO3.4_{jun}$. The positive correlation between NINO3.4 and $P_{set}$ was also found when comparing $P_{set}$ values for each NINO3.4 phase, especially for $NINO3.4_{may}$ and $NINO3.4_{jun}$ (Figure 8-e).

According with these results, the predictors with larger potential prediction power were: $NAOI_{jul}$ for the NAO index; $PDOI_{may}$ and $PDOI_{jun}$ for the PDO index; $SAMI_{may}$, $SAMI_{jun}$, $SAMI_{jul}$ and $SAMI_{aug}$ for the SAM index; and $NINO3.4_{may}$ and $NINO3.4_{jun}$ for the NINO3.4 index.

## ANN Model Setup

A multilayer feed-forward artificial neural network model (MFM) was used for developing the forecast model. As there is only one predictor variable, the monthly precipitation in September, the MFM output layer has one neuron. Several models were designed, trained, and assessed, varying the set of input variables, all of them predicting $P_{set}$. The number of neurons in the hidden layer was determined independently for each MFM configuration by a trial-and-error pruning process.

For all designed MFMs, the hidden layer consists of neurons with logistic sigmoid activation function, while the linear activation function was used in the output layer. The ANN training was carried out through the scale conjugate gradient method and considering the root mean square error as the performance measure.

The available records were split into three sets for training, testing, and validation. This division was made in order to use the cross-validation procedure. The data division was performed aiming to distribute precipitation events ($P_{set}$) of all intensities along the three sets as well as seeking similar average and standard deviation values among them. Among the 52 records of the available data set, 38 records (73%) were selected as the training set, 7 records (13%) were considered as the testing set and the same quantity for the validation set (Figure 9). The average precipitation is 67, 61 and 69 mm for the training, testing and validation sets, respectively, while the standard deviations are 52, 50 and 48 mm. All the MFMs input data were standardized in the range [0.1, 0.9] prior to the training procedure.

*Figure 9. Data division into training, testing and validation sets for ANN cross-validation process*

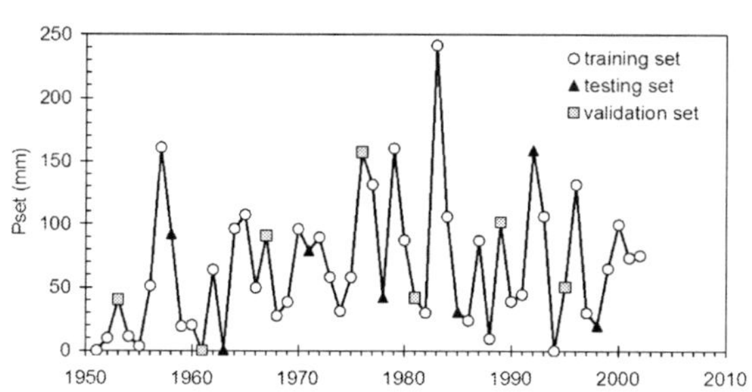


## Results

Among the designed and tested MFM configurations, we select the three with best performance to present here. All these three selected models have five predictors, one related to each of the NAO, PDO and NINO indices, and two predictors related with SAM index (Table 1). The variables $NAOI_{jul}$, $SAMI_{jul}$, $SAMI_{aug}$, and $NINO3.4_{may}$ are common predictors among the three models, the difference being restricted to the variable related to PDO index. Models 1, 2 and 3 use $PDOI_{may}$, $PDOI_{jun}$ and $PDOI_{jul}$ as predictors, respectively.

The RMSE, the Nash-Sutcliffe efficiency index (NS) and the mean-absolute error (MAE) were used to assess models performance (Table 2 and Figure 10). The scatter plot of observed and calculated $P_{set}$ values were also analyzed (Figure 11).

Relatively similar RMSE and MAE values were obtained for the three forecasting models considering the validation sets. The RMSE ranged from 33 to 36 mm, and the MAE was around 30 mm. The performance in terms of RMSE and MAE was reasonably similar among the training, testing and validation sets for each model.

In terms of NS coefficient, the MFM-2 (NS = 0.52) and MFM-3 (NS = 0.51) obtained slightly better skill at the validation set in comparison to MFM-1 (NS = 0.42). However, a more remarkable result is that both MFM-2 and MFM-3 maintained relatively similar NS scores among the training, testing and validation procedures (NS ranging between 0.45 and 0.59), what may be considered a good indication of model capability of generalization, while the MFM-1 showed a considerable decrease in this coefficient from the training (NS = 0.68) to the testing (NS = 0.30) and validation sets (NS = 0.43).

## FUTURE RESEARCH DIRECTIONS

Currently seasonal precipitation forecast is not widely used as a major input into agricultural

*Table 1. Description of the MFM configurations used to predict monthly precipitation in September*

| Model | Predictors | | | | Predictand | Architecture |
|---|---|---|---|---|---|---|
| | NAOI | PDOI | SAMI | NINO3.4 | | |
| MFM 1 | jul | may | jul, aug | may | | 5-4-1 |
| MFM-2 | jul | jun | jul, aug | may | $P_{set}$ | 5-4-1 |
| MFM-3 | jul | jul | jul, aug | may | | 5-3-1 |

*Table 2. Performance measures of MFM models in forecasting monthly precipitation in September for the training, testing and validation sets*

| Model | Training | | | Testing | | | Validation | | |
|---|---|---|---|---|---|---|---|---|---|
| | RMSE | MAE | NS | RMSE | MAE | NS | RMSE | MAE | NS |
| MFM-1 | 30.0 | 22.7 | 0.68 | 37.3 | 34.1 | 0.30 | 36.1 | 30.0 | 0.43 |
| MFM-2 | 35.7 | 28.4 | 0.54 | 32.8 | 30.0 | 0.46 | 33.2 | 30.5 | 0.52 |
| MFM-3 | 39.0 | 32.1 | 0.45 | 28.4 | 20.5 | 0.59 | 33.6 | 29.1 | 0.51 |

*Figure 10. Performance measures of MFM models in forecasting monthly precipitation in September for the training, testing and validation sets*

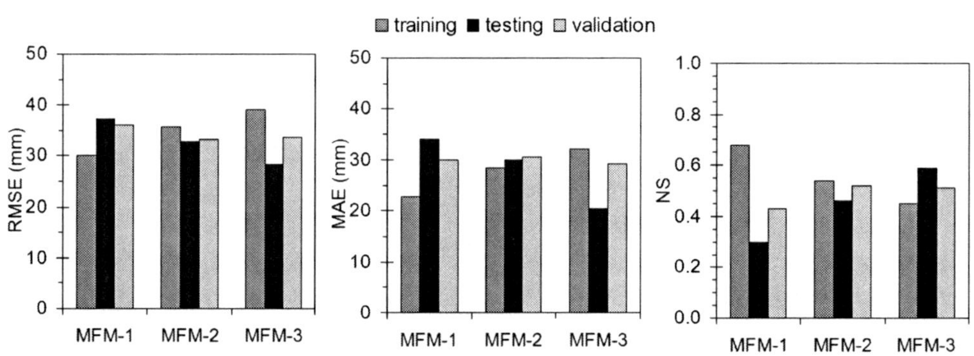


*Figure 11. Precipitation in September observed and calculated using the three models (MFM-1, MFM-2 and MFM-3) for the training, testing and validation sets (the 1:1 line is also shown)*

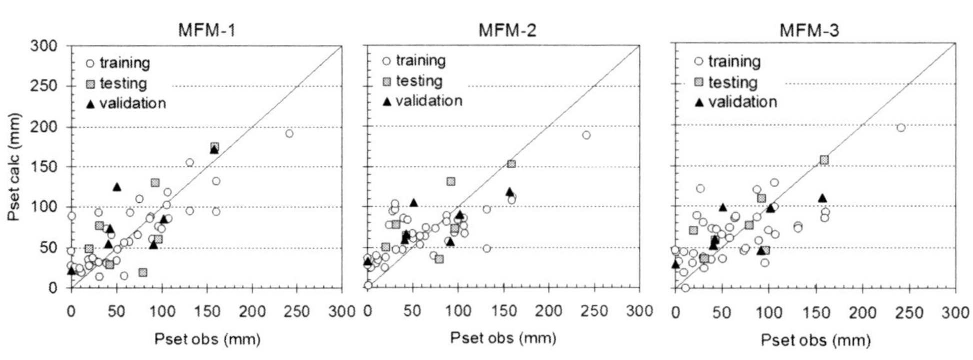

management (McIntosh et al., 2005). The main challenges are concerned to improving forecast skill and reliability, forecast communication to decision-makers and farmers, providing site-specific forecasts rather than regional information, and better understanding the agricultural systems in order to produce a more appropriate forecast (McIntosh et al., 2005, 2007; Ash et al., 2007; Goddard et al., 2001; Moeller et al., 2008). Additionally, there is a need to better evaluating the costs and benefits of weather and climate forecasts on whole society, which requires better understanding of the impacts of weather and climate themselves (Pielke and Carbone, 2002).

In this sense, it should be emphasized the importance of implementing seasonal precipitation forecasting systems focused on specific location and agricultural system. Even if the current seasonal forecast skill may not reach farmers expectance, practicing such a forecast system is an interesting way of learning how the forecast information is used and how to improve its value. Furthermore, operating the precipitation forecast system and analyzing its uncertainties may provide further insight to forecasters for the improvement of its performance as well as may increase reliability by farmers and other forecast users.

Skill improvement of site-specific statistical seasonal forecasts depends on deeper investigation of climate teleconnections and the effects of known modes of climate variability over the precipitation regime at the area of interest. Furthermore, the influence of land use change on the local climate needs also to be better understood, in order to separate the effects of climate variability and land use change.

In terms of methodological improvement, a promising approach is coupling GCM outputs to statistical methods such as ANN in order to produce local precipitation forecasts for agricultural management (Hansen et al., 2009). GCM may be suitable for simulating the effect of land use change over the climate, primarily if developed a two-way coupling with a hydrological model and with crop simulation models. However, an actual shortcoming of GCM is its inability to accurately reproduce modes of climate variability. Coupling GCM outputs with statistical models designed for a specific location may overcome this limitation and produce enhanced precipitation forecasts.

## SUMMARY AND CONCLUSION

Several agriculture practices are conditioned to precipitation conditions and thus dependent on climate variability. The knowledge of precipitation conditions within a season is a valuable information for agricultural management purposes, justifying the development and improvement of methods for precipitation forecasting. Climate predictions could lead decisions towards a strategic view, aiming at minimizing unwanted impacts and taking advantage of favorable conditions.

Data driven methods may be suitable for developing precipitation forecast models, as they allow to produce site-specific predictions and have low computational and modeling costs. One key step towards developing a data driven forecasting model is the selection of predictor variables. For seasonal precipitation forecast, climate indices are potential candidates for predictors, as they summarize modes of climate variability and teleconnections. The good results obtained in this experiment imply that in addition to physical basis correlation analysis may be a suitable technique for identifying the good predictor variables to be used as input to an ANN seasonal precipitation forecast model. Particularly, this approach is useful for finding the more appropriate lag between a variable candidate for predictor and the predictand.

The main issue behind a data driven model is the determination of a numerical relationship between predictors and predictands. In this sense, the Artificial Neural Networks technique is a powerful tool. However, despite of ANN has been applied in several problems of hydro-meteorological areas, the application of such technique for seasonal precipitation forecast is still rare. This chapter presented a successful example of ANN modeling aiming to forecast monthly precipitation in the onset of the wet season. A detailed discussion concerning the steps for developing an ANN forecast model was also presented, including: definition of the forecast variable, identification and selection of the predictors; definition of the ANN architecture; data handling; definition of ANN weights and bias; performance assessment. Although the ANN technique is labeled a black box model, it is necessary to understand (at least) the concepts presented in this chapter to design an ANN model and avoid common mismatches such as model overfitting.

The results obtained in the study case illustrate the high potential of predicting seasonal precipitation using ANN modeling and taking advantage of known modes of climate variability and teleconnections represented through climate indices. This approach can be successfully used to provide site-specific precipitation forecasts, valuable for certain agriculture-related decisions.

## REFERENCES

Agência Nacional de Energia Elétrica, A. N. E. E. L. (2005). *Atlas de Energia Elétrica do Brasil*, (2nd Ed.). Brasília, Brazil.

ASCE. (2000). Artificial neural networks in hydrology. I: preliminary concepts. *Journal of Hydrologic Engineering*, *5*(2), 115–123. doi:10.1061/(ASCE)1084-0699(2000)5:2(115)

Ash, A., McIntosh, P., Cullen, B., Carberry, P., & Smith, M. S. (2007). Constraints and opportunities in applying seasonal climate forecasts in agriculture. *Australian Journal of Agricultural Research*, *58*, 952–965. doi:10.1071/AR06188

Bebis, G., & Georgiopoulos, M. (1994). Feed-forward neural networks: why network size is so important. *IEEE Potentials*, 27-31.

Bishop, C. M. (1995). *Neural Network for Pattern Recognition*. Oxford, U.K: Clarendon.

Bowden, G. J., Dandy, G. C., & Maier, H. R. (2005). Input determination for neural network models in water resources applications. Part 1 - Background and methodology. *Journal of Hydrology (Amsterdam)*, *301*, 75–92. doi:10.1016/j.jhydrol.2004.06.021

Bravo, J. M., Collischonn, W., Pilar, J. V., & Tucci, C. E. M. (2009a). Otimização de regras de operação de reservatórios com incorporação da previsão de vazão. *Revista Brasileira de Recursos Hídricos*, *13*, 181–196.

Bravo, J. M., Paz, A. R., Collischonn, W., Uvo, C. B., Pedrollo, O. C., & Chou, S. C. (2009b). Incorporating forecasts of rainfall in two hydrologic models used for medium-range streamflow forecasting. *Journal of Hydrologic Engineering*, *13*(5), 435–445. doi:10.1061/(ASCE)HE.1943-5584.0000014

Bretherton, C. S., Smith, C., & Wallace, J. M. (1992). An intercomparison of methods for finding coupled patterns on climate data. *Journal of Climate*, *5*, 541–560. doi:10.1175/1520-0442(1992)005<0541:AIOMFF>2.0.CO;2

Brown, B. & co-authors (2009). Forecast Verification - Issues, Methods and FAQ. Retrieved May 12, 2009 from http://www.bom.gov.au/bmrc/wefor/staff/eee/verif/verif_web_page.html

Cabrera, V. E., Letson, D., & Podestá, G. (2007). The value of climate information when farm programs matter. *Agricultural Systems*, *93*, 25–42. doi:10.1016/j.agsy.2006.04.005

Casati, B., Wilson, L. J., Stephenson, D. B., Nurmi, P., Ghelli, A., & Pocernich, M. (2008). Forecast verification: current status and future directions. *Meteorological Applications*, *15*, 3–18. doi:10.1002/met.52

Chattopadhyay, S. (2007). Feed-forward artificial neural network model to predict the average summer-monsoon rainfall in India. *Geophysica Acta*, *55*(3), 369–382. doi:10.2478/s11600-007-0020-8

Chiew, F. H. S., Zhou, S. L., & McMahon, T. A. (2003). Use of seasonal streamflow forecasts in water resources management. *Journal of Hydrology (Amsterdam)*, *270*, 135–144. doi:10.1016/S0022-1694(02)00292-5

Coulibaly, P., Anctil, F., & Bobée, B. (2000). Daily reservoir inflow forecasting using artificial neural networks with stopped training approach. *Journal of Hydrology (Amsterdam)*, *230*, 244–257. doi:10.1016/S0022-1694(00)00214-6

Dawson, C. W., Harpham, C., Wilby, R. L., & Chen, Y. (2002). Evaluation of artificial neural network techniques for flow forecasting in the River Yangtze, China. *Hydrology and Earth System Sciences*, *6*(4), 619–926. doi:10.5194/hess-6-619-2002

Dawson, C. W., & Wilby, R. L. (1998). An artificial neural network approach to rainfall–runoff modelling. *Hydrological Sciences Journal*, *43*, 47–66. doi:10.1080/02626669809492102

Dawson, C. W., & Wilby, R. L. (2001). Hydrological modelling using artificial neural networks. *Progress in Physical Geography*, *25*(1), 80–108.

Everingham, Y. L., Muchow, R. C., Stone, R. C., Inman-Bamber, N. G., Singels, A., & Bezuidenhout, C. N. (2002). Enhanced risk management and decision-making capability across the sugarcane industry value chain based on seasonal climate forecasts. *Agricultural Systems*, *74*, 459–477. doi:10.1016/S0308-521X(02)00050-1

Fall, S., Niyogi, D., Mohanty, U. C., & Kumar, A. (2007). Application of weather prediction models for hazard mitigation planning: a case study of heavy off-season rains in Senegal. *Natural Hazards*, *41*, 227–243. doi:10.1007/s11069-006-9033-x

French, M. N., Krajewski, W. F., & Cuykendall, R. R. (1992). Rainfall forecasting in space and time using a neural network. *Journal of Hydrology (Amsterdam)*, *137*, 1–31. doi:10.1016/0022-1694(92)90046-X

Frenzel, S., & Pompe, B. (2007). Partial mutual information for coupling analysis of multivariate time series. *Physical Review Letters*, *99*(20), 204101. doi:10.1103/PhysRevLett.99.204101

Galacgac, E. S., & Balisacan, C. M. (2009). Traditional weather forecasting for sustainable agroforestry practices in Ilocos Norte Province, Philippines. *Forest Ecology and Management*, *257*, 2044–2053. doi:10.1016/j.foreco.2009.01.002

Goddard, L., Mason, S. J., Zebiak, S. E., Ropelewski, C. F., Basher, R., & Cane, M. A. (2001). Current approaches to seasonal-to-interannual climate predictions. *International Journal of Climatology*, *21*, 1111–1152. doi:10.1002/joc.636

Golden, R. M. (1996). *Mathematical Methods for Neural Network Analysis and Design*. Cambridge, UK: MIT Press.

Gong, D., & Wang, S. (1999). Definition of Antarctic Oscillation index. *Geophysical Research Letters*, *26*, 459–462. doi:10.1029/1999GL900003

Hammer, G. L., Hansen, J. W., Phillips, J. G., Mjelde, J. W., Hill, H., Love, A., & Potgieter, A. (2001). Advances in application of climate prediction in agriculture. *Agricultural Systems*, *70*, 515–553. doi:10.1016/S0308-521X(01)00058-0

Hansen, J. W. (2002). Realizing the potential benefits of climate prediction to agriculture: issues, approaches, challenges. *Agricultural Systems*, *74*, 309–330. doi:10.1016/S0308-521X(02)00043-4

Hansen, J. W., Mishra, A., Rao, K. P. C., Indeje, M., & Ngugi, R. K. (2009). Potential value of GCM-based seasonal rainfall forecasts for maize management in semi-arid Kenya. *Agricultural Systems*, *101*, 80–90. doi:10.1016/j.agsy.2009.03.005

Hastenrath, S. (1991). *Climate Dynamics of the Tropics*. Amsterdam: Kluwer Academic Publishers.

Hsu, K., Gupta, H. V., & Sorooshian, S. (1995). Artificial neural network modeling of the rainfall-runoff process. *Water Resources Research, 31*(10), 2517–2530. doi:10.1029/95WR01955

Hurrell, J. W. (1995). Decadal trends in the North Atlantic Oscillation: regional temperatures and precipitation. *Science, 269*, 676–679. doi:10.1126/science.269.5224.676

Hurrell, J. W., Kushnir, Y., Visbeck, M., & Ottersen, G. (2003). An overview of the North Atlantic Oscillation. In J. W. Hurrell, Y. Kushnir, G. Ottersen, & M. Visbeck (Eds), *The North Atlantic Oscillation: Climate Significance and Environmental Impact*s (pp. 1-35), Geophysical Monograph Series, 134. Washington, DC: American Geophysical Union.

Jeong, D.-I., & Kim, Y.-O. (2005). Rainfall-runoff models using artificial neural networks for ensemble streamflow prediction. *Hydrological Processes, 19*, 3819–3835. doi:10.1002/hyp.5983

Jones, C., & Carvalho, L. M. V. (2002). Active and break phases in the South American Monsoon System. *Journal of Climate, 15*, 905–914. doi:10.1175/1520-0442(2002)015<0905:AABPIT>2.0.CO;2

Jones, J. W., Hansen, J. W., Royce, F. S., & Messina, C. D. (2000). Potential benefits of climate forecasting to agriculture. *Agriculture Ecosystems & Environment, 82*, 169–184. doi:10.1016/S0167-8809(00)00225-5

Kalnay, E. (1996). The NCEP/NCAR Reanalysis 40-year Project. *Bulletin of the American Meteorological Society, 77*, 437–471. doi:10.1175/1520-0477(1996)077<0437:TNYRP>2.0.CO;2

Karamouz, M., Razavi, S., & Araghinejad, A. (2008). Long-lead seasonal rainfall forecasting using time-delay recurrent neural networks: a case study. *Hydrological Processes, 22*, 229–241. doi:10.1002/hyp.6571

Koutsouyiannis, D., Efstratiadis, A., Namassis, N., & Christofides, A. (2008). On the credibility of climate predictions. *Hydrological Sciences Journal, 53*(4), 671–684. doi:10.1623/hysj.53.4.671

Legates, D. R., & McCabe, G. J. (1999). Evaluating the use of "goodness-of-fit" measures in hydrologic and hydroclimatic model validation. *Water Resources Research, 35*(1), 233–241. doi:10.1029/1998WR900018

Liebmann, B., Camargo, S. J., Seth, A., Marengo, J. A., Carvalho, L. M. A., & Allured, D. (2007). Onset and end of the rainy season in South America in observations and the ECHAM 4.5 atmospheric general circulation model. *Journal of Climate, 20*, 2037–2050. doi:10.1175/JCLI4122.1

Liebmann, B., Kiladis, G. N., Marengo, J. A., Ambrizzi, T., & Glick, J. D. (1999). Submonthly convective variability over South America and the South Atlantic Convergence Zone. *Journal of Climate, 12*(7), 1877–1891. doi:10.1175/1520-0442(1999)012<1877:SCVOSA>2.0.CO;2

Luk, K. C., Ball, J. E., & Sharma, A. (2001). An application of Artificial Neural Networks for rainfall forecasting. *Mathematical and Computer Modelling, 33*, 683–693. doi:10.1016/S0895-7177(00)00272-7

Maier, H. R., & Dandy, G. C. (1997). Modelling cyanobacteria blue-green algae in the River Murray using artificial neural networks. *Mathematics and Computers in Simulation, 43*, 377–386. doi:10.1016/S0378-4754(97)00022-0

Maier, H. R., & Dandy, G. C. (2000). Neural networks for the prediction and forecasting of water resources variables: a review of modeling issues and applications. *Environmental Modelling & Software*, *15*(1), 101–124. doi:10.1016/S1364-8152(99)00007-9

Mantua, N. J., Hare, S. R., Zhang, Y., Wallace, J. M., & Francis, R. C. (1997). A pacific decadal climate oscillation with impacts on salmon. *Bulletin of the American Meteorological Society*, *78*, 1069–1079. doi:10.1175/1520-0477(1997)078<1069:APICOW>2.0.CO;2

Marengo, J. A., Soares, W. R., Saulo, C., & Nicolini, M. (2004). Climatology of the low-level jet east of the Andes as derived from NCEP-NCAR reanalysis - characteristics and temporal variability. *Journal of Climate*, *17*(12), 2261–2280. doi:10.1175/1520-0442(2004)017<2261:COTLJE>2.0.CO;2

McIntosh, P. C., Ash, A. J., & Smith, M. S. (2005). From oceans to farms: the value of a novel statistical climate forecast for agricultural management. *Journal of Climate*, *18*, 4287–4302. doi:10.1175/JCLI3515.1

McIntosh, P. C., Pook, M. J., Risbey, J. S., Lisson, S. N., & Rebbeck, M. (2007). Seasonal climate forecasts for agriculture: towards better understanding and value. *Field Crops Research*, *104*, 130–138. doi:10.1016/j.fcr.2007.03.019

Meinke, H., & Stone, R. C. (2005). Seasonal and inter-annual climate forecasting: the new tool for increasing preparedness to climate variability and change in agricultural planning and operations. *Climatic Change*, *70*, 221–253. doi:10.1007/s10584-005-5948-6

Minns, A. W., & Hall, M. J. (1996). Artificial Neural Networks as rainfall-runoff modes. *Hydrological Sciences Journal*, *41*(3), 399–417. doi:10.1080/02626669609491511

Moeller, C., Smith, I., Asseng, S., Ludwig, F., & Telcik, N. (2008). The potential value of seasonal forecasts of rainfall categories - case studies from the wheatbelt in Western Australia's Mediterranean region. *Agricultural and Forest Meteorology*, *148*, 606–618.

Moller, M. S. (1993). A scaled conjugate gradient algorithm for fast supervised learning. *Neural Networks*, *6*(4), 525–534. doi:10.1016/S0893-6080(05)80056-5

Moore, R. J., Bell, V. A., & Jones, D. A. (2005). Forecasting for flood warning. *Comptes Rendus Geoscience – External Geophysics. Climate and Enviroment*, *337*, 203–217.

Morid, S., Smakhtin, V., & Bagherzadeh, K. (2007). Drought forecasting using artificial neural networks and time series of drought indices. *International Journal of Climatology*, *27*, 2103–2111. doi:10.1002/joc.1498

Muluye, G. Y., & Coulibaly, P. (2007). Seasonal reservoir inflow forecasting with low frequency climatic indices: a comparison of data-driven methods. *Hydrological Sciences Journal*, *52*(3), 508–522. doi:10.1623/hysj.52.3.508

Murphy, A. H. (1993). What is a good forecast? An essay on the nature of goodness in weather forecasting. *Weather and Forecasting*, *8*, 281–293. doi:10.1175/1520-0434(1993)008<0281:WIAGFA>2.0.CO;2

Nan, S., & Li, J. (2003). The relationship between summer precipitation in the Yangtze River valley and the previous Southern Hemisphere Annular Mode. *Geophysical Research Letters*, *30*(24), 2266. doi:10.1029/2003GL018381

Nasrallah, H. A., Balling, R. C., Selovert, N. J., & Vose, R. S. (2001). Development of a seasonal forecast model for Kuwait winter precipitation. *Journal of Arid Environments*, *48*, 233–242. doi:10.1006/jare.2000.0746

Nguyen, T. D., Rosbjerg, D., Uvo, C. B., & Nguyen, K. Q. (2007b). Drought prediction in the Vietnamese central highlands. Changes in Water Resources Systems: Methodologies to Maintain Water Security and Ensure Integrated Management. In *Proceedings of Symposium HS3006 at IUGG2007*. Perugia, Italy: IAHS Publ.

Nguyen, T. D., Uvo, C. B., & Rosbjerg, D. (2007a). Relationship between tropical Pacific and Indian Ocean sea surface temperature and monthly precipitation over the central highland, Vietnam. *International Journal of Climatology*, *27*(11), 1439–1454. doi:10.1002/joc.1486

Nilsson, P., Uvo, C. B., & Berndtsson, R. (2006). Monthly runoff simulation: comparison of a conceptual model, neural networks and a combination of them. *Journal of Hydrology (Amsterdam)*, *321*, 344–363. doi:10.1016/j.jhydrol.2005.08.007

Nilsson, P., Uvo, C. B., Landman, W., & Nguye, T. D. (2008). Downscaling of GCM forecasts to streamflow over Scandinavia. *Hydrology Research*, *39*(1), 17–25. doi:10.2166/nh.2008.027

Nogués-Paegle, J., & co-authors. (2002). Progress in Pan American CLIVAR research: Understanding the South American Monsoon. *Meteorologica*, *27*, 3–30.

Ogallo, L. A., Boulahya, M. S., & Keane, T. (2000). Applications of seasonal to interannual climate prediction in agricultural planning and operations. *Agricultural and Forest Meteorology*, *103*, 159–166. doi:10.1016/S0168-1923(00)00109-X

Olsson, J., Uvo, C. B., Jinno, K., Kawamura, A., Nishiyama, K., & Koreeda, N. (2004). Neural Networks for rainfall forecasting by atmospheric downscaling. *Journal of Hydrologic Engineering*, *9*(1), 1–12. doi:10.1061/(ASCE)1084-0699(2004)9:1(1)

Pielke, R., & Carbone, R. E. (2002). Weather impacts, forecasts, and policy - an integrated perspective. *Bulletin of the American Meteorological Society*, 393–403. doi:10.1175/1520-0477(2002)083<0393:WIFAP>2.3.CO;2

Ramírez, M. C. V., Velho, H. F. C., & Ferreira, N. J. (2005). Artificial neural network technique for rainfall forecasting applied to the São Paulo region. *Journal of Hydrology (Amsterdam)*, *301*, 146–162. doi:10.1016/j.jhydrol.2004.06.028

Rasmusson, E. M., & Carpenter, T. H. (1982). Variations in tropical sea surface temperature and surface wind fields associated with the Southern Oscillation/El Niño. *Monthly Weather Review*, *110*(5), 354–384. doi:10.1175/1520-0493(1982)110<0354:VITSST>2.0.CO;2

Reboita, M. S., Ambrizzi, T., & Rocha, R. P. (2009). Relationship between the Southern Annular Mode and Southern Hemisphere Atmospheric Systems. *Revista Brasileira de Meteorologia*, *24*, 48–55. doi:10.1590/S0102-77862009000100005

Rumelhart, D. E., Hinton, G. E., & Williams, R. J. (1986). Learning internal representations by error propagation. In Rumelhart, D. E., & McClelland, J. L. (Eds.), *Parallel Distributed Processing* (*Vol. 1*, pp. 318–362). Cambridge, MA: MIT Press.

Silverman, D., & Dracup, J. A. (2000). Artificial Neural Networks and long-range precipitation prediction in California. *Journal of Applied Meteorology*, *39*, 57–66. doi:10.1175/1520-0450(2000)039<0057:ANNALR>2.0.CO;2

Sivakumar, M. V. K., Motha, R. P., & Das, H. P. (Eds.). (2005). *Natural disasters and extreme events in agriculture: impacts and mitigation*. The Netherlands: Springer. doi:10.1007/3-540-28307-2

Stokelj, T., Paravan, D., & Golob, R. (2002). Enhanced artificial neural network inflow forecasting algorithm for run-of-river hydropower plants. *Journal of Water Resources Planning and Management*, *128*(6), 415–423. doi:10.1061/(ASCE)0733-9496(2002)128:6(415)

Tantanee, S., Patamatamakul, S., Oki, T., Sriboonlue, V., & Prempree, T. (2005). Downscaled rainfall prediction model (DRPM) using a unit disaggregation curve (UDC). *Hydrology and Earth System Sciences Discussions*, *2*, 543–568. doi:10.5194/hessd-2-543-2005

Thompson, D. W. J., & Wallace, J. M. (2000). Annular modes in the extratropical circulation. Part I: Month-to-month variability. *Journal of Climate*, *13*, 1000–1016. doi:10.1175/1520-0442(2000)013<1000:AMITEC>2.0.CO;2

Tokar, A. S., & Johnson, P. A. (1999). Rainfall-runoff modeling using artificial neural networks. *Journal of Hydrologic Engineering*, *4*(3), 232–239. doi:10.1061/(ASCE)1084-0699(1999)4:3(232)

Trenberth, K. E. (1990). Recent observed interdecadal climate changes in the northern hemisphere. *Bulletin of the American Meteorological Society*, *71*, 988–993. doi:10.1175/1520-0477(1990)071<0988:ROICCI>2.0.CO;2

Trenberth, K. E., & Hurrell, J. W. (1994). Decadal atmosphere-ocean variations in the Pacific. *Climate Dynamics*, *9*, 303–319. doi:10.1007/BF00204745

Uvo, C. B., Olsson, J., Morita, O., Jinno, K., Kawamura, A., & Nishiyama, K. (2001). Statistical atmospheric downscaling for rainfall estimation in Kyushu Island, Japan. *Hydrology and Earth System Sciences*, *5*(2), 259–271. doi:10.5194/hess-5-259-2001

Vera, C., Higgins, W., Amador, J., Ambrizzi, T., Garreaud, R., & Gochis, D. (2006). Toward a unified view of the American Monsoon Systems. *Journal of Climate*, *19*(20), 4977–5000. doi:10.1175/JCLI3896.1

Wilby, R. L., Wedgbrow, C. S., & Fox, H. R. (2004). Seasonal predictability of the summer hydrometeorology of the River Thames, UK. *Journal of Hydrology (Amsterdam)*, *295*, 1–16. doi:10.1016/j.jhydrol.2004.02.015

Wilks, D. S. (2006). *Statistical methods in the atmospheric sciences*. Burlington, MA: Academic Press, Elsevier.

Yeh, W. W.-G., Becker, L., & Zettlemoyer, R. (1982). Worth of inflow forecast for reservoir operation. *Journal of Water Resources Planning and Management*, *108*(3), 257–269.

Zhang, Y., Wallace, J. M., & Battisti, D. S. (1997). ENSO-like interdecadal variability: 1900-93. *Journal of Climate*, *10*, 1004–1020. doi:10.1175/1520-0442(1997)010<1004:ELIV>2.0.CO;2

Zwiers, F. W., & von Storch, H. (2004). On the role of statistics in climate research. *International Journal of Climatology*, *24*, 665–680. doi:10.1002/joc.1027

## ADDITIONAL READING

Barnston, A. G., Mason, S. J., Goddard, L., Dewitt, D. G., & Zebiak, S. E. (2003). Multimodel ensembling in seasonal climate forecasting at IRI. *Bulletin of the American Meteorological Society*, *84*, 1783–1976. doi:10.1175/BAMS-84-12-1783

Cabrera, V. E., Breuer, N. E., & Hildebrand, P. E. (2006). North Florida dairy farmer perceptions toward the use of seasonal climate forecast technology. *Climatic Change*, *38*, 479–491. doi:10.1007/s10584-006-9053-2

Dettinger, M. D., Cayan, D. R., McCabe, G. M., & Marengo, J. A. (2000). Multiscale streamflow variability associated with El Niño/Southern Oscillation. In Diaz, H. F., & Markgraf, V. (Eds.), *El Niño and the Southern Oscillation - Multiscale Variability and Global and Regional Impacts* (pp. 113–146). Cambridge: Cambridge University Press. doi:10.1017/CBO9780511573125.005

Doblas-Reyes, F. J., Hagedorn, R., & Palmer, T. N. (2006). Developments in dynamical seasonal forecasting relevant to agricultural management. *Climate Research*, *33*, 19–26. doi:10.3354/cr033019

Fox, G., Turner, J., & Gillespie, T. (1999). The value of precipitation forecast information in winter wheat production. *Agricultural and Forest Meteorology*, *95*, 99–111. doi:10.1016/S0168-1923(99)00022-2

Fraisse, C. W., Breuer, N. E., Zierden, D., Bellow, J. G., Paz, J., & Cabrera, V. E. (2006). AgClimate: a climate forecast information system for agricultural risk management in the southeastern USA. *Computers and Electronics in Agriculture*, *53*, 13–27. doi:10.1016/j.compag.2006.03.002

Hansen, J. W. (2005). Integrating seasonal climate prediction and agricultural models for insights into agricultural practice. *Philosophical Transactions of the Royal Society B – Biological Sciences*, 360, 2037-2047.

Hansen, J. W., Challinor, A., Ines, A., Wheeler, T., & Moron, V. (2006). Translating climate forecasts into agricultural terms: advances and challenges. *Climate Research*, *33*, 27–41. doi:10.3354/cr033027

Harrison, M. (2005). The development of seasonal and inter-annual climate forecasting. *Climatic Change*, *70*, 201–220. doi:10.1007/s10584-005-5947-7

Hill, H. S. J., Park, J., Mjelde, J. W., Rosenthal, W., Love, H. A., & Fuller, S. W. (2000). Comparing the value of Southern Oscillation Index-based climate forecast methods for Canadian and US wheat producers. *Agricultural and Forest Meteorology*, *100*, 261–272. doi:10.1016/S0168-1923(99)00154-9

Huang, Y. (2009). Advances in artificial neural networks – methodological development and application. *Algorithms*, *2*, 973–1007. doi:10.3390/algor2030973

Jagtap, S. S., Jones, J. W., Hildebrand, P., Letson, D., O'Brien, J. J., & Podestá, G. (2002). Responding to stakeholder's demands for climate information: from research to applications in Florida. *Agricultural Systems*, *74*, 415–430. doi:10.1016/S0308-521X(02)00048-3

Kumar, D. N., Reddy, M. J., & Maity, R. (2007). Regional rainfall forecasting using large scale climate teleconnections and artificial intelligence techniques. *Journal of Intelligent Systems*, *16*(4), 307–322.

Lorenz, E. N. (1969). Three approaches to atmospheric predictability. *Bulletin of the American Meteorological Society*, *50*(5), 345–351.

Luk, K. C., Ball, J. E., & Sharma, A. (2000). A study of optimal model lag and spatial inputs to artificial neural network for rainfall forecasting. *Journal of Hydrology (Amsterdam)*, *227*, 56–65. doi:10.1016/S0022-1694(99)00165-1

Maqsood, I., Khan, M. R., & Abraham, A. (2004). An ensemble of neural networks for weather forecasting. *Neural Computing & Applications*, *13*, 112–122. doi:10.1007/s00521-004-0413-4

Mjelde, J. W., & Hill, H. S. J. (1999). The effect of the use of improved climate forecasts on variable costs, input usage, and production. *Agricultural Systems*, *60*, 213–225. doi:10.1016/S0308-521X(99)00033-5

Moura, A. D., & Hastenrath, S. (2004). Climate prediction for Brazil's Nordeste: performance of empirical and numerical modeling methods. *Journal of Climate*, *17*, 2667–2672. doi:10.1175/1520-0442(2004)017<2667:CPFBNP>2.0.CO;2

Nelson, R. A., Holzworth, D. P., Hammer, G. L., & Hayman, P. T. (2002). Infusing the use of seasonal climate forecasting into crop management practice in North East Australia using discussion support software. *Agricultural Systems*, *74*, 393–414. doi:10.1016/S0308-521X(02)00047-1

Paras, S. M., Kumar, A., & Chandra, M. (2008). A feature based neural network model for weather forecasting. *International Journal of Computational Intelligence*, *4*(3), 209–216.

Pezzi, L. P., & Kayano, M. T. (2009). An analysis of the seasonal precipitation forecasts in South America using wavelets. *International Journal of Climatology*, *29*, 1560–1573. doi:10.1002/joc.1813

Podestá, G., Letson, D., Messina, C., Royce, F., Ferreyra, R. A., & Jones, J. (2002). Use of ENSO-related climate information in agricultural decision making in Argentina: a pilot experience. *Agricultural Systems*, *74*, 371–392. doi:10.1016/S0308-521X(02)00046-X

Roebber, P. J., & Bosart, L. F. (1996). The complex relationship between forecast skill and forecast value: a real-world analysis. *Weather and Forecasting*, *11*, 544–559. doi:10.1175/1520-0434(1996)011<0544:TCRBFS>2.0.CO;2

Schneider, J. M., & Garbrecht, J. D. (2003). A measure of the usefulness of seasonal precipitation forecasts for agricultural applications. *Transactions of the ASAE. American Society of Agricultural Engineers*, *46*(2), 257–267.

Schneider, J. M., & Garbrecht, J. D. (2004). Dependability and effectiveness of seasonal forecasts for agricultural applications. *Transactions of the ASABE*, *49*(6), 1737–1753.

Schöngart, J., & Junk, W. J. (2007). Forecasting the flood-pulse in Central Amazonia by ENSO-indices. *Journal of Hydrology (Amsterdam)*, *335*, 124–132. doi:10.1016/j.jhydrol.2006.11.005

Smith, B. A., Hoogenboom, G., & McClendon, R. W. (2009). Artificial neural networks for automated year-round temperature prediction. *Computers and Electronics in Agriculture*, *68*, 52–61. doi:10.1016/j.compag.2009.04.003

Stone, R. C., Hammer, G. L., & Marcussen, T. (1996). Prediction of global rainfall probabilities using phases of the Southern Oscillation Index. *Nature*, *384*, 252–255. doi:10.1038/384252a0

Timilsena, J., Piechota, T., Tootle, G., & Singh, A. (2009). Associations of interdecadal/interannual climate variability and long-term Colorado river basin streamflow. *Journal of Hydrology (Amsterdam)*, *365*, 289–301. doi:10.1016/j.jhydrol.2008.11.035

Ziervogel, G., Bithell, M., Washington, R., & Downing, T. (2005). Agent-based social simulation: a method for assessing the impact of seasonal climate forecast applications among smallholder farmers. *Agricultural Systems*, *83*, 1–26. doi:10.1016/j.agsy.2004.02.009

## KEY TERMS AND DEFINITIONS

**Artificial Neural Network:** Computational approach based on analogies of the brain structure and brain functioning used in several practical problems.

**Climate Index:** A measure of intensity of modes of climate variability, which is calculated differently for each climate phenomena.

**Climate Variability:** Variability of climate conditions resultant of natural atmospheric and oceanic oscillations.

**Empirical Model:** Model defined by a statistical relationship developed using past values of input and output variables that may serve as a prediction model but not to explain or describe the functioning of the system.

**Predictand Variable:** Variable to be predicted by a forecast model.

**Predictor Variable:** Variable used as input for developing an empirical forecast model.

**Seasonal Precipitation Forecast:** Prediction of precipitation patterns for the seasonal rainy period, which usually comprise 1 to 6 months.

**Teleconnection:** A strong statistical relationship between the weather in different parts of the globe.

# Chapter 17
# Descriptive Methods and Compromise Programming for Promoting Agricultural Reuse of Treated Wastewater

**Hella Ben Brahim**
*University of Tunis - El Manar, Tunisia*

**Lucien Duckstein**
*University of Arizona, USA*

## ABSTRACT

*The present chapter reports on a practical case of decision making in agricultural TWW reuse. The studied site concerns the Cebala TWW irrigated perimeter in Tunisia, located in North Africa. Crucially though in this perimeter is that TWW is under-exploited. Thus, the proposed research dwells on the future of this perimeter which could oscillate between hope and deadlock. In case of hope, shall we improve the quality of the resource and possibly move to the tertiary treatment? And in case of deadlock, shall we give up the TWW reuse or substitute it by a conventional resource? To make an evaluation of this experience, the authors first gave a general overview of the descriptive study in a dashboard. Contingent valuation method was used to assess qualitative data and gauge the intangible effects. Then, the authors presented a systemic multicriteria method deployed to make decision-making process easier. Such a method or technique, which is grounded on Lp metrics, is called Compromise Programming.*

## INTRODUCTION

Reusing Treated WasteWater (TWW) in an agricultural perimeter is a profitable activity not only for farmers but further for concerned water resource management agencies. Relevant is that this activity may help tackle the scarcity of conventional resources in arid and semi-arid countries. It can also contribute to lessen the negative effects of TWW loading on the environment. Reusing TWW may present a multitude of disadvantages, however. Hence, when the wastewater resource is treated on the secondary stage, its reuse in agriculture is severally controlled. To this effect, international organizations such as WHO and local agencies

DOI: 10.4018/978-1-61692-871-1.ch017

set many restrictions over the irrigation of market gardening. This measure can partly explain the reluctance of a large number of farmers to reuse TWW. Further, most of the interviewed farmers reusing this type of resource expressed their resentment about the bad quality of the TWW they have access to. Though being quite cheap, many farmers refuse to pay advances to reuse TWW.

The management of TWW in agricultural perimeter entails many disadvantages for decision makers. The agricultural TWW reuse scheme needs an appropriate structure (appropriate irrigation and drainage system, roads, control of reuse practices, etc). So, governmental agencies are the most appropriate institution to control, supervise and offer this non-conventional resource. The investment implementation and the maintenance as well as operating fees are very expensive. Moreover, after 15 to 20 years of exploitation, this kind of schemes needs to be rehabilitated. Costly investments are needed as a result. This rehabilitation can be joined to the decision of improving the resource quality and then moved to the third level of treatment so as to authorize gardening market production. All those structural decisions must be carefully considered, studied and evaluated.

As the Tunisian experience in reusing TWW is important and dates from 1965, guidelines for exploitation and implementation schemes are widely developed. The need now is to evaluate this experience and to learn lessons from success and above all the cases of failure. The decision can be radical, such as deciding to give up the agricultural irrigation with TWW, to continue that experience or to substitute the resource with a conventional one (if it is possible). There is another situation that needs to be studied in case of continuing this experience. Such a situation can be expressed in the following query: Shall we improve the quality of the resource and may be reach the third level of treatment?.

The classical methods used to evaluate this experience are generally based on descriptive techniques. They give an idea about positive and negative effects, tangible and intangible effects. The difficulty lies in measuring the difference between those effects, especially in the case of qualitative ones. Many parameters can be considered in a dashboard that evaluates the reuse activity. If the positive and negative differences effects sum is negative, the general economic equilibrium will be less valuable.. This tool gives some priority to some factors. It helps offset certain weakness in the activity, but structural decision such us giving up the reuse cannot be made from the analysis using this dashboard.

The present study reports on a practical case of TWW agricultural reuse in Tunisia, in North Africa. Given the importance of the Tunisian experience in this sector, our focus will be on how we evaluate a pilot experience, using descriptive methods and multicriteria decision making approach. We will present in a first part most necessarily keys to prepare a useful survey for reliable data. In a second part, we present a systemic method to make decision. This multicriteria technique, which is based on Lp metrics, is called Compromise Programming (CP).

## DECISION MAKING AID FOR PROMOTING AGRICULTURAL REUSE OF TWW: THEORETICAL APPROACH AND DISCUSSION.

In this section, we start with presenting TWW characteristics as a waste to be loaded on the one hand and as a resource to be reused, on the other. This presentation qualifies this resource as an economic product to be promoted and valuated. In the meanwhile, it considers TWW as a negative externality of human and industrial activities that must be eliminated with respect of the restrictions of environmental preservation. Then, an overview of TWW reuse research and studies done so far the world over was given. We will focus on reusing TWW in agricultural activities. The third part of this section is devoted to current methods used for an ex-post evaluation of agricultural investment, such as irrigated schemes with TWW. The discussion of scientific researches in this field will be fully presented. Further, we will present descriptive analysis of data collected from surveys. By the end of this description, the strengths and weakness of each theoretical approach will be highlighted, and multicriteria decision aid methods will be introduced as an appropriate tool for the proposed topic. Scientific demonstration and contrary examples will be drawn upon to argue for our perspective. In the last sub-section of this study, we define the compromise programming (CP) method. Then, we outline developed scientific works linked or based on CP method. In so doing, the focus will be on its diverse uses in water resources management. Finally, we will prove and conclude that the success of multicriteria decision aid is based on the reliability of data analysis. We will also show that interactive decision-making methods depend on the understanding of human behavior and on a successful survey technique and sampling.

## TWW: A Waste or a Resource?

Within urban water use, about 75-85% of consumed water is being returned to the urban hydrologic system as a wastewater. This kind of waste is a mix of domestic and industrial water and storm water. The industrial wastewater quality is the most dangerous for human health and environment degradation. Indeed, it often contains metalloids, elevated levels of metals and suspended matters. It is important, however, to treat it at least in the secondary level. This can decrease negative externalities of loading TWW on the environment. In developing countries, wastewater is commonly discharged with little or no treatments (WHO, 2005, Qadir & al., 2008). The World Health Organization (WHO) and the Foundation of Agricultural Organization (FAO) set restrictions and guidelines on wastewater reuse and recycling. In some developing countries, such as Tunisia, environmental and agricultural government agencies created their own sanitation and set TWW reuse regulations and standards, based on international standards and guidelines (WHO., 2005, p. 53).

The fact of considering TWW as a waste limits its management to sanitation and health institutions. This perception is very acute, because in regions where water is in short supply, reusing wastewater helps tackle shortage of water resources. If we consider TWW as a resource, it becomes an economic product that must be exploited and valorized. In the same time, TWW management agencies must attribute a monetary value to this product. The price of this production input includes implementation, operating and system irrigation maintenance costs. In case of reusing TWW in agriculture, other costs can be considered, including the vulgarization of reuse guidelines and the health and the sanitary controls, among others.

Water resources are often considered as public goods that must be commonly shared between various sectors, users and regions. To avoid over-exploitation of those resources, economists advocate the definition of private water rights and water market. It remains difficult to speak about privatization, especially in case of exhaustible and non-renewable resource (Johansson et al., 2002). Privatization is more complex in case of agricultural TWW reuse. In fact, agricultural reuse must be severally controlled by governmental agencies, notably in case of wastewater secondary level treatment. In the last section which concerns the TWW resource farmer perception, focus will be on the discussion of other challenging problems.

## The TWW Reuse: Challenges, Controversies and Perspectives

Water reuse in agricultural irrigation is an ancient farming experience. Reusing rainwater in Europeans countries dates back to Minoan time (Angelakis et al., 1996). On the other hand, reusing wastewater goes back to Greek and Roman civilization (Angelakis et al., 2005). In this regard, Angelakis (2008) states that "*Wastewater has also been used by the Mediterranean civilisations, for example in the 14th and 15th centuries in the Milamese Marcites and in the Valencia Huearta and the North European ones, like in Great Britain, Germany, France and Poland*". Actually, the wastewater reuse experience has been generalized all over the world specially in developing countries (WHO., 2005., Angelakis & Durham, 2008., Bdour et al., 2009., Bahri., 2003., Al Nasir., 2008). The reuse targets vary from northern countries to southern ones. With respect to the northern countries, principal aims consist in environment preservation, eutrophication reduction and energy saving by selective treatment techniques (Angelakis & Durham., 2008., Tidăkar et al., 2007). In the case of Mediterranean Basin, the TWW reuse targets are principally linked to the scarcity of water resources. In this region, Israel is the TWW practices pioneer, followed by Cyprus, Jordan and Tunisia (Angelakis et al., 1998). The other agricultural world pioneer of TWW reuse experience is the Californian one. The wastewater is treated on the third level and it is reused on gardening and crops irrigation (Bahman, 1998).

The reuse of TWW, in some countries such as Brazil, starts late because of the ignorance of irrigation experience advantages and disadvantages (Da Fonesca et al., 2007). Other unknown drawbacks of this practice are the influence of tropical conditions parameters on irrigated soil characteristics (Leal et al., 2009). In this case, the sugarcane is irrigated with TWW and represents a renewable energy source.

The overview of TWW agricultural reuse experiences in emergent countries demonstrates an antagonistic fact: there are different treatment levels (if effluent is treated) and different resources qualities as well as different agricultural reuse practices; but influential reuse success factors are common and concerns farmers TWW acceptance and demand. The TWW demand is a key factor, for it can account for the sustainability of reuse schemes and the behavior of the farmer face to this resource. This demand depends on the water resources management policy – if it exists – and on the quality of the resource. Other factors can be influential, such as, legislative restrictions and guidelines, TWW price and prices of conventional water resources. The following Figure (Figure 1) is an illustration of the interplay of the different parameters underlying the discrepancy in the agricultural TWW reuse between the North and the South.

The most frequent user behavior (in our study, the farmer) face to this unconventional resource is some reluctance to reuse, suspicion and high risk perception (Wilson & Pfaff., 2008, Menegaki et al., 2007, Mutengu et al., 2007). Farmers

*Figure 1. Interaction between parameters explaining agricultural TWW demand increase from northern to southern countries*

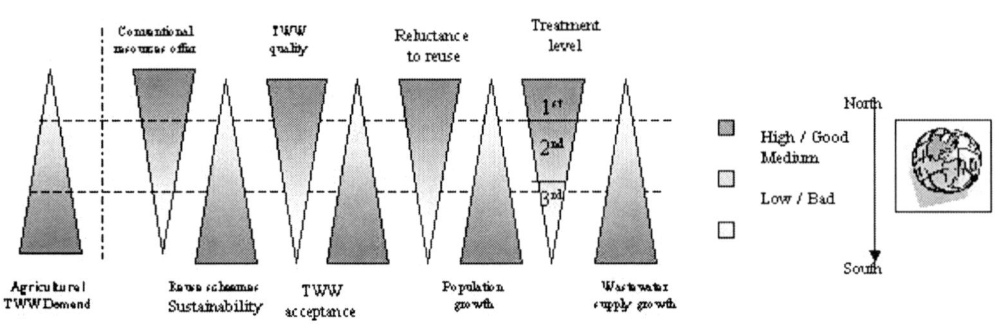

are also ignorant of potential benefits of TWW as nutrient sources for agricultural crop production (Yadav et al., 2002, Da Fonesca et al., 2007). Another common aspect concerns the indirect user of the resource behavior (the consumer of agricultural products irrigated with TWW). Face to the same product, one irrigated with conventional resource and the other with TWW, the choice is often made on the first product though being expensive. This behavior is different in case of industrial or fodder farming. It is considered in this case as an input and the price of the product which is irrigated with TWW is competitive compared to the same product irrigated with other conventional resource.

Other controversies that can be presented in this section are long-term human health effects and environmental degradation caused by TWW reuse. Depending on wastewater treatment level, the irrigation with TWW can have negative impacts on human health for both farmers and consumers in case of vegetable or gardening products. If the reuse is severely controlled, the risk is reduced and monitored. There must be no contact with the resource, hence the need to wear gloves and boots. The Drinking of TWW (treated at the most on the secondary level) is forbidden for either man or animals. Restrictions and guidelines are detailed on the work of international agencies concerned with the TWW management, such as WHO and FAO. The local agencies responsibility in application and adaptation of guidelines and restriction is an important factor to guarantee experience success and sustainability. They have other responsibilities such as popularization and sensitivity of health and sanitation rules. The role of farmers is, as well, important, especially in restrictions application and self-control.

The arbitrage between environmental reuse benefits and negative effects on soil degradation and groundwater contamination is intriguing. Reusing in agriculture is indeed an efficient solution to reduce TWW loading in natural environment on the one hand and in fertilizers consumption – TWW contains nutrients - in irrigated perimeters, on the other. However, mixed TWW (from many origins), may contain metal, metalloids, volatile or semi-volatile compounds and pathogens. In the long run, this kind of resource may have negative effects on soil and on groundwater from land-based activities (Leal et al., 2009, Bond, 1998). Eliminating or avoiding risks is far from easy. Hence, public agencies must choose the most propitious treatment and wastewater management policy that could minimize human health and environmental degradation risks. Those two parameters often help to establish TWW quality objectives. The revision of risk assessments is useful given the continuous scientific progress and the variation in public preferences. The implication of this is that the reuse management policy must be established in the long-term with consideration of such parameters.

In many scientific works, TWW is the future resource in agriculture in rare conventional resources context. The aim of this chapter is not to suggest guidelines, advices and recommendation to implement agricultural project irrigated with TWW. Rather, its objective is to evaluate this practice in pioneer countries such as Tunisia. In fact, to learn from this experience and present new perspectives, we must evaluate previous TWW irrigation projects. Such an evaluation provides new guidelines not only to have safe use but also to help decision maker (governmental agencies) to promote the reuse and implement successful new schemes. In some cases, it is hard to admit that the TWW reuse was not the appropriate solution for such or such perimeter. The problem is that this kind of investment is very expensive and the farmer behavior is not guaranteed. The evaluation of hazard is the most difficult part of scheme studies, especially in a context of information non-transparency. In this chapter, we will evaluate in a first step an agricultural perimeter irrigated with TWW. In a second step, we will present and choose solution(s) for this perimeter based on multicriteria method.

## Ex-Post Evaluation of Agricultural Irrigated Scheme with TWW

The opportunity cost of public funds is very high. We also notice the existence of other notable measures during the last decades, such as restrictions on agricultural public financing. Thus, economic and financial reliability are salient criteria in choosing to invest or not, as well as in selecting best investment (Loubier et al., 2001). The method of global cost is often used in this process of selection based on cost-effectiveness method principles. The aim of this method is to estimate an internal return rate (IRR) which must be higher than monetary market rate (MMR).

The most difficult part of this evaluation is to estimate externalities such as environmental and social effects. The pertinence of classical instruments such as references prices methods were criticized by Davarajan et al., (1995) and Billi (1996). Concrete application of optimal models of growth and collective offer of goods were disappointing. In this regard, the famous economist Arrow (1996) admits that:

- The use of this kind of models is not useful.
- Intuition and right direction are only the guarantor of good choice.
- All economists did not have the same intuition.

Focusing on the second point, we observe that it is subjective; it involves experience observation and good appreciation of actions and phenomena. Lessons must be drawn from both successful and unsuccessful experiences. This is possible by continuous control after investment implementation and by carrying out surveys among stockholders, in our case farmers. The third point shows that the decision maker perception and preferences must be considered in the decision maker aid system to guarantee its pertinence.

We can conclude that theoretical models based on classical tools are often not practical in real case studied, even in choosing agricultural irrigated scheme. This fact becomes more complicated in case of non-conventional resource, such as TWW. New criteria are introduced in the decision system aid such as health preservation and effects on environment. More than introducing new qualitative parameters – hard to evaluate – the implementation of Treated Wastewater Irrigated Perimeter (TWWIP) is often a public planned investment that presupposes a special context.

Implementing dams or laying out reservoirs are expensive public investments meant to resolve water resource scarcity in the long-term. The price of the resource is fixed by the balance of supply and demand. Farmers try to optimize the exploitation of the conventional resource by producing crops and other high added-value agricultural products. The farmer can use this new resource or exploit ground water or other conventional resource. TWW irrigated schemes are also costly, and they need pumping plant, regulation reservoir and irrigation and draining systems that cover the entire perimeter.

The size of the perimeter is estimated in advance and concerned farmers are obliged to respect restrictions. If wastewater is treated at the secondary level treatment, farming is limited to fodders, cereals, industrial and arboriculture. Farmer has no other choice than to reuse or to conduct by farming in dryness. In some region such as the northern Tunisia (ex-Cebala irrigated scheme), the price of farms irrigated with TWW is lower than others produced out of the perimeter. Another striking fact is that the land price value became lower than its neighbor's out of the perimeter. What results from this is that the implementing new investments or rehabilitating the project becomes suspicious. This fact is due to restrictions on farming products. The role of surveys and descriptive analysis methods to evaluate projects and help making are of vital important, for they help to decide whether the project is to be rehabilitated or not.

## Descriptive Analysis

The implementations of hydrologic agricultural schemes are often entrusted to private consulting with the task of studying the technical and financial and aspects, among others. Conducting a survey is important but it is often the most neglected part. The sample does not really reflect the real concerned population. Scientific research should be based on real data representing the population and the activity.

The success of data analysis process depends on many factors:

- The choice of reliable sample size.
- The conduct of survey.
- The Analysis of data (selection, cleaning)
- Calculation of indicators and comparison

**Statistical Methods for Sampling:**
There exist many statistical tools that can determine the sample frame for an ex-post evaluation survey. Given the goals and nature of our study, a stratified sample method is adopted. In an ex-post evaluation study of TWWIP, the size of the perimeter is limited. So, the size of the perimeter (in ha) is limited and the number of the population concerned is known in advance, along with the plot size and the TWW consumption.

In pursuit of choosing the representative sample, we first need to define important strata. Each stratum is divided into classes. The next step presents a post stratum divided into classes, too. The third step consists in splitting the population according to each stratum, as illustrated in the following example:

The stratum is S="farmers age" subdivided into five classes $C_i$ = [< 20 years[, [20, 30[, [30, 40[, [40, 50[ and [≥ 50 years [ with i= 1, 2, 3…5

The post stratum is PS = "parcel size" subdivided to Five classes $C_{ij}$= [< 5ha[, [5, 10[, [10, 20[,[20, 50[ and [≥ 50ha [ with j= 1, 2, 3…5

The population is subdivided first according to the strata. If the size of the population is n, $n_i$ represents the farmers' number of the population concerned. First evident equation can be presented as follows: $n_1 + n_2 + n_3 + n_4 + n_5 = n$.

The second step consists in subdividing each class of the first stratum into five other classes according to the post stratum.

$n_{ij}$ represents the member's number of farmers of class 1 of the stratum i and class j of the post stratum. ($n_{11}$ represents the number of farmers aged less than 20 and owns less than five hectares, $n_{11} + n_{12} + n_{13} + n_{14} + n_{15} = n_1$, Fig 2).

By and large, population is:

$$n = \sum_i \sum_j n_{ij}$$

We suppose that the size of the sample is p. "p" is calculated according to the enquiry needs. For example, direct interview is very costly; the size of the sample is sometimes reduced if the budget allocated to the survey cannot cover the entire perimeter. Of vital importance is that the sample should represent the population. To this end, the calculated $n_{ij}$ is first transformed into ratio, and then multiplied by the population size p.

$p_{ij}$ is the number of farmers, members of the sample according to class i of the stratum and class j of the post stratum. The $p_{ij}$ is calculated as follows:

$$p_{ij} = (n_{ij} / n) \times p$$

*Figure 2. Example of sample distribution*

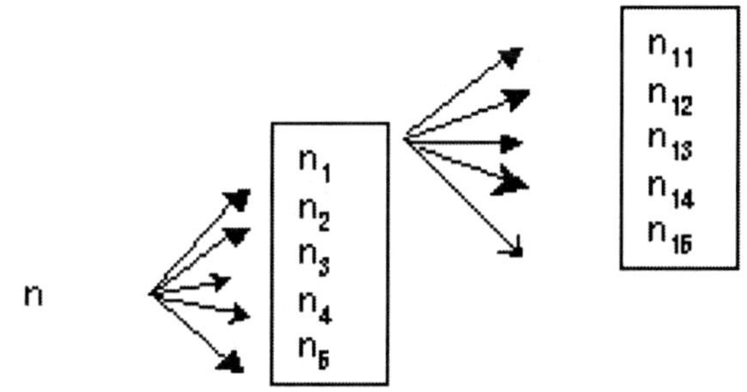

As indicated before, the sample is the key factor in guaranteeing reliable data. Concrete case is illustrated in (NAS, 2007) and in our case study.

### Design and Development of the Survey Questionnaire

The NAS (2007) defines the main phases of the questionnaire. We first distinguish the phase of screening survey. It gives an idea about the standing list of farms. The second phase concerns farming practices and it covers the use of chemicals fertilizers and pesticides. The last phase is about the evaluation of cost and returns.

It is arbitrary to design only one model of survey. The survey design depends generally on the aim of the investigation. One aim may need more than only one survey design in itself. To minimize potential false responses, it behooves us to start with defining the research problematic and then the survey aims.

Conducting interviews is deemed another important preliminary step. In emergent countries such as Tunisia, and above all in rural area, the number of illiterate framers is high. Worse still, information and communication technologies are not usually available and, if they exist, they are under-developed. As a result, relying on postal questionnaire or e-mail or phone communication is not so much productive. Thus, direct interview is endorsed to conduct a survey; this instrument, though being costly and time consuming, is considered the most reliable method to collect data. One implication of the use of this instrument is that the sample size singled out is often little.

The interviewer can add or correct farmers' responses by observing the farming activities or personal information about farmers such as living standards. He must first learn about the perimeter from technical reports. He can also carry out some surveys at governmental agencies managing the resource and the perimeter so as to ensure the reliability of the study.

In a survey with lengthy questionnaire on highly technical topics, answering items can be boring and interviewees may decide not to accomplish such. Thus, questions should be carefully designed and tested on a small sample of the population concerned to make any necessary modification

In a direct questionnaire, it is important to make the farmer feel at his ease and resume long questions. Farmers are not confident when questions concern their returns and incomes. Therefore, false data can be collected more than truth ones. To overcome this issue, we should learn about each topic of the survey and in some questions we need to supervise the farmer by suggesting an interval of value.

To evaluate qualitative effects such as TWW reuse effects on health or on environment, we need a special questionnaire based on new evaluation techniques. There exist many useful methods such as hedonic price method (Faucheux., 1995), Protection expenditures method (Desaigues and Point., 1993), contingent valuation method (OECD., 1994), which is focused upon in the present study. The CVM gives monetary value for qualitative service. For example, to evaluate the importance of having a safe environment or health, we propose supervised and progressive questionnaire. This questionnaire will help to evaluate the service (having safe environment) with yes/no questions, open questions, open-closed questions, etc. Questions are called Wiliness To Pay (WTP) or Wiliness To Accept (WTA) questions.

### Data Treatment and Analysis

Some data treatments are needed to eliminate false responses and make coherence. Often, the number of accepted responses for one or more parts of the survey is less than the sample size. This fact is observed when collecting data about farming incomes. The transparency of information is not always guaranteed. The right sense advocated by Arrow (1996) is also requested from the interviewer to minimize false responses. Codification of random and data is then requested. The simple selection of data gives an overview of the perimeter characteristics. Quantitative data are represented by central parameters (average, mode, median, etc.). Dispersion, concentration and forms data parameter can be measured by standard deviation, Gini index or symmetric coefficients. To understand qualitative data we can illustrate them via special graphics such as camembert. The analysis can be more developed by crossing two or more random and evaluating the correlation between them ($R^2$ = correlation rate). Econometric models can be also developed to show relations between random and make appropriate forecasting. The most challenging part of data analysis is the agricultural returns (farmer income) estimation. Farmers often tend to be coy about their real income, lest the questionnaires can be used by fiscal institute.

In emergent countries, agricultural activities are always subsidized. Financial institutes were sometimes obliged to favor farmers in paying credits. That is why, farmers refute to declare exactly their income. In TWWIP, this fact is easier to evaluate because farming is controlled by governmental agencies and the profit margin is somewhat known in advance. The case studied in section 5 gives an idea about the major difficulties experienced in carrying out a survey among farmers reusing TWW in the Cebala irrigated scheme.

## Shortcomings of Descriptive Methods

In an ex-post evaluation method of TWWIP, collecting data is very important. It helps to learn about the farming activities and returns of reusing a non-conventional water resource. Data are collected and then analyzed to have an overview of the whole activity and trends of evolution of farming activities. The problem with this method is that conclusions are too general. The weight of each studied topic is not exactly measured. For example, if we have a positive economic return and negative health impacts, it is difficult to measure how long the first effect will cover the second one. Calculating the general balance is often subjective, and needs sophistical computing method often unreliable.

The TWWIP has more than economic objectives. It is a project meant to preserve environment and enhance farmers' incomes. The objectives of implementing this kind of schemes are conflicting. The decision to give up the reuse of TWW is discussed in many studies. Surveys cannot determine whether we should give up reuse or not, because if economic returns of TWWIP is low, the quantity of loaded treated effluent in the environment decreases. So, the first economic objective was not reached but the environmental one is somehow reached because of TWW reuse. Appropriate decisions depend on decision maker objectives that surveys do not consider directly. That is why, we need for resolving the multi-objective problem a systemic method considering decision maker(s) aspirations and objectives.

## Multicriteria Analysis Methods (MCA)

The Multicriteria Decision Analysis (MCDA) is a systemic method. It is a framework for resolving a multiple-objective problem represented by criteria in different kinds of units (ratio, ordinal and cardinal). MCA helps to evaluate decision options. Many researchers were attracted by this analysis for its easiness and practical aspects. Reaching an ideal solution and an optimal one, which used to be beyond easiness or possibility, is now resolved due to the use of MCA. The practicality of this analysis helped to resolve water resources problems. Water planning decisions is characterized by intangible outcomes, including items such as recreation, environmental and human health (Hajkowicz and Higgins., 2008). Many scientific researchers (Dunning et al 2000, Joubert 2003, Glug et al., 2000 and Nayak & Panda., 2001) qualify MCA as an effective tool for water management helping to make structural, auditable, transparent and rigorous decisions. In the following, the objectives of using MCDA, its components and case studies are outlined.

Dodgson et al., (2000) states that MCDA can be used to identify the most preferred option, rank options, list a limited number of options, or discern acceptable from non-acceptable actions. A classification of MCDA problems was presented by Malcczweski (1999) as follows:

- Multi-Objective Decision Making (MODM) versus Multi-Attribute Decision Making (MADM).
- Individual versus group decision maker problems.
- Decision under certainty versus decision under uncertainty.

The evaluation criteria are standards of judgments or rules aimed to rank alternatives according to their desirability. Criteria distinguish MADM from MODM. Objectives are derived from a set of attributes as measurable quantities reflecting by their value the degree to which a particular objective is achieved.

Roy (1985) classifies MCDA into three several families:

- Multi-attribute methods
- Outclassing methods
- Interactive methods.

The last family of methods, which is called also, interactive local judgment approach (Caballero, 1996) is the focus of our study. This approach is constituted of techniques based on interaction between the decision maker and the analyst. The decision is made in parallel with the evolvement of the discussion between the two actors; and the discussion protocol must be done in many steps, that is, alternating discussion and calculation. In each step the reaction of decision maker is noted and the decision making process changes with the proposition of other solutions and criteria weights. Interactive methods are often chosen in resolving action choice problems. The most important characteristics that are also presented as limits of the method are:

- Limited to choice problem,
- The number of actions is high,
- Concerns only one actor, the decision- maker.

Interactive methods are, however, appreciated in resolving real problems because of its practicability. They were valorized by new computing techniques as well as by the emergency of a new concept called Decision Aid Interactive System (DAIS). One practical method, a member of the interactive decision aid approach, is the Compromise Programming which will be discussed in the next section.

## Compromise Programming Method (CP)

Compromise Programming (CP) is an interactive method used to resolve a multiple linear objective problem (Zelney, 1973, 1974). Goicoechea et al., (1982) state that: CP identifies solutions which are closest to the ideal solution as determined by some measure of distance. The *compromise solutions* are those identified as being closest to the ideal one and it constitutes the *compromise set.*

To understand the meaning of those two concepts, let's consider the concept of ideal solution and the appropriate used distance.

The vector $z^* = \{z_1^*, z_2^*, \ldots, z_i^*\}$ defines the ideal solution and the $z_i^*$ is the solution of the following problem:

Max $z_i$

Subject to

$x \in X$

$i = 1,2,3,.., p$

In general, the ideal solution x* is not feasible; however, it is considered as a standard for evaluation of the attainable non-dominated solution. The set $z^* (x^*) = \{z_1^*(x^*), z_2^*(x^*), \ldots, z_i^*(x^*)\}$ define the non- dominated set (in objective space) consisted of only one point. The evaluation of non-dominated point consists in measuring how these points come close to the ideal solution. The distance measure (the most frequently used) utilized in Compromise Programming is the family of $L_p$ - metrics. It is defined in two equivalent ways:

$$L_s = \left[ \sum_i^p \alpha_i^s (z_i^* - z_i(x))^s \right]^{1/s}$$

*or*

$$L_s = \sum_i^p \alpha_i^s (z_i^* - z_i(x))^s$$

*where*

$1 \le s \le \infty$

$\alpha_i = criteria\ weights$

If we note the compromise solution as $x_s^*$, this solution corresponds to the resolution of the following program:

Min $L_s(x) = L_s(x_s^*)$

Subject to

$x \in X$

The final expression of Lp metric used in this chapter (after transformations, see Goicoechea et al., 1982, 235-237p) is as follows:

$$\min \left\{ L_s(x) = \sum_{i=1}^p \alpha_i^s \left( \frac{z_i^* - z_i(x)}{z_i^* - z_i^{**}} \right)^s \right\} = L_s(x_s^*)$$

Subject to

$x \in X$

where $z_i^{**} = \min z_i$ ($x \in X$ and i= 1,2,.., p)

## Compromise Programming Method: A Bit of History

Goiccoechea et al., (1982) present illustrative examples for discrete and continuous setting. Many other scientific studies were based on CP method, especially in resolving environmental and water resource management problems. Chang et al., (1999) used the grey compromise programming method in exploiting and sharing water reservoir optimization problem. Shiau and Lee, (2005) used CP in measuring the distance between optimal and critical period of a reservoir supply. They first modeled the supply system in Shihmen reservoir (in Taiwan), then measured the gap between supplies period using Lp metrics and the CP method. The CP method helped to measure the utility for different attributes in different units. These measures were exploited in evaluating water quality investment using cost-utility analysis (Hajkowicz et al., 2008).

The CP method was also used in resolving different problems such as environment optimization (Tzeng et al., 1991; Wu & Chang., 2004), planning country (Marshall & Homans., 2004; Baja et al., 2007; Pongthanapanich., 2006) and economic problems (Ballestro., 2000, Tolga & Alptekin., 2007, Isikar et al., 2007, André et al., 2008).

The mathematical concept of the CP method was widely discussed in many scientific works. They were combined to other mathematical tools to create new techniques such as the fuzzy CP (Dadashzadeh & Nimse, 2006., Bilbao-terol., 2006, Karnib., 2004., Pradanovic & Simonovic., 2003.), the grey CP (Lin & Liu, 2004., Li, 2002., Wu & Chang., 2004, Chang & Tseng., 1999.), the hierarchical CP (Escobar & Moreno-Jiménez, 1997.,2002.) and the meta-heuristic CP (Gagné et al.,).

In the next section, a real evaluation and decision aid problem concerning TWW were developed. The descriptive analysis and the CP method were used to decide the best alternatives for the concerned case study.

## CASE STUDY: EX-POST EVALUATION OF CEBALA TWW IRRIGATED SCHEME AND DECISION MAKING ABOUT THE PERIMITER FUTURE

To illustrate and understand the theoretical aspects presented in the last section, we opt for the Cebala irrigated scheme case. This work is compounded of 4 steps:

- we present the historic evolution of the perimeter since its TWW reuse activities started. Data and information were collected from Ministry of Agriculture and Hydrologic Resources (MAHR), Ministry of Environment and Sustainable Development (MESD), and annual and technical reports.
- we present the chosen perimeter after 15 years of activities. Data were collected from a survey carried out among 103 farmers and covering 41% of the whole surface perimeter. The sample and topics were chosen according to the theoretical aspects presented in the first parts of the chapter.
- The multicriteria problem concerning the perimeter future and how to choose best alternatives for enhancing the economic and social perimeter returns are presented.
- A mix of results from descriptive and multicriteria decision analysis is presented and analyzed to deduce best alternatives for the perimeter.

### Cebala TWW Irrigated Scheme History

The Cebala TWW irrigated scheme is 18km far from Tunis, the capital of Tunisia, in North Africa. The perimeter was implemented in 1989. It is integrated to the big project of fitting out the Madjerda bass valley. The neighbor's perimeter to Cebala's, Kalaat Landalous and Henchir Tobias were also planned to be irrigated with Madjerda water resource. The area of Cebala TWWIP is about 3200ha and it is the most important perimeter irrigated with TWW. Technical studies proved that the perimeter has many chances of success out of its water resources shortage. Few ground water wells were listed and the quality of the resource is not appropriate for farming (high salinity and nitrate composition, more important than TWW salinity and nitrates characteristics).

The implementation of Cebala perimeter was a good opportunity to receive TWW produced by three treatment plants. Steps localization (not far from the perimeter) and their important treated effluent volumes encouraged the government to create the project. The study was integrated in the northern waters director plan (SCET & CAN., 1981).

The most challenging technical problem was the quality of the soil. Indeed, the soil is clayey with bad structure favoring fodders, cereal and industrial farming. This kind of crops is authorized to be irrigated with TWW in the Tunisian legislation. The technical problem linked with Cebala soil characteristics and aquifer location was the implementation of a drainage system. The aquifer groundwater is located at 1m far from the surface. This resource has a high salinity (between 1.24 to 7.10g/l). Intensive irrigation may cause the emergency of aquifer ground water causing the destruction of all crops thereby.

Before the implementation of Cebala TWWIP, farming consists of large crops [wheat (22%), barley (24%), fodders – oats (24%) and feve (12%)]. The soil occupation is about 82% and the rotations depend on soil quality. Few quantities of fertilizers are distributed (50 to 80% of recommended doses). Only wheat was treated with chemical weed by a few number of farmers. Crop yields in this perimeter were low and the economic situation for most farmers was bad. In the northern Cebala perimeter, some farmers were connected to Medjerda canal and produced irrigated crops such as fruits and gardening. Other form of farming is bovine and ovine breeding. With many precautions, this kind of farming was not considered in the project analysis. Tunisian authority judged in 1981 that the sanitary risks are very high with this kind of production. They authorized only the fattening of young bulls; and they judged that the health risks of eating beef is lower than drinking milk of cows reared in a perimeter irrigated with TWW. The most salient rule that must be respected is not to let cows or sheep directly graze in the perimeter. The contact risk of the animal with the resource is high and may cause health problems for both animal and human. Another constraint that caused reservation is the absence of legislative context for this kind of farming. The risk of having milk with bad odor or color is high, which accounts for the need to make experimentation before authorizing this kind of production.

In reality, the agrarian and economic-social structure of the perimeter varies from the north to the south. The southern of the perimeter (Zone 1, 2, 3, 4, 5, 6) is characterized by small areas, other zones are characterized by large parcels. The most important and rich farmers are localized in the northern of the perimeter. Farming is also different from the north to the south because of soil quality. The Cebala perimeter is constituted by registered private ownerships or by attributed ownerships after the abolition of state-owned property. In 1981, the result of the survey showed that the average size of parcels was about 26 ha (SCET & CNEA, 1981). The sample chosen by the research department do not represent the real structure of the perimeter. The choice of the average indicator is not appropriate, either, because of extreme value. Cebala population had more than 1400 peoples in 1981. The family size average is 6 members per household. 54% of the population has more than 20 years. The average number of workers was about 2 persons. The most common working activity was farming. Currently, it has more than 7500 people. The number of farmers has involved since the perimeter implementation. The structure of the family has changed and the size average is 5 members per house hold. We identified 437 farmers (owners and tenants).

The total project cost was evaluated about 26 million dinars (in 1981). The actualization of this amount (with consideration of credit rate along the 30 passed years) is equivalent to 32 millions $. The investment was very important and the infrastructure project establishment lasted more than 8 years. The investment is composed of four technical modifications. We note the building of an important regulation reservoir, the implementation of a draining system (underground and open drainage systems), irrigation network and roads. The initiation of farming activities was supervised by specialized trainers and encouraged by subsidization. Total costs and depreciation have been evaluated for 15 years. The technical-economic study stipulates that after this period the perimeter needs to be rehabilitated. Thus, many costly improvements must be implemented in the Cebala irrigated scheme. A call order was launched by the AWRM in 2005 to improve drainage and network irrigation system. This kind of investment is launched after the first planned period. This fact helps to guarantee the sustainability of the perimeter investment.

Our point of view is different from decision makers'. The IRR is calculated for a limited period (in this case for 15 years). After this period we must automatically verify if this rate was reached first. Otherwise; we must evaluate if intangible targets were fully or partly realized. Finally, if the general equilibrium could not be attainable within 15 years, we should ask this question: Shall new rehabilitation investments help to promote TWW reuse in Cebala perimeter? Other important queries should also be raised, such as the future of the perimeter:

1. Shall we continue the irrigation with TWW, give up this experience or irrigate with conventional water resource?
2. If we continue the reuse of TWW, shall we continue with the same TWW quality, or improve it or may be pass to the third level treated effluent?

Many alternatives are possible but we need first an ex-post evaluation of the perimeter based on descriptive analysis. We propose then an appropriate multi-criteria solution(s) to promote the Cebala irrigated perimeter activities and TWW reuse. Our approach is based on data collection and analysis. This step of work is the most important in the multicriteria paradigm. Our aim is to compare and exploit the result of the two approaches and give the appropriate solution for the proposed problem.

## Descriptive Data Analysis

### Sample Choice

The survey was carried out among 103 farmers, representing 20.6% of farmers' population. The area concerned covers 1339ha, that is, 40% of the total area of the perimeter.

The sample was chosen according to two criteria:

1. Owner or tenant: The number of farmers is about 437 farmers distributed between owners (80%) and tenants (20%). The sharecropping was not the most widespread. That is why, we focus on those two kinds of farmers.
2. Geographic zone: The total number of geographic zone is 12. The zone number 10 was not connected to the irrigation network because of technical constraints.

The stratified method was opted for to choose and determine the sample. Table 1 gives an idea about the distribution of the sample betweens zones and which was based on the two foregoing criteria.

### Questionnaire

The questionnaire is composed of 5 parts:

1. A background section: This part is primordial to know personal information about the interviewee and it is the first step to put him at ease. Often, interviewee thinks that this kind of interview is to know information about their income. Thus, the interviewer must give an idea about the principle aim of the survey and explain that the final return is to help involve farming in the studied perimeter. This part may contain informa-

*Table 1. Distribution of the sample between the visited zones*

| Zones | 1 | 2 | 3 | 4 | 5 | 6 | 7 | 8 | 9 | 11 | 12 | total |
|---|---|---|---|---|---|---|---|---|---|---|---|---|
| Area (ha) | 53 | 66,5 | 100,5 | 48 | 206 | 125,5 | 292,5 | 246 | 20 | 42 | 139 | 1339 |
| Parcels number | 10 | 6 | 12 | 24 | 66 | 53 | 66 | 8 | 1 | 5 | 6 | 257 |
| Average size (ha) | 5, 3 | 11 | 8, 3 | 2 | 3,12 | 2,3 | 4,4 | 30 | 20 | 8,4 | 23 | 5,21 |
| Interviewers number | 6 | 4 | 9 | 14 | 21 | 14 | 22 | 4 | 1 | 4 | 4 | 103 |
| Number of owner | 6 | 4 | 8 | 10 | 18 | 11 | 19 | 3 | 1 | 2 | 3 | 85 |
| Number of tenant | 0 | 0 | 1 | 4 | 3 | 3 | 3 | 1 | 0 | 2 | 1 | 18 |
| Total | 6 | 4 | 9 | 14 | 21 | 14 | 22 | 4 | 1 | 4 | 4 | 103 |

tion about age, sex, family situation, home location (if it is on the perimeter or not) and, agrarian structure.

2. Water resources: This part of the questionnaire concerns TWW resource and drainage system. Questions concern the network of irrigation system and drainage, irrigation material and technical constraints. Questions may concern also the price of the resource, the appreciation of its quality and problems of rental charges payment. In our survey, we tried to internalize some externalities such as the perception of resource quality by proposing scenarios concerning the enhancement of TWW price. This price evolution corresponds to technical improvement of wastewater treatment techniques; therefore, it implies resource quality improvement.
3. Farming production: This part of the survey is the trickiest. In fact, questions concern farming returns and farmers' incomes. Responses are often hedged and do not reflect the real situation in the perimeter. Some farmers try to prove that the resource is not profitable for the perimeter, so they decrease their income declaration. For others, more implicated in farming activities, they push up their incomes just to show that the resource is very useful. As intimated before, the interviewer must learn more about the perimeter and its production performances. Farming popularizers monitor continuously farmers and have a wide idea about their production and their incomes. The interviewer must learn about farming activities and work in collaboration with popularizers. Important information must be also collected concerning fertilizers, livestock and dead stock. What is hard to be evaluated is the value of the second kind of resources. It is difficult to verify in a survey the real dates of acquisition and we often suppose that it is totally depreciated.
4. Social information: This part of the survey often concerns labor resources and costs. In our survey, we asked about health problems that may be linked to TWW quality. Other questions were asked and concern guidelines and sanitary precautions respect.
5. Perspective: As the Cebala perimeter is located in peri-urban region and face to the urbanization problems, we deemed useful for the researcher to learn more about farmers' attachment to the agricultural land and farming activities. We proposed to the interviewees to speak freely about their well-being evolution after introducing the resource. We asked also about the possibility of selling their lands in case of urbanization and how much they estimate the value of their properties.

Data collected from the survey were treated and bad responses were eliminated. The final data were classified into 3 strata (land size, mechanized farming culture or not, kind of farming, that is, with or without bovine breeding).

**Results and Analysis**

The survey carried out among 103 farmers helped:

1. to understand the functioning of the perimeter and the interaction between farmers and governmental agencies, along with the perception and the acceptance of the non-conventional water resource by farmers,
2. to trace the history of the perimeter from collected information and to compare it to those cited in MAHR technical reports,
3. to approximate the income of farmers from irrigated farming and to compare it to that of non- irrigated farming.
4. to develop a dashboard throwing light on advantages and disadvantages of irrigating with TWW.

In this section, we will scrutinize the collected data to reach our first aim: helping DM to decide on the future of Cebala TWWIP.

The data analysis demonstrates that interviewees' age varies between 20 to 60 years. Educational attainment is very low (37% of interviewees are illiterate and 38% had a primary education). 57% of interviewees belong to large families. Only two women were interviewed in the perimeter because often man manages the farm and the wife helps in some tasks. The labor is for 77% of interviewees, familial. This kind of social fabric is a rural characteristic in most emergent countries.

After the implementation of the project, roads were significantly improved. However, some enquiry was not possible to be carried out because of road quality, especially in case of rainfall. The problem of drainage is also raised by interviewees and concerns stagnant water (71% of interviewees), blocked drain (19% of interviewees), drains spaced (57%).

The TWW resource was conducted from three treatment plants with different quality, the mix of the resource is the cause of health problems (rash and itching problems) for 22% of interviewees. The comparison of pumped and consumed TWW volumes shows that losses exceed 31 to 60%. The price of the resource has varied only for 7 years as shown in Figure 3.

Since 2002, the governmental agency (CRDA) has opted for a lump sum payment by speculation. The price depends on the irrigation system adopted (board irrigation, asperity irrigation). It varies between 150 to 170 $ per year and speculation. For some fodders like Lucerne (low added value crops), the price does not exceed 100$. Figure 4 compares the prices of conventional resources exploited in neighboring perimeters. TWW price also gives an idea about the covering rate of operating charges by the overall TWW sales.

Figure 5 shows that the relation between TWW demand and rainfall is very strong and antagonist. This fact is also verified for the whole year. In fact, the most important demand is in summer were the rainfall is lower.

The sample chosen has revealed an important fact about agrarian structure. In fact, in the perimeter study (SCET & CNEA 1981), the indicator chosen is the average size of properties, that is, 26ha per owner. This indicator is very large; thus, the sample gives a wide idea about the land distribution and the real land concentration. The survey demonstrated that 4% of interviewed persons possess more than 50 ha and exploit more 35% of the total surface. Per contra, 52% of interviewees owns less than 5ha and exploit 13% of the total perimeter. The land concentration is unequal and Gini rate calculated from survey data is about 59%. Thus, the concentration is relatively high. The most important form of farms is property. Tenants are often breeders and also owners of parcels in Cebala TWWIP. Renting

*Figure 3. Price evolution since Cebola irrigated scheme was implemented prices in DT ($1=1.2DT)*

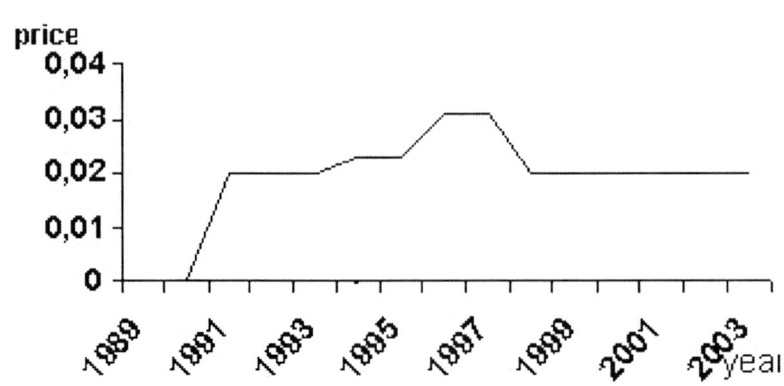

*Figure 4. Comparison of TWW price to other neighboring resources prices and its capability to cover exploitation charges*

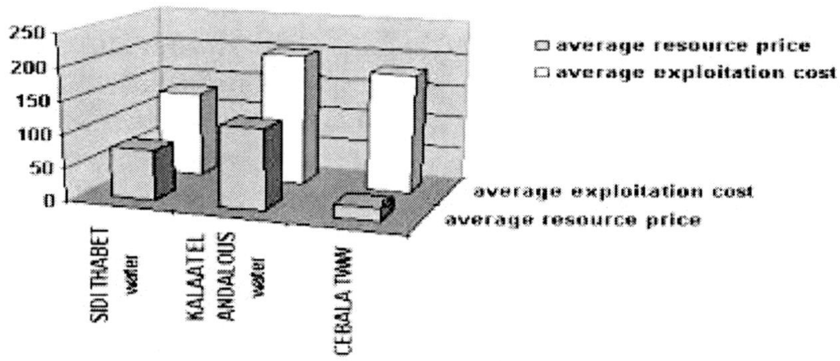


*Figure 5. Relation between rainfall and TWW consumption*

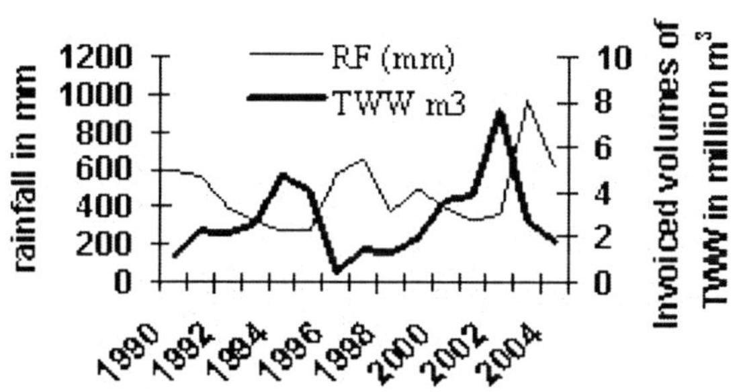


land helps them to satisfy their needs in fodders. Heirs represent 66% of interviewed farmers against 3% of new owners. The industrial crops (cotton and sugar beet) were tried in the Cebala TWWIP between 1991 and 1996 but problems of management and institutional context caused the experience to fail. The most important type of farming activity was the livestock which was the most unplanned. However, this form of farming is an ancient Cebala TWWIP practice. The government encouraged such by subsidizing farming activities. Table 2 presents farms typology.

*Table 2. Typology of the perimeter*

| *Strata* | Surface | Interviewee | interviewee kind | | Total |
|---|---|---|---|---|---|
| | | | owner | tenant | |
| Stratum 1 | < 5 ha | number | 38 | 10 | 48 |
| | | % | 36,9% | 9,7% | 46,6% |
| Stratum 2 | [5 ha, 10ha[ | number | 15 | 3 | 18 |
| | | % | 14,6% | 2,9% | 17,5% |
| Stratum 3 | [10 ha, 20 ha[ | number | 20 | 1 | 21 |
| | | % | 19,4% | 1,0% | 20,4% |
| Stratum 4 | [20 ha, 50 ha[ | number | 9 | 3 | 12 |
| | | % | 8,7% | 2,9% | 11,7% |
| Stratum 5 | >50 ha | number | 4 | | 4 |
| | | % | 3,9% | | 3,9% |
| Total | | number | 86 | 17 | 103 |
| | | % Surface | 83,5% | 16,5% | 100% |

As indicated before, data were classified into 3 strata: land size, mechanized farming culture or not, kind of farming (with or without bovine breeding). So, we distinguish 2 kind of crops system:

1. System of production 1: Cereals + Fodders.
2. System of production 2: Cereals + Fodders + bovine breeding.

Another important criterion explaining the differences in farmers' incomes is the owning of its dead stock or not and it is symbolized by $M_{ij}$ for the first case and by $NM_{ij}$ for the second situation. The index i corresponds to the strata number and the index j corresponds to the production system.

The sample becomes distributed as follow:

Number **4** in Table 6 means that we interviewed 4 farmers owning a parcel whose size is between 10 to 20ha. These farmers produce only cereals and fodders, and tend to rent their mechanical machine. The most valuable information about crops production and income concerns crops distribution between parcels. The selling price of farming products irrigated with TWW was for most responders under-valuated. Thus, to avoid any error in calculating the farmer income, we considered prices of crops production and input cost furnished by the popularizer in the Cebala perimeter. The approximated income calculated from crops offer and declared by interviewers is presented in intervals per hectare (ha) per year [Inc min, Inc max].

In the first part of the Table we calculate the income for interviewees with consideration of the new resource TWW. The size of the sub-sample is 37 interviewees. So, only 35% of the total interviewees are not interested by rearing activities. In that case, we consider its price as a cost and add new crops production such as sorghum or Lucerne. Other costs such as fertilizers are reduced. As can be seen, the last strata were eliminated. In fact, the contact with farmers owning more than 50ha was not easy; they refused to give us any idea about their farming system and we collected only complaints concerning TWW. The analysis of this Table shows that the resource caused an enhancement in farmers income, above all for little farms (less than 5 ha). This demonstrates that labor factor is more important than capital one. Thus, the theory of decreasing returns to scale (Ricardo, 1817) is verified. The negative value is often the cause underlying the giving up of farming in the Cebala perimeter or the lending of parcels. The main cause of farming abandon is living out of the perimeter and having other work. So, the presence of manpower is the most important factor guaranteeing the farming success. Another important factor which may cause such abandon is the non-possession of regular title deed. In that case, the access to governmental facilities and subsidizing is not possible; furthermore, financial credits for modernizing farming activities are prohibited. This fact is documented in the second kind of farming system which corresponds to producing cereals and fodders and

*Table 3. Sample distribution*

| Surfaces | System 1 | | | | System 2 | | | |
|---|---|---|---|---|---|---|---|---|
| | **M*** | **$n_i$** | **NM**** | **$n_i$** | **M** | **$n_i$** | **NM** | **$n_i$** |
| Strata 1 | M11 | 6 | NM11 | 11 | M12 | 8 | NM12 | 24 |
| Strata 2 | M21 | 6 | NM21 | 1 | M22 | 3 | NM22 | 11 |
| Strata 3 | M31 | 1 | NM31 | **4** | M32 | 10 | NM32 | 2 |
| Strata 4 | M41 | 5 | NM41 | 1 | M42 | 4 | NM42 | 2 |
| Strata 5 | M51 | 2 | NM51 | 0 | M52 | 2 | NM52 | 0 |

*Table 4. Farmers' incomes per strata, hectare and dead stock property (System1) (Updated Prices in $/ Year/ha)*

| **Strata** | **Surface (ha)** | **Intensification rate** | **Farmer income:System 1 (irrigated crops with TWW)** | | | | | |
|---|---|---|---|---|---|---|---|---|
| | | | **M** | **max** | **min** | **NM** | **max** | **min** |
| Strata 1 | 5 | 1,3 | M11 | 1330 | 1640 | NM11 | 1117 | 194 |
| Strata 2 | 10 | 1,1 | M21 | 1247 | 1538 | NM21 | 1062 | 202 |
| Strata 3 | 20 | 0,8 | M31 | 1034 | 1275 | NM31 | 890 | 168 |
| Strata 4 | 50 | 0,6 | M41 | 1008 | 1243 | NM41 | 865 | 134 |
| Strata | Surface (ha) | Intensification rate | Farmer income: System 1 (non irrigated crops) | | | | | |
| | | | M | max | min | NM | max | min |
| Strata 1 | 5 | 0.9 | M11 | 740 | 913 | NM11 | 385 | -30 |
| Strata 2 | 10 | 0.75 | M21 | 606 | 747 | NM21 | 469 | -48 |
| Strata 3 | 20 | 0,65 | M31 | 493 | 608 | NM31 | 375 | -69 |
| Strata 4 | 50 | 0,6 | M41 | 449 | 554 | NM41 | 339 | -70 |

breeding bovine. Three different kinds of breeders can be underscored:

1. Small breeder: owning at least 10 cows
2. Medium breeder: owning between 11 and 20 cows
3. Cattle rancher: Owning more than 20 cows

The third form of breeders is not the most common in the perimeter. That is why, we focused on the two first classes. The distribution of bovine heads per land area can be resumed in Figure 6 above.

As blatantly shown, the approximate income in the second system is higher than the first one, especially for small farmers. breeding bovine have enhanced their income and improved their well-being. The most important guarantee of the experience success is the labor factor. The technological progress (mechanization) is an important factor, too.

The growth of income can be twice if the farmer is the owner. Lending parcels is the only appeal for small farmers owning more than 10 cows. This fact helps to produce fodders to feed cows.

*Figure 6. Distribution of bovine heads per hectare*

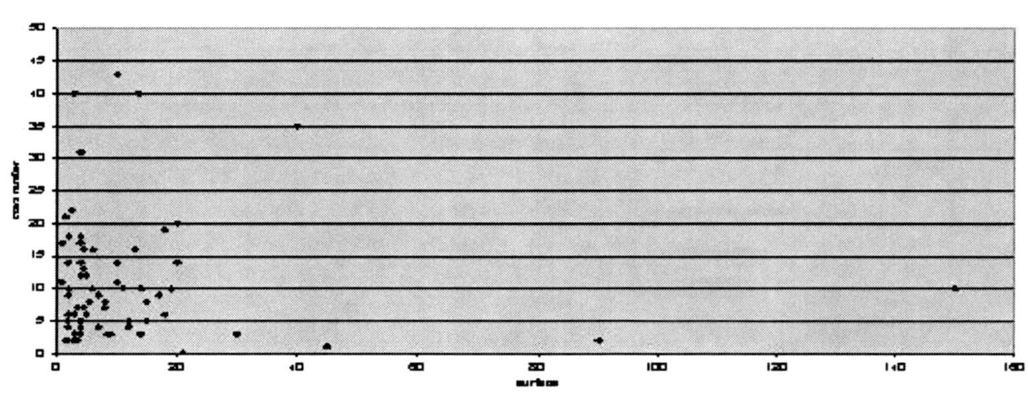

In 2003, drought has damaged fodder production in the southern Tunisia and the Cebala perimeter helped to alleviate this crisis by producing and selling fodders for the concerned Tunisian regions. It is noteworthy that our analysis was performed in a sectional instant. The continuous monitoring reduces the range of variation of the incomes (Table 5). Some costs are not considered in the evaluation process amid credit charges and mechanical material depreciation. Others outputs not considered in this study consist in animal fertilizer and feeder.

In this part of analysis we focus on positive and negative effects of introducing TWW in the Cebala perimeter. The evaluation of the spread between those effects is not easy to be measured and the total evaluation of effects that we call in macro economic analysis "the global equilibrium" cannot be gauged with accuracy.

Subjective factors require the weighting of some effects because of their importance for the decision maker. In case of governmental investment, social effects such as employment or equity are very important and widely explain the choice of investment with low economic returns but with high social returns.

Table 6 resumes negative and positive effects of TWW reuse. As we can observe the gap between the two kinds of effects cannot be measured. The evaluation of each effect is linked to subject appreciation from the DM.

In the last part of this section, we propose a systemic method based on multicriteria approach. This method provides a decision aid for the decision maker with respect to conflicting criteria and high level of aspiration

**Multicriteria Decision Making Analysis**

Three analytical steps are demarcated here, as illustrated below:

*First step*: we first present the problematic and the objective of this study and then we define all concerned actors (decision maker (DM), analyst and stakeholder) and their tasks. In this step we specify the decision maker objectives and then subdivide them to sub-objectives. The DM attributes a weight to each sub-objective and then chooses appropriate criteria (in different criteria) to each objective. The role of the analyst is to present attainable solutions for the problem and then evaluate the contribution of each solution for each criterion. The result of this step is presented

*Table 5. Farmers' incomes per strata, hectare and dead stock property (System2) (Updated Prices in $/ Year/ha)*

| **Strata** | **Surface (ha)** | **Intensifi-cation rate** | **Farmer income:System 2 + (small breeder)** | | | | | |
|---|---|---|---|---|---|---|---|---|
| | | | **M** | **max** | **min** | **NM** | **max** | **min** |
| Stratum 1 | 5 | 1,3 | M11 | 1841 | 407 | NM11 | 1686 | 220 |
| Stratum 2 | 10 | 1,1 | M21 | 1532 | 386 | NM21 | 1402 | 255 |
| Stratum 3 | 20 | 0,8 | M31 | 1489 | 310 | NM31 | 1489 | 349 |
| Stratum 4 | 50 | 0,6 | M41 | 1349 | 278 | NM41 | 1349 | 305 |
| Strata | Surface (ha) | Intensifica-tion rate | Farmer income: System 2 + (medium breeder) | | | | | |
| | | | M | max | min | NM | max | min |
| Stratum 1 | 5 | 0.9 | M11 | 2678 | 1025 | NM11 | 1348 | 618 |
| Stratum 2 | 10 | 0.75 | M21 | 2257 | 1228 | NM21 | * | * |
| Stratum 3 | 20 | 0,65 | M31 | 2382 | 1432 | NM31 | 2382 | 1394 |
| Stratum 4 | 50 | 0,6 | M41 | * | * | NM41 | * | * |

*Table 6. Negative and positive effects of TWW reuse*

| Characteristics | Details | Advantages | drawbacks |
|---|---|---|---|
| Location | | - Situated in peri-urban region, not far from sanitation network and treatment plants.<br>- Easy access of Cebala crops to great markets of the capital. | - Near to the capital: migration of farming labor to other economic sectors: industrial, commercial and service,<br>Close to a perimeter irrigated conventional re-source: farmers living in northern Cebala TWWIP are frustrated because they do not have access to this resource. |
| Physical re-sources | Climate | - in drought period, the resource has helped cereals and fodders to growth | - In wet period farmers drop the resource by the way a growth of perimeter charges management. |
| | Sol | | Clayly soil causing stagnation of TWW resource and in absence of drainage system, farming is not possible. Farmers confirm that in wet period, it is impossible to wear boats because of soil quality, which implies high health risks. In some zones of the perimeter, the soil has an equilibrated struc-ture. This kind of soil is convenient to produce gardening. Unfortunately, this kind of production is prohibited in the Cebala TWWIP. |
| | Hydro-geology | - The salinity of TWW is lower than aquifers one | - Irrigation favors aquifer rise than the accumulation of salt on surface. |
| Resources Physiques | Water resource | - TWW is rich in fertilizers<br>- Low salinity compared to medjerda resource salinity | - TWW with high bacterial charges<br>- Restrictions on crops production due to TWW quality<br>- Banned water wells because of TWW presence in the perimeter |
| | Humans | - Available family labor factor | - few number of qualified technician<br>- low motivation of farmers to involve and to grow their production in the perimeter,<br>- many farmers want to leave the perimeter and sell their parcels (especially young ones) |
| | Agrarian Structure | | - Great fragmentation of farming lands caused by inheritance.<br>- non-possession of regular title deed. |
| Economic factor | Resource demand | -The is strongly demanded in the southern perimeter. | - Demand linked to climatic vagaries. |
| | Resource offer | - Available resource, | - bad quality<br>- low flow in summer |
| | Resource price | - Low price (0,020DT) representing the 1/7 of conventional resource price | - low price of TWW causing management problems for the governmental agency because it cannot cover exploitation and load charges |
| | Vegetarian Production | - margin per ha is higher than before imple-menting TWWIP,<br>- High crops margin for small farmers | - Not diversified production, limited to cereals, and fodders<br>- Need of breeding system modernization. |
| | Bovine breeding | - Successful experience<br>- High beneficiary margin compared to sys-tems without breeding | - Breed bulls is not yet developed<br>- The number of rancher is low<br>- Need of modernizing the breeding system |

in an evaluation matrix. Such a result will be transformed into pay off matrix useful for computing analysis. The compromise programming (CP) ranks solutions that must be presented to the decision maker. If the DM judges that the solution is not conform to his aspiration and hopes, the entire decision making process should be revisited and criteria, weights and solutions can be modified.

*Second step*: generally, choosing to invest in agricultural project is relevant to governmental agencies. Their first aim is to enhance farmers' income and create economic dynamism in a deprived area. So, the objectives of farmers such as enhancing their income and resolving employment problems are respected and considered in the DM objectives. We propose in this section to consider the stakeholder (in our case the farmer) as a DM and restart the decision making process with him. Objectives, and therefore criteria and weights are also revisited. The first target of this process is to compare the solutions in both systems and then measure the agreement between the two actors.

*Third step*: In our decision making system, we consider two cases: with or without criteria weights. The result of this distinction is to evaluate the influence of DM preferences in a decision making system.

## First Decision Making System: AMHR Is The Decision Maker.

The logic of decision making in the proposed problem can be resumed in Figure 7.

Five solutions can be distinguished; they are built around the resource quality. The contact is with the chosen DM, in our case, a general director in the Tunisian Ministry of Agriculture and Hydrologic Resources. The DM contributes to the decision making system by choosing objectives, criteria and weights as presented in Table 7.

The work comprises three substantive steps:

1. Choice of objectives and sub-objectives: Indeed, the analyst proposed and discussed objectives with the DM, who accepted some and refuted others. He can also suggest other objectives which he sees pertinent and useful in the decision making process. Four objectives resulted from this interactive process, including:
   a. *Increase of the project profitability.*
   b. *Insurance of farmer well-being.*
   c. *Preservation of the environment.*
   d. *Guarantee of project sustainability*

Each objective is, in turn, subdivided into sub-objectives. This process contributes to the definition of the DM targets and the enhancement of the criteria selection.

Criteria selection: The analyst and the DM choose appropriate criteria to each sub-objective. For example, corresponds to:

*Figure 7. Logic of decision making*

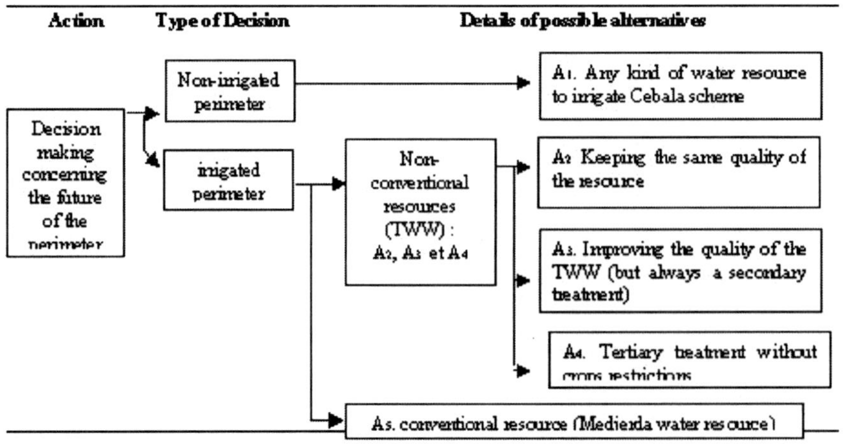

*Table 7. Objectives, criteria and weights of MAHR decision maker*

| Objectives | Sub-objectives | Criteria | Unites | Value | $w_1$ | W2 |
|---|---|---|---|---|---|---|
| *Increase the profitability of the project* | Maximize project return | 1. Internal return rate IRR. | % | [0,1] | 50 | 30 |
| | Maximize MAHR profit | 2. recovery rate | % | [0,1] | | 6 |
| | Maximize farmers' income | 3. farmers' income | (updated in $/year/ha) | Cardinal | | 10 |
| | Create new economic activities | 4. new labors | A-G | Ordinal | | 4 |
| *Insurance of farmer well-being* | Maximize the employment rate | 5. number of working days | (j/ ha/r an) | Cardinal | 26 | 8 |
| | Minimize sanitary risks | 6. sanitary risks | A-G | Ordinal | | 16 |
| | Maximizes the population levels | 7. urbanization | A-G | Ordinal | | 2 |
| *Preservation of the environment* | Minimize health risks | 8.illeness vectors (mosquitoes) | A-G | Ordinal | 10 | 2 |
| | Minimize bad odors | 9. bad odors | A-G | Ordinal | | 2 |
| | Maximize the preservation of environment from TWW loading | 10. loading TWW effects in the sce | A-G | Ordinal | | 6 |
| *Guarantee of project sustainability* | Minimize effects of reusing TWW in the long term | 11. soil Stalinization risk | A-G | Ordinal | 14 | 4 |
| | | 12. suspended matter | A-G | Ordinal | | 4 |
| | | 13. distribution network corrosion | A-G | Ordinal | | 6 |

$$\text{Recovery rate} = \frac{\text{farmers' charges (price of the TWW)}}{\text{operating and maintenance charges in the perimeter.}}$$

Against such a criterion we measure the contribution of farmers in paying the perimeter charges. Relevant in this criterion are its ratio scale and its value which varies in this interval [0, 1].

3. Weights attribution. We start first with ascribing a weight (W1) to each objective based on its importance. For example, the DM attributes the weight of 50% to the first objective – Increase of the project profitability – and only 10% to the environmental preservation objective. He judged that the first objective is the most important and its realization guarantees 50% of the overall output of the DM system. As we can observe, the attribution of weights is grounded on the DM personal perceptions. The sum of W1 (per percent) is equal to 1.

$$\sum_{i=1}^{4} W1_i = 1$$

Then, we consider each objective individually in such a way that its weight represents 100% ; thereafter, we subdivide it in sub-weights (W2). For example, we ask the DM what weight he would attribute to the first sub-objective per 50. This operation is replicated for the other sub-objectives; the sum of W2 (per percent) is equal to 1.

$$\sum_{j=1}^{13} W2_j = 1$$

Depending on the objectives (minimizing or maximizing) A, B,…,G corresponds to the best or the worst value. For example, the letter **A,** corresponding to the first solution and criterion number 11, is logical because the first alternative proposes to stop irrigation in the Cebala perimeter and in that case there is few risks of aquifer emergency. The farmer's income in DT/ha for the third solution (A3) will be **2671** equal to the income in the second solution. In fact, improving bacteriologic quality of TWW does not enable farmers to produce gardening than to increase his income.

Table 9 standardizes values in Table 8 to eliminate the effects of the different units. This standardization is performed by computing tool (computing program drawn on C++ universe and based on CP optimization program). This tool helped to calculate distances between possible solutions and ideal one for both cases, that is, with or without weights. Let's again observe the mathematical formula furnished in page 11. We start with giving negative values to objectives that must be minimized; then, we add two columns (maximum and minimal value for each criterion); next, we calculate for each action (solution) how far it is from the ideal solution which gives the best value for each criterion. We varied then the level of aspiration "s" and we rank solutions as illustrated in Table 10.

The marked values represent the most nearest action to the ideal one. We observe that if we do not consider the DM preferences (equal weights), the most preferred action is alternative A4 which corresponds to reusing tertiary TWW with no farming restrictions (possibility of producing gardening). With high level of aspiration (s > 5), the worst solution becomes stopping the irrigation (A1). This fact explains the importance of the resource in the perimeter, irrespective of its quality.

In case of considering different weights of criteria, the best solution becomes irrigation with a conventional resource (A5). In fact, DM attributed high weights for social and farmer's income, criteria Tertiary treated wastewater is ranked in the second place. Stopping irrigated farming remains the least satisfactory solution for the perimeter. The irrigation with Medjerda resource depends on technical feasibility of the solution and demand. Parcels near this conventional resource are often large (more than 10ha) and their owners prefer to produce cereals because of inputs charges and above all the absence of manpower in the perimeter if they choose irrigated crops production.

*Table 8. Evaluation matrix of MAHR decision maker*

| **Criteria** | **ALTERNATIVES** | | | | |
|---|---|---|---|---|---|
| | **A1** | **A2** | **A3** | **A4** | **A5** |
| 1. Internal return rate IRR. | 0 | 6,71 | 6,71 | 10 | 12 |
| 2. Recovery rate | 0 | 16 | 23 | 47,5 | 119 |
| 3. Farmers' income | 1320 | 2671 | **2671** | 5920 | 7153 |
| 4. New labors | E | D | C | B | A |
| 5. Number of working days | 7 | 10 | 10 | 60 | 60 |
| 6. Sanitary risks | A | G | D | B | A |
| 7. Urbanization | E | D | C | B | A |
| 8.Illeness vectors (mosquitoes) | A | G | F | B | B |
| 9. Bad odors | A | G | E | B | A |
| 10. Loading TWW effects on the sea | G | D | C | A | G |
| 11. Soil Salinization risk | **A** | B | B | D | D |
| 12. Suspended matter | A | E | D | C | B |
| 13. Distribution network corrosion | A | D | D | B | B |

*Table 9. Payoff matrix of MAHR decision maker*

| CRITERIA | ALTERNATIVES | | | | |
|---|---|---|---|---|---|
| | A1 | A2 | A3 | A4 | A5 |
| 1. Internal return rate IRR. | 0 | 28000 | 28000 | 42000 | 56000 |
| 2. Recovery rate | 0 | 6000 | 8500 | 18000 | 45000 |
| 3. Farmers' income | 10000 | 20000 | 20000 | 45000 | 54000 |
| 4. New labors | 8000 | 12000 | 16000 | 20000 | 24000 |
| 5. Number of working days | 5600 | 8000 | 8000 | 48000 | 48000 |
| 6. Sanitary risks | 16000 | 112000 | 64000 | 48000 | 32000 |
| 7. urbanization | 6000 | 8000 | 10000 | 12000 | 14000 |
| 8. Illness vectors (mosquitoes) | 2000 | 14000 | 12000 | 4000 | 4000 |
| 9. Bad odors | 2000 | 14000 | 10000 | 4000 | 2000 |
| 10. Loading TWW effects in the sec | 6000 | 24000 | 30000 | 42000 | 6000 |
| 11. Soil Stalinization risk | 6000 | 12000 | 12000 | 24000 | 24000 |
| 12. Suspended matter | 0 | 16000 | 12000 | 8000 | 4000 |
| 13. Distribution network corrosion | 0 | 18000 | 18000 | 12000 | 12000 |

*Table 10. CP analysis results in case of equal & different weights (1st DM)*

| Equal weights | | | | |
|---|---|---|---|---|
| s = 1 | s= 2 | s = 5 | s = 10 | s = ∞ |
| Action LP – DI | Action LP – DI | Action LP – DI | Action LP - DI | Action LP – DI |
| 5 3.25<br>4 4.38<br>1 7.00<br>3 8.44<br>2 10.47 | 4 1.56<br>5 1.60<br>3 2.46<br>1 2.64<br>2 2.99 | 4 1.04<br>5 1.16<br>3 1.26<br>1 1.47<br>2 1.48 | 4 1.20<br>3 1.10<br>5 1.07<br>2 1.21<br>1 1.21 | 4 1.00<br>3 1.00<br>5 1.02<br>2 1.05<br>1 1.06 |
| Different weights | | | | |
| s = 1 | s= 2 | s = 5 | s = 10 | s = ∞ |
| Action LP – DI | Action LP – DI | Action LP – DI | Action LP - DI | Action LP – DI |
| 5 18.00<br>4 30.65<br>3 61.47<br>1 66.00<br>2 74.31 | 5 8.73<br>4 11.80<br>3 22.12<br>2 26.58<br>1 34.00 | 5 6.03<br>4 7.90<br>3 15.36<br>2 17.96<br>1 30.04 | 5 6.02<br>4 7.52<br>3 15.00<br>2 16.69<br>1 30.00 | 5 6.00<br>4 7.50<br>3 15.00<br>2 16.04<br>1 30.00 |

## Second Decision Making System: Farmer Is The Decision Maker.

To learn more about the farmer acceptance of the TWW and the effect of DM decisions on stakeholders, we simulate the case that farmers can be a DM. Their objectives are represented by criteria. To each criterion, farmers attribute a weight. To this end, we choose a group of 20 farmers and we start the interactive process individually. All information about weights was aggregated then divided by 20 to obtain an average weight for each criterion. The final result of this investigation is resumed in Table 11.

Table 12 sketches out the contribution of each solution to each criterion. Some criteria are similar to those chosen by MAHR (maximizing income or minimizing illness vectors). The price

*Table 11. Objectives, criteria and weights of farmer DM*

| Objectives | Sub-objectives | Criteria | Units | Value | $W_1$ | $W_2$ |
|---|---|---|---|---|---|---|
| *Increase project profitability* | 1. low price of the resource | 1. Price | ($/m³) | Cardinal | 38 | 8 |
| | 2. Minimize fertilizer quantities | 2. Fertilizer quantities /ha | Kg/ ha | Cardinal | | 4 |
| | 3. Maximize farmer income | 3. Maximum income | $/year/ha | Cardinal | | 12 |
| | 4. Maximize land value | 4. Land value | M$/ha | Cardinal | | 14 |
| *Assure social well being* | 5. Maximize employment opportunity | 5. Number of working day | (j/ ha/ an) | Cardinal | 42 | 6 |
| | 6. Minimize sanitary risks | 6. Sanitary risks | A-G | Ordinal | | 14 |
| | 7. Minimize health risks | 7. Illness Vector mosquitoes | A-G | Ordinal | | 14 |
| | 8. Minimize bad odors | 8. Bad odors | A-G | Ordinal | | 8 |
| *project sustain-ability* | 9. maximize project sustainability | 9. Urbanization influence | A-G | Ordinal | 20 | |

*Table 12. Evaluation matrix of farmer DM*

| CRITERES | ALTERNATIVES | | | | |
|---|---|---|---|---|---|
| | A1 | A2 | A3 | A4 | A5 |
| 1. Price | 0 | 0,025 | 0,025 | 0,047 | 0,148 |
| 2. Fertilizing quantities /ha | 185 | 0 | 0 | 0 | 425 |
| 3. Maximum income | 1320 | 2671 | **2671** | 5920 | 7153 |
| 4. Land value | 62 | 39 | 39 | 62 | 148 |
| 5. Number of working day | 7 | 10 | 10 | 60 | 60 |
| 6. Sanitary risks | A | G | D | C | A |
| 7. Illness Vector mosquitoes | A | E | D | C | B |
| 8. Bad odors | A | G | D | C | A |
| 9. Urbanization influence | D | C | C | B | A |

of the resource TWW in action 4 (A4) is 0.038 DT/m³. It was calculated with the contingent valuation method using indirect questions (open-ended questions). The fourth criteria corresponds to the price of the hectare in Cebala irrigated perimeter and neighbors'. What is intriguing in this criterion is the depreciation of the hectare price after introducing the resource. In fact, the restrictions on crops selection are the cause of price decreasing. That is why, farmers presume that improving TWW quality and eliminating restrictions will enhance the price of the ha. Other factors that are found to impact on price are its localization in the perimeter near or far from principle roads and urbanization growth. This factor influences the sustainability of the project. Currently, the perimeter is protected by legislative laws but the growth of urbanization is important, mostly after the implementation of an hyper market close to Cebala perimeter.

The CP method based on LP metrics helped to rank solutions by calculating distances between the ideal solution and the proposed one. The final results are presented in Table 13.

As can seen, if we do not consider farmers' preferences (equal weights), the most preferred solution corresponds to reusing wastewater treated at the tertiary level. In fact, the price of the resource is lower compared to the conventional water one. In addition, the farmer's income

*Table 13. CP analysis results in case of equal & different weights (2nd DM)*

| Equal weights | | | | |
|---|---|---|---|---|
| s = 1 | s= 2 | s = 5 | s = 10 | s = ∞ |
| Action LP – DI | Action LP – DI | Action LP – DI | Action LP - DI | Action LP – DI |
| 5 2.33<br>4 3.02<br>1 4.26<br>3 5.52<br>2 5.65 | 4 1.26<br>5 1.45<br>1 1.96<br>3 2.10<br>2 2.22 | 4 0.87<br>5 1.14<br>3 1.26<br>1 1.27<br>2 1.32 | 4 0.82<br>5 1.07<br>3 1.10<br>1 1.12<br>2 1.13 | 4 0.81<br>5 1.02<br>3 1.02<br>1 1.03<br>2 1.03 |
| Different weights | | | | |
| s = 1 | s= 2 | s = 5 | s = 10 | s = ∞ |
| Action LP – DI | Action LP – DI | Action LP – DI | Action LP - DI | Action LP – DI |
| 5 16.67<br>4 39.91<br>1 51.19<br>2 67.21<br>3 68.34 | 5 10.09<br>4 17.44<br>1 26.71<br>3 27.39<br>2 27.43 | 5 8.15<br>4 12.29<br>3 17.37<br>2 17.40<br>1 20.53 | 5 8.00<br>4 11.56<br>3 15.42<br>2 15.42<br>1 20.02 | 5 8.00<br>4 11.42<br>3 14.31<br>2 14.31<br>1 20.00 |

is higher than the one obtained when the perimeter is irrigated with wastewater treated at the secondary level. The quality of the resource proposed in the fourth solution A4 satisfies health and sanitation standards. Obviously, the actual situation of the perimeter (irrigated with wastewater treated at the secondary level), which corresponds to action 2 (A2) is ranked at the last position. This situation is different in case of considering different weights. The worst solution becomes stopping the irrigation in the perimeter. In that case, economic criteria outweigh the other criteria. The attachment of farmers to their irrigated farming activities is great, especially for bovine breeders. Another important deduction is that farmers favor the exploitation of conventional resource if it exists. That is why, action 5 occupies the best position and it is the most preferred one for farmers.

The last Table (Table 14) demonstrates the convergence of solutions with consideration of the two scenarios (MAHR and farmers DM) and the two cases (with or without considering weights).

The DM chosen to represent MAHR interests knows perfectly farmers' preferences and problems; that is why, the final solution for both DM has converged for the same alternative. This is a logical result of decision making system in choosing governmental investment.

*Table 14. Ranking solution for both DM and different kinds of weights*

| | Equal weights | | | | | | Different weights | | | | | |
|---|---|---|---|---|---|---|---|---|---|---|---|---|
| | P = 1 | | P = 10 | | P = ∞ | | P = 1 | | P = 10 | | P = ∞ | |
| Ranking of alternatives | farmer | MA | farmer | MA | farmer | MA | farmer | MA | farmer | MA | farmer | MA |
| | 5 | 5 | 4 | 4 | 4 | 4 | 5 | 5 | 5 | 5 | 5 | 5 |
| | 4 | 4 | 5 | 3 | 5 | 3 | 4 | 4 | 4 | 4 | 4 | 4 |
| | 1 | 1 | 3 | 5 | 3 | 5 | 1 | 3 | 3 | 3 | 3 | 3 |
| | 3 | 3 | 1 | 2 | 1 | 2 | 2 | 1 | 2 | 2 | 2 | 2 |
| | 2 | 2 | 2 | 1 | 2 | 1 | 3 | 2 | 1 | 1 | 1 | 1 |

If the DM chooses to introduce a conventional resource, many parameters must be considered:

- The size of parcels near the Medjerda canal (in the northern of the perimeter) is large. Though their land or soil quality is better than parcels situated in the southern perimeter, farmers prefer to produce cereals and fodders because of manpower scarcity
- The irrigation experience in neighboring perimeters (Kalaat Landalous and Henchir Tobias) showed that farmers are reluctant to pay water resource charges. Governmental agencies often remove credits to encourage farmers to continue their farming activities. So, irrigating with conventional resource in Cebala perimeter does not entail that farmers will enhance their productivity and pay their duties in due time.
- The soil quality in the southern perimeter, where familial man-power is abundant, is not suitable to gardening. In addition, the aquifer situation (1m far from the soil) may cause saline water stagnation, thus crops destruction.
- The most profitable farming activity is dairy production. This activity does not require conventional resource, except for watering animals. Self-sufficiency in fodders is guaranteed by the TWW resource.

The treatment of wastewater at the tertiary level is costly. The Tunisian National Office of Sanitation (ONAS) within the ministry of environment and sustainable development is the only agency concerned with this kind of investment (to pass from the secondary to the tertiary treatment). The discussion with DM in this agency demonstrates that the actual quality of treated effluent conforms to the loading standards. So, this kind of investment is not urgent nor does it represent the first target. Another point raised is the TWW charges. Actually, farmers pay 0.020DT/$m^3$ to the MAHR. When the quality of the resource will be improved the question that could be raised concerns the payee, that is, MAHR or ONAS. Otherwise, what are the compensations to encourage ONAS to treat wastewater at the tertiary treatment? Decision making process in this kind of problem corresponds foremost to a group. We must stop considering farmer as a stakeholder but rather as a DM, and ONAS must be involved in the decision making system as one of the important actors.

## CONCLUSION

The agricultural reuse of treated wastewater helps to reduce the loaded TWW quantities and above all resolves the problem of water scarcity. This kind of resource is often managed by governmental agencies, and decisions concern not only implementation or rehabilitation of perimeters irrigated with TWW, but also stopping the TWW reuse. Within their decision making process, along with the economic and technical constraints, governmental agencies give much breed to social and environmental parameters which are hard to evaluate. The study presented in this chapter concerned the rehabilitation of a perimeter irrigated with TWW in an emergent country. The objective of our work was not to find the best solution to implement or to rehabilitate a perimeter but to discuss about its future. To this end, we used two approaches: the descriptive analysis and the multi-criteria decision making analysis. We simulate 4 cases as a possible evolution of an ancient perimeter which is suffering from under-exploitation and low returns. Those simulations plus the possibility to continue irrigation with the same quality of the resource are proposed as solutions to be ranked to help the DM to take decision. We focused first on the choice of the sample and on collecting data. The descriptive analysis presented a base for the decision making framework. An informative table was presented sketching out on the one hand tangible and intangible effects and on the other hand negative and positive aspects. The multicriteria tool used to attain our objective was the Compromise Programming (CP) which measures distances between ideal solution and proposed actions using the Lp metrics.

The analysis of results given by both procedures suggests that descriptive analysis and multicriteria methods are complementary. In fact, one of the most important steps of the multi-criteria decision making paradigm is the collection and the selection of data. Analyzing the descriptive data in-depth is not so much important in an interactive decision making system because we suppose that DM is well-informed about the problematic as well as all parameters.

Our experimental experience proved that the descriptive data analysis helps the analyst to understand better the perimeter functioning, and more importantly to discuss the choice of objectives, criteria and weights. Often, analysts compare between results obtained by using many multicriteria decision making methods and neglect the most important keys of success, i.e., sample choice, data collection and descriptive data analysis with statistical tools (if the case needs a survey).

The CP method is a systemic method meant to rank solutions. In a complex decision making problem that proposes a set of numerous solutions, this method is more appropriate than using the cost-benefit method which is the most used method in evaluating public projects. This method compares solution per twice and this procedure is long and complicated supposing the existence of a general equilibrium. The CP offers, in addition to its practicability, the possibility to compare and rank a set of non-dominated solutions. The number of conflicting criteria can be high, as well. The final solution can be discussed with the DM and the whole decision making process can be revisited. This method is flexible enough and invites actors (DM, analyst and stakeholder) to reflect about the convergence of their decisions, opinions and aspirations.

The selected solution must be quickly adopted to avoid problems of values actualization in evaluation matrixes. In our cases we noticed many divergences in stakeholders' behavior which can be explained by at least four important factors: the size of parcels, the type of activity, the labor factor and the possession of regular title deed. Therefore, we suggest for future work to limit the Cebala TWWIP to its southern part and to focus above all on how to promote and encourage the daily production. The concerned part of the perimeter can be defined by using the Geographical Information System (GIS). In fact, the motivation to reuse the resource of farmers farming in the north of the perimeter and owning large parcels is somewhere weak. The absence of labor factor in the southern perimeter explains this reluctance to reuse. We felt this need only via the direct contact with farmers. Our contact with the agency managing the TWW also informed us about the need to limit the agency charges because of the seasonal demand, the high operating and the maintenance charges. Decision-making process in choosing public investment should hence start with a direct evaluation of stakeholders needs by carrying out surveys. Thus, descriptive analysis methods based on statistical tools, albeit being time-consuming, are the only guarantees of decision making system success. Sample choice and questionnaire preparation should be given much care.

## ACKNOWLEDGMENT

Authors extends their thanks to the National Institute of Reseearch in Agricultural Engineering, Water, Forests in Tunisia and in particular to Mme Bahri for her help in carrying out the survey. They thank also, Dr Hemdane, a director in the MAHR for his acceptance de respond to our questionnaire as a decision maker. They dedicate this work to the memory of Professor François Brissaud, of the University of Montpellier II for his considerable participation in the descriptive analysis.

## REFERENCES

Al Nasir, F., & Batarseh, M. I. (2008). Agricultural reuse of reclaimed water and uptake of organic compounds: Pilot study at Mutah University wastewater treatment plant, Jordan. *Chemosphere*, *72*, 1203–1214. doi:10.1016/j.chemosphere.2008.01.064

André, F. J., Cardenete, M. A., & Romero, C. (2008). Using compromise programming for macroeconomic policy making in a general equilibrium framework: Theory and application to the Spanish economy. *The Journal of the Operational Research Society*, *59*, 875–883. doi:10.1057/palgrave.jors.2602415

Angelakis, A. N., & Durham, B. (2008). Water recycling and reuse in EUREAU countries: Trends and challenges. *Desalination*, *218*, 3–12. doi:10.1016/j.desal.2006.07.015

Angelakis, A. N., Koutsoyiannis, D., & Tchobanoglous, G. (2005). Urban wastewater and stormwater technologies in ancient Greece. *Water Resources*, *39*, 210–220.

Angelakis, A. N., Marecos Do Monte, M. H. F., Bontoux, L., & Asano, T. (1998). The status of wastewater reuse practice in the mediterranean Basin: need for guidelines. *Water Resources*, *33*, 2201–2217.

Angelakis, A. N., & Spyridakis, S. V. (1996). The status of water resources in Minoan times –A preliminary study. In Angelakis, A., & Issar, A. (Eds.), *Diachronic climatic impacts on water resources with emphasis on Mediterranean region*. Berlin: Springer-Verlag.

Arrow, K. J. (1995). Effet de serre et actualisation. *Revue de l'Énergie*, *471*, 631–636.

Bahman, S. (1998). Accounting for benfits of water reuse. In *Proc American Waterworks Assn 1998 Water Recycling and Reuse Symposium*.

Bahri, A. (2003). *Water reuse in the Middle East, North Africa and Mediterranean Countries*. Paper presented at Kyoto water forum, Kyoto, Japan.

Baja, S., Chapman, D. M., & Dragovich, D. (2007). Spatial based compromise programming for multiple criteria decision making in land use planning. *Environmental Modeling and Assessment*, *12*, 171–184. doi:10.1007/s10666-006-9059-1

Ballestro, E. (2000). Project finance: A multi-criteria approach to arbitration. *The Journal of the Operational Research Society*, *51*, 183–197.

Bdour, A. N., Hamdi, M. R., & Tarawneh, Z. (2009). Perspectives on sustainable wastewater treatment technologies and reuse in the urban area of the Mediterranean region. *Desalination*, *237*, 162–174. doi:10.1016/j.desal.2007.12.030

Belli, P. (1996). Is economic analysis of projects still useful? *World Bank Policy Research Working Paper 1689*.

Bilbao-terol, A., Perez-Gladish, B., & Antomil-Jbias, J. (2006). Selecting the optimum potfolio using fuzzy compromise programming and Sharpe's single-index model. *Applied Mathematics and Computation*, *182*, 644–664. doi:10.1016/j.amc.2006.04.028

Blumenthal, U., & Pearsey, A. (2002). *Critical Review of epidemiological evidence of the health effects of wastewater and excreta use in agriculture*. Geneva: WHO.

Bond, W. J. (1998). Effluent irrigation – an environmental challenge for soil science. *Australian Journal of Soil Research*, *36*, 543–555. doi:10.1071/S98017

Caballero, R., Ruiz, F., Rodriguez Uria, M. V., & Romero, C. (1996). Interactive meta-goal programming. *European Journal of Operational Research*, *175*, 135–154. doi:10.1016/j.ejor.2005.04.040

Chang, N.-B., Eh, S. C. Y., & Wu, G. C. (1999). Stability of Grey Compromise Programming and its application to watershed land-use planning. *International Journal of Systems Science*, *30*, 571–589. doi:10.1080/002077299292092

Chang, N.-B., & Tseng, C. C. (1999). Optimal evaluation of expansion alternatives for existing air quality monitoring network by grey compromise programming. *Journal of Environmental Management*, *56*, 61–77. doi:10.1006/jema.1999.0267

Da Fonesca, A. F., Herpin, U., Paula, A. M., Victoria, R. L., & Melfi, A. J. (2007). Agricultural use of treated sewage effluents: agronomical-environmental implications and perspectives for Brazil. *Sci. Agric*, *64*, 194–209.

Dadashzadeh, R., & Nimse, S. B. (2006). A compromise model for solving fuzzy linear programming problems. *Mathematica*, *48*, 134–148.

Davarajan, S., Squire, L., & Suthiwart-Narueput, S. (1995). Reviving project appraisal at the World Bank, *World Bank Policy Research Working Paper Policy working paper N°* 1496. Available at SSRN: http://ssrn.com/abstract=636145

Desaigues, B., & Point, P. (1993). *Economie du patrimoine naturel. La valorisation des bénéfices de protection de l'environnement*. Paris: Economica.

Dodgson, J., Spackman, M., Pearman, A., & Phillips, L. (2000). *Multi-Criteria analysis: An annual*. London: Department of the Environment, Transport and the Regions. Available at http://www.odpm.gov.uk/about/multicriteria/index.htm.

Dunning, D. J., Ross, Q. E., & Merkhofer, M. W. (2000). Multi attribute utility analysis for addressing section 316(b) of the clean water act. *Environmental Science & Policy*, *3*, 7–14. doi:10.1016/S1462-9011(00)00022-8

Escobar, M. T., & Moreno-Jiménez, J. M. (1997). The Hierarchical Compromise Programming. *Sociedad de Estadistica e InvestigaciSn Operativa*, *5*, 253–281.

Escobar, M. T., & Moreno-Jiménez, J. M. (2002). A linkage between the Analytic Hierarchy Process and the Compromise Programming Models. *Omega*, *30*, 359–365. doi:10.1016/S0305-0483(02)00053-1

Faucheux, S., & Noel, J.-F. (1995). *Economie des ressources naturelles et de l'environnement*. Paris: Armand Colin.

Flug, M., Seitz, H. L. H., & Scott, J. F. (2000). Multicriteria decision analysis applied to Glen Canyon Dam. *Journal of Water Resources Planning and Management*, *126*, 270–276. doi:10.1061/(ASCE)0733-9496(2000)126:5(270)

Gagné, C., Gravel, M., & Price, W. L. (April 2003). *Recherche de solutions de compromis en ordonnancement à l'aide de metaheuristiques*. Paper presented at the Conférence Francophone de MOdélisation et SIMulation, Toulouse (France).

Goicoechea, A., Hansen, D. R., & Duckstein, L. (1982). *Multiobjective decision analysis with engineering and business application*. New York: John Wiley and sons.

Hajkowicz, S., & Higgins, A. (2008). A comparison of multiple criteria analysis techniques for water resource management. *European Journal of Operational Research*, *184*, 255–265. doi:10.1016/j.ejor.2006.10.045

Hajkowicz, S., Spencer, R., Higgins, A., & Marion, A. (2008). Evaluating water quality investment using cost utility analysing. *Journal of Environmental Management*, *88*, 1601–1610. doi:10.1016/j.jenvman.2007.08.006

Isiklar, G., Alptekin, E., & Buyukozhan, G. (2007). Application of an hybrid intelligent decision support model in logistics outsourcing. *Computers & Operations Research*, *34*, 3701–3714. doi:10.1016/j.cor.2006.01.011

Johansson, R. C., Tsurb, Y., Roe, T. L., Doukkali, R., & Dinar, A. (2002). Pricing irrigation water: a review of theory and practice. *Water Policy*, *4*, 173–199. doi:10.1016/S1366-7017(02)00026-0

Joubert, A. R., Stewart, T. J., & Eberhard, R. (2003). Evaluation of water supply augmentation and water demand management option for the city of Cape Town. *Journal of Multi-criteria Decision Analysis*, *12*, 17–25. doi:10.1002/mcda.342

Leal, R. M. P., Herpin, U., Da Fonseca, A. F., Firme, L. P., Montes, C. R., & Melfi, A. J. (2009). Sodicity and salinity in a Brazilian oxisol cultivated with sugarcane irrigated with wastewater. *Agricultural Water Management*, *96*, 307–316. doi:10.1016/j.agwat.2008.08.009

Lin, Yi., & Liu, S. (2004). A historical introduction to grey systems theory. *Systems, man and cybernetics, 3*, 2403-2408.

Loubier, S., Terreaux, J.-P., & Gleyses, G. (2001). Sur les principes de l'évaluation économique des systèmes irrigués: Méthodes et applications. In *19th European Regional Conference of ICID,* Brno and Prague, Czech Republic, 4-8 June.

Luigi, P., Kramer, A., & Al Baz, I. (2007). The EMWater project – promoting efficient wastewater management and reuse in Mediterranean countries. *Desalinisation*, *215*, 56–63. doi:10.1016/j.desal.2006.10.027

Malczewski, J. (1999). *GIS and Multicriteria Decision Analysis*. New York: John Wiley and Sons.

Marshall, E.-P., & Homans, H.-A. (2004). A spatial analysis of the economic and ecological efficacy of land retirement. *Environmental Modeling and Assessment*, *9*, 65–75. doi:10.1023/B:ENMO.0000032097.99637.26

Menegaki, A. N., Hanley, N., & Tsagarakis, K. P. (2007). The social acceptability and valuation of recycled water in Crete: A study of consumers' and framers' attitude. *Ecological Economics*, *62*, 7–18. doi:10.1016/j.ecolecon.2007.01.008

Mutegu, S., Hoko, Z., & Makoni, F. S. (2007). An assessement of the public health hazard potential of wastewater reuse for crop production. A case of Bulawayo city, Zimbabwe. *Physics and Chemistry of the Earth*, *32*, 1195–1203.

National Academy of Sciences. (2007). *Understanding American Agriculture: Challenges for the agricultural Resource Management Survey*. National Academies Press. Retrieved from http://jbr.org/articles.html

Nayak, R. C., & Panda, R. K. (2001). Integrated management of a canal command in a river Delta using multi-objective techniques. *Water Resources Management*, *15*, 383–401. doi:10.1023/A:1015593417769

OECD. (1994). *Project and policy appraisal: integrating economic and environment.* Brussels: OECD publication.

Pongthanapanich, T. (2006). Optimal Coastal Land Use and Management in Krabi, Thailand: Compromise Programming Approach. *IME working paper no.67/06. SDU*, Denmark. Available online: http://econpapers.repec.org/

Prodanovic, P., & Simonovic, S. P. (2003). Fuzzy Compromise Programming for group decision making. *IEEE Transactions on Systems, Man, and Cybernetics. Part A, Systems and Humans*, *33*(3), 358–365. doi:10.1109/TSMCA.2003.817050

Qadir, A., & Wichelns, D. (2008). (in press). Raschid-Sally. L, McCornick, P. G., Drechsel. P., Bahri. A., & [The challenges of wastewater irrigation in developing countries. *Agricultural water management.*]. *Minhas. P. S.*

SCET International and CNEA. (1981). *Plan directeur des eaux du nord, 2ième tranche. Technical report*. MAHR.

Shiau, J. T., & Lee, H. C. (2005). Derivation of optimal hedging for a water supply reservoir through compromise programming. *Water Resources Management*, *19*, 111–132. doi:10.1007/s11269-005-1502-6

Tidăkar, P., Sjőberg, C., & Jőnsson, H. (2007). Local recycling of plant nutrients from small-scale wastewater systems to farmland – A Swedish scenario study. *Resources, Conservation and Recycling*, *49*, 388–405. doi:10.1016/j.resconrec.2006.05.004

Tolga, E., Alptekin, E. (2007). Product development process using a fuzzy compromise based goal programming approach (note de lecture). *Computer Science, Computational sciences and its applications, 1*, 832-845.

Tzeng, G.-H., Teng, Y.-J., & Hu, C.-P. (1991). Urban environmental evaluation and improvement: Application of multi-attribute utility and Compromise Programming. *Behaviometrika*, *29*, 83–98. doi:10.2333/bhmk.18.29_83

Wilson, Z., & Pfaff, B. (2008). Religious, philosophical and environmentalist perspectives on potable wastewater reuse in Durba, South Africa. *Desalinisation*, *228*, 1–9. doi:10.1016/j.desal.2007.07.022

World Health Organisation. (2005). *A regional overview of wastewater management and reuse in the Eastern Mediterranean*. Regional Office for the Estern Mediterranean. CEHA.

World Health Organisation. (2006). Guidelines for the safe use of wastewater, excreta and grey water. *Wastewater use in agriculture 2*. Genova.

Wu, C.-C., & Chang, N.-B. (2004). Corporate optimal production planning with varying environmental costs. A grey compromise programming approach. *European Journal of Operational Research*, *155*, 68–95. doi:10.1016/S0377-2217(02)00820-2

Yadav, R. K., Goyal, B., Sharma, R. K., Dubey, S. K., & Minhas, P. S. (2002). Post-irrigation impact of domestic sewage effluent on composition of soils, crops and ground water – a case study. *Environment International*, *28*, 481–486. doi:10.1016/S0160-4120(02)00070-3

Zeleny, M. (1973). Compromise Programming. In Cochrane, J. L., & Zeleny, M. (Eds.), *Multiple criteria decision making* (pp. 373–391). Columbia, SC: Univ. of South Carolina Press.

Zeleny, M. (1974). A concept of compromise solutions and method of displaced ideal. *Computers & Operations Research*, *1*(4), 479–496. doi:10.1016/0305-0548(74)90064-1

# APPENDIX

*Table of abbreviations*

| | |
|---|---|
| AMHR | Ministry of Agriculture and Hydrologic Resources |
| CP | Compromise programming |
| CRDA | Commissariat Régional de Développement Agricole (regional agency of agricultural development) |
| CVM | Contingent valuation Method |
| DAIS | Decision Aid Interactive System |
| DM | Decision Maker |
| FAO | Food and Agriculture Organization of the united nation |
| GIS | Geographic Information systems |
| IRR | Internal Return Rate |
| MADM | Multi-Attributes Decision Making |
| MCA | Multicriteria Analysis |
| MCDA | Multicriteria Decision Analysis |
| MESD | Ministry of Environment and Sustainable Development |
| MODM | Multi-Objectives Decision Making |
| OECD | Organization for Economic Co-operation Development |
| ONAS | Office National d'assainissement sanitaire |
| TWW | Treated Wastewater |
| TWWIP | Treated Wastewater Irrigated perimeter |
| WHO | World Health Organization |
| WTA | Wiliness To Accept |
| WTP | Wiliness To Pay |

# Chapter 18
# Towards Spatial Decision Support System for Animals Traceability

**Marcos Visoli**
*Embrapa Agriculture Informatics, Campinas, Brazil*

**Sandro Bimonte**
*Cemagref, TSCF, Clermont Ferrand, France*

**Sônia Ternes**
*Embrapa Agriculture Informatics, Campinas, Brazil*

**François Pinet**
*Cemagref, TSCF, Clermont Ferrand, France*

**Jean-Pierre Chanet**
*Cemagref, TSCF, Clermont Ferrand, France*

## ABSTRACT

*Animal traceability is a very important question for several government and private institutions from many points of view: economical, sanitary, etc. Traditional systems are able to memorize the main bovine movements, or to capture the geolocation of an animal using RFID. Now it should be possible to envisage a new generation of traceability systems in which the different locations are automatically recorded several times per day for each animal. These systems should also be coupled with analysis techniques to help decision-makers to take decisions, validate and/or reformulate their hypothesis. In this chapter the authors present a spatial decision support system dedicated to the animal geolocation acquisitions and analysis of possible sanitary problems. Indeed, in case of sanitary alerts, the system is able to determine the animals which have been in contact with a diseased animal exploiting historical trajectories of animals. It is applied to traceability of beef cattle using the Brazilian production system as a case study. OTAG focuses on improving methods and geotechnologies for recording reliable and accurate data on beef production.*

DOI: 10.4018/978-1-61692-871-1.ch018

## INTRODUCTION

The availability of technological solutions to guarantee a reliable bovine traceability is very important for any beef producer. In 2008, for example Brazil leaded the ranking of beef exporter in the world, with 2.2 million tons, corresponding to US$ 5.3 billion (ABIEC, 2009). These values represent 28% of the international marketing of beef and commercial relations with 170 countries. Nowadays, this country is the second biggest beef producer with 8.9 million tons in 2008, while United Sates maintain the first position with 12.1 million tons and Europe, the third one, produces 8.2 million tons (SINDICARNE, 2009). The Brazilian beef production is developed all over the territory and it includes approximately 225 million hectares distributed in 2.2 millions properties, according to the Brazilian Institute of Geography and Statistics, and a bovine population of 199.7 millions animals in 2007 (IBGE, 2009), which corresponds to 15% of bovine population in the world. This is one of the productive chains with more employees, being responsible for 7.2 millions of direct jobs. The bovine meat agribusiness has been growing annually in Brazil at the same time it is being structured in a competitive way. In the last decade, the medium growth was approximately 30%, while the growth of the exports was superior to 200%. The Brazilian beef cattle productive chain is characterized by the creation of animals under grazing, by semi-intensive system (grazing with supplementation) or by the intensive system (grazing with supplementation and feed-lot), in which the animal is fed by grains, being forbidden the use of animal protein (except dairy products). Considering the economic and health aspects of the beef cattle for producer and consumer countries, the continuous use of agricultural techniques becomes essential to contribute with the increase of the production and to guarantee the final product quality. Furthermore, it is necessary to use information technology aiming at the collection and treatment of information throughout the productive chain, for accelerating the decision making in any stage of the production process.

This chapter sums up the studies undertaken during a project supported by the European Community about geolocation at a large scale applied to bovine traceability. This proposal is issued from the project "Operational management and geo-decisional prototype to track and trace agricultural production" (OTAG), which is a Specific Support Action of the 6th Framework Programme of the European Community. The goal of OTAG is to provide conditions to know the relative risks concerning to a bovine traceability, in the context of Southern Cone Countries and the EU policies. Hence, the Brazilian beef cattle production system was taken as case study. The project is based on the existing knowledge in Europe and Canada concerning to information systems and geo-decisional tools, as well to the interaction among experts and user groups from South Cone, Canada and Europe. The following institutions are partners in the project: Cemagref (France), Brazilian Agricultural Research Corporation (Embrapa, Brazil), Université Laval (Canada) and Cooperative Program for the Technological Development of the Agro-food and Agroindustry in the Southern Cone (Procisur, Argentine).

The mainly goals of OTAG is to propose a technological solution and a global architecture for a new generation of animal traceability systems. Thus, we propose three-tier architecture composed of an *acquisition layer*, a *data management layer* and a *spatial decision support layer*. The acquisition layer allows measuring the animal geolocations in the pastures, using electronic devices embedded in necklaces carried out by animals. Then, these data are centralized in a spatial database (data management layer). This layer represents semantics of bovine traceability applications, including concepts of production system, animals and their movements. The richness of these database models allows consumers, researchers, farmers and government employers to manage, query and analyze data related to bovine production and movements. In particular, these records constitute a valuable information source for providing an efficient traceability of animals, and as a consequence, a better sanitary control and a reliable product quality. Indeed, using precise history of animals' geolocations we have developed a spatial decision system which allows decision-makers obtaining relatively exhaustive information about contacts between animals. In case of sanitary alerts (e.g. epidemic disease), they can deduce and define very precisely the animals and the geographical areas (farms, pastures, etc.) having potential risks. Consequently, our system permits answering classical trajectories queries like: "which are the farms where animals have been in a specific period?" or "how many time the animal had taken to make its trajectory?", but also contamination queries such as:

- which animals had been in contact during one specific period of time ?
- which animals had been in contact in one specific region, in one determined period of time ?
- which animals had been in contact with infected animals in one specific period of time ?
- which animals had entered in contact of infect animal in a particular period of time ?
- which are the animals not protected by vaccination that had been in contact with the animal a during a period of time ?
- which are the farms where animals have been in a specific period ?

The main problem to resolve the above queries is to find a model to represent trajectories covered by the animals and associated computation of contamination contacts which is a quite tedious task as it involves spatio-temporal data and relationships, contamination areas and time-windows. The proposed model allows answering queries in an acceptable time. The use of the concept of database for trajectories has been adopted as a possible solution. An important information that can be analyzed from the animal trajectories is the probability of contamination in accordance with the intensity of the contact between animals in a space of time. Then, a mathematical model to determine the probability of contamination for the intensity of contact between animals has been provided.

The chapter is structured as follows. Section 1 introduces the scientific background related to spatial databases and spatial decision support systems. Section 2 presents the computational architecture of the OTAG project. Future research directions and conclusions are presented in Sections 3 and 4, respectively.

## BACKGROUND

In this section, we present existing work about spatial databases and spatial decision support systems for trajectory data.

### Spatial Databases

In Brazil there are two information systems related to beef cattle production. SISBOV (traceability service of beef cattle production chain) contains general information about the animal life inside the farms from birth to death (SISBOV, 2009). The farmers who intend to export their production must use this system. To control the animal movement, Brazilian Ministry of Agriculture, Food and Livestock (MAPA) uses GTA system (Animal Traffic Guide). GTA is used to record places of origin and destination of any animal movement, as well as farm entry/exit data of animals. The information managed by SISBOV and GTA systems are complementary.

In Quebec, the Ministry of Agriculture has entrusted Agri-Traçabilité Quebec (ATQ) for developing and managing livestock identification and traceability system (AGROTRACABILITE,

2009). Quebec's livestock identification and traceability system allows knowing the main locations of each animal, its history, its transit and the animals which have been in contact with it. There are 3 types of information stored in the national ATQ database:

- the animal identification (thanks to an RFID ear tag);
- the site identification (an unique seven-digit number for each site, like farm, slaughterhouse and so on);
- the recording of each animal movement (and some additional events).

In Europe the regulation for animal identification, which is the responsibility of the Ministry of Agriculture of each individual EU country, is based on 4 principles:

- allocation of an unique national number for each animal;
- recording of the information regarding to the identity of every bovine into a cattle register held (recording made by producers for each livestock);
- communication of this information to a national computerized database;
- production of an identity document (cattle passport) to accompany every bovine travel.

These rules are the same in all the EU countries with some local adaptation. There are official procedures to make the equivalence of the data from different country.

In France, the Ministry of Agriculture delegates the practical running of the identification process to an official regional body (*Etablissement départemental de l'élevage*). The database for the registration of bovine in France is the BDNI (*Base de Données Nationale d'Identification*). It was created in 1999 by the Ministry. In January 2008, there was around 19,900,000 bovines registered and 246,000 farms. In France, there is also a project named GIEA which is focused on the data management of the farm (GIEA, 2009; Dufy et al., 2006; Brun et al., 2005; Poyet et al., 2003). A goal of this project is to propose a standard data format to facilitate information exchanges between agricultural software and French information systems. Once implemented in the main software applications used by farmers, this standard could be employed to encode data before their transmission between farms and national information systems. An UML modeling of the GIEA data format is available (Pinet et al., 2009 ; Booch et al., 1999).

Traditional systems are able to memorize the main bovine movements (e.g. from a farm to another one), or to capture the geolocation of an animal when a transporter scans its RFID tag with a specific device (see ATQ systems). Now it should be possible to envisage a new generation of bovine traceability systems in which the different locations are automatically recorded several times per day for each animal. Memorizing the precise history of animal geolocations will improve considerably the traceability especially for countries having very large pasture areas (such as Brazil). With these new systems, decision-makers will have relatively exhaustive information about contacts between animals. In case of sanitary alerts, they can determine very precisely the animals and the geographical areas (farms, pastures, etc.) having a potential risk. These kind of device can also improve farm management by knowing animal habit in function of different parameters (pastures quality, time period, forecast). These new systems can be considered as databases and decision support systems for trajectories data.

Definition of databases for trajectories data is an important research issues (Macedo et al., 2007). It deals with modeling and querying moving objects and their trajectories. Moving objects are geo-referenced objects whose position changes in time. Trajectory is the set of these positions in time. Several works have been defined on modeling and querying moving objects. On the other side, fully exploitation of trajectories is a recent new research area (Macedo et al., 2007; Spaccapietra et al., 2008). Preliminary works on databases for moving objects are found in Guting et al. (2000) and Guting and Schneider (2005). Guting et al. (2000) provide a formal theory for representing moving regions and points where geometries and /or their positions change in function of time. They propose an extension of Relational DBMS data and query model to take into account moving objects. Definition of spatio-temporal operators for moving objects has been addressed in several works. Sistla et al. (1997) define a data model which combines future temporal logic and topological operators. Classical relationships for moving objects are based on time instants and time intervals. By the same way spatial relationships are also interesting for moving objects. Then, concepts like direction, velocity and acceleration are mandatory for the analysis of moving objects. Su et al. (2001) extends the combination of spatial and temporal operators provided by Guting & Schneider (2005) to take into account these concepts using a data model based on linear constraint databases. This approach is representative of constraint models for moving objects. Indeed, trajectory of moving points can be represented as a discretized (finite) set of points (Guting & Schneider, 2005) or by a boolean combination of polynomial equalities and inequalities (Gertz, 2004). Moving objects into constrained network have been addressed into several works which define extensions of previous cited models (Vazirgiannis & Wolfson, 2001; Guting & Almeida, 2006). In those works two related models are defined: one for modeling moving objects and another one to define the topological network which constraints movement of objects. Handling spatio-temporal objects at physical level is also a hot topic. Indeed several indexing techniques have been developed to improve query performances (Saltenis et al., 2000; Lin et al., 2005; Pfser et al., 2000; Sun et al., 2004). They extend and adapt classical spatial and temporal techniques. As underlined in Macedo et al. (2007), an important research issue is the definition of spatial, temporal and spatio-temporal operators defined on the whole trajectory. Indeed users should be able to query spatial and temporal aspects of the trajectory (for example aggregating or comparing several trajectories (Baltzer et al., 2008; Meratnia & R. de By, 2002) extending spatio-temporal operators defined for moving objects (Guting and Schneider, 2005; Su et al., 2001). From a conceptual point of view, Spaccapietra et al. (2008) defines a trajectory as a set of moving spatial objects associated with a semantic component. The latter one describes trajectory as a set of movements (animals in paddocks) and stops (sleeping animals), the begin and the end of the trajectory. Thus, Spaccapietra et al. (2008) approach doesn't consider all points as belonging to the trajectory. Our approach is based on Spaccapietra et al. (2008) model which takes into account the complexity of real trajectory applications.

## Spatial Decision Support Systems

Spatial Decision Support Systems (SDSS) extends GIS functionalities (spatial data management, cartographic visualization and spatial analysis). They provide advanced spatial analysis methods and user-friendly and flexible interfaces, handling complex spatial data structures, and supporting aggregated and detailed data (Armstrong & Densham, 1990). In this way they allow decision-makers to make and validate hypothesis, and discover new trends and relations. Different kinds of SDSS have been developed, such as "Exploratory Spatial Data Analysis" systems (Andrienko et al., 2003) that integrate cartographic interactive visualization methods and spatial analysis techniques, or "Visual Spatial Data Mining" systems (Compieta et al., 2007) that provide spatial data mining techniques and cartographic visualization methods. Each approach addresses particular spatial decisional problems. Integration of valuable spatial data into data warehouses has become a challenging research area and leads to the definition of new kind of SDSS: Spatial OLAP (SOLAP). SOLAP systems allow analyzing huge volume of data, stored into a spatial data warehouse according to analysis axes called (spatial) dimensions, which are organized into hierarchies. The analysis subjects (measures) are aggregated using SQL functions (i.e. AVG, SUM, etc.). SOLAP has been defined by Bédard et al. (2001) as "a visual platform built especially to support rapid and easy spatio-temporal analysis and exploration of data following a multidimensional approach comprised of aggregation levels available in cartographic displays as well as in tabular and diagram displays". From SOLAP point of view, the inclusion of spatio-temporal data as subject analysis has been addressed in several studies (Bérniere et al., 2009; Baltzer et al., 2008; Gomez et al., 2009; Orlando et al., 2007; Tao and Papadias, 2005; Wan and Zeitouni, 2007). Once trajectory data are stored and analyzed, they have to be visualized. For that reason, ad-hoc visual techniques are necessary to interpret and understand results, formulate hypothesis and discover unknown spatial patterns and trends. Some proposals of visual techniques for trajectory data analysis have been developed by Andrienko et al. (2008).

To the best of our knowledge there is not a theoretical and implementation framework conceived expressly for contamination problem based on trajectory data model as described in Section "Spatial Decision Support System Layer".

## THE PROPOSED ARCHITECTURE

### Global Architecture

The OTAG architecture is organized in three layers as schematized in Figure 1.

Acquisition layer concerns the acquisition of georeferenced data in the agricultural pastures; the data are captured thanks to embedded electronic devices carried out by the animals. We have implemented an acquisition system based on GPS collar which is based on existing works (Butler et al., 2006; Ungara et al., 2009; Rodgers, 1996). In this chapter we do not detail this layer since it is not the focus of our study.

*Figure 1. Layers of the traceability architecture*

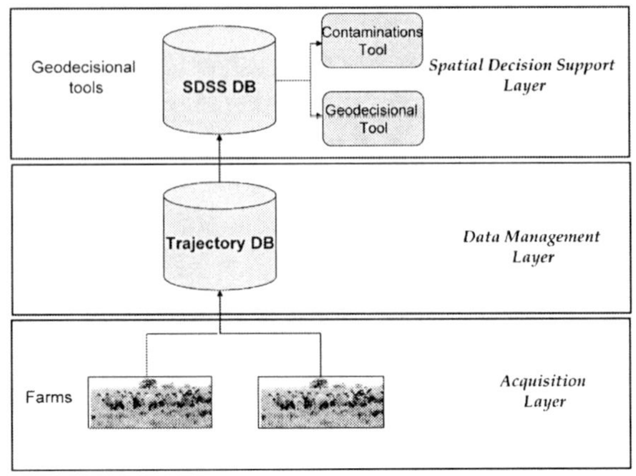


The second layer - Data Management - is responsible for storing collected data about animals (and their movements), farms, owner, etc.

The Spatial Decision Support layer is dedicated to the geo-decisional analysis; its main goal is to help public authorities making decisions (for instance in case of sanitary alerts). This last layer could be supported by different geo-decisional tools and by a data warehouse for integrating the database presented in Data Management layer with external sources of data about ground, pastures, climate, etc, or auxiliary information needful to geodecisional tools. Currently it provides a trajectory database of the Data Management layer with some auxiliary tables and some algorithms for computation of animals' contaminations.

The mainly use cases for the possible protagonists involved in the system are proposed bellow:

- **Government:** data about animal geolocations can be analysed by public authorities in order to make decisions in different situations when a sanitary alert occurs;

- **Consumers:** consumers are the domestic market of bovine meat. Our system could give information for those actors about production system practices and about the animal life and its origin. Those data contribute for increasing product quality;
- **Researchers:** researchers could get many information from the system, to compare the results of different production systems used in farms, to analyse data from different sources in order to study the real impact of grounds, pastures and climate in the bovine production systems and to analyse in a more realistic way the relation between animal movement and illness occurrence;
- **Farmers:** farmers are the most important actors in the system. The quality of information depends directly on the benefits that they could get by using the system.

## Data Management Layer

This section describes the conceptual model of the trajectory database (see Figure 1) using the UML formalism. The model is based on concepts like farm, farmer, cattle area, paddock, animal, sanitary event and geolocations. These concepts come from practices of Brazilian beef cattle production systems.

Figure 2 shows the global structure of the model which is based around four sub-models for the production systems, the animals, the moving management and trajectories.

Animals are temporally associated with cattle explorations through the class "Managing Event" (see Figure 3). Indeed, animals can move from different cattle explorations to change for example production systems, or to join an animal exposition, etc. By the same way, animals can be introduced into a cattle exploration following commercial transactions, etc. All this events are

*Figure 2. Conceptual model of the trajectory database*

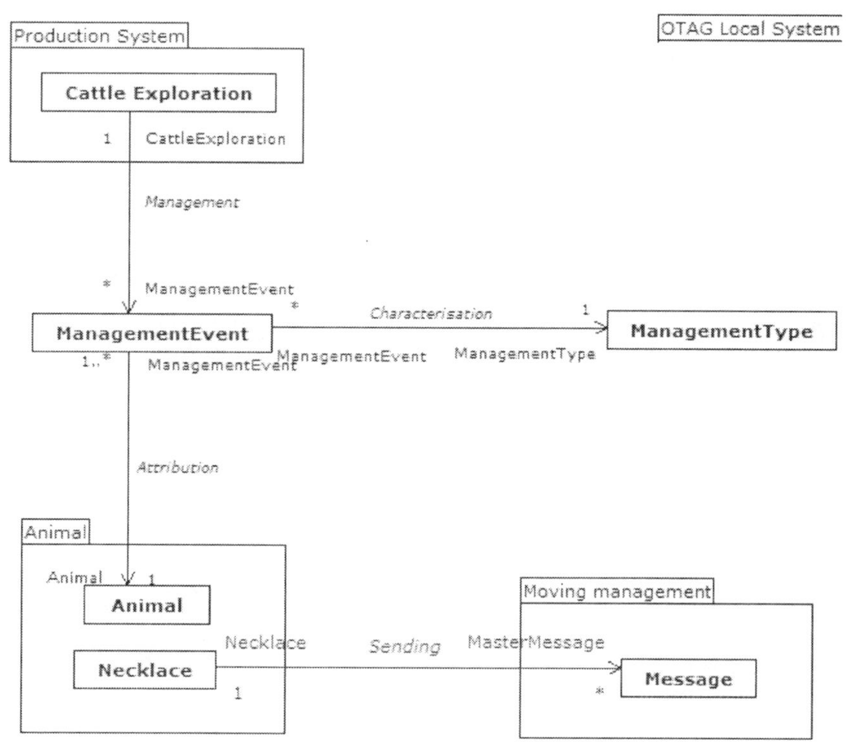

*Figure 3. Moving management sub-model*

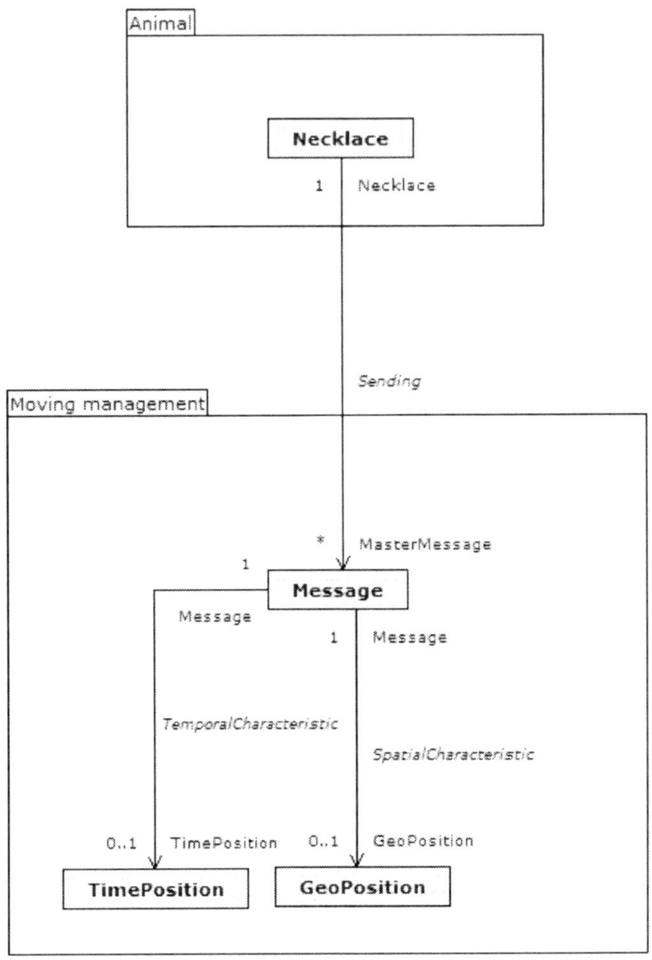


represented by class "ManagementType". The model "Moving management" is associated to "Animal" model as it represents the set of messages collected by the collars.

Each message is associated to a temporal class (TimePosition) that represents day and hour of acquisition, and to a spatial class (GeoPosition), that is a geographical coordinate (see Figure 3).

The "Animal" model is described in Figure 4. The animals receive vaccines periodically. Their weight and body temperatures are observed periodically too. Moreover, different fundamental events are recorded: birth, death, entrance and exit of the farm. The positioning of the animals is measured periodically thanks to necklaces. Necklace is an electronic device and an animal can use more than one necklace during its life. This information is represented by the classes "Animal event" and "Electronic device". It could be noted that in our model it is possible to introduce other observations, events, and electronic devices in a easy way by specializing these classes which makes the model extensible and flexible. Movements of animals (trajectories) are discussed in the next section that presents contamination calculation using trajectory data model.

The production system model is described in Figure 5. The farm is an agricultural establishment that belongs to one owner. Each cattle area is under the responsibility of a farmer that can be

*Figure 4. Animal sub-model*

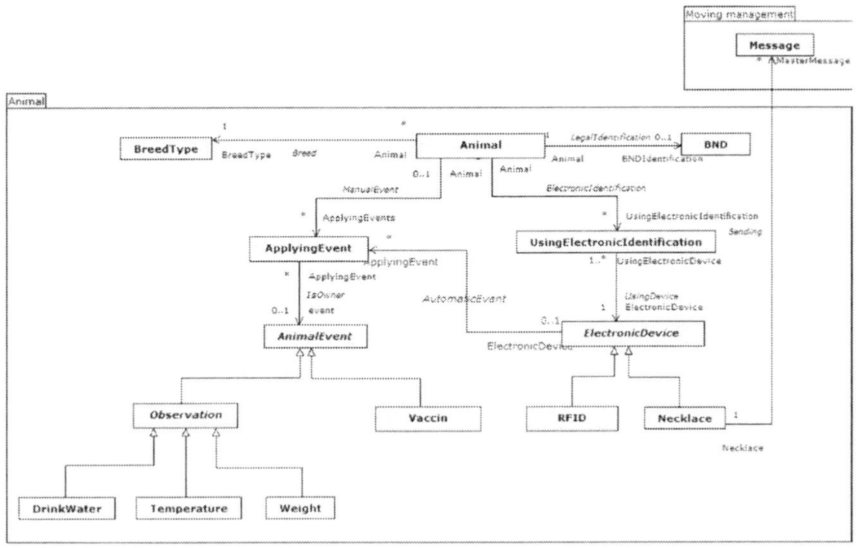


the owner or a leaseholder. Each cattle area corresponds to a geographical zone with one specific production system (extensive, semi-intensive and intensive), that can change periodically. This area is divided into paddocks; the paddock is associated with a feeding system, which can be modified throughout the time ("AppliedFeeding system" is a temporal class). A group of animals can be moved from a paddock to another one.

*Figure 5. Production system sub-model*

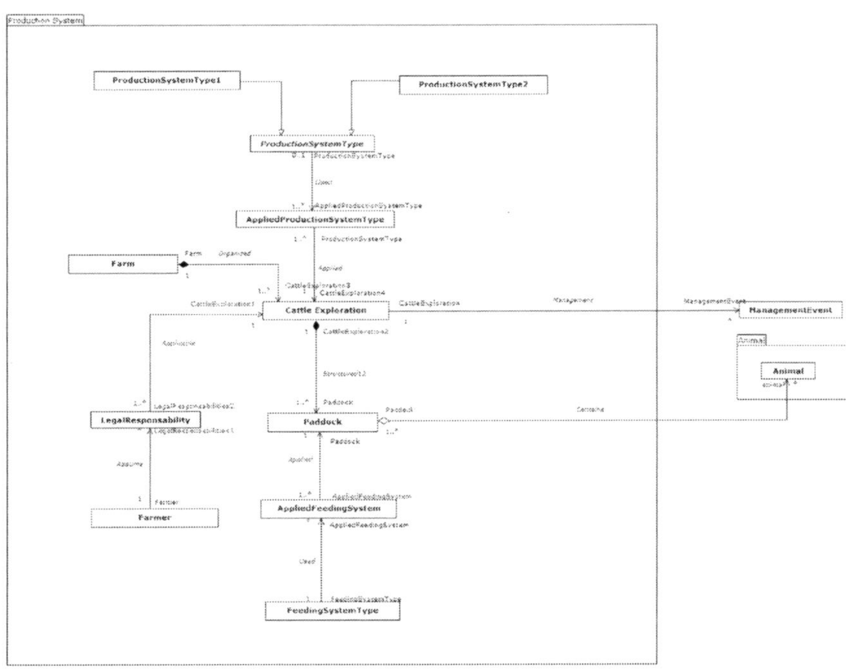

We note that different production systems can be applied to the same one cattle exploration as "Applied production system type" class is a temporal class. Cartographic representation of farms and paddock is shown in Figure 6.

We have implemented the data management layer using J2EE platform and PostgreSQL database system with its spatial extension (PostGIS). PostGIS provides a native supports for spatial data allowing storing and retrieving spatial data using particular indexes and spatial operators. A user-friendly visual interface has been developed for allowing the farmers to feed spatial databases using data concerning the production system model.

## Spatial Decision Support Layer

This section presents contributions of our work for the problems of bovine traceability and contact between animals using trajectory database concepts. First we describe delimitations of the problem and the representation of the movement of the animals as trajectories. Then, the algorithms for the identification of contacts between animals, and the calculation of the probability of contamination are presented. Since the Otag information system prototype is being tested, e.g., currently the data are being collected in the field; we have created a data generator of animal movements with geographic coordinates per hour, to carry through the computation of that probability.

*Figure 6. Cartographic representation of farms, paddocks ans points*

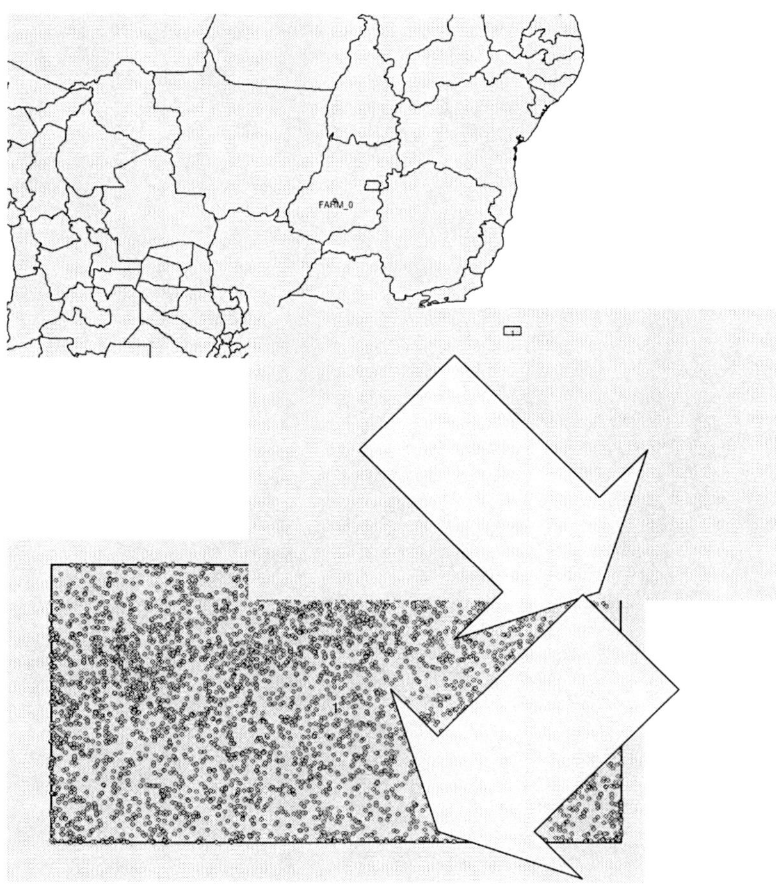

## Problem Delimitation

In this work we define contaminations algorithms considering the characteristics of Foot-and-Mouth Disease (FMD), although the general idea could be applied and adapted to any disease. FDM is a highly contagious disease who attacks all the animals of cleft hoof, mainly bovine. It occurs in all the ages, independent of sex, race, climate, etc. The virus is tied by direct contact of animals and by air, water and foods. The virus is present in great concentrations in the liquid of the vesicles which are formed in the mucosa of the tongue and around the nails. The blood contains great amounts of virus during the initial phases of the disease, when FMD is very contagious. Sometimes the infection is indirect: the virus is carried through foods, water, air and birds. Also the people who take care of sick animals lead the virus in their hands, clothes or the footwear, being capable of contaminate healthy animals. The virus incubation period varies from eighteen hours to three weeks.

In this work we consider two parameters for the calculation of contacts by using trajectories which are related to FMD but that are common for other diseases. The first one is *the number of days necessary to guarantee that there was no contamination problem: 90 days*. This period determines a time interval where the contacts must be calculated for posterior verification. The second factor is *the area that must be interdicted*. Initially it is possible to use these values to determine the size of an activity area (buffer) and posterior contact between animals. Finally, we make a realistic assumption on geographic location of animals' movements which allows improving contacts computation performance as it reduces the number of trajectories considered by our algorithm. Indeed, we consider scenarios where the farms use extensive and semi-extensive production systems. This means that animals can move between paddocks under the control of farmers. The movement of animals between farms or farms and slaughterhouse, which is carried through by some road, can be deduced and studied with the analysis of the existing road structure, but it isn't subject of our work. We are interested in bovines' movements in the farms, and specifically between paddocks.

## Bovine Trajectory Model

The data generator produces points that represent animal positions inside the paddock. For each point, geographic coordinates and time (day and hour) are collected. Thus, the following step is to transform these points into information that represent the movements of the animals in a format that allows carrying through more intelligent operations. For that goal, we consider that the displacement between two points is carried through straight-line. Thus, using this assumption, we proposed a model of trajectories according to the model proposed by Spaccapietra et al. (2008). In order to represent the trajectory of animals, we provide the definitions for point, segment, trajectory and travel.

Definition

- *Point: a point is a geolocalized acquisition represented by the geographic position and its associated time (day and hour);*
- *Segment: a segment is a line that connect two consequent points in a list of points ordered by time;*
- *Trajectory: a trajectory is a sequential set of segments, in accordance with the semantics defined by the decision-maker;*
- *Travel: set of trajectories.*

The semantics that determines the start and stop a trajectory depends on what it is being observed. For example, in our case study start and stop can be associated with the entry of the animal in the farm and its exit. Another example can correspond to the first and the last day of disease protection

*Figure 7. Segments and trajectories*

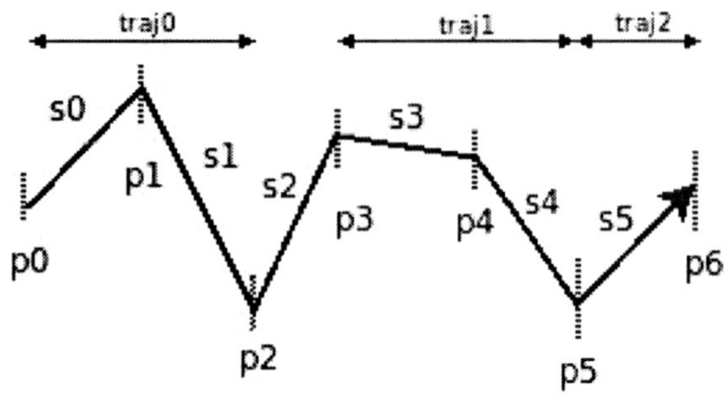


associated to some vaccination. An example of trajectories is shown in Figure 7.

The data model used to represent trajectories is shown in Figure 8. Three classes have been created: Travel, Trajectory and Segment. Segment is the simplest class and it represents collected points. Trajectory class is an aggregation of segments, and Travel class is an aggregation of trajectories.

## Contacts Definitions

As described in the previous section, two important variables for the algorithms of contacts between animals and calculation of the percentage of contamination have been considered: Activity area and IBC.

Definition

- *Activity area: is the buffer zone around an animal trajectory where there is a probability of contamination (seeFigure 9). Buffer represents a zone around the trajectory with an identical distance for each one of the sides;*
- *Interval between contacts (IBC): is a time range that represents the number of days where a risk of contamination exists (seeFigure 10).*

The FMD belongs to a group of diseases that are transmitted also being far from the contaminated animal. The distance from the trajectory can change depending on the wind, relief, or the rain occasion. In our work we consider a distance of 3 km.

In our case study the FMD IBC is 90 days. A contact between two animals occurs when an intersection of two trajectories exists and the

*Figure 8. Trajectory data model*

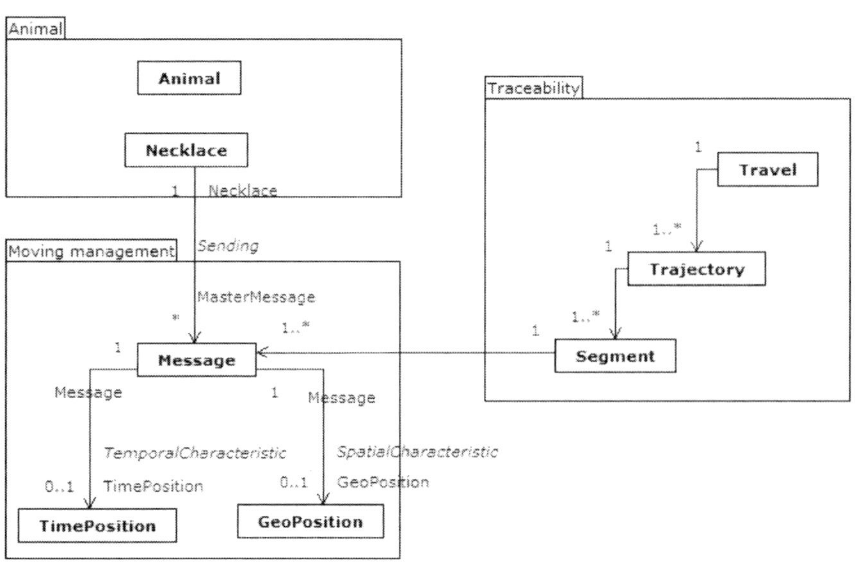

*Figure 9. Trajectory and buffer (activity area)*

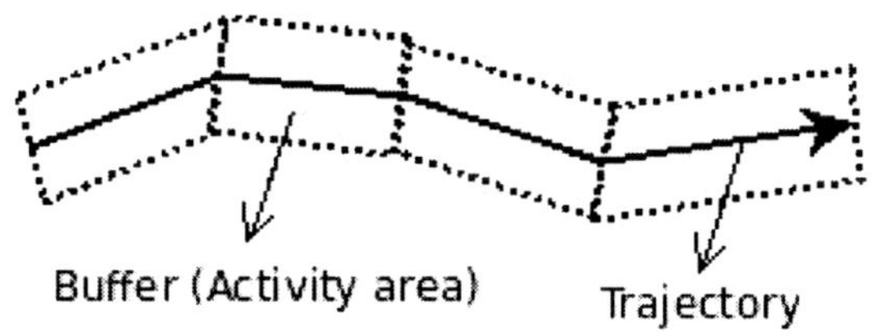


intersection occurs during the IBC. By this way, in the case of FMD we reduce the history of animal movements to trajectories defined on 90 days window. An example is shown in Figure 10 where using an IBC of 90 days, the animal C has no risk to be contaminated by animal A. On the other hand, animal B being inside the IBC has a contamination risk.

Finally, when the transmission of a disease such as the FMD is considered, it is not enough to verify only the intersection of the trajectories, but it is also necessary try to delimit an area around the trajectory (Activity area), and check the intersection of these areas with the trajectories of animals. Thus, before formalizing contamination contact that takes into account IBC and Activity Area, Temporal Contact and Spatial Contact concepts must be introduced.

Definition

- ***Temporal contact****: let trajA the trajectory of the animal A and IBCtrajA the subtrajectory of trajA that has starting and ending times equal to IBC bounds. Let trajB the trajectory of the animal B. A temporal contact exists when there is a temporal intersection (Allen, 1983) between IBCtrajA and trajB.*

*Figure 10. Trajectories and IBC*

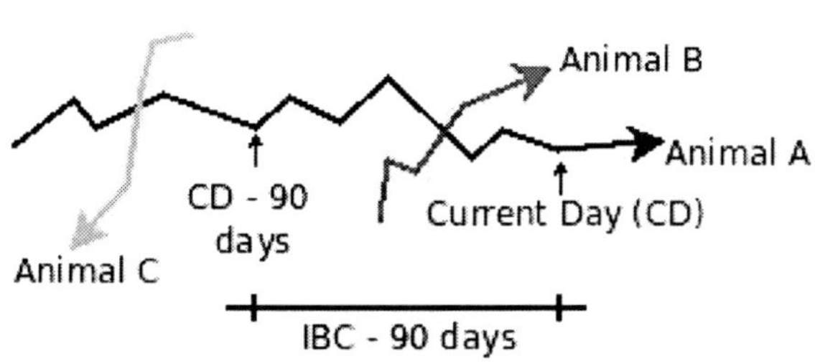


An example of temporal contact is shown in Figure 10. Here there is a temporal contact between animal A and animal B. In the other hand, there is no temporal contact between animal A and animal C, because animal C describes its trajectory outside the IBC bounds.

Definition

- ***Spatial contact****: let trajA the trajectory of the animal A and ACTtrajA the Activity Area of trajA. Let trajB the trajectory of the animal B. A spatial contact exists when there is a topological intersection between ACTtrajA and trajB.*

An example of spatial contact is shown in Figure 11. Here, there is a spatial contact between animal A and animal B, and between animal A and animal C. In the other hand, there is no spatial contact between animal A and animal D.

Now, with the definitions of Temporal Contact and Spatial Contact, we define the Spatial Temporal Contact, that represents the condition when an animal can be considered contaminated. Spatial Temporal Contact combines temporal and spatial contacts reflecting the two important parameters of FMD: IBC and Activity Area.

*Figure 11. Spatial contacts examples*

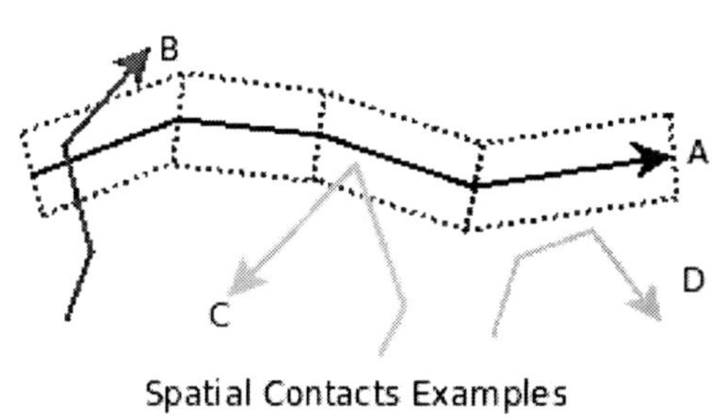

Definition

- ***Spatial temporal contact:****let trajA the trajectory of the animal A and let trajB the trajectory of the animal B. There is a spatial temporal contact if there is a temporal contact and a spatial contact.*

An example of spatial temporal contact is shown on Figure 12. In that Figure there is a spatial temporal contact between animal A and animal C, and between animal A and animal D. In the other hand, there is no spatial temporal contact between animal A and animal B, because a spatial contact exists but there is no temporal contact.

The following sections propose two algorithms aiming at calculating spatial temporal contacts between animals according the time granularity used to define segments: day and hour.

## Hourly Trajectories Contact Algorithm

The first algorithm considers each segment as a trajectory (Figure 13). Then, one segment is generated per hour. We call these trajectories *hourly trajectories.*

The first step of the algorithm is to create activity area using the spatial function of PostGIS ST_BUFFER (POSTGIS, 2009). Then, it verifies for each Activity Area if there is a topological intersection with other trajectories defined in the IBC (spatial temporal contact). Note that since topological intersection is a symmetric operator, we verify spatial temporal contact between two trajectories only once (line 4). To verify the topological intersection, the function PostGIS ST_INTERSECTS is used (POSTGIS, 2009). It verifies if there is an intersection between two geometric objects: line and polygon in our case. The complexity is O(m!*n!) where *m* is the number of days and *n* is the number of necklaces.

```
1 For each necklace (nck) {
2    For each trajectory (TrajA){
3       create an activity
area(ActTrajA)
4   For each trajectory (TrajB) of
necklaces with id greater than nck
5            verify spatial temporal
contact between ActTraj and Traj
6            insert TrajA and TrajB
into Contacts Table
}
```

In the following, we show an example of the algorithm using trajectories data of Table 1.

In the first step, the Activity area of the trajectory of necklace 10 is compared to all the other trajectories (necklaces 33, 47, 80 and 101). For each spatial temporal contact identified, one contact is registered. On the second step the Activity area of the trajectory of the necklace 33 is compared to the trajectories of necklaces 47, 80 and 101. The algorithm is executed until all the necklaces are analysed. The identified contacts

*Figure 12. Spatial temporal contacts examples*

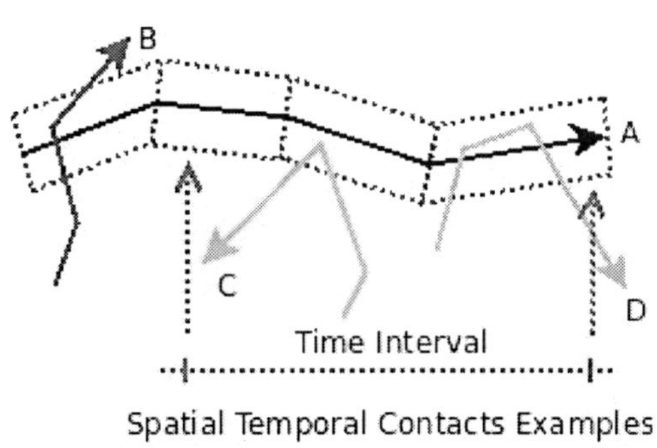


*Figure 13. Hourly trajectories: Each trajectory is equivalent to one segment*

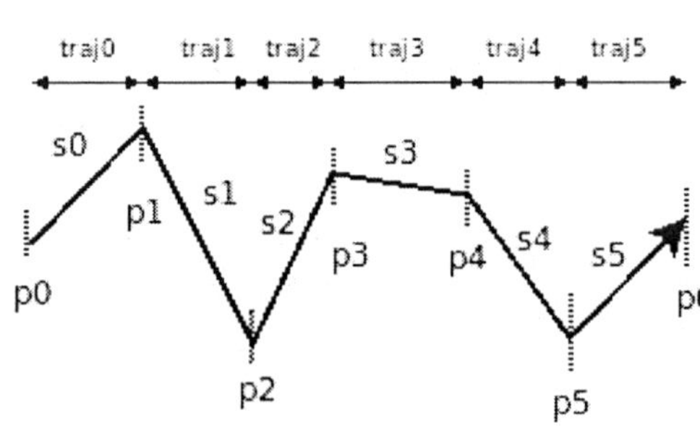

*Table 1. Sample trajectories data*

| Necklace id | Time | Trajectory (geometry) |
|---|---|---|
| 10 | 20080501 12:00 | t10 |
| 33 | 20080501 12:00 | t33 |
| 47 | 20080501 12:00 | t47 |
| 80 | 20080501 12:00 | t80 |
| 101 | 20080501 12:00 | t90 |

are stored in a table that contains the identifiers of the two intersected trajectories. Note that this table is stored into the SSDS database as it is used exclusively by the contamination tool of the SDSS layer. We have implemented the algorithms using PostGIS and Java Platform. An example of SQL query is presented in the Figure 14.

## Daily Trajectories Contact Algorithm

We provide an optimization in order to improve performance of previously proposed algorithm. This second version of the algorithm considers a trajectory as being all the segments of one day (24 hours): daily trajectories. Thus, a trajectory is represented by a geometric object called polyline. The number of comparisons is reduced (from 24 comparisons per day to only 1), and by consequence, also the number of contacts generated is less than the first algorithm. The algorithm is similar to the previous one. Here, the function ST_INTERSECTS uses the objects polylines and polygons.

**Experiments**

In this section we present experiments for both algorithms, considering a dataset with 2 farms, 9 paddocks, 395 animals and 3464940 positions. The experiments were executed on Intel Centrino processor 1.8 GHz with 2 GB ram memory. Figure 15 presents the results of the calculation of contacts for hourly and daily trajectories. The number of contacts is reduced substantially with the use of polylines.

Using daily trajectories, the identification of contacts loses the information of the segment where the contact happened. However, when this identification is necessary, a new query can be carried through in the database using necklaces identifiers.

The strategy of grouping segments by day also reduces the time for calculation of contacts (Figure 16). However, it implies a loss of precision about time when the contact happened, moving from hour to day scale precision.

*Figure 14. Example of the SQL query using PostGIS function ST_INTERSECTS*

```
SELECT
        a.animal_id as animal1, b.animal_id as animal2,
        a.time_position as time_position_1, b.time_position as   time_position_2,
        a.segment_id as segment1, b.segment_id as segment2
FROM
        animal_segments a, animal_segments b
WHERE
        (a.time_position >= timestamp ''' + day + '' 00:00') and
        (a.time_position < (timestamp ''' + day + '' 00:00' + interval '24 hours')) and
        (b.time_position < (timestamp ''' + day + '' 00:00' + interval '24 hours')) and
        (b.time_position >= (timestamp ''' + day + '' 00:00' - interval '90 days')) and
        (a.animal_id = '' + animalList.get(animal) + '') and
        (b.animal_id > '' + animalList.get(animal) + '') and
 ST_INTERSECTS(a.segment, b.segment)
```

*Figure 15. Number of contacts: segments by hour and segments grouped by day*

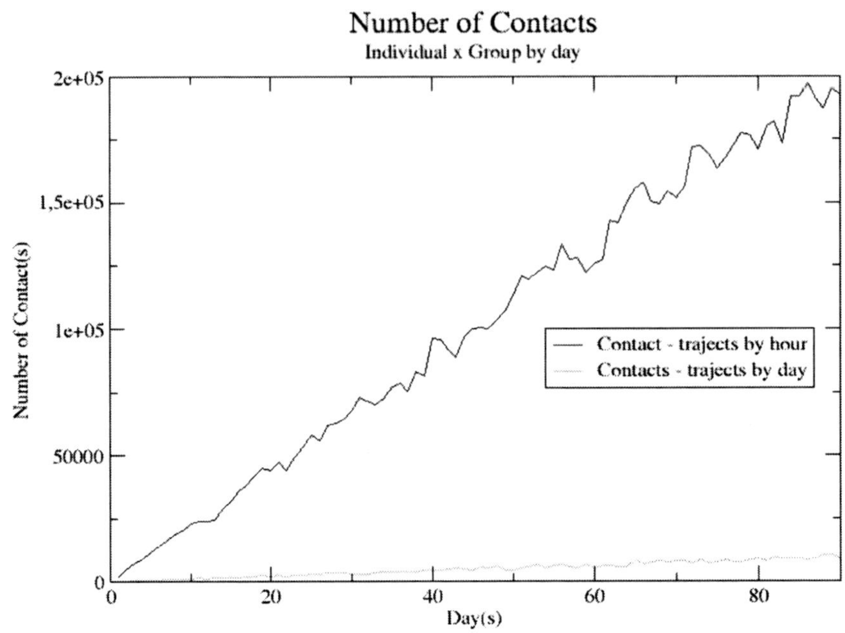


*Figure 16. Process time: segments by hour and segments grouped by day*

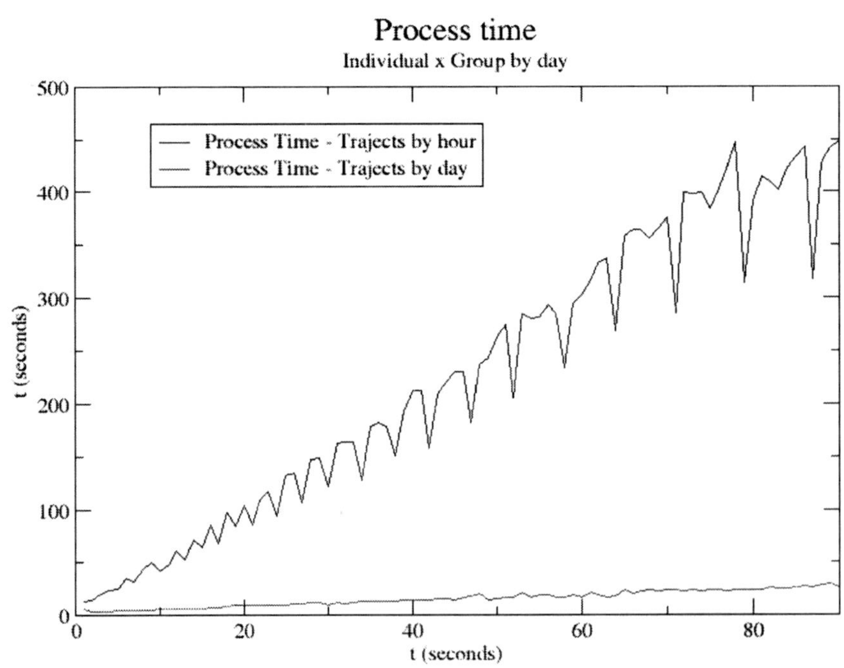

## Contamination Risk Calculation

In this section we investigate the use of contacts identification to calculate the contamination risk between two animals (AnimA and AnimB). It depends on two parameters: (i) the time spent by AnimB into the Activity Area of AnimA, and (ii) position of AnimB into the Activity Area of AnimA. We call these parameters respectively: *exposition time* and *spatial probability*.

Definition

Contamination Risk = *exposition time * spatial probability.*

We discuss two forms for the Contamination Risk calculation. The first approach considers only the use of the exposition time. In this case the spatial probability is uniform, equal to 1, implying that the power of contamination of an animal is identical in whole Activity Area. The second approach considers the use of spatial probability, meaning that the use of a specific function to represent the variation of the probability of contamination on the Activity Area.

## Use of Exposition Time and Uniform Spatial Probability

The exposition time is the time that an animal stays in the activity area of another animal. To calculate this time we need to identify the entry and exit points of the animal in the Activity Area. An example is shown in Figure 17. Here animal B enters in the Activity Area of animal A at the point *sc* and leaves the Activity Area at the point *ec*. Then, we calculate how long the animal takes to go from *sc* to *ec*. Hence we use the instants known at the bounds of the segment (*sp* and *fp*) and the length of the polyline identified by *sc* and *ec* (see Figure 17), supposing a constant velocity.

*Figure 17. Contamination calculation with uniform spatial probability*

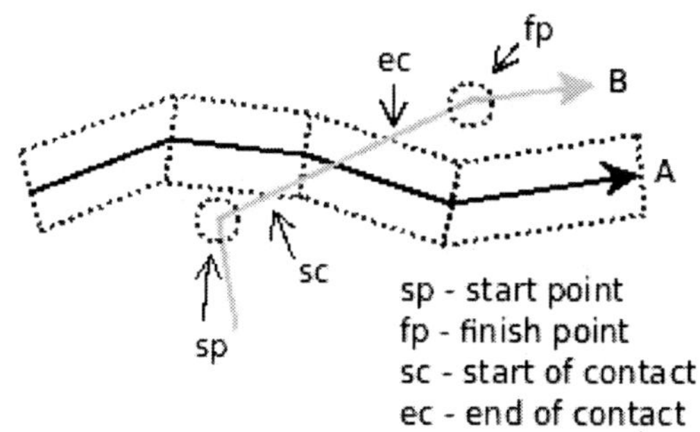


Formally, we can define the exposition time as:

*Exposition Time (B) = distance (sc (B), ec(B)) / uv(B)*

where:

*distance (sc(B), ec(B) is the length of the polyline inside the Activity Area that starts in sc and ending in ec ;*

*uv(B) is the Uniform Velocity of the animal B during the segment.*

In Figure 18 we show an example with two trajectories B and C that cross the same activity area of the trajectory A with the same velocity. Since the distance of animal C in the activity area is greater than the distance of animal B, then, according to our formula, the exposition time of C is greater than the exposition time of animal B.

*Figure 18. Example of exposition time*

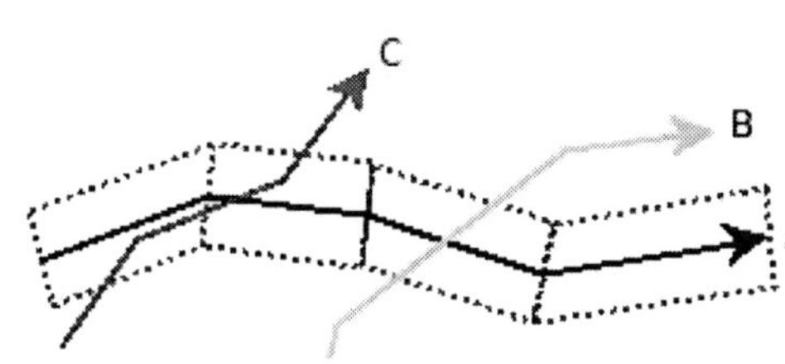

### Use of Exposition Time and Variable Spatial Probability

Here, we add the variable spatial probability for the computation of risk contamination. This means that we use a probability function that represents the position of the animal in the Activity Area. The contamination probability in the Activity Area is bigger for positions near to the trajectory. So, for the spatial probability, we use the Normal Distribution (or Gauss Distribution), with mean $\mu = 0$ and standard deviation $\sigma = 1$ as shown on Figure 19.

We propose to use a probability function that varies from 0 to 0.5 between a bound of the activity area and the trajectory, and from 0.5 to 0 between the trajectory and the other bound of activity area (see Figure 20). Consequently, in order to find the spatial probability, we map the initial and final positions of the intersected trajectory (*ec* and *sc*) in the Activity area on the x-axis of the Normalized Gaussian curve. Then using the Normal Distribution Table, which is a mathematical table for this function, we can calculate the probability of risk contamination (see Figure 20).

## FUTURE RESEARCH DIRECTIONS

Some works on introducing trajectory data into spatial data warehouse and SOLAP tools have been conducted in the last years. Baltzer et al. (2008) present a SQL extension for "grouping" similar trajectories through spatial data mining

*Figure 19. Normalized Gaussian curve with expected value $\mu = 0$ and variance $\sigma = 1$*

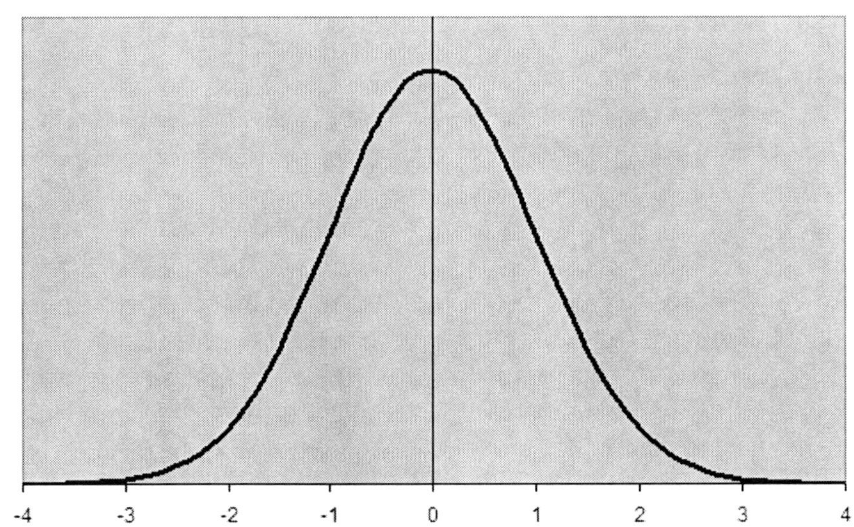


*Figure 20. Spatial probability: mapping segment intersection on the Normalized Gaussian curve*

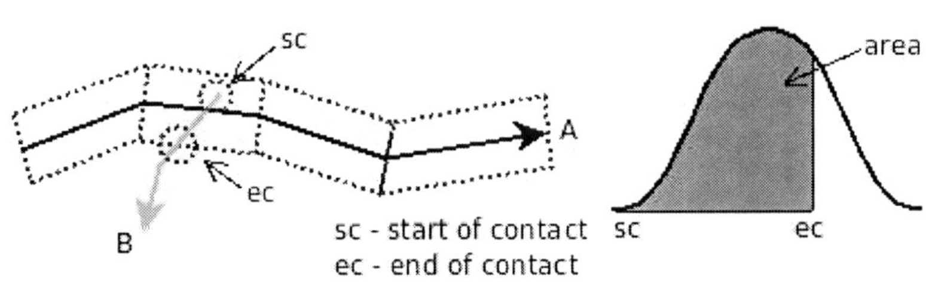

algorithms. Analyzed facts are described by numerical measures and/or spatial measures which is the average trajectory. Gomez et al. (2009) define a model which combines SOLAP and GIS data structures and operators. The model allows SOLAP queries filtered using spatial predicates and GIS queries using selection spatial predicates. This work defines geometric aggregation as the aggregation of numerical measures distributed on the surface of spatial dimension members. By the same way Orlando et al. (2007) define a spatial data warehouse with a spatial dimension representing a grid and a measure which is the number of trajectories per cell. An approximation technique is used to aggregate measures. Tao and Papadias (2005) provide an extension for classical spatial indexing techniques (i.e. R-Tree) to improve calculation of aggregation for numerical measures on spatio-temporal dimensions representing moving objects. A relational schema for huge volume of trajectory data is presented in Wan & Zeitouni (2007). Finally, introduction of trajectory data as analysis subject has not been investigated yet (Orlando et al., 2007b). Some works deal with knowledge discovery from trajectory data. Here, trajectories are defined as sequences of locations which are analyzed to discover patterns (Cao et al., 2007). Sometimes data are organized using graphs which allow using graph pattern mining methods to discover trajectory patterns (Lee et al., 2009).

Considering the promising works that could be found currently in the literature around SOLAP tools, this approach presents a great potential for analyzing data issued from the traceability of agricultural products by providing practical results. Nevertheless, they must be adapted in order to deal with analysis of a large volume of traceability data. In the near future we intend to work on the following topics:

- Definition of a structure to store the data;
- Development of efficient methods to access the data;
- Find ways to analyze the information in order to facilitate decisions;
- Improve the model of contamination, in the context of beef cattle traceability.

## CONCLUSION

The availability of technological solutions to guarantee a reliable bovine traceability is very important for producers and consumers countries. Traditional systems are able to memorize main bovine movements, or to capture their geolocations using RFID. Thus, it should be possible to envisage a new generation of traceability systems where the different locations are automatically recorded several times per day for each animal. These systems should also be coupled with analysis techniques to help decision makers to take decision, validate and reformulate their hypothesis.

In this chapter we presented a three-tier architecture for animal traceability information system that is composed by an acquisition layer, a data management layer, and a spatial decision support layer. The acquisition layer allows measuring the animal geolocations in the pastures, using electronic devices embedded in necklaces carried out by animals. In the data management layer, data are centralized in a spatial database that includes information about the production system, animals and their movements. The spatial database allows consumers, researchers, farmers and government employers to manage, query and analyze data related to bovine production and movements. In particular, these records constitute a valuable information source for providing an efficient traceability of animals, and retrieve all associated thematic information. As a consequence, the spatial database can be used as a support for a better sanitary control, and a reliable product quality system.

Indeed, using precise history of animals' geolocations, we have developed a spatial decision system which allows decision-makers obtaining relatively exhaustive information about contacts between animals. In this work, we discuss a contamination algorithm for foot-and-mouth disease (FMD) that considers animal trajectory data and the number of days to guarantee that it hasn't been contamination problem, that is, 90 days. This period determines a time interval where the contacts must be calculated for further verification. So, we have defined a contact as a spatial and temporal contact introducing the concept of Activity area that represents a region around the animal where a risk of contamination exists. Two algorithms have been proposed to improve this computation using different types of trajectories: daily and hourly trajectories. Then, we propose a mathematical model to determine the risk of contamination for the intensity of contact between two animals. For that, two parameters are considered. The first is the time spent by an animal in the Activity area, and the second is the distance between the animals.

The traceability information system presented here shows to have potential to increase product quality and trustworthiness for improving the precision of the traceability process. The proposal focuses on an important target that involves a large number of actors around an important subject for human health. This information system should be useful to governmental institutions and farmers in case of a recall, saving time, costs and collaborating to achieve the food security.

## REFERENCES


ABIEC. (2009). *Brasil lidera ranking de exportação da carne bovina*. Retrieved July, 2, 2009, from http://www.abiec.com.br/news_view.asp?id={CAACE975-B5D1-4337-9F3B-580E7118CB45}

AGROTRACABILITE. (2009). *Agi-traçabilité Québec*. Retrieved July, 2, 2009, from http://www.agri-tracabilite.qc.ca

Allen, J. (1983). Maintaining knowledge about temporal intervals. *Communications of the ACM*, *26*(11), 832–843. doi:10.1145/182.358434

Andrienko, G., & Andrienko, N. (2008). A Visual Analytics Approach to Exploration of Large Amounts of Movement Data. In *Proceedings of the* VISUAL 2008, (pp. 20-35). Berlin: Springer-Verlag.

Andrienko, N., Andrienko, G., & Gatalsky, P. (2003). Exploratory spatio-temporal visualization: an analytical review. *Journal of Visual Languages and Computing*, *14*(6), 503–541. doi:10.1016/S1045-926X(03)00046-6

Armstrong, M., & Densham, P. (1990). Database organization strategies for spatial decision support systems. *International Journal of Geographical Information Systems*, *4*(1), 3–20. doi:10.1080/02693799008941525

Baltzer, O., Dehne, F., Hambrusch, S. E., & Rau-Chaplin, A. (2008). OLAP for Trajectories. In *Proceedings of the* DEXA 2008, (pp. 340-347). Berlin: Springer-Verlag.

Bédard, Y., Merrett, T., & Han, J. (2001). Fundaments of Spatial Data Warehousing for Geographic Knowledge Discovery. In *Geographic Data Mining and Knowledge Discovery* (pp. 53–73). London: Taylor & Francis. doi:10.4324/9780203468029_chapter_3

Bernier, E., Gosselin, P., Badard, T., & Bédard, Y. (2009). Easier Surveillance Of Climate-Related Health Vulnerabilities Through A Web-Based Spatial Olap Application. *International Journal of Health Geographics*, *8*(18).

Booch, G., Rumbauch, J., & Jacobson, I. (1999). *The Unified Modeling Language User Guide*. Reading, MA: Addison-Wesley.

Brun, T., Poyet, P., Boop, M., & Vigier, F. (2005). Towards an agricultural ontology in France: Contributions of the farm Information Management Project (GIEA), In *Proceedings of the* EFITA/WCCA 2005 (pp. 1296-1302).

Butler, P., Corke, R., Peterson, P., & Rus, D. (2006). Dynamic Virtual Fences for Controlling Cows, in Experimental Robotics. *Springer Tracts in Advanced Robotics*, *21*, 513–522. doi:10.1007/11552246_49

Cao, H., Mamoulis, N., & Cheung, D. W. (2007). Discovery of Periodic Patterns in Spatiotemporal Sequences. *IEEE Transactions on Knowledge and Data Engineering*, *19*(4), 453–467. doi:10.1109/TKDE.2007.1002

Compieta, P., Di Martino, S., Bertolotto, M., Feerucci, F., & Kechadi, T. (2007). Exploratory spatio-temporal data mining and visualization. *Journal of Visual Languages and Computing*, *18*(3), 255–279. doi:10.1016/j.jvlc.2007.02.006

Dufy, L., Abt, V., & Poyet, P. (2006). GIEA: gestion des informations de l'exploitation agricole. Un projet au service de l'interopérabilité sémantique de la profession agricole. *Ingénieries - Eau, Agriculture. Territoires*, *48*, 27–36.

Gäuting, R. H., Almeida, V. T., & Ding, Z. (2006). Modeling and querying moving objects in networks. *The VLDB Journal*, *15*, 165–190. doi:10.1007/s00778-005-0152-x

Gäuting, R. H., Bäohlen, M. H., Erwig, M., Jensen, C. S., Lorentzos, N. A., Schneider, M., & Vazirgiannis, M. (2000). A foundation for representing and quering moving objects. *ACM Transactions on Database Systems*, *25*(1), 1–42. doi:10.1145/352958.352963

Gäuting, R. H., & Schneider, M. (2005). *Moving Objects Databases*. San Francisco: Morgan Kaufmann Publishers.

Geerts, F. (2004). Moving objects and their equations of motion. In *Proceedings of the First International Symposium on Applications of Constraint Databases,* (pp. 41–52). Berlin: Springer-Verlag.

GIEA. (2009). *Site Web GIEA*. Retrieved July, 2, 2009, from http://www.projetgiea.fr

Gomez, L. I., Kuijpers, B., Moelans, B., & Vaisman, A. A. (2009). A Survey of Spatio-Temporal OLAP. *International Journal of Data Warehousing and Mining*, *5*(3), 28–55.

IBGE. (2009). *Banco de Dados Agregados SIDRA – Tema: Pecuária*. Retrieved July 1, 2009, from http://www.sidra.ibge.gov.br/bda/pecua/

Lee, A., Liu, Y., Tsai, H., Lin, H., & Wu, H. (2009). Mining frequent patterns in image databases with 9D-SPA representation. *Journal of Systems and Software*, *82*(4), 603–618. doi:10.1016/j.jss.2008.08.028

Lin, D., Jensen, C. S., Ooi, B. C., & Saltenis, S. (2005). Efficient Indexing of the Historical, Present, and Future Positions of Moving Objects. In *Proceedings of MDM 2005*, (pp. 59-69). New York: ACM.

Macedo, J., Vangenot, C., Othman, W., Pelekis, N., Frentzos, E., & Kuijpers, B. (2008). Trajectory Data Models. In Giannoti, F., & Pedreschi, D. (Eds.), *Mobility, data mining, and Privacy – Geographic Discovery* (pp. 123–150). Berlin: Springer-Verlag. doi:10.1007/978-3-540-75177-9_6

Meratnia, N., & de By, R. (2002). Aggregation and comparison of trajectories. In *Proceedings of the 26th VLDB Conference*, (pp. 49-54). New York: ACM.

Oralndo, O., Orsini, R., Raffaeta, A., Roncato, A., & Silvestri, C. (2007). Spatio-Temporal Aggregations in Trajectory Data Warehouses. In [Berlin: Springer-Verlag.]. *Proceedings of the DaWa*, *K07*, 66–77.

Orlando, S., Orsini, R., & Raffaetà, A. (2007). Trajectory Data Warehouses: Design and Implementation Issues. *Journal of Computing Science and Engineering*, *1*(2), 1976–4677.

Pfoser, D., Jensen, C., & Theodoridis, Y. (2000). Novel Approaches to the Indexing of Moving Object Trajectories. In *Proceedings of the 26th VLDB Conference,* (pp. 395-406). New York: ACM.

Pinet, F., Roussey, C., Brun, T., & Vigier, F. (2009). The Use Of Uml As A Tool For The Formalisation Of Standards And The Design Of Ontologies. In *Agriculture. In Advances In Modeling Agricultural Systems*. Berlin: Springer-Verlag. doi:10.1007/978-0-387-75181-8_7

Poyet, P., & Brun, T. (2003). GIEA: gestion des informations de l`exploitation agricole - Vers des concepts et un langage communs pour les partenaires économiques et institutionnels de la profession agricole. *Ingénieries - Eau, Agriculture, Territoires, Technologies pour les agrosystèmes durables* (pp. 167 – 175).

Rodgers, A., Rempel, R., & Abraham, K. (1996). A GPS-Based Telemetry System. *Wildlife Society Bulletin*, *24*(3), 559–566.

Saltenis, S., Jensen, C. S., Leutenegger, S. T., & Lopez, M. A. (2000). Indexing the positions of continuously moving objects. In *Proc SIGMOD Conference*, (pp. 331-342). New York: ACM.

SINDICARNE. (2009). *Produção e exportação de carne bovina brasileira em 2009*. Retrieved July 2, 2009, from http://www.sindicarne.com.br/

SISBOV. (2009). *SISBOV web site*. Retrieved July 2, 2009, from http://www.localizaanimal.com.br/

Sistla, A. P., Wolfson, O., Chamberlain, S., & Dao, S. (1997). Modeling and querying moving objects. In *Proc. Int. Conf. on Data Engineering*, (pp. 422-432). Washington, DC, USA: IEEE.

Spaccapietra, S., Parent, C., Damiani, M. L., Macedo, J. A. F., Porto, F., & Vangenot, C. (2008). A conceptual view on trajectories. *Data & Knowledge Engineering*, *65*(1), 126–146. doi:10.1016/j.datak.2007.10.008

Su, J., Xu, H., & Ibarra, O. (2001). Moving Objects: Logical Relationships and Queries. In *Proceedings of the 7th Int. Symp. on Spatial and Temporal Databases*, (pp. 3-19). Berlin: Springer-Verlag.

Sun, J., Papadias, D., Tao, Y., & Liu, B. (2004). Querying about the Past, the Present, and the Future in Spatio-Temporal Databases. In *Proceedings of the 20th ICDE,* (pp. 202–213). Washington, DC: IEEE.

Tao, Y., & Papadias, D. (2005). Historical spatio-temporal aggregation. *ACM Transactions on Information Systems*, *23*(1), 61–102. doi:10.1145/1055709.1055713

Ungara, E., Henkinb, Z., Gutmana, M., Dolevc, A., Genizid, A., & Ganskopp, D. (2009). Inference of Animal Activity From GPS Collar Data on Free-Ranging Cattle. *Rangeland Ecology and Management*, *62*(4), 256–266.

Vazirgiannis, M., & Wolfson, O. (2001). A Spatiotemporal Model and Language for Moving Objects on Road Networks. In *Proceedings of the Advances in Spatial and Temporal Databases Conf.* (pp. 20-35). Berlin: Springer-Verlag.

Wan, T., & Zeitouni, K. (2007). An OLAP system for network-constraint moving objects, In *Proceedings of the Annual ACM Symposium on Applied Computing,* (pp. 13-18). New York: ACM.

## ADDITIONAL READING


Athanasios S. Voulodimos, Charalampos Z. Patrikakis, Alexander B. Sideridis, Vasileios A. Ntafis, Eftychia M. Xylouri. A complete farm management system based on animal identification using RFID technology Export. Computers and Electronics in Agriculture (01 September 2009)

Food Chain Evaluation Consortium (FCEC). Study on the introduction of electronic identification (EID) as official method to identify bovine animals within the European Union. Retrieved by November 2, 2009, from http://ec.europa.eu/food/animal/identification/bovine/elec_id_bovine_en.htm.

Ipema, A. H., Goense, D., Hogewerf, P. H., Houwers, H. W. J., & van Roest, H. (2008, November). Pilot study to monitor body temperature of dairy cows with a rumen bolus. *Computers and Electronics in Agriculture*, *64*(1), 49–52. doi:10.1016/j.compag.2008.05.009

McMeekin, T. A., & Baranyi, J. (2006). Information systems in food safety management. *International Journal of Food Microbiology*, *112*(3), 181–194. doi:10.1016/j.ijfoodmicro.2006.04.048

Pinto, D. B., Castro, I., & Vicente, A. A. (2006) The use of TIC's as a managing tool for traceability in the food industry. Food Research International, 39 (7). pp. 772-781. ISSN 0963-9969

Qu, X.-h. (2007). Zhuang, D.-f. Qiu, D.-s. Studies on GIS Based Tracing and Traceability of Safe Crop Product in China. *Agricultural Sciences in China*, *6*(6), 724–731. doi:10.1016/S1671-2927(07)60105-9


## KEY TERMS AND DEFINITIONS

**Animal Traceability:** The ability to verify the history, location, or application of an animal during its whole life, using documented and recorded identification.

**Spatial Decision Support System:** Interactive, computer-based system designed to support a user or group of users in achieving a higher effectiveness of decision making while solving a semi-structured spatial problem.

**Trajectory Databases:** Spatial database systems to manage trajectories. A trajectory is the description of the movement of geometry changing over time.

# Chapter 19

# Construction of Agri-Environmental Data Using Computational Methods:

## The Case of Life Cycle Inventories for Agricultural Production Systems

**Susumu Uchida**
*National Agriculture and Food Research Organization, Japan*

**Kiyotada Hayashi**
*National Agriculture and Food Research Organization, Japan*

**Masaei Sato**
*National Agriculture and Food Research Organization, Japan*

**Shingo Hokazono**
*National Agriculture and Food Research Organization, Japan*

## ABSTRACT

*Increased public focus on agri-environmental issues and recent policies on agricultural sustainability have necessitated the construction of a life cycle inventory (LCI) database for agricultural production systems. However, the current progress of LCI database construction is far from being complete in both developed and developing countries. In this chapter, an integrated view for a data-construction methodology for agri-environmental assessment is proposed. The applications of computational methods to the construction processes are also proposed, with a special focus on Japanese case studies. After discussing the methods for construction, the implications of LCI construction are presented, which includes the following issues: how to make a transition to sustainability, and how to achieve informed and science-based policy decisions by increasing the applicability of life cycle assessment and the level of preparedness. The Japanese case studies should support assessment and decisions in developed and developing countries.*

DOI: 10.4018/978-1-61692-871-1.ch019

## INTRODUCTION

Global food trends—such as intensified food–fuel competition and deforestation due to the increase in biofuel production—have necessitated informed decision making based on environmental and economic assessments. These include, for example, life cycle assessment (LCA), life cycle costing (LCC), environmental impact assessment (EIA) and strategic environmental assessment (SEA). The construction of a database is the first step towards achieving informed decision making processes, and mathematical models and computational methods are expected to play an important role.

However, at present, the life cycle inventory (LCI) data for agricultural production systems, which constitute a representative database for environmental assessment and are necessary for conducting LCAs of agricultural products, are generally inadequate, both quantitatively and qualitatively. This is applicable to Japan as well as to other countries. The reason for the inadequacy of LCI data is related to the characteristics of agricultural production, including the diversity of cultivation practices and the existence of a variety of statistical datasets.

In this chapter, we propose an integrated view of a data-construction methodology for agri-environmental assessment. Special attention is focused on the classification and characteristics of the computational methods for each stage of data construction and database development. The methods are broadly classified according to the degree of computation, as shown in Figure 1. Individual methods can be selected from Figure 1, or methods can often be combined depending on the data sources. The details of the application of such computational methods are described in the following subchapters. The following topics are discussed in terms of data construction: (a) life cycle assessment and life-cycle inventories, (b) foreground system data for agricultural production, (c) background system data for agricultural production and (d) LCI database management and development. Japanese cases are largely studied as instances of data construction in a country where systematized datasets in terms of agricultural production are not well established.

## BACKGROUND

As an analysis method, LCA has attracted considerable attention in recent years; it is regarded as the most comprehensive technique for quantitatively evaluating the environmental load associated with a product or project. It covers all the processes related to the subject under assessment; these include the extraction of natural resources, production, cultivation, manufacture, circulation, consumption, waste disposal and so on. The subject of analysis is extensive and exhaustive; there exist no one-dimensional data that can be directly

*Figure 1. Classification and characteristics of computational methods in LCI construction (Extracted from Hayashi, 2008)*

| Classification of computational method | Major data source | Data processing | Processing tool |
|---|---|---|---|
| Calculation | Field data<br>Experimental data | Simple calculation | Calculator |
| Estimation | Government statistics<br>Experimental data | Statistical processing | Spread sheet on computer |
| Simulation | Various | Construction or application of simulation model | Software for simulation or modeling |

applied to LCA because so many kinds of data are required in LCA. Therefore, LCA must always deal with diversity in the reliability and certainty of data. Unlike other methods such as multivariate analysis, LCA considers how to process the data. In other words, while other analyses process data by using already established methodologies, LCA allows us to choose the determination methodology for data processing. In this sense, LCA is characterized by a synergetic relationship between data and methodology.

LCI is the database construction component and assumes the basic and concrete contents of LCA. It serves as a useful database by itself and is applicable to other analyses such as multiple criteria decision making and mathematical programming. Since LCA is also the most fundamental application of LCI, it is important to describe it prior to LCI.

## LIFE CYCLE ASSESSMENT AND LIFE CYCLE INVENTORIES

ISO standards that establish LCA as including LCI are ISO 14040 (environmental management—life cycle assessment—principles and framework) and ISO 14044 (environmental management—life cycle assessment—requirements and guidelines). According to these standards, LCA is split into four stages: (a) goal and scope definition, (b) life cycle inventory analysis (LCI), (c) life cycle impact assessment (LCIA), and (d) life cycle interpretation. The structures of and relationships among these stages are shown in Figure 2. In the following section, we briefly outline these four stages.

### Goal and Scope Definition

Prior to assessment, the purpose and extent of the analysis must be clearly defined. The goal takes into account the intended audience. Two important items in the definition of scope are the *functional unit* and the *system boundary*. The functional unit is the basis of all analyses and is especially important when comparing two or more subjects. For agricultural production systems, the results of the comparison may vary between cases in which the area (hectare) or yield (ton) are chosen as functional units. Defining the system boundary involves considering what processes (e.g., quarrying for metallic resources, building of sheds for agricultural machines) are included in the assessment. The methodology and impact categories to be considered are also important in the goal and scope definition process. The latter,

*Figure 2. Stages in LCA (Adopted from The International Organization for Standardization, 2006)*

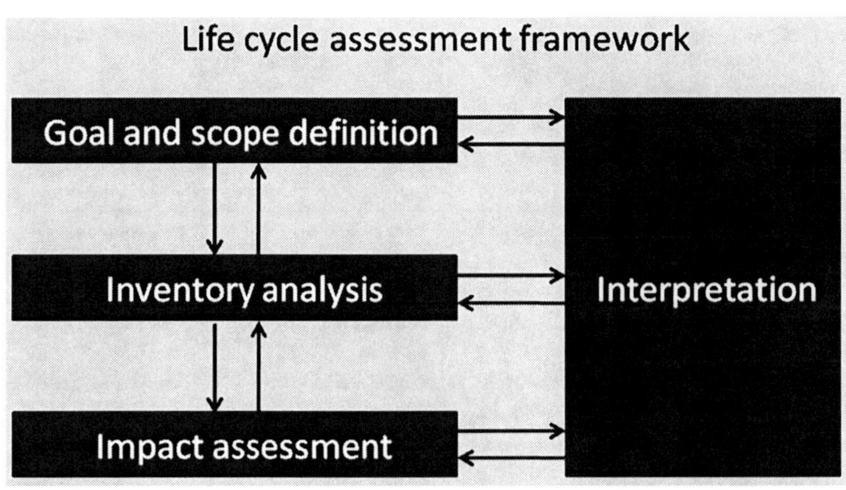

in particular, leads the discussion on what substances should be considered and is also related to defining the system boundary.

## Life Cycle Inventory Analysis (LCI)

This stage is the most practical and essential task in LCA. All resources that are input to the considered system and all emissions that exit the system are calculated in terms of the substances defined in the previous stage. In LCA studies, the data for direct environmental burdens are often called foreground system data; indirect environmental burdens, which are a sort of "hidden" data, are called background system data. For agricultural production systems, carbon dioxide ($CO_2$) emissions due to fossil fuel consumption during the operation of agricultural machines and the penetration of pesticides into the ground are examples of foreground system data. Background system data is what sets LCA apart from other assessment methods, and it includes emissions from the processes used to produce materials, manufacture machines, construct buildings, and so on. Inventory analysis utilizing these data must meet the requirements of the definitions for the goal and scope; a model of the system is usually constructed prior to data collection and calculation. When a flow in the system splits into multiple flows (e.g., by-products) the inventory data need to be allocated accordingly. Allocation is one of the most puzzling issues in LCA and requires certain knowledge on the subject.

## Life Cycle Impact Assessment (LCIA)

When the inventory data are established, they are sorted into impact categories and weighted according to their degree of contribution to the impact categories. This classification is a simple process of sorting based on what impact is related to each inventory. Impact categories denote the type of influence on the environment, human health, or socio-economic system, e.g., global warming, eutrophication, acidification, ozone depletion, ecotoxicity, land use, etc. The weighting in this stage is called characterization. Some substances have established coefficients in terms of their impact; these represent the degree of contribution to the impact category, and characterization in such cases simply means multiplying each coefficient by the corresponding inventory and summing the contributions of all the coefficients. The coefficient of global warming is the global warming potential (GWP), which is one of the most widely known indices. However if the coefficient has not been established or it would vary depending on the goal and scope of assessment, it must be estimated by LCA researchers themselves. Even the GWP varies in accordance with the time period of analysis, and it indicates the importance of the first step, i.e., the goal and definition of scope in LCA.

Normalization is an optional stage in LCIA, in which the ratios of summarized impacts are evaluated again relative to a reference value in the standards, such as the total impact from the target region. The results become dimensionless values, and they can be summed up according to the impact categories so that two or more subjects can be compared simply and directly. In some other cases, the summarized impacts are weighted based on other criterions and represented as values with the same dimension. The criterion is chosen depending on the goal, such as disability adjusted life year (DALY) and social cost; it is also possible to compare two or more subjects. This stage is called the endpoint, while each impact category is called a midpoint. Both cases are commonly encountered in LCA: the assessment is completed using both the midpoint and the endpoint, and the normalization and final weighting processes are skipped in the midpoint analysis. It also depends on the goal whether preferable method is a midpoint analysis or the endpoint analysis.

The entire LCIA process can be skipped in an LCI study; the next stage, i.e., life cycle interpretation, is conducted directly on the basis of the LCI results.

### Life Cycle Interpretation

The results of LCIA or LCI are finally evaluated qualitatively in accordance with the goal and the scope definition. The significant items that largely affect the results are identified, and checks for various issues, including sensitivity, are carried out. The uncertainty of the results must be evaluated in the sensitivity check to reflect the uncertainty of data and calculations with various levels of estimation and approximation; this comes from the fact that various kinds of data sources are used exhaustively in LCA. These considerations are then reported to the intended audience, which is the next step for LCA.

## FOREGROUND SYSTEM DATA FOR AGRICULTURAL PRODUCTION

Even though foreground system data, which are related to direct environmental burdens, are easier to obtain than background system data, the essential instability and inapplicability of agriculture in a quantitative manner means that special efforts are required to ensure data cogency. The methodologies for data collection and dataset construction are discussed in this subchapter.

### Measurement by Actual Field Survey and Experiments at Institutes

Of the data needed in LCIs of agricultural production systems, technical data such as working times, amounts of materials, and types of agricultural machines can only be obtained by direct measurement of the field being cultivated. If the subject crop is cultivated in a research institute as an experiment, cultivation data are usually collected. Therefore, data can be obtained if no problems with intellectual property exist for data offered by the data collector to LCI researchers and if the data content matches the aim of the study. These institute data are as valid as field data, however, an attention is necessary that the data may be abnormal because the experiment itself is often conducted under a special condition.

Cultivation data from farmers constitute actual field data. Such data are usually collected through individual interviews or questionnaires. In the EU, an abundance of accountancy data is accessible through the Farm Accountancy Data Network (FADN). In Japan, some of these questionnaires are implemented every 5 years by the government and compiled into statistical data as the Census of Agriculture and Forestry. However, no systematized databases such as FADN exist in Japan; researchers must conduct interviews by themselves in many cases. An actual survey of agricultural practices in fields is the surest way to obtain data in terms of certainty. However, a large number of surveys are required to obtain a statistically representative sample; many external factors greatly affect agricultural practices both qualitatively and quantitatively. Weather, temperature, and precipitation vary from one region to another; soil properties and material prices can be different in different areas; and manpower and dimension of the field vary farmer to farmer. It is very expensive to conduct surveys that cover all of these factors. Therefore, these types of surveys are usually conducted together with other macroscopic research.

Some farmers are subsidized in accordance with prescribed agricultural standards. In such cases, farmers have to submit their agricultural data regarding the standards. The data summarized by these farmers can be utilized as data sources, thereby leading to savings in terms of time and cost for both researchers and farmers. A typical example of such standards is GLOBALGAP, which certifies farmers as part of a good agricultural practice (GAP). ISO and the labeling system for carbon footprints are also related to agricultural production. In Japan, the Japanese Agricultural Standards (JAS) system has three standards involved with the production process: (a) specified JAS, (b) JAS with production information declaration, and (c) organic JAS. In addition, every prefectural government has an accreditation system for eco-farmers who employ sustainable agricultural production.

Apart from cultivation, field data such as greenhouse gas (GHG) emissions from soil are collected through measurements from researchers regardless of whether the field is in an institute or a farm; this is because this type of data has no direct relation to agricultural production and the measurement requires technical skills. In particular, measuring GHG emissions takes a very long time because it is conducted repeatedly at the same point over many years. Recently, some models predicting GHG emissions have been developed and applied to LCA, as mentioned in the following section. It is increasingly common to carry out measurements of GHG emissions; these measurements are conducted with the aim of aiding in the construction and modification of models, rather than serving as independent sources of data.

Although measurements in actual field surveys and experiments do not need complicated computation, simple statistical processing such as the derivation of average and standard deviation values must be introduced when many interviews or questionnaires are implemented. Nonetheless, for this stage, emphasis is laid on constructing a system and format for data collection rather than on developing computational methods.

Another important point for field surveys is the construction of a trusting relationship. Unlike nationwide researches, the outcome of field surveys is usually dominated by whether the subject farmers can establish sufficient trust with the researcher to volunteer information. Therefore, the researcher must spend a long time in building up communication and constructing good relationships. For efficiency, it is recommended to obtain the cooperation of local organizations or farming communities. In this regard, Japan Agricultural Cooperatives (JA) plays an important role in Japan.

## Estimation Using Literature Values and Simulation Models

Some data can be estimated using information from literature and simulation models without using direct surveys. In particular, for data obtained from purely physical or chemical processes, data estimation from the literature or from models is a common practice. The carbon content in a specific biomass is a typical example of information that can be accessed from the literature, and $CO_2$ emission from biomass incineration can be easily calculated using information on the mass ratio of carbon in $CO_2$. For instance, the guidelines from the Intergovernmental Panel on Climate Change (IPCC) (Houghton et al., 1996) provide substantial data on the LCI of GHG emissions and has been employed for developing inventories in many countries including Japan (Greenhouse Gas Inventory Office of Japan, 2009). Japanese inventory data by themselves are also a useful data source for estimating the LCI of GHG emissions in Japan.

For the foreground system of agricultural production, GHG emissions and nitrogen leaching from farmland soil are very important topics that have been covered in recent LCA studies, especially in terms of the environmental aspect. A number of studies have aimed to measure GHG emissions in various situations and derive the emission coefficient. For example, Tsuruta (1997) established emission coefficients of methane ($CH_4$) and dinitrogen monoxide ($N_2O$) in Japan, and Akiyama, Yagi, and Yan (2006) carried out research for developing the Japanese GHG inventory (Greenhouse Gas Inventory Office of Japan [GIO], 2009). $CO_2$ emissions from soil are strongly related to changes in soil organic carbon caused by turnover of soil organic matter. For upland fields, adequate oxygen in the air is supplied to the soil; only aerobic fermentation (and not anaerobic fermentation) takes place. Hence, when $CO_2$ is emitted to the atmosphere as gas, it is considered to comprise all the carbon components. In other words, $CO_2$ emissions are derived from the increase in soil organic carbon by subtracting it from the input of organic carbon included in manure compost and crop residue. Koga and Tsuruta (2006) conducted an LCI analysis in which $CO_2$ emission was estimated using this method, and Kusaba and Ishioka (2008) constructed a database for the measured soil properties.

However, the obtained data must be generally versatile with various soil conditions and agricultural operations. It is costly to measure the amount of emissions at each site in a survey; hence, model simulation is a useful tool for estimation. Several models have been developed to predict GHG emission amounts. The Rothamsted carbon (RothC) model (Coleman & Jenkinson, 1996) was originally developed to simulate the behavior of organic carbon in soil. The model only needs a small number of parameters as input and is easy to manage for researchers in fields other than soil science. Furthermore, this model is also highly accurate. Shirato, Hakamata, and Taniyama (2004) modified the RothC model for application to andosols, which are soil components of arable land in Japan. Andosols originate from volcanic ash and contain activated aluminum. The decomposition rate of humus in andosols was modified in the model because activated aluminum stabilizes humus. However, among the various GHG emissions, only $CO_2$ emissions from upland soil can be simulated; this is because the model does not aim to simulate GHG emissions.

The denitrification-decomposition (DNDC) model (Li, Frolking, & Frolking, 1992) aims to simulate GHG emissions from arable land. The model comprehensively addresses the behavior of both nitrogen and carbon; therefore, $CH_4$ and $N_2O$ emissions are evaluated in addition to $CO_2$. Furthermore, nitrate leaching into groundwater, which causes water pollution, can also be simulated. The drawback of this model is that many detailed parameters are needed for simulation and the model structure is complicated. One application of the model to Japanese agriculture was performed by the National Institute for Rural Engineering (2006).

The CENTURY model (Parton, Schimel, Cole, & Ojima, 1987; Parton, Stewart, & Cole, 1988) deals with the balances of more substances in the continental biosystem than the previous two models. These substances—carbon, nitrogen, phosphorus and sulfur—strongly affect the environmental burden. Carbon and nitrogen are integral components of GHG, phosphorus is a major water pollutant, and sulfur forms $SO_X$, which is a major air pollutant.

When investigating agricultural production alone, simulation models that predict crop growth are sometimes useful estimation tools. The International Consortium for Agricultural Systems Applications (ICASA) developed the software for a simulation model called the Decision Support System for Agrotechnology Transfer (DSSAT). This model simulates the growth of crops from parameters for soil, climate, and agricultural operations, including material input.

Since most of these simulation models are distributed in software packages, complicated computation is not necessary. However, some knowledge of data processing and MS-DOS commands is required to run some of the simulations smoothly.

## Utilization of Government Statistics and Data for Agricultural Extension

Government agencies often conduct nationwide researches, including a census, to figure out the state of the nation and utilize the obtained information for planning and decision making. Such comprehensive macro-data are the most reliable for obtaining average figures, although the specific factors underlying such data are ignored. For Japanese agriculture, there is an annual statistical survey of agricultural management as well as a census of agriculture and forestry every 5 years. The statistical survey is mainly concerned with financial data of agricultural management, while the census deals with the agricultural management style (kind of crop, scale, equipment, revenue and expenditure, etc.), personnel composition and average age, employment and commission, side jobs, etc. In addition, many statistical surveys of agriculture are implemented by the Ministry of Agriculture, Forestry and Fisheries of Japan. For dairy farming, the industry organization Japan Livestock Industry Association compiles data on farmer management.

Most of the data in these government surveys concern agricultural management and do not directly contribute LCI data. However, some surveys are conducted to determine the allocations of consumption, emission, and environmental burdens to a number of agricultural products or activities. Even though a farmer may produce multiple crops, irrespective of whether the crops are produced by rotation or by operating separate fields, we are often able to measure only the total environmental load from the entire farmland; in such cases, we must estimate the amount attributable to the production of a specific crop, and the allocation method used plays an important role in the analysis. The allocation methods are usually based on the field area, harvest yield, or proceeds of each crop. For example, when the total amount of material consumed in an area is given, the allocated amounts for each crop are estimated by multiplying a fraction of the abovementioned factors for each crop by the total amount. In another example, the lifetime of agricultural machines can be estimated by dividing the investment for the machines by its depreciated value.

More closely related to LCI are systematized techniques that are mainly published by prefectural governments with the aim of extending agriculture and agricultural technologies. Although the data are predetermined standard values that are not based on statistical surveys, they contain concrete procedures for cultivation of agricultural crops in terms of the kind and quantities of machines needed, fertilizers, agrichemicals, manpower, and time. Although the data represent standard technology for the subject prefecture only and their item and amount vary from one region to another, the numerical data greatly help in constructing LCI data for agricultural production systems.

When these data are utilized, some calculation or statistical processing is sometimes needed to modify the obtained data for applying data construction procedures, although these calculations are not technically complex.

## BACKGROUND SYSTEM DATA FOR AGRICULTURAL PRODUCTION

Background system data are related to indirect environmental burdens from the subject system. In general it is much more difficult to precisely evaluate background system data than foreground system data; various methods must be introduced to estimate background system data values.

### Analysis Using an Input-Output Table

One of the most commonly used methods in Japan for estimating background system data is analysis based on an input-output table. The table represents mutual values of input among sectors, including industries, public sectors and final consumption. Input-output analysis, which is a major application of these tables, is generally used to quantitatively evaluate the eventual multiplied effects of changes in activity for certain industries and to the whole economic system; this includes production for other industries and the final consumption of goods and services from other industries.

Let us consider the transportation vehicle manufacturing industry as an example. To produce vehicles, many parts such as engines, glass and tires must be produced in corresponding industries. However, these respective industries demand transportation vehicles to transport the products. Because of this circularity, the calculation seems to have no end when one wants to know how much glass does the transportation vehicle industry needs to produce their product. This problem can be cleared up using the calculations f an input-output matrix. When **X** and **C** represent column vectors whose elements consist of the production and final consumption, respectively, of industries, their relationship is as follows:

$$\mathbf{AX} + \mathbf{C} = \mathbf{X} \quad (1)$$

where **A** is a square matrix whose elements are the input of a product to another or the same industry divided by the production of the industry receiving the input. This proportion is called the input coefficient. Equation (1) represents the balance for the supply and demand, where **AX** denotes the products demanded by other industries as intermediate input for the production of their own products and on the right side is the total production. This total production as the supply meets all of the multiplied demands; the final consumption is on the left side of the formula, and its value is obtained by following transformation:

$$\mathbf{C} = (\mathbf{I} - \mathbf{A})\,\mathbf{X} \quad (2)$$

$$\mathbf{X} = (\mathbf{I} - \mathbf{A})^{-1}\mathbf{C} \quad (3)$$

where **I** is the unit matrix with all diagonal elements as one and other elements as zero. The inverse matrix $(\mathbf{I} - \mathbf{A})^{-1}$ is called the Leontief inverse matrix; its elements represent the produced value from industries induced by the final demand of another industry. In other words, it is the multiplied effect of the final demand from a certain product to other industries. This is the most basic and simple domestic origin–type of input-output analysis model. It has been developed and applied to various models by various researchers. As an example of the application, many regional input-output tables have been developed to analyze regional economies and utilized to obtain background system data in LCI of regional activities to mitigate environmental burdens.

If a value is given indicating some sort of direct environmental burden from each industry, the multiplied life cycle environmental burden is easily derived using the Leontief inverse matrix by multiplying it by a vector composed of the amount of burden per production value. The result is the ultimate impact, including all indirect effects caused by intermediate input. The National Institute for Environmental Studies (NIES) in Japan constructed such datasets for energy consumption and four environmental load substances: $CO_2$, nitrogen oxides ($NO_X$), sulfur oxides ($SO_X$) and suspended particulate matter (SPM) (Nansai, Moriguchi, & Touno, 2002). Needless to say, the first substance is a major GHG, while the others are major air pollutants. As a result of the analyses, the NIES established the coefficients of these substances for the production of each industry, including the indirect effects. This table is called 3EID and is one of the most popular tools for LCA and LCI studies in Japan.

These analyses are on a production value basis; they are predicated on the assumption that the price per weight and composition of products is homogeneous within each industry. The reliability of the result decreases when there is a distribution in the products, which is unfortunately so in most cases. Furthermore, input-output analysis has a problem with temporal change in the structure. The input-output table represents the static industrial structure every five years; the annual tables are extrapolated for more precise analysis. Therefore, dynamic structure changes in industries and the economy may cause errors when two or more databases at different points of time are used as the basis of analysis. The above extrapolated input-output table and 3EID database are often used; this is because 3EID is only published every five years, while no extrapolated tables are published. The types of structural change can be classified as technical and price changes. The former regards a change of production technologies: namely, changes in the amounts of material, labor, capital, etc. to produce a unit product. This change occurs usually within individual plants or firms due to internal and essential production factors, such as innovation, R&D, learning and strategic design change. Therefore, it is a gradual change that has very little effect on discrepancies between databases at different points of time and the errors when using them. On the other hand, price changes are often caused by external changes, such as changes in the supply-demand balance, amount of resources in the market and the global situation. These are rapid and large changes compared to technical changes, and they cause errors in analysis, especially when the price of particular goods or services is greatly changed. This is because the factor affecting analysis accuracy is not the absolute values of the price but their ratio. Although modified input-output tables have been released to prevent the influence of price changes on the analysis, they do not cover all of the instability.

Analysis based on input-output tables has some problems and needs to be addressed, as mentioned above. In spite of these troubles, input-output analysis is widely utilized because of its applicability and exhaustiveness. In many LCI and LCA studies, only the Leontief inverse matrix or established coefficient from input-output analysis is applied with simple calculations. However, the applied input-output model can be constructed individually for advanced analysis using mathematical software. The following are examples of such applied input-output analyses.

When a new technology needs to be evaluated for its comprehensive effect on the economic system, it can also be analyzed by regarding the technology as a new industry and adding it to the input-output table. For example, when a fermentation technology is developed that converts the starch from sweet potatoes to bioethanol fuel, the fermentation industry can be added to the input-output table. Although its input coefficients must be estimated and fixed—i.e. the amount of materials should be estimated in terms of monetary amounts that are input to produce a unit amount of bioethanol fuel—the change in production and demand for other industries by the change in demand for bioethanol fuel can be estimated.

Wastes have a significant relation with environmental burden. Most environmental load substances are emitted from wastes. It is important to consider the flow of wastes to assess environmental burdens from all kind of activity. Nakamura and Kondo (2002) developed a waste input-output table to estimate the waste flow. They classified wastes into 79 categories and set 13 waste treatment sectors. Through analysis using the waste input-output table, it is possible to predict the effect of a new waste treatment technology that reduces the final disposal products or converts wastes to energy sources. The multiplied effect to the entire socio-economic system can be analyzed by adding a sector of new technology to the waste input-output table as an industry, similar to the above example of fermentation. The effects on the amount of wastes can also be analyzed when the present industrial structure changes or a new technology is introduced into the waste input-output table as a new industry.

## Estimation Using Literature Values and Simulation Models

Estimation using information from the literature and simulation models is also useful when researching background and foreground system data. Datasets mentioned in the section on the foreground system, such as the Japanese GHG inventory (GIO, 2009), can be utilized also to obtain background system data. Although the datasets do not contain the multiplied effect, sometimes they provide better approximations than input-output analysis containing the multiplied effect; this can occur when the category of the obtained data covers a more specific area of the material. Furthermore, the input-output table can be modified further to make its classification finer. Afterwards, the obtained data can be applied to the modified input-output table. To give a concrete example, many reports have clarified the environmental burdens of the production process for agrichemicals and chemical fertilizers. Although the kinds of chemicals are too many to be covered by researchers, some of the data produced important information about the environmental burdens of agrichemicals and chemical fertilizers. A project by the Safety and Environmental Technology Group of Swiss Federal Institute of Technology is in progress to construct an estimation model for the environmental impact of chemical production. The model is expected to estimate inventory parameters of chemicals using basic information such as molecular weight and the number of functional groups (Wernet, Hellweg, Fischer, Papadokonstantakis, & Hungerbühler, 2008).

Input-output tables sometimes also act as a database. The main part of the table is described in monetary units, but the amounts of material or unit prices are given for some industries. In such sectors, the monetary data can be converted based on the mass, and the analysis accuracy can be improved by combining the information into the background system data.

A number of economic models have been developed to predict future activity, such as the amounts of production and consumption by companies and consumers. It is also possible to assess the change in environmental burdens when the economic circumstances or specific parameters vary, such as the demand or price of certain goods. These models treat the change in the coefficient of the inventory value per production value through changes in the supply-demand balance and whole industrial structure; it is assumed that they do not change the general input-output model and its applications, such as 3EID. These macroeconomic models are suitable for LCI analysis of new technology or products that are not yet introduced into an actual market but believed as causing a significant change in the economic structure rather than a simple appreciation of the input-output model. Macroeconomic models can also be constructed were the technology, or input coefficient, varies; this is done by assuming more complicated production functions and utility functions. The applications of the simulation results from these macroeconomic models to LCI studies, however, are rarely found; this is probably because the accuracy level of the models does not match that of LCI.

## Data by Industrial Associations

An input-output table classifies industries into tens to hundreds of categories. Even though the classification is fine enough for analysis, some categories include a wide variety of goods and services because many more of certain types go into circulation. In such categories, the reliability of the summarized coefficient decreases. By restricting the subject to a few factors, such as energy consumption or $CO_2$ emissions, many industrial associations release data reported by member companies of the association. If an assumption can be made that all products of the member company are for an industry with a different classification in the input-output table, the association data can be used as substitutional data. Major companies also publish environmental reports, including the amount of environmental burdens; this has been backed by an increase in society awareness of corporate social responsibility (CSR). For some impact categories, the laws stipulate that the amount of certain items must be reported. For example, the Law Concerning the Rational Use of Energy obligates companies which consume more than the threshold amount of energy to report annual energy consumption. The Pollutant Release and Transfer Register (PRTR) system stipulates companies as having to report the amount of chemicals emitted to nature. Data from individual companies sometimes help researchers analyze operations for agricultural production, and the data from monopolistic or oligopolistic companies can be regarded as representative for the whole industry. As an academic society, the Architectural Institute of Japan has released the database and tools for the LCA of buildings (Architectural Institute of Japan, 2006). Sometimes, the data of another industry can be utilized as a substitutional estimation; an example is using the vehicle industry as a substitute for the agricultural machine industry. However, the data should be noted as not being background system data, which should include all multiplied effects. Further analysis using the input-output table is needed to modify this data into strict background system data. Researchers must compare and judge reliabilities between the comprehensive background system data of general categories and the partial but specific data from individual sections or companies.

In contrast, material data is hard to obtain since it directly related to the technical information of companies, which is usually kept top secret. Analyzing the material flow for companies is difficult except for $CO_2$, which is the emission from combusting various fuels. This is the major reason why we must rely mainly on input-output analysis when background system data is being evaluated.

## LCI DATABASE MANAGEMENT AND DEVELOPMENT

Irrespective of the particularity of the subject of assessment or the extent of restriction of the scope, LCI ideally requires a database that covers all kinds of substances and products. Against the backdrop of increasing public concern for environmental problems and the necessity of project assessment, constructing such a comprehensive database set is an urgent challenge. Thus far, various organizations have produced LCA databases, as shown in Figure 3. Figure 4 shows a list of major LCI databases. Ecoinvent and IVAM LCA have the largest corpus of data. In particular, ecoinvent covers a wide area of the world. In addition, many industry organizations of various countries have also constructed databases of their own industries, as shown in Figure 5.

Although these databases have different file formats for data exchange, many of them are compatible with each other. Two major formats are becoming standard worldwide: the EcoSpold

*Figure 3. Global activity of construction of LCI databases (Extracted from Curran & Notten, 2006)*

| Level of Participation | Africa | APEC | Europe | Americas | Global |
|---|---|---|---|---|---|
| Multi-organisational and/or National or Multigovernment[1] | | Australia, Japan, Korea, Singapore, Taiwan | Denmark, Sweden, Switzerland | Canada, USA | |
| Consultants and Research Institutes (data made available) [2] | | | Denmark, Netherlands, Norway, Sweden, Switzerland, Germany, Italy | USA | Boustead, TEAM GaBi, SimaPro |
| Industry(data made available) [3] | | ISSF | PlasticsEurope (formerly APME), FEFCO, ISSF, Volvo, EPD-Norway, EAA | ISSF,APC | IISI, Nickel |
| Decentralized (academic, consultants etc.)[4] | Egypt, Mauritius, South Africa | China, India, Indonesia, Malaysia, Philippines, Thailand, Vietnam | Norway | Brazil, Chile, Colombia, Mexico | |

1.Co-ordinated effort to produce nationally representative and accessible database. Typically involves collaboration between several organisations and some degree of government funding.
2.Inventories produced by research organisations or consultants and made publicly available in a database, most often for a fee (e.g. databases included with LCA software).
3.Inventories produced and published under the auspices of a particular industry organisation. Includes cases where data made only partially available (e.g. for a fee, or only to parties with sufficient motivation for requesting the data). Most often data compiled by consultants, but includes cases where LCI development is done in-house, or by academic or other research organisations.
4.Includes inventories compiled by academic or other research organisations, made either partially or fully available on an ad-hoc basis (e.g., through journal publications). Countries may have some degree of information sharing (e.g. an LCA society), but no co-ordinated data gathering effort (i.e. studies are not organised into an accessible database).

*Figure 4. Existing national LCI databases (Extracted fromCurran & Notten, 2006)*

| Name | Availability | Language | Data focus (if any) | Geographic coverage | Number of datasets |
|---|---|---|---|---|---|
| Australian Life Cycle Inventory Data Project | Free | English | | Australia | >100 |
| BUWAL250 | Fee or included with SimaPro | German, English, French | Packaging materials | Switzerland | |
| Canadian Raw Materials Database | Free | English, French | Raw materials | Canada | >10 |
| DuboCalc | Upon request | top level in Dutch/underlying LCA data in English | Construction materials | Netherlands | >100 |
| Dutch Input Output | Licence fee | English | Input-output | Netherlands | >100 |
| ecoinvent | Licence fee | English, Japanese, German | | Global/ Europe/ Switzerland | >1000 |
| Eco-Quantum | | Dutch | | | |
| EDIP | Licence fee | Danish, English, German | | Denmark | >100 |
| Franklin US LCI | Available with SimaPro | English | | U.S.A | >10 |
| German Network on Life Cycle Inventory Data | On-going | German, English | | Germany | |
| ITRI Database | | Taiwanese, English | | | |
| IVAM LCA Data | Licence fee | Chinese, English | Construction, food, waste, etc. | Netherlands | >1000 |
| Japan National LCA Project | Fee | Japanese | | Japan | >600 |
| Korean LCI | On-going | | | | |
| LCA Food | Free | English | Food products | Denmark | |
| SPINE@CPM | Fee | English | - | Global | >100 |
| Swiss Agricultural Life Cycle Assessment Database (SALCA) | Free with contact | German | Agriculture | Switzerland | >100 |
| Thailand LCI Database Project | | Thai, English | | | |
| US LCI Database Project | Free with contact | English | | US | 73 |

format, which has been adopted for the ecoinvent database, and the ILCD format, which has been developed by the European Commission.

For the computation of LCA, software applications are used together with the databases. Figure 6 shows a list of the major LCA software;

*Figure 5. Existing industry organization LCI databases (Extracted from Curran & Notten, 2006)*

| Industry organisation | Availability | Product group or sector | Geographic coverage |
|---|---|---|---|
| American Iron and Steel Institute (AISI) | | Iron and steel | America |
| American Plastics Council (APC) | | Polymers | America |
| EDP-Norway | Free | Norwegian business (several sectors) | Norway and Europe |
| European Aluminium Association (EAA) | Free | Aluminium | Europe |
| European Copper Institute (ECI) | Free with contact | Copper | Europe |
| European Federation of Corrugated Board Manufacturers (FEFCO)<br>Groupement Ondulé, European Association of Makers of Corrugated Base Papers (GEO)<br>European Containerboard Organisation (ECO) | Free | Corrugated Board | Europe |
| International Iron and Steel Institute (IISI) | Free with contact | Steel | Global |
| ISSF International Stainless steel Forum (ISSF) | Free with contact | Stainless steel | Global |
| KCL(EcoData) | Fee | Pulp and paper | Finnish/Nordic |
| Nickel Institute | Free with contact | Nickel | Global |
| PlasticsEurope (formerly APME) | Free | Plastics | Europe |
| Volvo EPDs | Free | Trucks and busses | Europe |

in particular, SimaPro and GaBi are utilized worldwide. These LCA applications are characterized by the integration with databases, and the major formats, such as EcoSpold and ILCD, are supported by many applications for compatibility. Although standard personal computers have enough capacity to run the applications and users do not need to prepare any particular hardware, the calculation time depends on the specifications of the computers and the conditions of assessment. For instance, SimaPro takes less than a minute to run an assessment on a recent standard personal computer.

The Open LCA project has been organized by a German company, and is used for constructing open-source LCI databases and LCA software. This project started in 2006 and, at present, the beta 1.1.1 version is available. It is expected to be a new platform for LCA studies in the future.

The contents of these LCI databases have become increasingly rich; however, they still lack information in terms of both quality and quantity, compared to the enormous volume of required information. Therefore, projects to enrich and expand these databases are continuing.

## Project of JEMAI/JLCA

For Japan, there are few existing databases for detailed LCA, and this applies not only to the agricultural production system but also to almost all industries. The Japan Environmental Management Association for Industry (JEMAI) formed the Life Cycle Assessment Society of Japan (JLCA) and constructed the database for LCA together with the LCA software named "JEMAI-LCA Pro" in the project "Development of Assessment Technology of Life Cycle Environment Impacts of Products and so forth" from 1998 to 2003; this project was sponsored by Ministry of Economy, Trade and Industries in Japan. The main content of the database for the software and model are

*Figure 6. Existing LCA software (Extracted from Curran & Notten, 2006)*

| Name | Availability | Language | Data focus (if any) | Geographic coverage | Number of datasets[1] |
|---|---|---|---|---|---|
| BEES 3.0 | Free with contact | English | Building materials and products | USA | 200 |
| Boustead Model 5.0 | License fee | English | | Global | |
| CMLCA 4.2 | License fee only for commercial use | English | | Europe | |
| eiolca.net | Free | English | Input-Output | USA | >100 |
| EMIS | License fee | English, German | | Global | >1000 |
| Environmental Impact Estimator | License fee | English | Building materials and products | Canada, USA | >10 |
| GaBi | License fee | English, German, Japanese | | Global | >1500 |
| GEMIS | | English, German | | Europe | |
| GREET 1.7 | Free | English | Transportation sector, energy sector | USA | >20 |
| IDEMAT 2005 | License fee | English | Engineering | Netherlands | >100 |
| KCL-ECO 4.0 | License fee | English | | Global | |
| LCAiT | License fee | English | | | |
| MIET | | | | | |
| AIST-LCA (JEMAI-LCA) | License fee to JEMAI | Japanese | | Japan | >500 |
| Regis | License fee | English, German, Japanese | | Global | |
| SimaPro | License fee | English, Japanese | | Global | >1000[1] |
| TEAM | | | | Global | |
| Umberto | License fee | English, German, Japanese | | Europe | [1] |

1.Additional datasets are available for an additional fee (i.e. in a compatible format)

environmental burdens from industrial sectors; agricultural processes are not mentioned. Nonetheless, the data are helpful for analysis of background system data in agricultural production systems, especially for fuels, machines, construction material and transport. Another achievement of this project is the life-cycle impact assessment method based on endpoint modeling (LIME); this is a damage-calculated impact assessment method based on the constructed database set. LIME has recently been employed in various studies and assessments in Japan.

### Constructing Database for Japanese Agricultural Production Systems

The National Agriculture and Food Research Organization in Japan (NARO) started a project for constructing a database for agricultural production systems. The objective crops are rice, sugar beets, potatoes, sweet sorghum, sweet potato and sugarcane; these will be expanded in series. The computational method is based on the ecoinvent database and has been combined with an original method to take into account factors particular to Japan. The data format conforms to SimaPro, which is one of the most widely used LCA software and is easily accessible.

The scope of this database set includes materials, agricultural machines and transport machines input to the operation, and emission from fuel combustion, tyre abrasion and farmland soil. Constructions and labor are outside of the scope. Materials include fuels, seeds, fertilizers, agrichemicals and other materials such as plastics for mulching. The quantitative data in operations are based on interviews and data for agricultural extension. Details of the construction procedure for the dataset will be published together with the main dataset.

## FUTURE RESEARCH DIRECTIONS

LCA is used as a tool that supports decision making. When the magnitude of the scope in assessment is large enough, the effect of the subject for changing the surrounding socio-economic system cannot be neglected. However, the present LCA does not consider these consequences. Some studies deal with consequential LCA, which is a suitable tool for such type of assessment, compared to conventional attributional LCA. In attributional LCA, a part of the existing material flow and environmental burden is attributed to the subject of assessment using real average data and derived coefficients. In contrast, consequential LCA considers all the consequences of decision making, including changes in the socio-economic system, and therefore is usually based on scenario settings or economic models and marginal data of products, in terms of the material flow and environmental burden. Because it often relates to future prediction, spatial and time scales are important factors and need to be defined. Although much discussion is needed to establish its methodology, consequential LCA is expected to be one of the major technologies of LCA in the near future.

In terms of the database, the presence of spatial and time factors will be necessary in individual data for consequential LCA. Hence, the above discussion concerning database construction may substantially change. Although how they will change is uncertain, there is a correspondence with consequential LCA in some databases and formats. In the updated version of EcoSpold, for example, some meta-information items, including a macro-economic scenario have been added. This version corresponds to ecoinvent v3, which will be released in 2011, and to some extent these changes are considered to be more conscious regarding the aspect of consequential LCA.

## CONCLUSION

Now that we have discussed all of the important aspects of constructing LCI data, our next task is to consider the implications of data construction in decision making for the agricultural sector. Hence, in concluding this chapter, we provide future perspectives on agricultural decision making.

First, constructing LCI data implies an increase in the applicability of LCA. A remarkable recent trend in the application of LCA is its use in legislation. For example, to meet sustainability standards for biofuels, life cycle emissions of GHGs must be calculated. The development of an LCI database is necessary for the effective application of LCA to such legislation.

Second, the applicability of LCA has a major bearing on the level of preparedness for informed, science-based policy making, which can be related to research questions such as how to revitalize rural areas, how to realize a bio-economy and ultimately how to make a transition to sustainability.

Finally, decision analysis studies will play an important role in enabling the transition to sustainability on the basis of informed, science-based policy making. Indeed, LCA can be considered as a multi-criteria assessment. Thus, in order to make effective decisions, it is necessary to integrate multiple criteria. In integrating multiple dimensions, computational studies that include multi-criteria decision making, preference construction, and group decision making will be important.

The application of LCA in decision making will become increasingly important, especially in developing countries. However, these countries lack databases that are as detailed as those in Japan. Hence, constructing a method for Japanese LCI databases, as described in this chapter, can support further studies and assist decision makers in these developing countries.

## REFERENCES


Akiyama, H., Yagi, K., & Yan, X. (2006, March). *Direct $N_2O$ emissions and estimate of $N_2O$ emission factors from Japanese agricultural soils.* Paper presented at International Workshop on Monsoon Asia Agricultural Greenhouse Gas Emissions, Tsukuba, Japan.

Architectural Institute of Japan (Ed.). (2006). *Tatemono no LCA Shishin* (3rd ed.). Tokyo, Japan: Architectural Institute of Japan. (In Japanese)

Coleman, K., & Jenkinson, D. S. (1996). RothC-26.3---A model for the turnover of carbon in soil. In Powlson, D. S., Smith, P., & Smith, J. U. (Eds.), *Evaluation of Soil Organic Matter Models: Using Existing Long-Term Datasets* (pp. 237–246). Berlin, Germany: Springer.

Curran, M. A., & Notten, P. (2006). *Summary of global life cycle inventory data resources.* Report for Task force 1: Database Registry, SETAC/UNEP Life Cycle Initiative.

Greenhouse Gas Inventory Office of Japan. (2009). *National Greenhouse Gas Inventory Report of JAPAN.* Tokyo, Japan: National Institute for Environmental Studies, Japan.

Hayashi, K. (2008). Is organic farming environmentally superior? what agricultural LCA reveals. *Journal of Life Cycle Assessment, Japan*, *4*(2), 112–118.

Houghton, J. T., Meira Filho, L. G., Lim, B., Treanton, K., Mamaty, I., Bonduki, Y., et al. (Eds.). (1996). *Revised 1996 IPCC Guidelines for National Greenhouse Gas Inventories.* Paris, France: Organization for Economic Co-operation and Development.

Koga, N., & Tsuruta, H. (2006). Hokkaido no hatasaku nougyou kara hassei suru onshitsukouka gas no life cycle inventory bunseki. [in Japanese]. *Agriculture and Horticulture*, *81*(10), 1101–1109.

Kusaba, T., & Ishioka, G. (2008). Outline and characteristics of the Dojo Kankyo Kisochosa Kijunten Chosa Data Base: data base of long-term changes in soil properties and crops yields of arable field under different nutrient managements in Japan. [in Japanese]. *Miscellaneous Publication of the National Agricultural Research Center*, *8*, 1–87.

Leontief, W. W. (1941). *The structure of American economy, 1919-1929—An empirical application of equilibrium analyses*. Cambridge, MA: Harvard University Press.

Li, C., Frolking, S., & Frolking, T. A. (1992). A model of nitrous oxide evolution from soil driven by rainfall events: 1. Model structure and sensitivity. *Journal of Geophysical Research*, *97*(D3), 9759–9776.

Nakamura, S., & Kondo, Y. (2002). Input-Output Analysis of Waste Management. *Journal of Industrial Ecology*, *6*(1), 39–64. doi:10.1162/108819802320971632

Nansai, K., Moriguchi, Y., & Touno, S. (2002). *Embodied Energy and Emission Intensity Data for Japan Using Input-Output Tables*. Tsukuba, Japan: National Institute for Environmental Studies.

National Institute for Rural Engineering (Ed.). (2006). *Biomass Rikatuyou System no Sekkei to Hyouka*. Tsukuba, Japan: National Institute for Rural Engineering. (In Japanese)

Parton, W. J., Schimel, D. S., Cole, C. V., & Ojima, D. S. (1987). Analysis of factors controlling soil organic matter levels in Great Plains grasslands. *Soil Science Society of America Journal*, *51*, 1173–1179. doi:10.2136/sssaj1987.03615995005100050015x

Parton, W. J., Stewart, J. W. B., & Cole, C. V. (1988). Dynamics of C, N, P and S in grassland soils: a model. *Biogeochemistry*, *5*, 109–131. doi:10.1007/BF02180320

Shirato, Y., Hakamata, T., & Taniyama, I. (2004). Modified Rothamsted Carbon Model for Andosols and Its Validation: Changing Humus Decomposition Rate Constant with Pyrophosphate-Extractable Al. *Soil Science and Plant Nutrition*, *50*, 149–158.

The International Organization for Standardization. (2006). Environmental management -- Life cycle assessment -- Principles and framework. *International Standard ISO*, *14040*, 8.

Tsuruta, H. (1997). Emission rates of methane from rice paddy fields and nitrous oxide from fertilised upland fields estimated from intensive field measurement for three years (1992-1994) all over Japan. *Research Reports Division of Environmental Planning, National Institute of Agro-* [in Japanese]. *Environmental Sciences*, *13*, 101–131.

Wernet, G., Hellweg, S., Fischer, U., Papadokonstantakis, S., & Hungerbühler, K. (2008). Molecular-Structure-Based Models of Chemical Inventories using Neural Networks. *Environmental Science & Technology*, *42*(17), 6717–6722. doi:10.1021/es7022362

## ADDITIONAL READING

Akiyama, H., Yagi, K., & Yan, X. (2005). Direct $N_2O$ emissions from rice paddy fields: Summary of available data. *Global Biogeochemical Cycles*, *19*, GB1005..doi:10.1029/2004GB002378

Anton, A. (2004). *Utilización del análisis del ciclo de vida en la evaluación del impacto ambiental del cultivo bajo invernadero mediterráneo*. Unpublished doctoral dissertation, Universitat Politécnica de Catalunya (in Spanish).

Audsley, E., Alber, S., Clift, R., Cowell, S., Crettaz, P., & Gaillard, G. (1997). *Harmonisation of environmental life cycle assessment for agriculture. Final Report for Concerted Action AIR3-CT94-2028. European Commission DG VI Agriculture*. Silsoe, United Kingdom: Silsoe Research Institute.

Basset-Mens, C. (2005). *Propositions pour une adaptation de l'Analyse de Cycle de Vie aux systèmes de production agricole. Mise en œuvre pour l'évaluation environnementale de la production porcine.* Unpublished doctoral dissertation, Unité Mixte de Recherche INRA Agrocampus de Rennes-Quimper Sol Agronomie Spatialisation (in French).

Baumann, H., & Tillman, A. M. (2004). *The hitch hiker's guide to LCA*. Lund, Sweden: Studentlittertur.

Brentrup, F. (2003). *Life cycle assessment to evaluate the environmental impact of arable crop production*. Göttingen, Germany: Cuvillier Verlag.

Cowell, S. (1998). *Environmental life cycle assessment of agricultural systems: integration into decision-making*. Unpublished doctoral dissertation, University of Surrey.

Dalgaard, R. L. (2007). *The environmental impact of port production from a life cycle perspective*. Unpublished doctoral dissertation, Aarhus University and Aalborg University.

De Beaufort-Langeveld, A. S. H., Bretz, R., van Hoof, G., Hischier, R., Jean, P., Tanner, T., & Huijbregts, M. (Eds.). (2003). *Code of life-cycle inventory practice*. Raleigh, NC: Society of Environmental Toxicology and Chemistry (SETAC).

Elsayed, M. A., Matthews, R., & Mortimer, N. D. (2003). *Carbon and energy balances for a range of biofuels options*. Milano, Italy: Italian Thermotechnical Committee. Retrieved September 16, 2009, from http://www.cti2000.it/Bionett/BioDE-2003-101%20DTI%20biodiesel%20LCA%20study.pdf.

Fischer, G., Ermolieva, T., Ermoliev, Y., & van Velthuizen, H. T. (2006). Spatial recovering of agricultural values from aggregate information: Sequential downscaling methods. *International Journal of Knowledge and Systems Science*, *3*(1), 1–6.

Fischer, G., van Velthuizen, H., Shah, M., & Nachtergaele, F. O. (2002). *Global agro-economical assessment for agriculture in the 21st century: methodology and results*. Laxenburg, Austria: International Institute for Applied Systems Analysis.

Frischknecht, R. (1998). *Life cycle inventory analysis for decision-making: scope-dependent inventory system models and context-specific joint product allocation*. Unpublished doctoral dissertation, Swiss Federal Institute of Technology Zurich (ETH Zurich).

Frischknecht, R., Jungbluth, N., Althaus, H.-J., Doka, G., Heck, T., Hellweg, S., et al. (2007). *Overview and methodology: ecoinvent report No. 1*. Dübendorf, Switzerland: Swiss Centre for Life Cycle Inventories. Retrieved September 16, 2009, from http://www.ecoinvent.org/fileadmin/documents/en/01_OverviewAndMethodology.pdf

Fumoto, T., Kobayashi, K., Li, C., Yagi, K., & Hasegawa, T. (2008). Revising a process-based biogeochemistry model (DNDC) to simulate methane emission from rice paddy fields under various residue management and fertilizer regimes. *Global Change Biology*, *14*, 382–402. doi:10.1111/j.1365-2486.2007.01475.x

Hayashi, K. (2006). Environmental indicators for agricultural management: integration and decision making. *International Journal of Materials and Structural Reliability*, *4*(2), 115–127.

Hayashi, K. (2007). Dealing with multiple objectives in agriculture. In Weintraub, A., Romero, C., Bjørndal, T., & Epstein, R. (Eds.), *Handbook of Operations Research in Natural Resources* (pp. 17–31). Berlin, Germany: Springer-Verlag.

Hayashi, K., Gaillard, G., & Nemecek, T. (2006). Life cycle assessment of agricultural production Systems: current issues and future perspectives. *Proceedings of the International Seminar on Technology Development for Good Agricultural Practice in Asia and Oceania* (pp. 98–110). Taipei, Taiwan: Food and Fertilizer Technology Center.

Hayashi, K., & Nemecek, T. (2007). Ratio and additive models for measuring environmental-economic performance: case studies at field and farm levels. In, *Book of Proceedings: the 5th International Conference LCA in Foods* (pp.26–29). Gothenburg, Sweden: the Swedish Institute for Food and Biotechnology.

Hayashi, K., & Uchida, S. (2009, August). *Sustainable design of farming systems using the life cycle approach: the case of agricultural biomass production.* Paper presented at Farming Systems Design 2009: an International Symposium on Methodologies for Integrated Analysis of Farm Production Systems, Monterey, CA.

Heijungs, R., & Suh, S. (2002). *The computational structure of life cycle assessment.* Dordrecht, Netherlands: Kluwer Academic Publishers.

Hischier, R., Weidema, B., Althaus, H.-J., Bauer, C., Doka, G., Dones, R., et al. (2009). *Implementation of Life Cycle Impact Assessment Methods: ecoinvent report No. 3, v2.1.* St. Gallen, Switzerland: Swiss Centre for Life Cycle Inventories. Retrieved September 16, 2009, from http://www.ecoinvent.org/fileadmin/documents/en/03_LCIA-Implementation_v2.1.pdf

Hokazono, S., Hayashi, K., & Sato, M. (2009). Potentialities of organic and sustainable rice production in Japan from a life cycle perspective. *Agronomy Research*, 7(Special issue 1), 257-262.

Hunkeler, D., Lichtenvort, K., & Rebitzer, G. (Eds.). (2008). *Environmental life cycle costing.* Boca Raton, FL: CRC Press. doi:10.1201/9781420054736

Jungllluth, N. (2000). *Umweltfolgen des Nahrungsmittelkonsums: Beurteilung von Produkumerkmalen auf Grundlage einer modularen Ökobilanz.* Unpublished doctoral dissertation, Swiss Federal Institute of Technology Zurich (ETH Zurich) (in Germany).

Maruyama, K., Gocho, N., Moriya, T., & Hayashi, K. (in press). Life cycle assessment of super high-yield and conventional rice production systems: A comparison based on global warming potential and energy consumption. *Journal of Life Cycle Assessment, Japan.*

Matthews, R. B., & Stepens, W. (Eds.). (2002). *Crop-soil simulation models; Applications in developing countries.* Oxfordshire, UK: CABI Publishing. doi:10.1079/9780851995632.0000

Milà i Canals, L. (2003). *Contributions to LCA methodology for agricultural systems: site-dependency and soil degradation impact assessment.* Unpublished doctoral dissertation, Universitat Autònoma de Barcelona.

Miller, R. E., & Blair, P. D. (2009). *Input-Output Analysis: Foundations and Extensions* (2nd ed.). Cambridge, United Kingdom: Cambridge University Press. doi:10.1017/CBO9780511626982

Mouron, P. J. P. (2005). *Ecological-economic life cycle management of perennial tree crop systems: the case of Swiss fruit farms.* Unpublished doctoral dissertation, Swiss Federal Institute of Technology Zurich (ETH Zurich).

Nakamura, S., & Kondo, Y. (2009). *Waste Input-Output Analysis: Concepts and Application to Industrial Ecology.* New York, NY: Springer.

Nemecek, T., & Kägi, T. (2007). *Life cycle inventories of agricultural production systems: Ecoinvent report No. 15.* Dübendorf, Switzerland: Swiss Centre for Life Cycle Inventories.

(1999). InOrganization for Economic Co-operation and Development (Ed.). Environmental indicator for agriculture: *Vol. 1. Concepts and framework*. Paris, France: OECD Publishing.

(1999). InOrganization for Economic Co-operation and Development (Ed.). Environmental indicator for agriculture: *Vol. 2. Issues and Design*. Paris, France: OECD Publishing.

(2001). InOrganization for Economic Co-operation and Development (Ed.). Environmental indicator for agriculture: *Vol. 3. Method and Results*. Paris, France: OECD Publishing.

Papajorgji, P. J., & Pardalos, P. M. (Eds.). (2009). *Advances in modeling agricultural systems*. New York, NY: Springer.

Poppe, K. J., Beers, G., & de Putter, I. D. (1999). *PACIOLI6; Models for data and data for models workshop report*. Hague, Netherlands: Agricultural Economics Research Institute. Retrieved September 16, 2009, from http://www.lei.dlo.nl/publicaties/PDF/1999/6_xxx/6_99_01.pdf.

Rajagopal, D., & Zilberman, D. (2007). *Review of environmental, economic and policy aspects of biofuels*. Policy Research Working Paper. Berkeley, CA: The World Bank Development Research Group Sustainable Rural and Urban Development Team. Retrieved September 16, 2009, from http://are.berkeley.edu/~dwrh/CERES_Web/Docs/wps4341.pdf

Solomon, S., Qin, D., Manning, M., Chen, Z., Marquis, M., & Averyt, K. B. (Eds.). (2007). *Contribution of working group I to the fourth assessment report of the Intergovernmental Panel on Climate Change*. Cambridge, United Kingdom and New York, NY: Cambridge University Press.

Thornley, J. H. M., & France, J. (2007). *Mathematical models in agriculture; Quantitative methods for the plant, animal and ecological sciences* (2nd ed.). Oxfordshire, UK: CABI Publishing.

Uchida, S., Hayashi, K., Gau, M., Kajiyama, T., Shirasawa, S., Takahashi, H., et al. (2009, August) *Life cycle assessment of energy crop production for improving regional utilization of biomass*. Paper presented at 49th European Congress of the Regional Science Association International, Łódź, Poland.

Wegener Sleeswijk, A. W., Kleijn, R., van Zeijts, H., Reus, J. A. W. A., van Onna, M. J. G., Leneman, H., & Sengers, H. H. W. J. M. (1997). *Application of LCA to Agricultural Products*. Leiden, The Netherlands: Centre of Environmental Science, Leiden University.

Weidema, B., Hischier, R., Althaus, H.-J., Bauer, C., Doka, G., Dones, R., et al. (2009). *Code of practice: ecoinvent report No. 2*. St. Gallen, Switzerland: Swiss Centre for Life Cycle Inventories. Retrieved September 16, 2009, from http://www.ecoinvent.org/fileadmin/documents/en/02_CodeOfPractice_v2.1.pdf.

Weidema, B. P., & Meeusen, M. J. G. (Eds.). (2000). *Agricultural data for life cycle assessment*. Hague, Netherlands: Agricultural Economics Research Institute. Retrieved September 16, 2009, from http://www.lei.dlo.nl/publicaties/rapporten_engels.php3?id=55.

Weintraub, A., Romero, C., Bjørndal, T., & Epstein, R. (Eds.). (2007). *Handbook of operation research in natural resources*. New York, NY: Springer.

Williams, A. G., Audsley, E., & Sandars, D. L. (2006). *Determining the environmental burdens and resource use in the production of agricultural and horticultural commodities*, Main Report, Defra Research Project IS0205, Cranfield, UK: Cranfield University and Defra. Retrieved September 16, 2009, from http://randd.defra.gov.uk/Default.aspx?Menu=Menu&Module=More&Location=None&ProjectID=11442&FromSearch=Y&Publisher=1&SearchText=is0205&SortString=ProjectCode&SortOrder=Asc&Paging=10#Description.

## KEY TERMS AND DEFINITIONS

**Agri-Environmental Assessment:** An assessment of agriculture from the viewpoint of environmental adequacy. In particular, in agriculture, it is desirable to sufficiently consider not only the general environmental load such as global warming, eutrophication and human toxicity but also the biodiversity and quality of soil. These factors strongly depend on change in land use and soil organic matter, and the establishment of an agri-environmental production system with high sustainability. In LCA, such aspects are reflected in the goal and scope definition and the assessment is implemented along with this definition.

**Crop-Soil Simulation Model:** Models in which the balances of components of crops and changes in soil components are considered in detail. In terms of environmental issues, the elements to be considered are restricted to carbon, nitrogen and a few other elements when only the interaction of crops and soil are focused on. Several types of models have been developed and, in addition to the application of those models to assess the environmental load from soil, the recommended agricultural practices that should be followed in the interests of sustainability can also be clarified.

**Foreground and Background System:** Inclusive terms describing items having a direct and indirect causal relation to the subject process, respectively. While the foreground system is easy to determine, i.e. it can be 'visualized', since only the primary relationship must be checked, the background system includes all the multiplied, i.e. 'hidden', effects which require an infinitely repeating calculation. For the computation of the background system data, LCA software is usually required; input-output analysis is also a convenient method for the same.

**Impact Category:** Types of environmental impact are classified into globally significant issues such as global warming, human toxicity and eutrophication. The number of categories is around 10 in most LCA software packages. Generally, one impact category is related to some substances and each of them has different degrees of contribution; therefore, a characterization process is necessary for the contribution to be unified. When the scope of LCA includes the endpoint, each value of an impact category is again unified in the weighting process. In particular, in the assessment of agricultural production, the biodiversity and quality of soil are important impact categories, as mentioned in terms of agri-environmental assessment.

**Input-Output Table:** A matrix which was developed by Leontief (1941). This table indicates transaction values between two activities such as industries, public sector, consumption sector and so on. Each row of the table indicates an industry as a supplier, corresponding expenditure such as depreciation and tax, added value and so on. Each column of the table indicates an industry as a demander, public consumption, final consumption, import, export and so on. The values in a row indicate the distributed sectors and amounts of produced goods or services. On the other hand, the values in a columns indicate the purchased amounts of a sector from other (including the same) sectors and the amount of additional cost. The sum of the row and column of an industry indicates the total production. As an application, the table of input coefficients can also be utilized in input-output analysis; this table is derived from the transaction values by dividing them by the total production or consumption.

**Life Cycle Assessment (LCA):** A method to assess products or projects mainly from an environmental viewpoint. It covers all the processes related to the assessment, including mining of resources, production, cultivation, manufacture, circulation, consumption, waste disposal and so on. Therefore, as a methodology, LCA is related to various information processing theories and it is often regarded as an academic field rather than as an analytical tool.

**Life Cycle Inventory (LCI):** Aggregated datasets concerning certain substances which are inputted or outputted in a process that are mainly utilized in LCA. The data describes the amount of a substance and information about where it comes from and goes to as relationships among various processes. Processes are defined as a minimum unit of activity and the information belonging to it. In LCI analysis, all amounts of environmental load substances are evaluated as inputs to or outputs from all the processes related to a target process. Many projects have constructed exhaustive LCI databases by integrating the data of many processes and substances.

**Soil Organic Carbon:** Carbon component in soil organic matter. Its major sources are plant litter and compost input to arable land. The behaviour of soil organic carbon is modelled using the RothC model (Coleman & Jenkinson, 1996) which states that the decomposable components in the input material decay to humified organic matter, microbial biomass and $CO_2$. The first two among these components decay further and eventually, all soil organic carbon gradually changes to $CO_2$, except for a small amount of inert organic matter. In modern agriculture, the total amount of soil organic carbon is controlled by the balance between the input of organic matter and the rate of decay; this amount tends to gradually decrease, implying an unknown source of $CO_2$ emission. As a reverse process, attempts toward soil carbon sequestration are expected to be one of the promising measures for reducing GHG emissions by increasing the input of organic matter and introducing no tillage. In Japanese Andosols, active metal components such as aluminum and iron are considered to bind chemically with organic carbon, and therefore, it is necessary to modify the model. Inorganic carbon such as carbonates is also inert and it is not included in these considerations.

# Compilation of References


ABIEC. (2009). *Brasil lidera ranking de exportação da carne bovina*. Retrieved July, 2, 2009, from http://www.abiec.com.br/news_view.asp?id={CAACE975-B5D1-4337-9F3B-580E7118CB45}

Abramovay, R. (2006). Para una teoría de los estudios territoriales. In Manzanal, M., Neiman, G., & Lattuada, M. (Eds.), *Desarrollo rural – Organizaciones, instituciones y territorios* (pp. 51–70). Buenos Aires, Argentina: Ediciones Ciccus.

Abramovitz, J. N. (2001). *Unnatural Disasters*. Worldwatch Paper 158.

Acuña, R. S., Cruz, C. D., Do Vale, F. X. R., & Zambolim, L. (2005). PROGRESSA. In *Informática na agricultura, agroindústria e meio ambiente*. Universidade Federal De Viçosa (sd.UFV). Consultado em janeiro/2005 em http://www.ufv.br/ctq/home.htm#progressa

Addiscott, T. M. A., & Wagenet, R. J. (1985). Concepts of solute leaching in soils: a review of modeling approaches. *Journal of Soil Science*, *36*, 411–424. doi:10.1111/j.1365-2389.1985.tb00347.x

Adenáuer, M., Britz, W., Gocht, A., Gömann, H., Cristoiu, A., & Ratinger, T. (2006). *Modelling impacts of decoupled premiums: building-up a farm type layer within the EU-wide regionalised CAPRI model*. Selected paper presented at the 93rd EAAE Seminar. Prague, 22-23 September 2006.

Agência Nacional de Energia Elétrica, A. N. E. E. L. (2005). *Atlas de Energia Elétrica do Brasil*, (2nd Ed.). Brasília, Brazil.

Aggarwal, C. C. (2003). A framework for diagnosing changes in evolving data streams. In *Proceedings of the ACM SIGMOD*, 575-586.

Aggarwal, C. C., & Yu, P. S. (2008). LOCUST: An Online Analytical Processing Framework for High Dimensional Classication of Data Streams. In *Proceedings of the International Conference on Data Engineering*, (pp. 426-435).

Aggarwal, C. C., Han, J., Wang, J., & Yu, P. S. (2004). On demand classification of data streams. In *Proceedings of the ACM Knowledge Discovery & Data Mining Conference*, (pp. 503-508).

Aggarwal, C. C., Han, J., Wang, J., & Yu, P. S. (2004a). A framework for projected clustering of high dimensional data streams. In *Proceedings of the Very Large Data Base*, (pp. 852-863).

Agre, P. E., & Rosenschein, S. J. (Eds.). (1996). *Computational theories of interaction and agency*. Cambridge, MA: MIT Press.

AGROTRACABILITE. (2009). *Agi-traçabilité Québec*. Retrieved July, 2, 2009, from http://www.agri-tracabilite.qc.ca

Ahumada, O., & Villalobos, J. R. (2009). Application of planning models in the agri-food supply chain: A review. *European Journal of Operational Research*, *196*(1), 1–20. doi:10.1016/j.ejor.2008.02.014

Akiyama, H., Yagi, K., & Yan, X. (2006, March). *Direct $N_2O$ emissions and estimate of $N_2O$emission factors from Japanese agricultural soils*. Paper presented at International Workshop on Monsoon Asia Agricultural Greenhouse Gas Emissions, Tsukuba, Japan.

Aksoy, E. (2006, October) *Clustering with GIS: an attempt to classify turkish district data.* Paper presented at the XXIII FIG Congress. Munich, Germany.

Al Nasir, F., & Batarseh, M. I. (2008). Agricultural reuse of reclaimed water and uptake of organic compounds: Pilot study at Mutah University wastewater treatment plant, Jordan. *Chemosphere*, *72*, 1203–1214. doi:10.1016/j.chemosphere.2008.01.064

Alencar, C. E. R., Corrêa, R. F., & de Lima Neto, F. B. (2006). *Ferramenta para suporte à decisão de frentes de corte de cana-de-açúcar usando algoritmos genéticos.* Universidade de Pernambuco.

Allen, J. (1983). Maintaining knowledge about temporal intervals. *Communications of the ACM*, *26*(11), 832–843. doi:10.1145/182.358434

Almeida, F. (2003). o mundo dos negócios e o meio ambiente no século 21. In Trigueiro, A. (Ed.), *meio ambiente no século 21: 21 especialistas falam da questão ambiental nas suas áreas de conhecimento*, (pp. 123-141).

Altieri, M. A. (1995). *Agroecology: the science of sustainable agriculture.* Boulder, CO: Westview press.

Alves, E., Souza, G. S., & Garagorry, F. L. (1999). Evolution of the corn yield. *Brazilian Review of Agricultural Economics and Rural Sociology*, *37*(1), 183–202.

Alves, L. M., & Pesquero, J. F. (2009). Previsões de mudanças climáticas para a América do Sul: resultados da aplicação do "Earth Simulator". Boletim do Grupo de Pesquisa em Mudanças Climáticas –GPMC, São José dos Campos: CCST/INPe, n.7, abril de 2009, semestral, pg.2-4.

Ambrosano, G. M. B., Stimac, J. L., Silveira-Neto, S., Igue, T., & Nagai, V. (1996). Modelo matemático para simulação do controle biológico da broca-da-cana com o parasitóide *Trichogramma galloi. I. modelos conceituais. Bragantia, Campinas*, *55*(2), 371–382.

Anderberg, M. R. (1973). *Cluster analysis for applications.* New York: Academic Press.

André, F. J., Cardenete, M. A., & Romero, C. (2008). Using compromise programming for macroeconomic policy making in a general equilibrium framework: Theory and application to the Spanish economy. *The Journal of the Operational Research Society*, *59*, 875–883. doi:10.1057/palgrave.jors.2602415

Andrienko, N., Andrienko, G., & Gatalsky, P. (2003). Exploratory spatio-temporal visualization: an analytical review. *Journal of Visual Languages and Computing*, *14*(6), 503–541. doi:10.1016/S1045-926X(03)00046-6

Andrienko, G., & Andrienko, N. (2008). A Visual Analytics Approach to Exploration of Large Amounts of Movement Data. In *Proceedings of the* VISUAL 2008, (pp. 20-35). Berlin: Springer-Verlag.

Angelakis, A. N., & Durham, B. (2008). Water recycling and reuse in EUREAU countries: Trends and challenges. *Desalination*, *218*, 3–12. doi:10.1016/j.desal.2006.07.015

Angelakis, A. N., Koutsoyiannis, D., & Tchobanoglous, G. (2005). Urban wastewater and stormwater technologies in ancient Greece. *Water Resources*, *39*, 210–220.

Angelakis, A. N., Marecos Do Monte, M. H. F., Bontoux, L., & Asano, T. (1998). The status of wastewater reuse practice in the mediterranean Basin: need for guidelines. *Water Resources*, *33*, 2201–2217.

Angelakis, A. N., & Spyridakis, S. V. (1996). The status of water resources in Minoan times –A preliminary study. In Angelakis, A., & Issar, A. (Eds.), *Diachronic climatic impacts on water resources with emphasis on Mediterranean region.* Berlin: Springer-Verlag.

Anunciação, Y. M. T da, Liu, W. T. (1991) Estimativa da produtividade do trigo em campo experimental, utilizando o modelo fisiológico CERES trigo em condições de campo experimental. In *Congresso Brasileiro de Agrometeorologia, 7, 1991,* Viçosa, MG. Resumos..., (pp.22-25). Viçosa, Brazil: UFV.

Appa, G. M. (1973). The transportation problem and its variants. *Operational Research Quarterly*, *24*, 79–99.

Araujo, M. J., Corrêa, P. L. P., Sato, L. M., & Saraiva, A. M. (2007). A framework for species distribution modelling - A performance evaluantion approach. In *6th International Information and Telecommunication Technologies Symposium, 2007*, Brasilia, (I2TS'2007).

Architectural Institute of Japan (Ed.). (2006). *Tatemono no LCA Shishin* (3rd ed.). Tokyo, Japan: Architectural Institute of Japan. (In Japanese)

Arfini, F., & Paris, Q. (1995). A positive mathematical programming model for regional analysis of agricultural policies. In E, Sotte, (Ed.), *The Regional Dimension in Agricultural Economics and Policies, EAAE, Proceedings of the 40th Seminar, 26-28 June 1995, Ancona,* (pp. 17-35).

Armstrong, A. C., Portwood, A. M., Leeds-Harrison, P. B., Harris, G. L., & Catt, J. A. (1996). The validation of pesticide leaching models. *Pesticide Science*, *48*, 47–55. doi:10.1002/(SICI)1096-9063(199609)48:1<47::AID-PS434>3.0.CO;2-F

Armstrong, M., & Densham, P. (1990). Database organization strategies for spatial decision support systems. *International Journal of Geographical Information Systems*, *4*(1), 3–20. doi:10.1080/02693799008941525

Aronoff, S. (1991). *Geographic information systems: a management perspective*. Ottawa, Ontário, Canada: WDL.

Arriaza, M., & Gómez-Limón, J. A. (2003). Comparative performance of selected mathematical programming models. *Agricultural Systems*, *77*, 155–171. doi:10.1016/S0308-521X(02)00107-5

Arrow, K. J. (1995). Effet de serre et actualisation. *Revue de l'Énergie*, *471*, 631–636.

Asadi, M. E., & Clemente, R. S. (2003). Evaluation of CERES-MAIZE of DSSAT model to simulate nitrate leaching, yield and soil moisture content under tropical conditions. *Food. Agriculture & Environment*, *1*(3&4), 270–276.

ASCE. (2000). Artificial neural networks in hydrology. I: preliminary concepts. *Journal of Hydrologic Engineering*, *5*(2), 115–123. doi:10.1061/(ASCE)1084-0699(2000)5:2(115)

Ash, A., McIntosh, P., Cullen, B., Carberry, P., & Smith, M. S. (2007). Constraints and opportunities in applying seasonal climate forecasts in agriculture. *Australian Journal of Agricultural Research*, *58*, 952–965. doi:10.1071/AR06188

Assad, E. D., & Sano, E. E. (1998). *Sistemas de Informações Geográficas - Aplicações na Agricultura.* Brasília, DF: EMBRAPA.

Atamtürk, A., & Savelsbergh, M. (2005). Integer-programming software systems. *Annals of Operations Research*, *140*, 67–124. doi:10.1007/s10479-005-3968-2

Audsley, E. (1985). The development of operational research at the national institute of agricultural engineering. *The Journal of the Operational Research Society*, *36*(2), 117–123.

Audsley, E., Dumont, S., & Boyce, D. (1978). An economic comparison of methods of planting cereals, sugar beets and potatoes and their interactions with harvesting, timeliness and available labour by linear programming. *Journal of Agricultural Engineering Research*, *23*(3), 283–300. doi:10.1016/0021-8634(78)90102-6

Augustithis, S. P. (1995). *Atlas of textural patterns of ore minerals and metallogenic processes.* New York: de Gruyter.

Austin, M. P., Nicholls, A. O., Doherty, M. D., & Meyers, J. A. (1994). Determining species response functions to an environmental gradient by means of a beta-function. *Journal of Vegetation Science*, *5*, 215–228. doi:10.2307/3236154

Axtell, R. (2000). *Why agents? On the varied motivations for agent computing in the social sciences*. Working Paper 17, Center on Social and Economic Dynamics, Brookings Institution.

Azevedo Filho, A. J. B. V., & Neves, E. M. (1982). O uso de programação matemática na análise de investimento na pecuária de corte: técnicas intensivas X mercado de capitais. In *Anais do I Congresso Latino-Americano de Pesquisa Operacional e Engenharia de Sistemas,* (pp. 41-55). Rio de Janeiro, Brazil: Sobrapo.

Bação, F., Lobo, V., & Painho, M. (2005). The self-organizing map. the Geo-SOM, and relevant variants for geosciences. *Computers & Geosciences*, *31*, 155–163. doi:10.1016/j.cageo.2004.06.013

Bachinger, J., & Zander, P. (2006). ROTOR, a tool for generating and evaluating crop rotations for organic farming systems. *European Journal of Agronomy*, *26*, 130–143. doi:10.1016/j.eja.2006.09.002

Bacic, I. L. Z., Rossiter, D. G., & Mannaerts, C. M. (2008). Applicability of a distributed watershed pollution model in a data-poor environment in Santa Catarina State, Brazil. *Rev. Bras. Ciênc. Solo. Viçosa*, *32*(4), 1699–1712.

Bahman, S. (1998). Accounting for benfits of water reuse. In *Proc American Waterworks Assn 1998 Water Recycling and Reuse Symposium*.

Bahri, A. (2003). *Water reuse in the Middle East, North Africa and Mediterranean Countries*. Paper presented at Kyoto water forum, Kyoto, Japan.

Bailey, T. C., & Gatrell, A. C. (1995). *Interactive Spatial Data Analysis*. Longman.

Baja, S., Chapman, D. M., & Dragovich, D. (2007). Spatial based compromise programming for multiple criteria decision making in land use planning. *Environmental Modeling and Assessment*, *12*, 171–184. doi:10.1007/s10666-006-9059-1

Ballestro, E. (2000). Project finance: A multi-criteria approach to arbitration. *The Journal of the Operational Research Society*, *51*, 183–197.

Baltzer, O., Dehne, F., Hambrusch, S. E., & Rau-Chaplin, A. (2008). OLAP for Trajectories. In *Proceedings of the* DEXA 2008, (pp. 340-347). Berlin: Springer-Verlag.

Bandeira, P. S. (2006). Institucionalização de regiões no Brasil. *Revista da Sociedade Brasileira para o Progresso da Ciência, 58*(1), 34-37.

Barbará, D., & Chen, P. (2003). Using self-similarity to cluster large data sets. *Data Mining and Knowledge Discovery*, *7*(2), 123–152. doi:10.1023/A:1022493416690

Barbará, D., Chen, P., & Nazeri, Z. (2004). Self-similar mining of time association rules. *Proceedings of the Pacific-Asia Conference on Knowledge Discovery and Data Mining*, *3056*, 86–95.

Barkaoui, A., & Butault, J. P. (2003). *Révision à mi-parcours de la PAC: simulations de l'effect du découplage sur l'offre dans les régions françaises. Document de travail*. INRA-ESR Nancy.

Barkaoui, A., & Butault, J. P. (1999). *Positive Mathematical Programming and Cereals and Oilseeds Supply within EU under Agenda 2000*. Paper presented at the 9th European Congress of Agricultural Economics, Warsaw, August 1999.

Barnhart, C., Johnson, E. L., Nemhauser, G. L., Savelsbergh, M. W. P., & Vance, P. H. (1998). Branch and price: Column generation for solving huge integer programs. *Operations Research*, *46*(3), 316–329. doi:10.1287/opre.46.3.316

Barreto, G. A., & Araújo, A. F. R. (2001). Time in self-organizing maps: An overview of models. *International Journal of Computer Research*, *10*(2), 139–179.

Bassanezi, R. C., & Ferreira Junior, W. C. (1988). *Equações diferenciais com aplicações*. São Paulo, Brazil: Harbra Ltda. barioni, L. G., Beltrame, R. T., Quirino, C. R., Fernandes, D. R. (2007). Modelos determinista e estocástico em programas de transferência de embriões em bovinos. *Archivos Latinoamericanos de Producción Animal*, *15*, 107–113.

Bayless, E. R., Capel, P. D., Barbash, J. E., Webb, R. M. T., Hancock, T. C., & Lampe, D. C. (2008). Simulated fate and transport of metolachlor in the unsaturated zone, Maryland, USA. *Journal of Environmental Quality*, *37*, 1064–1072. doi:10.2134/jeq2006.0562

Bdour, A. N., Hamdi, M. R., & Tarawneh, Z. (2009). Perspectives on sustainable wastewater treatment technologies and reuse in the urban area of the Mediterranean region. *Desalination*, *237*, 162–174. doi:10.1016/j.desal.2007.12.030

Beale, R., & Jackson, T. (1990). *Neural Computing: An Introduction*. Bristol, UK: Institute of Physics Publishing.

Bebis, G., & Georgiopoulos, M. (1994). Feed-forward neural networks: why network size is so important. *IEEE Potentials*, 27-31.

Bédard, Y., Merrett, T., & Han, J. (2001). Fundaments of Spatial Data Warehousing for Geographic Knowledge Discovery. In *Geographic Data Mining and Knowledge Discovery* (pp. 53–73). London: Taylor & Francis. doi:10.4324/9780203468029_chapter_3

Belli, P. (1996). Is economic analysis of projects still useful? *World Bank Policy Research Working Paper 1689*.

Belmans, C., Wesseling, J. G., & Feddes, R. A. (1983). Simulation of the water balance of a cropped soil: SWATRE. *Journal of Hydrology (Amsterdam)*, *63*, 271–286. doi:10.1016/0022-1694(83)90045-8

Bergamasco, A. F., Silva, F. C. D. A., & Rodrigues, L. H. A. Trivelin, P.C.O. (2002). Aplicabilidade do modelo de balanço de Nitrogênio para previsão de resposta da cana-de-açúcar ao fertilizante nitrogenado. *Campinas: Embrapa Informação Agropecuária, dezembro, 9p. (Comunicado Técnico, 28)*.

Bernier, E., Gosselin, P., Badard, T., & Bédard, Y. (2009). Easier Surveillance Of Climate-Related Health Vulnerabilities Through A Web-Based Spatial Olap Application. *International Journal of Health Geographics*, *8*(18).

Beven, K., & Germann, P. (1981). Water flow in soil macropores. II. A combined flow model. *Journal of Soil Science*, *32*, 15–29. doi:10.1111/j.1365-2389.1981.tb01682.x

Beven, K., & Germann, P. (1982). Macropores and water flow in soils. *Water Resources Research*, *18*, 1311–1325. doi:10.1029/WR018i005p01311

Bilbao-terol, A., Perez-Gladish, B., & Antomil-Jbias, J. (2006). Selecting the optimum potfolio using fuzzy compromise programming and Sharpe's single-index model. *Applied Mathematics and Computation*, *182*, 644–664. doi:10.1016/j.amc.2006.04.028

Bishop, C. M. (1995). *Neural Network for Pattern Recognition*. Oxford, U.K: Clarendon.

Bixby, A., Downs, B., & Self, M. (2006). A scheduling and capable-topromise application for swift & company. *Interfaces*, *36*(1), 69–86. doi:10.1287/inte.1050.0184

Blanco, A., & Iglesias, E. (2005). *Modelling New EU Agricultural Policies: Global Guidelines, Local Strategies*. Selected paper presented at the 89th EAAE Seminar. 3-5 February, Parma, Italy.

Blum, C., & Roli, A. (2003). Metaheuristics in combinatorial optimization: Overview and conceptual comparison. *ACM Computing Surveys*, *35*(3), 268–308. doi:10.1145/937503.937505

Blumenthal, U., & Pearsey, A. (2002). *Critical Review of epidemiological evidence of the health effects of wastewater and excreta use in agriculture*. Geneva: WHO.

Boehlje, M., Fulton, J. & AP Gray and T. N. (2002). *Strategic issues in the changing agricultural industry*. Purdue University Cooperative Extension Service.

Boesten, J. J. T. I. (1994). Simulation of bentazon leaching in sandy loam soil from Mellby (Sweden) with the PESTLA model. *Journal of Environmental Science and Health Part A*, *29*, 1231–1253. doi:10.1080/10934529409376105

Boesten, J. J. T. I. (2000). From laboratory to field: uses and limitations of pesticide behaviour models for the soil/plant system. *Weed Research*, *40*, 123–138. doi:10.1046/j.1365-3180.2000.00158.x

Boesten, J. J. T. I., & Stroosnijder, L. (1986). Simple model for daily evaporation from fallow tilled soil under spring conditions in a temperate climate. *Netherlands Journal of Agricultural Science*, *34*, 75–90.

Boesten, J. J. T. I., & Van der Linden, A. M. A. (1991). Modeling the influence of sorption and transformation on pesticide leaching and persistence. *Journal of Environmental Quality*, *20*, 425–435. doi:10.2134/jeq1991.00472425002000020015x

Boesten, J. J. T. I. (1986). *Behaviour of herbicides in soil: simulation and experimental assessment*. Ph.D. thesis, Wageningen University and Research Centre, Wageningen.

Boman, M., & Holm, E. (2004). Multi-agent systems, time geography, and microsimulations. In Olsson, M.-O., & Sjöstedt, G. (Eds.), *Systems Approaches and their Application* (pp. 95–118). Amsterdam: Kluwer Academic.

Bond, W. J. (1998). Effluent irrigation – an environmental challenge for soil science. *Australian Journal of Soil Research*, *36*, 543–555. doi:10.1071/S98017

Bonham-Carter, G. F. (1994). *Geographic information systems for geoscientists*. Ontario, Canada: Pergamon.

Booch, G., Rumbauch, J., & Jacobson, I. (1999). *The Unified Modeling Language User Guide*. Reading, MA: Addison-Wesley.

Booltink, H. W. G., Hatano, R., & Bouma, J. (1993). Measurement and simulation of bypass flow in a structured clay soil: a physico-morphological approach. *Journal of Hydrology (Amsterdam)*, *148*, 149–168. doi:10.1016/0022-1694(93)90257-A

Borges Junior, J. C. F., Ferreira, P. A., Andrade, C. L. T., & Hedden-Dunkhorst, B. (2008). Computational modeling for irrigated agriculture planning. Part I: general description and linear programming. *Eng. Agríc.*, *28*(3), 471–482.

Bortoluzzi, E. C., Rheinheimer, D. S., Gonçalves, C. S., Pellegrini, J. B. R., Zanella, R., & Copetti, A. C. C. (2006). Contaminação de águas superficiais por agrotóxicos em função do uso do solo numa microbacia hidrográfica de Agudo, RS. *Revista Brasileira de Engenharia Agrícola e Ambiental*, *10*(4), 881–887.

Botelho, R. G. M. (1999). Planejamento ambiental em microbacias hidrográficas. In *Guerra, A.J.T., Silva, A.S., Botelho, R.G.M. (orgs.), Erosão e conservação do solo: conceitos, temas e aplicações* (pp. 269–300). Rio de Janeiro, Brazil: Bertrand Brasil.

Bouraoui, F. (2007). Testing the PEARL model in the Netherlands and Sweden. *Environmental Modelling & Software*, *22*, 937–950. doi:10.1016/j.envsoft.2006.06.004

Bowden, G. J., Dandy, G. C., & Maier, H. R. (2005). Input determination for neural network models in water resources applications. Part 1 - Background and methodology. *Journal of Hydrology (Amsterdam)*, *301*, 75–92. doi:10.1016/j.jhydrol.2004.06.021

Brasil. Ministério da Agricultura, Pecuária e Abastecimento. Secretaria de Produção e Agroenergia (2006). *Plano nacional de agroenergia 2006-2011*. (2. ed. rev. atual.). Brasília, DF: Embrapa Informação Tecnológica.

Brasil. Ministério da Agricultura, Pecuária e Abastecimento. Secretaria de Produção e Agroenergia. (2007). *Balanço nacional da cana-de-açúcar e de agroenergia 2007*. Brasília, DF: Secretaria de Política Agrícola.

Brasil. Ministério do Meio Ambiente (2001). *Memória das Reuniões de Discussões sobre Zoneamento Ecológico-Econômico no Brasil*. Brasília: MMA. 1 CD-rom.

Bravo, J. M., Collischonn, W., Pilar, J. V., & Tucci, C. E. M. (2009a). Otimização de regras de operação de reservatórios com incorporação da previsão de vazão. *Revista Brasileira de Recursos Hídricos*, *13*, 181–196.

Bravo, J. M., Paz, A. R., Collischonn, W., Uvo, C. B., Pedrollo, O. C., & Chou, S. C. (2009b). Incorporating forecasts of rainfall in two hydrologic models used for medium-range streamflow forecasting. *Journal of Hydrologic Engineering*, *13*(5), 435–445. doi:10.1061/(ASCE)HE.1943-5584.0000014

Bravo, C., Neto, J. J., Santana, F. S., & Saraiva, A. M. (2007). Towards an Adaptive Implementation of Genetic Algorithms. In *Latin American Conference on Informatics, Proceedings Latin American Conference on Informatics*, San Jose.

Bretherton, C. S., Smith, C., & Wallace, J. M. (1992). An intercomparison of methods for finding coupled patterns on climate data. *Journal of Climate*, *5*, 541–560. doi:10.1175/1520-0442(1992)005<0541:AIOMFF>2.0.CO;2

Briassoulis, H. (2000). Analysis of Land Use Change: Theoretical and Modeling Approaches. In S. Loveridge (Ed.), *The Web Book of Regional Science* (www.rri.wvu.edu/regscweb.htm). Morgantown, WV: Regional Research Institute, West Virginia University.

Briggs, G. G., Bromilow, R. H., & Evans, A. A. (1982). Relationships between lipophilicity and root uptake and translocation of non-ionised chemicals by barley. *Pesticide Science, 13*, 495–504. doi:10.1002/ps.2780130506

Bronswijk, J. J. B. (1988). Modeling of water balance, cracking and subsidence of clay soils. *Journal of Hydrology (Amsterdam), 97*, 199–212. doi:10.1016/0022-1694(88)90115-1

Brooke, A., Kendrick, D., & Meeraus, A. (1992). *GAMS: a user's guide, release 2.25*. San Francisco: The Scientific Press.

Brouwers, L., Danielson, M., Ekenberg, L., & Hansson, K. (2004). Multi-criteria decision-making of policy strategies with public-private re-insurance systems. *Risk Decision and Policy, 9*(1), 23–45. doi:10.1080/14664530490429535

Brouwers, L. (2002). *Flood risk management policy from an insurance perspective*. Licentiate Dissertation, Stockholm University, Department of Computer & Systems Sciences, Report No. 02-008.

Brouwers, L. (2005). *Microsimulation models for disaster policy making*. Doctoral Dissertation, Stockholm University, Department of Computer & Systems Sciences, Report No. 05-012.

Brown, B. & co-authors (2009). Forecast Verification - Issues, Methods and FAQ. Retrieved May 12, 2009 from http://www.bom.gov.au/bmrc/wefor/staff/eee/verif/verif_web_page.html

Brun, T., Poyet, P., Boop, M., & Vigier, F. (2005). Towards an agricultural ontology in France: Contributions of the farm Information Management Project (GIEA), In *Proceedings of the* EFITA/WCCA 2005 (pp. 1296-1302).

Brunini, O. (2007). *Zoneamento de culturas bioenergéticas no Estado de São Paulo*. Retrived October 9, 2007, from http://www.infobibos.com/Artigos/2007_3/zoneamento/Index.htm.

Burrough, P. A. (1986). *Principles of geographical information systems*. New York: Oxford. Brasil. Ministério da Agricultura. Divisão de Pesquisa Pedológica (1973). *Levantamento de reconhecimento dos solos do Estado do Rio Grande do Sul*. Recife, Brazil: MA/DPP – SA/DRNR (Boletim técnico, 30).

Busby, J. R. (1986). A biogeographical analysis of Nothofagus cunninghamii (Hook.) Oerst. in southeastern Australia. *Australian Journal of Ecology, 11*, 1–7. doi:10.1111/j.1442-9993.1986.tb00912.x

Butler, P., Corke, R., Peterson, P., & Rus, D. (2006). Dynamic Virtual Fences for Controlling Cows, in Experimental Robotics. *Springer Tracts in Advanced Robotics, 21*, 513–522. doi:10.1007/11552246_49

Buysse, J., & Van Huylenbroeck, G. (2005). *Impact of alternative implementations of the Agenda 2000 Mid Term Review*. Selected Paper presented at the 11 th Congress of the EAAE. Copenhagen, Denmark, 24-27 August, 2005.

Buysse, J., Fernagut, B., Harmignie, O., Henry De Frahan, B., Lauwers, L., Polomé, P., et al. *(2004)*. *Modelling the impact of sugar reform on Belgian Agriculture*. Selected paper presented at the International Conference on Policy Modelling, Paris, 30 June- 2 July, 2004.

Caballero, R., Ruiz, F., Rodriguez Uria, M. V., & Romero, C. (1996). Interactive meta-goal programming. *European Journal of Operational Research, 175*, 135–154. doi:10.1016/j.ejor.2005.04.040

Cabrera, V. E., Letson, D., & Podestá, G. (2007). The value of climate information when farm programs matter. *Agricultural Systems, 93*, 25–42. doi:10.1016/j.agsy.2006.04.005

Caldwell, S., & Keister, L. A. (1996). Wealth in America: Family stock ownership and accumulation 1960-95. In Clark, G. P. (Ed.), *Microsimulation for urban and regional policy analysis*. Pion.

Caliari, M., Júnior, M. S. S., & Serra, G. E. (2004). Programação matemática aplicada à colheita de cana-de-açúcar com enfase na recuperação máxima de açúcar. *Pesquisa Agropecuária Tropical., 35*, 1–8.

Camara, G., Souza, R. C. M., Freitas, U. M., & Garrido, J. (1996). SPRING: Integrating remote sensing and GIS by object-oriented data modeling. *Computers & Graphics, 20*(3), 395–403. doi:10.1016/0097-8493(96)00008-8

Câmara, G., Davis, C., & Monteiro, A. M. V. (2001). *Introdução à ciência da geoinformação*. São José dos Campos, Brazil: INPE.

Campanhola, C., & Valarini, P. J. (2001). A agricultura orgânica e seu potencial para o pequeno agricultor. *Cadernos de Ciência & Tecnologia, 18*(3), 69–101.

Canhos, V. (2005). *A framework for species distribution modeling. Thematic Research Project – Supported by FAPESP*. São Paulo, Brazil: CRIA/EPUSP/INPE.

Canhos, V. P. (2003). Biodiversity Informatics: patterns, protocols and tools. *Science and Culture, 55*(2), 45-47. Retrieved August 27, 2009, from http://cienciaecultura.bvs.br/pdf/cic/v55n2/15530.pdf. (in Portuguese).

Cao, H., Mamoulis, N., & Cheung, D. W. (2007). Discovery of Periodic Patterns in Spatiotemporal Sequences. *IEEE Transactions on Knowledge and Data Engineering, 19*(4), 453–467. doi:10.1109/TKDE.2007.1002

Capra, F. (2003). Alfabetização ecológica: o desafio para a educação do século 21. In *Trigueiro, A. (coord) (2003). Meio ambiente no século 21: 21 especialistas falam da questão ambiental nas suas áreas de conhecimento* (pp. 19–33). Rio de Janeiro: Sextante.

CAPRI. (2000). *Common Agricultural Policy Regional Impact (Fair3-CT96-1849). Final Consolidated Report, Institute for Agricultural Policy, University of Bonn, Department of Economics, University College Galway, Institut Agronomique Meditérranéen de Montpellier, Departamento di Economia, Sociologia y Politica Agraria, Universidad Politécnica de Valencia, Università degli Studi di Bologna, Dipartamento di Protezione e Valorizzazione Agro-Alimentare (DIPROVAL)*. Sezione Economía.

Carpenter, G., Gillison, A. N., & Winter, J. (1993). DOMAIN: a flexible modeling procedure for mapping potential distributions of plants, animals. *Biodiversity and Conservation, 2*, 667–680. doi:10.1007/BF00051966

Carsel, R. F., Mulkey, L. A., Lorber, M. N., & Baskin, L. B. (1985). The pesticide root zone model (PRZM), a procedure for evaluating pesticide leaching threats to ground water. *Ecological modelling. Amsterdan, 30*, 49–69.

Casati, B., Wilson, L. J., Stephenson, D. B., Nurmi, P., Ghelli, A., & Pocernich, M. (2008). Forecast verification: current status and future directions. *Meteorological Applications, 15*, 3–18. doi:10.1002/met.52

Castellazzi, M. S., Wood, G. A., Burgess, P. J., Morris, J., Conrad, K. F., & Perry, J. N. (2008). A systematic representation of crop rotations. *Agricultural Systems, 97*, 26–33. doi:10.1016/j.agsy.2007.10.006

Castro, N. R. A., Rigitano, R. L. O., Lima, J. M., & Scorza Júnior, R. P. (2006). Avaliação do simulador MACRO (versão 5.0) para predição da lixiviação do inseticida tiametoxam em duas classes de solos. *Pesticidas: Revista Brasileira de Ecotoxicologia e Meio Ambiente, 16*, 1–12.

Cerdeira, A. L., Gomes, M. A. F., Pessoa, M. C. P. Y., Bonato, P. S., & Lanchote, V. L. (Maggio. (2000). Tebuthiuron In Soil And Groudwater In Sugarcane Area In Brazil. *Bollettino Dei Chimici Igiesnisti, 51*(4), 51–57.

Cerdeira, A. L., Pessoa, M. C. P. Y., Santos, N. A. G., & Lanchote, V. L. (2005d). Lixiviação De Atrazina Em Solo Em Área De Recarga Do Aqüífero Guarani. *Revista Brasileira De Herbicidas, 2*, 92–101.

Cerdeira, A. L., Santos, N. A. G., Pessoa, M. C. P. Y., Gomes, M. A. F., & Lanchote, V. L. (2005b). Herbicide Leaching On A Recharge Area Of The Guarany Aquifer In Brazil. *Journal of Environmental Science and Health. Part. B, Pesticides, Food Contaminants, and Agricultural Wastes, 40*(1), 159–165. doi:10.1081/PFC-200034294

Cerdeira, A. L., Santos, N. A. G., Ueta, J., Shuhama, I. K., Pessoa, M. C. P. Y., Smith, S. Jr, & Lanchote, V. L. (2004, July). Atrazine In Water And Biodegradation In A Recharge Area Of Guarany Aquifer In Brazil. *Bulletin of Environmental Contamination and Toxicology*, *73*(1), 117–124..doi:10.1007/s00128-004-0402-4

Cerdeira, A. L., Souza, M. D., Bolonhezi, D., Queiroz, S. C. N., Ferracini, V. L., & Ligo, M. A. V. (2005c). Effects Of Sugar Cane Mechanical Harvesting Followed By No-Tillage Crop Systems On Leaching Of Triazine Herbicides In Brazil. *Bulletin of Environmental Contamination and Toxicology*, *75*(4), 805–812. doi:10.1007/s00128-005-0822-9

Cerdeira, A. L., Souza, M. D., Queiroz, S. C. N., Ferracini, V. L., Bolonhezi, D., & Gomes, M. A. F. (2007a). Leaching and half-life of the herbicide tebuthiuron on a recharge area of Guarany aquifer in sugarcane fields in Brazil. *Journal of Environmental Science and Health*, *42*, 635–639. doi:10.1080/03601230701465593

Cerdeira, A. L., Pessoa, M. C. P. Y., Bolonhezi, D., Gomes, M. A. F., Spadotto, C. A., Souza, M. D., et al. (2007b) Effects of agriculture on ground water in a recharge area of Guarany Aquifer in Brazil. In *37th Annual Mississipi Water Resources Conference,* April 24-25, Jackson, MS, (pp.120-125).

Cerdeira, A. L., Pessoa, M. C. P. Y., Bolonhezi, D., Gomes, M. A. F., Spadotto, C. A., Souza, M. D., et al. (2007c). Agriculture and ground water quality in a sugarcane area in São Paulo State, Brazil. In *Annual Mississipi Water Resources Conference, 37*, Jackson, MS, (p. 39).

Cerdeira, A. L., Souza, M. D., Bolonhezi, D., Queiroz, S. C. N., Ferracini, V. L., Ligo, M. A. V., et al. (2005a, April) Soil Tillage Systems And Herbicide Leaching In Brazil, In *35th Annual Mississipi Water Resources Conference,* April 26-27.

Cerdeira, A. L., Souza, M. D., Ferracini, V. L., Queiroz, S. C. N., Bolonhezi, D., Gomes, M. A. F., et al. (2006, April). Effect Of Sugarcane Coverage On The Behavior Of Tebuthiuron In Soil In Brazil. In *36th Annual Mississippi Water Resources Conference*, Jackson, Mississippi, 25 A 26/04/2006, (pp. 83-85).

Cerdeira, A. L., Souza, M. D., Pessoa, M. C. P. Y., Queiroz, S.C.N., Ferracini, V. L., Lanchote, V. L., Queiroz, R. H. C., Bolonhezi, D., Farjani Neto, C. (Outubro/2005e). Efeito De Sistemas De Cultivo Com Cana-De-Açúcar Sobre A Lixiviação Dos Herbicidas Do Grupo Das Triazinas Em Região De Recarga Do Aqüífero Guarani. *Jaguariúna/Sp.: Embrapa Meio Ambiente, Pp.1-25 (Boletim De Pesquisa E Desenvolvimento, Nº 32).*

Cerdeira, A.L., Pessoa, M. C. P. Y., Gomes, M. A. F., Bolonhezi, D., Souza, M. D., Farjani Neto, C. (2007d). Proposta de Boas Práticas Agrícolas para as Áreas de Afloramento do Aqüífero Guarani em Ribeirão Preto, SP. *Jaguariúna/SP: Embrapa Meio Ambiente, agosto, 65,* 86.

Chakrabarti, D., & Faloutsos, C. (2002). F4: large-scale automated forecasting using fractals. In *Proceedings of the International Conference on Information and Knowledge Management*, McLean, VA - EUA, (Vol. 1, pp. 2-9).

Chang, N.-B., Eh, S. C. Y., & Wu, G. C. (1999). Stability of Grey Compromise Programming and its application to watershed land-use planning. *International Journal of Systems Science*, *30*, 571–589. doi:10.1080/002077299292092

Chang, N.-B., & Tseng, C. C. (1999). Optimal evaluation of expansion alternatives for existing air quality monitoring network by grey compromise programming. *Journal of Environmental Management*, *56*, 61–77. doi:10.1006/jema.1999.0267

Chattopadhyay, S. (2007). Feed-forward artificial neural network model to predict the average summer-monsoon rainfall in India. *Geophysica Acta*, *55*(3), 369–382. doi:10.2478/s11600-007-0020-8

Chiew, F. H. S., Zhou, S. L., & McMahon, T. A. (2003). Use of seasonal streamflow forecasts in water resources management. *Journal of Hydrology (Amsterdam)*, *270*, 135–144. doi:10.1016/S0022-1694(02)00292-5

Chopra, P. (2002). *A very simple example of Neural Networks using back propagation.* Retrieved July 15, 2008, from http://paraschopra.com

Claessen, M. E. C. (1997). *Manual de métodos de análise de solo.* Rio de Janeiro, Brazil: Centro Nacional de Pesquisa de Solos.

Clarke, H. R. (1989). Combinatorial aspects of cropping pattern selection in agriculture. *European Journal of Operational Research*, *40*, 70–77. doi:10.1016/0377-2217(89)90273-7

Claval, P. (2008). Espaces et territoire: les bifurcations de la science régionale. *Géographie, Économie. Société*, *10*, 157–184.

Cobuci, J. A., Euclydes, R. F., & Verneque, R. S. (2000). Curva De Lactação Na Raça Guzerá. *Revista Brasileira de Zootecnia*, *29*(5), 1332–1339. doi:10.1590/S1516-35982000000500011

Cochran, W. G. (1977). *Sampling techniques*. (3ed.). New York: John Wiley & Sons.

Cohen, S. Z., Wauchope, R. D., Klein, A. W., Eadsporth, C. V., & Grancy, R. (1995). Offsite Transport Of Pesticides In Water – Mathematical Models Of Pesticide Leaching And Runoff. *Pure and Applied Chemistry*, *67*, 2109–2148. doi:10.1351/pac199567122109

Cohen, S. Z., Creeger, S. M., Carsel, R. F., & Enfield, C. G. (1984). Treatment And Disposal Of Pesticide Wastes. In Kruger, R. F., & Seiber, J. N. (Eds.), *Acs Symp. Ser. 259* (pp. 297–325). Washington, DC: Amer. Chem. Soc.

Coleman, K., & Jenkinson, D. S. (1996). RothC-26.3---A model for the turnover of carbon in soil. In Powlson, D. S., Smith, P., & Smith, J. U. (Eds.), *Evaluation of Soil Organic Matter Models: Using Existing Long-Term Datasets* (pp. 237–246). Berlin, Germany: Springer.

Collares, J. E. R., Lauria, C. A., & Carrilho, M. M. (1993). Pesquisa de previsão e acompanhamento de safras baseada em painéis de amostras de áreas. In Instituto Nacional de Pesquisas Espaciais (Ed.), *VII Simpósio Brasileiro de Sensoriamento Remoto,* (Vol. 4, pp. 450-453). São José dos Campos, Brazil: INPE.

Comisión Económica para América Latina (Ed.). (1983). *Expansión de la frontera agropecuaria y medio ambiente en América Latina*. Madrid, Spain: Centro Internacional de Formación en Ciencias Ambientales.

Compieta, P., Di Martino, S., Bertolotto, M., Feerucci, F., & Kechadi, T. (2007). Exploratory spatio-temporal data mining and visualization. *Journal of Visual Languages and Computing*, *18*(3), 255–279. doi:10.1016/j.jvlc.2007.02.006

Coppin, P., Jonckheere, I., Nackaerts, K., & Muys, B. (2004). Digital change detection methods in ecosystem monitoring: a review. *International Journal of Remote Sensing*, *25*(9), 1565–1596. doi:10.1080/0143116031000101675

Corrêa, P. L. P., Saraiva, A. M., Cugnasca, C. E., & Hirakawa, A. R. (2006). An architecture of systems for integration of Biodiversity distributed databases. *Journal of the Brazilian Society of Agroinformatics*, *6*(1), 80–100.

Costa Neto, P. L. de O. (2002). *Estatística*. São Paulo, Brazil: Editora Edgard Blücher Ltda.

Costa, A. M., Santos, L. M. R., Alem, D. J., & Santos, R. H. S. (2009, July). Sustainable Vegetable Crop Supply Problem with Perishable Stocks. Paper presented at the EURO Summer Institute, Lleida, Spain.

Costa, J. A. F. (1999) *Classificação automática e análise de dados por redes neurais auto-organizáveis*. Unpublished doctoral dissertation, University of Campinas, Campinas, São Paulo.

Costa, J. A. F., & Andrade Netto, M. L. (2003). *Segmentação do SOM Baseada em Particionamento de Grafos*. Paper presented at the VI Brazilian Congress on Neural Networks, São Paulo, (pp. 451-456).

Coulibaly, P., Anctil, F., & Bobée, B. (2000). Daily reservoir inflow forecasting using artificial neural networks with stopped training approach. *Journal of Hydrology (Amsterdam)*, *230*, 244–257. doi:10.1016/S0022-1694(00)00214-6

Coutinho, H. L. C., Leite, A.A.S., Paraiba, L.C., Cerdeira, A., Filizola, H., Silva, E.F., Gomes, M.A.F. (n.d.). *Análise Espacial Do Potencial De Lixiviação De Herbicidas Em Áreas De Recarga Do Aqüífero Guarani Na Alta Bacia Do Rio Taquari*. Retrieved from Http://Www.Riosvivos.Org.Br/Arquivos/1934241170.Pdf 20p. (Accessed On September, 2009)

Cromer, R. N., Cameron, D. M., Rance, S. J., Ryan, P. A., & Brown, M. (1993). Response to nutrients in E. grandis. Nitrogen acumulation. *Forest Ecology and Management*, *62*(1-4), 231–243. doi:10.1016/0378-1127(93)90052-O

Cruz, C. D. (1990). *Aplicação De Algumas Técnicas Multivariadas No Melhoramento De Plantas. Piracicaba: Escola Superior De Agronomia Luiz De Queiroz/Usp.* Tese De Doutorado.

Cruz, F. M. (2006). *Avaliação Do Potencial De Contaminação Dos Recursos Hídricos Por Pesticidas Utilizando O Software Agroscre.* Belém: Ufpa/Ccnt, 2006. 86p. (Monografia, Engenharia Ambiental. Curry, R.B., Peart, R.M., Jones, J.W., Boote, K.J., Allen Jr., L.H. (1990). Simulation As A Tool For Analyzing Crop Response To Climate Change. *Transactions of the ASAE. American Society of Agricultural Engineers*, *33*, 981–990.

Cruz, C. D., & Regazzi, A. J. (1994). *Modelos Biométricos Aplicados Ao Melhoramento Genético*. Viçosa, Brazil: Ufv.

Cunha, G. R., & Assad, E. D. (2001). Uma visão geral do número especial da revista brasileira de agrometeorologia sobre zoneamento agrícola no brasil. *Revista Brasileira de Agrometeorologia*, *9*, 377–385.

Cunha, N.G., Silvira, R.J.C., Severo, C.R.S. (2006) *Solos e terras do planalto Sul-Rio-Grandense e planícies costeiras*. Pelotas: Embrapa Clima Temperado. (Circular técnica, 55).

Curran, M. A., & Notten, P. (2006). *Summary of global life cycle inventory data resources*. Report for Task force 1: Database Registry, SETAC/UNEP Life Cycle Initiative.

Da Fonesca, A. F., Herpin, U., Paula, A. M., Victoria, R. L., & Melfi, A. J. (2007). Agricultural use of treated sewage effluents: agronomical-environmental implications and perspectives for Brazil. *Sci. Agric*, *64*, 194–209.

da Mota, F. S., Beirsdorf, M. I. C., Acosta, M. J. C., Motta, W. A., & West, S. L. (1974). *Zoneamento agroclimático do Rio Grande do Sul e Santa Catarina*. Pelotas, Brazil: IPEAS (Circular, 50).

Dadashzadeh, R., & Nimse, S. B. (2006). A compromise model for solving fuzzy linear programming problems. *Mathematica*, *48*, 134–148.

Dantzig, G. B., & Wolfe, P. (1960). Decomposition principle for linear programs. *Operations Research*, *8*, 101–111. doi:10.1287/opre.8.1.101

Darrow, W. K. (1994). The effect of drought en eucalypt species growing on shallow soil in South Africa: effect on mortality and growth. *ICFR Bulletin Series*, *7*, 1–24.

Davarajan, S., Squire, L., & Suthiwart-Narueput, S. (1995). Reviving project appraisal at the World Bank, *World Bank Policy Research Working Paper Policy working paper N°* 1496. Available at SSRN: http://ssrn.com/abstract=636145

Davis, J. (1986). *Statistics and data analysis in geology*. New York: Wiley.

Dawson, C. W., Harpham, C., Wilby, R. L., & Chen, Y. (2002). Evaluation of artificial neural network techniques for flow forecasting in the River Yangtze, China. *Hydrology and Earth System Sciences*, *6*(4), 619–926. doi:10.5194/hess-6-619-2002

Dawson, C. W., & Wilby, R. L. (1998). An artificial neural network approach to rainfall–runoff modelling. *Hydrological Sciences Journal*, *43*, 47–66. doi:10.1080/02626669809492102

Dawson, C. W., & Wilby, R. L. (2001). Hydrological modelling using artificial neural networks. *Progress in Physical Geography*, *25*(1), 80–108.

De Juana, E. (Ed.). (1990). *Áreas importantes para las aves en España*. Madrid: Monografías de la Sociedad Española de Ornitología, n.º 3.

De Paula, K. R. (2005). *Fatores Genéticos E Ambientais Na Produção De Sementes De Acácia Negra*. Curitiba: Engenharia Florestal/Universidade Federal Do Paraná. 77p. (Dissertação De Mestrado).

De Siqueira, O. J. F. (1991). Exemplo Utilizando O Sistema De Suporte De Decisão Dssat-Nitrogênio X Trigo. In *Reunião Nacional De Pesquisa De Trigo, 16, Resumos... Dourados: Embrapa/Uepae*. Avanços Na Informática Aplicados Às Relações Solo-Planta-Clima.

De Siqueira, O. J. F. Berg, M. Van Den. (1991). Validação Do Modelo De Simulação Soygro-Soja – Passo Fundo, Rs. In *Soja: Resultados De Pesquisa 1990-1991,* (pp. 139-144). Passo Fundo: Embrapa/Cnpt.

De Wit, C. T. (1993). Fundamentals of dynamic simulation: philosophy and terminology. In Leffelaar, P. A. (Ed.), *On systems analysis and simulation of ecological processes: with examples in CSMP and FORTRAN* (pp. 3–9). Dordrecht: Kluwer.

Desaigues, B., & Point, P. (1993). *Economie du patrimoine naturel. La valorisation des bénéfices de protection de l'environnement*. Paris: Economica.

Desaulniers, G., Desrosiers, J., & Solomon, M. M. (Eds.). (2005). *Column generation*. Berlin: Springer. doi:10.1007/b135457

Detlefsen, N., & Jensen, A. L. (2007). Modelling optimal crop sequences using network flows. *Agricultural Systems*, *94*, 566–572. doi:10.1016/j.agsy.2007.02.002

Dias, J. R. (2003). *Aplicação Do Modelo Hidrologico Agnps2001 Utilizando Dados Observados Na Baica Do Arroio Vacacaí-Mirim*. Santa Maria,Rs/Ufsm-Ppge, Dissertação De Mestrado.

Dias, C. T. S. (2009). *Regressão Segmentada Como Modelo Na Determinação De Exigências Nutricionais De Peixes*. Retrieved from http://Br.Monografias.Com/Trabalhos/Regressao-Segmentada-Exigencias-Nutricionais-Peixes/Regressao-Segmentada-Exigencias-Nutricionais-Peixes.Shtml (Accesed On 1st September, 2009).

Díaz, J. A., & Pérez, I. G. (2000). Simulation and Optimization of sugar cane transportation in harvest season. *Proceedings of the Simulation Conference, Winter, 2*, 1114-1117.

Diniz, C. C. (1995). *A dinâmica regional recente da economia brasileira e suas perspectivas*. Brasília, DF: Instituto de Pesquisa Econômica Aplicada. (Texto para discussão Nº 375.)

Dodgson, J., Spackman, M., Pearman, A., & Phillips, L. (2000). *Multi-Criteria analysis: An annual*. London: Department of the Environment, Transport and the Regions. Available at http://www.odpm.gov.uk/about/multicriteria/index.htm.

Dogliotti, S., Rossing, W. A. H., & Van Ittersum, M. K. (2003). ROTAT, a tool for systematically generating crop rotations. *European Journal of Agronomy*, *19*, 239–250. doi:10.1016/S1161-0301(02)00047-3

Dossa, D. (1994). Programação Linear Na Gestão Da Propriedade Rural: Um Enfoque Alternativo. *Teoria E Evidencia Econômica,* (4), 33-60.

Dourado-Neto, D., Teruel, D.A., Reichardt, K., Nielsen, D.R., Frizzone, J.A., Bacchi, O.O.S. (1998) Principles Of Crop Modeling And Simulation: I. Uses Of Mathematical Models In Agricultural Science. *Scientia Agricola, Piracicaba, 55*(Número Especial), 46-50.

Dubus, I. G., Beulke, S., & Brown, C. D. (2000). *A briefing document on the application of inverse modeling techniques to pesticide leaching models*. Silsoe, UK: Cranfield University.

Dufy, L., Abt, V., & Poyet, P. (2006). GIEA: gestion des informations de l'exploitation agricole. Un projet au service de l'interopérabilité sémantique de la profession agricole. *Ingénieries - Eau, Agriculture. Territoires*, *48*, 27–36.

Dunning, D. J., Ross, Q. E., & Merkhofer, M. W. (2000). Multi attribute utility analysis for addressing section 316(b) of the clean water act. *Environmental Science & Policy*, *3*, 7–14. doi:10.1016/S1462-9011(00)00022-8

Earsterling, W.E.Iii, Crosson, R.P., Rosenberg, N.J., Mc Kenney, M.S., Katz, L.A., Lemon, K.M. (1993). Agricultural Impacts And Responses To Climate Change In The Missouri_Iowa_Nebraska_Kansas (Mink) Region. *Climatic Change*, *24*, 23–61. doi:10.1007/BF01091476

Efron, B. (1979). Bootstrap Methods: Another Look at the Jackknife. *Annals of Statistics*, *7*(1), 1–26. doi:10.1214/aos/1176344552

Eldridge, K. G., Davidson, J., Hardwood, C. E., & van Wyk, G. (1994). *Eucalypt domestication and breeding*. Oxford, UK: Clarendon Press.

Elith, J., Graham, C. H., Anderson, R. P., Dudík, M., Ferrier, S., & Guisan, A. (2006). Novel methods improve prediction of species' distributions from occurrence data. *Ecography*, *29*, 129–151. doi:10.1111/j.2006.0906-7590.04596.x

El-Nazer, T., & McCarl, B. A. (1986). The choice of crop rotation: A modeling approach and case study. *American Journal of Agricultural Economics*, *68*(1), 127–136. doi:10.2307/1241657

Embrapa. (1997). *Sistema brasileiro de classificação de solos*. Rio de Janeiro, Brazil: Centro Nacional de Pesquisa de Solos.

Environmental Systems Research Institute. (1999). *ArcGIS 9.1: handbook*. Redlands, CA: Environmental Systems Research Institute.

Epiphanio, J. C. N., Luiz, A. J. B., & Formaggio, A. R. (2002). Estimativa de áreas agrícolas municipais, utilizando sistema de amostragem simples sobre imagens de satélite. *Bragantia*, *61*(2), 187–197. doi:10.1590/S0006-87052002000200012

Epiphanio, R. D. V., Formaggio, A. R., Rudorff, B. F. T., Maeda, E. E., & Luiz, A. J. B. (2010). Estimating soybean crop areas using spectral-temporal surfaces derived from MODIS images in Mato Grosso, Brazil. *Pesquisa Agropecuaria Brasileira*, *45*(1), 72–80.

Epiphanio, J. C. N., Barros Neto, O. O., Luiz, A. J. B., & Formaggio, A. R. (2001). Sistema de amostragem em imagem como base para estimativa de áreas agrícolas no município de Ipuã-SP. In Instituto Nacional de Pesquisas Espaciais (Ed.), *X Simpósio Brasileiro de Sensoriamento Remoto*, (pp. 59-66). São José dos Campos, Brazil: INPE.

Equi, L., Gallo, G., Marziale, S., & Weintraub, A. (1997). A combined transportation and scheduling problem. *European Journal of Operational Research*, *97*(1), 94–104. doi:10.1016/S0377-2217(96)00085-9

Ermolieva, T., Ermoliev, Y., Fischer, G., & Galambos, I. (2003). The role of financial instruments in integrated catastrophic flood management. *Multinational Finance*, *7*(3&4), 207–230.

Ermolieva, T. (1997). *The design of optimal insurance decisions in the presence of catastrophic risks*. Interim Report IR-97-068. IIASA, Laxenburg, Austria.

Escobar, M. T., & Moreno-Jiménez, J. M. (1997). The Hierarchical Compromise Programming. *Sociedad de Estadistica e InvestigaciSn Operativa*, *5*, 253–281.

Escobar, M. T., & Moreno-Jiménez, J. M. (2002). A linkage between the Analytic Hierarchy Process and the Compromise Programming Models. *Omega*, *30*, 359–365. doi:10.1016/S0305-0483(02)00053-1

Esquerdo, J. C. D. M., Antunes, J. F. G., Baldwin, D. G., Emery, W. J., & Zullo, J. Jr. (2006). An automatic system for AVHRR land surface product generation. *International Journal of Remote Sensing*, *27*(18), 3925–3942. doi:10.1080/01431160600763956

Everingham, Y. L., Muchow, R. C., Stone, R. C., Inman-Bamber, N. G., Singels, A., & Bezuidenhout, C. N. (2002). Enhanced risk management and decision-making capability across the sugarcane industry value chain based on seasonal climate forecasts. *Agricultural Systems*, *74*, 459–477. doi:10.1016/S0308-521X(02)00050-1

Extoxnet/Pip. (2008). *Extension Toxicology Network/Pesticide Information Profiles*. http://Extoxnet.Orst.Edu/Pips/Hexazin.Htm (Accesed On 18th August, 2008).

Fall, S., Niyogi, D., Mohanty, U. C., & Kumar, A. (2007). Application of weather prediction models for hazard mitigation planning: a case study of heavy off-season rains in Senegal. *Natural Hazards*, *41*, 227–243. doi:10.1007/s11069-006-9033-x

Faloutsos, C., & Kamel, I. (1994, 1994). Beyond Uniformity and Independence: Analysis of R-trees Using the Concept of Fractal Dimension. In *Proceedings of the ACM Symposium on Principles of Database Systems*, Minneapolis, MN, (pp. 4-13).

FAO. (2008). Major Food and Agricultural Commodities and Producers, Food and Agriculture Organization of the United Nations, The Statistics Division, Economic and Social Department. Retrieved October, 2008 from http://www.fao.org/es/ess/top/country.html?lang=en.

Faucheux, S., & Noel, J.-F. (1995). *Economie des ressources naturelles et de l'environnement.* Paris: Armand Colin.

Feddes, R. A., Kowalik, P. J., Kolinska-Malinka, K., & Zaradny, H. (1976). Simulation of field water uptake by plants using a soil water dependent root extraction function. *Journal of Hydrology (Amsterdam)*, *31*, 13–26. doi:10.1016/0022-1694(76)90017-2

Feddes, R. A., Kowalik, P. J., & Zaradny, H. (1978). *Simulation of field water use and crop yield.* Wageningen, The Netherlands: PUDOC.

Fernandes, E. N. (1997). *Sistema Inteligente De Apoio Ao Processo De Avaliação De Impactos Ambientais De Atividades Agropecuárias. Tese (Doutorado).* Universidade Federal De Viçosa, Viçosa, Mg.

Fernandes, E. N., & Silva, C. A. B. (1995). Agroflo: Sistema Especialista Para Planejamento De Sistemas Agroflorestais. In *Agrosoft 95 - Feira E Congresso De Informática Aplicada À Agropecuária E Agroindústria, Juiz De Fora.* Anais... Disponível Em: http://Www.Agrosoft.Org.Br/Trabalhos/Ag95/Doc35.Htm>. Acesso Em: 12 Out. 2004.

Fernandes, E. N., Silva, E., Fernandes Filho, E. I., & Silva, C. A. B. (2003). Erosys: Sistema Integrado Para Avaliação De Impactos Ambientais De Atividades Agropecuárias. In João Fernando Marques, Ladislau Araujo Skorupa, Jose Maria Gusman Ferraz. (Org.), *Indicadores De Sustentabilidade Em Agroecossistemas*, (pp. 251-273). Jaguariuna: Embrapa Meio Ambiente.

Ferracini, V. L., & Pessoa, M. C. P. Y. (2008). *Aspectos Toxicológicos E Ambientais Dos Agrotóxicos Aplicados Na Cultura Do Melão Produção Integrada De Melão*, (pp. 237 - 247).

Ferracini, V. L., Pessoa, M. C. P. Y., Oliveira, F. C., & Silva, A. De S. (2005). *Potenciais Contaminantes Ambientais E Subsídios Para Produção Com Qualidade Na Citricultura Citros.* Campinas, Brazil: Instituto Agronômico E Fundag.

Ferracini, V. L., Queiroz, S. C. N., Gomes, M. A. F., Cerdeira, A. L., Pereira, A. S., Souza, M. D., Santos, G. L. (2006, June). *Monitoramento Do Herbicida Tebutiuron Em Água Subterrânea Na Microbacia Do Córrego Espraiado, Região De Ribeirão Preto*, (pp. 1-10). Sp Jaguariúna/Sp: Embrapa Meio Ambiente (Documentos, Nº 54).

Ferracini, V.L., Pessoa, M.C.P.Y., Silva, A.S., Spadotto, C.A. (2001). Análise De Risco De Contaminação De Águas Subterrâneas E Superficiais Da Região De Petrolina (Pe) E Juazeiro (Ba). *Pesticidas: Revista De Ecotoxicologia E Meio Ambiente, 11*, 1-16.

Ferrer, J.-C., Cawley, A. M., Maturana, S., Toloza, S., & Vera, J. (2008). An optimization approach for scheduling wine grape harvest operations. *International Journal of Production Economics*, *112*(2), 985–999. doi:10.1016/j.ijpe.2007.05.020

Ferrer-Troyano, F., Aguilar-Ruiz, J. S., & Riquelme, J. C. (2006). Data streams classification by incremental rule learning with parameterized generalization. In *Proceedings of the ACM Symposium of Applied Computing*, (pp. 657-661).

Ferrier, S., Watson, G., Pearce, J., & Drielsma, M. (2002). Extended statistical approaches to modelling spatial pattern in Biodiversity in northeast New South Wales. 1. Species-level modeling. *Biodiversity and Conservation*, *11*, 2275–2307. doi:10.1023/A:1021302930424

Ferrier, S., & Watson, G. (1997). *An evaluation of the effectiveness of environmental surrogates and modelling techniques in predicting the distribution of biological diversity.* Canberra, Australia: NSW National Parks and Wildlife Service. Retrieved August 27, 2009, from http://www.environment.gov.au/Biodiversity/publications/technical/surrogates/pubs/surrogates.pdf

Fialho, F.B., & Ledur, M.C. (Setembro. (2000b). *Estimador Da Produção De Ovos*. Concórdia: Embrapa - Suínos E Aves. *Comunicado Técnico*, *257*, 1–2.

Fialho, F.B., Ledur, M.C. (Setembro, 2000a). *Modelo Matemático Para Curvas De Produção De Ovos*. Concórdia: Embrapa - Suínos E Aves, Comunicado Técnico N. 256, pp.1-3.

Fielding, A. H., & Bell, J. F. (1997). A review of methods for the assessment of prediction errors in conservation presence/absence models. *Environmental Conservation*, *24*, 38–49. doi:10.1017/S0376892997000088

Figueiredo, R. O., Watrin, O. S., Gerhard, P., Kato, O. R., Souza, E. L., Oliveira, F. A., et al. (2007). Watershed Studies In A Region Mainly Occupied By Small Holder Farms In The Eastern Amazon. In *Lba-Eco 11th Science Team Meeting, 2007*, Salvador, Book Of Abstracts.

Flexer, A. (2001). On the use of self-organizing maps for clustering and visualization. *Intelligent Data Analysis*, *5*, 373–384.

Flores, C. A., Filippini Alba, J. M., & Wrege, M. S. (Eds.). (2009). *Zoneamento agroclimático do eucalipto para o Estado do Rio Grande do Sul e edafoclimático na região do Corede Sul – RS*. Pelotas, Brazil: Embrapa Clima Temperado.

Flores, C. A., Garrastazu, M. C., Hasenack, H., & Weber, E. (2006). *Zoneamento edáfico para as culturas da mamona, soja, girassol e canola na região Noroeste do Estado do Rio Grande do Sul*. Pelotas, Brazil: Embrapa Clima Temperado (Documentos, 174).

Flores, M. (2004). Desenvolvimento territorial rural: uma proposta de estudo para apoio à formulação de políticas públicas. In Lages, et al (Ed.), *Territórios em movimento: cultura e identidade como estratégia de inserção competitiva*. Brasília, DF: SEBRAE.

Flug, M., Seitz, H. L. H., & Scott, J. F. (2000). Multicriteria decision analysis applied to Glen Canyon Dam. *Journal of Water Resources Planning and Management*, *126*, 270–276. doi:10.1061/(ASCE)0733-9496(2000)126:5(270)

Flury, M. (1996). Experimental evidence of transport of pesticides through field soils – a review. *Journal of Environmental Quality*, *25*, 25–45. doi:10.2134/jeq1996.00472425002500010005x

Flury, M., Flüher, H., Jury, W. A., & Leuenberger, J. (1994). Susceptibility of soils to preferential flow of water: a field study. *Water Resources Research*, *30*, 1945–1954. doi:10.1029/94WR00871

FOCUS. (1997). *Surface water models and EU registration of plant protection products: Final report of the work of the Regulatory Modelling Working Group on Surface Water*. EC document reference 6476/VI/96.

FOCUS. (2000). *FOCUS groundwater scenarios in the EU review of active substances*: *Report of the FOCUS groundwater scenarios workgroup*. EC document reference Sanco/312/2000 rev.2.

FOCUS. (2001). *FOCUS surface water scenarios in the EU evaluation process under 91/414/EEC: Report of the FOCUS Working Group on Surface Water Scenarios*. EC Document Reference SANCO/4802/2001-rev.2.

Fontoura Junior, J. A. S., Menezes, L. M., Corrêa, M. N., & Dionello, N. J. L. (2007). Utilização De Modelos De Simulação Em Sistemas De Produção De Bovinos De Corte. *Veterina-ria e Zootecnia*, *14*(1), 19–30.

Fontoura Junior, J. A. S., Siewerdt, F., Dionello, N. J. L., & Correa, M. N. (2009). Modelo De Simulação Do Desempenho Reprodutivo De Fêmeas Bovinas De Corte Com Base No Escore De Condição Corporal. *Revista Brasileira de Zootecnia*, *38*(8), 1627–1635. doi:10.1590/S1516-35982009000800030

Food and Agriculture Organization of the United Nations – FAO. (1996). *Multiple frame agricultural surveys: volume 1 current survey based on area and list sampling methods*. Rome: FAO. (FAO Statistical Development Series, 7).

Food and Agriculture Organization of the United Nations – FAO. (1999). *Multiple frame agricultural surveys: volume 2 agricultural survey programmes based on area frame or dual frame (area and list) sample design*. Rome: FAO. (FAO Statistical Development Series, 10).

Fortaleza, E. A. T., & Brasil, B. D. N. D. Cap.20, Ferracini, V. L., Queiroz, S. C. N., Souza, M. D., Gomes, M. A. F., Cerdeira, A. L., Rampazzo, P., Balderrama, O. (May, 2007). Tebuthiuron Monitoring In Sandy Soils Of The Guarany Aquifer Recharge Zone In Southeastern Brazil. In *1st Latin American Pesticide Residue Workshop (Laprw 2007), Santa Maria, Rs, Brazil - May 6-11, 2007, Book Of Abstratcts*, (pp. 65).

France, J., & Thornley, J. (2007). *Mathematical Models in Agriculture*. Wallingford, UK: Cabi Publishing.

Freijer, J. I., Tiktak, A., Hassanizadeh, S. M., & Van der Linden, A. M. A. (1996). PESTRAS: *Vol. 3. 1.: a one dimensional model for assessing leaching, accumulation and volatilization of pesticides in soil*. Bilthoven, The Netherlands: RIVM.

French, M. N., Krajewski, W. F., & Cuykendall, R. R. (1992). Rainfall forecasting in space and time using a neural network. *Journal of Hydrology (Amsterdam)*, *137*, 1–31. doi:10.1016/0022-1694(92)90046-X

Frenzel, S., & Pompe, B. (2007). Partial mutual information for coupling analysis of multivariate time series. *Physical Review Letters*, *99*(20), 204101. doi:10.1103/PhysRevLett.99.204101

Gagné, C., Gravel, M., & Price, W. L. (April 2003). *Recherche de solutions de compromis en ordonnancement à l'aide de metaheuristiques*. Paper presented at the Conférence Francophone de MOdélisation et SIMulation, Toulouse (France).

Gala, P. Y. L., Lyneb, P., Meyerb, E., & Solerc, L. G. (2008). Impact of sugarcane supply scheduling on mill sugar production: A South African case study. *Agricultural Systems*, *96*(1-3), 64–74. doi:10.1016/j.agsy.2007.05.006

Galacgac, E. S., & Balisacan, C. M. (2009). Traditional weather forecasting for sustainable agroforestry practices in Ilocos Norte Province, Philippines. *Forest Ecology and Management*, *257*, 2044–2053. doi:10.1016/j.foreco.2009.01.002

Gallego, F. J. (2004). Remote sensing and land cover area estimation. *International Journal of Remote Sensing*, *25*(15), 3019–3047. doi:10.1080/01431160310001619607

Gama, J., Medas, P., & Rodrigues, P. (2005). Learning decision trees from dynamic data streams. In *Proceedings of the ACM Symposium of Applied Computing*, (pp. 573-577).

Gao, J. (2002). Integration of GPS with remote sensing and GIS: reality and prospect. *Photogrammetric Engineering and Remote Sensing*, *68*(5), 447–453.

Garagorry, F. L., & Chaib Filho, H. (2008a). *Elementos de agrodinâmica*. Retrieved June 29, 2009, from http://www.embrapa.br/publicacoes/tecnico/RELFIN1.pdf

Garagorry, F. L., & Chaib Filho, H. (2008b). Avaliação conjunta da agrodinâmica de 20 produtos no Cerrado. In *Anais do 2º Simpósio Internacional de Savanas Tropicais: Desafios e estratégias para o equilíbrio entre sociedade, agronegócio e recursos naturais*. Brasília, DF: Embrapa Cerrados. 1 CD.

Garagorry, F. L., & Chaib Filho, H. (2008c). Mudanças no perfil agrícola das microrregiões do Cerrado. In *Anais do 2º Simpósio Internacional de Savanas Tropicais: Desafios e estratégias para o equilíbrio entre sociedade, agronegócio e recursos naturais*. Brasília, DF: Embrapa Cerrados. 1 CD.

Garrastazú, M. C., da Silva, M. B., & de Siqueira, O. J. W. (2006). *Estruturação da base cartográfica digital para o extremo sul do RS*. Pelotas, Brazil: Embrapa Clima Temperado (Comunicado técnico, 142).

Gäuting, R. H., Almeida, V. T., & Ding, Z. (2006). Modeling and querying moving objects in networks. *The VLDB Journal*, *15*, 165–190. doi:10.1007/s00778-005-0152-x

Gäuting, R. H., Bäohlen, M. H., Erwig, M., Jensen, C. S., Lorentzos, N. A., Schneider, M., & Vazirgiannis, M. (2000). A foundation for representing and quering moving objects. *ACM Transactions on Database Systems*, *25*(1), 1–42. doi:10.1145/352958.352963

Gäuting, R. H., & Schneider, M. (2005). *Moving Objects Databases*. San Francisco: Morgan Kaufmann Publishers.

Gazzoni, D. L., Pedroso Junior, M., Garagorry, F., & Moscardi, F. (1998). Mathematical Simulation Model Of The Velvetbean Caterpillar. *Pesquisa Agropecuária Brasileira, Brasília, 33*(4), 385–396.

Gebler, L. (2000). *Aplicabilidade Da Rotina Przm Do Modelo Matemático Przm3 Em Uma Situação Ambiental Do Oeste De Santa Catarina. Florianópolis: Universidade Federal De Santa Catarina*. Tese De Mestrado.

Geerts, F. (2004). Moving objects and their equations of motion. In *Proceedings of the First International Symposium on Applications of Constraint Databases*, (pp. 41–52). Berlin: Springer-Verlag.

Genovez, P. (2002). *Território e Desigualdades: análise espacial intra-urbana no estudo da dinâmica de exclusão/inclusão social no espaço urbano em São José dos Campos-SP*. Unpublished master thesis, National Institute for Space Research, São José dos Campos, São Paulo, Brazil.

Gerke, H. H., & Van Genuchten, M. Th. (1993). Evaluation of a first-order water transfer term for variably-saturated dual-porosity flow models. *Water Resources Research, 29*, 1225–1238. doi:10.1029/92WR02467

Gerke, H. H., & Van Genuchten, M. Th. (1996). Macroscopic representation of structural geometry for simulating water and solute movement in dual-porosity media. *Advances in Water Resources, 19*, 343–357. doi:10.1016/0309-1708(96)00012-7

Giannotti, M. A. (2001). *Geotecnologias na análise de impactos sócios-ambientais: o caso da queima da cana-de-açúcar na região de Piracicaba. Unpublished master's theses*. São José dos Campos, Brazil: Instituto Nacional de Pesquisas Espaciais.

Gimblett, H. R. (Ed.). (2002). *Integrating geographic information systems and agent-based modeling techniques for simulating social and ecological processes*. Oxford, UK: Oxford University Press.

Giuliano, A.D. (Agosto 1995). *Verificação Experimental De Um Modelo De Simulação Do Movimento De Solutos Em Um Meio Poroso*. São Carlos: If/Usp, Dissertação De Mestrado, Departamento De Física E Ciência Dos Materiais.

Glen, J. J. (1987). Mathematical models in farm planning: a survey. *Operations Research, 35*(5), 641–666. doi:10.1287/opre.35.5.641

Gliessman, S. R. (2000). *Agroecology: ecological processes in sustainable agriculture*. Chelsea, UK: Ann Arbor.

Gocht, A. (2005). *Assessment of simulation behaviour of different mathematical programming approaches*. Selected paper presented at the 89th EAAE symposium on Modelling agricultural policies: state of the art and new challenges, Parma, 3-5 February.

Goddard, L., Mason, S. J., Zebiak, S. E., Ropelewski, C. F., Basher, R., & Cane, M. A. (2001). Current approaches to seasonal-to-interannual climate predictions. *International Journal of Climatology, 21*, 1111–1152. doi:10.1002/joc.636

Gohin, A., & Chantreuil, F. (1999). La programmation mathématique positive dans les modèles d'exploitation agricole: principes et importance du calibrage. *Cahiers d'Économie et Sociologie Rurales., 52*, 59–78.

Goicoechea, A., Hansen, D. R., & Duckstein, L. (1982). *Multiobjective decision analysis with engineering and business application*. New York: John Wiley and sons.

Golden, R. M. (1996). *Mathematical Methods for Neural Network Analysis and Design*. Cambridge, UK: MIT Press.

Golfari, L., Caser, R. L., & Moura, V. P. G. (1978). *Zoneamento ecológico esquemático para o reflorestamento no Brasil*. Belo Horizonte, Brazil: Centro de Pesquisa Florestal da Região do Cerrado (Série técnica, 11).

Gomes, M. A. F., Filizola, H. F., & Spadotto, C. A. (2006a). Classificação Das Áreas De Recarga Do Sistema Aqüífero Guarani No Brasil Em Domínios Pedomorfoagroclimáticos - Subsídio Aos Estudos De Avaliação De Risco De Contaminação Das Águas Subterrâneas. *Revista Do Departamento De Geogragia (Usp), 18*, 67–74.

Gomes, M. A. F. (1995) Impacto Da Agricultura Intensiva De Cana-De-Açúcar Sobre As Propriedades Do Solo E Da Água Subterrânea - Estudo De Caso Em Ribeirão Preto, Sp. In *Encontro Nacional Da International Association For Impact Assessment - Iaia. 4*, (pp. 383-387). Belo Horizonte, Anais... Belo Horizonte: Gráfica E Editora Cultura.

Gomes, M. A. F., Filizola, H. F., Spadotto, C. A., Pereira, A. S (Junho/2006b). *Caracterização Das Áreas De Afloramento Do Aqüífero Guarani No Brasil - Base Para Uma Proposta De Gestão Sustentável*, Jaguariúna/Sp.: Embrapa Meio Ambiente,. Pp.1-20 (Documentos, Nº 52).

Gomes, M. A. F., Queiroz, S. C. N., Ferracini, V. L., Pessoa, M. C. P. Y., Pereira, A. S., Filizola, H. F., Souza, M. D. (Junho, 2004). *Ocorrência Dos Herbicidas Tebutiuron E Hexazinone E Implicações Na Qualidade Da Água Subterrânea - Estudo De Caso Na Microbacia Do Córrego Espraiado*. Ribeirão Preto/Sp. In.: I Congresso De Meio Ambiente, Paulínia/Sp, 21 A 25/6/2004, Anais... 2004, Pp.150-151.

Gomes, M. A. F., Queiroz, S. C. N., Ferracini, V. L., Pereira, A. S., Santos, G. L. (Setembro/2006c). *Monitoramento Do Herbidicida Diuron Em Água Subterrânea Na Microbacia Do Córrego Espraiado*. Região De Ribeirão Preto/Sp, Jaguariúna/Sp: Embrapa Meio Ambiente, Pp 1-11(Documentos, Nº 55).

Gomes, M. A. F., Spadotto, C. A., & Pessoa, M. C. P. Y. (2002). Avaliação Da Vulnerabilidade Natural Do Solo Em Áreas Agrícolas: Subsídio À Avaliação Do Risco De Contaminação Do Lençol Freático Por Agroquímicos. *Pesticidas: Revista De Ecotoxicologia E Meio Ambiente, Centro De Pesquisa E Processamento De Alimentos, Curitiba, 12*(Jan./Dez.), 169-179.

Gomez, L. I., Kuijpers, B., Moelans, B., & Vaisman, A. A. (2009). A Survey of Spatio-Temporal OLAP. *International Journal of Data Warehousing and Mining*, *5*(3), 28–55.

Gómez, J., & Rodríguez, O. (2000). Effects of vinasse on sugarcane (Saccharum officinarum) productivity (in spanish). [LUZ]. *Rev. Fac. Agron.*, *17*, 318–326.

Gomory, R. E. (1958). Outline of an algorithm for integer solutions to linear programs. *Bulletin of the American Mathematical Society*, *64*, 275–278. doi:10.1090/S0002-9904-1958-10224-4

Gonçalves, F. A. (2008). *Validação Do Modelo Wepp Na Predição De Erosão Hídrica Para Condição Edafoclimática Da Região De Viçosa, Mg. Viçosa, Mg/ Ufv-Ppgea*. Tese De Doutorado.

Gong, D., & Wang, S. (1999). Definition of Antarctic Oscillation index. *Geophysical Research Letters*, *26*, 459–462. doi:10.1029/1999GL900003

Goss, D. W. (1992). Screening Procedure For Soils And Pesticides For Potential Water Quality Impacts. *Weed Technology*, *6*(3), 701–708.

Greenhouse Gas Inventory Office of Japan. (2009). *National Greenhouse Gas Inventory Report of JAPAN*. Tokyo, Japan: National Institute for Environmental Studies, Japan.

Grodski, L., Caramori, P. H., Bootsma, A., de Oliveira, D., & Gomes, J. (1996). Riscos de ocorrência de geada no Estado do Paraná. *Revista Brasileira de Agrometeorologia*, *4*(1), 93–99.

Guahyaba, A. Da S. (2004). Modelagem Matemática Em Avicultura: Uma Questão De Sobrevivência. *Aveworld, 2(9*, Abril/Maio), 36-38.

Guha, S., Meyerson, A., Mishra, N., Motwani, R., & O'Callaghan, L. (2003). Clustering Data Streams: Theory and Practice. *IEEE Transactions on Knowledge and Data Engineering*, *15*(3), 515–528. doi:10.1109/TKDE.2003.1198387

Gurgel, H. C., Ferreira, N. J., & Luiz, A. J. B. (2003) Estudo da variabilidade do NDVI sobre o Brasil, utilizando-se a análise de agrupamentos. *Revista brasileira de engenharia agrícola e ambiental, 7*(1), 85-90. doi: 10.1590/S1415-43662003000100014.

Gustafson, D. I. (1989). Groundwater Ubiquity Score: A Simple Method For Assessing Pesticide Leachability. *Environmental Toxicology and Chemistry*, *8*(4), 339–357. doi:10.1002/etc.5620080411

Haese, K., & Goodhill, G. J. (2001). Auto-SOM: Recursive Parameter Estimation for Guidance of Self-Organizing Feature Maps. *Neural Computation, 13*, 595–619. doi:10.1162/089976601300014475

Hägerstrand, T. (1975). Survival and arena: On the life-history of individuals in relation to their geographical environment. *Monadnock, 49*, 9–29.

Hajkowicz, S., & Higgins, A. (2008). A comparison of multiple criteria analysis techniques for water resource management. *European Journal of Operational Research, 184*, 255–265. doi:10.1016/j.ejor.2006.10.045

Hajkowicz, S., Spencer, R., Higgins, A., & Marion, A. (2008). Evaluating water quality investment using cost utility analysing. *Journal of Environmental Management, 88*, 1601–1610. doi:10.1016/j.jenvman.2007.08.006

Halkidi, M., & Vazirgiannis, M. (2002). *Clustering validity assessment using multi representatives.* Paper presented at the SETN Conference, Thessaloniki, Greece.

Hammer, G. L., Hansen, J. W., Phillips, J. G., Mjelde, J. W., Hill, H., Love, A., & Potgieter, A. (2001). Advances in application of climate prediction in agriculture. *Agricultural Systems, 70*, 515–553. doi:10.1016/S0308-521X(01)00058-0

Haneveld, W. K., & Stegeman, A. W. (2005). Crop succession requirements in agricultural production planning. *European Journal of Operational Research, 166*, 406–429. doi:10.1016/j.ejor.2004.03.009

Hansen, J. W. (2002). Realizing the potential benefits of climate prediction to agriculture: issues, approaches, challenges. *Agricultural Systems, 74*, 309–330. doi:10.1016/S0308-521X(02)00043-4

Hansen, J. W., Mishra, A., Rao, K. P. C., Indeje, M., & Ngugi, R. K. (2009). Potential value of GCM-based seasonal rainfall forecasts for maize management in semi-arid Kenya. *Agricultural Systems, 101*, 80–90. doi:10.1016/j.agsy.2009.03.005

Hansson, K., Danielson, M., & Ekenberg, L. (2008). A framework for evaluation of flood management strategies. *Environmental Management, 86*, 465–480. doi:10.1016/j.jenvman.2006.12.037

Hao, Q., Meng, F., Zhou, Y., & Wang, J. (2004). Determining the optimal selective harvest strategy for mixed-species stands with a transition matrix growth model. *New Forests, 29*(3), 207–219. doi:10.1007/s11056-005-5656-0

Harms, S. K., & Deogun, J. S. (2004). Sequential association rule mining with time lags. *Journal of Intelligent Information Systems, 22*(1), 7–22. doi:10.1023/A:1025824629047

Hastenrath, S. (1991). *Climate Dynamics of the Tropics.* Amsterdam: Kluwer Academic Publishers.

Hayashi, K. (2008). Is organic farming environmentally superior? what agricultural LCA reveals. *Journal of Life Cycle Assessment, Japan, 4*(2), 112–118.

Haykin, S. (1998). *Neural Networks: A Comprehensive Foundation* (2nd ed.). New York: Prentice Hall.

Heady, E. O. (1954). Simplified presentation and logical aspects of linear programming technique. *Journal of Farm Economics.*

Heckelei, T., & Wolff, H. (2003). Estimation of Constrained Optimisation Models for Agricultural Supply Analysis Based on Generalised Maximum Entropy. *European Review of Agriculture Economics, 30*(1), 27–50. doi:10.1093/erae/30.1.27

Heckelei, T., & Britz, W. (1999). *Maximum entropy specification of PMP.* CAPRI Working Paper 08199.

Heckelei, T., & Britz, W. (2005). *Models Based on Positive Mathematical Programming: State of the Art and Futher Extensions.* Plenary paper presented at the 89th EAAE Seminar, 3-5 February, Parma, (pp. 48-73).

Helming, J. F. M., Peeters, L., & Veendendaal, P. J. J. (2001). Assessing the Consequences of Environmental Policy Scenarios in Flemish Agriculture. In Heckelei, T., Witzke, H. P., & Henrichsmeyer, W. (Eds.), *Agricultural Sector Modelling and Policy Information Systems*. Kiel, Germany: Vank Verlag.

Henningsson, M., Karlsson, J., & Rönnqvist, M. (2006). *Optimization models for forest road upgrade planning*. International Series In Operations Research, Handbook of Operations Research In Natural Resources.

Henry de Frahan, B., Buysse, J., Polomé, P., Fernagut, B., Harmignie, O., & Lauwers, L. (2007). Positive mathematical programming for agricultural and environmental policy analysis: review and practice. In Weintraub, A., Bjorndal, T., Epstein, R., & Romero, C. (Eds.), *Management of Natural Resources: A Hanbook of Operations Research Models, Algorithms and Implementations*. Berlin: Springer, International Series in Operations Research and Management Science.

Hernandes, F. B. T., Sousa, S. A. V., Zocoler, J. L., & Frizone, J. A. (2000) Simulação E Efeito De Veranicos Em Culturas Desenvolvidas Na Região De Palmeira D`Oeste, Estado De São Paulo. In *Xxix Congresso Brasileiro De Engenharia Agrícola (Conbea), Fortaleza-Ceará, 4 A 7 De Julho De 2000*. Available At: Http://Www.Agr.Feis.Unesp.Br/Veranico.Htm (24/10/2009)

Higgins, A. (2002). Australian Sugar Mills Optimize Harvester Rosters to Improve Production. *Interfaces*, *32*(3), 15–25. doi:10.1287/inte.32.3.15.41

Higgins, A. (2006). Scheduling of road vehicles in sugarcane transport: A case study at an Australian sugar mill. *European Journal of Operational Research*, *170*(3), 987–1000. doi:10.1016/j.ejor.2004.07.055

Higgins, A., Antony, G., Sandell, G., Davies, I., Prestwidge, D., & Andrew, B. (2004). A framework for integrating a complex harvesting and transport system for sugar production. *Agricultural Systems*, *82*(2), 99–115. doi:10.1016/j.agsy.2003.12.004

Higgins, A., & Davies, I. (2005). A simulation model for capacity planning in sugarcane transport. *Computers and Electronics in Agriculture*, *47*(2), 85–102. doi:10.1016/j.compag.2004.10.006

Higgins, A. & Laredo, L. (2006). Improving harvesting and transport planning within a sugar value chain. *Journal of the Operational Research Society, 57* (4), 367–376 (10).

Higgins, A., Thorburn, P., Archer, A. & Jakku, E (2007). Opportunities for value chain research in sugar industries. *Agricultural Systems, 94*, 611-621 (3).

Hijmans, R. J., Cameron, S. E., Parra, J. L., Jones, P. G., & Jarvis, A. (2005). Very high resolution interpolated climate surfaces for global land areas. *International Journal of Climatology*, *25*, 1965–1978. doi:10.1002/joc.1276

Hildreth, C. G., & Reiter, S. (1951). On the choice of a crop rotation plan. In Activity Analysis of Production and Allocation, T.C. Koopmans (Ed.), *Proceedings of the Conference on Linear Programming*, (pp. 177-188).

Holben, B. N. (1986). Characteristics of maximum value composite images from temporal AVHRR data. *International Journal of Remote Sensing*, *7*, 1417–1435. doi:10.1080/01431168608948945

Homem de Melo, F. (1983). La producción de biomasa y el crecimiento de la superficie de cultivo en el Brasil. In Comisión Económica para América Latina (Ed.), *Expansión de la frontera agropecuaria y medio ambiente en América Latina* (pp. 105-138). Madrid, Spain: Centro Internacional de Formación en Ciencias Ambientales.

Honda, R., & Konishi, O. (2001). Temporal rule discovery for time-series satellite images and integration with RDB. In *Proceedings of the European Conference on Principles and Practice of Knowledge Discovery in Databases*, Freiburg, Germany, (pp. 204-215).

Houghton, J. T., Meira Filho, L. G., Lim, B., Treanton, K., Mamaty, I., Bonduki, Y., et al. (Eds.). (1996). *Revised 1996 IPCC Guidelines for National Greenhouse Gas Inventories*. Paris, France: Organization for Economic Co-operation and Development.

Howitt, R. E. (1995). Positive Mathematical Programming. *American Journal of Agricultural Economics*, *77*(2), 329–342. doi:10.2307/1243543

Hsu, K., Gupta, H. V., & Sorooshian, S. (1995). Artificial neural network modeling of the rainfall-runoff process. *Water Resources Research*, *31*(10), 2517–2530. doi:10.1029/95WR01955

Hurrell, J. W. (1995). Decadal trends in the North Atlantic Oscillation: regional temperatures and precipitation. *Science*, *269*, 676–679. doi:10.1126/science.269.5224.676

Hurrell, J. W., Kushnir, Y., Visbeck, M., & Ottersen, G. (2003). An overview of the North Atlantic Oscillation. In J. W. Hurrell, Y. Kushnir, G. Ottersen, & M. Visbeck (Eds), *The North Atlantic Oscillation: Climate Significance and Environmental Impacts* (pp. 1-35), Geophysical Monograph Series, 134. Washington, DC: American Geophysical Union.

Ibáñez, J., & Pérez, C. (1999). Impactos de la reforma de la PAC de 1992 sobre el subsector agrícola español. *Estudios Agrosociales y Pesqueros*, *185*, 9–30.

Instituto Brasileiro de Geografia e Estatística. (2009). *Sistema IBGE de Recuperação Automática*. Retrieved June 29, 2009, from http://www.sidra.ibge.gov.br

Instituto Brasileiro de Geografia e Estatística – IBGE. (2003a). *CIDADES*. Retrieved May 20, 2003, from http://www.ibge.gov.br/cidadesat

Instituto Brasileiro de Geografia e Estatística – IBGE. (2003b). *SIDRA*. Retrieved April 22, 2003, from http://www.sidra.ibge.gov.br

Instituto Geográfico Nacional. (2003). *Generación de la Base de Datos de cambios Corine Land Cover y estadísticas por Km2. Proyecto Image & Corine 2000*. Madrid: Ministerio de Fomento, Área de Teledetección del Instituto Geográfico Nacional.

Intergovernmental Panel On Climate Change (Ipcc). (2001). *Climate Change 2001: The Scientific Basis: Contribution Of Working Group I To The Third Assessment Report Of The Intergovernmental Panel On Climate Change, (Edited By J. T. Houghton Et Al.)*. New York: Cambridge Univ. Press.

IPCC. (2007). *Climate Change 2007: Synthesis Report*. Retrieved March 07, 2009, from http://www.ipcc.ch/pdf/assessment-report/ar4/syr/ar4_syr.pdf

Ipt -,Instituto De Pesquisas Tecnológicas. (1994). Caracterização Do Potencial De Contaminação Das Águas Subterrâneas. Por Agroquímicos. *Hidrogeologia Da Microbacia, Relatório, 32*(605).

Isiklar, G., Alptekin, E., & Buyukozhan, G. (2007). Application of an hybrid intelligent decision support model in logistics outsourcing. *Computers & Operations Research*, *34*, 3701–3714. doi:10.1016/j.cor.2006.01.011

Ivey, F. J. (1999). Desenvolvimento E Aplicação De Modelos De Crescimento Para Frango De Corte. In *Simpósio Internacional Acav-Embrapa Sobre Nutrição De Aves, 1999, Concórdia, Santa Catarina, Brasil*, (pp. 22 –35).

Jarvis, N. J. (1989). A simple empirical model for root water uptake. *Journal of Hydrology (Amsterdam)*, *107*, 57–72. doi:10.1016/0022-1694(89)90050-4

Jarvis, N. J. (2007). A review of non-equilibrium water flow and solute transport in soil macropores: principles, controlling factors and consequences for water quality. *European Journal of Soil Science*, *58*, 523–546. doi:10.1111/j.1365-2389.2007.00915.x

Jarvis, N. J., Bergström, L. F., & Brown, C. D. (1995). Pesticide leaching models and their use for management purposes. In Roberts, T. R., & Kearney, P. C. (Eds.), *Environmental behaviour of agrochemicals* (pp. 185–220). New York: John Wiley.

Jayet, P. A., Marzochi, E., Hofstetter, A., & Donati, M. (2000). *Modular approach of agricultural supply modelling through linear programming*. EUROTOOLS (Proyect FAIR 5-CT97-3403), Working Paper Series n.26.

Jaynes, E. T. (1957a). Information theory and statistical mechanics. *Physical Review*, *106*, 620–630. doi:10.1103/PhysRev.106.620

Jaynes, E. T. (1957b). Information theory and statistical mechanics II. *Physical Review*, *108*, 171–190. doi:10.1103/PhysRev.108.171

Jaynes, E. T. (1982). On the rationale of maximum entropy methods. *Proceedings of the IEEE*, *70*, 939–952. doi:10.1109/PROC.1982.12425

Jena, S. D. & Poggi de Aragão, M. V. (2009). Sugar cane cultivation and harvesting: MIP approach and valid inequalities. *Anais do XLI Simpósio Brasileiro de Pesquisa Operacional.*, 969-980.

Jena, S. D. (2009). *A Mixed Integer Programming approach for sugar cane cultivation and harvest planning*. Master thesis, Pontifical Catholic University of Rio de Janeiro, Brazil.

Jensen, J. R. (2000). *Remote sensing of the environment: an earth resource perspective*. Upper Saddle River: Prentice Hall.

Jeong, D.-I., & Kim, Y.-O. (2005). Rainfall-runoff models using artificial neural networks for ensemble streamflow prediction. *Hydrological Processes*, *19*, 3819–3835. doi:10.1002/hyp.5983

Jin, C., Qian, W., Sha, C., Yu, J. X., & Zhou, A. (2003). Dynamically maintaining frequent items over a data stream. *Proceedings of the ACM Conference on Information and Knowledge Management*, 287-294.

Johansson, R. C., Tsurb, Y., Roe, T. L., Doukkali, R., & Dinar, A. (2002). Pricing irrigation water: a review of theory and practice. *Water Policy*, *4*, 173–199. doi:10.1016/S1366-7017(02)00026-0

Johnson, N. L., & Kotz, S. (1969). *Discrete Distributions*. New York: John Wiley & Sons.

Johnson, E. L., Nemhauser, G. L., & Savelsbergh, M. W. P. (2000). Progress in Linear Programming-Based Algorithms for Integer Programming: An Exposition. *INFORMS Journal on Computing*, *12*(1), 2–23. doi:10.1287/ijoc.12.1.2.11900

Jones, J., Hoogenboom, G., Porter, C., Boote, K., Batchelor, W., & Hunt, L. (2003). The DSSAT cropping system model. *European Journal of Agronomy*, *18*(3), 235–265. doi:10.1016/S1161-0301(02)00107-7

Jones, C., & Carvalho, L. M. V. (2002). Active and break phases in the South American Monsoon System. *Journal of Climate*, *15*, 905–914. doi:10.1175/1520-0442(2002)015<0905:AABPIT>2.0.CO;2

Jones, J. W., Hansen, J. W., Royce, F. S., & Messina, C. D. (2000). Potential benefits of climate forecasting to agriculture. *Agriculture Ecosystems & Environment*, *82*, 169–184. doi:10.1016/S0167-8809(00)00225-5

Joubert, A. R., Stewart, T. J., & Eberhard, R. (2003). Evaluation of water supply augmentation and water demand management option for the city of Cape Town. *Journal of Multi-criteria Decision Analysis*, *12*, 17–25. doi:10.1002/mcda.342

Jovanovic, T., & Booth, T. (2002). *Improved species climatic profiles*. Australia: Union Offset Printing.

Júdez, L., Chaya, C., Martínez, S., & González, A. A. (2001). Effects of the Measures Envisaged in "Agenda 2000" on Arable Crop Producers and Beef and Veal Producers: An Application of Positive Mathematical Programming to Representative Farms of a Spanish Region. *Agricultural Systems*, *67*, 121–138. doi:10.1016/S0308-521X(00)00051-2

Júdez, L., de Miguel, J. M., Mas, J., & Bru, R. (2002). Modeling Crop Regional Production Using Positive Mathematical Programming. *Mathematical and Computer Modelling*, *35*, 77–89. doi:10.1016/S0895-7177(01)00150-9

Júdez, L., Fuentes-Pila, J., de Miguel, J. M., & Martínez, S. (2001a). Programación Matemática Positiva: Un análisis de los efectos de la Agenda 2000 en algunas explotaciones tipo españolas. Agricultura en Sao Paulo. *Journal of Agricultural Economics*, *48*(1), 99–118.

Júdez, L., de Andrés, R., Ibáñez, M., de Miguel, J. M., & Urzainqui, E. (2005). *The PROMAPA.G Model. Document of GENEDEC (Project FP6 502184)*. 31 March.

Júdez, L., de Andrés, R., Ibáñez, M., & Urzainqui, E. (2008). *A method for including in PMP models activities non –existent in the baseline situation*. Selected poster presented at the XII th Congress EAAE, Ghent, Belgium, 26-29 August.

Júdez, L., de Miguel, J. M., Piniés, M., Legorburu, I. G., & Miguel, J. L. (2003). *Effects of Mid-Term review measures on some representative Spanish farm: an application of the PROMAPA model*. Selected paper presented 80 th EAAE Seminar, Ghent, 24-26 September 2004.

Júdez, L., Ibáñez, M., de Andrés, R., & Urzainqui, E. (2006). *Single Farm Payment, Modulation and Dual Values of Land*. Document of GENEDEC (Project FP6 502184). Presented at Rethymno meeting 12-13 May.

Júdez, L., Martínez, S., & Fuentes-Pila, J. (1998): Positive Mathematical Programming Revisited. EUROTOOLS (Proyect FAIR 5-CT97-3403), Working Paper Series n.2.

Julea, A., Méger, N., & Trouvé, E. (2006). Sequential patterns extraction in multitemporal satellite images. In *Proceedings of the European Conference on Principles and Practice of Knowledge Discovery in Databases*, Berlin, Germany, (pp. 96-99).

Kalnay, E. (1996). The NCEP/NCAR Reanalysis 40-year Project. *Bulletin of the American Meteorological Society*, *77*, 437–471. doi:10.1175/1520-0477(1996)077<0437:TNYRP>2.0.CO;2

Kantorovich, L. (1960). Mathematical methods of organizing and planning production (translated from a report in Russian, dated 1939). *Management Science*, *6*, 366–422. doi:10.1287/mnsc.6.4.366

Karamouz, M., Razavi, S., & Araghinejad, A. (2008). Long-lead seasonal rainfall forecasting using time-delay recurrent neural networks: a case study. *Hydrological Processes*, *22*, 229–241. doi:10.1002/hyp.6571

Karlsson, J., Rönnqvist, M., & Bergström, J. (2003). Short-term harvest planning including scheduling of harvest crews. *International Transactions in Operational Research*, *10*(5), 413–431. doi:10.1111/1475-3995.00419

Karlsson, J., Rönnqvist, M., & Bergström, J. (2004). An optimization model for annual harvest planning. *Canadian Journal of Forest Research*, *34*(8), 1747–1754. doi:10.1139/x04-043

Kaschner, K., Ready, J. S., Agbayani, E., Rius, J., Kesner-Reyes, K., Eastwood, P. D., et al. (2008). *AquaMaps: Predicted range maps for aquatic species*. Retrieved August 27, 2009, from http://www.aquamaps.org. Version 10/2008.

Kaski, S., & Kohonen, T. (1996). Exploratory data analysis by the self-organizing map: structures of welfare and poverty in the world. In Refenes, A.-P. N., Abu-Mostafa, Y., Moody, J., & Weigend, A. (Ed.), *Third International Conference on Neural Networks in the Capital Markets* (pp. 498-507). Singapore: World Scientific.

Keller, A. A. (1982). *Development And Analysis Of An Irrigation Scheduling Program With Emphasis On Forecasting Consumptive Use*. Ms Thesis, Agricultural And Irrigation Engineering Department, Utah State University, Logan.

Kifer, D., Ben-David, S., & Gehrke, J. (2004). Detecting change in data streams. In *Proceedings of the Very Large Data Base*, (pp. 180-191).

King, R. B. (2002). Land cover mapping principles: a return to interpretation fundamentals. *International Journal of Remote Sensing*, *23*(18), 3525–3545. doi:10.1080/01431160110109606

Kleinberg, J. M. (2003). Bursty and hierarchical structure in streams. *Data Mining and Knowledge Discovery*, *7*(4), 373–397. doi:10.1023/A:1024940629314

Koga, N., & Tsuruta, H. (2006). Hokkaido no hatasaku nougyou kara hassei suru onshitsukouka gas no life cycle inventory bunseki. [in Japanese]. *Agriculture and Horticulture*, *81*(10), 1101–1109.

Koua, E. L. (2003). *Using self-organizing maps for information visualization and knowledge discovery in complex geospatial datasets*. Paper presented at the XXI International Cartographic Conference (ICC), (pp. 1694-1702).

Koutsouyiannis, D., Efstratiadis, A., Namassis, N., & Christofides, A. (2008). On the credibility of climate predictions. *Hydrological Sciences Journal*, *53*(4), 671–684. doi:10.1623/hysj.53.4.671

Kroes, J. G., & Van Dam, J. C. (Eds.). (2003). *Reference manual SWAP version 3.0.3*. Wageningen, The Netherlands: Alterra.

Krol, M. S., Jaeger, A., & Bronstert, A. (2003). Integrated Modeling Of Climate Changes Impact In Northearstern Brazil. In *Global Change And Regional Impacts, Water Availability, Vulnerability Of Ecosystems And Society In The Semi-Arid Northeast Of Brazil, Gaiser, Krol, Frischkorn, And Araújo*. Berlin: Springer. Disponivel Em Http://Documente.Nl/44592/1/2003_Krol_Et_Al_Waves_Book_Springer.Pdf

Kuepker, B., & Kleinhanss, W. (2006). *Comparative Assessment of National and Alternative Decoupling Schemes in EU Member States*. Selected paper presented at the 93 rd EAAE Seminar, Prague, 22-23 September 2006.

Kusaba, T., & Ishioka, G. (2008). Outline and characteristics of the Dojo Kankyo Kisochosa Kijunten Chosa Data Base: data base of long-term changes in soil properties and crops yields of arable field under different nutrient managements in Japan. [in Japanese]. *Miscellaneous Publication of the National Agricultural Research Center*, *8*, 1–87.

Laabs, V., Amelung, W., Pinto, A., Wantzen, M., Silva, C. J., & Zech, W. (2002). Pesticides in surface water, sediment, and rainfall of the northeastern Pantanal basin, Brazil. *Journal of Environmental Quality*, *31*, 1636–1648. doi:10.2134/jeq2002.1636

Lal, H., Hoogenboom, G., Calixte, J. P., Jones, J. W., & Beinroth, F. H. (1993). Using Crop Simulation Models And Gis For Regional Productivity Analysis. *Transactions of the ASAE. American Society of Agricultural Engineers*, *36*(1).

Land, A. H., & Doig, A. G. (1960). An Automatic Method for Solving Discrete Programming Problems. *Econometrica*, *28*, 497–520. doi:10.2307/1910129

Lane, L. J., & Nearing, M. A. (Eds.). (1980). *Usda Water Erosion Prediction Project: Hill Slope Profile Model Documentation*, (2ed.). West Lafayette: Usda: Sea, (Usda-Sea Research Report, 26).

Larsbo, M., & Jarvis, N. J. (2003). *MACRO 5.0. A model of water flow and solute transport in macroporous soil: technical description*. Uppsala, Sweden: Swedish University of Agricultural Sciences.

Larsson, M. H., & Jarvis, N. J. (1999). Evaluation of a dual-porosity model to predict field-scale solute transport in a macroporous soil. *Journal of Hydrology (Amsterdam)*, *215*, 153–171. doi:10.1016/S0022-1694(98)00267-4

Lavorenti, A., Prata, F., & Regitano, J. B. (2003). Comportamento de pesticidas em solos – fundamentos. *Tópicos em Ciência do Solo*, *3*, 291–334.

Leal, R. M. P., Herpin, U., Da Fonseca, A. F., Firme, L. P., Montes, C. R., & Melfi, A. J. (2009). Sodicity and salinity in a Brazilian oxisol cultivated with sugarcane irrigated with wastewater. *Agricultural Water Management*, *96*, 307–316. doi:10.1016/j.agwat.2008.08.009

Lee, A., Liu, Y., Tsai, H., Lin, H., & Wu, H. (2009). Mining frequent patterns in image databases with 9D-SPA representation. *Journal of Systems and Software*, *82*(4), 603–618. doi:10.1016/j.jss.2008.08.028

Legates, D. R., & McCabe, G. J. (1999). Evaluating the use of "goodness-of-fit" measures in hydrologic and hydroclimatic model validation. *Water Resources Research*, *35*(1), 233–241. doi:10.1029/1998WR900018

Leistra, M. (1971). Diffusion of 1,3-dichloropropene from a plane source in soil. *Pesticide Science*, *2*, 75–79. doi:10.1002/ps.2780020206

Leistra, M., & Dekkers, W. A. (1976). Computed leaching of pesticides from soil under field conditions. *Water, Air, and Soil Pollution*, *5*, 491–500. doi:10.1007/BF00280848

Leistra, M., Van der Linden, A. M. A., Boesten, J. J. T. I., Tiktak, A., & Van den Berg, F. (2002). *PEARL model for pesticide behaviour and emissions in soil-plant systems: description of the processes in FOCUS PEARL v.1.1.1*. Wageningen, The Netherlands: Alterra.

Leite, L. F. C., Madari, B. E., Machado, P. L. O. A., & Doraiswamy, P. C. (2007) Modelagem Da Dinâmica Da Matéria Orgânica Do Solo Em Sistema Integração Lavoura-Pecuária Usando Cqestr. In *Anais Congresso Brasileiro De Ciência Do Solo. Gramado, Ufrgs.*

Leonard, R.A, Knisel, W.G, Still, D. A. (1987). Gleams: Groundwater Loading Effects Of Agricultural Management Systems. *Transactions Of Asae, St. Joseph, 30* (5), 1403-1418.

Leontief, W. W. (1941). *The structure of American economy, 1919-1929—An empirical application of equilibrium analyses*. Cambridge, MA: Harvard University Press.

Levy, T. L., Mattice, J. D., & Kochenderfer, J. N. (1989). Hexazinone Persistence And Mobility Of A Steep Forested Watershed. *Journal of Environmental Quality, 18*, 507–514. doi:10.2134/jeq1989.00472425001800040021x

Lewis, G. N. (1901). The Law Of Physico-Chemical Change. *Proceedings of the American Academy of Arts and Sciences, 37*, 49–69.

Li, C., Frolking, S., & Frolking, T. A. (1992). A model of nitrous oxide evolution from soil driven by rainfall events: 1. Model structure and sensitivity. *Journal of Geophysical Research, 97*(D3), 9759–9776.

Liebmann, B., Camargo, S. J., Seth, A., Marengo, J. A., Carvalho, L. M. A., & Allured, D. (2007). Onset and end of the rainy season in South America in observations and the ECHAM 4.5 atmospheric general circulation model. *Journal of Climate, 20*, 2037–2050. doi:10.1175/JCLI4122.1

Liebmann, B., Kiladis, G. N., Marengo, J. A., Ambrizzi, T., & Glick, J. D. (1999). Submonthly convective variability over South America and the South Atlantic Convergence Zone. *Journal of Climate, 12*(7), 1877–1891. doi:10.1175/1520-0442(1999)012<1877:SCVOSA>2.0.CO;2

Lima, L. M., Watrin, O. S., Pessoa, M. C. P. Y., Pereira, A. S., Campinas, D. S. N., Figueiredo, R. O., & Costa, F. R. (Abril, 2007). Simulação Da Movimentação De Agrotóxicos No Solo Com O Uso De Geotecnologias, Como Instrumento Para Avaliação Do Potencial De Risco De Contaminação De Águas Subterrâneas Na Amazônia Oriental. In *Anais Xiii Simpósio Brasileiro De Sensoriamento Remoto*, Florianópolis, Brasil, 21 A 26 Abril 2007, Inpe, (pp. 3397-3404).

Lin, D., Jensen, C. S., Ooi, B. C., & Saltenis, S. (2005). Efficient Indexing of the Historical, Present, and Future Positions of Moving Objects. In *Proceedings of MDM 2005*, (pp. 59-69). New York: ACM.

Lin, Yi., & Liu, S. (2004). A historical introduction to grey systems theory. *Systems, man and cybernetics, 3*, 2403-2408.

Linderoth, J. T., & Ralphs, T. K. (2005). Noncommercial Software for Mixed-Integer Linear Programming. In *Integer Programming: Theory and Practice* (pp. 253-303). Boca Raton, FL: CRC Press Operations Research Series.

Linnerooth-Bayer, J. (2001). *An insurance system for Hungary: Three policy scenarios*. Working paper, IIASA, Laxenburg, Austria.

Liu, W. T. (1989). Application Of Ceres-Maize Model To Yield Prediction Of A Brazilian Maixe Hybrid. *Agricultural and Forest Meteorology, 45*, 299–312. doi:10.1016/0168-1923(89)90050-6

Lorena, A. C., Siqueira, M. F., Giovanni, R., Carvalho, A. C. P. L. F., & Prati, R. C. (2008). Potential Distribution Modelling Using Machine Learning Classifiers. In *The Twenty First International Conference on Industrial, Engineering & Other Applications of Applied Intelligent Systems*, 2008, Wroclaw, *Proceedings The Twenty First International Conference on Industrial, Engineering & Other Applications of Applied Intelligent Systems* (LNAI 5027, pp. 255-264).

Lorenzi, H., Souza, H. M., Costa, J. T. M. C., & Ferreira, E. (2004). *Brazilian end exotic palm cultivated*. Editora Plantarum. (in Portuguese)

Loubier, S., Terreaux, J.-P., & Gleyses, G. (2001). Sur les principes de l'évaluation économique des systèmes irrigués: Méthodes et applications. In *19th European Regional Conference of ICID,* Brno and Prague, Czech Republic, 4-8 June.

Lourencetti, C., Spadotto, C. A., Santiago-Silva, M., & Ribeiro, M. (2005) Avaliação Do Potencial De Águas Subterrâneas Por Pesticidas: Comparação Entre Métodos De Previsão Por Lixiviação. *Pesticidas: Revista De Ecotoxicologia E Meio Ambiente, Curitiba, 15*, 1-14.

Lowe, T. J., & Preckel, P. V. (2004). Decision technologies for agribusiness problems: A brief review of selected literature and a call for research. *Manufacturing and Service Operations Management, 6*(3), 201–208. doi:10.1287/msom.1040.0051

Lu, D., Mausel, P., Brondizios, E., & Moran, E. (2004). Change detection techniques. *International Journal of Remote Sensing, 25*(12), 2365–2407. doi:10.1080/0143116031000139863

Lübbecke, M. E., & Desrosiers, J. (2005). Selected topics in column generation. *Operations Research, 53*(6), 1007–1023. doi:10.1287/opre.1050.0234

Lucas, M. T., & Chhajed, D. (2004). Applications of location analysis in agriculture: a survey. *The Journal of the Operational Research Society, 55*(6), 561–578. doi:10.1057/palgrave.jors.2601731

Luigi, P., Kramer, A., & Al Baz, I. (2007). The EMWater project – promoting efficient wastewater management and reuse in Mediterranean countries. *Desalinisation, 215*, 56–63. doi:10.1016/j.desal.2006.10.027

Luiz, A. J. B. (2002). *Sensoriamento remoto agrícola.* São José dos Campos, Brazil: INPE.

Luiz, A. J. B., Oliveira, J. C., Epiphanio, J. C. N., & Formaggio, A. R. (2002). Auxílio das imagens de satélite aos levantamentos por amostragem em agricultura. *Agricultura em São Paulo, 49*(1), 41–54.

Luiz, A. J. B. (2003). *Estatísticas agrícolas por amostragem auxiliadas pelo sensoriamento remoto.* Unpublished doctoral dissertation, Instituto Nacional de Pesquisas Espaciais, São José dos Campos, Brazil.

Luiz, A. J. B., & Epiphanio, J. C. N. (2001). Amostragem por pontos em imagens de sensoriamento remoto para estimativa de área plantada por município. In Instituto Nacional de Pesquisas Espaciais (Ed.), *X Simpósio Brasileiro de Sensoriamento Remoto*, (pp. 111-118). São José dos Campos, Brazil: INPE.

Luiz, A. J. B., & Gürtler, S. (2003). Aleatorização de pontos no território de um município, usando o SPRING, para a estimativa de área agrícola por amostragem. In Sociedade Brasileira de Informática Aplicada à Agropecuária e Agroindústria (Ed.), *IV Congresso Brasileiro da Sociedade Brasileira de Informática Aplicada à Agropecuária e Agroindústria SBI-Agro*. (pp. 9-12). Lavras, Brazil: UFLa.

Luk, K. C., Ball, J. E., & Sharma, A. (2001). An application of Artificial Neural Networks for rainfall forecasting. *Mathematical and Computer Modelling, 33*, 683–693. doi:10.1016/S0895-7177(00)00272-7

Lyon, S. (2006). Evaluating fair trade consumption: politics, defetishization and producer participation. *International Journal of Consumer Studies, 30*, 452–464. doi:10.1111/j.1470-6431.2006.00530.x

Macedo, J., Vangenot, C., Othman, W., Pelekis, N., Frentzos, E., & Kuijpers, B. (2008). Trajectory Data Models. In Giannoti, F., & Pedreschi, D. (Eds.), *Mobility, data mining, and Privacy – Geographic Discovery* (pp. 123–150). Berlin: Springer-Verlag. doi:10.1007/978-3-540-75177-9_6

Machado, R. E., & Vettorazzi, C. A. (2003). Simulação Da Produção De Sedimentos Para A Microbacia Hidrográfica Do Ribeirão Dos Marins (Sp). *Revista Brasileira de Ciencia do Solo, 27*(4), 735–741. doi:10.1590/S0100-06832003000400018

Machado, R. E., Vettorazzi, C. A., & Cruciani, D. E. (2003, Março). Simulação De Escoamento Em Uma Microbacia Hidrográfica Utilizando Técnicas De Modelagem E Geoprocessamento. *Rev. Bras. Recursos Hídricos.*, *8*(1), 147–155.

Mackay, D. (1993). Toxic Substances Modeling: Fugacity Concepts. In Systems, C. E. O. E. (Ed.), *Young, P.C* (pp. 611–615). New York: Pergamon Press Ltd.

Magnol, D., Oliveira, G., & Menegotto, R. (2001). *Cenário gaúcho: representações históricas e geográficas*. São Paulo, Brazil: Moderna.

Maienfisch, P., Angst, M., Brandl, F., Fischer, W., Hofer, D., & Kayser, H. (2001). Chemistry and biology of thiamethoxam: a second generation neonicotinoid. *Pest Management Science*, *57*, 906–913. doi:10.1002/ps.365

Maier, H. R., & Dandy, G. C. (1997). Modelling cyanobacteria blue-green algae in the River Murray using artificial neural networks. *Mathematics and Computers in Simulation*, *43*, 377–386. doi:10.1016/S0378-4754(97)00022-0

Maier, H. R., & Dandy, G. C. (2000). Neural networks for the prediction and forecasting of water resources variables: a review of modeling issues and applications. *Environmental Modelling & Software*, *15*(1), 101–124. doi:10.1016/S1364-8152(99)00007-9

Makatouni, A. (2002). What motivates consumers to buy organic food in the UK? Results from a qualitative study. *British Food Journal*, *104*, 345–352. doi:10.1108/00070700210425769

Malczewski, J. (1999). *GIS and Multicriteria Decision Analysis*. New York: John Wiley and Sons.

Maluf, J. R. T., Westphalen, S. L., Matznauer, R., & Maluf, D. E. (2007). *Zoneamento agroclimático da cana-de-açúcar no Estado do Rio Grande do Sul, visando a produção de açúcar e álcool.* Paper presente at the XV Brasilian Congress of Agrometeorology, Aracajú, Brazil.

Manjhi, A., Shkapenyuk, V., Dhamdhere, K., & Olston, C. (2005). Finding (recently) frequent items in distributed data streams. In *Proceedings of the IEEE International Conference on Data Engineering*, (pp. 767-778).

Mantua, N. J., Hare, S. R., Zhang, Y., Wallace, J. M., & Francis, R. C. (1997). A pacific decadal climate oscillation with impacts on salmon. *Bulletin of the American Meteorological Society*, *78*, 1069–1079. doi:10.1175/1520-0477(1997)078<1069:APICOW>2.0.CO;2

Marengo, J. A., Soares, W. R., Saulo, C., & Nicolini, M. (2004). Climatology of the low-level jet east of the Andes as derived from NCEP-NCAR reanalysis - characteristics and temporal variability. *Journal of Climate*, *17*(12), 2261–2280. doi:10.1175/1520-0442(2004)017<2261:COTLJE>2.0.CO;2

Marshall, E.-P., & Homans, H.-A. (2004). A spatial analysis of the economic and ecological efficacy of land retirement. *Environmental Modeling and Assessment*, *9*, 65–75. doi:10.1023/B:ENMO.0000032097.99637.26

Martella, D. L., Eldon, A., & Weintraub, A. (1998). Forest management challenges for operational researchers. *European Journal of Operational Research*, *104*, 1–17. doi:10.1016/S0377-2217(97)00329-9

Martello, S., & Toth, P. (1990). *Knapsack problems: Algorithms and Computer Implementations*. Chichester, UK: John Wiley and Sons.

Martín Lou, M. A., & Martínez Vega, J. (1998). *Situación socioeconómica de la ZEPA Encinares del río Alberche y río Cofio (Comunidad de Madrid)*. Madrid: Instituto de Economia y Geografía (CSIC).

Masini, G. (2003). *Supply chain planning optimization in the fruit industry*. CEPAC Working Paper.

Mast, J. N., Veblent, T. T., & Hodgson, M. E. (1997). Tree invasion within a pine/grassland ecotone: An approach with historic aerial photography and GIS modelling. *Forest Ecology and Management*, *93*(3), 181–194. doi:10.1016/S0378-1127(96)03954-0

May, P., Veiga Neto, F. C., & Cheves Pozo, O. V. (2000). *Biodiversity economic valuation*. Brasília: Ministério do Meio Ambiente. (in Portuguese)

McCarl, B. A., Candler, W. V., Doster, D. H. & Robbins, P., R. (1978). Experience with mass audience linear programming for farm planning. *Mathematical Programming Studies, 9.*

McIntosh, P. C., Ash, A. J., & Smith, M. S. (2005). From oceans to farms: the value of a novel statistical climate forecast for agricultural management. *Journal of Climate, 18*, 4287–4302. doi:10.1175/JCLI3515.1

McIntosh, P. C., Pook, M. J., Risbey, J. S., Lisson, S. N., & Rebbeck, M. (2007). Seasonal climate forecasts for agriculture: towards better understanding and value. *Field Crops Research, 104*, 130–138. doi:10.1016/j.fcr.2007.03.019

Medeiros, S. L. P., Gosse, G., Chartier, M., & Manfron, P. A. (2000). Ajuste Do Módulo De Produção De Matéria Seca No Modelo Ceres-Sorghum Para Simulação Da Cultura Do Sorgo Sacarino. *Revista Brasileira De Agrometeorologia. Santa Maria, 8*(2), 223–229.

Medeiros, H. R. (2003). *Avaliação De Modelos Matemáticos Desenvolvidos Para Auxiliar A Tomada De Decisão Em Sistemas De Produção De Ruminantes Em Pastagens.* Piracicaba: Esalq/Usp 98p.

Meinke, H., & Stone, R. C. (2005). Seasonal and interannual climate forecasting: the new tool for increasing preparedness to climate variability and change in agricultural planning and operations. *Climatic Change, 70*, 221–253. doi:10.1007/s10584-005-5948-6

Mello, L. M. R., Garagorry, F. L., & Chaib Filho, H. (2008). Spatial dynamics of grapes production in Brazil between 1975 and 2005. In *31st World Congress of Vine and Wine* (file PIII02.pdf, pp. 1-7).

Mendonça, L. R. C. (2004) *Simulador De Cenários Bioeconômicos Para Suporte A Decisão No Gerenciamento De Fazendas Produtoras De Gado De Corte.* Goiânia: Universidade Federal De Goiás- Escola De Engenharia Elétrica E De Computação, 116p. (Dissertação De Mestrado).

Menegaki, A. N., Hanley, N., & Tsagarakis, K. P. (2007). The social acceptability and valuation of recycled water in Crete: A study of consumers' and framers' attitude. *Ecological Economics, 62*, 7–18. doi:10.1016/j.ecolecon.2007.01.008

Meratnia, N., & de By, R. (2002). Aggregation and comparison of trajectories. In *Proceedings of the 26th VLDB Conference*, (pp. 49-54). New York: ACM.

Merz, J. (1991). Microsimulation – A survey of principles, developments and applications. *Forecasting, 7*, 77–104. doi:10.1016/0169-2070(91)90035-T

Miklós, A. A. W., & Gomes, M. A F. (1996). *Levantamento Semidetalhado Dos Solos Da Bacia Hidrográfica Do Córrego Do Espraiado*, Ribeirão Preto- Sp. (Relatório De Consultoria).

Millington, R. J., & Quirk, J. P. (1961). Permeability of porous solids. *Transactions of the Faraday Society, 57*, 1200–1207. doi:10.1039/tf9615701200

Ministério da Agricultura. Pecuária e Abastecimento. (2009). *Projeções do agronegócio: Brasil 2008/2009 a 2018/19*. Brasília, DF: Ministério da Agricultura, Pecuária e Abastecimento.

Ministério do Desenvolvimento Agrário – MDA. (2005). *Referências para a gestão social de territórios rurais.* Brasília, DF: MDA. Série Documentos nº 3.

Minns, A. W., & Hall, M. J. (1996). Artificial Neural Networks as rainfall-runoff modes. *Hydrological Sciences Journal, 41*(3), 399–417. doi:10.1080/02626669609491511

Minoti, R. T. E. Crestana, S. (2004). Perda De Solo Como Indicador De Vulnerabilidade De Bacias Hidrográficas: Artigos Publicados Na Revista Brasileira De Ciência Do Solo E Em Congressos Brasileiros De Ciência Do Solo, Manejo E Conservação. In Espíndola, E.L.G., Schalch, V. (Orgs.), *Bacia Hidrográfica: Diversas Abordagens Em Pesquisa,* (pp. 47-54). São Carlos: Rima.

Minoti, R.T. (2006). *Abordagens Qualitativa E Quantitativa Das Micro-Bacias Hidrográficas E Áreas Alagáveis De Um Compartimento Do Médio Mogi-Superior/Sp*. Tese Doutorado Em Ciências Da Engenharia Ambiental, São Carlos: Usp/Eesc.

Mitchell, S. A. (2004). *Operational forest harvest scheduling optimisation: a mathematical model and solution strategy*. Doctoral dissertation, University of Auckland, New Zealand.

Mitja, D., & Ferraz, I. (2001). Establishment of babaçu in pastures in Pará, Brazil. *Palms*, *45*(3), 138–147.

Mitja, D., Martins, E. D., Silva, J. C. S., & Miranda, I. S. (2007). Biodiversité: les Palmiers, Babaçu, exploitation d'une invasion. *Science Studies*, *38*, 10.

Moeller, C., Smith, I., Asseng, S., Ludwig, F., & Telcik, N. (2008). The potential value of seasonal forecasts of rainfall categories - case studies from the wheatbelt in Western Australia's Mediterranean region. *Agricultural and Forest Meteorology*, *148*, 606–618.

Moller, M. S. (1993). A scaled conjugate gradient algorithm for fast supervised learning. *Neural Networks*, *6*(4), 525–534. doi:10.1016/S0893-6080(05)80056-5

Monteith, J. L. (1981). Evaporation and surface temperature. *Quarterly Journal of the Royal Society*, *107*, 1–27. doi:10.1256/smsqj.45101

Moore, R. J., Bell, V. A., & Jones, D. A. (2005). Forecasting for flood warning. *Comptes Rendus Geoscience – External Geophysics. Climate and Enviroment*, *337*, 203–217.

Morid, S., Smakhtin, V., & Bagherzadeh, K. (2007). Drought forecasting using artificial neural networks and time series of drought indices. *International Journal of Climatology*, *27*, 2103–2111. doi:10.1002/joc.1498

Mualem, Y. (1976). A new model for predicting the hydraulic conductivity of unsaturated porous media. *Water Resources Research*, *12*, 513–522. doi:10.1029/WR012i003p00513

Mueller, C. C. (1983). El Estado y la expansión de la frontera agrícola en la Amazonia. In Comisión Económica para América Latina (Ed.), *Expansión de la frontera agropecuaria y medio ambiente en América Latina* (pp. 37-78). Madrid, Spain: Centro Internacional de Formación en Ciencias Ambientales.

Mueller, C. C., & Martha Júnior, G. B. (2008) A agropecuária e o desenvolvimento socioeconômico recente do Cerrado. In F. G. Faleiro & A. L. Farias Neto (Eds.), *Savanas: Desafios e estratégias para o equilíbrio entre sociedade, agronegócio e recursos naturais* (pp. 105-169). Planaltina, DF: Embrapa Cerrados.

Muluye, G. Y., & Coulibaly, P. (2007). Seasonal reservoir inflow forecasting with low frequency climatic indices: a comparison of data-driven methods. *Hydrological Sciences Journal*, *52*(3), 508–522. doi:10.1623/hysj.52.3.508

Muñoz, M. E. S., Giovanni, R., Siqueira, M. F., Sutton, T., Brewer, P., & Pereira, R. S. (2009). openModeller: a generic approach to species' potential distribution modeling. *GeoInformatica*. doi:.doi:10.1007/s10707-009-0090-7

Murakami, E., Santana, F. S., Stange, R. L., & Saraiva, A. M. (2009). The integration of open standards for Enterprise Service Bus in a solution for agribusiness and environmental applications development. In *Proceedings of the Joint International Agricultural Conference*, Wageningen. Holand.

Murphy, A. H. (1993). What is a good forecast? An essay on the nature of goodness in weather forecasting. *Weather and Forecasting*, *8*, 281–293. doi:10.1175/1520-0434(1993)008<0281:WIAGFA>2.0.CO;2

Murray, I., Blàzquez, M., & Rullan, O. (2005). Los cambios en la cobertura de la Tierra. Una revisión bibliográfica desde la geografía. *Biblio 3W, Revista Bibliográfica de Geografía y Ciencias Sociales, Universidad de Barcelona, 10*, 571. Retrieved Mars 15 2005, from http://www.ub.es/geocrit/b3w-571.htm

Mutegu, S., Hoko, Z., & Makoni, F. S. (2007). An assessement of the public health hazard potential of wastewater reuse for crop production. A case of Bulawayo city, Zimbabwe. *Physics and Chemistry of the Earth*, *32*, 1195–1203.

Myers, N., Mittermeier, R. A., Mittermeier, C. G., Fonseca, G. A. B., & Kent, J. (2000). Biodiversity hotspots for conservation priorities. *Nature*, *403*, 853–858. doi:10.1038/35002501

Nakamura, S., & Kondo, Y. (2002). Input-Output Analysis of Waste Management. *Journal of Industrial Ecology*, *6*(1), 39–64. doi:10.1162/108819802320971632

Nan, S., & Li, J. (2003). The relationship between summer precipitation in the Yangtze River valley and the previous Southern Hemisphere Annular Mode. *Geophysical Research Letters*, *30*(24), 2266. doi:10.1029/2003GL018381

Nansai, K., Moriguchi, Y., & Touno, S. (2002). *Embodied Energy and Emission Intensity Data for Japan Using Input-Output Tables*. Tsukuba, Japan: National Institute for Environmental Studies.

Nasrallah, H. A., Balling, R. C., Selovert, N. J., & Vose, R. S. (2001). Development of a seasonal forecast model for Kuwait winter precipitation. *Journal of Arid Environments*, *48*, 233–242. doi:10.1006/jare.2000.0746

National Academy of Sciences. (2007). *Understanding American Agriculture: Challenges for the agricultural Resource Management Survey*. National Academies Press. Retrieved from http://jbr.org/articles.html

National Institute for Rural Engineering (Ed.). (2006). *Biomass Rikatuyou System no Sekkei to Hyouka*. Tsukuba, Japan: National Institute for Rural Engineering. (In Japanese)

Nayak, R. C., & Panda, R. K. (2001). Integrated management of a canal command in a river Delta using multi-objective techniques. *Water Resources Management*, *15*, 383–401. doi:10.1023/A:1015593417769

Nemhauser, G. L., & Wolsey, L. A. (1988). *Integer and Combinatorial Optimization*. New York: Wiley-Interscience Publication.

Nguyen, T. D., Uvo, C. B., & Rosbjerg, D. (2007a). Relationship between tropical Pacific and Indian Ocean sea surface temperature and monthly precipitation over the central highland, Vietnam. *International Journal of Climatology*, *27*(11), 1439–1454. doi:10.1002/joc.1486

Nguyen, T. D., Rosbjerg, D., Uvo, C. B., & Nguyen, K. Q. (2007b). Drought prediction in the Vietnamese central highlands. Changes in Water Resources Systems: Methodologies to Maintain Water Security and Ensure Integrated Management. In *Proceedings of Symposium HS3006 at IUGG2007*. Perugia, Italy: IAHS Publ.

Nilsson, P., Uvo, C. B., & Berndtsson, R. (2006). Monthly runoff simulation: comparison of a conceptual model, neural networks and a combination of them. *Journal of Hydrology (Amsterdam)*, *321*, 344–363. doi:10.1016/j.jhydrol.2005.08.007

Nilsson, P., Uvo, C. B., Landman, W., & Nguye, T. D. (2008). Downscaling of GCM forecasts to streamflow over Scandinavia. *Hydrology Research*, *39*(1), 17–25. doi:10.2166/nh.2008.027

Nix, H. (1986). A biogeographic analysis of Australian elapid snakes. In *Atlas of Elapid Snakes of Australia* (pp. 4–15). Canberra, Australia: Australian Government Publishing Service.

Nofziger, D. L., & Hornsby, A. G. (1994). [- *Chemical Movement In Layered Soils*. Gainesville, FL: University Of Florida.]. *Cellular and Molecular Life Sciences*, 94.

Nogués-Paegle, J., & co-authors. (2002). Progress in Pan American CLIVAR research: Understanding the South American Monsoon. *Meteorologica*, *27*, 3–30.

O'Sullivan, D., & Haklay, M. (2000). Agent-based models and individualism – Is the world agent-based? *Environment & Planning A*, *32*, 1409–1425. doi:10.1068/a32140

Odum, E. P. (Trad. Gomes, A.M.A.) (2004). *Fundamentos De Ecologia,* (7ª Ed.). Lisboa: Fundação Calouste Gulbenkian.

Offermann, F., Kleinhanss, W., Huettel, S., & Kuepker, B. (2005). Assessing the 2003 CAP reform impacts on German Agriculture using the farm group model FARMIS. In F. Arfini, (ed.), *Modelling agricultural policies: state of the art and new challenges. Proceedings of the 89 th EAAE Seminar,* Parma, Italy.

Ogallo, L. A., Boulahya, M. S., & Keane, T. (2000). Applications of seasonal to interannual climate prediction in agricultural planning and operations. *Agricultural and Forest Meteorology*, *103*, 159–166. doi:10.1016/S0168-1923(00)00109-X

Oliveira, F. R., Pacheco, D. F., Leonel, A., & de Lima Neto, F. B. (2006). Intelligent Support Decision in Sugar Cane Harvest. *IV World Congress on Computers in Agriculture (WCCA 2006)*, Orlando, (pp. 456-462).

Olsson, J., Uvo, C. B., Jinno, K., Kawamura, A., Nishiyama, K., & Koreeda, N. (2004). Neural Networks for rainfall forecasting by atmospheric downscaling. *Journal of Hydrologic Engineering*, *9*(1), 1–12. doi:10.1061/(ASCE)1084-0699(2004)9:1(1)

Openshaw, S., & Turton, I. (1996). A parallel Kohonen algorithm for the classification of large spatial datasets. *Computers & Geosciences*, *22*(9), 1019–1026. doi:10.1016/S0098-3004(96)00040-4

Oralndo, O., Orsini, R., Raffaeta, A., Roncato, A., & Silvestri, C. (2007). Spatio-Temporal Aggregations in Trajectory Data Warehouses. In [Berlin: Springer-Verlag.]. *Proceedings of the DaWa*, *K07*, 66–77.

Orcutt, G. H. (1957). A new type of socio-economic system. *The Review of Economics and Statistics*, *58*, 773–794.

Orcutt, G. H., & Greenberger, M. Korbel, J. & Rivlin, A. (1961). *Microanalysis of socioeconomic systems: A simulation study*. New York: Harper & Row.

Orlando, S., Orsini, R., & Raffaetà, A. (2007). Trajectory Data Warehouses: Design and Implementation Issues. *Journal of Computing Science and Engineering*, *1*(2), 1976–4677.

Oseki, H. (1998). *Economic Models For Brazilian Farm Environment*. Available At Http://Www.Jircas.Affrc.Go.Jp/Kankoubutsu/News/Newsletter/Nl1998/No.15/03ozeki.Htm (Accessed On August 2009).

Osterburg, B., Offermann, F., & Kleinhanss, W. (2001). A sector consistent farm group model for German agriculture. In Heckelei, T., Witzke, H.P. & Henrichsmeyer, W. (eds.), *Agricultural Sector Modelling and Policy Information Systems*. Kiel, Germay: Wissenschaftsverlag, Vauk.

Pacheco, D. F., Regueira, F. S., & Lima Neto, F. B. (2005). Utilização de Redes Neurais Artificiais em colheitas de cana-de-açúcar para predição de PCC, TCH e fibra. *ALCOOLbrás*, *90*, 60–63.

Pacheco, D. F., & Lima Neto, F. B. (2008). *A Fair Comparison of Representations, Operators and Algorithms for the Sugarcane Harvest Problem*. IEEE International Conference on Systems, Man, and Cybernetics.

Pacheco, D. F., Oliveira, F. R., & Lima Neto, F. B. (2008). *Including Multiobjective Abilities in the Hybrid Intelligent Suite for Decision Support*. International Joint Conference on Neural Networks-IJCNN (World Congress on Computational Intelligence-WCCI2008).

Pacheco, D. F., Pontes, T. D., & Lima Neto, F. B. (2007). *How to Obtain Fair Managerial Decisions in Sugar Cane Harvest Using NSGA II*. 7th International Conference on Hybrid Intelligent systems (HIS 07).

Papadimitriou, S., Brockwell, A., & Faloutsos, C. (2004). Adaptive, unsupervised stream mining. *Very Large Data Base Journal*, *13*(3), 222–239.

Papajcsik, I. A., & Bodero, J. (1988). Modeling Lactation Curves Of Friesian Cow In A Subtropical Climate. *Animal Production*, *47*, 201–207. doi:10.1017/S0003356100003275

Paraíba, L. C. (2007). Pesticide Bioconcentration Modelling For Fruit Trees. *Chemosphere*, *66*, 1468–1475. doi:10.1016/j.chemosphere.2006.09.017

Paraíba, L. C., Cerdeira, A. L., Silva, E. F., Martins, J. S., & Coutinho, H. L. C. (2003). Evaluation Of Soil Temperature Effect On Herbicide Leaching Potential Into Groundwater In The Brazilian Cerrado. *Chemosphere*, *53*, 1087–1095. doi:10.1016/S0045-6535(03)00594-0

Paraíba, L. C., Plese, L. P. M., Foloni, L. L., & Carrasco, J. M. (2007). Simulation Of The Fate Of The Insecticide Carbofuran In A Rice Field Using A Level Iv Fugacity Model. *Spanish Journal of Agricultural Research*, *5*(1), 43–50.

Paraíba, L. C., & Spadotto, C. A. (2002). Soil Temperature Effect In Calculating Attenuation And Retardation Factors. *Chemosphere*, *48*, 905–912. doi:10.1016/S0045-6535(02)00181-9

Paraiba, L. C., Boeira, R. C., Jonsson, C. M., & Carrasco, J. M. (2006). Fator De Bioconcentração De Poluentes Orgânicos De Lodo Em Frutos De Laranjeiras. *Pesticidas: Revista De Ecotoxicologia E Meio Ambiente, Curitiba, 16*(Jan./Dez), 125-134.

Paraíba, L.C., Pulino, P. (2003). *Simulação Numérica Da Dispersão-Advecção De Pesticidas No Solo Sob Efeito Da Temperatura*. Jaguariúna: Embrapa Meio Ambiente (Embrapa Meio Ambiente. Documentos, 35).

Paranhos, S. B. (1987). *Cana-De-Açúcar - Cultivo E Utilização*. Campinas: Fundação Cargil.

Paris, Q., & Howitt, R. E. (1998). An Analysis of Ill-posed Production Problems Using Maximum Entropy. *American Journal of Agricultural Economics*, *80*(1), 124–138. doi:10.2307/3180275

Paris, Q., Montresor, E., Arfini, F., & Mazzocchi, M. (2000). An Integrated Multi-Phase Model for Evaluating Agricultural Policies Through Positive Information. In Heckelei, T., Witzke, H.P. & Henrichsmeyer, W. (eds.), *Agricultural Sector Modelling and Policy Information Systems.Proceedings of the 65 th EAAE Seminar*, March 29-31, 2000 at Bonn University, (pp. 100-110). Kiel, Germany: Vauk Verlag.

Parker, D. C., Manson, S. M., Janssen, M. A., Hoffmann, M. J., & Deadman, P. (2003). Multi-agents systems for the simulation of land-use and land-cover change: A review. *Annals of the Association of American Geographers. Association of American Geographers*, *93*(2), 314–337. doi:10.1111/1467-8306.9302004

Parton, W. J., Schimel, D. S., Cole, C. V., & Ojima, D. S. (1987). Analysis of factors controlling soil organic matter levels in Great Plains grasslands. *Soil Science Society of America Journal*, *51*, 1173–1179. doi:10.2136/sssaj1987.03615995005100050015x

Parton, W. J., Stewart, J. W. B., & Cole, C. V. (1988). Dynamics of C, N, P and S in grassland soils: a model. *Biogeochemistry*, *5*, 109–131. doi:10.1007/BF02180320

Pausnitz, J. N. (1969). *Molecular Thermodynamics Of Fluid-Phase Equilibria*. Englewood Cliffs, NJ: Prentice-Hall.

Pearson, K. (1896). Mathematical contributions to the theory of evolution. III regression, heredity and panmixia. *Philos Trans Royal Society London Ser A*, *187*, 253–318. doi:10.1098/rsta.1896.0007

Peña, J., Poveda, R. M., Bonet, A., Bellot, J., & Escarré, A. (2005). Cartografía de las coberturas y usos del suelo de la Marina Baixa (Alicante) para 1956, 1978 y 2000. *Investigaciones Geográficas*, *37*, 93–107.

Pereira, R. S., & Peterson, A. T. (2001). The use of modeling in the definition of strategies for conservation of Biodiversity. *Biodiversity - economic and social value*. (in Portuguese). Retrieved August 27, 2009, from http://www.comciencia.br/reportagens/biodiversidade/bio18.htm

Peres, F.C. (1990). *Confinamento De Gado De Corte: Um Modelo De Analise De Decisão*. Tese De Livre Docência Em Administração Rural, Piracicaba: Usp/Esalq.

Perkins, L. B., Yarborough, D. E., & Guthrie, K. E. (2006). Detection Of Hexazinone In Maine Groundwater- A Nine Year Study. *Acta Hort. (Ishs)*, *715*, 329-336. Accessed At: Http://Www.Actahort.Org/Books/715/715_49.Htm

Persona, L., Corrêa, P. L. P., & Saraiva, A. M. (2003). Environmental Niche Modeling in Biodiversity with Genetic Algorithms. (in Portuguese). In *Proceedings of the IEEE - I2TS'2003 2nd International Information and 68 Telecommunication Technologies Symposium*, Florianópolis.

Pessoa, M. C. P. Y. (1994). *Simulação E Inteligência Artificial Aplicadas Ao Estudo Da Dinâmica Populacional Do Bicudo Do Algodoeiro Na Região De Campinas/Sp*. Campinas: Densis-Fee/Unicamp, Tese De Doutorado.

Pessoa, M. C. P. Y., Gomes, M. A. F., Souza, M. D., Cerdeira, A. L., Nicolella, G., & Monitcelli, A. (1999). Simulação Do Movimento De Herbicidas Utilizados No Monocultivo De Cana-De-Açúcar Em Latossolos Da Área De Recarga Do Aquífero Botucatu (Guarani) Em Ribeirão Preto, Sp. *Revista Científica Rural. Bagé, 4*(1), 15–24.

Pessoa, M. C. P. Y., Chaim, A., Capalbo, D. M. F., Hamada, E., Tambasco, F. J., Ferraz, J. M. G., et al. (2006b). *Boas Práticas Agrícolas E Meio Ambiente - Elementos De Apoio Para As Boas Práticas Agrícolas E O Sistema Appcc/Pas Campo*. - 2.Ed. Rev., Atual. - Brasília, Df: Embrapa, (pp. 13-33) (Série Qualidade E Segurança Dos Alimentos).

Pessoa, M. C. P. Y., Ferracini, V. L., Chaim, A., & Scramin, S. (Dezembro/2004b). *Software Agroscre - Apoio À Avaliação De Tendências De Transporte De Princípios Ativos De Agrotóxicos*. Jaguariúna/Sp: Embrapa Meio Ambiente, (pp. 1-23). (Boletim De Pesquisa E Desenvolvimento, Nº 26).

Pessoa, M. C. P. Y., Filizola, H. F., & Gomes, M. A. F. (2005). *Simulação Da Movimentação Vertical De Produtos Aplicados Em Milho, Milho Safrinha E Soja Em Latossolos Vermelhos E Neossolos Quartzarênicos De Mineiros*, Jaguariúna: Embrapa Meio Ambiente, 27p. (Embrapa Meio Ambiente. Boletim De Pesquisa E Desenvolvimento, 31).

Pessoa, M. C. P. Y., Gomes, M. A. F., Neves, M. C., Cerdeira, A. L., & Souza, M. D. (2003a). Identificação De Áreas De Exposição Ao Risco De Contaminação De Águas Subterrâneas Pelos Herbicidas Atrazina. Diuron E Tebutiuron. *Pesticidas: Revista De Ecotoxicologia E Meio Ambiente, Curitiba, 13* (Jan./Dez.), 111-122.

Pessoa, M. C. P. Y., Gomes, M. A. F., Nicolella, G., Souza, M. D., Cerdeira, A. L., & Monticelli, A. (1997a). Simulação Do Movimento Dos Herbicidas Hexazinone, Diuron, Atrazina, Ametrina E Simazina Aplicados Na Cultura De Cana-De-Açúcar Em Solos Da Microbacia Do Córrego Do Espraiado. Ribeirão Preto,Sp. In *Congresso Brasileiro De Ciências De Solo. 26, Rio De Janeiro, Resumos Do Congresso Brasileiro De Ciências De Solo, 26, Cdrom.*

Pessoa, M. C. P. Y., Sá, L. A. N., Kodaira, J. Y., Wilcken, C. F., & Almeida, G. R. (2008). *Simulação Da Dinâmica Populacional Do Psilideo-De-Concha, Glycaspis Brimblecombei (Hemiptera: Psyllidae) E Identificação De Estratégias Para A Criação Laboratorial De Seu Parasitoide Psyllaephagus Bliteus (Hymenoptera: Encyrtidae)*. Jaguariúna: Embrapa Meio Ambiente, 32p. (Embrapa Meio Ambiente. Boletim De Pesquisa E Desenvolvimento, 49).

Pessoa, M. C. P. Y., & Scramin, S. (2004a). Modelagem Matemática E Simulação Da Exposição Do Ambiente À Aplicação De Agrotóxicos: Apoio À Avaliação De Risco Ambiental. In: Silva, C.M.M.S, Fay, E. F. (Eds.), *Agrotóxicos E Ambiente. Embrapa Informação Tecnológica*, (pp. 319-364). Brasília/Df.

Pessoa, M.C.P.Y., Chaim, A., Gomes, M.A.F., Silva, A. S., Soares, J. M. (2003b). Simulação Do Movimento De Aldicarb E Tebuthiuron Em Solos Sob Cultivos De Banana E Cana-De-Açúcar No Semi-Árido Brasileiro. *Rev. Bras. De Eng. Agrícola E Ambiental (Agriambi), 7*(2), 297-302.

Pessoa, M.C.P.Y., Gomes, M.A.F., Scramin, S., Silva, A. De S., Gussakov, K. C. (2006a). Vulnerabilidade Natural Das Grandes Bacias Hidrográficas Brasileiras À Tendência De Contaminação De Águas Por Agrotóxicos Em Função Dos Tipos De Solos Predominantes. Pesticidas: Revista De *Ecotoxicologia E Meio Ambiente, 16*(Jan/Dez), 39-52.

Pessoa, M.C.P.Y., Luchiari Junior, A., Fernandes, E.N., Lima, M.A. De. (1997b). *Principais Modelos E Simuladores Usados Em Análise De Impacto Ambiental Da Agricultura*. Jaguariúna: Embrapa-Cnpma, 87p., (Documentos, N.8).

Pessoa, M.C.P.Y., Scramin, S., Chaim, A., Ferracini, V.L. (Junho/2007). *Avaliação Do Potencial De Transporte De Agrotóxicos Usados No Brasil Por Modelos Screening E Planilha Eletrônica*. Jaguariúna/Sp: Embrapa Meio Ambiente. Boletim De Pesquisa E Desenvolvimento, Nº 44.

Pessoa. M. C. P. Y., Gomes. M. A. F., Souza. M. D., Cerdeira. A. L., Nicolella. G., Monticelli, A. (1998). Simulação Do Movimento De Herbicidas Utilizados No Monocultivo De Cana-De-Açúcar, Em Areia Quartzosa Da Área De Recarga Do Aquífero Guarani (Antigo Botucatu) Em Ribeirão Preto,Sp. *Revista Científica Rural. Bagé/Rs, 3* (2), 11-19.

Pfoser, D., Jensen, C., & Theodoridis, Y. (2000). Novel Approaches to the Indexing of Moving Object Trajectories. In *Proceedings of the 26th VLDB Conference,* (pp. 395-406). New York: ACM.

Phillips, S. J., Anderson, R. P., & Schapire, R. E. (2006). Maximum entropy modeling of species geographic distributions. *Ecological Modelling, 190*, 231–259. doi:10.1016/j.ecolmodel.2005.03.026

Phillips, S. J., Dudík, M., & Scaphire, R. E. (2009). Retrieved August 27, 2009, from http://www.cs.princeton.edu/~schapire/maxent

Phillips, S. J., Dudík, M., & Schapire, R. E. (2004). A Maximum Entropy Approach to Species Distribution Modeling. In *Proceedings of the 21st International Conference on Machine Learning*, Banff, Canada.

Picini, A. G., Valeriano, M. De M., Lombardi Neto, F. And Zullo Junior, J. (2005) Ajuste Dos Simuladores De Dados Meteorológicos Do Modelo Epic Para Diferentes Locais Do Estado De São Paulo. *Revista Brasileira De Engenharia Agrícola E Ambiental., 9*(2), 239-246, 1415-4366.

Pielke, R., & Carbone, R. E. (2002). Weather impacts, forecasts, and policy - an integrated perspective. *Bulletin of the American Meteorological Society*, 393–403. doi:10.1175/1520-0477(2002)083<0393:WIFAP>2.3.CO;2

Piñeiro, R., Aguilar, J. F., Munt, D. D., & Feliner, G. N. (2007). Ecology matters: Atlantic-Mediterranean disjunction in the sand-dune shrub Armeria pungens (Plumbaginaceae). *Molecular Ecology, 16*, 2155–2171. doi:10.1111/j.1365-294X.2007.03280.x

Pinet, F., Roussey, C., Brun, T., & Vigier, F. (2009). The Use Of Uml As A Tool For The Formalisation Of Standards And The Design Of Ontologies. In *Agriculture. In Advances In Modeling Agricultural Systems*. Berlin: Springer-Verlag. doi:10.1007/978-0-387-75181-8_7

Pino, F. A. (1999). Estatísticas agrícolas para o século XXI. *Agricultura em São Paulo, 46*(2), 71–105.

Plec, D. F., Pessoa, M. C. P. Y., Lima, M. A., & Vilella, O. V. (Agosto, 2007). Avaliação Do Fluxo De Emissão De Gases Estufa Provenientes De Fontes Agrícolas Por Simulação De Sistemas: Estudo De Caso De Emissão De Metano Por Arroz Irrigado Por Inundação Em Pindamonhangaba, Sp. In *Anais... Congresso Interinstitucional De Iniciação Científica (Pibic/Cnpq)*, Campinas/Sp: Instituto Agronômico De Campinas, 1 A 2 De Agosto De 2007. Cd-Rom.

Plese, L. P. M., Silva, C. L., & Foloni, L. L. (2009). Distribuição Nos Compartimentos Ambientais Dos Herbicidas Utilizados Nas Culturas De Algodão, Café E Citros. *Planta Daninha. Viçosa-Mg, 27*(1), 123–132.

Polhill, J. G., Gotts, N. M., & Law, A. N. R. (2001). Imitative versus nonimitative strategies in a land-use simulation. *Cybernetics and Systems, 32*, 285–307. doi:10.1080/019697201300001885

Polomé, P., Harmignie, O., Henry de Frahan, B., Buysse, J., Van Huylenbroeck, G., Van Meensel, J., et al. (2005). *Farm –level Multi-output Multi-input Cost Function: a Comparison of Least Squares and Entropy Estimators*. Selected paper presented at the 11 th EAAE Congress, Copenhagen, 23-27 August 2005.

Pongthanapanich, T. (2006). Optimal Coastal Land Use and Management in Krabi, Thailand: Compromise Programming Approach. *IME working paper no. 67/06. SDU*, Denmark. Available online: http://econpapers.repec.org/

Pontius, R. G. Jr, Shusas, E., & McEachern, M. (2004). Detecting important land changes while accounting for persistence. *Agriculture Ecosystems & Environment*, *100*, 251–268. doi:10.1016/j.agee.2003.09.008

Pontius, R. G., Jr., & Malizia, N. R. (2004). Effect of category aggregation on map comparison. In Egenhofer, M.J., Freksa, C., & Miller, H.J. (Ed.), *GISscience 200*, (LNCS 3234, pp. 251-268). Berlin Heildelberg: Springer-Verlag.

Poyet, P., & Brun, T. (2003). GIEA: gestion des informations de l`exploitation agricole - Vers des concepts et un langage communs pour les partenaires économiques et institutionnels de la profession agricole. *Ingénieries - Eau, Agriculture, Territoires, Technologies pour les agrosystèmes durables* (pp. 167 – 175).

Poyton, R. J. (1979). *Tree planting in Southern Africa: the eucalyptus*. Pretoria, South Africa: Department of Forestry.

Pradhan, S. (2001). Crop area estimation using GIS, remote sensing and area frame sampling. *JAG: International Journal of Applied Earth Observation and Geoinformation*, *3*(1), 86–92. doi:10.1016/S0303-2434(01)85025-X

Praseres, A. L. (1976). *Combinação De Atividades Pela Programação Linear Em Empresas Agrícolas Na Microrregião De Imperatriz, Maranhão*. Piracicaba: Usp/Esalq, 150p. (Dissertação De Mestrado).

Prodanovic, P., & Simonovic, S. P. (2003). Fuzzy Compromise Programming for group decision making. *IEEE Transactions on Systems, Man, and Cybernetics. Part A, Systems and Humans*, *33*(3), 358–365. doi:10.1109/TSMCA.2003.817050

Qadir, A., & Wichelns, D. (2008). (in press). Raschid-Sally. L, McCornick, P. G., Drechsel. P., Bahri. A., & [The challenges of wastewater irrigation in developing countries. *Agricultural water management.*]. *Minhas. P. S.*

Queiroz, S. C. N., Ferracini, V. L., Rosa, M. A., & Cerdeira, A. L. (Maio, 2006) Validação De Método Por Clae Para Determinação De Hexazinone E Diuron Em Solo. In *29ª Reunião Anual Da Sociedade Brasileira De Química. Águas De Lindóia/Sp, 19 A 22/05/2006.*

Queiroz, S.C.N., Ferracini, V.L., Rosa, M. A., Cerdeira, A.L. (2007, Dezembro). Método Para Determinação De Hexazinona E Diurom Em Solo. *Boletim De Pesquisa E Desenvolvimento,* (47).

Quenouille, M. H. (1956). Notes on Bias in Estimation. *Biometrika*, *43*(3/4), 353–360. doi:10.2307/2332914

Quirino, T. R. (1990). *Construção Do Conhecimento Científico*. Brasília: Embrapa Sede.

Raidl, J. R., & Puchinger, J. (2008). Combining (Integer) Linear Programming Techniques and Metaheuristics for Combinatorial Optimization. [SCI]. *Studies in Computational Intelligence*, *114*, 31–62. doi:10.1007/978-3-540-78295-7_2

Ramírez, M. C. V., Velho, H. F. C., & Ferreira, N. J. (2005). Artificial neural network technique for rainfall forecasting applied to the São Paulo region. *Journal of Hydrology (Amsterdam)*, *301*, 146–162. doi:10.1016/j.jhydrol.2004.06.028

Rao, P. S. C., Hornsby, A. G., & Jessup, R. E. (1985). Indices For "Ranking" The Potential For Pesticide Contamination Of Groundwater. *Soil And Crop Science Society Of Florida*, *44*, 1–8.

Rasmusson, E. M., & Carpenter, T. H. (1982). Variations in tropical sea surface temperature and surface wind fields associated with the Southern Oscillation/El Niño. *Monthly Weather Review*, *110*(5), 354–384. doi:10.1175/1520-0493(1982)110<0354:VITSST>2.0.CO;2

Ratnaparkhi, A. (1997). *A Simple Introduction to Maximum Entropy Models for Natural Language Processing.* Institute for Research in Cognitive Science - IRCS Report 97–08 at University of Pennsylvania.

Ray, S. S., & Pokharna, S. S., & Ajai. (1999). Cotton yield estimation using agrometeorological model and satellite-derived spectral profile. *International Journal of Remote Sensing*, *20*(14), 2693–2702. doi:10.1080/014311699211741

Reboita, M. S., Ambrizzi, T., & Rocha, R. P. (2009). Relationship between the Southern Annular Mode and Southern Hemisphere Atmospheric Systems. *Revista Brasileira de Meteorologia*, *24*, 48–55. doi:10.1590/S0102-77862009000100005

Reenberg, A. (2001). Agricultural land use pattern dynamics in the Sudan–Sahel: towards an event-driven framework. *Land Use Policy*, *18*(4), 309–319. doi:10.1016/S0264-8377(01)00020-5

Renard, K. G., Foster, G. R., Weesies, G. A., & Mccool, D. K. (Eds.). (1991). *Predicting Soil Erosion By Water - A Guide To Conservation Planning With The Revised Universal Soil Loss Equation (Rusle)*. Washington, DC: Usda-Ars.

Resende, M., & Almeida, J. A. (1985). Modelos De Predição De Perda De Solo: Uma Ferramenta Para Manejo E Conservação Do Solo. *Informe Agropecuário, Belo Horizonte*, *11*(128), 38–54.

Ribeiro, M. X., Balan, A. G. R., Felipe, J. C., Traina, A. J. M., & Traina, C., Jr. (2005), Mining Statistical Association Rules to Select the Most Relevant Medical Image Features. In *First International Workshop on Mining Complex Data, in conjunction with ICDM'05*, Houston, TX, (pp. 91-98). Washington, DC: IEEE Computer Society.

Rich, J. L. (1999). *A computational study of vehicle routing applications*. Doctoral dissertation, Houston, TX, USA.

Richardson, C. W., & Wright, D. A. (1984). *Wgen: A Model For Generating Daily Weather Variables*. Washington, DC: Agricultural Research Service, Usda.

Rickman, R. W., Douglas, C. L. Jr, Albrecht, S. L., Bundy, L. G., & Berc, J. L. (2001). Cqestr: A Model To Estimate Carbon Sequestration In Agricultural Soils. *J. Soil Cons.*, *56*, 237–242.

Rodgers, A., Rempel, R., & Abraham, K. (1996). A GPS-Based Telemetry System. *Wildlife Society Bulletin*, *24*(3), 559–566.

Rodrigues, P., Gama, J., & Pedroso, J. P. (2008b). Hierarchical Clustering of Time-Series Data Streams. *IEEE Transactions on Knowledge and Data Engineering*, *20*(5), 615–627. doi:10.1109/TKDE.2007.190727

Rodrigues, L. C. L. (1997). *Planejamento Agropecuário Através De Um Modelo De Programação Linear Não-Desterminista. Piracicaba: Usp/Esalq*. Dissertação De Mestrado.

Rodrigues, F. A., Avilla, A. O., Rodrigues, E. S. C., Corrêa, P. L. P., Saraiva, A. M., & Rocha, R. L. A. (2009). Species distribution modeling with neural networks. In *e-Biosphere 2009*, 1-3 June 2009, London-UK. Retrieved August 27, 2009, from http://www.e-biosphere09.org/assets/files/e-Biosphere%20Abstracts%20Volume%20-%20FINAL.pdf

Rodrigues, F. A., Rodrigues, E. S. C., Sato, L. M., Midorikawa, E. T., Corrêa, P. L. P., & Saraiva, A. M. (2008). Parallelization of the Jackknife Algorithm Applied to a Biodiversity Modeling System. In *7th International Information and Telecommunication Technologies Symposium, 2008, Foz do Iguaçu. Proceedings of the 7th International Information and Telecommunication Technologies Symposium*, (pp. 58–65).

Rodrigues, P. P., Gama, J., & Lopes, L. M. B. (2008a). Clustering Distributed Sensor Data Streams. In *Proceedings of the European Conference on Principles and Practice of Knowledge Discovery in Databases*, (pp. 282-297).

Röhm, O., & Dabbert, S. (2003). Integrating Agri-Environmental Programs into Regional Production Models: An Extension of Positive Mathematical Programming. *American Journal of Agricultural Economics*, *85*(1), 254–265. doi:10.1111/1467-8276.00117

Röhm, O., & Dabbert, S. (1999). Modelling regional production and Income Effects. In Huylenbroeck, G., & Witby, M. (Eds.), *Countryside Stewardship:Farmers, Policies and Markets*. Amsterdam: Pergamon.

Romani, L. A. S., Ávila, A. M. H. d., Zullo, J., Jr., Traina, C., Jr., & Traina, A. J. M. (2009). Mining Climate and Remote Sensing Time Series to Discover the Most Relevant Climate Patterns. In *Proceedings of the XXIV Simpósio Brasileiro de Banco de Dados*, Fortaleza, CE, Brasil.

Romani, L. A. S., Sousa, E. P., Ribeiro, M. X., Zullo, J., Jr., Traina, C., Jr., & Traina, A. J. M. (2009). Employing Fractal Dimension to Analyze Climate and Remote Sensing Data Streams. In *Proceedings of the SIAM Multimedia Data Mining Workshop (SDM)*, Sparks, Nevada, 1-15.

Rondon, E. O. O., Murakami, A. E., & Sakaguti, E. S. (2002). Modelagem Computacional Para Produção E Pesquisa Em Avicultura. *Revista Brasileira De Ciência Avícola, 4*(1), 9. Available At: Http://Www.Scielo.Br/Scielo.Php?Pid=S1516-635x2002000100001&Script=Sci_Arttext (25/10/2009)

Rosseti, L. (2001). Zoneamento agrícola em aplicações de crédito e securidade rural no brasil. *RBAgro*, *9*(3), 386–399.

Roy, D. N., Konar, S. K., Charles, D. A., Feng, J. C., Prasad, F., & Campbell, R. A. (1989). Determination Of Persistence. Movement And Degradation Of Hexazinone In Selected Canadian Boreal Forest Soils. *Journal of Agricultural and Food Chemistry*, *37*, 443–447. doi:10.1021/jf00086a039

Rumelhart, D. E., Hinton, G. E., & Williams, R. J. (1986). Learning representations by back-propagating errors. *Nature*, *323*, 533–536. doi:10.1038/323533a0

Rumelhart, D. E., Hinton, G. E., & Williams, R. J. (1986). Learning internal representations by error propagation. In Rumelhart, D. E., & McClelland, J. L. (Eds.), *Parallel Distributed Processing* (*Vol. 1*, pp. 318–362). Cambridge, MA: MIT Press.

Ryerson, R. A., Curran, P. J., & Stephens, P. R. (1997). Agriculture. In Philipson, W. R. (Ed.), *Manual of Photographic Interpretation* (pp. 285–397). Bethesda, MD: ASPRS.

Sabourin, E., & Teixeira, O. A. (2002). *Planejamento e desenvolvimento dos territórios rurais*. Brasília: Embrapa Informação Tecnológica.

Sakurai, Y., Faloutsos, C., & Yamamuro, M. (2007). Stream Monitoring under the Time Warping Distance. In *Proceedings of the IEEE International Conference on Data Engineering*, (pp. 1046-1055).

Salle, C. T., Guahyba, A. S., Wald, V. B., Silva, A. B., Salle, F. O., & Fallavena, L. C. B. (2001). Uso De Redes Neurais Artificiais Para Estimar Parâmetros De Produção De Galinhas Reprodutoras Pesadas Em Recria. *Revista Brasileira de Ciência Avícola*, *3*, 257–264. doi:10.1590/S1516-635X2001000300008

Salle, C. T. P., Cé, M. C., Lorenzini, G., Sfoggia, M. V. B., Guahyba, A. S., Moraes, H. L. S., & Nascimento, V. P. (1998a). Correlation Between Aflatoxin And Ocratoxin Levels With Production Parameters In *A Poultry Company. Abstracts Of The Iv Asia-Pacific Poultry Health Conference. Melbourne - Australia,* (pp. 131-131).

Salle, C. T. P., Cé, M. C., Santos, C. H. C., Guahyba, A. S., & Nascimento, V. P. (1998b). Use Of Statistical Technique On The Interpretation Of Routine Serologic Data Produced By Poultry Industry. *Abstracts Of The Iv Asia-Pacific Poultry Health Conference. Melbourne - Australia*, (pp. 148-148).

Saltenis, S., Jensen, C. S., Leutenegger, S. T., & Lopez, M. A. (2000). Indexing the positions of continuously moving objects. In *Proc SIGMOD Conference*, (pp. 331-342). New York: ACM.

Santana, F. S., Siqueira, M. F., Saraiva, A. M., & Corrêa, P. L. P. (2008). A reference business process for ecological niche modelling. [INSERT FIGURE 001]. *Ecological Informatics*, *3*, 75–86. doi:10.1016/j.ecoinf.2007.12.003

Santos, L. M. R., Costa, A. M., Arenales, M. N., & Santos, R. H. S. (2010). Sustainable vegetable crop supply problem. *European Journal of Operational Research*, *204*, 639–647. doi:10.1016/j.ejor.2009.11.026

Santos, L. M. R., Michelon, P., Arenales, M. N., & Santos, R. H. S. (2008). Crop rotation scheduling with adjacency constraints. *Annals of Operations Research*. doi:.doi:10.1007/s10479-008-0478-z

Santos, L. M. R. (2009). *Programação de Rotação de culturas – modelos e métodos de solução. Doctoral dissertation*, ICMC/USP - University of São Paulo, Brazil.

Saraiva, A. M., Correa, P. L. P., Sato, L. M., Rodrigues, F. A., Santana, F. S., Rodrigues, E. S. C., et al. (2009) A Service-Based Framework for Species Distribution Modeling. In *Conference Abstracts Of E-Biosphere Conference,* London, 1-3, june, 2009.

Sawyer, D. R. (1983). Ocupación y desocupación de la frontera agrícola en el Brasil: Un ensayo de interpretación estructural y espacial. In Comisión Económica para América Latina (Ed.), *Expansión de la frontera agropecuaria y medio ambiente en América Latina* (pp. 79-104). Madrid, Spain: Centro Internacional de Formación en Ciencias Ambientales.

SCET International and CNEA. (1981). *Plan directeur des eaux du nord, 2ième tranche. Technical report*. MAHR.

Schroeder, M. (1991). *Fractals, Chaos, Power Laws* (6 ed.). New York: W. H. Freeman.

Scorza Júnior, R. P., & Boesten, J. J. T. I. (2005). Simulation of pesticide leaching in a cracking clay soil with the PEARL model. *Pest Management Science, 61*, 432–448. doi:10.1002/ps.1004

Scorza Júnior, R. P., Jarvis, N. J., Boesten, J. J. T. I., Van der Zee, S. E. A. T. M., & Roulier, S. (2007). Testing MACRO (version 5.1) for pesticide leaching in a Dutch clay soil. *Pest Management Science, 63*, 1011–1025. doi:10.1002/ps.1434

Secretaria de Planejamento e Gestão – RS. (2007). *Atlas socioeconômico do Rio Grande do Sul.* Retrieved July 1st, 2007, from http://www.scp.rs.gov.br/.

Severini, S., & Cortignani, R. (2008). *Introducing deficit irrigation crop techniques derived by crop growth models into a Positive Mathematical Programming model.* Selected paper presented at the XII th Congress EAAE, Ghent, Belgium, 26-29 August.

Seyfang, G. (2006). Ecological citizenship and sustainable consumption: Examining local organic food networks. *Journal of Rural Studies, 22*, 383–395. doi:10.1016/j.jrurstud.2006.01.003

Sharpley, A. N., & Willians, J. R. (Eds.). (1990). *Epic-Erosion Productivity Impact Calculator: I. Model Documentation.* Springfield, IL: Ntis, (Usda. Technicall Bulletin, 1768).

Shiau, J. T., & Lee, H. C. (2005). Derivation of optimal hedging for a water supply reservoir through compromise programming. *Water Resources Management, 19*, 111–132. doi:10.1007/s11269-005-1502-6

Shirato, Y., Hakamata, T., & Taniyama, I. (2004). Modified Rothamsted Carbon Model for Andosols and Its Validation: Changing Humus Decomposition Rate Constant with Pyrophosphate-Extractable Al. *Soil Science and Plant Nutrition, 50*, 149–158.

Silva, C. L., Foloni, L.L., Paraíba, L.C., Plese, L.P.M. (2007). Previsão Ambiental Da Distribuição Dos Pesticidas Aplicados Em Arroz. Pesticidas: R. Ecotoxicol. E Meio Ambiente, *17*(Jan./Dez.), 75-86.

Silva, F. A. M. (2004). Parametrização E Modelagem Do Balance Hídrico Em Sistemas De Plantio Direto No Cerrado Brasileiro. Campinas,Sp: Unicamp-Feagri, Tese Doutorado.

Silva, F. C., & Bergamasco, A. F. (2001). *Levantamento De Modelos Matemáticos Descritos Para A Cultura Da Cana-De-Açúcar*. Campinas: Embrapa Informática Agropecuária. Também Disponível Em Http://Www.Unitau.Br/Prppg/Publica/Biocienc/Downloads/Levantamentomodelos-N1-2001.Pdf Consulted In 12/11/2004

Silva, M. A. S., Monteiro, A. M. V., & de Medeiros, J. S. (2004a). *Visualization of geospatial data by component planes and u-matrix*. Paper presented at the VI *Brazilian Symposium on GeoInformatics*, Campos do Jordão, Brazil.

Silva, M. A. S., Monteiro, A. M. V., & de Medeiros, J. S. (2004b). *Semi-Automatic Geospatial Data Clustering by Self-Organizing Maps*. Paper presented at the VI Brazilian Symposium on Neural Networks, 2004, São Luis, MA.

Silverman, D., & Dracup, J. A. (2000). Artificial Neural Networks and long-range precipitation prediction in California. *Journal of Applied Meteorology*, *39*, 57–66. doi:10.1175/1520-0450(2000)039<0057:ANNALR>2.0.CO;2

Simon, M. F., & Garagorry, F. L. (2005). The expansion of agriculture in the Brazilian Amazon. *Environmental Conservation*, *32*(3), 203–212. doi:10.1017/S0376892905002201

SINDICARNE. (2009). *Produção e exportação de carne bovina brasileira em 2009*. Retrieved July 2, 2009, from http://www.sindicarne.com.br/

Siqueira, O. J. F., Farias, J. R. B., & Sans, L. M. A. (1994). Potential Effects Of Global Climate Change For Brazilian Agriculture And Adaptive Strategies For Wheat, Maize And Soybean. Revista *Brasileira De Agrometeorologia. Santa Maria*, *2*, 115–129.

Siqueira, M. F. (2005). *Use of Fundamental Niche Modeling in the Pattern Evaluation of Vegetal Species Geographic Distribution*. PhD Thesis. Department of Ambient Engineering of University of Sao Carlos. Sao Carlos, Sao Paulo, Brazil. (In Portuguese).

Sistla, A. P., Wolfson, O., Chamberlain, S., & Dao, S. (1997). Modeling and querying moving objects. In *Proc. Int. Conf. on Data Engineering*, (pp. 422-432). Washington, DC, USA: IEEE.

Sivakumar, M. V. K., Motha, R. P., & Das, H. P. (Eds.). (2005). *Natural disasters and extreme events in agriculture: impacts and mitigation*. The Netherlands: Springer. doi:10.1007/3-540-28307-2

Soares, R. C. B. (1995). *Avaliação Da Resposta Imunológica Em Reprodutoras De Perus (Meleagridis Gallopavo) Diante Do Vírus Vacinal Da Doença De Newcastle Através De Modelos Matemáticos*. Porto Alegre: Ufrgs/ Faculdade De Veterinária, Dissertação De Mestrado.

Soares, F. U. (2008). *Zoneamento agroclimático para cana-de-açúcar (Sacharum ssp.) no Rio Grande do Sul por meio de geoprocessamento: estudo de caso no município de Jaguari*. Unpublished master dissertation, Federal University of Santa Maria, Brazil.

Sousa, E. P. M. d., Traina, C. Jr, Traina, A. J. M., Wu, L., & Faloutsos, C. (2007b). A Fast and Effective Method to Find Correlations among Attributes in Databases. [DMKD]. *Data Mining and Knowledge Discovery*, *14*(3), 367–407. doi:10.1007/s10618-006-0056-4

Sousa, S. A. V. (1999). Programa Computacional Para Simulação Da Ocorrência De Veranicos E Queda De Produção. [P. Tese ] [Doutorado] [ – Escola Superior De Agricultura "Luiz De Queiroz" - Universidade De São Paulo.]. *Piracicaba*, *1999*, 124.

Sousa, E. P. M. d., Traina, A. J. M., Traina Jr., C., & Faloutsos, C. (2007a). Measuring Evolving Data Streams' behavior through their Intrinsic Dimension. *New Generation Computing Journal*, 25(Special Issue on Knowledge Discovery from Data Streams), 33-59.

Souto, A. R., Crestana, S. (2000) Identificação Das Áreas Potenciais De Produção De Sedimentos Com O Modelo Agnps E Técnicas De Sig Em Uma Microbacia Hidrográfica. *Rev. Bras. Eng. Agríc. Ambient., Campina Grande*, *4*(3), Dez.

Souza, R. F. P. (2007). Economia Do Meio Ambiente E Responsabilidade Social: Os Métodos De Valoração Econômica E Controle Ambiental. In *Anais... Xlv Congresso Da Sociedade Brasileira De Economia, Administração E Sociologia Rural*, (pp.1-11). 22 A 25 De Julho 2007, Londrina: Uel.

Spaccapietra, S., Parent, C., Damiani, M. L., Macedo, J. A. F., Porto, F., & Vangenot, C. (2008). A conceptual view on trajectories. *Data & Knowledge Engineering*, *65*(1), 126–146. doi:10.1016/j.datak.2007.10.008

Spadotto, C. A. (2002). Screening Method For Assessing Pesticide Leaching Potential. *Pesticidas: Revista De Ecotoxicologia E Meio Ambiente, Curitiba*, *12*, 69-78.

Spielman, S. E., & Thill, J.-C. (2008). Social area analysis, data mining, and GIS. *Computers, Environment and Urban Systems, 32*, 110–122. doi:10.1016/j.compenvurbsys.2007.11.004

Steele, G. V., Johnson, H. M., Sandstrom, M. W., Capel, P. D., & Barbash, J. E. (2009). Occurrence and fate of pesticides in four contrasting agricultural settings in the United States. *Journal of Environmental Quality, 37*, 1116–1132. doi:10.2134/jeq2007.0166

Steenhuis, T. S., & Walter, M. F. (1980). Closed form solution for pesticide loss in runoff water. *Transactions of the ASAE. American Society of Agricultural Engineers, 23*, 615–620.

Stöckle, C. O., Donatelli, M., & Nelson, R. (2003). CropSyst, a cropping systems simulation model. *European Journal of Operational Research, 18*, 289–307.

Stockwell, D., & Peters, D. (1999). The GARP modelling system: problems and solutions to automated spatial prediction. *International Journal of Geographical Information Science, 13*(2), 143–158. doi:10.1080/136588199241391

Stockwell, D. R. B., & Noble, I. R. (1991). Induction of sets of rules from animal distribution data: a robust and informative method of data analysis. *Mathematics and Computers in Simulation, 32*, 249–254. doi:10.1016/0378-4754(90)90247-G

Stockwell, D. R. B. (1992). *Machine learning and the problem of prediction and explanation in ecological modelling*. Doctoral Thesis, Australian National University, Australia.

Stokelj, T., Paravan, D., & Golob, R. (2002). Enhanced artificial neural network inflow forecasting algorithm for run-of-river hydropower plants. *Journal of Water Resources Planning and Management, 128*(6), 415–423. doi:10.1061/(ASCE)0733-9496(2002)128:6(415)

Streck, N. A., & Alberto, C. M. (2006). Simulação Do Impacto Da Mudança Climática Sobre A Água Disponível Do Solo Em Agroecossistemas De Trigo, Soja E Milho Em Santa Maria, Rs. *Ciência Rural, 36*, 424–433.

Su, J., Xu, H., & Ibarra, O. (2001). Moving Objects: Logical Relationships and Queries. In *Proceedings of the 7th Int. Symp. on Spatial and Temporal Databases*, (pp. 3-19). Berlin: Springer-Verlag.

Suárez, F., Oñate, J., Malo, J., & Peco, B. (1997). Las políticas agroambientales y de conservación de la naturaleza en España. *Economía Agraria, 179*, 267–296.

Suji, E. R., Tigano, M. S., & Sosa-Gomes, D. (2002). Simulação Do Impacto Do Fungo *Nomuraea Rileyi* Em Populações Da Lagarta Da Soja. *Anticarsia Gemmatalis* Pesquisa. *Agropecuária Brasileira, Brasília, 37*(11), 1551–1558.

Suji. E. R. (1994). *Padrão De Distribuição Das Populações Anuais E Modelo Fenológico Para O Manejo Da Cigarrinha-Das-Pastagens, Deois Flavopicta (Homoptera: Cercopidae)*. Dissertação Mestrado. Campinas/Sp: Ib-Unicamp.

Sun, J., Papadias, D., Tao, Y., & Liu, B. (2004). Querying about the Past, the Present, and the Future in Spatio-Temporal Databases. In *Proceedings of the 20th ICDE*, (pp. 202–213). Washington, DC: IEEE.

Sutton, T., Giovanni, R., & Siqueira, M. F. (2007). Introducing openModeller – a fundamental niche modeling framework. *OSGEO Journal, 1*. Retrieved August 27, 2009, from http://www.osgeo.org/ojs/index.php/journal/article/viewFile/109/90

Tantanee, S., Patamatamakul, S., Oki, T., Sriboonlue, V., & Prempree, T. (2005). Downscaled rainfall prediction model (DRPM) using a unit disaggregation curve (UDC). *Hydrology and Earth System Sciences Discussions, 2*, 543–568. doi:10.5194/hessd-2-543-2005

Tao, Y., & Papadias, D. (2005). Historical spatio-temporal aggregation. *ACM Transactions on Information Systems, 23*(1), 61–102. doi:10.1145/1055709.1055713

Tashizawa, T., & Rezende, W. (2000). *Estratégia Empresarial*. São Paulo: Makron Books.

Tavares, J. N. (2004). *Free Course on Theory of the Field - Entropy. Measures of Gibbs. Mathematic Center of University of Porto*. Physics Center of Porto. (in Portuguese)

Tedeschi, L. O., Fox, D. G., Sainz, R. D., Medeiros, S. R., Boin, C., & Barioni, L. G. (2005). Mathematical Models In Ruminant Nutrition. *Scientia Agricola. Piracicaba, Brasil*, *62*(1), 76–91.

Teixeira, O. A., Melo, R. L. de & França, V. L. A. (2007). *Estado de Sergipe: uma proposta de territorialização para o planejamento*. SEPLAN Report.

Ternes, S., Fernandes, J. F. R., & Yang, H. M. (2005). O Uso De Modelagem Matemática E Simulação Na Avaliação Do Controle Biológico Da Minadora Do Citros. *Revista Brasileira De Agroinformática*, *7*(2), 1–16.

Terry, A. S., Carter, A. D., Humphrey, R. L., Capri, E., Grua, B., & Panagopoulous, A. C. (2008). A monitoring programme for 1,3-dichloropropene and metabolites in groundwater in five EU countries. *Pest Management Science*, *64*, 923–932. doi:10.1002/ps.1585

Tese De Doutorado. Medeiros, G.A, Perini, M., Lucarelli, J.R.F., Daniel, L.A. (2001) Influência Do Impacto Da Compactação Do Solo Sobre A Condutividade Hidráulica Saturada Para Um Solo Argiloso. *Revista Ecossistema*, *26*(2), Agosto-Dezembro.

The International Organization for Standardization. (2006). Environmental management -- Life cycle assessment -- Principles and framework. *International Standard ISO*, *14040*, 8.

*The World Factbook 2009*. (2009). Washington, DC: Central Intelligence Agency. Retrieved February 9, 2010, from https://www.cia.gov/library/publications/the-world-factbook/index.html

Thenkabail, P. S., Ward, A. D., Lyon, J. G., & Merry, C. J. (1994). Thematic mapper vegetation indices for determining soybean and corn growth parameters. *Photogrammetric Engineering and Remote Sensing*, *60*(4), 437–442.

Thompson, D. W. J., & Wallace, J. M. (2000). Annular modes in the extratropical circulation. Part I: Month-to-month variability. *Journal of Climate*, *13*, 1000–1016. doi:10.1175/1520-0442(2000)013<1000:AMITEC>2.0.CO;2

Thornthwaite, C. W. (1948). An approach toward a rational classification of climate. *Geographical Review*, *38*, 55–94. doi:10.2307/210739

Tidåkar, P., Sjőberg, C., & Jönsson, H. (2007). Local recycling of plant nutrients from small-scale wastewater systems to farmland – A Swedish scenario study. *Resources, Conservation and Recycling*, *49*, 388–405. doi:10.1016/j.resconrec.2006.05.004

Tiktak, A., Van den Berg, F., Boesten, J. J. T. I., Van Kraalingen, D., Leistra, M., & Van der Linden, A. M. A. (2002). *Manual of FOCUS PEARL version 1.1.1*. Bilthoven, The Netherlands: RIVM/Alterra.

Tiktak, A., Van der Linden, A. M. A., & Swartjes, F. A. (1994). *PESTRAS: A one dimensional model for assessing leaching and accumulation of pesticides in soil*. Bilthoven, The Netherlands: RIVM.

Tillman, D., Cassman, K. G., Matson, P. A., Naylor, R., & Polask, S. (2002). Agricultural sustainability and intensive production practices. *Nature*, *418*, 671–677. doi:10.1038/nature01014

Tojo-Soler, C., Folegatti, M. V., & Faria, R. T. (2001). Uso Do Modelo Ceres-Maize Para Identificação De Características Genéticas Desejáveis E De Praticas Adequadas De Manejo De Milho "Safrinha". *Revista Brasileira De Agrometeorologia. Santa Maria*, *9*(2), 339–346.

Tokar, A. S., & Johnson, P. A. (1999). Rainfall-runoff modeling using artificial neural networks. *Journal of Hydrologic Engineering*, *4*(3), 232–239. doi:10.1061/(ASCE)1084-0699(1999)4:3(232)

Tolga, E., Alptekin, E. (2007). Product development process using a fuzzy compromise based goal programming approach (note de lecture). *Computer Science, Computational sciences and its applications, 1*, 832-845.

Tomazoni, J.C. & Guimarães, E. (2005). A Sistematização Dos Fatores Da Eups Em Sig Para Quantificação Da Erosão Laminar Na Bacia Do Rio Jirau. *Revista Brasileira De Cartografia, N°57*(03), 235-244.

Traina, C. Jr, Sousa, E. P. M., & Traina, A. J. M. (2005). Using Fractals in Data Mining. In Kantardzic, M. M., & Zurada, J. (Eds.), *New Generation of Data Mining Applications*. New York: Wiley/IEEE Press.

Traina, A. J. M., Traina, C., Jr., Papadimitriou, S., & Faloutsos, C. (2001, August 26-29, 2001). Tri-Plots: Scalable Tools for Multidimensional Data Mining. In *Proceedings of the ACM International Conference on Knowledge Discovery and Data Mining*, San Francisco, CA, (pp. 184-193).

Traina, C., Jr., Traina, A. J. M., Wu, L., & Faloutsos, C. (2000, october 2-4, 2000). Fast feature selection using fractal dimension. *Proceedings of the Brazilian Symposium on Databases*, João Pessoa, PB, 158-171.

Trenberth, K. E. (1990). Recent observed interdecadal climate changes in the northern hemisphere. *Bulletin of the American Meteorological Society*, *71*, 988–993. doi:10.1175/1520-0477(1990)071<0988:ROICCI>2.0.CO;2

Trenberth, K. E., & Hurrell, J. W. (1994). Decadal atmosphere-ocean variations in the Pacific. *Climate Dynamics*, *9*, 303–319. doi:10.1007/BF00204745

Trigueiro, A. (Coord). (2003). *Meio Ambiente No Século 21: 21 Especialistas Falam Da Questão Ambiental Nas Suas Áreas De Conhecimento*. Rio De Janeiro: Sextante.

Trindade Neto, M. A. (n.d.). *Fluxo Biológico Do Fósforo No Metabolismo De Suínos Alimentados Com Dietas Contendo Fitase*. Http://Br.Monografias.Com/Trabalhos/Fluxo-Biologico-Fosforo-Metabolismo-Suinos/Fluxo-Biologico-Fosforo-Metabolismo-Suinos.Shtml. (Accessed On 01/09/2009).

Tsuji, G. Y., Uehara, G., & Balas, S. (Eds.). (1994). *Dssat: A Decision Support System For Agrotechnology Transfer. Version 3. V.1,2 And 3*. University Of Hawaii, Honolulu, HI.

Tsunechiro, A., & de Freitas, B. B. (2001). Os cinqüenta municípios brasileiros maiores produtores de milho e soja. *Informações Econômicas*, *31*(7), 53–58.

Tsuruta, H. (1997). Emission rates of methane from rice paddy fields and nitrous oxide from fertilised upland fields estimated from intensive field measurement for three years (1992-1994) all over Japan. *Research Reports Division of Environmental Planning, National Institute of Agro-* [in Japanese]. *Environmental Sciences*, *13*, 101–131.

Tzeng, G.-H., Teng, Y.-J., & Hu, C.-P. (1991). Urban environmental evaluation and improvement: Application of multi-attribute utility and Compromise Programming. *Behaviometrika*, *29*, 83–98. doi:10.2333/bhmk.18.29_83

Ungara, E., Henkinb, Z., Gutmana, M., Dolevc, A., Genizid, A., & Ganskopp, D. (2009). Inference of Animal Activity From GPS Collar Data on Free-Ranging Cattle. *Rangeland Ecology and Management*, *62*(4), 256–266.

União da Industria da Cana de Açucar. (n.d.). *Portal Única*. Retrieved in October 30, 2007 from http://www.portalunica.com.br/portalunica/

United States. Department of Agriculture. Natural Resources Conservation Service. Soil Survey Staff. (1999). *Soil taxonomy: a basic system of soil classification for making and interpreting soil surveys*. Washington, DC: USDA.

Urzedo, A. P. F. M., Rigitano, R. L. O., Guerreiro, M. C., & Castro, N. R. A. (2006). Dissipação do inseticida tiametoxam em solos da região de Lavras-MG. *Pesticidas: Revista de Ecotoxicologia e Meio Ambiente*, *16*, 31–38.

Uvo, C. B., Olsson, J., Morita, O., Jinno, K., Kawamura, A., & Nishiyama, K. (2001). Statistical atmospheric downscaling for rainfall estimation in Kyushu Island, Japan. *Hydrology and Earth System Sciences*, *5*(2), 259–271. doi:10.5194/hess-5-259-2001

Valocchi, A. J. (1990). Use of temporal moment analysis to study reactive transport in aggregated porous media. *Geoderma*, *46*, 233–247. doi:10.1016/0016-7061(90)90017-4

Van Dam, J. C., Groenendijk, P., Hendriks, R. F. A., & Kroes, J. G. (2008). Advances of modeling water flow in variably saturated soils with SWAP. *Vadose Zone Journal*, *7*(2), 640–653. doi:10.2136/vzj2007.0060

Van den Berg, F., & Boesten, J. J. T. I. (1998). *Pesticide leaching and accumulation model (PESTLA) version 3.3. Description and user's guide*. Wageningen, The Netherlands: SC-DLO.

Van Genuchten, M. Th. (1980). A closed form equation for predicting the hydraulic conductivity of unsaturated soils. *Soil Science Society of America Journal*, *44*, 892–898. doi:10.2136/sssaj1980.03615995004400050002x

Van Genuchten, M. Th., & Dalton, F. N. (1986). Models for simulating salt movement in aggregated field soils. *Geoderma*, *38*, 165–183. doi:10.1016/0016-7061(86)90013-3

Van Genuchten, M. Th., Leij, F. J., & Yates, S. R. (1991). *The RETC code for quantifying the hydraulic functions of unsaturated soils*. Riverside, CA: US Salinity Laboratory.

Van Ness, H. C., & Abbott, M. M. (1982). *Classical Thermodinamics Of Non-Electrolyse Solutions*. New York: Mcgraw-Hill. Veloso, R. F., Chaib Filho, H. (1999). *Uma Proposta De Modelagem De Sistemas Agrícolas Nos Cerrados*. Juiz De Fora: Agrosoft. Retrieved from http://Agrosoft.Softex.Br/Trabalhos/Ag99/Artigo56.Htm

Van Ommen, H. C., Van Genuchten, M. Th., Van der Molen, W. H., Dijksma, R., & Hulshof, J. (1989). Experimental and theoretical analysis of solute transport from a diffuse source of pollution. *Journal of Hydrology (Amsterdam)*, *105*, 225–251. doi:10.1016/0022-1694(89)90106-6

Van Oort, P. A. J. (2006). Interpreting the change detection error matrix. *Remote Sensing of Environment*, *108*(1), 1–8. doi:10.1016/j.rse.2006.10.012

Vanclooster, M., Boesten, J. J. T. I., Trevisan, M., Brown, C. D., Capri, E., & Eklo, O. M. (2000). A European test of pesticide-leaching models: methodology and major recommendations. *Agricultural Water Management*, *44*, 1–19. doi:10.1016/S0378-3774(99)00081-5

Vanderbei, R. J. (2008). *Linear programming: foundations and extensions* (3rd ed.). New York: Springer.

Vapinik, V. (1995). *The Nature of Statistical Learning Theory*. Berlin: Springer-Verlag.

Vári, A. (1999). *Floor control development in Hungary: Willingness-to-pay study*. Unpublished report, IIASA, Laxenburg, Austria.

Varsta, M. (2002). *Self-organizing maps in sequence processing*. Unpublished doctoral dissertation, Helsinki University of Technology. Espoo, Finland.

Vazirgiannis, M., & Wolfson, O. (2001). A Spatiotemporal Model and Language for Moving Objects on Road Networks. In *Proceedings of the Advances in Spatial and Temporal Databases Conf.* (pp. 20-35). Berlin: Springer-Verlag.

Veldkamp, A., & Lambin, E. F. (2001). Predicting land-use change. *Agriculture Ecosystems & Environment*, *85*(1-3), 1–6. doi:10.1016/S0167-8809(01)00199-2

Veloso, R. F., Gomes, E. G., Barioni, L. G., Fernandes, F. B., & Chaib Filho, H., Oliveira. A. J. De, Silva, A.P. Da, Coloca, A. (2007). Programação Matemática Para Avaliação De Viabilidade Econômica Da Produção De Leite Bovino Em Pequena Escala No Sul De Goiás. *Engevista*, *9*(1), 57–81.

Vera, C., Higgins, W., Amador, J., Ambrizzi, T., Garreaud, R., & Gochis, D. (2006). Toward a unified view of the American Monsoon Systems. *Journal of Climate*, *19*(20), 4977–5000. doi:10.1175/JCLI3896.1

Verburg, P. H., & Veldkamp, A. (2005). Introduction to the special issue on spatial modeling to explore land use dynamics. *International Journal of Geographical Information Science*, *19*(2), 99–102. doi:10.1080/13658810410001713362

Verburg, P. H., Rounsevell, M. D. A., & Veldkamp, A. (2006). Scenario-based studies of future land use in Europe. *Agriculture, Ecosystems & Environment*, *114*(***), 1-6.

Verneque, R. S. (2009). *Aspectos Genéticos E Ambientais Da Curva De Lactação De Vacas Da Raça Guzerá*. Retrieved from http://Www.Br.Monografias.Com/Trabalhos/Aspectos-Geneticos-Ambientais/Aspectos-Geneticos-Ambientais.Shtml (Accessed On 01/09/2009)

Vesanto, J. (1999). SOM based data visualization methods. *Intelligent Data Analysis*, *3*, 111–126. doi:10.1016/S1088-467X(99)00013-X

Vesanto, J., & Alhoniemi, E. (2000). Clustering of the Self-Organizing Map. *IEEE Transactions on Neural Networks*, *11*(3), 586–600. doi:10.1109/72.846731

Vesanto, J. (1997). *Data Mining Techniques Based on the Self-Organizing Map*. Unpublished master thesis, Helsinki University of Technology.

Vieira, S.R. (1998). Permeâmetro: novo aliado na avaliação de manejo do solo. *O Agronômico, 47/50*, 32-33.

Vilcarromero, A. C. S. (1996). *Análise De Modelos Determinísticos Em Fitopatologia*. Dissertação De Mestrado, Imecc/Unicamp, Campinas/Sp.

Vissotto Júnior, D. (2003). *Modelagem numérica para interações superfície-atmosfera no Estado do Paraná*. Master dissertation, Federal University of Paraná, Curitiba, PR.

Vogel, H. J., Cousin, I., Ippisch, O., & Bastian, P. (2006). The dominant role of structure for solute transport in soil: experimental evidence and modeling of structure and transport in a field experiment. *Hydrology and Earth System Sciences*, *10*, 495–506. doi:10.5194/hess-10-495-2006

Vryzas, Z., Vassiliou, G., Alexoudius, C., & Papadopoulou-Mourkidou, E. (2009). Spatial and temporal distribution of pesticide residues in surface waters in northeastern Greece. *Water Research*, *43*(1), 1–10. doi:10.1016/j.watres.2008.09.021

Wagenet, R. J., & Hutson, J. L. (1989). *Leachm: Leaching Estimation And Chemistry Model*. Ithaca, NY: Cornell University Center For Environment Resource, V.2, Version 2.

Wagner, A. B. A. Shiraishi. J. N. & Alvino, B. M. (2001). *Babaçu economy, preliminary data survey*. MIQCB, Balaios Typographia, Sao Luis-MA, 294p. (in Portuguese)

Walker, A. (1974). A simulation model for prediction of herbicide persistence. *Journal of Environmental Quality*, *3*, 396–401. doi:10.2134/jeq1974.00472425000300040021x

Wan, T., & Zeitouni, K. (2007). An OLAP system for network-constraint moving objects, In *Proceedings of the Annual ACM Symposium on Applied Computing,* (pp. 13-18). New York: ACM.

Weast, R. C. (1974). *Handbook of chemistry and physics*. Cleveland, OH: CRC Press.

Weaver, W. (1948). Science and complexity. *American Scientist*, *36*, 536–544.

Weber, E., Hasenack, H., & Ferreira, C. J. S. (2005) *Adaptação do modelo digital de elevação do SRTM para o sistema de referência oficial brasileiro e recorte por unidade da federação*. Porto Alegre, Brazil: UFRGS. Retrieved Dezember 01, 2005 from http://www.ecologia.ufrgs.br/labgeo

Weintraub, A., & Navon, D. (1976). A forest management planning model integrating silvicultural and transportation activities. *Management Science*, *22*, 1299–1309. doi:10.1287/mnsc.22.12.1299

Wernet, G., Hellweg, S., Fischer, U., Papadokonstantakis, S., & Hungerbühler, K. (2008). Molecular-Structure-Based Models of Chemical Inventories using Neural Networks. *Environmental Science & Technology*, *42*(17), 6717–6722. doi:10.1021/es7022362

Wesseling, J. G. (1991). *Meerjarige simulaties van grondwateronttrekking voor verschillende bodemprofielen, grondwatertrappen en gewassen met het model SWATRE*. Wageningen, The Netherlands: SC-DLO.

Wilby, R. L., Wedgbrow, C. S., & Fox, H. R. (2004). Seasonal predictability of the summer hydrometeorology of the River Thames, UK. *Journal of Hydrology (Amsterdam)*, *295*, 1–16. doi:10.1016/j.jhydrol.2004.02.015

Wilkinson, G. G. (1996). A review of current issues in the integration of GIS and remote sensing data. *International Journal of Geographical Information Systems*, *10*(1), 85–101.

Wilks, D. S. (2006). *Statistical methods in the atmospheric sciences*. Burlington, MA: Academic Press, Elsevier.

Williams, J. R. (1975). Sediment Yield Prediction With Universal Equation Using Runoff Energy Factor. In *Present Prospective Technology For Predicting Sediment Yields And Source,* (pp. 244-252). Washington, DC: Usda: Ars.

Wilson, Z., & Pfaff, B. (2008). Religious, philosophical and environmentalist perspectives on potable wastewater reuse in Durba, South Africa. *Desalinisation, 228*, 1–9. doi:10.1016/j.desal.2007.07.022

Winter, K., & Hewitson, B. (1994). Self organizing maps - applications to census data. In Hewitson, B., & Crane, R. (Eds.), *Neural nets: applications in geography*. Amsterdam: Kluwer.

Wischmeier, W. H., & Smith, D. D. (1978). *Predicting Rainfall Erosion And Losses: A Guide Of Conservation Planning*. Washington, DC: Usda.

Wooldridge, M. (2002). *An introduction to multiagent systems*. Chichester, UK: John Wiley & Sons.

World Health Organisation. (2005). *A regional overview of wastewater management and reuse in the Eastern Mediterranean*. Regional Office for the Estern Mediterranean. CEHA.

World Health Organisation. (2006). Guidelines for the safe use of wastewater, excreta and grey water. *Wastewater use in agriculture 2*. Genova.

Wrege, M. S., Caramori, P. H., Gonçalves, A. C. A., Bertonha, A., Ferreira, R. C., & Caviglione, J. H. (2005). Regiões potenciais para cultivo da cana-de-açúcar no Paraná, com base na análise do risco de geadas. *Revista Brasileira de Agrometeorologia, 13*(1), 113–122.

Wrege, M. S. (2004). *Delimitação de regiões com potencial de cultivo econômico da cana-de-açúcar (Saccharum spp.), no estado do Paraná por meio da análise de riscos climáticos.* Unpublished Doctoral Dissertation, State University of Maringa, Maringa, Brazil.

Wu, T., Song, G., Ma, X., Xie, K., Gao, X., & Jin, X. (2008). Mining geographic episode association patterns of abnormal events in global earth science data. *Science in China, 51*, 155–164. doi:10.1007/s11431-008-5008-3

Wu, S., & Chow, T. W. (2004). Clustering of the self-organizing map using a clustering validity index based on inter-cluster and intra-cluster density. *Pattern Recognition, 37*, 175–188. doi:10.1016/S0031-3203(03)00237-1

Wu, C.-C., & Chang, N.-B. (2004). Corporate optimal production planning with varying environmental costs. A grey compromise programming approach. *European Journal of Operational Research, 155*, 68–95. doi:10.1016/S0377-2217(02)00820-2

Xiao, X., Boles, S., Liu, J., Zhuang, D., Frolking, S., Li, Ch., & Salas, W., & MOORE, B. (2005). Mapping paddy rice agriculture in southern China using multi-temporal MODIS images. *Remote Sensing of Environment, 95*, 480–492. doi:10.1016/j.rse.2004.12.009

Yadav, R. K., Goyal, B., Sharma, R. K., Dubey, S. K., & Minhas, P. S. (2002). Post-irrigation impact of domestic sewage effluent on composition of soils, crops and ground water – a case study. *Environment International, 28*, 481–486. doi:10.1016/S0160-4120(02)00070-3

Yee, T. W., & Mitchell, N. D. (1991). Generalized additive models in plant ecology. *Journal of Vegetation Science, 2*, 587–602. doi:10.2307/3236170

Yeh, W. W.-G., Becker, L., & Zettlemoyer, R. (1982). Worth of inflow forecast for reservoir operation. *Journal of Water Resources Planning and Management, 108*(3), 257–269.

Young, R., Onstad, C., Bosch, D., & Anderson, W. (1986). *Agricultural Nonpoint Source Pollution Model: A Watershed Analysis Tool, Model Documentation.* Morris, MN: Agricultural Research Service, U.S. Department Of Agriculture.

Zaki, M. J. (2001). Spade: An efficient algorithm for mining frequent sequences. *Machine Learning, 42*(1-2), 31–60. doi:10.1023/A:1007652502315

Zeleny, M. (1974). A concept of compromise solutions and method of displaced ideal. *Computers & Operations Research, 1*(4), 479–496. doi:10.1016/0305-0548(74)90064-1

Zeleny, M. (1973). Compromise Programming. In Cochrane, J. L., & Zeleny, M. (Eds.), *Multiple criteria decision making* (pp. 373–391). Columbia, SC: Univ. of South Carolina Press.

Zhang, Y., Wallace, J. M., & Battisti, D. S. (1997). ENSO-like interdecadal variability: 1900-93. *Journal of Climate*, *10*, 1004–1020. doi:10.1175/1520-0442(1997)010<1004:ELIV>2.0.CO;2

Zhaorong, J., Higgins, A., & Prestwidge, D. B. (2005). An integrated statistical and optimisation approach to increasing sugar production within a mill region. *Computers and Electronics in Agriculture*, *48*(2), 170–181. doi:10.1016/j.compag.2005.03.004

Zhu, Y., & Shasha, D. (2003). Efficient elastic burst detection in data streams. In *Proceedings of the ACM Knowledge Discovery & Data Mining*, (pp. 336-345).

Zwiers, F. W., & von Storch, H. (2004). On the role of statistics in climate research. *International Journal of Climatology*, *24*, 665–680. doi:10.1002/joc.1027

Zwillinger, D., & Kokoska, S. (2000). *CRC Standard probability and statistics tables and formulae*. Boca Raton, FL: Chapman & Hall/CRC.

# About the Contributors

**Hércules Antonio do Prado** is a researcher in Computer Science, focusing his work on data/text mining, neural networks, knowledge-based systems, and knowledge management. In 1984 he joined the Brazilian Agricultural Research Corporation and, in 1992, the Catholic University of Brasília as an assistant professor. He received his D.Sc. in Computer Science at the Federal University of Rio Grande do Sul, Brazil (2001) his M.Sc. in Systems Engineering from the Federal University of Rio de Janeiro (1989). In 1999 he stayed in the Information Sciences Department of University of Pittsburgh as a Visitor Scholar, developing research for his doctoral program. He undergraduated in Computer Systems at the Federal University of São Carlos, Brazil (1976).

**Alfredo José Barreto Luiz** is a researcher in Applied Statistics at the Brazilian Agricultural Research Corporation (Embrapa Environment) and a post-doctoral fellow at Brazilian National Institute of Space Research (INPE). He received his D.Sc. in Remote Sensing at INPE and his Master degree in Statistics and Quantitative Methods from the University of Brasília (UnB). He did his undergraduate course in Agronomy at Federal University of Lavras (UFLa). His research interest includes the study of scenarios and the spatial and temporal dynamics of agriculture, beside data analysis in biometrics and environmetrics.

**Homero Chaib Filho** was born in 1954 in Belém-PA, Brazil. In 1970 was awarded the diploma of merit, by the National Council of nuclear energy of Brazil, for the monograph on the development of nuclear energy in Brazil. In 1977 has taken his bachelor's degree in Mathematics at Universidade de Brasília, Brazil. In 1983, took his master in Production Engineering, with emphasis on Operation Research, at Universidade de Santa Catarina, Brazil. At Escuela superior de Ingenieros agrónomos of the Universidad de Madri, took his Doctoral degree in Applied Mathematics, in 1990. Since 1977, he was a researcher at Embrapa, where he worked in the areas of data analysis and computational methods. In 2009 came to retire at Embrapa to seek new horizons.

* * *

**Geraldo da Silva e Souza** finished his Ph.D. in Statistics in 1979 at North Carolina State University with honors of Phi-Kappa-Phi. He pursued a postdoctoral in Econometrics at the University of North Carolina (1993) and in Statistics at the University of Brasilia (2002). He has a Master of Science degree in Systems Engineering from COPPE-UFRJ (1974). He has a BS degree in Mathematics (UFRJ, 1972) and a BS in Economics (UERJ, 1972). He was assistant professor of statistics at the University of Illinois at Chicago, Department of Information and Decision Sciences, and visiting professor at the University of North Carolina and North Carolina State University. Currently, he is a full professor of Statistics at the University of Brasilia and a senior researcher at Embrapa. His main research interests are in Probability, Statistics, Econometrics, and Operation Research.

**Eliseu Roberto de Andrade Alves** is an agronomist from the Federal University of Viçosa, Brazil, and MS and Ph.D. in Agricultural Economics from Purdue University. He was creator, founder, and CEO of Embrapa (1979-1985). Currently, he is a researcher of Embrapa, working with agricultural policy, institutional development, and economics of production. He received, among many honors, the award for International Distinguish Alumnus by the National Association of the State Lands Grant Colleges and Universities in the USA, and received the important distinction of the National Order of Scientific Merit, Brazil and from Purdue University the distinction Doctor Honoris Causa.

**Fernando Luis Garagorry** was born in 1938 in Montevideo, Uruguay. He obtained a degree in Mathematical Statistics from the Inter-American Statistics Training Center (CIENES, University of Chile, Santiago). Later on he got an MS and a PhD at Cornell University, with a major in Operations Research and minors in Mathematics and Statistics. He was Professor of Statistics at the University of Concepción, Chile, and then moved to Embrapa, where he has been a Senior Researcher since 1975. He has been a long-time member of the American Statistical Association, the Institute for Operations Research and the Management Sciences (INFORMS, U.S.A.) and the Brazilian Operations Research Society (SOBRAPO).

**Carlos Pérez-Hugalde** is Associate Professor in the Department of Statistics and Management Science in Agriculture at the Universidad Politécnica de Madrid. He has a Master of Science in Agrofood Systems and Management by the International Centre for Advanced Mediterranean Agronomic Studies and a PhD in Agricultural Engineering by the Universidad Politécnica de Madrid. His research interest includes explorative multivariate analysis, econometric modeling of agricultural policies, and application of statistics in agricultural and environmental sciences.

**Raúl Romero-Calcerrada** is Senior Lecturer at the Rey Juan Carlos University of Madrid. He has a degree in Geography and a PhD in Cartography, Geography and Remote Sensing Systems from the Universidad de Alcalá. In broad terms, his areas of scientific interest are centered on territory planning in rural areas and on Natural Protected Areas, from a Sustained Development perspective. He is especially interested in the study of land cover changes as a means to understand the interactions between socio-economic transformations and environmental alterations (such as Forest Fires), and to establish more efficient formulas of land use planning and rural development.

**Patricia Delgado-Pérez** has a degree in Agricultural Engineering by the Universidad Politécnica de Madrid. Her professional experience is related to environmental management.

**Luciana A. S. Romani** received the B.Sc. degree in Computer Science from the Federal University of São Carlos, Brazil, in 1993 and the M.Sc. degree in Computer Science from State University of Campinas, Brazil in 2000. She is currently a Ph.D. student in Computer Science at ICMC - USP São Carlos, Brazil. She is researcher of the Brazilian Agricultural Research Corporation since 1994. Her research interests include remote sensing applied to agriculture, data mining, information visualization and human-computer interaction.

**Elaine P. M. de Sousa** received the B.Sc., M.Sc. and Ph.D. degrees in Computer Science from the University of São Paulo, Brazil, in 1998, 2000 and 2006, respectively. She is currently an assistant professor with the Computer Science Department at the University of Sao Paulo in Sao Carlos, Brazil. Her research interests include database, fractals, data mining and data stream mining.

**Marcela X. Ribeiro** received the B.Sc. degree in Computer Engineering and the M.Sc. in Computer Science from the Federal University of São Carlos, Brazil, in 2002 and 2004, respectively. She received the Ph.D. degree in computer science at the Mathematics and Computer Science Institute of the University of São Paulo at São Carlos, Brazil, in 2008. She is currently a professor with the Computer Science Department of the Federal University of São Carlos, São Carlos, Brazil. Her research interests include multimedia data mining, visual data mining, visualization, computer-aided diagnosis (CAD) and content-based image retrieval (CBIR).

**Ana Maria H. de Avila** received the B.Sc. degree in Meteorology from the Federal University of Pelotas, and the M.Sc. in Agrometeorology from Federal University of Rio Grande do Sul. She received the Ph.D. degree in Agricultural Engineering at UNICAMP, Brazil in 2006. Since July 2006, she has been working at Center of Research Meteorological and Climatological applied to Agricultural (CEPAGRI) at UNICAMP. She is currently Director of CEPAGRI and group member of Climate Risk Zoning in Brazil. Her research interests include agrometeorology and climatology.

**Jurandir Zullo Jr.** received the M.Sc. degree in Applied Mathematics and the Ph.D. degree in Electrical Engineering from the University of Campinas (UNICAMP), Brazil, in 1990 and 1994, respectively. He received the B.Sc. in Agricultural Engineering (1987) and Applied Mathematics (1985) at the University of Campinas (Unicamp). Since 1987, he has been working at Center of Research Meteorological and Climatological applied to Agricultural (CEPAGRI) at UNICAMP. His research interests include agrometeorology, remote sensing, climatology and mathematical modeling.

**Caetano Traina Jr.** received the Ph.D. in Computational Physics, MSc in Computer Science and BSc in Electrical Engineering respectively in 1986, 1982, 1977 from the University of Sao Paulo - Brazil. From 1998 to 2000 he spent a sabbatical period with the Computer Science Department of Carnegie Mellon University - USA, working with multimedia databases regarding similarity queries. He is currently a Full Professor with the Computer Science Department at University of São Paulo at São Carlos. His research interests include multimedia databases, similarity queries on complex data, data mining and metric access methods. He has supervised 45 graduate students in these fields.

**Agma J. M. Traina** received the Ph.D. in Computational Physics, MSc in Computer Science and BSc in Computer Science respectively in 1991, 1987, 1983 from the University of Sao Paulo - Brazil. From 1998 to 2000 she spent a sabbatical period with the Computer Science Department of Carnegie Mellon University - USA, working with multimedia databases and information visualization. She is currently a Full Professor with the Computer Science Department at University of São Paulo at São Carlos. Her research interests include image databases, image mining, visual data mining, content-based image retrieval (CBIR), indexing methods for multidimensional data, information visualization, image processing for agroclimate applications and complex data. She has supervised 30 graduate students in these fields.

**Antonio Roberto Formaggio** is Agronomist by Escola Superior de Agricultura Luiz de Queiroz of University of São Paulo (ESALQ-USP), with Master degree in Remote Sensing of Environment from the brazilian National Institute for Space Research (INPE) and with Doctor degree in Remote Sensing and Agriculture at ESALQ-USP. He is researcher and post-graduation teacher/advisor in Remote Sensing and Environment and Agriculture since 1989, in INPE. His research interests include agricultural statistics, impacts of agriculture in the environment, hyperspectral remote sensing, estimation of biophysical variables of crop canopies, environmental modeling, beside radiative transfer models for the retrieval biophysical variables of vegetation.

**José Carlos Neves Epiphanio** is a senior scientist and professor at the Brazilian National Institute for Space Research (INPE) since 1979. His background and main research interests are in agriculture and remote sensing. He has a D.Sc. from University of São Paulo (USP) and a post-doc at University of Arizona. Crop classification, multitemporal analysis and interactions between crops and environment are the main focus of his research. He has been being the organizer of the Brazilian Symposium on Remote Sensing in the last 10 years. Currently he is involved with the program of development of China-Brazil Earth Resources Satellite (CBERS), in which he is in charge of the Application Program.

**Fernando Ulhmann Soares** is a geographer from Federal University of Pelotas, 2001. M. Sc. in Geomatics, Federal University of Santa Maria, Brazil, 2008. He was engaged as Professor at the Federal Institute of Education, Science and technology of Goias in 2008 working with Geomatics, Regional Planning, GIS, Environment and Agroenergy.

**José Maria Filippini Alba** holds a Bachelor degree in Chemistry, University of the Republic, Uruguay, 1988. M.Sc in Geochemistry, State University of Sao Paulo, Brazil, 1993. Dr. Sc. in Geochemistry and Remote Sensing, State University of Sao Paulo, Brazil, 1998, with collaboration of State University of Campinas, Brasil; University of Åbo, Finland and Geological Survey of Finland. Post - Doctorate in Geochemistry and GIS, State University of Campinas, Brasil, 2003 - 2005. He was engaged as researcher by the Brazilian Agricultural Research Corporation in 2005, working with environmental monitoring, remote sensing imagery and GIS modeling.

**Elódio Sebem** has a Bachelor degree in Forestry, Federal University of Santa Maria, 1998. M. Sc. in Agricultural Engineering, Federal University of Santa Maria, Brazil, 2000. Dr. Sc. in Photogrammetry, Remote Sensing and Graphics Engineering, Politechnic University of Madrid, Spain, 2006. He was engaged as Professor at Federal University of Santa Maria in 2006 working with remote sensing, technical cadastre, data banks, GIS and development of software.

**Marcos Silveira Wrege** holds a Bachelor degree in Agronomy, State University "Julio de Mesquita Filho", Brazil, 1991. M. Sc. in Agriculture (Irrigation and Drainage), the State University "Julio de Mesquita Filho", Brazil, 1995. D. Sc. in Agronomy (Crop Production) by State University of Maringa, Brazil, 2004. Researcher of the Brazilian Agricultural Research Corporation since 2001, his main research area is Agrometeorology with contributions about climate risk reduction, age of planting, water balance, agricultural modeling, climate change, GIS, forestry and bioenergy.

**Marcos Aurélio Santos da Silva** is a researcher in Computer Science at the Brazilian Agricultural Research Corporation (Embrapa Coastal Tablelands). He received his M.Sc. in Applied Computer at the National Institute for Space Research, Brazil (2004). He is undergraduated in Computer Science at the Federal University of Sergipe, Brazil (1998). His research interests include spatial analysis, neural networks and territorial studies.

**Edmar Ramos de Siqueira** is a researcher in Forestry Engineering at the Brazilian Agricultural Research Corporation (Embrapa Coastal Tablelands). He received his D.Sc. in Forestry Engineering at the Federal University of Paraná, Brazil (1988) and his M.Sc. in Forestry Science at the Federal University of Viçosa (1978). He is undergraduated in Forestry Engineering at the Federal University of Viçosa, Brazil (1975). His research interests include territorial development, agroforestry systems and sustainable development.

**Olívio Alberto Teixeira** is a full professor at the Federal University of Sergipe. He has a D.Sc. in Sociology from University of Paris X, Nanterre, France (2000), a M.Sc. in Rural Sociology and Economy from Federal University of Paraíba, Brazil (1990) and undergraduated in Economics at the Federal University of Santa Catarina, Brazil (1985). His research interests include rural development, interdisciplinary studies, territorial development and family farming.

**Maria Geovania Lima Manos** is an analyst in Business Management at the Brazilian Agricultural Research Corporation (Embrapa Coastal Tablelands). She received her M.Sc. in Economics at the Federal University of Sergipe, Brazil (2009). She is undergraduated in Economics at the Federal University of Sergipe, Brazil (2004). Her research interests include regional development, territorial development, credit, technological innovation and rural productive chains.

**Antonio Miguel Vieira Monteiro** has a degree in Electrical, a Master of Science degree in Applied Computer Science from the National Institute for Space Research (INPE – Brazil) and a DPhil in Electronic Engineering and Control/Computer Science at the Space Science Centre, School of Engineering and Applied Sciences, The University of Sussex at Brighton, October 1993. Miguel works at INPE and his research interests aims at building sensitive geographic indicators and discussing the use of geotechnologies and spatial analysis methods applied to urban and public health problems. His main interest areas include remote sensing, geotechnologies, geographic databases and software engineering in open source environments.

**Carlos Alberto Flores** holds a Bachelor degree in Agronomy, Federal University of Santa Maria, Brazil, 1973. M. Sc. in Soil Management, Federal University of Santa Maria, Brazil, 2001. He was engaged as researcher by the Brazilian Agricultural Research Corporation in 1990, working with soil mapping and agricultural zoning.

**Marilice Cordeiro Garrastazu** holds a Bachelor degree in Forestry, Federal University of Santa Maria, Brazil, 1999. M. Sc. in Agricultural Engineering (emphasis in Remote Sensing), Federal University of Santa Maria, Brazil, 2001. She was engaged as researcher by the Brazilian Agricultural Research Corporation in 2005, working with environmental monitoring, GIS and Agricultural Zoning.

**Sanjay Dominik Jena** was born in 1980 in Cologne, Germany. He completed an apprenticeship in Mathematics and Information Technology at AXA Germany (2002). Afterwards he graduated in Computer Science at the University of Applied Sciences Cologne, Germany (2007), while working in industry as a software developer. He then obtained a Master degree at the Pontifical Catholic University of Rio de Janeiro, Brazil, in Computer Science focused on Operations Research (2009). In his thesis, he investigated the use of optimization methods for the harvest planning in sugar cane industry. In the same year, he joined a Ph.D. program at the University of Montreal, Canada, where he is now performing his research at the research center CIRRELT, focusing on or methods applied to the forestry sector.

**Marcus V.S. Poggi de Aragao** is an associate professor at the Pontificia Universidade Catolica do Rio de Janeiro (PUC-Rio). He holds a Ph.D. in Applied Mathematics from Ecole Polytechnique de Montreal, Canada. He graduated in Electrical Engineering at PUC-Rio, where he also obtained his M.Sc. in Systems Engineering. Dr. Poggi main interests range from decomposition techniques for mixed-integer programming, and heuristics, metaheuristics and matheuristics for combinatorial optimization problems, to applications in logistics, telecommunications and production planning.

**Lucinio Júdez Asensio** is Full Professor of the Department of Statistics and Management Science in Agriculture (ETS Agricultural Engineers) at the Universidad Politécnica de Madrid (UPM) and director of the Associated Unit of "Statistical" Methods at the CSIC (Consejo Superior de Investigaciones Cientificas). His fields of work are Statistics, Operational Research and Econometrics applied to Agriculture. In recent years he has worked on different aspects of FADN and PMP models, developed in the context of several national and European research projects and contracts with the Spanish Ministry of Agriculture.

**Rosario de Andrés Gómez de Barreda** is a Tenured Scientist in the Economics Department at the Institute of Economics, Geography and Demography of the Consejo Superior de Investigaciones Cientificas (CSIC). She is also a member of the Associated Unit known as "Statistical Methods". This Unit, under the direction of Professor Lucinio Júdez, is a joint research initiative between CSIC and the Universidad Politecnica de Madrid (UPM) where she has been working on different agrarian economic analyses, in particular those related to the impact of the EU Common Agricultural Policy (CAP) on the Spanish Agricultural Sector. She is author of 26 book chapters. She has published more than 30 research papers in academic journals and attended more than 30 national and international conferences. She has worked on 17 competitive research projects in Spain and also on two European projects funded by the Fifth and the Sixth Framework EU Programmes.

**Miguel Angel Ibáñez Ruiz** is a former student of ETS Agricultural Engineers, (Spain), and holds a Doctorate Degree from the Universidad Politécnica de Madrid. He is currently a Professor at the Department of Statistics and Management Science in Agriculture (ETS Agricultural Engineers) at the Universidad Politécnica de Madrid (UPM). His fields of work are Statistics, Statistical Computing, Operational Research and Econometrics applied to Agriculture. In recent years he has participated in two European research projects related to the impact of the EU Common Agricultural Policy (CAP) on the Spanish Agricultural Sector.

**José-Luis Miguel de Diego** is Technical Director of the Coordinadora de Organizaciones de Agricultores y Ganaderos (COAG). He is a specialist in European agricultural policy and in agricultural economics, having worked in these areas for the last 15 years. He has a degree in Agricultural Engineering from the Universidad Politécnica de Madrid (UPM) and an MBA from the Instituto de la Empresa de Madrid.

**Elvira Urzainqui Miqueleiz** is a Tenured Scientist in the Economics Department at the Institute of Economics, Geography and Demography of the Consejo Superior de Investigaciones Cientificas (CSIC). She is also a member of the CSIC/UPM Associated Unit of "Statistical Methods". In recent years she has been working on analyses of the impact of the EU Common Agricultural Policy (CAP) on the Spanish Agricultural Sector. Since 1990, she has participated in two European projects funded by the Fifth and the Sixth Framework EU Programmes and in several national and regional projects. In the context of the Associated Unit, she is co-author of 5 books and 25 book chapters and has published several research papers in academic journals and attended many national and international conferences.

**Maria Conceição Peres Young Pessoa** is graduated in Applied Mathematics at IMECC/State University of Campinas (UNICAMP) (Brazil, 1985), MSc in Electrical Engineering (Electronic and Communication) at DT/FEE-Unicamp (Brazil, 1989), and PhD. in Electrical Engineering (Automation) at DENSIS/FEE-Unicamp (Brazil, 1994). Since 1989, she has been a researcher of Embrapa (Brazilian Agricultural Research Corporation linked to the Brazilian Ministry of Agriculture, Livestock and Food Supply), acting on computational-mathematic applied for environmental impact assessment. As main applications, she has been working with mathematical-modeling simulation applied for risk assessment of pesticide dynamics; integrated pest management (with emphasis on biological control and population dynamic of agricultural insects), greenhouse gas emission from Brazilian paddy-rice for the national inventory reference report; as well as, other systemic aspects related to environmental impacts assessment of agriculture like good agricultural practices of production systems.

**Elizabeth Nogueira Fernandes** is graduated in Forestry Engineering at Federal University of Viçosa (Brazil , 1979), MSc in Forest Science at Federal University of Viçosa (Brazil , 1993), PhD in Forest Science at Federal University of Viçosa (Brazil, 1997), and Post-Doctorate in Environmental Economy at State University of Campinas (Unicamp) (Brazil, 2005). Actually, she is researcher of Embrapa (Brazilian Agricultural Research Corporation linked to the Brazilian Ministry of Agriculture, Livestock and Food Supply), who acting at Embrapa Dairy Cattle, and professor of Urban Environmental Management Course of the Vianna Junior Institute. She has experience on the area of Forestry Resources and Forestry Engineering, with emphasis on Environmental Impact Assessment on the following issues: agrosilvi-pastoral systems, organic milk, emergy analysis of production systems, and natural resources valuation analysis.

**Sonia Claudia** do Nascimento de Queiroz is a researcher from the Brazilian Agricultural Research Corporation - Ministry of Agriculture, Livestock and Food Supply-Embrapa. She received Ph.D. degree in Analytical Chemistry at the University of Campinas- Unicamp (Brazil, 2001). Nowadays she is developing activities at Embrapa Environment, in Jaguariúna - São Paulo State. She has experience on the following topics: chromatography coupled to mass spectrometry, development and validation of chromatographic methods; residues and contaminants in food and environmental samples; behavior of pesticides and contaminants in the environment and water quality.

**Vera Lúcia Ferracini** is a researcher in Chemistry Science at the Brazilian Agricultural Research Corporation - Ministry of Agriculture, Livestock and Food Supply-Embrapa. She received her D.Sc. in Organic Chemistry Science at the University of Campinas of Campinas, Brazil (1995) and her master degree in Analytical Chemistry from the Federal University of Brasília, Brazil (1978). In 1988 she joined the Department of Biology at University of Texas at Austin as a Visitor Scholar, developing research in Natural Products program. She is graduated in Chemistry at the Federal University of Brasília, Brazil (1975). Her research interest includes natural products and pesticides residues analyses. Vera Lúcia Ferracini was born in 1951 in Penápolis, State of São Paulo - Brazil. In 1978-1979 she became an assistant professor of the Chemistry Department at University of Brasilia. In period of 1979-1984 she was a research in fuel ethanol of Research Center and Tecnological Analysis at Petroleum National Agency- Brasília/DF. Since 1984 she is a researcher at Embrapa Environment - Jaguariuna, SP - Brasil.

**Marco Antonio Ferreira Gomes** is graduated in Geology at Federal University of Mato Grosso (Brazil, 1981), MSc in Agronomy (Soil and Plant Nutrition) at Federal University of Viçosa (Brazil, 1987), and PhD in Agronomy (Soil and Plant Nutrition) at Federal University of Viçosa (Brazil, 1993). Nowadays is researcher of Embrapa (Brazilian Agricultural Research Corporation linked to the Brazilian Ministry of Agriculture, Livestock and Food Supply), acting at Embrapa Environment, and coordinator of the Environmental Engineering Course of the Jaguariúna Faculty (FAJ). He has experience on the area of sustainable use and occupation of the rural space with emphasis on groundwater and soil qualities, erosive process and pesticide contamination risks.

**Manoel Dornelas de Souza** is graduated in Agricultural Engineering at the University of Brasilia (Brazil, 1981), MSc in Management and Soil Conservation at the Federal Unicersity of Rio Grande do Sul (Brazil, 1985), and PhD. in Soil and Plant Nutrition at the School of Agriculture Luiz de Queiroz - University of São Paulo(Brazil, 1993). Researcher of Embrapa Environment, which is one of 40 Units of the Brazilian Agricultural Research Corporation (Embrapa) - linked to the Brazilian Ministry of Agriculture, Livestock and Food Supply, he has experience in agronomy, with emphasis on Soil Physics, acting on the following topics: dynamics of solutes in soil quality, organic wastes and recovery of degraded areas.

**Lana M. R. dos Santos** is a full professor at the Universidade Federal de Viçosa. She has a D.Sc. and M.Sc. in Computicional Mathematics. Her research interests include algorithms and applications of optimization in different sustainable production models.

**Marcos N. Arenales** is a full professor at University of São Paulo, in Brazil. He is undergraduated in Mathematics in São Paulo State University (1976), and has Master and Doctorate degrees in Applied Mathematics and Operations Research from the State Universtiy of Campinas (1978, 1984). His interests include linear and combinatorial optimization and its applications.

**Alysson M. Costa** obtained his Ph.D. in Administration (Quantitative Methods) at HEC Montreal - University of Montreal in 2006. He is Assistant Professor of Operations Research at ICMC - University of Sao Paulo at Sao Carlos. His main research interests are: (mixed-integer linear) optimization; assembly line balancing; sustainable agriculture; and exact and approximate algorithms and applications of optimization in different fields.

**Ricardo H. S. Santos** is a full professor at the Universidade Federal de Viçosa. He is an Agronomist and has a D.Sc. and s M.Sc. in Plant Science. He teaches the Agroecolgy course at Graduate and Undergraduate level. His research interests include organic agriculture, leguminous green manure and coffee crop.

**Rômulo Penna Scorza Júnior** is a researcher in Mathematical Modeling and Simulation at Embrapa Western Region Agriculture , in Dourados, MS, Brazil, which is one of the research units of the Brazilian Agricultural Research Corporation - Embrapa. He received his Ph.D. in Environmental Sciences at Wageningen University and Research Centre, Wageningen, The Netherlands, in 2002 and his M.Sc. in Entomology at the Federal University of Lavras, Lavras, Brazil, in 1997. His research interests are mainly on pesticide behavior in soils, mathematical modeling and simulation of pesticide leaching in soils and risk assessment of water resources contamination by pesticides.

**Renê Luís de Oliveira Rigitano** is a Full Professor at the Federal University of Lavras - Brazil, with Ph.D. on Entomology/Insecticides by the University of London - Imperial College of Science and Technology - England. His research interests are mainly on pesticides residues in agricultural products, translocation of pesticides in plants and behavior of pesticides in soils.

**Jos Boesten** is a senior researcher at Alterra which is part of Wageningen University and Research Centre. He received his M. Sc. in Soil Science in 1980 at Wageningen University and his Ph. D. on simulation of herbicide behaviour in soil in 1986 (also at Wageningen University). He is since 2003 a member of the Panel of Plant protection products and their Residues of the European Food Safety Agency (EFSA). His research interests are development of methodologies for exposure assessment of pesticides in soil, groundwater and surface water.

**Lisa Brouwers** is a lecturer at the Royal Institute of Technology (KTH) in Stockholm, Sweden, and holds a Doctorate Degree in Computer and systems sciences. She designs and implements individual-based simulation models for policy support. Currently she shares her time between KTH, the Ministry of Health and Social Affairs, and the Swedish Institute for Infectious Disease Control. At both governmental organizations she leads policy oriented research projects where simulation models are the main tools used for forecasting the effect of various policy interventions. Her main expertise lies within the area of policy modelling for health care and public health, with a special interest in epidemiology.

**Magnus Boman** has a doctoral degree from the Department of Computer & Systems Sciences at Stockholm University. In 2003, he was appointed full professor at KTH, in Intelligent Software Services. Since 2001, he shares his time equally between KTH and the Swedish Institute of Computer Science (SICS). Dr Boman has published more than 40 journal papers and book chapters on topics in computer science, logic, economics, and systems theory, as well as several books, and has been an invited speaker in academics and industry in nine countries. He is on the editorial board of the Robotics and Autonomous Systems journal. In 2002, he co-founded the international series of Mobility Roundtables, yearly events for industry and academia covering all aspects of mobile information technology. He is a founding member of the Stockholm Group on Epidemic Modelling (www.s-gem.se), an alumni of IIASA, and a member of TED.

**Pedro Luiz Pizzigatti Corrêa** is Professor at School of Engineering of the University of São Paulo (USP), at the Computer and Digital Systems Engineering Department, Brazil. He holds degrees on Computer Science (USP, 1987) with PhD on Electrical Engineering/Digital Systems (USP, 2002). He is researcher at the Agricultural Automation (Laboratory at School of Engineering - USP), since 2003, and his research interests include distributed database and information systems applied to agribusiness and to the environment in a broad sense. He worked as researcher at Brazilian Agricultural Research Corporation (EMBRAPA 1990-1993), as consulting of software development to Brazilian Financial Institutions based on EDI (Electronic Data Interchange) (1993-1997), and as consulting of the UNDP (United Nations Development Programme) in an electronic government project of São Paulo State - Brazil (1997-2002).

**Mariana Aparecida Carvalhaes** is researcher in Natural Resources at the Brazilian Agricultural Research Corporation (EMBRAPA). She received her D.Sc. in Environmental Engineering Science (2005) and her master degree in Ecology (1997) at the University of São Paulo (USP), Brazil. She obtained her degree in Biological Science at the São Paulo State University Júlio de Mesquita Filho, Brazil (1991). Her research interests include is prospection of native vegetal species of economic interest.

**Antonio Mauro Saraiva** is Full Professor at School of Engineering of the University of São Paulo (USP), at the Computer and Digital Systems Engineering Department. He holds degrees on Electrical Engineering (USP, 1980) and on Agronomy (ESALQ-USP, 1987) with PhD on Electrical Engineering/ Digital Systems (USP, 1998). He is the coordinator of the Agricultural Automation Laboratory (Laboratory at School of Engineering - USP) since 1989, and his research interests include instrumentation and control systems, and information systems applied to agribusiness and to the environment in a broad sense. He was the president of the Brazilian Society of Informatics applied to Agribusiness, SBIAgro.

**Fabrício Augusto Rodrigues** is a regular doctorate student at the School of Engineering of the University of São Paulo, Brazil. He received his master degree in Computer Science from the Federal University of Campina Grande, Brazil (2002) and he is graduated in Information Systems from the Potiguar University, Brazil (2000). He started his teaching activities in 2002, at the Faculty of Technology and Sciences in Feira de Santana, Brazil, where he acted until 2005. In the same year, he moved to College ÁREA1 in Salvador, Brazil, where he taught until 2007, besides acting as the coordinator of the Computation Nucleus. His research interest includes biodiversity informatics, machine learning and performance analysis of algorithms.

**Elisângela Silva da Cunha Rodrigues** is a regular doctorate student at the School of Engineering of the University of São Paulo, Brazil. She received her master degree in Computer Science from the Federal University of Campina Grande - Brazil (2002) and she is graduated in Informatics from the Federal University of Pelotas, Brazil (2000). She started her teaching activities in 2002, at the Faculty of Technology and Sciences of Feira de Santana, Brazil, where she acted until 2005. In the same year, she moved to College ÁREA1 in Salvador, Brazil, where she acted until 2007. In 2006, she participated as researcher (supported by Petrobrás) of the Automated Well Management System Project (SGPA), at the Federal University of Bahia - Brazil. Her research interest includes theory of computation, biodiversity informatics, machine learning and adaptive technology.

**Ricardo Luis de Azevedo da Rocha** is Professor at School of Engineering of the University of São Paulo (USP), at the Computer and Digital Systems Engineering Department, Brazil. He obtained his degree in Electronics Engineering, focus on Digital and Computer Systems, from the Pontifical Catholic University of Rio de Janeiro, Brazil (1982). He was granted his MSc (1995) and PhD (2000) in Electrical Engineering/Digital Systems from the School of Engineering of the University of São Paulo, Brazil. He is currently a professor at the Engineering School, and his research interests are Adaptive Techniques and Technologies, Computational Models, Intelligent Systems, Kolmogorov Complexity and Foundations of the Theory of Computation. Dr. Rocha has been a member of the Association for Computing Machinery - ACM since 2001, and also of the Brazilian Computing Society - SBC since 2000.

**Adriano Rolim da Paz** is a Professor at the Department of Civil and Environmental Engineering of the Federal University of Paraíba, Brazil. From the beginning of 2009 to September of 2010 he was a Researcher at the Brazilian Agricultural Research Corporation (Embrapa Satellite Monitoring). He received his D.Sc. (2010) and M.Sc. (2003) in Water Resources and Environmental Sanitation from the Institute of Hydraulic Research of the Federal University of Rio Grande do Sul, Brazil. He is Undergraduated in Civil Engineering at the Federal University of Paraíba (2001). His research interest includes hydrologic modeling, streamflow and precipitation forecasting, GIS applications for water resources, and environmental monitoring.

**Cintia Bertacchi Uvo** is a full professor at Lund University, Sweden. She has a Ph.D. in Water Resources Engineering from this same University (1998), a M.Sc. in Climatology from the National Space Research Institute, Brazil (1989) and a B.Sc. in Meteorology from the University of São Paulo, Brazil (1983). Considering her broad background in meteorology, climatology and water resources engineering, her research focuses on the interactions of climate and surface hydrology in diverse time and space scales, from daily to decadal. Her expertise includes the development of data driven models for analysis and forecast of hydrological variables, from climate conditions. She has a large international experience that comprises collaborations with scientific institutions in 13 countries.

**Juan Martín Bravo** is a Professor at the Department of Hydraulic Structures of the Federal University of Rio Grande do Sul, Brazil. He has a D.Sc. (2010) and a M.Sc. (2006) in Water Resources and Environmental Sanitation at the Institute of Hydraulic Research of the Federal University of Rio Grande do Sul, Brazil and undergraduated in Civil Engineering at the National University of Nordeste, Argentina (2003). His research interests include evolutionary algorithms, artificial neural networks, short-term streamflow forecasting, long-range precipitation forecast, hydrologic modeling, multi-purpose reservoir operation, and agent-based modeling.

**Walter Collischonn** is a Hydrology professor at the Federal University of Rio Grande do Sul (Brazil). He has a Doctorate degree in Water Resources and Environmental Sanitation from the Institute of Hydraulic Research of the Federal University of Rio Grande do Sul, Brazil (2001). His main research interests are large-scale hydrologic modeling and GIS applications for water resources.

**Humberto Rocha** is an engineer, PhD and his expertise is hydrometeorology. He helds a chair of Professor Titular in the Department of Atmospheric Sciences at the University of São Paulo, where his scientific research comprehends the water and carbon cycle of Brazilian land terrestrial ecosystems, using field observations with automated flux towers and simulations with computational numerical modelling. He has advised several grad students and authors over 40 international articles in per-review journals. In the Large-Scale Biosphere-Atmosphere Interaction Experiment in Amazonia (LBA) he was the South American P.I. for several projects sponsored by NASA, and is the lead scientist of projects addressed to data collection at multiple flux towers and simulation of river discharge, rainfall and crop production, which includes catchment studies over areas with Sugar cane, Eucalyptus, Savanna, Amazonian forest and Atlantic forest, in cooperation with private, NGO and government institutions.

**Hella Ben Brahim** was born in 1971 in Tunis, Tunisia. She is an economist with emphasis on decision support systems applied to environmental problems, has a Masters in statistics and a PhD in economic. University Tunis el Manar. In 2002, she became a member of European Group on Multicriteria Decision Aid. Her research interest includes multicriteria decision support, hydrological problems and wastewater reuse.

**Lucien Duckstein**. After finishing a graduate degree in Engineering in Toulouse (France), Lucien Duckstein was granted a Ph.D in Civil Engineering in 1962 at Colorado State University and then joined the Systems Engineering Faculty at the University of Arizona in Tucson. He taught courses in Systems Design, risk and uncertainty analysis in Water Resources Engineering. His research dealt with Bayesian decision theory, multicriterion analysis, fuzzy rule-based modeling, with applications in hydrology and hydrometeorology. This research led to the authorship or co-authorship of 5 books and many professional papers. He became professor emeritus in 1997, and then joined as an adjunct professor the ENGREF (école nationale du génie rural des eaux et des forêts) in Paris where he lives now.

**Marcos Cezar Visoli** is a researcher in computer science at the Brazilian Agricultural Research Corporation – Embrapa, specifically at Embrapa Informatics Agriculture, one of research unit of Embrapa. He has a B.S. in Computer Science (1989) from the Federal University of Santa Catarina, Brazil, and a M.Sc. in Computer Science (2009) from University Blaise Pascal, Clermont-Ferrand, France. His research interest includes agroenvironmental spatial-temporal databases and spatial data warehouses. He is the manager of Free Software Laboratory - LabSoL, with focus on open source technologies to systems integration and data interoperability.

Born in 1978, **Sandro Bimonte** is researcher at CEMAGREF, and more exactly he is at TSCF. He obtained his PhD at INSA-Lyon, France (2004-2007). From 2007-2008, he carried out researches at IMAG, France. He is Managing Editor de Journal of Decision Systems, Editorial Board member of International Journal of Decision Support System Technology, and International Journal of Data Mining, Modelling and Management and member of the Commission on GeoVisualization of the International Cartographic Association. His research activities concern Spatial Data Warehouses and Spatial OLAP, Visual Languages, Geographic Information Systems, Spatio-temporal Databases and GeoVisualization.

**Sônia Ternes** is a researcher in Computational Mathematics at the Brazilian Agricultural Research Corporation (Agricultural Informatics Unit). She holds a B.S. in Statistics (1984) and a B.S. in Applied

Mathematics (1985), both from the Institute of Mathematics, Statistics and Scientific Computing at the State University of Campinas (Unicamp), Brazil, a M.Sc. in Software Engineering from the Faculty of Electrical Engineering of the State University of Campinas (Unicamp) (1992) and a D.Sc. in Mathematical Modeling and Simulation from the Faculty of Electrical Engineering of the State University of Campinas (Unicamp) (2001). She also developed post-doctoral research in Agent Based Models and Microsimulation at the Laboratoire d'Ingénierie pour les Systèmes Complexes from Cemagref, France (2010). Since April 2010 she is responsible for leading the Laboratory of Computational Mathematics team at Embrapa Agricultural Informatics.

**François Pinet** received his M.Sc in Computer Science in 1997 (ENS and University of Lyon). He received his PhD in Computer Science in 2002 (INSA Lyon, France). He is currently a first class researcher at the French Institute for Agricultural and Environmental Engineering. His field of research is in environmental information systems and geomatics. He belongs to several scientific committees of different conferences and journals in these fields. Dr. Pinet has coauthored over 50 scientific papers and is involved in numerous national and international IT projects. He has been a co-organizer for several workshops on environmental systems. In France, he also teaches several graduate courses (M.Sc and PhD degrees) on IT techniques applied to environmental sciences. He is editor-in-chief of the International Journal on Agricultural and Environmental Information Systems (IGI Global).

**Jean-Pierre Chanet** is an electronic engineer and he has a Master degree in robotic and a PhD thesis in computer science from University Blaise Pascal (Clermont-Fd - France). He is the manager of Information system group of the French institute for environment: Cemagref. His main research topic is the management of information in wireless sensor networks. He is involved in different European research projects, like NETADDED, OTAG, Eranet ICT & Robotic. He is a reviewer for several international journals and conferences on ICT.

**Susumu Uchida** is a postdoctoral fellow with the National Agriculture and Food Research Organization (Japan). He received his first master's degree in engineering from the University of Tokyo (1990). Following this, he worked with the Mitsubishi Materials Corporation as a researcher and an engineer. Later, he received his second master's degree in environmental science (2005) and PhD (2008) from the Graduate School of Life and Environmental Sciences, University of Tsukuba. He is currently conducting research on the life cycle assessment of energy crops and is engaged in the construction of a related database.

**Kiyotada Hayashi** is the Leader of the Environmental Assessment and Management Research Team at the National Agricultural Research Center, National Agriculture and Food Research Organization in Tsukuba, Japan. He specializes in life cycle assessment and decision analysis in agriculture. He has been a Guest Research Scholar at the International Institute for Applied Systems Analysis (IIASA) and an OECD Fellow at the Swiss Federal Institute of Technology (ETH) Zurich. His recent research activities include sustainability assessment of agricultural production and biomass utilization systems. Recently, he organized an international conference on the transition to sustainability in agriculture under the OECD framework of biological resource management for sustainable agricultural systems.

**Masaei Sato** is a researcher with the Decision Support System Research Team at the National Agricultural Research Center, National Agriculture and Food Research Organization. He received his master's degree in agricultural management and information science from the Tottori University (1999). Since then and up until 2008, he was engaged in research with the National Agricultural Research Center for Tohoku Region and the Agriculture, Forestry and Fisheries Finance Corporation (which was founded by the Japanese government). His research interests include agricultural production economics, risk management, decision support systems, and business intelligence.

**Shingo Hokazono** is a former student of the Graduate School of Bioresources, Mie University, and holds a doctorate degree in arts and sciences (2008). His research interests include farm management, life cycle assessment, and sustainability in organic farming. He is currently a postdoctoral fellow with the National Agriculture and Food Research Organization, and is engaged in the development of the Life Cycle Inventories of Organic Farming in Japan.

# Index

## V



## W



## Z

Breinigsville, PA USA
23 November 2010
249730BV00003BA/2/P